T0180294

Lecture Notes in Computer Science 2059

Edited by G. Goos, J. Hartmanis and J. van Leeuwen

Springer
Berlin
Heidelberg
New York
Barcelona
Hong Kong
London
Milan
Paris
Singapore
Tokyo

Carlo Arcelli   Luigi P. Cordella
Gabriella Sanniti di Baja (Eds.)

Visual Form 2001

4th International Workshop on Visual Form, IWVF4
Capri, Italy, May 28–30, 2001
Proceedings

 Springer

Series Editors

Gerhard Goos, Karlsruhe University, Germany
Juris Hartmanis, Cornell University, NY, USA
Jan van Leeuwen, Utrecht University, The Netherlands

Volume Editor

Carlo Arcelli
Gabriella Sanniti di Baja
Istituto di Cibernetica, CNR
Via Toiano 6, 80072 Arco Felice (Naples), Italy
E-mail: {car/gsdb}@imagm.cib.na.cnr.it
Luigi P. Cordella
Università di Napoli "Frederico II"
Dipartimento di Informatica e Sitemistica
Via Claudio 21, 80125 Naples, Italy
E-mail: cordel@unina.it

Cataloging-in-Publication Data applied for

Die Deutsche Bibliothek - CIP-Einheitsaufnahme

Visual form 2001 : proceedings / 4th International Workshop on Visual
Form, IWVF-4, Capri, Italy, May 28 - 30, 2001. Carlo Arcelli ...
(ed.). - Berlin ; Heidelberg ; New York ; Barcelona ; Hong Kong ;
London ; Milan ; Paris ; Singapore ; Tokyo : Springer, 2001
    (Lecture notes in computer science ; Vol. 2059)
    ISBN 3-540-42120-3

CR Subject Classification (1998): I.3-5, H.2.8, G.1.2

ISSN 0302-9743
ISBN 3-540-42120-3 Springer-Verlag Berlin Heidelberg New York

Springer-Verlag Berlin Heidelberg New York
a member of BertelsmannSpringer Science+Business Media GmbH

http://www.springer.de

© Springer-Verlag Berlin Heidelberg 2001
Printed in Germany

Typesetting: Camera-ready by author, data conversion by PTP-Berlin, Stefan Sossna
Printed on acid-free paper   SPIN: 10781585   06/3142   5 4 3 2 1 0

Preface

This proceedings volume includes papers accepted for presentation at the 4th International Workshop on Visual Form (IWVF4), held in Capri, Italy, 28–30 May 2001. IWVF4 was sponsored by the International Association for Pattern Recognition (IAPR), and organized by the Department of Computer Science and Systems of the University of Naples "Federico II" and the Institute of Cybernetics of the National Research Council of Italy, Arco Felice (Naples). The three previous IWVF were held in Capri in 1991, 1994, and 1997, organized by the same institutions.

IWVF4 attracted 117 research contributions from academic and research institutions in 26 different countries. The contributions focus on theoretical and applicative aspects of visual form processing such as shape representation, analysis, recognition, modeling, and retrieval. The reviewing process, accomplished by an international board of reviewers, listed separately, led to a technical program including 66 contributions. These papers cover important topics and constitute a collection of recent results achieved by leading research groups from several countries. Among the 66 accepted papers, 19 were selected for oral presentation and 47 for poster presentation. All accepted contributions have been scheduled in plenary sessions to favor as much as possible the interaction among participants. The program was completed by seven invited lectures, presented by internationally well known speakers: Alfred Bruckstein (Technion, Israel), Horst Bunke (University of Bern, Switzerland), Terry Caelli (University of Alberta, Canada), Sven Dickinson (University of Toronto, Canada), Donald Hoffman (University of California, Irvine, USA), Josef Kittler (University of Surrey, UK), and Shimon Ullman (The Weizmann Institute of Sciences, Israel). A panel on the topic *State of the Art and Prospects of Research on Shape at the Dawn of the Third Millennium* has also been scheduled to conclude IWVF4.

IWVF4 and this proceedings volume would not have been possible without the financial support of the universities, research institutions, and other organizations that contributed generously. We also wish to thank contributors, who responded to the call for papers in a very positive manner, invited speakers, all reviewers and members of the Scientific and Local Committees, as well as all IWVF4 participants, for their scientific contribution and enthusiasm.

Carlo Arcelli
March 2001   Luigi P. Cordella
Gabriella Sanniti di Baja

IWVF4 Organization

The 4th International Workshop on Visual Form (IWVF4) was organized jointly by the Department of Computer Science and Systems of the University of Naples "Federico II", and by the Institute of Cybernetics of the National Research Council of Italy. IWVF4 was sponsored by the International Association for Pattern Recognition.

IWVF4 Chairs

Carlo Arcelli   Istituto di Cibernetica, CNR
Luigi P. Cordella   Università di Napoli "Federico II"
Gabriella Sanniti di Baja   Istituto di Cibernetica, CNR

Scientific Committee

Keiichi Abe (Japan)
Josef Bigün (Sweden)
Aurelio Campilho (Portugal)
Leila De Floriani (Italy)
Alberto Del Bimbo (Italy)
Gregory Dudek (Canada)
Gösta Granlund (Sweden)
Richard Hall (USA)
Benjamin Kimia (USA)
Gerard Medioni (USA)
Annick Montanvert (France)
Petra Perner (Germany)
Hanan Samet (USA)
Gabor Szekely (Switzerland)

Sergey V. Ablameyko (Belarus)
Gunilla Borgefors (Sweden)
Virginio Cantoni (Italy)
Koichiro Deguchi (Japan)
Dov Dori (Israel)
Giovanni Garibotto (Italy)
Erik Granum (Denmark)
Rangachar Kasturi (USA)
Walter Kropatsch (Austria)
Dimitris Metaxas (USA)
Nabeel Murshed (Brazil)
Maria Petrou (UK)
Pierre Soille (Italy)
Albert Vossepoel (The Netherlands)

Local Committee

Claudio De Stefano (Università del Sannio)
Pasquale Foggia (Università di Napoli "Federico II")
Maria Frucci (Istituto di Cibernetica, CNR)
Angelo Marcelli (Università di Salerno)
Giuliana Ramella (Istituto di Cibernetica, CNR)
Carlo Sansone (Università di Napoli "Federico II")
Francesco Tortorella (Università di Cassino)

Publicity

Salvatore Piantedosi (Istituto di Cibernetica, CNR)

Referees

K. Abe	S. V. Ablameyko	M. G. Albanesi
J. Alves Silva	S. Antani	J. Assfalg
S. Berretti	M. Bertini	M. Bertozzi
J. Bigün	G. Boccignone	G. Borgefors
E. Bourque	A. Bruckstein	H. Bunke
T. Caelli	A. Campilho	V. Cantoni
C. Colombo	E. Conte	D. Crandall
L. De Floriani	K. Deguchi	A. Del Bimbo
D. De Menthon	V. di Gesù	S. Dickinson
D. Dori	G. Dudek	M. Ferraro
A. Fusiello	G. Garibotto	R. Glantz
G. Granlund	E. Granum	R. Hall
E. R. Hancock	D. D. Hoffman	A. Imiya
M. Jenkin	R. Kasturi	B. Kimia
R. Kimmel	N. Kiryati	J. Kittler
W. Kropatsch	F. Leymarie	L. Lombardi
P. Magillo	J. Mahoney	V. Y. Mariano
G. Medioni	A. M. Mendonça	C. Montani
A. Montanvert	M. Monti	V. Murino
N. Murshed	G. Nagy	H. Nakatani
A. Narasimhamurthy	J. Neumann	H. Nishida
I. Nyström	P. Pala	J. H. Park
P. Perner	M. Petrou	P. Rosin
H. Samet	E. Saund	S. Sclaroff
T. Sebastian	D. Shaked	R. Sim
A. Soffer	P. Soille	T. Sugiyama
S. Svensson	G. Szekely	F. Tortorella
S. Ullman	U. Vallone	M. Vento
A. Verri	S. Vetrella	A. Vossepoel
Y. Yacoob	T. Zhang	

Sponsors

Azienda Autonoma Soggiorno e Turismo di Capri
Consiglio Nazionale delle Ricerche
Dipartimento di Informatica e Sistemistica, Università di Napoli "Federico II"
Elsag
Gruppo Nazionale di Cibernetica e Biofisica, CNR
Istituto di Cibernetica, CNR
Regione Campania
Università degli Studi di Napoli "Federico II"
Università degli Studi di Salerno

Table of Contents

Chapter III. Analysis

Chapter IV. Recognition

Chapter V. Modelling and Retrieval

Chapter VI. Applications

Invariant Recognition and Processing of Planar Shapes

Alfred M. Bruckstein

Ollendorff Professor of Science
Computer Science Department,
Technion - IIT, Haifa 32000 Israel
freddy@cs.technion.ac.il

Abstract. This short paper surveys methods for planar shape recognition and shape smoothing and processing invariant under viewing distortions and possibly partial occlusions. It is argued that all the results available on these problems implicitly follow from considering two basic topics: invariant location of points with respect to a given shape (i.e. a given collection of points) and invariant displacement of points with regard to the given shape.

1   Introduction

Vision is a complex process aimed at extracting useful information from images: the tasks of recognizing three-dimensional shapes from their two-dimensional projections, of evaluating distances and depths and spatial relationships between objects are tantamount to what we mean by seeing. In spite of promises, in the early 60's, that within a decade computers will be able "to see", we are not even close today to having machines that can recognize objects in images the way even the youngest of children are capable to do. As a technological challenge, the process of vision has taught us a lesson in modesty: we are indeed quite limited in what we can accomplish in this domain, even via deep mathematical results and the deployment ever-more-powerful electronic computing devices. In order to address some practical technological image analysis questions and in order to appreciate the complexity of the issues involved in "seeing" it helps to consider simplified vision problems such as "character recognition" and other "model-based planar shape recognition" problems and see how far our theories ("brain-power") and experiments ("number-crunching power") can take us toward working systems that accomplish useful image analysis tasks. As a result of such efforts we do have a few vision systems that work and there is a vast literature in the "hot" field of computer vision dealing with representation, approximation, completion, enhancement, smoothing exaggeration/characterization and recognition of planar shapes. This paper surveys methods for planar shape recognition and processing (smoothing, enhancement, exaggeration etc.) invariant under distortions that occur when looking at the planar shapes from various points of view. These distortions are Euclidean, Similarity, Affine and Projective maps

C. Arcelli et al. (Eds.): IWVF4, LNCS 2059, pp. 3–10, 2001.

of the plane to itself and model the possible viewing projections of the plane where a shape resides, into the image plane of a pinhole camera, capturing the shape from arbitrary locations. A further problem one must often deal with when looking at shapes is occlusion. If several planar shapes are superimposed in the plane or are floating in 3D-space they can and will (fully, or partially) occlude each other. Under full occlusion there is of course no hope for recognition, but how about partial occlusion? Can we recognize a planar shape from a partial glimpse of its contour? Is there enough information in a portion of the projection of a planar shape to enable its recognition? We shall here address such questions too. The main goal of this paper will be to point out that all the proposed methods to address the above mentioned topics implicitly require the solution of the following two basic problems: distortion-invariant location of points with respect to given planar shape (which for our purposes can be a planar region with curved or polygonal boundaries or in fact an arbitrary set of points) and invariant displacement, motion or relocation of points with respect to the given shape.

2   Invariant Point Locations and Displacements

A planar shape S, for our purpose, will be a set of points in R^2 points that usually specify a connected a planar region with a boundary that is either smooth or polygonal. The viewing distortions are classes of transformations $V_\phi : R^2 \to R^2$ parameterized by a set of values ϕ, and while the class of transformations is assumed to be known to us, the exact values of the parameters is not. The classes of transformations considered are continuous groups of transformations modeling various imaging modalities, the important examples being:

- The Euclidean motions (parameterized by a rotation angle and a two-dimensional translation vector, i.e. ϕ has 3 parameters).
- Similarity transformations (Euclidean motions complemented by uniform scaling transformations, i.e. $|\phi| = 4$ parameters).
- Equi-Affine and Affine Mappings (parameterized by 2×2 matrix - 4 parameters - or 3 if the matrix has determinant 1 - and a translation vector, i.e. $|\phi| = 6$ or 5 parameters).
- Projective Transformations (modeling the perspective projection with $|\phi| = 8$ parameters).

Given a planar shape $S \subset R^2$ and a class of viewing distortions $V_\phi : R^2 \to R^2$ we consider the following problem:

Two observers A and B look at $S_A = V_{\phi_A}(S)$ and at $S_B = V_{\phi_B}(S)$ respectively without knowing ϕ_A and ϕ_B. In other words A and B look at S from different points of view and the details of their camera location orientation and settings are unknown to them. Observer A chooses a point P_A in its image plane R^2, and wants to describe its location w.r.t. $V_{\phi_A}(S)$ to observer B, in order to enable him to locate the corresponding point $P_B = V_{\phi_B}(V_{\phi_A}^{-1}(P_A))$. A knows that B looks at $S_B = V_{\phi_B}(S) = V_{\phi_B}(V_{\phi_A}^{-1}(S_A))$, but this is all the information available to A and B. How should A describe the location of P_A w.r.t. S_A to B?

Solving this problem raises the issue of characterizing a position (P_A) in the plane of S_A in a way that is invariant to the class of transformations V_Φ.

To give a very simple example: Let S be a set of indistinguishable points in the plane $\{P_1, P_2, \ldots, P_N\}$ and V_ϕ be the class of Euclidean motions. A new point $\tilde{P}$ should be described to observers of $V_\phi\{P_1, \ldots, P_N\} = \{V_\phi(P_1), V_\phi(P_2) \ldots V_\phi(P_N)\}$ so that they will be able to locate $V_\phi(\tilde{P})$ in their "images". How should we do this? Well, we shall have to describe $\tilde{P}$'s location w.r.t. $\{P_1, P_2, \ldots, P_N\}$ in an Euclidean-invariant way. We know that Euclidean motions preserve length and angles between line segments so there are several ways to provide invariant coordinates in the plane w.r.t. the shape S. The origin of an invariant coordinate system could be the Euclidean-invariant (in fact even Affine-invariant) centroid of the points S, i.e. $O_S = \left(\sum_{i=i}^N P_i\right)/N$. As one of axes (say the x-axis) of the "shape-adapted invariant" coordinate system, one may choose the longest or shortest (or closest in length to the "average" length) vector among $\{\overline{OP_i}\}$ for $i = 1, 2, \ldots, N$. This being settled, all one has to do is to specify $\tilde{P}$ in this adapted and Euclidean-invariant coordinate system with origin at O_s and orthogonal axes with the x-axis chosen as described above. Note that other solutions are possible. We here assumed that the points of S are indistinguishable, but otherwise the problem would be much simpler. Note also that ambiguous situations can and do arise. In case all the points of S form a regular N-gon, there are N equal length vectors $\{OP_i\}$ $i = 1, 2, \ldots, N$ and we can not specify uniquely an x-axis. But, a moment of thought will reveal that in this case the location of any point in the plane is inherently ambiguous up to rotations of $2\pi/N$.

Contemplating the above-presented simple example one realizes that solving the problem of invariant point location is heavily based on the invariants of the continuous group of transformations V_ϕ. The centroid of S, O_S, an invariant under V_ϕ enabled the description of $\tilde{P}$ using a distance $d(O_S, \tilde{P})$, the length of vector $O_S\tilde{P}$ (again a V_ϕ-invariant), up to a further parameter that locates $\tilde{P}$ on the circle centered at O_S with radius $d(O_S P)$, and then the "variability" or inherent "richness" of the geometry of S enables the reduction of the remaining ambiguity.

Suppose next that we want not only to locate points in ways that are invariant under V_ϕ but we also want to perform invariant motions. This problem is already completely solved in the above presented example, once an "S-shape-adapted" coordinate system becomes available. Any motion can be defined with respect to this coordinate system and hence invariantly reproduced by all viewers of S. In fact, when we establish an adapted frame of references we implicitly determine the transformation parameters, ϕ, and effectively undo the action of V_ϕ.

To complicate the matters further consider the possibility that the shape S will be partially occluded in some of its views. Can we, in this case, establish the location of $\tilde{P}$ invariantly and perform some invariant motions as before? Clearly, in the example when S is a point constellation made of N indistinguishable points, if we assume that occlusion can remove arbitrarily some of the points, the situation becomes rather hopeless. However, if the occlusion is restricted to wiping out only points covered by a disk of radius limited to some R, or alternatively, we can assume that we shall always see all points within a certain radius around an (unknown) center point in the plane, the prospects of being able to solve the problem, at least in certain lucky instances, are much better. Indeed, returning to our simple example, assume that we have many indistinguishable landmark points (forming a "reference" shape S in the plane), and that a mobile robot navigates in the plane, and has a radius of sensing or visibility of R. At each location $\tilde{P}$ of the robot in the plane it will see all points of S whose distance from $\tilde{P}$ is less than R, up to its own arbitrary rotation. Hence, the question of being able to specify $\tilde{p}$ from this data becomes the problem of robotic self location in this context. So given a reference map (showing the "landmark" points of S in some "absolute" coordinate system), we want the robot to be able to determine its location on this map from what it sees (i.e. a portion of the points of S translated by $\tilde{P}$ and seen in an arbitrary rotated coordinate system). To locate itself the robot can (and should) do the following:
Using the arbitrarily rotated constellation of points of S within its radius of sensing, i.e. $\tilde{S}(\tilde{P}, R) = \{\Omega_\theta(P_i - \tilde{P})/P_i \in S, \ d((P_i\tilde{P}) \leq R\}$ when Ω_θ is a rotation matrix 2×2 about $\tilde{P}$, "search" in S a similar constellation by checking various center points (2 parameters: $x_{\tilde{p}}, y_{\tilde{p}}$) and rotations (1 parameter: θ).

As stated above, this solution involves a horrendous 3-dimensional search and it must be avoided by using various tricks like signatures and (geometric) hashing based on "distances" from $\tilde{P}$ to $\Omega_\theta(P_i - \tilde{P})$ and distances between the P_i's seen from $\tilde{P}$. This leads to more efficient Hough-Transform like solutions for the self location problem.

It would help if the points of S would be ordered on a curve, say a polygonal boundary of a planar region, or would be discrete landmarks on a continuous but visible boundary curve in the plane. Fortunately for those addressing shape analysis problems this is most often the case.

3 Invariant Boundary Signatures for Recognition under Partial Occlusions

If the shape S is a region of R^2 with a boundary curve $\partial S = C$ that is either smooth or polygonal, we shall have to address the problem of recognizing the shape S from V_ϕ-distorted portions of its boundary. Our claim is that if we can effectively solve the problem of locating a point $\tilde{P}$ on the curve C in a V_ϕ-invariant way based on the local behavior of C in a neighborhood of $\tilde{P}$, then we shall have a way to detect the possible presence of the shape S from a portion of its boundary. How can we locate $\tilde{P}$ on C in V_ϕ-invariant ways? We shall have to associate to $\tilde{P}$ a set of numbers ("co-ordinates" or "signatures") that are invariant under the class of V_ϕ-transformations. To do so, one again has to rely on known geometric invariants of the group of viewing transformation assumed to act on S to produce its image. The fact that we live on a curve C makes life a bit easier.

As an example, consider the case where C is a polygonal curve and V_ϕ is the group of Affine-transformations. Since all the viewing transformations map lines into lines and hence the vertices of the poly-line C into vertices of a transformed poly-line $V_\phi(C)$ we can define the local neighborhood of each vertex $C(i)$ of C, as the "ordered" constellation of $2n+1$ points $\{C(i-n), \ldots, C(i-1), C(i), C(i+1), \ldots, C(i+n)\}$ and associate to $C(i)$ invariants of V_ϕ based on this constellation of points. Affine transformations are known to scale areas by the determinant of their associated 2×2 matrix of "shear and scale" parameters, hence we know that ratios of corresponding areas will be affine invariant. Therefore we could consider the areas of the triangles $\Delta_1 = [(C(i-1)C(i)C(i+1)], \Delta_2 = [C(i-2)C(i)C(i+2)] \cdots \Delta_n = [C(i-n), C(i), C(i+n)]$ and associate to $C(i)$ a vector of ratios of the type $\{\Delta_k/\Delta_l | k, l = \{1, 2, \ldots, n\}, k \neq l\}$. This vector will be invariant under the affine group of viewing transformation and will (hopefully) uniquely characterize the point $C(i)$ in an affine-invariant way.

The ideas outlined so far provide us a procedure for invariantly characterizing the vertices of a poly-line, however, we can use similar ideas to also locate intermediate points situated on the line segments connecting them. Note that the number n in the example above is a locality-parameter : smaller n's imply more local characterization in terms of the size of neighborhoods on the curve C. Contemplating the foregoing example we may ask how to adapt this method to smooth curves where there are no vertices to enable us to count "landmark" points to the left and to the right of the chosen vertex in view-invariant ways. There is a beautiful body of mathematical work on invariant differential geometry providing differential invariants associated to smooth curves and surfaces, work that essentially carried out Klein's Erlangen program for differential geometry, and is reported on in books and papers that appeared about 100 years ago. The differential invariants enable one to determine a V_ϕ-invariant metric, i.e. a way to measure "length" on the curve C invariant with respect to the viewing distortion, similar to the way one finds, rather easily, the Euclidean-invariant arclength on smooth curves. If we have an invariant metric, we claim that our problem of invariant point characterizations on C can be readily put in the same framework as in the example of a poly-line. Indeed we can now use the invariant metric to locate to the left and right of $\tilde{P}$ on C - if we define $\tilde{P} \triangleq C(0)$, and describe C as $C(\mu)$ where μ is the invariant metric parameterization of C about $C(0) = \tilde{P}$ - the points $\{C(0 - n\Delta), \ldots, C(0 - \Delta), C(0 + \Delta), \ldots, C(0 + n\Delta)\}$, and these $2n + 1$ points will form an invariant constellation of landmarks about $\tilde{P} = C(0)$. Here Δ is arbitrarily chosen as a small "invariant" distance in terms of the invariant metric. It is very beautiful to see that letting $\Delta \searrow 0$ one often recovers, from the global invariant quantities that were defined on the constellation of points about $C(0) = \tilde{P}$, differential invariant quantities that correspond to known "generalized invariant curvatures" (that generalize the classical curvature obtained if V_ϕ is the simplest, Euclidean viewing distortion). Therefore to invariantly locate a point $\tilde{P}$ on C, we can use the existing V_ϕ invariant metrics on C (if C is a polygon - the ordering of vertices is an invariant metric!) to determine about $\tilde{P}$ an invariant constellation of "landmark" points on the boundary curve and use global invariants of V_ϕ to associate to $\tilde{P}$ an "invariant signature vector" $I_{\tilde{P}}(\Delta)$. If $\Delta \searrow 0$ this vector yields, for "good" choices of invariant quantities "generalized invariant curvatures" for the various viewing groups of transformations V_ϕ.

We however do not propose to let $\Delta \searrow 0$. Δ is a locality parameter (together with n) and we could use several small but finite values for Δ to produce (what we called) a "scale-space" of invariant signature vectors $\{I_{\tilde{P}}\}_{\Delta_i \in Range}$.

This freedom allows us to associate to a curve $C(\mu)$, parameterized in terms of its "invariant metric or arclength", a vector valued scale space of signature functions $\{I_P(\Delta_i j\mu)\}_{\Delta_i \in Range}$, that will characterize it in a view-invariant way. This characterization is local (its locality being in fact under our control via Δ and n) and hence useful to recognize portions of boundaries in scenes where planar shapes appear both distorted and partially occluded.

4   Invariant Smoothing and Processing of Shapes

Smoothing and other processes of modifying and enhancing planar shapes involves moving their points to new locations. In the spirit of the discussion above, we want to do this in "viewing-distortion-invariant" ways. To do so we have to locate, i.e. invariantly characterize the points of a shape S (or of its boundary $C = \delta S$) and then invariantly move them to new locations in the plane. The discussions of the previous sections showed us various ways to invariantly locate points in the plane of S (or on S). Moving points around is not much more difficult. We shall have to associate to each point (of S, or in the plane of S) a vector whose direction and length have been defined so as to take us to another point, in a way that is V_ϕ-invariant. In the example of S being a constellation of points with a robot using the points of S to locate itself at $\tilde{P}$, we also want it to determine a new place to go, i.e. to determine a point $\tilde{P}_{new}$, so as to have the property that from $V_\phi(\tilde{P})$ a robot using the points $\{V_\phi(P_1) \ldots V_\phi(P_N)\}$ will be able to both locate itself and move to $V_\phi(\tilde{P}_{new})$. Of course on shapes we shall have to do motions that achieve certain goals like smoothing the shape or enhancing it in desirable ways. To design view-distortion invariant motions, we can (and indeed must) rely on invariant point characterizations. Suppose we are at a point $\tilde{P}$ on the boundary $C = \delta S$ of a shape S, and we have established a constellation of landmark points about $\tilde{P}$. We can use the invariant point constellation about $\tilde{P}$ to define a V_ϕ-invariant motion from $\tilde{P}$ to $\tilde{P}_{new}$.

Let us consider again a simple example: if V_ϕ is the Affine group of viewing transformations, the centroid of the point constellations about $\tilde{P}$ is an invariantly defined candidate for $\tilde{P}_{new}$. Indeed it is an average of points around $\tilde{P}$ and the process of moving $\tilde{P}$ to such a $\tilde{P}_{new}$ or, differentially, toward such a new position can (relatively easily) be proved to provide an affine invariant shape smoothing operation. If S is a polygonal shape, i.e. $\partial S = C$ is a poly-line, then moving the vertices according to such a smoothing operation can be shown to shrink any shape into a polygonal ellipse, the affine image of a regular polygon with the same number of vertices as the original shape. In fact ellipses and polygonal ellipses are the results of many reasonably defined invariant averaging processes.

5   Concluding Remarks

The main point of this paper is the thesis that in doing "practical" view-point invariant shape recognition or shape processing for smoothing or enhancement, one has to rely on the interplay between global and local (or even differential) invariants of the group of viewing transformations.

Invariant reparameterization of curves based on "adapted metrics" enables us to design generalized local (not necessarily differential) signatures for partially occluded recognition. These signatures have many incarnations - they can be scalars, vectors or even a scale-space of values associated to each point on shape boundaries. They are sometimes quite easy to derive, and generalize the differential concept of "invariant curvature" in meaningful ways. A study of the interplay between local and global invariances of viewing transformations is also very useful for shape smoothing, generating invariant scale-space shape representations, and leads to various invariant shape enhancement operations.

Many students, collaborators and academic colleagues and friends have helped me develop the point of view exposed in this paper. I am grateful to all of them for the many hours of discussions and debates on these topics, for agreeing and disagreeing with me, for sometimes fighting and competing, and often joining me on my personal journey into the field of applied invariance theory. The list of papers provided below are our contributions to this area and further extremely relevant contributions to invariant shape signatures and shape processing by Weiss, Cyganski, VanGool, Brill, Morel, Faugeras, Olver, Adler, Cipolla and their colleagues can easily be located in the literature.

References

1. Bruckstein, A.M., Katzir, N., Lindenbaum, M. and Porat, M.: Similarity Invariant Signatures for Partially Occluded Planar Shapes. International Journal of Computer Vision, Vol. 7/3, (1992) 271–285 (work done in 1990)
2. Bruckstein, A.M. and Netravali, A.: On Differential Invariants of Planar Curves and Recognizing Partially Occluded Planar Shapes. Annals of Mathematics and Artificial Intelligence, Vol. 13, (1995) 227–250 (work done in 1990)
3. Bruckstein, A.M., Holt, R.J., Netravali, A and Richardson, T.J.: Invariant Signatures for Planar Shape Recognition under Partial Occlusion. CVGIP: Image Understanding, Vol. 58/1 (1993) 49–65 (work done in 1991)
4. Bruckstein, A.M., Sapiro, G. Shaked, D.: Afine Invariant Evolutions of Planar Polygons. International Journal of Pattern Recognition and Artificial Intelligence, Vol. 9/6 (1996) 991–1014 (work done in 1992)
5. Bruckstein, A.M. and Shaked, D.: On Projective Invariant Smoothing and Evolutions of Planar Curves and Polygons. Journal of Mathematical Imaging and Vision, Vol. 7 (1997) 225–240 (work done in 1993)
6. Bruckstein, A.M. and Shaked, D.: Skew-Symmetry Detection via Invariant Signatures. Pattern Recognition, Vol. 31/2 (1998) 181–192 (work done in 1994)
7. Bruckstein, A.M., Rivlin, E. and Weiss, I.: Scale-Space Semi-Local Invariants. Image and Vision Computing, Vol. 15 (1997) 335–344 (work done in 1995)

Recent Advances in Structural Pattern Recognition with Applications to Visual Form Analysis

Horst Bunke

Department of Computer Science, University of Bern
Neubrückstr.10, CH-3012 Bern, Switzerland
`bunke@iam.unibe.ch`

Abstract. Structural pattern recognition is characterized by the representation of patterns in terms of symbolic data structures, such as strings, trees, and graphs. In this paper we review recent developments in this field. The focus of the paper will be on new methods that allow to transfer some well established procedures from statistical pattern recognition to the symbolic domain. Examples from visual form analysis will be given to demonstrate the feasibility of the proposed methods.

Keywords: structural pattern recognition; string, tree and graph matching; edit distance; median, generalized median, and weighted mean computation; clustering; self-organizing map.

1 Introduction

Pattern recognition is based on the concept of similarity. If the objects under consideration are represented by means of feature vectors from an n-dimensional feature space, then similarity can be measured by means of distance functions, such as Euclidean or Mahalanobis distance. These categories of distance measures belong to the statistical approach to pattern recognition [1]. In the present paper we focus on structural pattern recognition [2,3]. This approach is characterized by the use of symbolic data structures, such as strings, trees, or graphs for pattern representation. Symbolic data structures are more powerful than feature vectors, because they allow a variable number of features to be used. Moreover, not only unary properties of the patterns under study can be represented, but also contextual relationships between different patterns and subpatterns.

In order to compute the similarity of symbolic data structures, suitable measures, for either similarity or distance, must be provided. One widely used class of distance functions in the domain of strings is string edit distance [4]. This distance function can be extended to the domain of graphs [5,6,7,8]. For other distance functions on trees and graphs see [9,10,11].

Although symbolic data structures are more powerful than feature vectors for pattern representation, one of the shortcomings of the structural approach is the lack of a rich set of basic mathematical tools. As a matter of fact, the vast

C. Arcelli et al. (Eds.): IWVF4, LNCS 2059, pp. 11–23, 2001.
© Springer-Verlag Berlin Heidelberg 2001

majority of all recognition procedures used in the structural domain are based on nearest neighbor classification using one of the distance measures mentioned above. By contrast, a large set of methods have become available in statistical pattern recognition, including various types of neural networks, decision theoretic methods, machine learning procedures, and clustering algorithms [1].

In this paper we put particular emphasis on novel work in the area of structural pattern recognition that aims at bridging the gap between statistical and structural pattern recognition in the sense that it may yield a basis for adapting various techniques from the statistical to the structural domain. Especially the topic of clustering symbolic structures will be addressed.

In the next section, basic concepts from the symbolic domain will be introduced. Then in Section 3, median and generalized median of a set of symbolic structures will be presented and computational procedures discussed. The topic of Section 4 is weighted mean of symbolic data structures. In Section 5 it will be shown how the concepts introduced in Sections 2 to 4 can be used for the purpose of graph clustering. Application examples with an emphasis on visual form analysis will be given in Section 6. Finally a summary and conclusions will be presented in Section 7.

2   Basic Concepts in Structural Matching

Due to space limitations, we'll explicitly mention in this section only concepts and algorithms that are based on graph representations. The corresponding concepts and algorithms for strings and trees can be derived as special cases.

In a graph used in pattern recognition, the nodes typically represent objects or parts of objects, while the edges describe relations between objects or object parts. Formally, a graph is a 4-tuple, $g = (V, E, \mu, \nu)$ where V is the set of nodes, $E \subseteq V \times V$ is the set of edges, $\mu : V \to L_V$ is a function assigning labels to the nodes, and $\nu : E \to L_E$ is a function assigning labels to the edges. In this definition, L_V and L_E is the set of node and edge labels, respectively.

If we delete some nodes from a graph g, together with their incident edges, we obtain a *subgraph* $g' \subseteq g$. A graph *isomorphism* from a graph g to a graph g' is a bijective mapping from the nodes of g to the nodes of g' that preserves all labels and the structure of the edges. Similarly, a *subgraph isomorphism* from g' to g is an isomorphism from g' to a subgraph of g. Another important concept is *maximum common subgraph*. A maximum common subgraph of two graphs, g and g', is a graph g'' that is a subgraph of both g and g' and has, among all possible subgraphs of g and g', the maximum number of nodes. Notice that the maximum common subgraph of two graphs is usually not unique.

Graph matching is a generic term that denotes the computation of any of the concepts introduced above, as well as graph edit distance (see further below). Graph isomorphism is a useful concept to find out if two objects are the same, up to invariance properties inherent to the underlying graph representation. Similarly, subgraph isomorphism can be used to find out if one object is part of another object, or if one object is present in a group of objects. Maximum common subgraph can be used to measure the similarity of objects even if there exists no graph or subgraph isomorphism between the corresponding graphs. Clearly, the larger the maximum common subgraph of two graphs is, the greater is their similarity.

Real world objects are usually affected by noise such that the graph representation of identical objects may not exactly match. Therefore it is necessary to integrate some degree of error tolerance into the graph matching process. A powerful alternative to maximum common subgraph computation is *error-tolerant graph matching* using *graph edit distance*. In its most general form, a graph *edit operation* is either a deletion, insertion, or substitution (i.e. label change). Edit operations can be applied to nodes as well as to edges. They are used to model the errors and distortions that may change an ideal graph into a distorted version. In order to enhance the modeling capabilities, a cost is usually assigned to each edit operation. The costs are real non-negative numbers. The higher the cost of an edit operation is, the less likely is the corresponding error to occur. The costs are application dependant and must be defined by the system designer based on knowledge from the underlying domain.

The concepts of *error-tolerant graph matching* and *graph edit distance* will be described only informally. For formal treatments, see [7], for example. An error-tolerant graph matching can be understood as a sequence of edit operations that transform graph g_1 into g_2 such that the accumulated cost of all edit operations needed for this transformation is minimized. The cost associated with such a sequence of edit operations is called the *graph edit distance* of g_1 and g_2, and is written as

$$d(g_1, g_2) = \min_S \{c(S) | S \text{ is a sequence of edit operations that} \qquad (1)$$
$$\text{transform } g_1 \text{ into } g_1\}$$

where $c(S)$ is the accumulated cost of all edit operations in sequence S. Clearly, if $g_1 = g_2$ then no edit operation is needed and $d(g_1, g_2) = 0$. On the other hand, the more g_1 and g_2 differ from each other, the more edit operations are needed, and the larger is $d(g_1, g_2)$.

The concepts that correspond to graph isomorphism, subgraph isomorphism, maximum common subgraph and graph edit distance in the domain of strings are identity of strings, subsequence, longest common subsequence and string edit distance, respectively. Notice, however, that the corresponding algorithms in the string domain are of lower complexity. For example, string edit distance is quadratic in the length of the two strings under comparison [4], while graph edit distance is exponential [7].

3   Median and Generalized Median of Symbolic Data Structures

Clustering is a key concept in pattern recognition. While a large number of clustering algorithms have become available in the domain of statistical pattern recognition, relatively little attention has been paid to the clustering of symbolic structures, such as strings, trees, or graphs [12,13,14]. In principle, however, given a suitable similarity (or dissimilarity) measure, for example, edit distance, many of the clustering algorithms originally developed in the context of statistical pattern recognition, can be applied in the symbolic domain.

In this section we review work on a particular problem in clustering, namely, the representation of a set of similar objects through just a single prototype. This problem typically occurs after a set of objects has been partitioned into clusters. Rather than storing all members of a cluster, only one, or a few, representative elements are being retained.

Assume that we are given a set $P = \{p_1, \cdots, p_n\}$ of patterns and some distance function $d(p_1, p_2)$ to measure the dissimilarity between patterns p_1 and p_2. A straightforward approach to capturing the essential information in set P is to find a pattern $\bar{p}$ that minimizes the average distance to all patterns in P, i.e.,

$$\bar{p} = \arg \min_p \frac{1}{n} \sum_{i=1}^{n} d(p, p_i) \tag{2}$$

Let's call pattern $\bar{p}$ the *generalized median* of P. If we constrain the representative to be a member of the given set P, then the resultant pattern

$$\hat{p} = \arg \min_{p \in P} \frac{1}{n} \sum_{i=1}^{n} d(p, p_i) \tag{3}$$

is called the *median* of P.

In the context of this paper we consider the case where the patterns are represented by means of symbolic data structures, particulary strings or graphs. The task considered in the following is the computation of the median and generalized median of a set of strings or graphs.

First, we notice that the computation of the median of symbolic structures is a straightforward task. It requires just $O(n^2)$ distance computations. (Notice, however, that each of these distance computations has a high computational complexity, in general.) But the median is restricted in the sense that it can't really generalize from the given patterns represented by set P. Therefore, generalized median is the more powerful and interesting concept. However, the actual computational procedure for finding a generalized median of a given set of symbolic structures is no longer obvious.

The concept of median and generalized median of a set of strings was introduced in the pattern recognition literature in [15]. An optimal procedure for computing the median of a set of strings was proposed in [16]. This procedure is an extension of the algorithm for string edit distance computation [4]. However, this extension suffers from a high computational complexity, which is exponential in the number of strings in set P. Hence its applicability is restricted to rather small sets and short strings. A suboptimal version of the same algorithm and its application to the postprocessing of OCR results is described in [17]. For another version of the method that uses positional information to prune the high-dimensional search space and its application to handwriting recognition see [18]. Other suboptimal procedures for the computation of generalized median of a set of strings are described in [19,20,21,22]. In [23] genetic algorithms are studied.

The high computational complexity of generalized median computation becomes even more severe if graphs rather than strins are used for pattern representation. Nevertheless, optimal algorithms and suboptimal methods based on genetic search are investigated in [24,25]. An application to graphical symbol recognition is discussed in [26].

4   Weighted Mean of Symbolic Data Structures

Consider two patterns, p_1 and p_2. We call pattern p a *weighted mean* of p_1 and p_2 if, for some real number α with $0 \leq \alpha \leq d(p_1, p_2)$, the following two conditions hold:

$$d(p_1, p) = \alpha, \tag{4}$$

$$d(p_1, p_2) = \alpha + d(p, p_2). \tag{5}$$

Here $d(., .)$ denotes again some distance function. Clearly, if p_1 and p_2 are represented in terms of feature vectors then weighted mean computation can be easily solved by means of vector addition. In [23,27] a procedure for the computation of weighted mean in the domain of strings has been proposed. This procedure is based on the 'classical' algorithm for string edit distance computation [4]. For a pair of strings, p_1 and p_2, it first computes the edit distance. This yields an edit matrix with an optimal path. If one takes a subset, S, of the edit operations on the optimal path and applies them on p_1, a new string, p, is obtained. It can be proven that string p is a weighted mean of p_1 and p_2, obeying equations (4) and (5) with α being equal to the accumulated cost of all edit operations in subset S. Furthermore, it can be proven that the procedure given in [27] is complete, i.e., there is no weighted mean of two strings, p_1 and p_2, that can't be generated by means of the given procedure.

Weighted mean is a useful tool for tasks where a pattern, p_1, has to be changed so as to make it more similar to another pattern, p_2. Intuitively speaking, the weighted mean, p, of a pair of patterns, p_1 and p_2, is a structure that is located between p_1 and p_2 in the symbolic pattern space.

In [28] weighted mean has been extended from the domain of strings to the domain of graphs. A concrete application of weighted mean in the graph domain will be described in the next section.

5   Clustering of Symbolic Data Structures

Self organizing map (*som*) is a very well established method in the area of statistical pattern recognition and neural networks [29]. A pseudo code description of the classical *som*-algorithm is given in Fig. 1. The algorithm can serve two

som-algorithm
(1) **input:** a set of patterns, $X = \{x_1, \ldots, x_N\}$
(2) **output:** a set of prototypes, $Y = \{y_1, \ldots, y_M\}$
(3) **begin**
(4)   initialize $Y = \{y_1, \ldots, y_M\}$ randomly
(5)   **repeat** select $x \in X$ randomly
(7)   find y^* such that $d(x, y^*) = \min\{d(x, y)|y \in Y\}$
(8)   **for** all $y \in N(y^*)$ **do**
(9) $y = y + \alpha(x - y)$
(10)   reduce learning rate α
(11)   **until** termination condition is true
(12) **end**

Fig. 1. The som-algorithm

purposes, either clustering or mapping a high-dimensional pattern space to a lower-dimensional one. In the present paper we focus on its application to clustering. Given a set of patterns, X, the algorithm returns a prototype y_i for each cluster i. The prototypes are sometimes called *neurons*. The number of clusters, M, is a parameter that must be provided a priori. In the algorithm, first each prototype y_i is randomly initialized (line 4). In the main loop (lines 5-11) one randomly selects an element $x \in X$ and determines the neuron y^* that is nearest to x. In the inner loop (lines 8,9) one considers all neurons y that are within a neighborhood $N(y^*)$ of y^*, including y^*, and updates them according to the formula in line 9. The effect of neuron updating is to move neuron y closer to pattern x. The degree by which y is moved towards x is controlled by the parameter α, which is called the *learning rate*. It has to be noted that α is dependent on the distance between y and y^*, i.e. the smaller this distance is the larger is the change on neuron y. After each iteration through the repeat-loop, the learning rate α is reduced by a small amount, thus facilitating convergence of the algorithm. It can be expected that after a sufficient number of iterations the y_i's have moved into areas where many x_j's are concentrated. Hence each y_i can be regarded a cluster center. The cluster around center y_i consists of exactly those patterns that have y_i as closest neuron.

In the original version of the *som*-algorithm all x_j and y_i are feature vectors. In this section, its adaption to the graph domain is discussed, see [30,31]. (The algorithm can also be adapted to the string domain; see [32].) To make the algorithm applicable in the graph domain, two new concepts are needed. First, a graph distance measure has to be provided in order to find graph y^* that is closest to x (see line 7). Secondly, a graph updating procedure implementing line 9 has to be found. If we use graph edit distance, and the weighted mean graph computation procedure (see Section 4) as updating method, the algorithm in Fig. 1 can in fact be applied in the graph domain. For further implementational details see [30,31].

Using the concepts of edit distance and generalized median, also the well-known k-means clustering algorithm [33] can be transferred from the domain

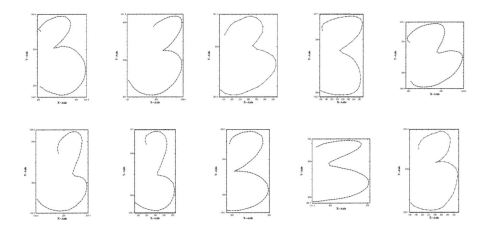

Fig. 2. Ten different instances of digit 3 (from [36])

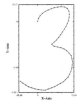

Fig. 3. Generalized median of the digits in Fig. 2

of feature vectors to the symbolic domain. In this case no updating procedure as given in Section 4 is needed. Instead, cluster centers are computed using the generalized median as described in Section 3. In fact, k-means clustering can be understood as a batch version of *som*, where neuron updating is done only after a complete cycle through all input patterns, rather than after presentation of each individual pattern [34].

Both *som* and k-means clustering require the number of clusters being known beforehand. In [35] the application of validation indices was studied in order to find the most appropriate number of clusters automatically.

6   Application Examples

In this section we'll first show an example of generalized median computation for the domain of strings. In the example online handwritten digits from a subset [36] of UNIPEN database [37] are used. Each digit is originally given as a sequence $s = (x_1, y_1), \ldots, (x_n, y_n)$ of points in the $x - y$-plane. In order to transform such

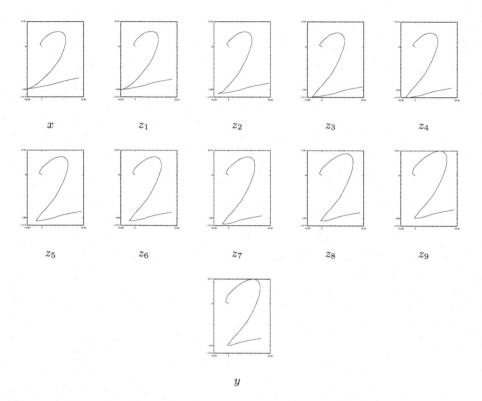

Fig. 4. A sequence of 11 instances of digit 2: $x, z_1, z_2, \ldots, z_9, y$. The first and last digits are from the database [36]. The other digits are weighted means computed for various values of α

a sequence of points into a string, we first resample the given data points such that the distance between any consecutive pair of points has a constant value, Δ. That is, s is transfomed into sequence $s' = (\overline{x}_1, \overline{y}_1), \ldots, (\overline{x}_m, \overline{y}_m)$ where $|(\overline{x}_{i+1}, \overline{y}_{i+1}) - (\overline{x}_i, \overline{y}_i)| = \Delta$ for $i = 1, \ldots, m - 1$. Then from sequence s' a string $z_1 \ldots z_{m-1}$ is generated where z_i is the vector pointing from $(\overline{x}_i, \overline{y}_i)$ to $(\overline{x}_{i+1}, \overline{y}_{i+1})$.

The costs of the edit operations are defined as follows: $c(z \to \epsilon) = c(\epsilon \to z) = |z| = \Delta$, $c(z \to z') = |z - z'|$. Notice that the minimum cost of a substitution is equal to zero (if and only if $z = z'$), while the maximum cost is 2Δ. The latter case occurs if z and z' are parallel and have opposite direction.

Ten different instances of digit 3 are shown in Fig. 2. Their generalized median obtained with the algorithm described in [23] is presented in Fig. 3. Intuitively speaking this instance of digit 3 represents the characteristic features of the samples in Fig. 2 very well. It suitably captures the variation in shape among the different patterns in the given set.

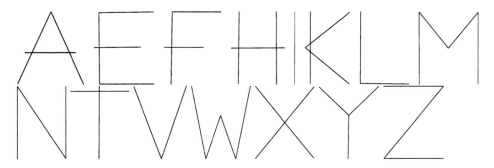

Fig. 5. 15 characters each representing a different class

Next we show an example of weighted mean computation [27]. The same string representation, edit costs and database as for generalized median computation were used in this example. Fig. 4 shows a sequence of 11 instances of digit 2. The first and last instance of this sequence, x and y, are taken from the database [36]. All other digits, $z_1, z_2, \ldots, z_9$ are generated under the procedure described in Section 4. String z_i corresponds to the weighted mean of x and y for $\alpha = \frac{i}{10} d(x, y)$; $i = 1, \ldots, 9$. It can be clearly observed that with an increasing value of α the characters represented by string z_i are becoming more and more similar to y.

Finally, an example of graph clustering using the *som* algorithm introduced in Section 5 will be given [31]. In this example, graph representations of capital characters were used. In Fig. 5, 15 characters are shown, each representing a different class. The characters are composed of straight line segments. In the corresponding graphs, each line segment is represented by a node with the coordinates of the endpoints in the image plane as attributes. No edges are included in this kind of graph representation. The edit costs are defined as follows. The cost of deleting or inserting a line segment is proportional to its length, while substitution costs correspond to the difference in length of the two considered line segments.

For each of the 15 prototypical characters shown in Fig. 5, ten distorted versions were generated. Examples of distorted A's and E's are shown in Fig. 6 and 7, respectively. The degree of distortion of the other characters is similar to Fig. 6 and Fig. 7. As a result of the distortion procedure, a sample set of 150 characters were obtained. Although the identity of each sample was known, this information was not used in the experiment described below.

The clustering algorithm described in Section 5 was run on the set of 150 graphs representing the (unlabeled) sample set of characters, with the number of clusters set to 15. As the algorithm is non-deterministic, a total of 10 runs were executed. The cluster centers obtained in one of these runs are shown in Fig. 8. Obviously, all cluster centers are correct in the sense that they represent meaningful prototypes of the different character classes. In all other runs similar results were obtained i.e., in none of the runs an incorrect prototype was gen-

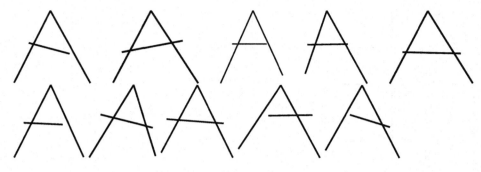

Fig. 6. Ten distorted versions of character A

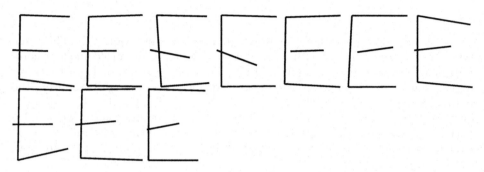

Fig. 7. Ten distorted versions of character E

erated. Also all of the 150 given input patterns were assigned to their correct cluster center.

From these experiments it can be concluded that the new graph clustering algorithm is able to produce a meaningful partition of a given set of graphs into clusters and find an appropriate prototype of each cluster.

7   Conclusions

In this paper, some recent developments in structural pattern recognition are reviewed. In particular the median and generalized median of a set as well as the weighted mean of a pair of symbolic structures are discussed. These concepts are interesting in their own right. But their combination, together with the edit distance of symbolic structures, allows to extend clustering procedures, such as *k*-means or *som*, from vectorial pattern representations into the symbolic domain. A number of examples from the area of shape analysis are given to demonstrate the applicability of the proposed methods.

From the general point of view, the procedures discussed in this paper can be regarded a contribution towards bringing the disciplines of statistical and structural pattern recognition closer together. In the field of statistical pattern

Fig. 8. Cluster centers obtained in one of the experimental runs

recognition, a rich set of methods and procedures have become available during the past decades. On the other hand, the representational power of feature vectors is limited when compared to symbolic data structures used in the structural domain. But structural pattern recognition suffers from the fact, that only a limited repository of mathematical tools have become available. In fact, most recognition procedures in the structural approach follow the nearest-neighbor paradigm, where an unknown pattern is matched against the full set of prototypes. The present paper shows how clustering procedures that were originally developed in statistical pattern recognition, can be adapted to the symbolic domain. It can be expected that a similar adaption will be possible for other types of pattern recognition algorithms.

Acknowledgment. Part of the material presented in this paper is based on the Diploma Theses [23,30]. Thanks are due to Dr. X. Jiang for many stimulating discussions on various aspects of symbolic matching.

References

1. Jain, A., Duin, R. and Mao, J.: Statistical Pattern Recognition: A Review, IEEE Trans. PAMI 22, 2000, 4–37.
2. Fu, K. *Syntactic Pattern Recognition and Applications*, Prentice Hall, 1982.
3. Bunke, H. and Sanfeliu, A. editors. *Syntactic and Structural Pattern Recognition - Theory and Applications*, World Scientific Publ. Co., 1990.
4. Wagner, R. and Fischer, M. The string-to-string correction problem, Journal of the ACM 21, 1974, 168–173.
5. Sanfeliu, A., and Fu, K.: A distance measure between attributed relational graphs for pattern recognition, IEEE Trans. SMC 13, 1998, 353–362.
6. Wang, J., Zhang, K. and Chirn, G. Algorithms for approximate graph matching, Information Sciences, 82, 1995, 45–74.
7. Messmer, B., and Bunke, H.: A new algorithm for error-tolerant subgraph isomorphism detection, IEEE Trans. PAMI 20, 1998, 493–504.
8. Myers, R., Wilson, R., and Hancock, E.: Bayesian graph edit distance, IEEE Trans. PAMI 22, 2000, 628–635.

9. Christmas, W.J., Kittler, J., and Petrou, M.: Structural matching in computer vision using probabilistic relaxation, IEEE Trans. PAMI 8, 1995, 749–764.
10. Pelillo, M., Siddiqi, K. and Zucker, S.: Matching hierarchical structures using associated graphs, IEEE Trans. PAMI 21, 1999, 1105–1120.
11. Cross, A., Wilson, R., and Hancock, E.: Genetic search for structural matching, In B. Buxton, R. Cipolla (eds.): Computer Vision - ECCV '96, LNCS 1064, Springer Verlag, 1996, 514–525.
12. Englert, R. and Glanz,R. Towards the clustering of graphs. Proc. 2nd IAPR-TC-15 Workshop on Graph Based Representations, 125–133, 2000.
13. Lu, S.-Y. A tree-to-tree distance and its application to cluster analysis, IEEE Trans. PAMI 1, 1979, 219–224.
14. Seong, D., Kim, D. and Park, K. Incremental clustering of attributed graphs, IEEE Trans. SMC 23, 1993, 1399–1411.
15. Kohonen, T.: Median strings, Pattern Recognition Letters 3, 1985, 309–313.
16. Kruskal, J.B.: An overview of sequence comparison: Time warps, string edits and macromolecules, SIAM Review 25, 1983, 201–237.
17. Lopresti, D., Zhou, J.: Using consensus sequence voting to correct OCR errors, Comp. Vision and Image Understanding 67, 1997, 39–47.
18. Marti, U.-V., Bunke, H.: Use of positional information in sequence alignment for multiple classifier combination, submitted.
19. Casacuberta, F., de Antonio, M. D.: A greedy algorithm for computing approximate median strings, Proc. of National Symposium on Pattern Recognition and Image Analysis, Barcelona Spain, 1997, 193–198.
20. Kruzslicz, F.: Improved Greedy Algorithm for Computing Approximate Median Strings, Acta Cybernetica 14, 1999, 331–339.
21. Jiang, X., Schiffmann, L., Bunke, H.: Computation of median shapes, Proc. 4th Asian Conference on Computer Vision, Taipei, Taiwan, 2000, 300–305.
22. Martinez - Hinarejos, C., Juan, A., Casacuberta, F.: Use of median string for classification, Proc. 15th Int. Conf. on Pattern Recognition, Barcelona, 2000, Vol.2, 907–910.
23. Abegglen, K.: Computing the Generalized Median of a Set of Strings, MS Thesis, Department of Computer Science University of Bern, forthcoming (in German).
24. Bunke, H., Münger, A. and Jiang,X. Combinatorial search versus genetic algorithms: a case study based on the generalized mean graph problem, Pattern Recognition Letters 20, 1999, 1271–1277.
25. Jiang, X., Münger, A. and Bunke,H. Computing the generalized median of a set of graphs, Proc. 2nd IAPR-TC-15 Workshop on Graph Based Representations, 2000, 115–124.
26. Jiang, X., Münger, A., Bunke, H.: Synthesis of representative Graphical symbols by computing generalized means, A. Chhabra, D. Dori (eds.); Graphics Recognition, Springer Verlag, LNCS 1941, 2000, 183–192.
27. Bunke, H., Jiang, X., Abegglen, K., Kandel, A.: On the weighted mean of a pair of strings, submitted.
28. Bunke, H., Günter, S.: Weighted mean of a pair of graphs, sumbitted.
29. Kohonen, T. *Self-Organizing Map*, Springer Verlag, 1997.
30. Günter, S., Graph Clustering Using Kohonen's Method, MS Thesis, Department of Computer Science University of Bern, 2000.
31. Günter, S., Bunke, H.: Self-organizing feature map for clustering in the graph domain, submitted.
32. Kohonen, T., Somervuo, P.: Self-organizing maps of symbol strings, Neurocomputing 21, 1998, 19–30.

33. Jain, A.K., Murty, M.N. and Flynn, P.J. Data Clustering: A Review, ACM Computing Surveys 31, 1999, 264–323.
34. Kohonen, T.: The self-organizing map, Neurocomputing 21, 1998, 1–6.
35. Günter, S., Bunke, H.: Validation indices for graph clustering, submitted.
36. Alpaydin, E., Alimoglu, F.: Pen-Based Recognition of Handwritten Digits, Department of Computer Engineering, Bogazici University, Istanbul Turkey,
ftp://ftp.ics.uci.edu/pub/machine-learning-databases/pendigits/
37. Guyon, I., Schomaker, L., Plamondon, R., Liberman, M., Janet, S.: UNIPEN project of on-line data exchange and recognizer benchmarks, Proc. 12th Int. Conf. on Pattern Recognition, Jerusalem, 1994, 29–33.

On Learning the Shape of Complex Actions

Terry Caelli, Andrew McCabe, and Gordon Binsted

Department of Computing Science, Research Institute for Multimedia Systems
(RIMS),The University of Alberta, Edmonton, Alberta, CANADA T6G 2E9

Abstract. In this paper we show how the shape and dynamics of
complex actions can be encoded using the intrinsic curvature and torsion
signatures of their component actions. We then show how such invariant
signatures can be integrated into a Dynamical Bayesian Network which
compiles efficient recurrent rules for predicting and recognizing complex
actions. An application in skill analysis is used to illustrate our approach.

Keywords: Differential Geometry, Invariance, Dynamical Bayesian Networks, hidden Markov models, learning complex actions.

1   Introduction

There is an ever increasing number of tasks where it would be useful to be able to
have machines encode, predict and recognize complex spatio-temporal patterns
defined by the dynamics and trajectories of interacting components in 3D over
time. These include gesture recognition, robot skill acquisition, prosthetics and
skill training for human expertise. Current work in this area is characterized by
two quite different approaches. For those consistent with a long tradition of Kinematics, Robotics and Biomedical Engineering, the problem is typically posed in
terms of deterministic control models involving solutions to forward (dynamics
to trajectories) or inverse (trajectories to dynamics) kinematics[12]. The other
exclusively behavioural approach (which predominates the gesture recognition
literature) uses Machine Learning and Pattern Recognition approaches to the
recognition of complex actions[9,3,1,11]. The benefits of this latter perspective
is that it is inherently concerned with recognition within the context of variability. The benefits of the former is that it allows for more detailed modeling of
underlying processes. In this paper we endeavor, more or less, to integrate both
perspectives into a single, invariant, stochastic model - in terms of Differential
Geometry and Dynamical Bayesian Networks (DBN).

New types of active sensors, like Magnetic Field Sensors (MFS), have now
evolved to provide fully 3D encoding of position and pose changes of moving
objects (see, for example, the MIT GANDALF program[1]). They provide us with
more reliably ways of extracting shape trajectories (given an adequate calibration
procedure) and they have already been used in the area of virtual reality HCI,

---

[1] http://gn.www.media.mit.edu/groups/gn/projects

C. Arcelli et al. (Eds.): IWVF4, LNCS 2059, pp. 24–39, 2001.

tele-robotics, film industry and, more recently, for more detailed studies of human kinematics[2]. MFS have also been used to recognize Combat Signals[2] using 3D signature correlation measures. However, there are still a number of open questions of interest to this paper.

1. **Encoding the shape and dynamics of complex actions.** How to uniquely encode the trajectory shape and dynamics of complex actions invariant to the absolute position and orientation of the action?
2. **Descriptions of complex actions.** How to decompose complex actions into basic components which can apply in a consistent and complete fashion while also providing a rich symbolic description?
3. **Learning, recognizing and predicting complex actions.** How to compile rules which define the ways in which many different types of complex actions are performed, recognized and predicted?

Of particular interest is to study the last question, thus posing our research in terms of how to recognize and/or predict complex 3D action trajectories by simply recording their total instantaneous velocity and acceleration dynamics. This is equivalent to learning forward kinematics, in contrast to the learning of dynamics from spatial trajectories - inverse kinematics. In this work we have explored these issues using MFS and, in particular, the Polhemus System[3]. However, the following discussion, treatment and algorithms equally apply to 3D feature data collected via passive vision sensors.

2   Trajectory Shapes and Dynamics

In our case we have, for each sensor and action, i, a recorded 3D trajectory defined by:

$$C_i(t) = (x_i(t), y_i(t), z_i(t)) \tag{1}$$

where x, y, z correspond to the cartesian coordinates of the sensor position, relative to the transmitter origin, over time, t. It is well known that in order to compute derivatives of such data it is necessary to regularize them using multiscaled operators. Past approaches have focused on gaussian pyramids (scale-space) methods for the encoding of contour features[7].

Although such an approach produces useful computations of derivatives, it is not adaptive to the signal's inherent local variations. More importantly, it does not, per se, offer a best-fitting approximation to the space curve, at a given scale. This is particularly relevant when there is a need to encode the trajectory and its dynamics at a given scale using physically referenced quantities like velocity and acceleration. For these reasons we have used a multi-scaled least-squares filter specifically designed to track the curve at a given scale: the Savitzky-Golay (SG) filter[8]. The filter is derived as follows.

---

[2] http://www.hitl.washington.edu/scivw/JOVE/ Articles/dsgbjsbb.txt
[3] http://www.polhemus.com/

We first denote the digital convolution of a signal $f(t)$ with a filter c, as:

$$g_t = \sum_{n=n_L}^{n_R} c_n f_{t+n} \tag{2}$$

where n_L and n_R correspond to point to the "left" and " right" of the position t. A moving window averaging would correspond to $c_n = 1/(n_L + n_R + 1)$, a constant over all positions. However this "0-moment" filter, by definition, does not preserve higher order moments of the function. The SG filter is designed to preserve such higher order moments by approximating the underlying function within a moving window by a polynomial. Since the process of least squares fitting involves linear matrix inversion the coefficients of a fitted polynomial are linearly combined with the data values. This implies that the coefficients can be computed only once and applied by convolution procedures as defined above.

The coefficients, a, are derived by (least squares) fitting a polynomial of degree M in t, namely, $a_0 + a_1 t + a_2 t^2 + .. + a_M t^M$, to the values $f_{-n_L}, ..., f_{n_R}$. The design coefficient matrix, A, for this problem is defined by:

$$A_{ij} = i^j; i = -n_L, .., n_R; j = 0, .., M \tag{3}$$

and the solution for the vector of a_j's is:

$$a = (A^T A)^{-1} A^T f \tag{4}$$

where A^T and A^{-1} correspond to the transpose and inverse of the matrix, A, respectively.

Due to the linearity of this solution we can project the $(A^T A)^{-1} A^T$ component onto unit orthogonal vectors $e_n = (0, 0, .., 1, 0, .., 0)$ with unity only in the nth position which corresponds to the nth window position in c_n (Eqn. (2)). Applying Eqn (4) to each basic vector results in the generation of what we term "SG filter" kernels which can be applied as standard convolution operators on the data. That is,

$$c_n = (A^T A)^{-1} A^T e_n. \tag{5}$$

The result is a set of polynomial SG least squares smoothing filters which can be applied to each of the $x(t), y(t), z(t)$ recordings of the form:

$$X(t : n, m) = SG(n, m) * x(t) \tag{6}$$
$$Y(t : n, m) = SG(n, m) * y(t) \tag{7}$$
$$Z(t : n, m) = SG(n, m) * z(t) \tag{8}$$

where $*$ denotes convolution.

The most important benefit of such polynomial approximations is that higher-order derivatives can be determined algebraically from the derived polynomial coefficients. For the case of quartic polynomials, the first, second and third derivative forms can be directly computed as each defines a polynomial design matrix whose terms are shifted one to the left of the former - due to the properties of polynomial differentiation. Such filters are well-known to preserve, to some extent, discontinuities and variations in the degrees of smoothness of the data.

The magnitudes of velocity, $V(t)$, acceleration, $A(t)$ and displacement, $D(t)$, for 3D motion can then be computed using these coefficients for each position parameter, $(X(t), Y(t), Z(t))$, as:

$$V(t) = \sqrt{(X_t(t:n,m))^2 + (Y_t(t:n,m))^2 + (Z_t(t:n,m))^2} \qquad (9)$$

$$A(t:n,m) = \sqrt{(X_{tt}(t:n,m))^2 + (Y_{tt}(t:n,m))^2 + (Z_{tt}(t:n,m))^2}. \qquad (10)$$

In the following section we will see how they can also be used to compute the curvature and torsion values of a curve.

In all, then, the SG filter satisfies a number of constraints required for the computation of dynamics and spatio-temporal trajectories by providing least squares filter for smoothing and differentiation in terms of polynomial filter kernels at any number of scales defined by the window size and order of polynomial.

2.1   Invariant Signatures: $\kappa\tau - va$ Spaces

There are many cases where it is necessary to encode action trajectories in ways which are invariant to their absolute position and pose. Consider, for example, situations where finger movements are used to communicate invariant to the pose and position of the hand(s): situations where only the relative motion of the movement trajectories are important. Current measures which use joint angles, relative displacements and feature ratios do not necessarily guarantee uniqueness, invariance and an implicit ability to reconstruct a given action. For such reasons we have explored intrinsic shape descriptors from Differential Geometry: curvature (κ) and torsion (τ). These features have already been used by others for invariant shape descriptions [7]. The curvature measures the amount of arc-rate of change of the tangent vector. From the unit tangent vector

$$\boldsymbol{T}(t) = \boldsymbol{C}_t(t)/\|\boldsymbol{C}_t(t)\| \qquad (11)$$

at a position on the trajectory, t, we can compute the curvature vector, $T_t(t)$, whose magnitude is the curvature

$$\kappa(t) = \|\boldsymbol{T}_t(t)\|. \qquad (12)$$

Torsion, $\tau(u)$, measures the degree to which the curve departs from a planar path as represented by the plane defined by the tangent and normal vectors (the osculating plane[4]). The normal to a curve is the unit curvature vector:

$$\boldsymbol{N}(t) = \boldsymbol{T}_t(t)/\|\boldsymbol{T}_t(t)\| \qquad (13)$$

and the binormal vector, the vector orthogonal to the tangent and normal is defined by:

$$\boldsymbol{B}(t) = \boldsymbol{C}_t(t) \times \boldsymbol{N}(t). \qquad (14)$$

Torsion can then be computed as the scalar product (projection) of the arc-rate of change of the binormal and the normal vectors (the *second curvature*):

$$\tau(t) = -\boldsymbol{B}_t(t)\boldsymbol{N}(t). \tag{15}$$

These measures can be directly computed from the derivatives using the SG filters discussed above[4].

The magnitude of curvature, at scale (n, m), is more readily computed as (dropping the scale parameters, n, m):

$$\|\kappa(t)\| = \frac{Y_t(t)Z_{tt}(t) - Z_t(t)Y_{tt}(t) - X_t(t)Z_{tt}(t) - Z_t(t)X_{tt}(t) + X_t(t)Y_{tt}(t) - Y_t(t)X_{tt}(t)}{\|C_t(t)\|^3} \tag{16}$$

Torsion, τ, at scale (n, m), is defined by (again, dropping the scale parameters, n, m):

$$\tau(t) = \frac{\begin{vmatrix} X_t(t) & Y_t(t) & Z_t(t) \\ X_{tt}(t) & Y_{tt}(t) & Z_{tt}(t) \\ X_{ttt}(t) & Y_{ttt}(t) & Z_{ttt}(t) \end{vmatrix}}{E^2(t) + F^2(t) + G^2(t)} \tag{17}$$

where

$$E(t) = \begin{vmatrix} Y_t(t) & Z_t(t) \\ Y_{tt}(t) & Z_{tt}(t) \end{vmatrix} \tag{18}$$

$$F(t) = \begin{vmatrix} Z_t(t) & X_t(t) \\ Z_{tt}(t) & X_{tt}(t) \end{vmatrix} \tag{19}$$

$$G(t) = \begin{vmatrix} X_t(t) & Y_t(t) \\ X_{tt}(t) & Y_{tt}(t) \end{vmatrix} \tag{20}$$

where $|Z|$ denotes the determinant of matrix Z.

Once computed these values determine the Serret-Frenet equations of a curve which defines how the trihedral vectors $(T(t), N(t), B(t))$ at position t changes (are transformed into their values at $t + 1$) by:

$$\boldsymbol{T}_t(t) = \kappa\boldsymbol{N}(t) \tag{21}$$
$$\boldsymbol{N}_t(t) = -\kappa(t)\boldsymbol{T}(t) + \tau(t)\boldsymbol{B}(t) \tag{22}$$
$$\boldsymbol{B}_t(t) = -\tau(t)\boldsymbol{N}(t). \tag{23}$$

The Serret-Frenet equations provide proof that the *shape* of a curve can be uniquely determined up to its absolute position and orientation in 3D. That is, κ and τ define the transformation matrix which carries the moving trihedron from one point on a curve to the next.

In all, then, we first define the SG-derived multi-scaled $\kappa\tau(t)$ curves as the locus of points in $\kappa\tau(t)$ space defining the values of $\kappa(t)$ and $\tau(t)$ as a function of time, t, as shown in Figure 1. The dynamics are then defined by a similar plot of acceleration, $a(t)$, and velocity, $v(t)$, over the same temporal index - as also

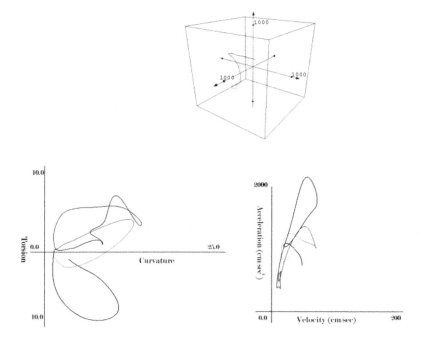

Fig. 1. Complex action statics and dynamics. Top: shows sample 3D component action trajectory recorded from one sensor. The three axes are calibrated in mms. $\kappa\tau$ (lower left) and va (lower right)- fully 3D velocity and acceleration signatures over time (scale: n=40,m=4). The grey level on each signature indicates the temporal evolution of the action.

shown in Figure 1: curves in $va(t)$ space. The benefits of this representation is that we can encapsulate the shape and dynamics of fully 3D actions via 2 simple contours in two two-dimensional spaces - invariant to their absolute position and pose in 3D. The additional benefits of such plots are that they visually demonstrate correlations and other properties between the components of shape and dynamics as well as identify "critical" points such as those where $va(t)$ or $\kappa\tau(t)$ curves change their directions.

Complex actions can then be uniquely encoded up to their absolute (spatio-temporal) position and orientation by:

1. each component's (i) invariant $\kappa\tau(t), va(t)$ signatures;
2. the relative spatial ($\delta\boldsymbol{x}_{ij}$) and temporal ($\delta t_{ij}$) positions for the initial state ($t = 0$) of each pair (i, j) of action components;
3. the relative direction of the initial tangent vectors (δT_{ij}).

Accordingly, a complex action composed of $n \geq 2$ components can be uniquely defined, up to their absolute position and pose, by n $\kappa\tau, va$ signatures and $8n(n-$

$1)/2 = 4n(n-1)$ additional initial relational spatio-temporal position(six) and direction(two, a unit vector) values.

Figure 2 shows the degree to which the computations of $\kappa\tau$ and va curves are invariant to the position and pose of the actions. In this case we have used the SG filter with a window size defined by $n_L = n_R = 20$ and a fourth-oder polynomial fit, with sampling at 120 Hz.

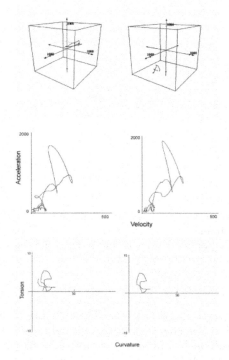

Fig. 2. Top row: same action in different positions and orientations. Middle and bottom rows: shows va and $\kappa\tau$ curves for each action. Note the invariance of their signatures (See text for details).

Screw Decomposition. Although the above derivations provide invariant descriptions of the two components of complex actions (trajectory shape and dynamics), they do not provide, as such, a symbolic representation or taxonomy of the action components which make practical sense to those who need to analyze and describe actions. To these ends we have developed, what we term, a "screw decomposition model" (SDM) to generate such descriptions. This approach is related to screw kinematics theory as developed over the past two centuries to encode mechanical motions in general[12]. This model differs from past work in so far as we use Screw Theory to encode motions. It follows simply from the observation that the trajectory of a "screw action" is helical and that a helix has constant curvature and torsion everywhere[4]. Specifically, for a helix defined by:

$$C(t) = (a \cos t, a \sin t, bt) \tag{24}$$

it can be shown that

$$\kappa = \frac{a}{a^2 + b^2}; \tau = \frac{b}{a^2 + b^2} \tag{25}$$

and

$$a = \frac{\kappa}{\kappa^2 + \tau^2}; b = \frac{\tau}{\kappa^2 + \tau^2}. \tag{26}$$

These relations show how changes in curvature and torsion values define the types of instantaneous (or prolonged if constant for a period of time) screw actions (for example, "left-handed" and "right-handed") which can approximate a given action component. Equally, temporally contiguous points which are close together on a $\kappa\tau$ curve can be approximated by a single screw action or helix. In particular, we define a "fixed point" in $\kappa\tau$ space is one that does not change for a given time interval over the evolution of the action trajectory and representing the trajectory of nearby points in $\kappa\tau$ space. That is, contiguity in local shape and time lie at the basis of our approach to encoding such actions. Consequently, our aim was to:

- describe action components in terms of a sequence of screws (hashed curvature and torsion values) and so changes in screw "states";
- determine the relationships between different types of screws and how they are performed - the screw dynamics (in va space);
- explore how to compile rules (Machine Learning) which define the execution of complex motions via the above invariant decomposition model over all component actions.

3   Learning, Recognizing, and Predicting Actions

Since this work is not concerned with innovation models but, rather, encoding, recognition and prediction processes, we adopt a Machine Learning perspective for learning the relationships between the dynamics and trajectory of individual action components, and their interdependencies. We explore how Dynamical Bayesian Networks (DBN), consisting of sets of coupled hidden Markov models (HMMs) and initial position and pose information for each sensor, can be used to encode, predict and recognize complex actions. Each component HMM is used to model the relationship between the sensor's invariant trajectory signature(curvature and torsion) and its dynamics (velocity and acceleration). More formally, in this case we define our complex action DBN as

$$\Lambda = \{\Pi, A, B\} \tag{27}$$

where

$$\mathbf{\Pi} = \{\pi_u^i = p(S_u^i); i = 1, .., N; u = 1, .., N_i\} \tag{28}$$

where N corresponds to the number of HMMs (nodes) in the network, N_i to the number of states in the i'th HMM. The generalized state transmission matrix, $\mathbf{A}$, is defined by

$$\mathbf{A} \equiv a_{uv}^{ij} = p(S_v^j(t+1)/S_u^i(t)) \tag{29}$$

where i, j correspond to a pair of HMMs and u, v to their states. When $i = j$ the state transitions are *within* a given HMM while when $i \neq j$ the state transitions apply *between* a pair of HMM states. In both cases we have used a single unit lag model (t to $t + 1$). However, in general, different types of lags can be used. It should be noted that $\mathbf{A}$ can also encode restricted interactive models by not allowing state transitions between specific HMMs and even specific states within and between HMM's as well as unidirectional or bidirectional dependencies between t and $t + 1$. In other words, A is the "causal model matrix" for the DBN.

The final component of the DBN is the matrix $\mathbf{B}$ defined by

$$\mathbf{B} \equiv \mathbf{b}_u^i(o_k^i) = p(o_k^i/S_u^i). \tag{30}$$

This corresponds to, for a given HMM, i, the probability of a given observation, k, that is o_k^i, given the state u, S_u^i, of HMM i. Finally, we define $b_u(o_t^i) = p(o_t^i/S_u^i)$ as the probability of a given observation for HMM i, at time, t. This term is necessary in developing the proposed Generalized Baum Welch (GBW) model (see below).

Discrete HMM and DBN models assume a finite number of states and observations and, as we will see, there are complications in using such discrete models as their defining characteristics affect performance in terms of encoding, generalizations, prediction and discrimination. Consequently, in this work (in contrast to most reported studies) we have explored how the number of states and observations affect encoding and discrimination performance. Our preference is to develop continuous action mixture models - a topic under investigation in a related project. Albeit, we use each discrete HMM to encode the dependency of trajectory shape, or the type of screw action ($\kappa\tau$), on (observed) dynamics (va). In this case we have also combined velocity and acceleration into a single discrete dynamical observable variable.

Another difference between this DBM model and past approaches is that, in this case, the states are not really "hidden". Here the HMMs are being used to encode the relationships between two classes of observable variables, and we have training data where states and observations can be extracted in parallel during the training phase. It is therefore possible for us to generate initial estimates for each HMM using a standard moving window method. Each process in this initial estimation scheme is defined as follows.

1. $\kappa\tau$ quantization into screw (action) states: $S_1, .., S_u, .., S_{N_i}$ using percentile-based attribute quantization over κ and τ values.

2. va quantization into dynamic "symbols": $o(va_1) \equiv o_1, .., o(va_k), .., o(va_{M_i}) \equiv o_{M_i}$, again, using percentile-based attribute quantization of v and a values.
3. Using the moving window method we obtain, from the training data, initial estimates of each type of action component HMM, from $w = 1, .., N_i; k = 1, .., M_i$:

$$p(S_u^i) = \sum_t^T I_w(S_u^i(t))/T \tag{31}$$

$$p(S_v^j(t)/S_u^i(t-1)) = \sum_{r=1}^T I_w(S_v^j(t+r))/I_j(S_u^i(t+r-1))/T \tag{32}$$

$$p(o_k^i/S_u^i) = \sum_t^T I_w(o_k^i(t)/S_u^i(t))/T \tag{33}$$

where T corresponds to the length of the observation sequence, and

$$I_i(z) = \begin{cases} 1 & \text{iff} \quad z \equiv i \\ 0 & \text{otherwise} \end{cases} \tag{34}$$

These initial estimates are then used as input to the Baum-Welch procedure which, via Expectation-Maximization, adjusts the complete DBN model to fit the training data and the inherent non-stationary performance characteristics of such models.

3.1   Estimation Update and Prediction

Estimation: The Generalized Baum Welch Algorithm. As a generalization of the single HMM case[10] we define the generalized forward operator as:

$$\alpha_1(S_u^i) = p(S_u^i)p(o_1^i/S_u^i) = \pi_u^i b_u(o_1^i) \tag{35}$$

$$\alpha_{t+1}(S_u^i) = \sum_{i=j} \sum_v \alpha_t(S_u^i)a_{uv}^{ij}b(o_{t+1}^i/S_v^i) + \sum_{i \neq j} \sum_w \alpha_t(S_w^j)a_{uw}^{ij}b(o_t^j/S_w^j) \tag{36}$$

The first component of the right hand side of Equation (36) corresponds to the intra-HMM forward operator while the second to the inter-HMM components. Together they can be represented in a single matrix form as:

$$\alpha_{t+1} = \alpha_t \mathbf{A} \mathbf{B}_t \tag{37}$$

which encodes all possible inter and intra state transitions within the given model as a function of the causal model. In this case the diagonal block submatrices of A correspond to the intra-HMM state transitions while the outer rectangular blocks correspond to state transition weights between each HMM. Accordingly we can model the DBN by blocking specific types of inter and intra state transitions. It is for this reason that we term A the "causal model" matrix.

In a similar way to the forward operator, the generalized backward operator can be defined, in matrix form, as:

$$\boldsymbol{\beta}_t = \boldsymbol{\beta}_{t+1} \mathbf{A}' \boldsymbol{b}_t \tag{38}$$

where $\mathbf{A}'$ corresponds to the transpose of $\mathbf{A}$ and, $\forall ij,\ \beta_T(S_j^i) \equiv 1$.

As with the single HMM Baum Welch (EM) procedure we first estimate the expected state transitions between any two states within a given model. As defined above this incorporates influences of other HMM states within and between the HMMs according to the DBN causal model. The net result, however, has the same format as the single HMM case, where, at a given time point, we have:

$$\boldsymbol{\Phi}_t(i, j; u, v) = \alpha_t \mathbf{A} \beta_{t+1} \tag{39}$$

Integrating over t we obtain:

$$\hat{\mathbf{A}} \equiv \hat{\mathbf{a}}_{uv}^{ij} = \sum_{t=1}^{T} \boldsymbol{\Phi}_t(i, j; u, v)/T \tag{40}$$

$$\hat{\boldsymbol{\Pi}} \equiv \{\hat{\pi}_u^i\} = \{\sum_{j,v} \hat{\mathbf{a}}_{uv}^{ij} / \sum_j N_j\} \tag{41}$$

and

$$\hat{\psi}_u^i(t) = \sum_{j,v} \boldsymbol{\Phi}_t(i, j; u, v) / \sum_j N_j \tag{42}$$

and so

$$\hat{\mathbf{B}} \equiv \hat{\mathbf{b}}_u^i(o_k^i) = \sum_{t=1}^{T} p(S_u^i(t)/o_k^i(t)/(T\hat{\psi}_u^i(t)) \tag{43}$$

The above formulation allows us to consider a number of algorithms as a function of the degree to which the intra- and inter- state dependencies are estimation in parallel or sequentially. The fully parallel Generalized Baum Welch algorithm is as follows:

Parallel GBW Estimation Method

1. Select a causal model for $\mathbf{A}$ by excluding possible dependencies.
2. Generate initial estimates of $\boldsymbol{\Lambda}$ from the training data using the moving window method (see above).
3. Re-estimate DBN as $\hat{\boldsymbol{\Lambda}} = \{\hat{\boldsymbol{\Pi}}, \hat{\mathbf{A}}, \hat{\mathbf{B}}\}$
4. If $\hat{\boldsymbol{\Lambda}} \simeq \boldsymbol{\Lambda}$ STOP
5. Set $\boldsymbol{\Lambda} = \hat{\boldsymbol{\Lambda}}$ and GoTo 2.

However, we have used an equally viable, and computationally less demanding, sequential ("residual") implementation of the Generalized Baum Welch operators - while continuing to investigate efficient ways of implementing the fully parallel version.

Residual GBW Estimation Method

1. Select a causal model for **A** by excluding possible dependencies.
2. For each sensor, i:
   - Generate initial estimates for each sensor HMM λ^i from the training data using the moving window method.
   - Re-estimate the sensor's HMM parameters till convergence using the standard Baum Welch algorithm.
3. Then compute the inter-HMM state dependencies in accord with the causal model (off–diagonal blocks of A_{uv}^{ij}) and an associated predictor method).

Prediction: The Generalized Viterbi Algorithm. Analogous to the single HMM Viterbi algorithm, the generalized method predicts the optimal *set* of state sequences which fits the observations. It is based upon computing the most likely state sequence for each of the component HMMs. This is well-known to be computationally expensive and so less optimal solutions are typically used, in particular, where only the previous best set of states are retained in determining the appropriate state at a given time. That is:

Algorithm

Initialization

$$For\ 1 \leq i \leq N,\ 1 \leq u \leq N_i$$
$$\delta_1(S_u^i) = \pi b_u(o_1^u)$$
$$\Phi_1(S_u^i) = 0$$

Recursion

$$For\ 2 \leq t \leq T,\ 1 \leq j \leq N,\ 1 \leq v \leq N_j$$
$$\delta_t(S_v^j) = max_{i,j;u,v}[\delta_{t-1}(S_u^i)a_{uv}^{ij}]b_v(o_t^j)$$
$$\Phi_t(S_v^j) = argmax_{i,j;u,v}[\delta_{t-1}(S_u^i)a_{uv}^{ij}]$$

Using the more efficient residual DBN estimation model, we simply used a corresponding sequential state estimation approach. That is, the Viterbi algorithm was applied to each individual sensor and then a second Viterbi procedure was used to estimate the degree to which one most likely sensor state sequence could predict another from the off-diagonal blocks of the causal model matrix, A ($i \neq j$ components of Eqn (36)). This was implemented by simply defining the predictor state sequence as the "inter-HMM" state sequences and the dependent state sequence as corresponding to the equivalent "observation sequence" (see below).

4   Assessing DBN Performance

The Generalized Viterbi algorithm defines the most likely state sequence for each node of the DBN in terms of the final (joint) posterior maximum likelihood (log) probability of the state sequences given the observation sequences. This latter measure is a reasonable measure for pattern recognition but not necessarily the most representative way of measuring how well the model encodes, predicts or can track the training data. For these reasons we have used the Viterbi-derived posterior maximum likelihood (log) probability for classifying actions and a different measure to represent how well the action is encoded via a given DBN. This latter measure is based upon computing the Hamming distance between observed and predicted observation sequences using a Monte Carlo method. That is, for each sensor and for each task/participant, we generate predicted observation sequences by randomly selecting observations according to the estimated DBN model probabilities and the Viterbi-estimated state sequences. This is computed a number of times to result in a mean and standard deviation of the similarity between observed and predicted observation sequences using the reversed normalized Hamming distance defined by:

$$\varsigma(\hat{o}, o) = \sum_{t}^{T} I_{o(t)}(\hat{o}(t))/T \qquad (44)$$

where

$$I_{o(t)}(\hat{o}(t)) = \begin{cases} 1 & \text{iff} \quad \hat{o}(t) \equiv o(t) \\ 0 & \text{otherwise.} \end{cases} \qquad (45)$$

This constitutes a direct measure of the likelihood that the particular set of observations sequences (one per sensor or node of the DBN) matches those predicted from the complete model using the Viterbi solutions for each of the node's predicted state sequences. Comparing these values indicates the uniqueness of the Viterbi solutions over the complete DBN. This measure is useful as it provides a direct estimate (μ, σ) of how well the model can encode the training data. However, it does not provide a way of determining the discriminatory power of the DBN model for a given action in predicting the data from the model. For this reason we have also introduced an additional component to the measure - how well a given DBN model can predict observations from data known *not* to arise from the model ($\neg o$) - that is, a different task or participant, etc. This comparison results in the prediction-discrimination function (PDF) which clearly varies as a function of the number of states(n_s) and observations(n_o):

$$PDF(n_s, n_o) = \varsigma_{n_s, n_o}(\hat{o}, o) - \varsigma_{n_s, n_o}(\hat{o}, \neg o) \qquad (46)$$

This is analogous to the "detectability" score in signal detection theory: p("Hit") - p("False Alarm") for the case of the ideal detector[6]. Again, $\neg o$ corresponds to observations not from the training set. This measure of "encoding-relative-to-discrimination" penalizes the Hamming Distance measure as a function of the degree to which the HMM can equally track observation sequences not generated from the target.

5   Experiments and Results

Although the above model defines an invariant stochastic model for encoding, learning and predicting complex actions it's discrete formulation presents a number of empirical issues. To study its performance we have considered four "assembly" and "disassembly" tasks as illustrated in Figure 3 below. In all, three of the authors performed each task 10 times using the 4 sensors attached to the forearms and hands as shown in Figure 3. Each task was performed using the same initial resting position of the arms and hands and at a fixed sitting position at a table upon which the parts were laid out in identical positions. Each participant was allowed some rehearsal trails before commencing the formal recordings. All data was collected for each participant over one recording session. This resulted in: 4(tasks) x 4(sensors) x 10(trials) x 3(participants) = 480 data steams sampled at 120Hz for approximately 20 seconds each - approximately 2,000 x 480 data points or 2 Megabytes of data. This data was smoothed and sub-sampled to 12 Hz for a number of reasons. One, the shear amount of data needed to be reduced for real-time computational purposes. Two, since our model involves *differential* invariants, it was more likely for the state transition matrices to be less redundant as the sampling decreased.

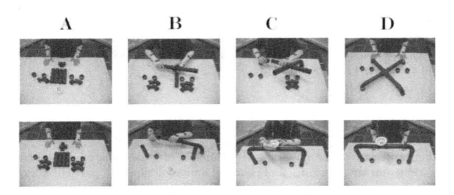

Fig. 3. Shows 4 construction tasks for each of the two construction tasks. The remaining two tasks consisted of disassembling each object. All tasks were initiated from the same hand and arm resting positions shown in column one.

Half of the trials were used for training the model, half for testing on unseen examples.

In order to ascertain the optimal numbers of states and symbols over all tasks, trials and participants, we generated a large number of DBN solutions varying in the numbers of clusters in va and $\kappa\tau$ attributes. In both cases the clustering "resolution" was determined in terms of a generalized percentile splitting method on each attribute resulting in a binning or quantization of the respective spaces based upon equal frequencies. This type of binning is efficient though not necessarily optimal. However, 16 states of $\kappa\tau$ and 9 discrete observation values of va were found to optimize the the prediction-discrimination function (*PDF*).

With these numbers of observations and states, we then determined, for a given task and participant, the degrees to which each sensor observations could be predicted from their own training data using only each sensor's individual HMM model. Over all four tasks, participants and sensors results showed that we could correctly and consistently predict observed sensor dynamics on average 85% (0.85) of the time with a standard deviation of $\pm 10\%$ (0.1) using the Viterbi generated optimal state sequences.

We then measured the degree to which each arm sensor states could predict each hand sensor states by the following procedure. The 16 × 16 sensor state dependencies do not provide a direct measure of the degree to which one sensor state sequence could predict those of another - particularly when there is no "ground truth". Consequently we adopted an alternate strategy to examine such correlations. Since we were concerned with analyzing the possible dependencies of the hand sensor states on the arm sensor states, we defined the Viterbi-predicted hand state sequence as dependent state "observations" relative to the arm sensor state sequence. We then estimated an equivalent inter sensor HMM based on the off-diagonal estimates of the causal model matrix A. From its Viterbi solution we could predict, using the Monte Carlo method and the same reverse Hamming distances, the degree to which each arm state sequence could predict each hand state sequence at four different time lags of $t = 0, -1, -2, -3$. From this analysis we found the best prediction occurred with zero lag, with a 55% prediction rate between the sensors. This is highly significant since the random performance would be at 6%(1/16) performance level.

6   Discussion

In this project we have investigated a number of issues related to the encoding, prediction and recognition of complex human actions. Of particular interest has been how to formulate the recognition of complex actions, their forward and inverse kinematics in terms of sets of hidden Markov models or Dynamical Bayesian Networks. These initial investigations show that this type of model has significant potential in so far as it incorporates variability in action performance, estimation, prediction and classification all within the same framework. Open issues still include those pertaining to optimal estimation of the complete DBN model and how to implement the generalized Baum Welch and Viterbi-type methods. Past work on estimation[5] illustrates the complexity of this problem even with discrete models and in this work the residual analysis approach to the problem provided useful insights into the types of dependencies existing between limb segments.

What is also needed, for this approach to be more robust, is to replace the current discrete state and observation variable models with mixture or related models. The clustering methods currently used for generating discrete state and observation values have demonstrable positive and negative affects on HMM performance as a function of the number and types of states and observations.

The use of Differential Geometry for uniquely encoding action trajectories in invariant ways and in a regularized fashion has proved useful and also has potential integrating more formally with Screw Theory in Kinematics[12]. The SG filters have proven useful as they not only provide efficient ways for computing derivatives but also provide filters which can reproduce the physical values required to maintain validity of velocities, acceleration, etc.

In all, then, this paper proposes an approach to the analysis, learning and prediction of the shape of complex human actions. We have explored a unique but invariant coding scheme and a method for encapulating the variations and interactions which occur in complex actions via Dynamical Bayesian Networks. Together they illustrate how Stochastic Differential Geometry can be a powerful tool for the future analysis of complex kinematic systems.

References

1. Pentland A. Bobick A. and Poggio T. Vsam at the media laboratory and cbcl: Learning and understanding action in video imagery. MIT Media Lab. Report, 1998.
2. R. Boulic. Human motion capture driven by orientation measurements. *Presence: Teleoperations and virtual Environments*, 8(2):187–203, 1999.
3. S. Bryson. Measurement and calibration of static distortion of position data from 3d trackers. Technical report, NASA Ames Research Center, 1992.
4. M. Do Carmo. *Differential Gemoetry of Curves and Surfaces*. Prentice Hall, New Jersey, 1976.
5. Z. Ghahramani and M. Jordon. Factorial hidden markov models. *Machine Learning*, 29:245–275, 1997.
6. D. Green and J. Swets. *Signal Detection Theory and Psychophysics*. Wiley, New York, 1966.
7. F. Mokhtarian. A theory of multiscale, torsion-based shape representation for space curves. *Computer Vision and Image Understanding*, 68(1):1–17, 1997.
8. W. Press, S. Teukolsky, W. Vetterling, and B. Flannery, editors. *Numerical Recipes in C*. Cambridge University Press, Cambridge, UK, 1996.
9. F. Raab, E. Blood, T. Steiner, and H. Jones. Magnetic position and orientation tracking system. *IEEE Transactions on Areospace and Electronic Systems*, (5):709–717, 1979.
10. L. Rabiner. A tutorial on hidden markov models and selected applications in speech recognition. *Proceedings of the IEEE*, 77(2):257–286, 1989.
11. A. Wilson and A. Bobick. Hidden markov models for modeling and recognizing gesture under variation. *International Journal of Pattern Recognition and Artificial Intelligence*, 2001.
12. V. Zatsiorsky. *Kinematics of Human Motion*. Human Kinetics, Windsor, Canada, 1998.

Mereology of Visual Form

Donald D. Hoffman

Department of Cognitive Science, University of California, Irvine 92697 USA

Abstract. Visual forms come in countless varieties, from the simplicity of a sphere, to the geometric complexity of a face, to the fractal complexity of a rugged coast. These varieties have been studied with mathematical tools such as topology, differential geometry and fractal geometry. They have also been examined, largely in the last three decades, in terms of mereology, the study of part-whole relationships. The result is a fascinating body of theoretical and empirical results. In this paper I review these results, and describe a new development that applies them to the problem of learning names for visual forms and their parts.

1 Introduction

From the anatomy and physiology of the retina, we know that the processes of vision begin from a source that is at once rich and impoverished: photon quantum catches at 5 million cones and 120 million rods in each eye [1]. This source is rich in the sheer number of receptors involved, the dynamic range of lighting over which they operate, and the volume of data they can collect over time. This source is impoverished in its language of description. The language can only state how many quanta are caught and by what receptors. It can say nothing about color, texture, shading, motion, depth or objects, all of which are essential to our survival. For this reason we devote precious biological resources—hundreds of millions of neurons in the retina and tens of billions of neurons in the cerebral cortex—to construct richer languages and more adaptive descriptions of the visual world.

A key criterion for these more adaptive descriptions is that they allow us to predict, with economy of effort, future events that can affect our survival. We construct a world of objects and their actions, because carving the world this way lets us quickly learn important predictions. Running toward a rabbit leads to predictably different results than running toward a lion. These are important object-specific properties that cannot be learned in the language of quantum catches.

We carve the world more finely still, dividing objects themselves into parts. Parts aid in the recognition of objects. Parts also allow more refined predictions: If, for instance, one is fighting a conspecific it might be critical to attend to certain parts, such as arms or legs or jaws, and relatively safe to ignore other parts such as ears. Moreover, some parts of shapes are better remembered than others [2–4]. The centrality of parts to human vision can be seen in the following

C. Arcelli et al. (Eds.): IWVF4, LNCS 2059, pp. 40–50, 2001.

six figures, each of which can be explained, as we will see shortly, by three rules for computing the parts of objects.

In Figure 1 you probably see hill-shaped parts with dashed lines in the valleys between them. But if you turn the figure upside down, you will see a new set of hills, and now the dashed lines lie on top of the new hills [5].

In Figure 2, which of the two half moons on the right looks most similar to the half moon on the left? In controlled experiments almost all subjects say that the bottom looks more similar to the half moon on the left —even though the top half moon, not the bottom, has the same bounding curve as the half moon on the left [6].

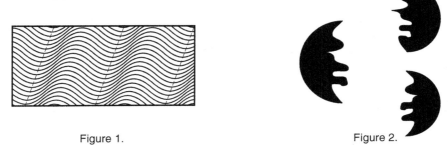

Figure 1.   Figure 2.

In Figure 3, most observers say the staircase on the right looks upside down, whereas the one on the left can be seen either as right side up or as upside down [7].

In Figure 4, the display on the left looks transparent, but the one on the right does not [8]. The luminances in the two cases are the same.

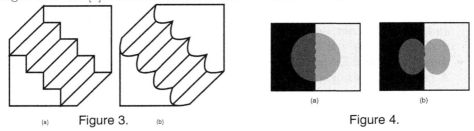

(a)   Figure 3.   (b)   Figure 4.

In Figure 5, the symmetry of the shape on the left is easier to detect than the repetition of the shape on the right [9–12].

In Figure 6, a heart shape pops out among a set of popcorn-shaped distractors, as shown on the left, but not vice versa, as shown on the right [13,14].

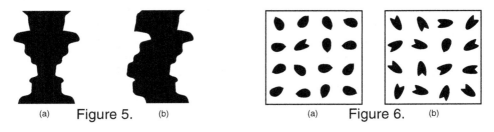

(a)   Figure 5.   (b)   (a)   Figure 6.   (b)

Figure 6 suggests that we begin to construct parts, or at least boundaries between parts, early in the stream of visual processing. Empirical evidence in support of this suggestion comes from experiments using visual search [13,14] with stimuli like Figure 6. In one experiment conducted by Hulleman, te Winckel and Boselie [13], subjects searched for a heart-shaped target amidst popcorn-shaped distractors, and vice versa. The heart had a concave (i.e., inward pointing) cusp which divided it into two parts, a left and right half. The popcorn had a convex (outward pointing) cusp and no obvious parts. Its convex cusp was chosen to have the same angle as the concave cusp of the heart. The data indicate that subjects search in parallel for the heart targets, but search serially for the popcorn targets. It appears that parts are important enough to the visual system that it devotes sufficient resources to search in parallel for part boundaries. This early and parallel construction of part boundaries explains why parts affect the perception of visual form in the wide variety of ways illustrated in Figures 1 through 6.

2   The Minima Rule

Recent experiments suggest that human vision divides shapes into parts by the coordinated application of three geometric rules: the *minima rule*, the *short-cut rule*, and the *part salience rule*. The rules are as follows:

- **Minima Rule (for 3D Shapes):** *All concave creases and negative minima of the principal curvatures (along their associated lines of curvature) form boundaries between parts* [5,7].
- **Minima Rule (for 2D Silhouettes):** *For any silhouette, all concave cusps and negative minima of curvature of the silhouette's bounding curve are boundaries between parts* [5,7].
- **Short-cut rule:** *Divide silhouettes into parts using the shortest possible cuts. A cut is (1) a straight line which (2) crosses an axis of local symmetry, (3) joins two points on the outline of a silhouette, such that (4) at least one of the two points has negative curvature. Divide 3D shapes into parts using minimal surfaces* [15].
- **Salience rule:** *The salience of a part increases as its protrusion, relative area, and strength of part boundaries increases.* [7].

Together these rules can explain each visual effect illustrated in Figures 1–6. In Figure 1, the straight lines in the valleys are negative minima of the principal curvatures and therefore, according to the minima rule, they are part boundaries. When you turn the illustration upside down this reverses figure and ground, so that negative minima and positive maxima of curvature reverse places. Therefore, according to the minima rule, you should see new part boundaries and new parts [5]. A quick check of the illustration will confirm this prediction.

In Figure 2, the half moon on the top right has the same contour as the half moon on the left. Yet most observers pick the half moon on the bottom right as more similar to the half moon on the left, even though its contour is mirror reversed and two of the minima parts have switched positions. This is explained by the minima rule because this rule carves the bottom half moon into parts at the same points as the half moon on the left, whereas it carves the top half moon into different parts [6]. Apparently shape similarity is computed part by part, not point by point.

In Figure 3 the staircase on the right looks inverted, because the parts defined by the minima rule for the inverted interpretation have more salient part boundaries (sharper cusps) than for the upright interpretation. Other things being equal, human vision prefers that choice of figure and ground which leads to the most salient minima-rule parts [7].

In Figure 4 we see transparency on the left but not on the right, even though all the luminances are identical. The reason is that on the right there are minima-rule part boundaries aligned with the luminance boundaries, so we interpret the different grays as different colors of different parts rather than as effects of transparency [8].

In Figure 5 we detect the symmetry on the left more easily than the repetition on the right, an effect noted long ago by Mach [12]. The minima rule explains this because in the symmetric shape the two sides have the same parts, whereas in the repetition shape the two sides have different parts [9]. Again it appears that shapes are compared part by part, not point by point.

In Figure 6 the heart pops out on the left, but the popcorn does not pop out on the right, even though the two have cusps with identical angles [13]. The minima rule explains this because the concave cusp in the heart is a part boundary whereas the convex cusp on the popcorn is not. Minima part boundaries are computed early in the flow of visual processing since parts are critical to the visual representation of shape.

The minima rule makes precise a proposal by Marr and Nishihara [16] that human vision divides shapes into parts at "deep concavities". The minima rule has strong ecological grounding in the principle of transversality from the field of differential topology. This principle guarantees that, except for a set of cases whose total measure is zero, minima part boundaries are formed whenever two separate shapes intersect to form a composite object, or whenever one shape grows out of another [5,7].

3   Other Part Rules

Given the central role of parts in object perception, it is no surprise that several theories of these parts have been proposed. One class of theories claims that human vision uses certain basic shapes as its definition of parts. Proponents of basic-shape theories have studied many alternatives: polyhedra [17–19], generalized cones and cylinders [16,20], geons [21,22], and superquadrics [23]. Of these, the geon theory of Biederman is currently most influential. Geons are a special class of generalized cylinders, and come in 24 varieties. The set of geons is derived from four nonaccidental properties [21,24,25]. These properties are whether (a) the cross section is straight or curved; (b) the cross section remains constant, expands, or expands and contracts; (c) the cross section is symmetrical or asymmetrical; and (d) the axis is straight or curved. These properties are intended to make recognition of geons viewpoint invariant, although some experiments suggests that geon recognition may nevertheless depend on viewpoint [26].

A second class of theories claims that human vision defines parts not by basic shapes but by rules which specify the boundaries between one part and its neighbors. The minima rule is one such theory. Another is the theory of "limbs" and "necks" developed by Siddiqi and Kimia [27], building on their earlier work [28]. They define a limb as "a part-line going through a pair of negative curvature minima with co-circular boundary tangents on (at least) one side of the part-line" ([27], p. 243). Their "part-line" is what I call a "part cut." Two tangents are "co-circular" if and only if they are both tangent to the same circle ([29], p. 829). Siddiqi and Kimia define a neck as "a part-line which is also a local minimum of the diameter of an inscribed circle" ([27], p. 243).

4   Naming Parts

Figure 7a shows a *peen*. After looking at the figure, you know that a peen is the part of a hammer that is shaded gray in Figure 7c. You also are sure that a peen is not the part shaded gray in Figure 7b. This exercise in ostensive definition is easy for us; we discover the meaning of *peen* quickly and without conscious effort. Yet in several respects our performance is striking. When we view the hammer in Figure 7a and guess the meaning of peen, we discard countless possibilities since, as is hinted by Figure 7b, there are countless ways to partition the hammer or any other shape. But despite these countless possibilities we all pick the same part. Deductive logic alone does not compel us to choose a unique part from the figure. Our choice must be constrained by rules in addition to the rules of deductive logic. In principle many different rules can yield a unique choice. But since we all pick the *same* part, it is likely we all use the same rules. What rules then do we use that make us all pick the same part when someone points and names?

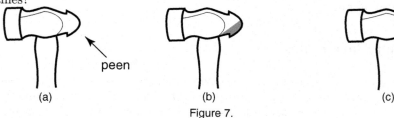

Figure 7.

We propose the following hypothesis:

Minima-Part Bias. When human subjects learn, by ostensive definition, names for parts of objects, they are biased to select parts defined by the minima and short-cut rules.

The motivation for this hypothesis is that the units of representation computed by the visual system and delivered to higher cognitive processes are natural candidates to be named by the language system. I propose that parts defined by the minima rule are such units. As an example, in Figure 8a is a curve, which is ambiguous as to which side is figure and which is ground. If we take the ground to be on the left, as in Figure 8b, then the minima rule gives us the part boundaries that are indicated by the short line segments. As you can see in Figure 8b, each part boundary lies in a region of the curve that is concave with respect to the figure (or, equivalently, convex with respect to the ground), and each part boundary passes through the point with highest magnitude of curvature within its region. If we take the ground to be on the right, as in Figure 8c, then the minima rule gives us a completely different set of part boundaries, indicated by the new set of short line segments. The reason is that switching figure and ground also switches what is convex and concave, and the minima rule says to use only the concave cusps and concave minima of curvature as part boundaries.

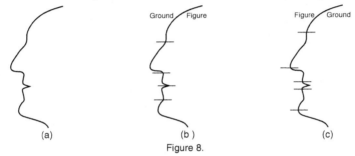

(a)   (b)   (c)

Figure 8.

We can use the curve in Figure 8a to create the well-known face goblet illusion, as shown in Figure 9a.

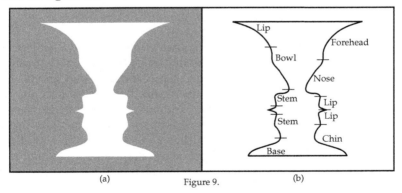

(a)   Figure 9.   (b)

One can see this either as a goblet in the middle or two faces on the sides. If we see the faces as figure, then the minima rule gives the part boundaries shown on the right side of Figure 9b. These part boundaries divide the face, from top to bottom, into a forehead, nose, upper lip, lower lip, and chin, as labeled in the figure. If instead we see the goblet as figure, then the minima rule gives the part boundaries shown on the left side of Figure 9b. These part boundaries divide the goblet, from top to bottom, into a lip, bowl, stem (with three parts), and base. Note that the parts defined by the minima rule are the parts we name. Other parts are not named. For instance, take the bowl part of the goblet and ask what we would call this same part on the face (where it is not a part defined by the minima rule). We would say it is "the lower part of the forehead and the upper part of the nose." This is not a name but a complex description of an unnamed part of the face. Thus the parts that we name on the face and the goblet are precisely the parts derived from the minima rule.

The outline shape of an object can play an important role in its recognition across depth rotations [30]. To test the minima-part bias with such outline shapes, Rodriguez, Nilson, Singh and Hoffman generated five random silhouettes each having five minima parts [31]. On each trial the observer saw one silhouette with an arrow pointing to it. Orientations of the silhouettes were changed from trial to trial. The observer was instructed that the arrow pointed to "a dax" on the shape. The syntax of count nouns was used in these instructions to direct the observer's attention to shape rather than substance interpretations of "a dax" [32–35]. On each trial the observer was given three choices, displayed in random order, for the dax: (1) a minima part, (2) a maxima part, and (3) a convex part cut at inflections. The arrow was placed so that it pointed towards the inflection, in order to minimize possible biases due to the position of the arrow. They found that minima parts were chosen as the dax about 75% of the time, far more frequently than the maxima or inflections.

To further control for possible biases of the arrow position, a second experiment eliminated the arrow. Instead observers were instructed that the shape had a "dax" near its top. On each trial the silhouette was rotated so that the 3 parts of interest (minima, maxima, inflections) were near the top, but never precisely vertical. They again found that minima parts were chosen about 75% of the time.

The minima-part bias makes a striking prediction. Suppose a subject can see a shape undergo reversals of figure and ground. Then the parts of that shape that the subject will name should also change each time figure and ground reverse. The reason is that the minima rule defines part boundaries only at concave regions of an object. Thus when figure and ground reverse so also do concave and convex, so that subjects should see a new set of part boundaries. This prediction was illustrated in Figures 8 and 9. Rodriquez et al. tested this prediction of the minima-part bias, using a simple method to induce a figure-ground reversal: global reversal of contrast together with enclosure [31]. Subjects viewed a shape, with an arrow pointing toward a "dax" on the shape, and picked the part that looked most natural to be the "dax". On a different trial they saw precisely the same shape, with the arrow pointing in exactly the same way, but with reversed contrast. This reversed contrast induced subjects to reverse figure and ground. In this case the minima-part bias predicted that they would pick a different part, one with the new minima of curvature for its part boundaries, as the most natural "dax". The shapes used were five random curves each having three or four minima parts. Observers chose between minima, maxima, and inflection options as before. The results were again as predicted by the minima-rule bias.

Rodriguez et al. also devised at test in which the predictions of the geon theory differ dramatically from those of the minima-rule bias [31]. They used a two-alternative forced-choice paradigm in which subjects saw two objects side by side and were told that one of the objects had a dax on it. Subjects had to choose which of the two objects had the dax on it. For most pairs of objects that were shown to the subjects, the minima rule and geon theory predicted opposite choices. Each of the objects was composed of two shapes. The first, the base shape, was either an elongated box or a cylinder. The second was one of twelve shapes, six of which were geons and six of which were nongeons. The geons were (1) a curved box or cylinder, (2) a tapered box or cylinder, and (3) a curved and tapered box or cylinder. The six nongeons were created from the geons by smoothly changing the cross section from square to circular or vice versa as it sweeps along the axis. Each of these twelve shapes was attached to it base shape in one of two ways: (1) with a minima part boundary at the attachment or (2) with no minima boundary at the attachment. This led to a total of 18 composite objects. Thus there were four types of objects, defined by the type and attachment of the second shape. These were (1) geons with minima, (2) geons without minima, (3) nongeons with minima, and (4) nongeons without minima. We can label these, respectively, +G+M, +G-M, -G+M, and -G-M. In a two-alternative forced-choice paradigm there are $\binom{4}{2} = 6$ ways of presenting pairs of these objects. These ways are listed below, together with the predictions of the geon theory and the minima rule as to which of the two objects is most likely to be chosen as the one having the dax.

CASE	OBJECT 1	OBJECT 2	GEON	MINIMA
1	+G+M	+G-M	same	object 1
2	+G+M	-G+M	object 1	same
3	+G+M	-G-M	object 1	object 1
4	+G-M	-G+M	object 1	object 2
5	+G-M	-G-M	object 1	same
6	-G+M	-G-M	same	object 1

As you can see from this table, in five of the six cases the minima rule and the geon theory make different predictions. Only in the third case do their predictions agree. Rodriquez et al. used all cases in a two-alternative forced-choice paradigm to test which theory correctly predicts observers' choices. They found that where the predictions disagreed, each subject chose overwhelmingly in accord with the minima bias and not in accord with the geon theory.

In a final experiment, Rodriguez et al. found that subjects also use the minima-part bias in generalizing the names of parts. On each trial they showed a subject an object with a single attached part, and told the subject that the object had a "dax" on it. Then they showed the subject that same object, but with the part transformed either by translation, scaling, or both translation and scaling. On some trials a minima part was transformed, on others a maxima part, and on others an inflection part. Subjects had to decide if this new object also had a "dax" on it. Subjects did generalize the part name to transformed minima parts, but not to transformed maxima or inflection parts. This indicates that the minima-rule bias guides both the initial attachment of names to parts, and the generalization of part names to new parts.

5   Conclusion

Visual form is not available at the retina, but must be constructed by tens of billions of neurons in the visual system. The description of visual form requires the visual system to carve the visual world into objects, and to carve these objects even more finely into parts. Human vision apparently does this by constructing minima-rule part boundaries early in the course of visual processing. These boundaries, together with geometric rules for constructing part cuts which use these boundaries, leads to an articulation of objects into parts. The potential shapes of these parts are countless, and are not limited to a preordained set such as geons or generalized cylinders. These parts are among the fundamental units of visual description that are delivered to higher cognitive processes, including language. As a result, in language acquisition humans employ the minima-part bias when learning the names of parts by ostensive definition. The minima-part bias has so far only been tested in adults. It will be of interest to see if young children also employ this bias when learning names for parts.

References

1. Tyler, C.W.: Analysis of Human Receptor Density. In: Lakshminarayanan, V. (ed.): Basic and Clinical Applications of Vision Science. Kluwer Academic, Norwell, Massachusetts (1997) 63–71
2. Bower, G.H., Glass, A.L.: Structural Units and the Redintegrative Power of Picture Fragments. J. Exper. Psych. **2** (1976) 456–466
3. Braunstein, M.L., Hoffman, D.D., Saidpour, A.: Parts of Visual Objects: An Experimental Test of the Minima Rule. Perception **18** (1989) 817–826
4. Palmer, S.E.: Hierarchical Structure in Perceptual Representation. Cognit. Psychol. **9** (1977) 441–474
5. Hoffman, D.D., Richards, W.A.: Parts of Recognition. Cognition **18** (1984) 65–96
6. Hoffman, D.D.: The Interpretation of Visual Illusions. Sci. Am. **249** 6 (1983) 154–162
7. Hoffman, D.D., Singh, M.: Salience of Visual Parts. Cognition **63** (1997) 29–78
8. Singh, M., Hoffman, D.D.: Part Boundaries Alter the Perception of Transparency. Psych. Sci. **9** (1998) 370–378
9. Baylis, G.C., Driver, J.: Obligatory Edge Assignment in Vision: The Role of Figure and Part Segmentation in Symmetry Detection. J. Exp. Psych.: Hum. Percept. Perform. **21** (1995) 1323–1342
10. Bruce, V., Morgan, M.J.: Violations of Symmetry and Repetitions in Visual Patterns. Perception **4** (1975) 239–249

11. Hoffman, D.D.: Visual Intelligence: How We Create What We See. W.W. Norton, New York (1998)
12. Mach, E.: The Analysis of Sensations, and the Relation of the Physical to the Psychical. Dover, New York (1885/1959) (Translated by Williams, C.M.)
13. Hulleman, J., te Winkel, W., Boselie, F.: Concavities as Basic Features in Visual Search: Evidence from Search Asymmetries. Percept. Psychophys. **62** (2000) 162–174
14. Wolfe, J.M., Bennett, S.C.: Preattentive Object Files: Shapeless Bundles of Basic Features. Vis. Res. **37** (1997) 24–43
15. Singh, M., Seyranian, G., Hoffman, D.D.: Parsing Silhouettes: The Short-Cut Rule. Percept. Psychophys. **61** (1999) 636–660
16. Marr, D., Nishihara, H.K.: Representation and Recognition of Three-Dimensional Shapes. Proc. Roy. Soc. Lond. B, **200** (1978) 269–294
17. Roberts, L.G.: Machine Perception of Three-Dimensional Solids. In: Tippett, J.T. et al. (eds.): Optical and Electrooptical Information Processing. MIT Press, Cambridge, Massachusetts (1965) 211–277
18. Waltz, D.: Generating Semantic Descriptions from Drawings of Scenes with Shadows. In: Winston, P. (ed.): The Psychology of Computer Vision. McGraw-Hill, New York (1975) 19–91
19. Winston, P.A.: Learning Structural Descriptions from Examples. In: Winston, P. (ed.): The Psychology of Computer Vision. McGraw-Hill, New York (1975) 157–209
20. Brooks, R.A.: Symbolic Reasoning among 3-D Models and 2-D Images. Artific. Intell. **17** (1981) 205–244
21. Biederman, I.: Recognition-By-Components: A Theory of Human Image Understanding. Psych. Rev. **94** (1987) 115–147
22. Biederman, I.: Higher Level Vision. In: Osherson, D., Kosslyn, S., Hollerbach, J. (eds.) An Invitation to Cognitive Science, Volume 2. MIT Press, Cambridge, Massachusetts (1990) 41–72
23. Pentland, A.P.: Perceptual Organization and the Representation of Natural Form. Artific. Intell. **28** (1986) 293–331
24. Lowe, D.: Perceptual Organization and Visual Recognition. Kluwer, Amsterdam (1985)
25. Witkin, A.P., Tenenbaum, J.M.: On the Role of Structure in Vision. In: Beck, J., Hope, B., Rosenfeld, A. (eds.) Human and Machine Vision. Academic Press, New York (1983) 481–543
26. Tarr, M.J., Williams, P., Hayward, W.G. Gauthier, I.: Three-Dimensional Object Recognition Is Viewpoint Dependent. Nature Neurosci. **1** (1998) 275–277
27. Siddiqi, K., Kimia, B.B.: Parts of Visual Form: Computational Aspects. IEEE PAMI **17** (1995) 239–251
28. Kimia, B.B., Tannenbaum, A.R., Zucker, S.W.: Entropy Scale-Space. Plenum Press, New York (1991) 333–344
29. Parent, P., Zucker, S.W.: Trace Inference, Curvature Consistency, and Curve Detection. IEEE PAMI **11** (1989) 823–839
30. Hayward, W.G., Tarr, M.J., Corderoy, A.K.: Recognizing Silhouettes and Shaded Images Across Depth Rotation. Perception **28** (1999) 1197–1215
31. Rodriguez, T., Nilson, C., Singh, M., Hoffman, D.D.: Word and Part. (under review).
32. Imai, M., Gentner, D.: A Cross-Linguistic Study of Early Word Meaning: Universal Ontology and Linguistic Influence. Cognition **62** (1997) 169–200
33. Landau, B., Smith, L.B., Jones, S.: Syntactic Context and the Shape Bias in Children's and Adults' Lexical Learning. J. Mem. Lang. **31** (1992) 807–825

34. Soja, N.N., Carey, S., Spelke, E.S.: Ontological Categories Guide Young Children's Inductions of Word Meanings: Object Terms and Substance Terms. Cognition **38** (1990) 179–211
35. Subrahmanyam, K., Landau, B., Gelman, R. Shape, Material, and Syntax: Interacting Forces in Children's Learning in Novel Words for Objects and Substances. Lang. Cognit. Proc. **14** (1999) 249–281

On Matching Algorithms for the Recognition of Objects in Cluttered Background

Josef Kittler and Alireza Ahmadyfard

Centre for Vision, Speech and Signal Processing
University of Surrey
Guildford, GU2 7XH, UK
J.Kittler@surrey.ac.uk
A.Ahmadyfard@surrey.ac.uk

Abstract. An experimental comparative study of three matching methods for the recognition of 3D objects from a 2D view is carried out. The methods include graph matching, geometric hashing and the alignment technique. The same source of information is made available to each method to ensure that the comparison is meaningful. The experiments are designed to measure the performance of the methods in different imaging conditions. We show that matching by geometric hashing and alignment is very sensitive to clutter and measurement errors. Thus in realistic scenarios graph matching is superior to the other methods in terms of both recognition accuracy and computational complexity.

1   Introduction

Object recognition is one of the crucial tasks in computer vision. A practical solution to this problem would have numerous applications and would greatly impact on the field of intelligent robotics. In this paper we are concerned with the problem of finding instances of 3D objects using a single 2D image of the scene. A frontal image of each object is used as the object model.

In a model based object recognition there are two major interrelated problems, namely that of object representation and the closely related problem of object matching. A number of representation techniques have been proposed in the computer vision literature which can be broadly classified into two categories: feature based, and holistic (appearance based). We shall not be dismissive of the appearance based approaches as they possess positive merits and no doubt can play a complementary role in object recognition. However, our motivation for focusing on feature based techniques is their natural propensity to cope better with occlusion and local distortion.

The matching process endeavours to establish the correspondence between the features of an observed object and of a hypothesised model. This invariably involves the determination of the object pose. The various object recognition techniques proposed in the literature differ in the way the models are invoked and verified. The techniques range from the alignment methods [4] [5] where the hypothesised interpretation of image data and the viewing transformation is

C. Arcelli et al. (Eds.): IWVF4, LNCS 2059, pp. 51–66, 2001.
© Springer-Verlag Berlin Heidelberg 2001

based on the correspondence of a minimal set of features. The candidate interpretation and pose are then verified using other image and model features. The
other end of the methodological spectrum is occupied by geometric hashing[7]
[6] or hough transform methods [10] where all the scene features are used jointly
to index into a model database. This approach is likely to require a smaller
number of hypothesis verifications. However its success is largely dependent on
the ability reliably to extract distinctive features. The verification stage in both,
alignment and geometric hashing methods involves finding a global transformation between scene object and model.

In contrast, the philosophy behind graph matching methods is to focus on local consistency of interpretation [1][2]. In an earlier work [3] we developed a
recognition method in which test and model image regions are represented in
the form of relational attributed graphs. The graph matching is accomplished
by the probabilistic relaxation labelling method [11] which has been adapted for
this application.

The aim of this paper is carry out an extensive experimental comparative study
of the geometric hashing, alignment and graph matching methods. The same
source of information is made available to each method to ensure that the comparison is meaningful. The experiments are designed to measure the performance
of the methods in different imaging conditions. The experimental results show
that matching by geometric hashing and alignment is very sensitive to clutter
and measurement errors. We shall argue that the success of the graph matching
approach stems from the fact that the model and scene images are compared by
considering local matches. This prevents the propagation of errors throughout
the image. In contrast, error propagation plagues the alignment and geometric
hashing methods which perform the scene/image model comparison in a global
coordinate system. Thus we shall show that, in realistic scenarios, graph matching is superior to the other methods in terms of both recognition accuracy and
computational complexity.

We begin by reviewing the methods to be compared. In section 3 we describe
the experiments designed for the comparison of the methods. The results of the
experiments are reported in section 4. Finally we draw the paper to conclusion
in the last section.

2   Methodology

In order to make a comparative study of matching approaches meaningful it is
essential that all methods use the same information for the object comparison.
Thus in our study all methods deploy identical object representation. In this
regard we consider an object image, which either serves as an object model or
captures a scene containing unknown object(s) to be interpreted, as a collection
of homogeneous planar regions obtained by segmenting the input image. In the
former case we refer to it as the model image whereas in the latter we call it a test
image. In our representation each region of the image is described using region
colour (in the YUV system), region area, region centroid and a number of high

curvature points extracted from the region boundary. Furthermore the boundary of each region is used to characterise its shape. In the following subsections we explain how each recognition method employs the given information.

2.1   Geometric Hashing and Alignment

In geometric hashing and alignment the recognition task is considered as the problem of finding the best transformation which matches object features in the test image to the corresponding features in the object model. The transformation parameters are computed using a minimum number of corresponding features in the model and test image planes. Assuming the transformation is affine, it can be determined from the knowledge of three corresponding points in the respective test and model image planes. The two methods differ in the way the model-test triplets are selected to generate transformation hypotheses.

Alignment Method: In the alignment method the test image is aligned against a candidate model image using all the possible pairs of model-test triplets. Each of these pairs defines a transformation between the two planes associated with the triplets[9]. The validity of each generated hypothesis is initially assessed at a coarse level by quantifying the goodness of match between a small set of features of the candidate model and the test image. A large number of hypotheses will be eliminated at this stage. In the fine verification stage the transformation, which exhibits the highest degree of match between all the model image features and the test image features, is considered as the solution.

It is apparent that because of the potentially large number of possible combinations of test and model image triplets the process is very time consuming. We take advantage of our region-based representation to reduce the number of candidate triplets. For this purpose, we consider only those combinations of interest points which are associated with the same region. A further reduction of the number of model-test triplets is achieved by filtering out those pairs belonging to regions with considerable colour difference.

In the coarse verification stage we measure the average Euclidean distance between the transformed centroid coordinates of the corresponding model and test image regions. Furthermore, the number of region correspondences is registered as another matching criterion.

We consider a pair of model-test image regions to match if the differences in the colour and area measurements and also in the centroid coordinates of the regions fall within a predetermined threshold. Note that for this evaluation we map the model features to the test image. Any candidate transformation which provides a matching distance below the predetermined threshold, with the number of corresponding regions exceeding the required minimum, passes this pruning stage.

In the fine verification stage for each hypothesised transformation we measure the average Euclidean distance between the boundary samples of a transformed model region and the corresponding test image region. This measurement is made by searching for the minimum distance between the two strings of samples. The search involves considering all the possible shifts between the strings.

The transformation between the test image and a candidate model, which yields a sufficiently high number of matches, will be accepted as a solution; otherwise the procedure continues by selecting another triplet from the test image. In case no such transformation between the test image and a candidate model is found other models would be considered in turn.

Geometric Hashing Method: Geometric hashing paradigm is based on an intensive off-line model processing stage, where model information is indexed into a hash-table using a minimum number of transformation invariant features referred to as a basis. In the recognition phase the features in the test image are expressed in a similar manner using bases picked from the test image. The comparison of the invariant coordinates of model and test feature points in the hash table leads to finding the best pair of bases which provides the maximum number of matches between the model and test feature points.

Let us consider the geometric hashing method[8] in more detail. Taking the three points, $\mathbf{e}_{00}$, $\mathbf{e}_{01}$ and $\mathbf{e}_{10}$, as a basis, the affine coordinates (ζ, η) of an arbitrary point, $\mathbf{v}$, can be expressed using the following formula:

$$\mathbf{v} = \zeta(\mathbf{e}_{10} - \mathbf{e}_{00}) + \eta(\mathbf{e}_{01} - \mathbf{e}_{00}) + \mathbf{e}_{00}$$

One of the important features of such a coordinate system is its invariance to affine transformations. Consider an arbitrary point and a basis from the same plane which are transformed by an affine transformation. The following expression shows that the affine coordinate of the point remains invariant:

$$\mathbf{Tv} = \zeta(\mathbf{Te}_{10} - \mathbf{Te}_{00}) + \eta(\mathbf{Te}_{01} - \mathbf{Te}_{00}) + \mathbf{Te}_{00}$$

Using this property the model information can be represented in an invariant form. For each model image and for each ordered non-collinear triple of the feature points the coordinates of all other model points are computed taking this triplet as an affine basis of the 2D plane. Each such coordinate is used as an entry to a hash table, where the identity of the basis in which the coordinate was obtained and the identity of the model (in case of more than one model) are recorded. In the recognition phase the given test image is represented in the same way. An arbitrary triplet of test image feature points is chosen to define a basis and affine coordinates of all the other scene interest points are computed based on this basis. For each coordinate the appropriate entry of the hash-table is checked and a vote for the basis and the model identity in the corresponding entry is noted. After the voting process the hash table is inspected to find the model-bases with a large number of votes. Such model-bases define hypothesised transformations between the test image and a specific model image. In the verification stage the model features are verified against the test image features by applying each of the candidate transformations. If the verification fails the algorithm will continue by picking another triplet from the test image. We implemented a recognition system based on the above geometric hashing algorithm. In this system we apply the same pruning strategy as in the alignment method to reduce the number of image bases. The verification of a hypothesised transformation is also carried out in the same way as in the fine verification stage of the alignment method.

2.2   Attributed Relational Graph Matching

In the graph-based method [3] image information is represented in the form of an attributed relational graph. In particular all the information associated with the object models is represented in a model graph and test image information is represented in a test graph. In these graphs each node characterises a region described in terms of measurements referred to as unary measurements. The affine invariant representation is achieved by presenting the regions in their normalised form. Furthermore, we measure the relation between a pair of regions using binary measurements and represent them as links between graph nodes. In the recognition phase we match the test graph against the model graph using the probabilistic relaxation labelling technique. In the following, we explain the adopted representation and the recognition process in more detail.

Graph Representation: We start by addressing the problem of finding an affine transform that maps the region R to region r in a normalised space where the transformed regions should be comparable.
In order to define such a transform uniquely we impose the following constraints on it:

1. the reference points (x_0, y_0) and (x_1, y_1) of the region R are to be mapped to points $(1, 0)$ and $(0, 0)$ of r respectively.
2. the normalised region r is to have a unit area and the second order cross moment equal to zero.

To simplify the transformation task we split it into two sub-tasks. First, using a similarity transformation matrix $\mathbf{T}_s$, the reference points (x_0, y_0) and (x_1, y_1) of R are mapped to points $(1, 0)$ and $(0, 0)$ in the new region R_0 respectively. The matrix $\mathbf{T}_s$ can readily be shown to be:

$$\mathbf{T}_s = \frac{1}{k_n} \begin{pmatrix} x_1 - x_0 & y_0 - y_1 & 0 \\ y_1 - y_0 & x_1 - x_0 & 0 \\ x_0^2 + y_0^2 - y_0 y_1 - x_0 x_1 & x_0 y_1 - y_0 x_1 & k_n \end{pmatrix} \tag{1}$$

where $k_n = (x_1 - x_2)^2 + (y_1 - y_2)^2$ is the distance of the two reference points. Second, we determine the affine transform, $\mathbf{T}_a$, that modifies the new region, R_0, to the normalised region, r. Such a matrix can be calculated taking into account the relations between the second order moments of R_0 and r :

$$\mathbf{T}_a = \begin{pmatrix} 1 & 0 & 0 \\ -u_{1,1}/u_{0,2} & k & 0 \\ 0 & 0 & 1 \end{pmatrix} \tag{2}$$

where $u_{1,1}, u_{0,2}$ are the second order moments of the normalised region r and $k = k_n/(Area\ of\ region\ R)$. The transformation matrix that maps region R to the normalised region, r, will then be given as $\mathbf{T} = \mathbf{T}_s\mathbf{T}_a$.

Let us define $\mathbf{B} = \mathbf{T}^{-1}$. The affine invariant coordinates of an arbitrary point P of region R can be defined as :

$$C_{\mathbf{B}}(P) = P\mathbf{B}^{-1} \tag{3}$$

It has been shown that matrix $\mathbf{B_{ij}} = \mathbf{B}_i\mathbf{B}_j^{-1}$, associated with a pair of regions R_i and R_j, is an affine invariant measurement[3]. We refer to this matrix as the binary relation matrix.

In order to normalise a region we consider its centroid as one of the required reference points. The way the second reference point is selected is different for the test image and the model. In the case of the model the highest curvature point on the boundary of the region is chosen as the second reference point, while in the test image for each region a number of points of high curvature are picked and consequently a number of representations for each test region are provided. The selection of more than one point on the boundary is motivated by the fact that an affine transformation may change the ranking and distort the position of high curvature points.

We represent all of the model images in a common graph and refer to it as the model graph. Suppose that the model graph contains M nodes (normalised regions). Then $\Omega = \{\omega_1, \omega_2, \cdots, \omega_M\}$ denotes the set of labels for the nodes which define their identity. Each node ω_i is characterised by a measurement vector $\breve{x}_i$. $\breve{\mathbf{X}} = \{\breve{x}_1, \breve{x}_2, \cdots, \breve{x}_M\}$ is the set of these unary measurement vectors. Let $\breve{\mathbf{N}}_i$ be the index set of nodes neighbouring node i. The set of relational measurements between a pair of the graph nodes is denoted $\breve{\mathbf{A}} = \{\breve{A}_{ij} | (i,j) i \in \{0, \cdots, M\}, j \in \breve{N}_i\}$. As a unary attribute vector for node, ω_i, we take vector $\breve{x}_i = \left[S^T, C^T \right]^T$ where the components of vector S are the coordinates of a number of equally spaced samples on the boundary of the ith normalised region. Vector C is a representative of the region colour. The binary measurement vector $\breve{A}_{ij}$ associated with the pair of nodes, ω_i, and, ω_j, is defined as follows:

$$\breve{A}_{ij} = \left[b_{ij_{mn}}(m = 1,2; n = 1,2), ColorRalation_{ij}{}^T, AreaRalation_{ij} \right]^T$$

where $b_{ij_{mn}}$ is the (m,n)th element of the binary relation matrix $\mathbf{B_{ij}}$ associated with the region pair. In this vector $ColorRelation_{ij}$ and scalar $AreaRelation_{ij}$ express the colour relations and area ratios of the region pair respectively.

Similarly, we capture the test image information in a test graph where $\mathbf{a} = \{a_1, a_2, \cdots, a_N\}$ is the set of test graph nodes and, $\mathbf{X}$, and, $\mathbf{A}$, denote the set of unary and binary measurements respectively. The identity of each graph node, a_i, is denoted by θ_i. Recall that the representation of the test image differs from that of the model in the sense that more than one bases are provided for each test region. The multiple representation for each test node is defined in terms of a set of unary measurement vectors, x_i^k, with index k indicating that the vector is associated with the kth representation of the ith node. To each neighbouring pair of regions, a_i, a_j, we associate binary relation vectors A_{ij}^{kl} with semantically identical components to those of the model binary vectors. The multiple unary measurement vectors x_i^k and binary relations A_{ij}^{kl} constitute the combined unary and binary relation representation $\mathbf{X}_i = \{x_i^k | k \in \{1, \cdots, L\}\}$ and $\mathbf{A}_{ij} = \{A_{ij}^{kl} | k, l \in \{1, \cdots, L\}\}$ where, L, denotes the number of representations used for the test regions.

Graph Matching: For matching we have adopted the relaxation labelling technique of [11] and adapted it to our task. The problem considered here is much more complex than in previous applications of relaxation methods due to a large number of nodes in the model graph. Similarly to [11] we add a null label to the label set to reduce the probability of incorrect labelling but we try to neutralise the support of neighbouring objects when they take the null label. As we discuss later this allows us better to cope with the matching problem.

We divide the matching process into two stages: first, finding the best representation of object under a particular label hypothesis and second, updating the label probabilities by incorporating contextual information. In the first stage, we compare the unary attribute measurements of each object, a_i, of the test image with the same measurements for all its admissible interpretations and construct a list, $\mathbf{L}_i$, containing the labels which can be assigned to the object. Simultaneously for each label in this list we find the best representation. For assessing each representation we measure the mean square distance between the normalised region boundary points stored in a vector, $\mathbf{S}$, and the unary attribute vector of the hypothesised label. In other words the merit of kth representation of object a_i in the context of the assignment of label ω_α to that object is evaluated as:

$$E(\theta_i = \omega_\alpha) = \min_k (x_i^k[S] - \breve{x}_\alpha[S])^2$$

In ideal conditions, for the correct basis, the above measurement for the corresponding regions would be zero. However, due to errors in the extraction of the reference points and as a result of the segmentation noise affecting the boundary pixel positions this measurement is subject to errors. Thus criterion function E is compared against a predefined threshold. A label is entered in the list of hypotheses only if the measurement value is less than the threshold. The best representation basis, k, is also recorded in the node label list. At the end of this process we have a label list for each object with the best representation for each label in the list. Hence we do not need to distinguish between different representations by superscript indices on the unary and binary vectors. Instead we index them with a star to indicate that the best representation is being considered.

In the second stage, we consider the possible label assignments for each object, a_i, and iteratively update the probabilities using their previous values and supports provided by the neighbouring objects. In addition to the above pruning measures, at the end of each iteration we eliminate the labels the related probabilities of which drop below a threshold value. This will make our relaxation method faster and more robust as well. Indeed the updating of probabilities of unlikely matches not only takes time but also increases the probability of incorrect assignment due to increased entropy of the interpretation process which is a function of not only probability distribution but also of the number of possible interpretations.

Returning to our problem of assigning labels from label set Ω to the set of objects in the test graph, let $p(\theta_i = \omega_{\theta_i})$ denote the probability of label ω_{θ_i} being the correct interpretation of object a_i. Christmas et al[11] developed a theoretical underpinning of probabilistic relaxation using a Bayesian framework. They derived a formula which iteratively updates the probabilities of different labels an object can take using contextual information measured in a neighbourhood of the object. To overcome the problem of inexact matching a null label ω_0 is added to the label set for assigning to the objects for which no other labels are appropriate[11] .

The probability that the object a_i takes label ω_{θ_i} is updated in the $(n + 1)$ st step of the algorithm using the support provided from the neighbouring objects:

$$p^{(n+1)}(\theta_i = \omega_{\theta_i}) = \frac{p^{(n)}(\theta_i = \omega_{\theta_i})Q^{(n)}(\theta_i = \omega_{\theta_i})}{\sum_{\omega_\lambda \in \mathbf{L}_i} p^{(n)}(\theta_i = \omega_\lambda)Q^{(n)}(\theta_i = \omega_\lambda)} \tag{4}$$

$$Q^{(n)}(\theta_i = \omega_\alpha) = \prod_{j \in N_i} \sum_{\omega_\beta \in \mathbf{L}_j} p^{(n)}(\theta_i = \omega_\beta)p(A_{ij}^*|\theta_i = \omega_\alpha, \theta_j = \omega_\beta) \tag{5}$$

where function Q quantifies the support that assignment $(\theta_i = \omega_\alpha)$ receives at the nth iteration step from the neighbours of object i in the test.

The summation in the support function measures the average consistency of the labelling of object a_j (a neighbour object of a_i) in the context of the assignment, $(\theta_i = \omega_{\theta_i})$, on object a_i. In each term of this summation the consistency of a particular label on object a_j is evaluated. The compatibility of the labels is measured in terms of the value of the distribution function of binary relation measurements between a_i and a_j. In [11] the consistency of the neighbouring object a_j even when it takes the null label is involved in the support computation. As a result if the consistency of other labels on a_j is low(which will frequently happen) the contribution to the support provided by the null label will be dominant and cause an undesirable effect. Since the assignment of the null label to a_j does not provide any relevant information for the labelling of object a_i its support should be neutral. The undesirable behaviour of the null label has been confirmed in all experiments. It is frequently manifest in incorrect labelling. To avoid this problem we set the support term related to the null label to a constant value. Because of the product operator in the definition of support function Q this constant cannot be set to zero.

In the first step of the process the probabilities are initialised based on the unary measurements. Denote by $p^{(0)}(\theta_i = \omega_{\theta_i})$ the initial label probabilities evaluated using the unary attributes as:

$$p^{(0)}(\theta_i = \omega_{\theta_i}) = p(\theta_i = \omega_{\theta_i}|x_i^*) \tag{6}$$

Applying the Bayes theorem we have :

$$p^{(0)}(\theta_i = \omega_{\theta_i}) = \frac{p(x_i^*|\theta_i = \omega_{\theta_i})p(\theta_i = \omega_{\theta_i})}{\sum_{\omega_\alpha \in \mathbf{L}_i} p(x_i^*|\theta_i = \omega_\alpha)p(\theta_i = \omega_\alpha)} \tag{7}$$

with the normalisation carried out over labels in the label list $\mathbf{L}_i$. Let ζ be the proportion of test nodes that will assume the null label. Then the prior label probabilities will be given as :

$$p(\theta_i = \omega_\lambda) = \begin{cases} \zeta & \lambda = 0 \ (null \ label) \\ \frac{1-\zeta}{M} & \lambda \neq 0 \end{cases} \tag{8}$$

where M is the number of labels (model nodes).

We shall assume that the errors on unary measurements are statistically independent and their distribution function is Gaussian i.e.

$$p(x_i^*|\theta_i = \omega_\alpha) = \mathcal{N}_{x_i^*}(\breve{x}_\alpha, \Sigma_u) \tag{9}$$

where Σ_u is a diagonal covariance matrix for measurement vector x_i^*. In the support function, Q, the term $p(A_{ij}^*|\theta_i = \omega_\alpha, \theta_j = \omega_\beta)$ behaves as a compatibility coefficient in other relaxation methods. In fact it is the density function for the binary measurement A_{ij}^* given the matches $\theta_i = \omega_\alpha$ and $\theta_j = \omega_\beta$. Similarly, the distribution function of binary relations is centred on the model binary measurement $\breve{A}_{\alpha\beta}$. It is assumed that deviations from this mean are also modelled by a Gaussian. Thus we have:

$$p(A_{ij}^*|\theta_i = \omega_\alpha, \theta_j = \omega_\beta) = \mathcal{N}_{A_{ij}^*}(\breve{A}_{\alpha\beta}, \Sigma_b) \tag{10}$$

where Σ_b is the covariance matrix of the binary measurement vector A_{ij}^* .

The iterative process will be terminated in one of the following circumstances:

1. If in the last iteration none of the probabilities changed by more than threshold ϵ.
2. If the number of iterations reached some specified limit

3   Experiments

In this section we describe the experiments carried in order to compare the geometric hashing, alignment and graph-based methods. The aim of the experiments is to compare the recognition methods from the point of view of recognition performance and the processing time. We test the matching methods on two different databases: groceries database and traffic signs database. In each database a number of images containing the objects of interest are used as the test images and a frontal image of each object is used as the object model.

In the first experiment we test the methods on the groceries database. The database contains images of 24 objects taken under controlled conditions. Each object is imaged from 20 different viewing angles ranging from -90 to $+90$ degrees. The set of 20 images for one of the objects is shown in Fig 1. Fig 2 shows

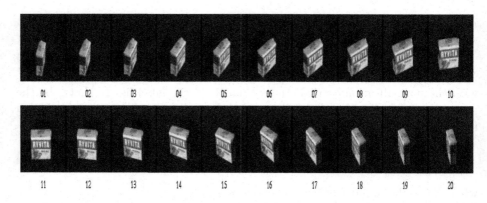

Fig. 1. The images of an object of the database taken from different viewing angels

object models for this database. Note that the resolution of the model images is twice that of the test images. This difference in resolution is designed to assess the ability of the methods to match objects under considerable scaling. The recognition rate is measured for the viewing ranges of 45 (pose 9, 10, 11) and 90 degrees (pose 6 to 15) separately.

The second experiment is designed to compare the performance of the matching methods in sever clutter conditions, considerable illumination changes and scaling. We have chosen a traffic sign database for this purpose. It consists of sixteen outdoor images of traffic signs shown in Fig 3 which are regarded as the test images. The object models are shown in Fig 4. The performance is measured in terms of recognition rate.

In each experiment in order to extract the required information, the test and model images are first segmented using a region growing method. For each segmented region, colour, area, centroid coordinates and the coordinates of the boundary points are extracted. The region feature points are obtained by measuring the curvature along the boundary. The curvature points are recorded in the decreasing order of curvature. In the first experiment on the traffic signs database we extract seven interest points for each region. The experiment has shown that even in the case of sever affine transformation this number of features is sufficient to obtain a model-test triplet match between the corresponding regions. In any case to find out how the number of feature points affects the performance of the individual methods we have experimented with different values of this parameter on the groceries database.

4   Results

The results of the first experiments are reported in Fig 5. Fig 5(a) shows the recognition rate versus the number of region feature points. For each method at a specific number of feature points we provide two measures: the recognition

Fig. 2. Boxes database

rate for pose 9 to 11 and the recognition rate for pose 6 to 15. We draw the attention to the most important points which emerged from these results.

As expected the alignment method performs better than geometric hashing from the recognition rate point of view. In contrast to geometric hashing the method does not miss any proper transformation during the coarse verification stage. However the graph matching method performs best.

By increasing the number of feature points the performance of the geometric hashing and alignment methods improves faster than that of the graph-based

Fig. 3. The test images

Fig. 4. The traffic signs as the library of model images

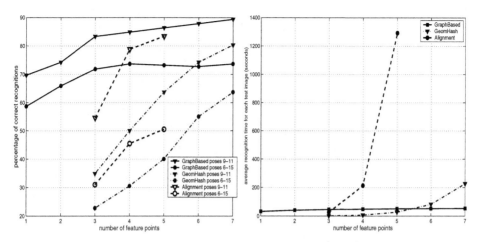

(a) Recognition rates versus the number of feature points for each region

(b) Processing times versus the number of feature points for each region

Fig. 5. The performances of the methods versus the number of feature points

Table 1. The result of the experiments on the traffic sign objects for different recognition methods

Performance	GeomHashing	Alignment	Graph-Based
Correct Recog	50%	62.5%	88.89%

method. In fact the performance of the geometric hashing and alignment methods totally depends on the accuracy of the feature points coordinates. Using more feature points increases the probability of obtaining more accurate model-test triplet pairs and consequently the likelihood of recognising the object. For two reasons the graph matching method is not affected as much as the other methods by this problem. First of all, we use feature points to normalise each region individually hence the errors in feature point coordinates are not propagated throughout the test image. Second, we use contextual information to moderate local imperfections caused by inaccurate region normalisation.

As Fig 5(b) shows the recognition time for the alignment and geometric hashing method is an exponential function of the number of feature points. The alignment recognition time is even worse than the geometric hashing method due to the lack of an effective candidate pruning strategy. In the graph-based method the number of feature points affects only the first part of the algorithm in which the best representation for each region is selected. In other words an increase in the number of feature points does not increase the graph complexity and consequently the matching time remains constant.

We applied the recognition methods to the traffic sign image database. The correct recognition rate is given in Table 1. There are two main reasons for the poor performance of geometric hashing: an excessively high count of false votes generated by the clutter, and different complexity of the objects of interest. The clutter in an image produces a large number of features which potentially increase the number of false votes for each model-triplet. Equally, since objects may be of very different complexity there are considerable differences in the number of feature points among the object models. Thus during hashing it is more likely to give a false vote to a complex object than to simple ones.

To illustrate this problem we plot feature points of a test image in a hash table for two different candidates of model-test triplets. In the first case the triplet by which the test feature points are normalised has chosen the corresponding triplet of the relevant model Fig 6(a). Each square represents a bin in the hash table. It is centred on a normalised feature of the model and each dot represents a normalised test feature point. The bin size is determined by error tolerance for the corresponding feature points. In the second case we consider test image feature points against the feature points of an irrelevant model. The test and model images are normalised using an arbitrary pair of model-test triplets. As can be seen in Fig 6(a) the number of feature points (squares) in the relevant model is considerably lower than for the irrelevant model shown in Fig 6(b). As a result the number of votes given to the incorrect triplet-model(case 2) is more than two times the number of votes given for the correct triplet-model(case 1). In the alignment method this problem with the selection of candidates does not arise but we spend much time verifying a large number of candidates. Although we use colour information to reduce the number of possible model-test triplets, because of the considerable illumination changes in the images of the database the colour constraints cannot effectively be applied. The results show that the recognition rate for the alignment system is better than for geometric hashing but it is not comparable to graph-based method.

5   Discussion and Conclusion

An extensive experimental comparative study of three distinct matching methods for the recognition of 3D objects from a 2D view was carried out. The methods investigated include graph matching, geometric hashing and the alignment technique.

The experiments were conducted on different databases in order to test the methods under different conditions. The results of the experiments demonstrated the superiority of the graph matching method over the other two methods in terms of both recognition accuracy and the speed of processing. Most probably the success of the graph matching approach owes to the fact that the model and test images are compared by considering local matches between the normalised image regions. This is in contrast to the alignment and geometric hashing methods which measure the distance between the features of the two images in a global coordinate system. A crucial advantage of the graph-based matching method is that the global match between object and the model is defined in terms of lo-

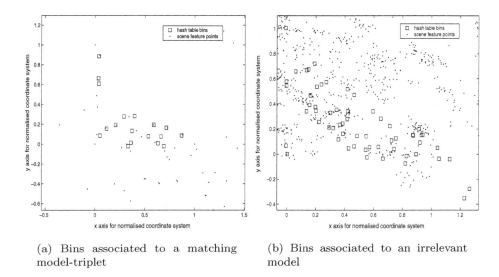

(a) Bins associated to a matching model-triplet

(b) Bins associated to an irrelevant model

Fig. 6. Location of normalised test image features with respect to hash table bins in the two different cases

cal consistencies. This prevents the propagation of errors throughout the image. Moreover the aggregative nature of the graph matching process makes it resilient to imperfect local matches caused by occlusion and measurement errors.

From the computational point of view the complexity of the graph matching method allows for the matching task to be completed in a reasonable time. Moreover it is independent from the number of features used for representation. This is not the case for the alignment and geometric hashing methods whose processing times depend on the number of region feature points exponentially. The alignment and geometric hashing are accomplished faster than the graph matching only when a very few feature points are involved in matching. However for this case the alignment and geometric hashing recognition rates are unacceptably poor. Reasonable performance rates are delivered by the alignment method only when five feature points or more are used for region representation. However at this point the alignment method ceases to be computationally feasible. The geometric hashing as tested in the complex scenes of our database failed to perform satisfactorily on all counts. Thus graph based matching proved to be the most promising and will be developed further by incorporating appearance information in the representation.

References

[1] Lopez A. and Pla F.: Dealing with segmentation errors in region-based stereo matching. Pattern Recognition. **33**(8) (2000) 1325–1338.

[2] Shokoufandeh A., Marsic I., and Dickinson S. J.: View-based object recognition using saliency maps. Image and Vision Computing. **17**(5) (1999) 445–460.

[3] Ahamdyfard A. and Kittler J.: Region-based object recognition: Pruning multiple representations and hypotheses. British Machine Vision Conference. Bristol,UK.(2000) 745–754.

[4] Huttenlocher D.P. and Ullman S.: Object recognition using alignment. First international conference on computer vision.London.(1987) 102–111.

[5] Huttenlocherr D.P and Ullman S.: Recognizing solid objects by alignment with an image. International Journal of Computer Vision.(1990) 195–212.

[6] Wolfson H.J.: Model-based object recognition by geometric hashing. International conference on computer vision.(1990) 526–536.

[7] Wolfson H.J. and Lamdan Y.: Geometric hashing: A general and efficient model-based recognition scheme. Second International Conference on Computer Vision.(1988) 238–249.

[8] Wolfson H.j., Lamdan Y., and Schwartz J.T.: On recognition of 3-d object from 2-d images. Robotics and Automation.(1988) 1407–1413.

[9] Nagao K. and Grimson W. Eric L.: Object recognition by alignment using invariant projections of planar surfaces. Technical report, A.I.Memo **1463**,Artificial Intelligence Lab,MIT. (1994).

[10] Lee M., Kittler J., and Wong K.C.: Generalised hough transform in object recognition. 11th International Conference on Pattern Recognition. **3**(1992) 285–289.

[11] Christmas W.J., Kittler J., and Petrou M.: Structural matching in computer vision using probabilistic relaxation. IEEE Transactions on Pattern Analysis and Machine Intelligence.(1995) 749–764.

A Unified Framework for Indexing and Matching Hierarchical Shape Structures

Ali Shokoufandeh[1] and Sven Dickinson[2]

[1] Department of Mathematics and Computer Science
Drexel University
3141 Chestnut Street
Philadelphia, PA USA 19104-2875
[2] Department of Computer Science
University of Toronto
6 King's College Rd.
Toronto, Ontario Canada M5S 3G4

Abstract. Hierarchical image structures are abundant in computer vision, and have been used to encode part structure, scale spaces, and a variety of multiresolution features. In this paper, we describe a unified framework for both indexing and matching such structures. First, we describe an indexing mechanism that maps the topological structure of a directed acyclic graph (DAG) into a low-dimensional vector space. Based on a novel eigenvalue characterization of a DAG, this topological signature allows us to efficiently retrieve a small set of candidates from a database of models. To accommodate occlusion and local deformation, local evidence is accumulated in each of the DAG's topological subspaces. Given a small set of candidate models, we will next describe a matching algorithm that exploits this same topological signature to compute, in the presence of noise and occlusion, the largest isomorphic subgraph between the image structure and the candidate model structure which, in turn, yields a measure of similarity which can be used to rank the candidates. We demonstrate the approach with a series of indexing and matching experiments in the domains of 2-D and (view-based) 3-D generic object recognition.

1   Introduction

The indexing and matching of hierarchical (e.g., multiscale or multilevel) image features is a common problem in object recognition. Such structures are often represented as rooted trees or directed acyclic graphs (DAGs), where nodes represent image feature abstractions and arcs represent spatial relations, mappings across resolution levels, component parts, etc [40,14]. The requirements of matching include computing a correspondence between nodes in an image structure and nodes in a model structure, as well as computing an overall measure of distance (or, alternatively, similarity) between the two structures. Such matching problems can be formulated as *largest isomorphic subgraph* or *largest isomorphic subtree* problems, for which a wealth of literature exists in the graph algorithms

C. Arcelli et al. (Eds.): IWVF4, LNCS 2059, pp. 67–84, 2001.
© Springer-Verlag Berlin Heidelberg 2001

community. However, the nature of the vision instantiation of this problem often precludes the direct application of these methods. Due to occlusion and noise, no significant isomorphisms may exist between two DAGs or rooted trees. Yet, at some level of abstraction, the two structures (or two of their substructures) may be quite similar.

The matching procedure is expensive and must be used sparingly. For large databases of object models, it is simply unacceptable to perform a linear search of the database. Therefore, an indexing mechanism is essential for selecting a small set of candidate models to which the matching procedure is applied. When working with hierarchical image structures, in the form of graphs, indexing is a challenging task, and can be formulated as the fast selection of a set of candidate models that share a subgraph with the query. But how do we test a given candidate without resorting to subgraph isomorphism? If there were a small number of subgraphs shared among many models, representing a vocabulary of *object parts*, one could conceive of a two-stage indexing process, in which image structures were matched to the part vocabulary, with parts "voting" for candidate models [6]. However, we're still faced with the complexity of subgraph isomorphism, albeit for a smaller database (vocabulary of parts).

In this paper, we present a unified solution to the problems of indexing and matching hierarchical structures. Drawing on techniques from the domain of eigenspaces of graphs, we present a technique that maps any rooted hierarchical structure, i.e., DAG or rooted tree, to a vector in a low-dimensional space. The mapping not only reduces the dimensionality of the representation, but does so while retaining important information about the branching structure, node distribution, and overall structure of the graph – information that is critical in distinguishing DAGs or rooted trees. Moreover, the technique accommodates both noise and occlusion, meeting the needs of an indexing structure for vision applications. Armed with a low-dimensional, robust vector representation of an input structure, indexing can be reduced to a nearest-neighbor search in a database of points, each representing the structure of a model (or submodel).

Once a candidate is retrieved by the indexing mechanism, we exploit this *same* eigen-characterization of hierarchical structure to compute a node-to-node correspondence between the input and model candidate hierarchical structures. We therefore unify our approaches to indexing and matching through a novel representation of hierarchical structure, leading to an efficient and effective framework for the recognition of hierarchical structures from large databases. In this paper, we will review our representation, described in [33,35], including a new analysis on its stability. We then describe the unifying role of our representation in the indexing and matching of hierarchical structures. Finally, we demonstrate the approach on two separate object recognition domains.

2   Related Work

Eigenspace approaches to shape description and indexing are numerous. Due to space constraints, we cite only a few examples. Turk and Pentland's eigenface approach [38] represented an image as a linear combination of a small number of basis vectors (images) computed from a large database of images. Nayar and Murase extended this work to general 3-D objects where a dense set of views was acquired for each object [17]. Other eigenspace methods have been applied to higher-level features, offering more potential for generic shape description and matching. For example, Sclaroff and Pentland compute the eigenmodes of vibration of a 2-D region [26], while Shapiro and Brady looked at how the modes of vibration of a set of 2-D points could be used to solve the point correspondence problem under translation, rotation, scale, and small skew [28].

In an attempt to index into a database of graphs, Sossa and Horaud use a small subset of the coefficients of the d_2-polynomial corresponding to the Laplacian matrix associated with a graph [36], while a spectral graph decomposition was reported by Sengupta and Boyer for the partitioning of a database of 3-D models, where nodes in a graph represent 3-D surface patches [27]. Sarkar [25] and Shi and Malik [31] have formulated the perceptual grouping and region segmentation problems, respectively, as graph partitioning problems and have used a generalized eigensystem approach to provide an efficient approximation. We note that in contrast to many of the above approaches to indexing, the representation that we present in this paper is independent of the contents of the model database and uses a uniform basis to represent all objects.

There have been many approaches to object recognition based on graph matching. An incomplete list of examples include Sanfeliu and Fu [24], Shapiro and Haralick [30,29], Wong et al. [41,22], Boyer and Kak [2] (for stereo matching), Kim and Kak [11], Messmer and Bunke [16], Christmas et al. [4], Eshera and Fu [7], Pellilo et al. [20], Gold and Rangarajan [10], Zhu and Yuille [42], and Cross and Hancock [5]. Although many of these approaches handle both noise and occlusion, none unify both indexing and matching through a single spectral mechanism.

3   Indexing Hierarchical Structures

We make the assumption that if a DAG[1] has rich structure in terms of depth and/or branching factor, its topology alone may serve as a discriminating index into a database of model structures. Although false positives (e.g., model DAGs that have the same structure, but whose node labels are different) may arise, they may be few in number and can be pruned during verification. As stated in Section 1, we seek a reduced representation for a DAG that will support efficient indexing and matching. An effective topological encoding of a DAG's structure should: **1)** map a DAG's topology to a point in some low-dimensional space; **2)** capture local topology to support matching/indexing in the presence of occlusion; **3)** be invariant to re-orderings of the DAG's branches, i.e., re-orderings which do not affect the parent-child relationships in the DAG; **4)** be

---

[1] Although a hierarchical structure can take the form of a DAG or rooted tree, we will henceforth limit our discussion to DAGs, since a rooted tree is a special case of a DAG.

as unique as possible, i.e., different DAGs should have different encodings; **5)** be stable, i.e., small perturbations of a DAG's topology should result in small perturbations of the index; and **6)** should be efficiently computed.

3.1 An Eigen-Decomposition of Structure

To describe the topology of a DAG, we turn to the domain of eigenspaces of graphs, first noting that any graph can be represented as a symmetric $\{0, 1, -1\}$ adjacency matrix, with 1's (-1's) indicating a forward (backward) edge between adjacent nodes in the graph (and 0's on the diagonal). The eigenvalues of a graph's adjacency matrix encode important structural properties of the graph. Furthermore, the eigenvalues of a symmetric matrix A are invariant to any orthonormal transformation of the form $P^t A P$. Since a permutation matrix is orthonormal, the eigenvalues of a graph are invariant to any consistent re-ordering of the graph's branches. However, before we can exploit a graph's eigenvalues for indexing purposes, we must establish their stability under minor topological perturbation, due to noise, occlusion, or deformation.

We will begin by showing that any structural change to a graph can be modeled as a two-step transformation of its original adjacency matrix. The first step transforms the graph's original adjacency matrix to a new matrix having the same spectral properties as the original matrix. The second step adds a noise matrix to this new matrix, representing the structural changes due to noise and/or occlusion. These changes take the form of the addition/deletion of nodes/arcs to/from the original graph. We will then draw on an important result that relates the distortion of the eigenvalues of the matrix resulting from the first step to the magnitude of the noise added in the second step. Since the eigenvalues of the original matrix are the same as those of the transformed matrix (first step), the noise-dependent eigenvalue bounds therefore apply to the original matrix. The result will establish the insensitivity of a graph's spectral properties to minor topological changes.

Let's begin with some definitions. Let $A_G \in \{0, 1, -1\}^{m \times m}$ denote the adjacency matrix of the graph G on m vertices, and assume H is an n-vertex graph obtained by adding $n - m$ new vertices and a set of edges to the graph G. Let $\Psi : \{0, 1, -1\}^{m \times m} \rightarrow \{0, 1, -1\}^{n \times n}$, be a *lifting operator* which transforms a subspace of $R^{m \times m}$ to a subspace of $R^{n \times n}$, with $n \geq m$. We will call this operator *spectrum preserving* if the eigenvalues of any matrix $\mathcal{A} \in \{0, 1, -1\}^{m \times m}$ and its image with respect to the operator $(\Psi(\mathcal{A}))$ are the same up to a degeneracy, i.e., the only difference between the spectra of $\mathcal{A}$ and $\Psi(\mathcal{A})$ is the number of zero eigenvalues ($\Psi(\mathcal{A})$ has $n - m$ more zero eigenvalues then $\mathcal{A}$).

As stated above, our goal is to show that any structural change in graph G can be represented in terms of a spectrum preserving operator and a noise matrix. Specifically, if A_H denotes the adjacency matrix of the graph H, then there exists a spectrum preserving operator $\Psi()$ and a noise matrix $E_H \in \{0, 1, -1\}^{n \times n}$ such that:

$$A_H = \Psi(A_G) + E_H \tag{1}$$

We will define $\Psi()$ as a lifting operator consisting of two steps. First, we will add $n - m$ zero rows and columns to the matrix A_G, and denote the resulting matrix by A'_G. Next, A'_G will be pre- and post-multiplied by a permutation matrix P and its transpose P^t, respectively, aligning the rows and columns corresponding to the same vertices in A_H and $\Psi(A_G)$. Since the only difference between the eigenvalues of A'_G and A_G is the number of zero eigenvalues, and $P A'_G P^t$ has the same set of eigenvalues as the matrix A'_G, $\Psi()$ is a spectrum preserving operator. As a result, the noise matrix E_H can be represented as $A_H - \Psi(A_G) \in \{0, 1, -1\}^{n \times n}$.

Armed with a spectrum-preserving lifting operator and a noise matrix, we can now proceed to quantify the impact of the noise on the original graph's eigenvalues. Specifically, let λ_k for $k \in \{1, ..., n\}$ denote the kth largest eigenvalue of the matrix A. A seminal result of Wilkinson [39] (see also Stewart and Sun [37]) states that:

Theorem 1. *If A and $A + E$ are $n \times n$ symmetric matrices, then:*

$$\lambda_k(A) + \lambda_k(E) \le \lambda_k(A + E) \le \lambda_k(A) + \lambda_1(E), \quad \text{for all } k \in \{1, ..., n\} \quad (2)$$

For H and G, we know that $\lambda_1(E_H) = ||E_H||_2$. Therefore, using the above theorem, for all $k \in \{1, ..., n\}$:

$$|\lambda_k(A_H) - \lambda_k(\Psi(A_G))| = |\lambda_k(\Psi(A_G) + E_H) - \lambda_k(\Psi(A_G))|$$

$$\le \max\{|\lambda_1(E_H)|, |\lambda_n(E_H)|\} \quad (3)$$

$$= ||E_H||_2.$$

The above chain of inequalities gives a precise bound on the distortion of the eigenvalues of $\Psi(A_G)$ in terms of the largest eigenvalue of the noise matrix E_H. Since $\Psi()$ is a spectrum preserving operator, the eigenvalues of A_G follow the same bound in their distortions.

The above result has several important consequences for our application of a graph's eigenvalues to graph indexing. Namely, if the perturbation E_H is small in terms of its complexity, then the eigenvalues of the new graph H (e.g., the query graph) will remain close to their corresponding non-zero eigenvalues of the original graph G (e.g., the model graph), independent of where the perturbation is applied to G. The magnitude of the eigenvalue distortion is a function of the number of vertices added to the graph due to the noise or occlusion. Specifically, if the noise matrix E_H introduces k new vertices to G, then the distortion of every eigenvalue can be bounded by $\sqrt{k-1}$ (Neumaier [18]). This bound can be further tightened if the noise matrix has simple structure. For example, if E_H represents a simple path on k vertices, then its norm can be bounded by $(2 \cos \pi/(k+1))$ (Lovász and Pelikán [15]). In short, large distortions are due to the introduction/deletion of large, complex subgraphs to/from G, while small structural changes will have little impact on the higher order eigenvalues G. The eigenvalues of a graph are therefore stable under minor perturbations in graph structure.

3.2   Formulating an Index

Having established the stability of a DAG's eigenvalues under minor perturbation of the graph, we can now proceed to define an index based on the eigenvalues. We could, for example, define a vector to be the sorted eigenvalues of a DAG, with the resulting index used to retrieve nearest neighbors in a model DAG database having similar topology. However, for large DAGs, the dimensionality of the index (and model DAG database) would be prohibitively large. Our solution to this problem will be based on eigenvalue sums rather than on the eigenvalues themselves.

Specifically, let T be a DAG whose maximum branching factor is $\Delta(T)$, and let the subgraphs of its root be $T_1, T_2, \ldots, T_S$. For each subgraph, T_i, whose root degree is $\delta(T_i)$, we compute the eigenvalues of T_i's submatrix, sort the eigenvalues in decreasing order by absolute value, and let S_i be the sum of the $\delta(T_i) - 1$ largest absolute values. The sorted S_i's become the components of a $\Delta(T)$-dimensional vector assigned to the DAG's root. If the number of S_i's is less than $\Delta(T)$, then the vector is padded with zeroes. We can recursively repeat this procedure, assigning a vector to each nonterminal node in the DAG, computed over the subgraph rooted at that node. The reasons for computing a description for each node, rather than just the root, will become clear in the next section.

Although the eigenvalue sums are invariant to any consistent re-ordering of the DAG's branches, we have given up some uniqueness (due to the summing operation) in order to reduce dimensionality. We could have elevated only the largest eigenvalue from each subgraph (non-unique but less ambiguous), but this would be less representative of the subgraph's structure. We choose the $\delta(T_i) - 1$ largest eigenvalues for two reasons: 1) the largest eigenvalues are more informative of subgraph structure, and 2) by summing $\delta(T_i) - 1$ elements, we effectively normalize the sum according to the local complexity of the subgraph root.

To efficiently compute the submatrix eigenvalue sums, we turn to the domain of semidefinite programming. A symmetric $n \times n$ matrix A with real entries is said to be positive semidefinite, denoted as $A \succeq 0$, if for all vectors $x \in R^n$, $x^t A x \geq 0$, or equivalently, all its eigenvalues are non-negative. We say that $U \succeq V$ if the matrix $U - V$ is positive semidefinite. For any two matrices U and V having the same dimensions, we define $U \bullet V$ as their inner product, i.e., $U \bullet V = \sum_i \sum_j U_{i,j} V_{i,j}$. For any square matrix U, we define $\text{trace}(U) = \sum_i U_{i,i}$. Let I denote the identity matrix having suitable dimensions. The following result, due to Overton and Womersley [19], characterizes the sum of the first k largest eigenvalues of a symmetric matrix in the form of a semidefinite convex programming problem:

Theorem 2 (Overton and Womersley [19]). *For the sum of the first k eigenvalues of a symmetric matrix A, the following semidefinite programming* *characterization holds:*

$$\lambda_1(A) + \ldots + \lambda_k(A) = \max A \bullet U$$
$$\text{s.t.} \quad \text{trace}(U) = k \tag{4}$$
$$0 \preceq U \preceq I,$$

The elegance of Theorem (2) lies in the fact that the equivalent semidefinite programming problem can be solved, for any desired accuracy ϵ, in time polynomial in $O(n\sqrt{n}L)$ and $\log \frac{1}{\epsilon}$, where L is an upper bound on the size of the optimal solution, using a variant of the Interior Point method proposed by Alizadeh [1]. In effect, the complexity of directly computing the eigenvalue sums is a significant improvement over the $O(n^3)$ time required to compute the individual eigenvalues, sort them, and sum them.

3.3   Properties of the Index

Our topological index satisfies the six criteria outlined in Section 1. The eigen-decomposition yields a low-dimensional (criterion 1) vector assigned to each node in the DAG, which captures the local topology of the subgraph rooted at that node (criterion 2 – this will be more fully explained in Section 3.4). Furthermore, a node's vector is invariant to any consistent re-ordering of the node's subgraphs (criterion 3). The components of a node's vector are based on summing the largest eigenvalues of its subgraph's adjacency submatrix. Although our dimensionality-reducing summing operation has cost us some uniqueness, our partial sums still have very low ambiguity (criterion 4). From the sensitivity analysis in Section 3.1, we have shown our index to be stable to minor perturbations of the DAG's topology (criterion 5). As shown in Theorem 2, these sums can be computed even more efficiently (criterion 6) than the eigenvalues themselves. The vector labeling of all DAGs isomorphic to T not only has the same vector labeling, but spans the same subspace in $R^{\Delta(T)-1}$. Moreover, this extends to any DAG which has a subgraph isomorphic to a subgraph of T.

3.4   Candidate Selection

Given a query DAG corresponding to an image, our task is to search the model DAG database for one or more model DAGs which are similar to the image DAG. If the number of model DAGs is large, a linear search of the database is intractable. Therefore, the goal of our indexing mechanism is to quickly select a small number of model candidates for verification. Those candidates will share coarse topological structure with the image DAG (or one of its subgraphs, if it is occluded or poorly segmented). Hence, we begin by mapping the topology of the image DAG to a set of indices that capture its structure, discounting any information associated with its nodes. We then describe the structure of our model database, along with our mechanism for indexing into it to yield a small set of model candidates. Finally, we present a local evidence accumulation procedure that will allow us to index in the presence of occlusion.

A Database for Model DAGs. Our eigenvalue characterization of a DAG's topology suggests that a model DAG's topological structure can be represented as a vector in δ-dimensional space, where δ is an upper bound on the degree of any vertex of any image or model DAG. If we could assume that an image DAG represents a properly segmented, unoccluded object, then the vector of eigenvalue sums, which we will call the *topological signature vector (or TSV)*, computed at the image DAG's root, could be compared with those topological signature vectors representing the roots of the model DAGs. The vector distance between the image DAG's root TSV and a model DAG's root TSV would be inversely proportional to the topological similarity of their respective DAGs, as finding two subgraphs with "close" eigenvalue sums represents an approximation to finding the largest isomorphic subgraph.

Unfortunately, this simple framework cannot support either cluttered scenes or large occlusion, both of which result in the addition or removal of significant structure. In either case, altering the structure of the DAG will affect the TSV's computed at its nodes. The signatures corresponding to the roots of those subgraphs (DAGs) that survive the occlusion will not change. However, the signature of a root of a subgraph that has undergone any perturbation will change which, in turn, will affect the signatures of any of its ancestor nodes, including the root of the entire DAG. We therefore cannot rely on indexing solely with the root's signature. Instead, we will exploit the local subgraphs that survive the occlusion.

We can accommodate such perturbations through a local indexing framework analogous to that used in a number of geometric hashing methods, e.g., [13,8]. Rather than storing a model DAG's root signature, we will store the signatures of *each* node in the model DAG, along with a pointer to the object model containing that node as well as a pointer to the corresponding node in the model DAG (allowing access to node label information). Since a given model subgraph can be shared by other model DAGs, a given signature (or location in δ-dimensional space) will point to a list of (model object, model node) ordered pairs. At runtime, the signature at each node in the image DAG becomes a separate index, with each nearby candidate in the database "voting" for one or more (model object, model node) pairs. Nearby candidates can be retrieved using a nearest neighbor retrieval method. In our implementation, model points were stored in a Voronoi database, whose off-line construction (decomposition) is $O((kn)^{\lfloor (\delta+1)/2 \rfloor +1}) + O((kn)^{\lfloor (\delta+1)/2 \rfloor} \log(kn))$ ([21]), and whose run-time search is $O(\log^{\delta}(kn))$ for fixed δ [21]; details are given in [33].

Accumulating Local Evidence. Each node in the image DAG will generate a set of (model object, model node) votes. To collect these votes, we set up an accumulator with one bin per model object. Furthermore, we can weight the votes that we add to the accumulator. For example, if the label of the model node is not compatible with the label of its corresponding image node, then the vote is discarded, i.e., it receives a zero weight. If the nodes are label-compatible, then we can weight the vote according to the distance between their respective TSV's – the closer the signatures, the more weight the vote gets.

We can also weight the vote according to the complexity of its corresponding subgraph, allowing larger and more complex subgraphs (or "parts") to have higher weight. This can be easily accommodated within our eigenvalue framework, for the richer the structure, the larger its maximum eigenvalue:

Theorem 3 (Lovász and Pelikán [15]). *Among the graphs with n vertices, the star graph $(K_{1,n-1})$, has the largest eigenvalue $(\sqrt{n-1})$, while the path on n nodes (P_n) has the smallest eigenvalue $(2\cos\pi/(n+1))$.*

Since the size of the eigenvalues, and hence their sum, is proportional to both the branching factor as well as the number of nodes, the magnitude of the signature is also used to weight the vote. If we let u be the TSV of an image DAG node and v the TSV of a model DAG node that is sufficiently close, the weight of the resulting vote, i.e., the local evidence for the model, is computed as (we use $p = 2$):

$$W = \frac{||u||_p}{1 + ||v - u||_p} \tag{5}$$

Once the evidence accumulation is complete, those models whose support is sufficiently high are selected as candidates for verification. The bins can, in effect, be organized in a heap, requiring a maximum of $O(\log k)$ operations to maintain the heap when evidence is added, where k is the number of non-zero object accumulators. Once the top-scoring models have been selected, they must be individually verified according to some matching algorithm.

4   Matching Hierarchical Structures

Each of the top-ranking candidates emerging from the indexing process must be verified to determine which is most similar to the query. If there were no clutter, occlusion, or noise, our problem could be formulated as a graph isomorphism problem. If we allowed clutter and limited occlusion, we would search for the largest isomorphic subgraphs between query and model. Unfortunately, with the presence of noise, in the form of the addition of spurious graph structure and/or the deletion of salient graph structure, large isomorphic subgraphs may simply not exist. It is here that we call on our eigen-characterization of graph structure to help us overcome this problem.

Each node in our graph (query or model) is assigned a TSV, which reflects the underlying structure in the subgraph rooted at that node. If we simply discarded all the edges in our two graphs, we would be faced with the problem of finding the best correspondence between the nodes in the query and the nodes in the model; two nodes could be said to be in close correspondence if the distance between their TSVs (and the distance between their domain-dependent node labels) was small. In fact, such a formulation amounts to finding the maximum cardinality, minimum weight matching in a bipartite graph spanning the two sets of nodes. At first glance, such a formulation might seem like a bad idea (by throwing away all that important graph structure!) until one recalls that the graph structure is really encoded in the node's TSV. Is it then possible to reformulate a noisy, largest isomorphic subgraph problem as a simple bipartite matching problem?

Unfortunately, in discarding all the graph structure, we have also discarded the underlying hierarchical structure. There is nothing in the bipartite graph matching formulation that ensures that hierarchical constraints among corresponding nodes are obeyed, i.e., that parent/child nodes in one graph don't match child/parent nodes in the other. This reformulation, although softening the overly strict constraints imposed by the largest isomorphic subgraph formulation, is perhaps too weak. We could try to enforce the hierarchical constraints in our bipartite matching formulation, but no polynomial-time solution is known to exist for the resulting formulation. Clearly, we seek an efficient approximation method that will find corresponding nodes between two noisy, occluded DAGs, subject to hierarchical constraints.

Our algorithm, a modification to Reyner's algorithm [23], combines the above bipartite matching formulation with a greedy, best-first search in a recursive procedure to compute the corresponding nodes in two rooted DAGs. As in the above bipartite matching formulation, we compute the maximum cardinality, minimum weight matching in the bipartite graph spanning the two sets of nodes. Edge weight will encode a function of both topological similarity as well as domain-dependent node similarity. The result will be a selection of edges yielding a mapping between query and model nodes. As mentioned above, the computed mapping may not obey hierarchical constraints. We therefore greedily choose only the best edge (the two most similar nodes in the two graphs, representing in some sense the two most similar subgraphs), add it to the solution set, and recursively apply the procedure to the subgraphs defined by these two nodes. Unlike a traditional depth-first search which backtracks to the next statically-determined branch, our algorithm effectively recomputes the branches at each node, always choosing the next branch to descend in a best-first manner. In this way, the search for corresponding nodes is focused in corresponding subgraphs (rooted DAGs) in a top-down manner, thereby ensuring that hierarchical constraints are obeyed.

Before formalizing our algorithm, some definitions are in order. Let $G = (V_1, E_1)$ and $H = (V_2, E_2)$ be the two DAGs to be matched, with $|V_1| = n_1$ and $|V_2| = n_2$. Define d to be the maximum degree of any vertex in G and H, i.e., $d = \max(\delta(G), \delta(H))$. For each vertex v, we define $\chi(v) \in R^{d-1}$ as the unique topological signature vector (TSV), introduced in Section 3.2.[2] Furthermore, for any pair of vertices u and v, let $C(u, v)$ denote the domain dependent node label distance between vertices u and v. Finally, let $\Phi(G, H)$ (initially empty) be the set of final node correspondences between G and H, representing the solution to our matching problem.

---

[2] Note that if the maximum degree of a node is d, then excluding the edge from the node's parent, the maximum number of children is $d - 1$. Also note that if $\delta(v) < d$, then then the last $d - \delta(v)$ entries of χ are set to zero to ensure that all χ vectors have the same dimension.

The algorithm begins by forming a $n_1 \times n_2$ matrix $\Pi(G, H)$ whose (u, v)-th entry has the value $C(u, v)\|\chi(u) - \chi(v)\|_2$, assuming that u and v are compatible in terms of their node labels, and has the value ∞ otherwise. Next, we form a bipartite edge weighted graph $\mathcal{G}(V_1, V_2, E_{\mathcal{G}})$ with edge weights from the matrix $\Pi(G, H)$.[3] Using the scaling algorithm of Goemans, Gabow, and Williamson [9], we then find the maximum cardinality, minimum weight matching in $\mathcal{G}$. This results in a list of node correspondences between G and H, called $\mathcal{M}_1$, that can be ranked in decreasing order of similarity.

From $\mathcal{M}_1$, we choose (u_1, v_1) as the pair that has the minimum weight among all the pairs in $\mathcal{M}_1$, i.e., the first pair in $\mathcal{M}_1$. (u_1, v_1) is removed from the list and added to the solution set $\Phi(G, H)$, and the remainder of the list is *discarded*. For the rooted subgraphs G_{u_1} and H_{v_1} of G and H, rooted at nodes u_1 and v_1, respectively, we form the matrix $\Pi(G_{u_1}, H_{v_1})$ using the same procedure described above. Once the matrix is formed, we find the matching $\mathcal{M}_2$ in the bipartite graph defined by weight matrix $\Pi(G_{u_1}, H_{v_1})$, yielding another ordered list of node correspondences. The procedure is recursively applied to (u_2, v_2), the edge with minimum weight in $\mathcal{M}_2$, with the remainder of the list discarded.

This recursive process eventually reaches the bottom of the DAGs, forming a list of ordered correspondence lists (or matchings) $\{\mathcal{M}_1, \ldots, \mathcal{M}_k\}$. In backtracking step i, we remove any subgraph from the graphs G_i and H_i whose roots participate in a matching pair in $\Phi(G, H)$ (we enforce a one-to-one correspondence of nodes in the solution set). Then, in a depth-first manner, we recompute $\mathcal{M}_i$ on the subgraphs rooted at u_i and v_i (with solution set nodes removed). As before, we choose the minimum weight matching pair, and recursively descend. Unlike a traditional depth-first search, we are dynamically recomputing the branches at each node in the search tree. Processing at a particular node will terminate when either rooted subgraph loses all of its nodes to the solution set. The precise algorithm is given in Figure 1; additional details and examples are given in [35,32].

In terms of algorithmic complexity, observe that during the depth-first construction of the matching chains, each vertex in G or H will be matched at most once in the forward procedure. Once a vertex is mapped, it will never participate in another mapping again. The total time complexity of constructing the matching chains is therefore bounded by $O(n^2\sqrt{n}\log\log n)$, for $n = \max(n_1, n_2)$ [9]. Moreover, the construction of the $\chi(v)$ vectors will take $O(n\sqrt{n}L)$ time, implying that the overall complexity of the algorithm is $\max(O(n^2\sqrt{n}\log\log n), O(n^2\sqrt{n}L)$. The above algorithm therefore provides, in polynomial time better than $O(n^3)$ an approximate optimal solution to the largest isomorphic subgraph problem in the presence of noise.

---

[3] $G(A, B, E)$ is a weighted bipartite graph with weight matrix $W = [w_{ij}]$ of size $|A| \times |B|$ if, for all edges of the form $(i, j) \in E$, $i \in A$, $j \in B$, and (i, j) has an associated weight $= w_{i,j}$.

procedure isomorphism(G,H)

    $\Phi(G, H) \leftarrow \emptyset$

    $d \leftarrow \max(\delta(G), \delta(H))$

    for $u \in V_G$ compute $\chi(u) \in R^{d-1}$ (see Section 3.2)

    for $v \in V_H$ compute $\chi(v) \in R^{d-1}$ (see Section 3.2)

    call match(root(G),root(H))

    return(cost($\Phi(G, H)$))

end

procedure match(u,v)

    do

        {

        let $G_u \leftarrow$ rooted subgraph of G at u

        let $H_v \leftarrow$ rooted subgraph of H at v

        compute $|V_{G_u}| \times |V_{H_v}|$ weight matrix $\Pi(G_u, H_v)$

        $\mathcal{M} \leftarrow$ max cardinality, minimum weight bipartite matching

            in $\mathcal{G}(V_{G_u}, V_{H_v})$ with weights from $\Pi(G_u, H_v)$ (see [9])

        $(u', v') \leftarrow$ minimum weight pair in $\mathcal{M}$

        $\Phi(G, H) \leftarrow \Phi(G, H) \cup \{(u', v')\}$

        call match(u',v')

        $G_u \leftarrow G_u - \{x | x \in V_{G_u} \text{ and } (x, w) \in \Phi(G, H)\}$

        $H_v \leftarrow H_v - \{y | y \in V_{H_v} \text{ and } (w, y) \in \Phi(G, H)\}$

        }

    while ($G_u \neq \emptyset$ and $H_v \neq \emptyset$)

Fig. 1. Algorithm for Matching Two Hierarchical Structures.

5   Demonstration

In this section, we briefly illustrate our unified approach to indexing and matching on two different object recognition domains.

5.1   2-D Generic Object Recognition

To demonstrate our approach to indexing, we turn to the domain of 2-D object recognition [33,35]. We adopt a representation for 2-D shape that is based on a coloring of the shocks (singularities) of a curve evolution process acting on simple closed curves in the plane [12]. Any given 2-D shape gives rise to a rooted *shock tree*, in which nodes represent parts (whose labels are drawn from four qualitatively-defined classes) and arcs represent relative time of formation (or relative size). Figure 2 illustrates a 2-D shape, along with its corresponding shock tree.

We demonstrate our indexing algorithm on a database of 60 object silhouettes. In Figure 3, query shapes are shown in the left column, followed by the top ten database candidates (based on accumulator scores), ordered left to right. The

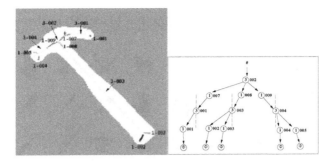

Fig. 2. An illustrative example taken from [35]. The labels on the shocks of the hammer (left) correspond to vertices in the derived shock graph (right).

candidate in the box is the closest candidate, found by computing the distance (using the matcher) between the query and each database shape (linear search). For unoccluded shapes, the results are very encouraging, with the correct candidate ranking very highly. With increasing occlusion, high indexing ambiguity (smaller unoccluded subtrees are less distinctive and "vote" for many objects that contain them) leads to slightly decreased performance, although the target object is still ranked highly. We are investigating the incorporation of more node information, as well as the encoding of subtree relations (currently, each subtree votes independently, with no constraints among subtrees enforced) to improve indexing performance.

5.2 View-Based 3-D Object Recognition

For our next demonstration, we turn to the domain of view-based object recognition, in which salient blobs are detected in a multiscale wavelet decomposition of an image [3]. Figure 4 shows three images of an origami figure, along with their computed multiscale blob analyses (shown inverted for improved visibility). Each image yields a DAG, in which nodes correspond to blobs, and arcs are directed from blobs at coarser scales to blobs at finer scales if the distance between their centers does not exceed the sum of their radii. Node similarity is a function of the difference in saliency between two blobs.

In Figure 5, we show the results of matching the first and second, and first and third images, respectively, in Figure 4. The first and third images are taken from similar viewpoints (approx 15° apart), so the match yields many good correspondences. However, for the first and third images, taken from different viewpoints (approx 75° apart), fewer corresponding features were found. Note that since only the DAG structure is matched (and not DAG geometry), incorrect correspondences may arise when nodes have similar saliency and size but different relative positions. A stronger match can be attained by enforcing geometric consistency (see [34]).

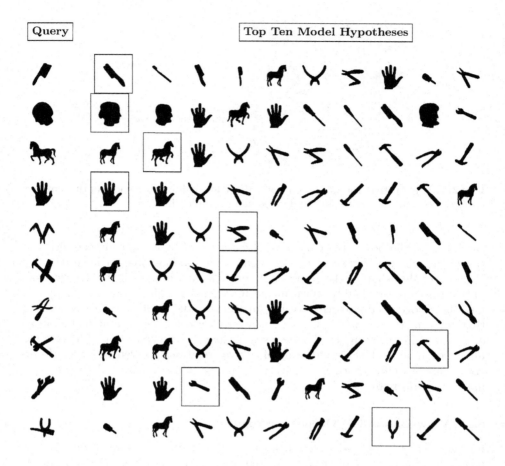

Fig. 3. Indexing Demonstration. Shape on left is query shape, followed by top ten candidates in decreasing score from left to right. Boxed candidate is closest to query (using a linear search based on matcher).

6   Conclusions

We have presented a unified solution to the tightly-coupled problems of indexing and matching hierarchical structures. The structural properties of a DAG are captured by the eigenvalues of its corresponding adjacency matrix. These eigenvalues, in turn, can be combined to yield a low-dimensional vector representation of DAG structure. The resulting vectors can be used to retrieve, in the presence of noise and occlusion, structurally similar candidates from a database using efficient nearest-neighbor searching methods. Moreover, these same vectors contribute to the edge weights in a recursive bipartite matching formulation that computes an approximation to the largest isomorphic subgraph of two graphs (query and model) in the presence of noise and occlusion. Our formulation is gen-

Fig. 4. Top row contains original images, while bottom row (shown inverted for improved visibility) contains corresponding multiscale blob analyses (see text).

eral and is applicable to the indexing and matching of any rooted hierarchical structure, whether DAG or rooted tree. The only domain-dependent component is the node label distance function, which is used in conjunction with the topological distance function to compute a bipartite edge weight. We have tested the approach extensively on the indexing and matching of shock graphs, and have only begun to test the approach on other domains, including the preliminary results reported in this paper in the domain of multiscale blob matching.

Acknowledgements. The authors gratefully acknowledge the support of NSERC, NSF, and ARO for their support of this work. The authors would also like to thank our collaborators, Kaleem Siddiqi, Steven Zucker, and Ee-Chien Chang, for their kind permission to use the figures shown in Section 5.

References

1. F. Alizadeh. Interior point methods in semidefinite programming with applications to combinatorial optimization. *SIAM J. Optim.*, 5(1):13–51, 1995.
2. K.L. Boyer and A.C. Kak. Structural stereopsis for 3-D vision. *IEEE Transactions on Pattern Analysis and Machine Intelligence*, 10(2):144–166, March 1988.
3. E. Chang, Stephane Mallat, and Chee Yap. Wavelet foveation. *Journal of Applied and Computational Harmonic Analysis*, 9(3):312–335, October 2000.
4. W. J. Christmas, J. Kittler, and M. Petrou. Structural matching in computer vision using probabilistic relaxation. *IEEE Transactions on Pattern Analysis and Machine Intelligence*, 17:749–764, August 1995.
5. A.D. Cross and E.R. Hancock. Graph matching with a dual-step em algorithm. *IEEE Transactions on Pattern Analysis and Machine Intelligence*, 20(11):1236–1253, November 1998.

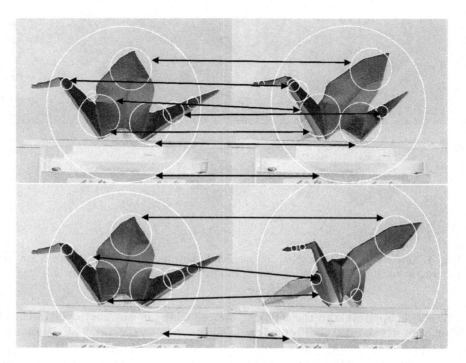

Fig. 5. Results of matching the first two, and first and third images, respectively, in Figure 4. The first two images are taken from similar viewpoints (approx 15° apart), and hence many correspondences were found. However, for the first and third images, taken from different viewpoints (approx 75° apart), fewer corresponding features were found.

6. S. Dickinson, A. Pentland, and A. Rosenfeld. From volumes to views: An approach to 3-D object recognition. *CVGIP: Image Understanding*, 55(2):130–154, 1992.

7. M. A. Eshera and K. S. Fu. A graph distance measure for image analysis. *IEEE Trans. SMC*, 14:398–408, May 1984.

8. P. Flynn and A. Jain. 3D object recognition using invariant feature indexing of interpretation tables. *CVGIP:Image Understanding*, 55(2):119–129, March 1992.

9. H. Gabow, M. Goemans, and D. Williamson. An efficient approximate algorithm for survivable network design problems. *Proc. of the Third MPS Conference on Integer Programming and Combinatorial Optimization*, pages 57–74, 1993.

10. Steven Gold and Anand Rangarajan. A graduated assignment algorithm for graph matching. *IEEE PAMI*, 18(4):377–388, 1996.

11. W. Kim and A. C. Kak. 3d object recognition using bipartite matching embedded in discrete relaxation. *IEEE Transactions on Pattern Analysis and Machine Intelligence*, 13(3):224–251, 1991.

12. B. B. Kimia, A. Tannenbaum, and S. W. Zucker. Shape, shocks, and deformations I: The components of two-dimensional shape and the reaction-diffusion space. *International Journal of Computer Vision*, 15:189–224, 1995.

13. Y. Lamdan, J. Schwartz, and H. Wolfson. Affine invariant model-based object recognition. *IEEE Transactions on Robotics and Automation*, 6(5):578–589, October 1990.

14. Tony Lindeberg. Detecting Salient Blob–Like Image Structures and Their Scales With a Scale–Space Primal Sketch—A Method for Focus–of–Attention. *International Journal of Computer Vision*, 11(3):283–318, December 1993.

15. L. Lovász and J. Pelicán. On the eigenvalues of a tree. *Periodica Math. Hung.*, 3:1082–1096, 1970.

16. B. T. Messmer and H. Bunke. A new algorithm for error-tolerant subgraph isomorphism detection. *IEEE Transactions on Pattern Analysis and Machine Intelligence*, 20:493–504, May 1998.

17. H. Murase and S. Nayar. Visual learning and recognition of 3-D objects from appearance. *International Journal of Computer Vision*, 14:5–24, 1995.

18. A. Neumaier. Second largest eigenvalue of a tree. *Linear Algebra and its Applications*, 46:9–25, 1982.

19. M. L. Overton and R. S. Womersley. Optimality conditions and duality theory for minimizing sums of the largest eigenvalues of symmetric matrices. *Math. Programming*, 62(2):321–357, 1993.

20. M. Pelillo, K. Siddiqi, and S. Zucker. Matching hierarchical structures using association graphs. *IEEE Transactions on Pattern Analysis and Machine Intelligence*, 21(11):1105–1120, November 1999.

21. F. Preparata and M. Shamos. *Computational Geometry*. Springer-Verlag, New York, NY, 1985.

22. Recognition and shape synthesis of 3D objects based on attributed hypergraphs. A. wong and s. lu and m. rioux. *IEEE Transactions on Pattern Analysis and Machine Intelligence*, 11:279–290, 1989.

23. S. W. Reyner. An analysis of a good algorithm for the subtree problem. *SIAM J. Comput.*, 6:730–732, 1977.

24. A. Sanfeliu and K. S. Fu. A distance measure between attributed relational graphs for pattern recognition. *IEEE Transactions on Systems, Man, and Cybernetics*, 13:353–362, May 1983.

25. S. Sarkar. Learning to form large groups of salient image features. In *IEEE CVPR*, Santa Barbara, CA, June 1998.

26. S. Sclaroff and A. Pentland. Modal matching for correspondence and recognition. *IEEE Transactions on Pattern Analysis and Machine Intelligence*, 17(6):545–561, June 1995.

27. K. Sengupta and K. Boyer. Using spectral features for modelbase partitioning. In *Proceedings, International Conference on Pattern Recognition*, Vienna, Austria, August 1996.

28. L. Shapiro and M. Brady. Feature-based correspondence: an eigenvector approach. *Image and Vision Computing*, 10(5):283–288, June 1992.

29. L. G. Shapiro and R. M. Haralick. Structural descriptions and inexact matching. *IEEE Transactions on Pattern Analysis and Machine Intelligence*, 3:504–519, 1981.

30. L. G. Shapiro and R. M. Haralick. A metric for comparing relational descriptions. *IEEE Transactions on Pattern Analysis and Machine Intelligence*, 7:90–94, January 1985.

31. J. Shi and J. Malik. Normalized cuts and image segmentation. In *IEEE Conference on Computer Vision and Pattern Recognition*, San Juan, Puerto Rico, June 1997.

32. A. Shokoufandeh and S. Dickinson. Applications of bipartite matching to problems in object recognition. In *Proceedings, ICCV Workshop on Graph Algorithms and Computer Vision (web proceedings: http://www.cs.cornell.edu/iccv-graph-workshop/papers.htm)*, September 1999.
33. A. Shokoufandeh, S. Dickinson, K. Siddiqi, and S. Zucker. Indexing using a spectral encoding of topological structure. In *IEEE Conference on Computer Vision and Pattern Recognition*, pages 491–497, Fort Collins, CO, June 1999.
34. A. Shokoufandeh, I. Marsic, and S. Dickinson. View-based object recognition using saliency maps. *Image and Vision Computing*, 17:445–460, 1999.
35. K. Siddiqi, A. Shokoufandeh, S. Dickinson, and S. Zucker. Shock graphs and shape matching. *International Journal of Computer Vision*, 30:1–24, 1999.
36. H. Sossa and R. Horaud. Model indexing: The graph-hashing approach. In *Proceedings, IEEE CVPR*, pages 811–814, 1992.
37. G.W. Stewart and J.-G. Sun. *Matrix Perturbation Theory*. Academic Press, San Diego, 1990.
38. M. Turk and A. Pentland. Eigenfaces for recognition. *Journal of Cognitive Neuroscience*, 3(1):71–86, 1991.
39. J. Wilkinson. *The Algebraic Eigenvalue Problem*. Clarendon Press, Oxford, England, 1965.
40. A. Witkin. Scale space filtering. In Alex Pentland, editor, *From Pixels to Predicates*. Ablex, Norwood, NJ, 1986.
41. A. K. C. Wong and M. You. Entropy and distance of random graphs with application to structural pattern recognition. *IEEE Transactions on Pattern Analysis and Machine Intelligence*, 7:599–609, September 1985.
42. S. Zhu and A. L. Yuille. Forms: a flexible object recognition and modelling system. *International Journal of Computer Vision*, 20(3):187–212, 1996.

A Fragment-Based Approach to Object Representation and Classification

Shimon Ullman, Erez Sali, and Michel Vidal-Naquet

The Weizmann Institute of Science
Rehvot 76100, Israel
Shimon@wisdom.weizmann.ac.il

Abstract. The task of visual classification is the recognition of an object in the image as belonging to a general class of similar objects, such as a face, a car, a dog, and the like. This is a fundamental and natural task for biological visual systems, but it has proven difficult to perform visual classification by artificial computer vision systems. The main reason for this difficulty is the variability of shape within a class: different objects vary widely in appearance, and it is difficult to capture the essential shape features that characterize the members of one category and distinguish them from another, such as dogs from cats.

In this paper we describe an approach to classification using a fragment-based representation. In this approach, objects within a class are represented in terms of common image fragments that are used as building blocks for representing a large variety of different objects that belong to a common class. The fragments are selected from a training set of images based on a criterion of maximizing the mutual information of the fragments and the class they represent. For the purpose of classification the fragments are also organized into types, where each type is a collection of alternative fragments, such as different hairline or eye regions for face classification. During classification, the algorithm detects fragments of the different types, and then combines the evidence for the detected fragments to reach a final decision. Experiments indicate that it is possible to trade off the complexity of fragments with the complexity of the combination and decision stage, and this tradeoff is discussed.

The method is different from previous part-based methods in using class-specific object fragments of varying complexity, the method of selecting fragments, and the organization into fragment types. Experimental results of detecting face and car views show that the fragment-based approach can generalize well to a variety of novel image views within a class while maintaining low mis-classification error rates. We briefly discuss relationships between the proposed method and properties of parts of the primate visual system involved in object perception.

1 Introduction

The general task of visual object recognition can be divided into two related, but somewhat different tasks – classification and identification. Classification is concerned with the general description of an object as belonging to a natural class of similar objects, such as a face or a dog. Identification is a more specific level of

C. Arcelli et al. (Eds.): IWVF4, LNCS 2059, pp. 85–100, 2001.
© Springer-Verlag Berlin Heidelberg 2001

recognition, that is, the recognition of a specific individual within a class, such as the face of a particular person, or the make of a particular car. For human vision, classification is a natural task: we effortlessly classify a novel object as a person, dog, car, house, and the like, based on its appearance. Even a three-year old child can easily classify a large variety of images of many natural classes. Furthermore, the general classification of an object as a member of a general class such as a car, for example, is usually easier than the identification of the specific make of the car [25]. In contrast, current computer vision systems can deal more successfully with the task of identification compared with classification. This may appear surprising, because specific identification requires finer distinctions between objects compared with general classification, and therefore the task appears to be more demanding.

The main difficulty faced by a recognition and classification system is the problem of variability, and the need to generalize across variations in the appearance of objects belonging to the same class. Different dog images, for example, can vary widely, because they can represent different kinds of dogs, and for each particular dog, the appearance will change with the imaging conditions, such as the viewing angle, distance, and illumination conditions, with the animal's posture, and so on. The visual system is therefore constantly faced with views that are different from all other views seen in the past, and it is required to generalize correctly from past experience and classify correctly the novel image. The variability is complex in nature: it is difficult to provide, for instance, a precise definition for all the allowed variations of dog images. The human visual system somehow learns the characteristics of the allowed variability from experience. This makes classification more difficult for artificial systems than individual identification. In performing identification of a specific car, say, one can supply the system with a full and exact model of the object, and the expected variations can be described with precision. This is the basis for several approaches to identification, for example, methods that use image combinations [29] or interpolation [22] to predict the appearance of a known object under given viewing conditions. In classification, the range of possible variations is wider, since now, in addition to variations in the viewing condition, one must also contend with variations in shape of different objects within the same class.

In this paper we outline an approach to the representation of object classes, and to the task of visual classification, that we call a fragment-based representation. In this approach, images of objects within a class are represented in terms of class-specific fragments. These fragments provide common building blocks that can be used, in different combinations, to represent a large variety of different images of objects within the class. Following a brief review of related approaches, we discuss the problem of selecting a set of fragments that are best suited for representing a class of related objects, given a set of example images. We then illustrate the use of these fragments to perform classification. Finally, we conclude with some comments about similarities between the proposed approach and aspects of the human visual system.

2   A Brief Review of Related Past Approaches

A variety of different approaches have been proposed in the past to deal with visual recognition, including the tasks of general classification and the identification of individual objects. We will review here briefly some of the main approaches, focusing in particular on methods that are applicable to classification, and that are related to the approach developed here.

A popular framework to classification is based on representing object views as points in a high-dimensional feature space, and then performing some partitioning of the space into regions corresponding to the different classes. Typically, a set of n different measurements are applied to the image, and the results constitute an n-dimensional vector representing the image. A variety of different measures have been proposed, including using the raw image as a vector of grey-level values, using global measures such as the overall area of the object's image, different moments, Fourier coefficients describing the object's boundary, or the results of applying selected templates to the image. Partitioning of the space is then performed using different techniques. Some of the frequently used techniques include nearest-neighbor classification to class representatives using, for example, vector quantization techniques, nearest-neighbor to a manifold representing a collection of object or class views [17], separating hyperplanes performed, for example, by Perceptron-type algorithms and their extensions [15], or, more optimally, by support vector machines [30]. The vector of measurements may also serve as an input to a neural network algorithm that is trained to produce different outputs for inputs belonging to different classes [21].

More directly related to our approach are methods that attempt to describe all object views belonging to the same class using a collection of some basic building blocks and their configuration. One well-known example is the Recognition By Components (RBC) method [3] and related schemes using generalized cylinders as building blocks [4, 12, 13]. The RBC scheme uses a small number of generic 3-D parts such as cubes, cones, and cylinders. Objects are described in terms of their main 3-D parts, and the qualitative spatial relations between parts.

Other part-based schemes have used 2-D local image features as the underlying building blocks. These building blocks were typically small simple image features, together with a description of their qualitative spatial relations. Examples of such features include local image patches [1, 18], corners, direct output of local receptive fields of the type found in primary visual cortex [7], wavelet functions [26], simple line or edge configurations [23], or small texture patches [32].

The eigenspace approach [28] can also be considered as belonging to this general approach. In this method, a collection of objects within a class, such as a set of faces, are constructed using combinations of a fixed set of building blocks. The training images are described as a set of grey-level vectors, and the principle components of the training images are extracted. The principal components are then used as the building blocks for describing new images within the class, using linear combination of the basic images. For example, a set of 'eigenfaces' is extracted and used to represent a large space of possible faces. In this approach, the building blocks are global rather than local in nature. As we shall see in the next section, the building blocks selected by our method are intermediate in complexity: they are considerably more complex than simple local features used in previous approaches, but they still correspond to partial rather than global object views.

3   The Selection of Class-Based Fragments

The classification of objects using a fragment-based representation raises two main problems. The first is the selection of appropriate fragments to represent a given class, based on a set of image examples. The second is performing the actual classification based on the fragment representation. In this section we deal with the first of these problems, the selection of fragments that are well-suited for the classification task. Subsequent sections will then deal with the classification process.

Our method for the selection and use of basic building blocks for classification is different from previous approaches in several respects. First, unlike other methods that use local 2-D features, we do not employ universal shape features. That is, the set of basic building blocks used as shape primitives are not the same for all classes, as used, for instance, in the RBC approach. Instead, we use object fragments that are specific to a class of objects, taken directly from example views of objects in the same class. As a result, the shape fragments used to represent faces, for instance, would be different from shape fragments used to represent cars, or letters in the alphabet. These fragments are then used as a set of common building blocks to represent, by different combinations of the fragments, different objects belonging to the class. Second, the fragments we use as building blocks are extracted using an optimization process that is driven directly by requirements of the classification task. This is in contrast with other scheme where the basic building elements are selected on the basis of other criteria, such as faithful reconstruction of the input image. Third, the fragments we detect are organized into equivalence sets that contain views of the same general region in the objects under different transformations and viewing conditions. As we will see later, this novel organization plays a useful role in performing the classification task.

The use of the combination of image fragments to deal with intra-class variability is based on the notion that images of different objects within a class have a particular structural similarity - they can be expressed as combinations of common substructures. Roughly speaking, the idea is to approximate a new image of a face, say, by a combination of images of partial regions, such as eyes, hairline and the like of previously seen faces. In this section we describe briefly the process of selecting class-based fragments for representing a collection of images within a class. We will focus here mainly on computational issues, possible biological implications will be discussed elsewhere. In the following sections, we describe the use of the fragment representation for performing classification tasks.

3.1 Selecting Informative Fragments

Given a set of images that represent different objects from a given class, our scheme selects a set of fragments that are used as a basis for representing the different shapes within the class. Examples of fragments for the class of human faces (roughly frontal) and the class of cars (sedans, roughly side views) are illustrated in Figure 1. The fragments are selected using a criterion of maximizing the mutual information I(C,F) between a class C and a fragment F. This is a natural measure to employ, because it measures how much information is added about the class once we know whether the fragment F is present or absent in the image. In the ensemble of natural images in general, prior to the detection of any fragment, there is an a-priori probability p(C) for the appearance of an image of a given class C. The detection of a fragment F adds information and reduces the uncertainty (measured by the entropy) of the image. We select fragments that will increase the information regarding the presence of an image from the class C by as much as possible, or, equivalently, reduce the uncertainty by as much as possible. This depends on p(F|C), the probabilities of detecting the fragment F in images that come from the class C, and on p(F|NC) where NC is the complement of C.

A fragment F is highly representative of the class of faces if it is likely to be found in the class of faces, but not in images of non-faces. This can be measured by the likelihood ratio p(F|C) / p(F|NC). Fragments with a high likelihood ratio are highly distinctive for the presence of a face. However, highly distinctive features are not necessarily useful fragments for face representation. The reason is that a fragment can be highly distinctive, but very rare. For example, a template depicting an individual face is highly distinctive: its presence in the image means that a face is virtually certain to be present in the image. However, the probability of finding this particular fragment in an image and using it for making classification is low. On the other hand, a simple local feature, such as a single eyebrow, will appear in many more face images, but it will appear in non-face images as well. The most informative features are therefore fragments of intermediate size. In selecting and using optimal fragments for classification, we distinguish between what we call the `merit' of a fragment and its `distinctiveness'. The merit is defined by the mutual information

$$I(C,F) = H(C) - H(C/F) \tag{1}$$

where I is the mutual information, and H denotes entropy [6]. The merit measures the usefulness of a fragment F to represent a class C, and the fragments with maximal merit are selected as a basis for the class representation. The distinctiveness is defined by the likelihood ratio above, and it is used in reaching the final classification decision, as explained in more detail below. Both the merit and the distinctiveness can be evaluated given the estimated value of p(C), p(F|C), p(F|NC). In summary, fragments are selected on the basis of their merit, and then used on the basis of their distinctiveness.

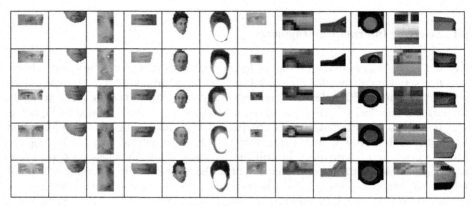

Fig. 1. Examples of face and car fragments

Our procedure for selecting the fragments with high mutual information is the following. Given a set of images, such as cars, we start by comparing the images in a pairwise manner. The reason is that a useful building block that appears in multiple car images must appear, in particular, in two or more images, and therefore the pairwise comparison can be used as an initial filter for identifying image regions that are likely to serve as useful fragments. We then perform a search of the candidate fragments in the entire database of cars, and also in a second database composed of many natural images that do not contain cars. In this manner we obtain estimations for $p(F|C)$ and $p(F|NC)$ and, assuming a particular $p(C)$, we can compute the fragment's mutual information. For each fragment selected in this manner, we extend the search for optimal fragments by testing additional fragments centered at the same location, but at different sizes, to make sure that we have selected fragments of optimal size. The procedure is also repeated for searching an optimal resolution rather than size.

We will not describe this procedure in further detail, except to note that large fragments of reduced resolution are also highly informative. For example, a full-face fragment at high resolution is non-optimal because the probability of finding this exact high-resolution fragment in the image is low. However, at a reduced resolution, the merit of this fragment is increased up to an optimal value, at which it starts to decrease. In our representation we use fragments of intermediate complexity in either size or resolution, and it includes full resolution fragments of intermediate size, and larger fragments of intermediate resolution.

4   Performing Classification

The set of fragments extracted from the training images during the learning stage are then used to classify new images. In this section we consider the problem of performing object classification based on the fragment representation described above.

In performing classification, the task is to assign the image to one of a known set of classes (or decide that the image does not depict any known class). In the following discussion, we consider a single class, such as a face or a car, and the task is to decide whether or not the input image belongs to this class. This binary decision can also be extended to deal with multiple classes. We do not assume that the image contains a single object at a precisely known position, consequently the task includes a search over a region in the image. We can therefore view the classification task also as a detection task, that is, deciding whether the input image contains a face, and locating the position of the face if it is detected in the image.

The algorithm consists of two main stages: detecting fragments in the image, followed by a decision stage that combines the results of the individual fragment detectors. In the first stage, fragments are detected by comparing the image at each location with stored fragment views. The comparison is based on gray-level similarity, that is insensitive to small geometric distortions and gray level variations. As for the combination stage, we have compared two different approaches: a simple and a complex scheme. The simple scheme essentially tests that a sufficient number of basic fragments have been detected. The more complex scheme is based on a more complete probability distribution of the fragments, and takes into account dependencies between pairs of fragments. We found that, using the fragments extracted based on informativeness, the simple scheme is powerful and works almost as well as the more elaborate scheme. This finding is discussed later in this section.

In the following sections we describe the algorithm in more details. We begin by describing the similarity measure used for the detection of the basic fragments.

5 Detecting Individual Fragments

The detection of the individual fragment is based on a direct gray-level comparison between stored fragments and the input image. To allow some distortions and scale changes of a fragment, the comparison is performed first on smaller parts of the fragments, that were taken in the implementation to be patches of size 5*5 pixels. We describe first the gray-level similarity measure used in the comparison, and then how the comparisons of the small patches are used to detect individual fragments.

We have evaluated several gray-level comparison methods, both known and new, to measure similarity between gray level patches in the stored fragment views and patches in the input image. Many of the comparison methods we tested gave satisfactory results within the subsequent classification algorithm, but we found that a method that combined qualitative image based representation suggested by Bhat and Nayar [2] with gradient and orientation measures gave the best results. The method measured the qualitative shape similarity using the ordinal ranking of the pixels in the regions, and also measured the orientation difference using gradient amplitude and direction. For the qualitative shape comparison we computed the ordinal ranking of the pixels in the two regions, and used the normalized sum of displacements of the pixels with the same ordinal ranking as the measure for the regions similarity.

The similarity measure $D(F,H)$ between an image patch H and a fragment patch F is a weighted sum of their sum of these displacements d_i, the absolute orientation difference of the gradients $|\alpha_F - \alpha_H|$ and their absolute gradient difference $|G_F - G_H|$:

$$D(F,H) = k_1 \sum_i d_i + k_2 |\alpha_F - \alpha_H| + k_3 |G_F - G_H| \qquad (2)$$

This measure appears to be successful because it is mainly sensitive to the local structure of the patches and less to absolute intensity values.

For the detection of fragments in the images we first compared local 5x5 gray level patches in each fragment to the image, using the above similarity measure. Only regions with sufficient variability were compared, since in flat-intensity regions the gradient, orientation and ordinal-order have little meaning. We allowed flexibility in the comparison of the fragment view to the image by matching each pixel in the fragment view to the best pixel in some neighborhood around its corresponding location. Most of the computations of the entire algorithm are performed at this stage. To detect objects at different scale in the image, the algorithm is performed on the image at several scales. Each level detects objects at scale differences of ±35%. The combination of several scales enables the detection of objects under considerable changes in their size.

6   Combining the Fragments and Making a Decision

Following the detection of the individual fragments, we have a set of 'active' fragments, that is, fragments that have been detected in the image. We next need to combine the evidence from these fragments and reach a final decision as to whether the class of interest is present in the image.

In this section we will consider two alternative methods of combining the evidence from the individual fragments and reaching a decision. Previous approaches to visual recognition suggest there is a natural trade-off between the use of simple visual features that require a complex combination scheme, and the use of more complex features, but with a simpler combination scheme.

A number of recognition and classification schemes have used simple local image features such as short oriented lines, corners, Gabor or wavelet basis functions, or local texture patches. Such features are generic in nature, that is, common to all visual classes. Consequently, the combination scheme must rely not only on the presence in the image of particular features, but also on their configurations, for example, their spatial relations, and pair-wise or higher statistical interdependencies between the features. A number of schemes using this approach [1, 14, 26, 33], therefore employ, in addition to the detection of the basic features, additional positional information, and probability distribution models of the features. In contrast, a classifier that uses more complex, class-specific visual features could employ a simpler combination scheme because the features themselves already provide good evidence about the presence of the class in question. In the next section, we first formulate the classification as a problem of reaching optimal decision using probabilistic evidence. We then compare experimentally the classification performance of a fragment-based classifier that uses a 'simple' combination method, and one that uses a more complex scheme.

6.1 Probability Distribution Models

We can consider in general terms the problem of reaching a decision about the presence of a class in the images based on some set of measurements denoted by X. Under general conditions, the optimal decision is obtained by evaluating the likelihood ratio defined as:

$\dfrac{P(X \mid C_1)}{P(X \mid C_0)}$, where $P(X|C_0)$, $P(X|C_1)$ are the conditional probabilities of X within

and outside the class. The elements of X express, in the fragment-based scheme, the fragments that have been detected in the image.

The direct use of this likelihood ratio in practice raises computational problems in learning and storing the probability functions involved. A common solution is to use restricted models using assumptions about the underlying probability distribution of the feature vector. In such models, the number of parameters used to encode the probability distribution is considerably smaller than for a complete look-up table representation, and these parameters can be estimated with a higher level of confidence.

A popular and useful method for generating a compact representation of a probability distribution is the use of Belief-Networks, or Bayesian-Networks. A Bayesian-Network is a directed graph where each node represents one of the variables used in the decision process. The directed edges correspond to dependency relationships between the variables, and the parameters are conditional probabilities between inter-connected nodes. A detailed description of Bayesian-Networks can be found in [19]. The popularity of Bayesian-Network methods is due in part to their flexibility and ability to represent probability distributions with dependencies between the variables. There are several methods that enable the construction of a network representation from training data, and algorithms that efficiently compute the probability of all the variables in the network given the observed values of some of the variables. In the following section, we use this formalism to compare two methods for combining the evidence from the fragments detected in the first stage.

One is a simple scheme sometimes called 'naïve Bayesian' method, and the second a more elaborate scheme using a Bayesian-Network type method.

6.2 Naïve-Bayes

The assumption underlying the naive-Bayes classifier is that the entries of the feature vector can be considered independent when the computing the likelihood ratio $\dfrac{P(X \mid C_1)}{P(X \mid C_0)}$. In this case, the class-conditional probabilities can be expressed by the product:

$$P(X_1, \ldots X_N \mid C) = \prod_{i=1}^{N} P(X_i \mid C) \tag{3}$$

The values of the single probabilities are directly measured from the training data. In practice, this means that we first measure the probability of each fragment X_i within and outside the class. To reach a decision we simply multiply the relevant probabilities together. This method essentially assumes independence between the different fragments. (More precisely, it assumes conditional independence.) The actual computation in our classification scheme was performed using the fragment types, rather than the fragments themselves. This means that for each fragment type (such as a hairline or eye region), the best-matching fragment was selected. The combination then proceeds as above, by multiplying the probabilities of these fragments, one from each type.

6.3 Dependence-Tree Combination

The Dependence-tree model is a simple Bayesian-Network that describes a probability distribution which incorporates some relevant pairwise dependencies between variable, unlike the independence assumptions used in the naïve-Bayes scheme. The features are organized into a tree structure that represents statistical dependencies in a manner that allows an efficient computation of the probability of an input vector. The tree structure permits the use of some, but not all, of the dependencies between features. An optimal tree representation can be constructed from information regarding pairwise dependencies in the data [5]. The probability of an input vector is computed by multiplying together the probabilities of each node given the value of its parent. More formally:

$$P(X_1, \ldots X_N \mid C) = P(X_1 \mid C) \times \prod_{i=2}^{N} P(X_i \mid X_{j(i)}, C) \tag{4}$$

where j(i) represents the parent of node i. X_1 is the root of the tree (which represent the class variable), that does not have a parent. The conditional probabilities are estimated during the learning phase directly from the training data.

6.4 The Trade-Off between Feature Complexity and Combination Complexity

We have implemented and compared the two schemes outlined above. This allows us to compare a simple combination scheme that is based primarily on the presence or absence of fragments in the image, and a more elaborate scheme that uses a more refined model of the probability distribution of the fragments. Figure 3 shows the performance of a fragment-based classifier trained to detect side views of cars in low-resolution images, using both combination schemes.

The results are presented in the form of Receiver Operating Characteristic (ROC) curves [9]. Each point in an ROC curve represents a specific pair of false-alarms and hit-rate of the classifier, for a given likelihood ratio threshold. The efficiency of a classifier can be evaluated by the 'height' of its ROC curve: for a given false-alarm rate, the better classifier will be the one with higher hit probability. The overall performance of a classifier can be measured by the area under the performance curve in the ROC graph.

The curves for both methods are almost identical, showing that including pairwise dependencies in the combination scheme, rather than using the information of each feature individually, has a marginal effect on the classifier performance. This suggests that most of the useful information for classification is encoded in the image fragments themselves, rather than their inter-dependencies. This property of the classifier depends on the features used for classification. When simple generic features are used, the dependencies between features at different locations play an important role in the classification process. However, when more complex features are used, such as the ones selected by our information criterion, then a simpler combination scheme will suffice.

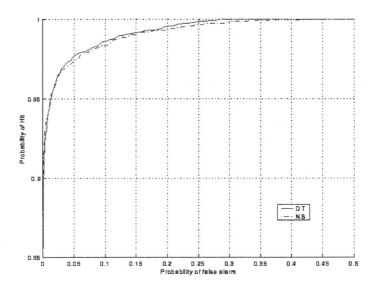

Fig. 2. Receiver Operating Characteristic curves for both classifiers.   NB: Naïve-Bayes classifier. DT: Dependence-Tree classifier.

7 Experimental Results

We have tested our algorithm on face and car views. A number of similar experiments were performed, using somewhat different databases and different details of the fragment extraction and combination procedures. In an example experiment, we tested face detection, using a set of 1104 part views, taken from a set of 23 male face views under three illuminations and three horizontal rotations. The parts were grouped into eight fragment types – eye pair, nose, mouth, forehead, low-resolution view, mouth and chin, single eye and face outline. For cars, we used 153 parts of 6 types. Figure 5 shows several examples. Note that although the system used only male views in few illuminations and rotations, it detects male and female face views under various viewing conditions.

The results of applying the method to these and other images indicate that the fragment-based representation generalizes well to novel objects within the class of interest. Using face fragments obtained from a small set of examples it was possible to classify correctly diverse images of males and females, in both real images and drawings, that are very different from the faces in the original training set.   This was achieved while maintaining low false alarm rates on images that did not contain faces. Using a modest number of informative fragments in different combinations, appears to have an inherent capability to deal with shape variability within the class.   The fragment-based scheme was also capable of obtaining significant position invariance, without using explicit representation of the spatial relationships between fragments. The insensitivity to position as well as to other viewing parameters was obtained primarily by the use of a redundant set of overlapping fragments, including fragments of intermediate size and higher resolution, and fragments of larger size and lower resolution.

8 Some Analogies with the Human Visual System

In visual areas of the primate cortex neurons respond optimally to increasingly complex features in the input. Simple and complex cells in the primary visual area (V1) responds best to a line or edge at a particular orientation and location in the visual field [10]. In higher-order visual areas of the cortex, units were found to respond to increasingly complex local patterns. For example, V2 units respond to collinear arrangements of features [31], some V4 units respond to spiral, polar and other local shapes [8], TE units respond to moderately complex features that may resemble e.g. a lip or an eyebrow [27], and anterior IT units often respond to complete or partial object views [11, 24]. Together with the increase in the complexity of their preferred stimuli, units in higher order visual areas also show increased invariance to viewing parameters, such as position in the visual field, rotation in the image plane, rotation in space, and some changes in illumination [11, 20, 24, 27].

Fig. 3. Examples of face and car detection. The images are from the Weizmann image database, from the CMU face detector gallery, and the last two from www.motorcities.com.

The preferred stimuli of IT units are highly dependent upon the visual experience of the animal. In monkeys trained to recognize different wire objects, units were found that respond to specific full or partial views of such objects [11]. In animals trained with fractal-like images, units were subsequently found that respond to one or more of the images in the training set [16].

These findings are consistent with the view that the visual system uses object representations based on class related fragments of intermediate complexity, constructed hierarchically. The preferred stimuli of simple and intermediate complexity neurons in the visual system are specific 2-D patterns. Some binocular information can also influence the response, but this additional information, which adds 3-D information associated with a fragment under particular viewing conditions, can be incorporated in the fragment based representation. The lower level features are simple generic features. The preferred stimuli of the more complex units are dependent upon the family of training stimuli, and appear to be class-dependent rather than, for example, a small set of universal building blocks. Invariance to viewing parameters such as position in the visual field or spatial orientation appears gradually, possibly by the convergence of more elementary and less invariant fragments onto higher order units. From this theory we can anticipate the existence of two types of intermediate complexity units that have not been reported so far. First, for the purpose of classification, we expect to find units that respond to different types of partial views. As an example, a unit of this kind may respond to different shapes of hairline, but not to a mouth or nose regions. Second, because the invariance of complex shapes to different viewing parameters is inherited from the invariance of the more elementary fragment, we expect to find intermediate complexity units, responding to partial object views at a number of different spatial orientations and perhaps different illumination conditions.

Acknowledgement. This research was supported in part by the Israel Ministry of Science under the Scene Teleportation Research Project. We acknowledge the use of material reported in the Proceedings of the Korean BMCV meeting, Seoul, 1999.

References

1. Amit, Y.,Geman, D., Wilder, K.: Joint Induction of Shape Features and Tree Classifiers. IEEE Trans. on Pattern Analysis and Machine Intelligence, Vol. 19, 11 (1997) 1300-1306
2. Bhat, D., Nayar, K. S.: Ordinal Measures for Image Correspondence. IEEE Trans. on Pattern Analysis and Machine Intelligence, Vol. 20, 4 (1998) 415-423
3. Biederman, I.: Human image understanding: recent research and theory. Computer Vision, Graphics and Image Processing 32 (1985) 29-73
4. Binford, T. O.: Visual perception by computer. IEEE conf. on systems and control, Vol. 94, 2 (1971) 115-147

5. Chow, C.K., Liu, C.N.: Approximating Discrete Probability Distributions with Dependence Trees. IEEE Transactions on Information Theory, Vol. 14, 3 (1968) 462-467
6. Cover, T.M. & Thomas, J.A.: Elements of Information Theory. Wiley Series in Telecommunication, New York (1991)
7. Edelman, S.: Representing 3D objects by sets of activities of receptive fields. Biological cybernitics 70 (1993) 37-45
8. Gallant, J.L., Braun, J., Van Essen, D.C.: Selectivity for polar, hyperbolic, and cartesian gratings in macaque visual cortex. Science, 259 (1993) 100-103
9. Green, D. M., Swets, J. A.: Signal Detection Theory and Psychophysics. Wiley, Chichester New York Brisbane Toronto (1966). Reprinted by Krieger, Huntingdon, New York (1974)
10. Hubel, D. H., Wiesel, T. N.: Receptive fields and functional architecture of monkey striate cortex. Journal of physiology 195 (1968) 215-243
11. Logothetis, N. K., Pauls J., Bülthoff H. H., Poggio T.: View-dependent object recognition in monkeys. Current biology, 4 (1994) 401-414
12. Marr, D.: Vision. W.H. Freeman, San Francisco (1982)
13. Marr, D., Nishihara, H. K.: Representation and recognition of the spatial organization of three dimensional structure. Proceedings of the Royal Society of London B, 200 (1978) 269-294
14. Mel, W. B.: SEEMORE: Combining color, shape and texture histogramming in a neurally inspired approach to visual object recognition. Neural computation 9 (1997) 777-804
15. Minsky, M., Papert, S.: Perceptrons. The MIT Press, Cambridge, Massachusetts (1969)
16. Miyashita, Y.,Chang, H.S.: Neuronal correlate of pictorial short-term memory in the primate temporal cortex. Nature, 331, (1988) 68-70
17. Murase, H.,Nayar, S.K.: Visual learning and recognition of 3-D objects from appearance. International J. of Com. Vision, 14 (1995) 5-24
18. Nelson, C. R., Selinger A.: A Cubist approach to object recognition. International Conference on Computer Vision '98 (1998) 614-621
19. Pearl, J.: Probabilistic Reasoning in Intelligent Systems: Networks of Plausible Inference. Morgan Kaufman Publishers, San Mateo, California (1988)
20. Perret, D. I., Rolls, E. T., Caan W.: Visual neurons responsive to faces in the monkey temporal cortex. Experimental brain research, 47 (1982) 329-342
21. Poggio, T., Sung, K.: Finding human faces with a gaussian mixture distribution-base face model. Computer analysis of image and patterns (1995) 432-439
22. Poggio, T., Edelman, S.: A network that learns to recognize three-dimensional objects. Nature, 343 (1990) 263-266
23. Riesenhuber, M.,Poggio, T.: Hierarchical models of object recognition in cortex. Nature Neuroscience, Vol. 2, 11 (1999) 1019-1025
24. Rolls, E. T.: Neurons in the cortex of the temporal lobe and in the amygdala of the monkey with responses selective for faces. Human neurobiology, 3 (1984) 209-222
25. Rosch, E. Mervis, C.B., Gray, W.D., Johnson, S.M., Boyes-Braem, P.: Basic objects in natural categories. Cognitive Psychology, 8 (1976) 382-439
26. Schneiderman, H., Kanade. T.: Probabilistic modeling of local appearance and spatial relationships for object recognition. Proc. IEEE Comp. Soc. Conference on Computer Vision and Pattern Recognition, CVPR98, (1998) 45-51
27. Tanaka, K.: Neural Mechanisms of Object Recognition. Science, Vol. 262 (1993) 685-688.

28. Turk M., Pentland A.: "Eigenfaces for recognition", Cognitive Neuroscience, 3 (1990) 71-86
29. Ullman, S., Basri, R.: Recognition by Linear Combination of Models. IEEE Trans. on Pattern Analysis and Machine Intelligence, Vol. 13, 10 (1991) 992-1006
30. Vapnik, V.: The Nature of Statistical Learning Theory. Springer, New York (1995)
31. von der Heydt, R., Peterhans, E., Baumgartner G.: Illusory Contours and Cortical Neuron Responses. Science, 224 (1984) 1260-1262
32. Weber, M, Welling M. & Perona, P.: Towards Automatic Discovery of Object Categories. Proc. IEEE Comp. Soc. Conference on Computer Vision and Pattern Recognition, CVPR2000, 2, (2000) 101-108
33. Wiskott, L., Fellous J. M., Krüger N., von der Malsburg, C.: Face Recognition by Elastic Bunch Graph Matching. Intelligent Biometric Techniques in Fingerprint and Face Recognition, 11 (1999) 355-396

Minimum-Length Polygons in Approximation Sausages

Tetsuo Asano[1], Yasuyuki Kawamura[1], Reinhard Klette[2], and Koji Obokata[1]

[1] School of Information Science, JAIST
Asahidai, Tatsunokuchi, 923-1292 Japan
t-asano@jaist.ac.jp
[2] CITR, University of Auckland, Tamaki Campus, Building 731
Auckland, New Zealand
r.klette@auckland.ac.nz

Abstract. The paper introduces a new approximation scheme for planar digital curves. This scheme defines an approximating sausage 'around' the given digital curve, and calculates a minimum-length polygon in this approximating sausage. The length of this polygon is taken as an estimator for the length of the curve being the (unknown) preimage of the given digital curve. Assuming finer and finer grid resolution it is shown that this estimator converges to the true perimeter of an r-compact polygonal convex bounded set. This theorem provides theoretical evidence for practical convergence of the proposed method towards a 'correct' estimation of the length of a curve. The validity of the scheme has been verified through experiments on various convex and non-convex curves. Experimental comparisons with two existing schemes have also been made.

Keywords: Digital geometry, digital curves, multigrid convergence, length estimation.

1 Introduction and Preliminary Definitions

Approximating planar digital curves is one of the most important topics in image analysis. An approximation scheme is required to ensure convergence of estimated values such as curve length toward the true length assuming a digitization model and an increase in grid resolution. For example, the *digital straight segment approximation method* (DSS method), see [3,8], and the *minimum length polygon approximation method* assuming one-dimensional grid continua as boundary sequences (MLP method), see [9], are methods for which there are convergence theorems when specific convex sets are assumed to be the given input data, see [6,7,10]. This paper studies the convergence properties of a new minimum length polygon approximation method based on so-called *approximation sausages* (AS-MLP method).

Motivations for studying this new technique are as follows: the resulting DSS approximation polygon depends upon starting point and the orientation of the

C. Arcelli et al. (Eds.): IWVF4, LNCS 2059, pp. 103–112, 2001.

boundary scan, it is not uniquely defined, but it may be calculated for any given digital object. The MLP approximation polygon is uniquely defined, but it assumes a one-dimensional grid continua as input which is only possible if the given digital object does not have cavities of width 1 or 2. The new method leads to a uniquely defined polygon, and it may be calculated for any given digital object.

Let r be the *grid resolution* defined as being the number of grid points per unit. We consider *r-grid points* $g_{i,j}^r = (i/r, j/r)$ in the Euclidean plane, for integers i, j. Any r-grid point is assumed to be the center point of an *r-square* with *r-edges* of length $1/r$ parallel to the coordinate axes, and *r-vertices*.

The digitization model for our new approximation method is just the same as that considered in case of the DSS method, see [4,5,6]. That is, let S be a set in the Euclidean plane, called *real preimage*. The set $C_r(S)$ is the union of all those r-squares whose center point $g_{i,j}^r$ is in S. The boundary $\partial C_r(S)$ is the *r-frontier* of S. Note that $\partial C_r(S)$ may consists of several non-connected curves even in the case of a bounded convex set S. A set S is *r-compact* iff there is a number $r_S > 0$ such that $\partial C_r(S)$ is just one (connected) curve, for any $r \geq r_0$. This definition of r-compactness has been introduced in [6] in the context of showing multigrid convergence of the DSS method.

The validity of the proposed scheme has been verified through experiments on various curves, which are described in Section 5. It has also been compared with the existing schemes in convergence and computation time.

2   Approximation Scheme

Given a connected region S in the Euclidean plane and a grid resolution r, the r-frontier of S is uniquely determined. We consider r-compact sets S, and grid resolutions $r \geq r_S$ for such a set, i.e. $\partial C_r(S)$ is just one (connected) curve. In such a case the r-frontier of S can be represented in the form $P = (v_0, v_1, \ldots, v_{n-1})$ in which the vertices are clockwise ordered so that the interior of S lies to the right of the boundary. Note that all arithmetic on vertex indices is modulo n.

Let δ be a real number between 0 and $1/(2r)$. For each vertex of P we define forward and backward shifts: The *forward shift* $f(v_i)$ of v_i is the point on the edge (v_i, v_{i+1}) at the distance δ from v_i. The *backward shift* $b(v_i)$ is that on the edge (v_{i-1}, v_i) at the distance δ from v_i.

For example, in the approximation scheme as detailed below we will replace an edge (v_i, v_{i+1}) by a line segment $(v_i, f(v_{i+1}))$ interconnecting v_i and the forward shift of v_{i+1}, which is referred to as the *forward approximating segment* and denoted by $L_f(v_i)$. The *backward approximating segment* $(v_i, b(v_{i-1}))$ is defined similarly and denoted by $L_b(v_i)$. Refer to Fig. 1 for illustration. Now we have three sets of edges, original edges of the r-frontier, forward and backward approximating segments. Let $0 < \delta \leq .5/r$. Based on these edges we define a connected region $A_r^\delta(S)$, which is homeomorphic to the annulus, as follows:

Given a polygonal circuit P describing an r-frontier in clockwise orientation. By reversing P we obtain a polygonal circuit Q in counterclockwise order. In

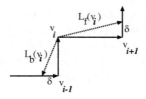

Fig. 1. Definition of the forward and backward approximating segments associated with a vertex v_i.

the initialization step of our approximation procedure we consider P and Q as the *external* and *internal* bounding polygons of a polygon P_B homeomorphic to the annulus. It follows that this initial polygon P_B has area contents zero, and as a set of points it coincides with $\partial C_r(S)$.

Now we 'move' the external polygon P 'away' from $C_r(S)$, and the internal polygon Q 'into' $C_r(S)$ as specified below. This process will expand P_B step by step into a final polygon which contains $\partial C_r(S)$, and where the Hausdorff distance between P and Q becomes non-zero. For this purpose, we add forward and backward approximating segments to P and Q in order to increase the area contents of the polygon P_B.

To be precise, for any forward or backward approximating segment $L_f(v_i)$ or $L_b(v_i)$ we first remove the part lying in the interior of the current polygon P_B and updating the polygon P_B by adding the remaining part of the segment as a new boundary edge. The direction of the edge is determined so that the interior of P_B lies to the right of it.

Definition 1. *The resulting polygon P_B^δ is referred to as the* approximating sausage *of the r-frontier and denoted by $A_r^\delta(S)$.*

The width of such an approximating sausage depends on the value of δ. It is easy to see that as far as the value of δ is at most half of the grid size, i.e., less or equal $1/(2r)$, the approximating sausage $A_r^\delta(S)$ is well defined, that is, it has no self-intersection. It is also immediately clear from the definition that the Hausdorff distance from the r-frontier $\partial C_r(S)$ to the boundary of the sausage $A_r^\delta(S)$ is at most $\delta \leq 1/(2r)$.

We are ready to define the final step in our AS-MLP approximation scheme for estimating the length of a digital curve. Our method is similar to that of the MLP as introduced in [9].

Definition 2. *Assume a region S having a connected r-frontier. An AS-MLP curve for approximating the boundary of S is defined as being a shortest closed curve $\gamma_r^\delta(S)$ lying entirely in the interior of the approximating sausage $A_r^\delta(S)$, and encircling the internal boundary of $A_r^\delta(S)$.*

It follows that such an AS-MLP curve $\gamma_r^\delta(S)$ is uniquely defined, and that it is a polygonal curve defined by finitely many straight segments. Note that this curve depends upon the choice of the approximation constant δ. An example of such a shortest closed curve $\gamma_r^\delta(S)$ is given in Fig. 2, with $\delta = .5/r$.

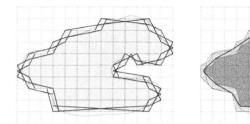

Fig. 2. Left: construction of approximating sausage. Right: approximation by shortest internal path.

3   Properties of the Digital Curve

We discuss some of the properties of the approximating polygonal curve $\gamma_r^\delta(S)$ defined above, assuming that $\partial C_r(S)$ is a single connected curve.

Non-selfintersection: The AS-MLP curve $\gamma_r^\delta(S)$ is defined as being a shortest closed curve lying in the approximating sausage. Since it is obvious from the definition that the sausage has no self-intersection, so does the curve.

Controllability: The width of an approximating sausage can be controlled by selecting a value of δ, with $0 < \delta \le .5/r$.

Smoothness: Compared with the other two approximation schemata DSS and MLP, our approximating curve is 'more smooth' in the following sense: the angle associated with a corner of an approximating polygon is the smaller one of its internal angle and external angle. We consider the minimum angle of all these angles associated with a corner of the AS-MLP curve. Similarly, such minimum angles may be defined for approximating DSS and MLP curves. It holds that the minimum AS-MLP angle is always greater than or equal to the minimum DSS or minimum MLP angle, if a convex set S has been digitized. Note that 'no small angle' means 'no sharp corner'.

Linear complexity: Due to the definition of our curve $\gamma_r^\delta(S)$ the number of its vertices is at most twice that of the r-frontier.

Computational complexity: Assuming that a triangulation of an approximating sausage is given, linear computation time suffices to find a shortest closed path: we can triangulate an approximating sausage in linear time since the vertices of the sausage can be calculated only using nearby segments. So, linear time is enough to triangulate it. Then, we can construct an adjacency graph, which is a tree, representing adjacency of triangles again in linear time. Finally, we can find a shortest path in linear time by slightly modifying the linear-time algorithm for finding a shortest path within a simple polygon.

Figure 3 gives visual comparisons of the proposed AS-MLP method with two existing schemes DSS and MLP.

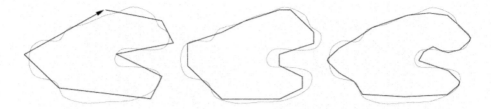

Fig. 3. Original region with DSS (left), MLP (center), and proposed approximation using $\delta = .5/r$ (right).

4   Convergence Theorem

In this section we prove the main result of this paper about the multigrid convergence of the AS-MLP curve based length estimation of the perimeter of a given set S.

Theorem 1. *The length of the approximating polygonal curve $\gamma_r^\delta(S)$ converges to the perimeter of a given region S if S is a r-compact polygonal convex bounded set and $0 < \delta \leq .5/r$.*

We sketch a proof of this theorem with an investigation of geometric properties of the r-frontier of a convex polygonal region S.

   We first classify r-grid points into interior and exterior ones depending on whether they are located inside of the region S or not. Then, CH_{in} is defined to be the convex hull of the set of all interior r-grid points. CH_{out} is the convex hull of the set of those exterior r-grid points adjacent horizontally or vertically to interior ones. See Fig. 4 for illustration.

Lemma 1. *The difference between the lengths of CH_{in} and CH_{out} is exactly $4\sqrt{2}/r$.*

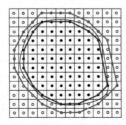

Fig. 4. Interior r-grid points (filled circles) and exterior points (empty circles) with the convex hulls CH_{in} of a set of interior points and CH_{out} of a set of exterior points adjacent to interior ones.

Now, we are ready to state the following lemma which is of crucial importance for proving the convergence theorem.

Lemma 2. *Given an r-compact polygonal convex bounded set S, the approximating polygonal curve $\gamma_r^\delta(S)$ is contained in the region bounded by CH_{in} and CH_{out}, for $0 < \delta \leq .5/r$.*

Let CH be the convex hull of the set of vertices of the approximating polygonal curve $\gamma_r^\delta(S)$. The convex hull CH is also bounded by CH_{in} and CH_{out}. Obviously, the vertices of CH are all intersections of approximating segments. Furthermore, exterior intersections do not contribute to CH, where external (internal, resp.) intersections are those on the external (internal, resp.) boundary of the approximating sausage. Therefore, we can evaluate the perimeter of CH. An increase in distance of an internal intersection from the boundary of CH_{in} corresponds to an increase in length of an approximating segment, and a decrease of distance of its associated intersection to the inner hull CH_{in}. Thus, such an intersection is farthest at a corner defined by two unit edges. Thus, the maximum distance from CH_{in} to CH is bounded by $\sqrt{2}/6$, which implies that the perimeter of CH is bounded by $\sqrt{2}\pi/3$.

Lemma 3. *Let CH be the convex hull of all internal intersections defined above. Then, the approximating polygonal curve $\gamma_r^\delta(S)$ lies between the two convex hulls CH_{in} and CH. The maximum gap between CH_{in} and CH is bounded by $\sqrt{2}/6$, and for their perimeter we have*

$$Perimeter(CH) \leq Perimeter(CH_{in}) + 4\sqrt{2}/r. \tag{1}$$

So, if the approximating polygonal curve $\gamma_r^\delta(S)$ is convex, then we are done. Unfortunately, it is not always convex. In the remaining part of this section we evaluate the largest possible difference on lengths between $\gamma_r^\delta(S)$ and CH.

Lemma 4. *The approximating polygonal curve $\gamma_r^\delta(S)$ is concave when two consecutive long edges of lengths d_{i-1} and d_i with intervening unit edge satisfy $d_i > 3d_{i-1} + 1$.*

By analysis of the possible differences from the convex chain, we obtain the following theorem.

Theorem 2. *Let S be a bounded, convex polygonal set. Then, there exists a grid resolution r_0 such that for all $r \geq r_0$ it holds that any AS-MLP approximation of the r-frontier $\partial C_r(S)$, with $0 < \delta \leq .5/r$, is a connected polygon with a perimeter l_r and*

$$|Perimeter(S) - l_r| \leq (4\sqrt{2} + 8 * 0.0234)/r = 5.844/r. \tag{2}$$

5   Experimental Evaluation

We have seen above that the perimeter estimation error by AS-MLP is bounded in theory by $5.8/r$ for a grid resolution r, for convex polygons. To illustrate

Fig. 5. Experimental objects.

its practical behavior we report on experiments on various curves, which are described below. Although we have restricted ourselves to convex objects in the preceding proof, we took non-convex curves as well in these experiments. Figure 6 illustrates a set of objects used for experiments as suggested in [5].

CIRCLE: the equation of the circle is

$$(x - 0.5)^2 + (y - 0.5)^2 = 0.4^2.$$

YINYANG: the lower part of the yinyang symbol is composed by arcs of 3 half circles: the lower arc is a part of CIRCLE, and the upper arcs are parts of circles whose sizes are half of CIRCLE.

LUNULE: this object is the remainder of two circles, where the distance between both center points is 0.28.

SINC: the sinc equation corresponding to the upper curve is

$$y = sin\left(\frac{\pi x}{4\pi x}\right).$$

SQUARE: the edges of the isothetic SQUARE are of length 0.8.

5.1 Two Existing Approximation Schemes

We sketch both existing schemes which are used for comparisons, where the DSS and MLP implementation reported in [4] has been used for experimental evaluation. First, the digital straight segment (DSS) algorithm traces an r-frontier, i.e. vertices and edges on $\partial C(S)$, i.e. a boundary of $C(S)$, and detects a consecutive sequence of maximum length DSSs. The sum of the lengths of these DSS is used as DSS curve length estimator. The DSS algorithm runs in linear time.

The minimum-length polygon (MLP) approximation needs two boundaries, of set $I(S)$ and of set $O(S)$, as input. Roughly saying, $I(S)$ is the union of r-squares that are entirely included in S, in other words, all four r-vertices of such a square are included in a convex set S; and $O(S)$ is obtained by 'expanding' $I(S)$ by a dilation using one r-square as structuring element. The MLP algorithm calculates the shortest path in the area $O(S) \setminus I(S)$ circumscribing the boundary of $I(S)$. The length of such a shortest path is used as MLP curve length estimator. The MLP algorithm also takes linear time.

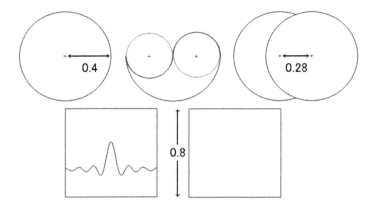

Fig. 6. Test sets drawn in unit size.

In the experiments we computed the errors of three approximation schemes for the specified objects digitized in grid resolutions $r = 32 \sim 1024$. For DSS and AS-MLP, $C(S)$ was used as a digitized region, where $C(S)$ is a set of pixels whose midpoints are included in S. For MLP, $I(S)$ and its expansion was used.

5.2   Experiments

Following the given implementations of DSS, and MLP, also our new AS-MLP scheme has been implemented in C++ for comparisons. We have computed the curve length error in percent compared to the true perimeter of a given set S. The error E_{DSS} of the DSS estimation scheme is defined by

$$E_{DSS} = \frac{P(S) - P(DSS_S)}{P(S)}$$

where $P(S)$ is the true perimeter of S and $P(DSS_S)$ is the perimeter of the approximation polygon given by the DSS scheme. E_{MLP} and E_{ASMLP} are analogously defined.

Figure 7 shows the errors for all five test curves, the boundaries of CIRCLE, YINYANG, LUNULE, SINC, and SQUARE in that order, from top to bottom. The diagrams for DSS, MLP, and AS-MLP are arranged from left to right, in each row of the figure. The graphs illustrate that AS-MLP has smaller errors in general than MLP has, but DSS is the best among the three.

6   Conclusion

We proposed a new approximation scheme for planar digital curves and analyzed its convergence to the true curve length by stating a theorem for convex sets. To

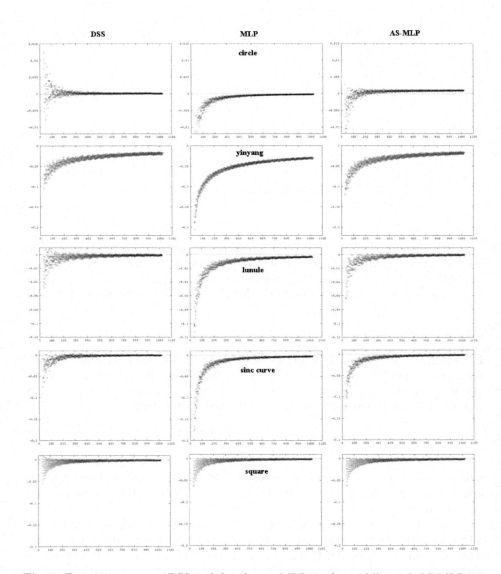

Fig. 7. Estimation errors: DSS in left column, MLP in the middle, and AS-MLP on the right; top row for circle, followed by lower part of yinyang, lunule, sinc curve, and square.

verify its practical performance we have implemented this scheme and tested it on various curves including non-convex ones. The results reflected the theoretical analysis of the three schemes, that is, DSS is the best in accuracy, and our scheme is in the middle. The AS-MLP approximation curves are smoother (see our definition above) than the MLP or DSS curves,

Acknowledgment. The used C++ programs for DSS and MLP are those implemented for and described in [4], and the test data set has been copied from [5].

References

1. T. Asano, Y. Kawamura, R. Klette, and K. Obokata. A new approximation scheme for digital objects and curve length estimations. in *Proc. of Image and Vision Computing New Zealand*, Hamilton, pages 26–31, 2000.
2. T. Buelow and R. Klette. Rubber band algorithm for estimating the length of digitized space-curves. in *Proc. IEEE Conf. ICPR*, Barcelona, Vol. III, pages 551-555, 2000.
3. A. Hübler, R. Klette, and K. Voss. Determination of the convex hull of a finite set of planar points within linear time. *EIK*, 17:121–139, 1981.
4. R. Klette, V. Kovalevsky, and B. Yip. On the length estimation of digital curves. in *Proc. Vision Geometry VIII*, Denver, SPIE-3811, pages 117–129, 1999.
5. R. Klette and Ben Yip. The length of digital curves. *Machine GRAPHICS & VISION*, 9:673–703, 2000.
6. R. Klette and J. Žunić. Multigrid convergence of calculated features in image analysis. *J. Mathem. Imaging and Vision*, 173–191, 2000.
7. V. Kovalevsky and S. Fuchs. Theoretical and experimental analysis of the accuracy of perimeter estimates. in *Robust Computer Vision* (W. Förstner, S. Ruwiedel, eds.), Wichmann, Karlsruhe, pages 218–242, 1992.
8. A. Rosenfeld. Digital straight line segments. *IEEE Trans. Comp.*, 28:1264–1269, 1974.
9. F. Sloboda, B. Zaťko, and P. Ferianc. Minimum perimeter polygon and its application. in *Theoretical Foundations of Computer Visi/-on* (R. Klette, W.G. Kropatsch, eds.), Mathematical Research **69**, Akademie Verlag, Berlin, pages 59-70, 1992.
10. F. Sloboda, B. Zaťko, and J. Stoer. On approximation of planar one-dimensional continua. in *Advances in Digital and Computational Geometry*, (R. Klette, A. Rosenfeld and F. Sloboda, eds.) Springer, pages 113–160, 1998.

Optimal Local Distances for Distance Transforms in 3D Using an Extended Neighbourhood

Gunilla Borgefors and Stina Svensson

Centre for Image Analysis, Swedish University of Agricultural Sciences
Lägerhyddvägen 17, SE-75237 Uppsala, SWEDEN
{gunilla|stina}@cb.uu.se

Abstract. Digital distance transforms are useful tools for many image analysis tasks. In the 2D case, the maximum difference from Euclidean distance is considerably smaller when using a 5×5 neighbourhood compared to using a 3×3 neighbourhood. In the 3D case, weighted distance transforms for neighbourhoods larger than $3 \times 3 \times 3$ has almost not been considered so far. We present optimal local distances for an extended neighbourhood in 3D, where we use the three weights in the $3 \times 3 \times 3$ neighbourhood together with the $(2, 1, 1)$ weight from the $5 \times 5 \times 5$ neighbourhood. A good integer approximation is shown to be $\langle 3, 4, 5, 7 \rangle$.

1 Introduction

Digital distance transforms have been used for computing distances in images since the 60s, [6,7]. The basic idea is to, by using propagation of local distances, approximate the Euclidean distance in a computationally convenient way. The reason the Euclidean distance is approximated it that it is rotation independent (up to digitization effects). The distance transform can be computed during two scans, one forward and one backward, over the image, considering only a small neighbourhood around each pixel/voxel.

Compared to the earliest approach, which was based on the computation of the number of steps in a minimal path, the Euclidean distance can be better approximated by using different weights for steps representing different neighbour relations. For 2D images, optimal local distances in neighbourhoods up to 13×13 have been investigated [1,8]. The maximum difference to the Euclidean distance was considerably smaller when using a 5×5 neighbourhood compared to a 3×3 neighbourhood, 1.41% compared to 4.49% for optimal weights.

In a distance transform certain regularity criteria for the weights need to be satisfied. In fact, only rather limited intervals for the different weights are allowed. In [4], necessary conditions for an nD distance transform to be a metric are presented. In 3D, optimization of local distances, in a $3 \times 3 \times 3$ neighbourhood, fulfilling these conditions is presented in [3]. It was shown that there are two different valid cases, for which the distance functions, and hence the optimizations, are different.

C. Arcelli et al. (Eds.): IWVF4, LNCS 2059, pp. 113–122, 2001.
© Springer-Verlag Berlin Heidelberg 2001

In this paper, we will compute the optimal local distances for an extended neighbourhood in 3D, where we use the three weights in the $3 \times 3 \times 3$ neighbourhood together with the $(2, 1, 1)$ weight from the $5 \times 5 \times 5$ neighbourhood. As for the $3 \times 3 \times 3$ neighbourhood, different cases occur, more precisely eight difference cases. We will illustrate all cases and optimize for the most straightforward of them.

An early attempt to find optimal local distances in a $5 \times 5 \times 5$ neighbourhood can be found in [9]. However, nothing is mentioned about the validity of the resulting distance transforms with respect to the above mentioned regularity criteria. Weights for 3D distance transforms in extended neighbourhoods have recently also been treated in [5]. There, regularity and optimization criteria different from ours are used, and only integer weights are considered.

2   Geometry and Equations

Consider a 3D bi-level image, consisting of object and background voxels. In the corresponding distance image, or a distance transform (DT), each voxel in the object can be labelled with the distance to the closest voxel in the background. A good underlying concept for all digital distance, introduced in [10], is:

Definition 1. *The distance between two points $\boldsymbol{x}$ and $\boldsymbol{y}$ is the length of the shortest path connecting $\boldsymbol{x}$ to $\boldsymbol{y}$ in an appropriate graph.*

Each voxel has three types of neighbours in its $3 \times 3 \times 3$ neighbourhood: 6 face neighbours, 12 edge neighbours, and 8 point neighbours. A path between two voxels in a 3D image can thus include steps in 26 directions, if only steps between immediate neighbours are allowed. In a $5 \times 5 \times 5$ neighbourhood, each voxel has 124 neighbours, belonging to one of six different types, i.e., the types in the $3 \times 3 \times 3$ neighbourhood plus three new types d, e, and f. See Fig. 1 for the various possible steps in a $5 \times 5 \times 5$ neighbourhood (each of course occur in several rotations). These six possible steps are usually called *prime steps* or *local steps*. The corresponding distances are called *local distances*.

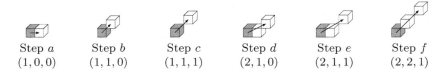

| Step a | Step b | Step c | Step d | Step e | Step f |
| $(1, 0, 0)$ | $(1, 1, 0)$ | $(1, 1, 1)$ | $(2, 1, 0)$ | $(2, 1, 1)$ | $(2, 2, 1)$ |

Fig. 1. Local distances in a $5 \times 5 \times 5$ neighbourhood of a voxel (grey).

Optimizing for all six local distances is complicated and may not be necessary to discover new useful weighted distance transforms that are considerably better than the ones using only a $3 \times 3 \times 3$ neighbourhood. Here, we add only one of the three possible steps in the $5 \times 5 \times 5$ neighbourhood, the e step. Adding this step can be assumed to make the most difference to the $3 \times 3 \times 3$ weighted DT, as its Euclidean length, $\sqrt{6}$, provides the largest decrease compared to using two a, b, c steps, $a + c = 1 + \sqrt{3}$ (cf. $d = a + b$, $f = b + c$). Expressed in another way: the "corner" cut off by this new step is the sharpest of the three d, e, or f corners. The new distance is denoted $\langle a, b, c, e \rangle$.

Let $DT(i, j, k)$ for an object voxel $v(i, j, k)$ be the minimal length of a path connecting v to any voxel w in the background using only the local steps $\langle a, b, c, e \rangle$. Due to symmetry, when computing optimal local distances, it is enough to consider distances from the origin to a voxel (x, y, z), where $0 \leq z \leq y \leq x \leq M$ and M is the maximum dimension of the image.

The length of a minimal path is independent of the order in which the steps are taken. We can, therefore, assume that the minimal path consists of a number of straight line segments, equal to the number of different directions used.

Not all combinations of local distances a, b, c, and e are useful. The following property is desirable:

Definition 2. *Consider two voxels that can be connected by a straight line, i.e., by using one type and direction of local step. If that line defines a distance between the voxels, i.e., is a minimal path, then the resulting DT is semi-regular. If there are no other minimal paths, then the DT is regular.*

In [4], necessary conditions for weighted distance transforms in dimensions ≥ 2 to be a metric, i.e., the corresponding distance function is positive definite, symmetric and fulfils the triangle inequality, are presented. The result is summarized by Theorem 1.

Theorem 1. *A distance transform in $\mathbf{Z}^n$ that is a metric is semi-regular. A semi-regular distance transform in $\mathbf{Z}^2$ is a metric.*

Theorem 1 implies for the 2D case that a weighted DT is a metric if and only if it is semi-regular, and for the 3D case that a metric DT is semi-regular. Thus, 3D DTs should be semi-regular, even though it is not a sufficient condition for being metric.

For the $\langle a, b, c, e \rangle$, we can have four types of straight paths (Def. 2). We need to investigate these straight lines and find the ways they can be approximated using other steps, to find conditions for the straight path to be the shortest. By such investigation (see [3] for the procedure) we find that the semi-regularity criteria are given by the inequalities in (1), which defines a hyper-volume in a, b, c, e parameter space.

$$ e \leq a + c \qquad e \leq 2b \qquad b \leq c \qquad 2e \geq 3b \qquad 4c \leq 3e \qquad (1) $$

The regularity criteria are not sufficient to determine unique expressions for the weighted DTs. We have found that the hyper-volume (1) is divided into eight subspaces, originating eight different cases of semi-regular DTs. (For the $3 \times 3 \times 3$ neighbourhood, there are two different cases, [3].) The eight Cases occur in the following hyper-volumes:

$$\text{Case I:} \begin{cases} e \geq a+b \\ e \geq \frac{b}{2}+c \end{cases} \qquad \text{Case II:} \begin{cases} e \leq a+b \\ e \geq \frac{b}{2}+c \\ e \geq a+2b-c \end{cases}$$

$$\text{Case III:} \begin{cases} e \geq \frac{b}{2}+c \\ e \leq a+2b-c \end{cases} \qquad \text{Case IV:} \begin{cases} e \leq a+b \\ e \leq \frac{b}{2}+c \\ e \geq a+2b-c \\ e \geq -a+b+c \end{cases}$$

$$\text{Case V:} \begin{cases} e \geq a+b \\ e \leq \frac{b}{2}+c \\ e \leq -a+b+c \end{cases} \qquad \text{Case VI:} \begin{cases} e \geq -a+b+c \\ e \leq b+c \\ e \leq a+2b-c \end{cases}$$

$$\text{Case VII:} \begin{cases} e \geq a+b \\ e \leq -a+b+c \end{cases} \qquad \text{Case VIII:} \begin{cases} e \leq a+b \\ e \leq -a+b+c \end{cases}$$

The differences between two of the possibilities, Case I and Case VIII, are illustrated by path maps for all voxels $0 \leq z \leq y \leq x \leq 4$ in Fig. 2. From the path maps we can see that, e.g., the minimal path from $(0,0,0)$ to $(2,2,1)$ consists of a combination of one b- and one c-step in Case I and by one a- and one e-step in Case VIII. Examples of $\langle a, b, c, e \rangle$-balls computed for Case I-VIII with strict inequalities are shown in Fig. 3. They are all polyhedra, but with different configurations of vertices and faces.

In the following, we will focus on the first of the eight cases. This case can perhaps be said to be the "natural" one, in the sense that it is the case where $\langle 1, \sqrt{2}, \sqrt{3}, \sqrt{6} \rangle$ is placed, i.e., where the Euclidean distances are used as local distances, and where the equations are the "expected" ones. The parameter space is limited by the inequalities (2) (from (1) and Case I).

$$e \leq a+c \qquad e \geq \frac{b}{2}+c \qquad e \geq a+b \qquad e \leq 2b \qquad b \leq 2a \qquad (2)$$

The weighted distance between the origin and (x, y, z) in Case I becomes

$$D(x,y,z) = \begin{cases} z(c-b)+y(c+b-e)+x(e-c), & \text{if } x \leq y+z, \\ z(e-b-a)+y(b-a)+ax, & \text{if } x \geq y+z. \end{cases} \qquad (3)$$

3   Optimality Computations, Case I

Many different optimality criteria have been used in the literature, see, e.g., [3,5, 9]. The results from optimizations using different criteria are often very similar, see discussion in [2]. Optimality is here defined as minimizing the maximum difference between the weighted distance and the Euclidean distance in a cubic image of size $M \times M \times M$.

For the optimization, the same Lemma as was used for the $3 \times 3 \times 3$ neighbourhood, [3, Lemma 1], will be very useful.

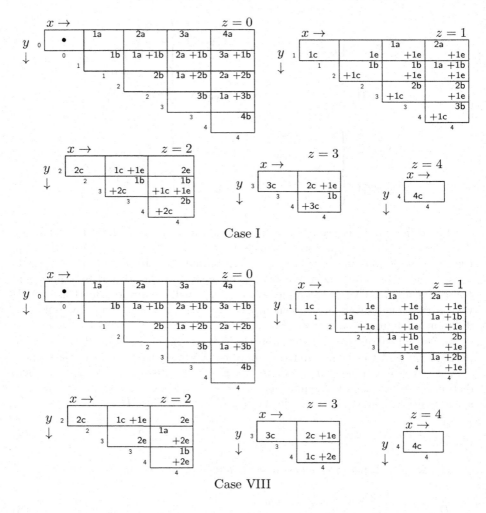

Fig. 2. Path maps for all voxels $0 \le z \le y \le x \le 4$ (Case I, top, and Case VIII, bottom).

Lemma 1. *The function*

$$f(\eta) = \alpha\eta + \gamma\xi + \beta M - \sqrt{M^2 + \xi^2 + \eta^2},$$

where $|\alpha| < 1$ and $M > 1$ has the maximum value

$$f_{\max} = \beta M + \gamma\xi - \sqrt{M^2 + \xi^2} \cdot \sqrt{1 - \alpha}, \quad \textit{for } \eta = \frac{\alpha\sqrt{M^2 + \xi^2}}{\sqrt{1 - \alpha^2}}$$

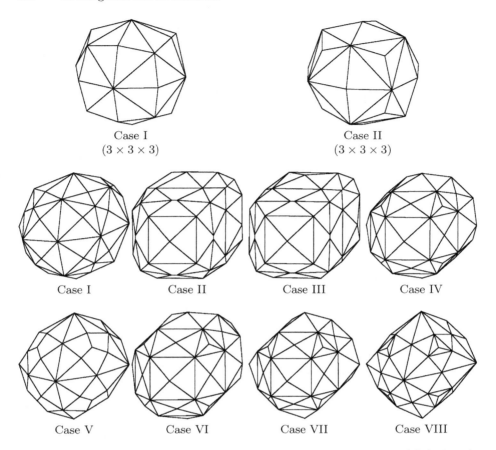

Fig. 3. $\langle a, b, c \rangle$-balls fulfilling Case I and II for a $3 \times 3 \times 3$ neighbourhood, [3]. $\langle a, b, c, e \rangle$-balls fulfilling Case I-VIII, see text.

The difference between the computed distance (see Eq. (3)) and the Euclidean distance is

$$
\begin{aligned}
\text{Diff}_\mu &= (c - b)z + (c + b - e)y + (e - c)x - \sqrt{x^2 + y^2 + z^2}, &&\text{if } x \leq y + z, \\
\text{Diff}_\kappa &= (e - b - a)z + (b - a)y + ax - \sqrt{x^2 + y^2 + z^2}, &&\text{if } x \geq y + z.
\end{aligned}
\tag{4}
$$

This difference is to be minimized for an $M \times M \times M$ image. If we put the origin in the corner of the image, we can assume that the maximum difference, maxdiff, occurs for $x = M$ (remember $0 \leq z \leq y \leq x$). We have $0 \leq y \leq M$ and $0 \leq z \leq y$. Due to the nature of the distance function, we must separate the cases $M \leq y + z$ and $M \geq y + z$. The interesting areas in the y, z-plane can be seen in Fig. 4. The maximum difference can occur within the two triangular areas (μ_2 and κ_2), on the boundaries ($\mu_1, \mu_3, \kappa_1, \kappa_3,$ and λ_2), or at the corners

$(\mu_4, \kappa_4, \lambda_1, \text{ and } \lambda_3)$. Starting with the two areas, μ_2 and κ_2, the extrema occur for $\frac{\partial}{\partial z}(\text{Diff}) = 0$. By using Lemma 1, we have

$$\text{Diff}^{\max}_{\mu_2} = (e - c)M + (c + b - e)y - \sqrt{M^2 + y^2}\sqrt{1 - (c - b)^2}, \quad \text{if } 0 < z < M - y,$$
$$0 < y < M,$$
$$\text{Diff}^{\max}_{\kappa_2} = aM + (b - a)y - \sqrt{M^2 + y^2}\sqrt{1 - (e - b - a)^2}, \quad \text{if } M - y < z < y,$$
$$\tfrac{M}{2} < y < M.$$

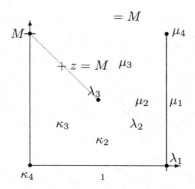

Fig. 4. The different areas, lines, and points where the maximal differences from the Euclidean distance can occur in Case I.

On the lines in Fig. 4, we have the following expressions for Diff:

$$\text{Diff}_{\kappa_1} = (b - a)y + aM - \sqrt{M^2 + y^2}, \qquad \text{if } 0 < y < M$$
$$\text{Diff}_{\kappa_3} = (e - 2a)y + aM - \sqrt{M^2 + 2y^2}, \qquad \text{if } 0 < y < \tfrac{M}{2}$$
$$\text{Diff}_{\mu_1} = (c - b)z + bM - \sqrt{2M^2 + z^2}, \qquad \text{if } 0 < z < M$$
$$\text{Diff}_{\mu_3} = (2c - e)y + (e - c)M - \sqrt{M^2 + 2y^2}, \qquad \text{if } \tfrac{M}{2} < y < M$$
$$\text{Diff}_{\lambda_2} = (2b - e)y + (e - b)M - \sqrt{M^2 + y^2 + (M - y)^2}, \qquad \text{if } \tfrac{M}{2} < y < M.$$

The maxima of all these expressions can be found by using Lemma 1 with $\xi = 0$. This, together with the values at the corner points, see Fig. 4, gives the following possibilities for the maximum difference in Case I

$$E_{\kappa_1} = \left(a - \sqrt{1 - (b - a)^2}\right)M, \qquad \text{for } (M, y^{\kappa_1}_{\max}, 0),$$
$$E_{\kappa_2} = \left(a - \sqrt{1 - (e - b - a)^2 - (b - a)^2}\right)M, \text{for } (M, y^{\kappa_2}_{\max}, 0),$$
$$E_{\kappa_3} = \left(a - \sqrt{1 - \tfrac{1}{2}(e - 2a)^2}\right)M, \qquad \text{for } (M, y^{\kappa_3}_{\max}, y^{\kappa_3}_{\max}),$$
$$E_{\kappa_4} = (a - 1)M, \qquad \text{for } (M, 0, 0),$$
$$E_{\lambda_1} = (b - \sqrt{2})M, \qquad \text{for } (M, M, 0),$$
$$E_{\lambda_2} = \tfrac{1}{2}\left(e - \sqrt{3}\sqrt{2 - (2b - e)^2}\right)M, \qquad \text{for } (M, y^{\lambda_2}_{\max}, y^{\lambda_2}_{\max}),$$
$$E_{\lambda_3} = \tfrac{1}{2}\left(e - \sqrt{6}\right)M, \qquad \text{for } (M, \tfrac{M}{2}, \tfrac{M}{2}),$$

$$E_{\mu_1} = \left(b - \sqrt{2}\sqrt{1 - (c-b)^2}\right) M, \qquad\qquad \text{for } (M, M, z_{\max}^{\mu_1}),$$

$$E_{\mu_2} = \left(e - c - \sqrt{1 - (c-b)^2 - (c+b-e)^2}\right) M, \qquad \text{for } (M, y_{\max}^{\mu_2}, z_{\max}^{\mu_2}),$$

$$E_{\mu_3} = \left(e - c - \sqrt{1 - \tfrac{1}{2}(2c-e)^2}\right) M, \qquad\qquad \text{for } (M, y_{\max}^{\mu_3}, y_{\max}^{\mu_3}),$$

$$E_{\mu_4} = \left(c - \sqrt{3}\right) M, \qquad\qquad\qquad\qquad\qquad \text{for } (M, M, M).$$

Numerical experiments shows that the optimum occurs for $E_{\kappa_4} = E_{\lambda_1} = E_{\mu_4}$. This implies that

$$b = \sqrt{2} + a - 1$$
$$c = \sqrt{3} + a - 1$$

We use this to limit our problem to two variables, a and e. Numerical experiments also show that at the optimum $E_{\kappa_2} = -E_{\kappa_4}$ and that the optimum occurs on the boundary $e = \frac{b}{2} + c$ of the Case I hyper-volume. This leads to the solution

$$a_{\text{opt}} = \tfrac{1}{17}\left(7 + 2\sqrt{3} - \sqrt{2} + 2\sqrt{16\sqrt{6} + 24\sqrt{3} + 22\sqrt{2} - 99}\right) \approx 0.9545,$$
$$b_{\text{opt}} = \sqrt{2} + a - 1 \qquad\qquad\qquad\qquad\qquad\qquad\qquad \approx 1.3687,$$
$$c_{\text{opt}} = \sqrt{3} + a - 1 \qquad\qquad\qquad\qquad\qquad\qquad\qquad \approx 1.6865,$$
$$e_{\text{opt}} = \tfrac{3}{2}a + \sqrt{3} + \tfrac{1}{\sqrt{2}} - \tfrac{3}{2} \qquad\qquad\qquad\qquad\qquad \approx 2.3709,$$
with maxdiff $= (1 - a) M \approx 0.0455M.$

For the computation of good integer local distances, we also need to find the optimal solution for $a \equiv 1$. This is due to the fact that for the integer solution a becomes a scale factor, i.e., in reality $= 1$. Thus, we solve $E_{\kappa_1}^* = E_{\kappa_2}^* = -E_{\lambda_1}^* = -E_{\mu_4}^*$, where the star denotes that $a = 1$ has been substituted in the expressions. The solution is given by

$$a_{opt}^* \qquad\qquad\qquad\qquad = 1,$$
$$b_{opt}^* = \tfrac{1}{\sqrt{2}} + \sqrt{\sqrt{2} - 1} \qquad \approx 1.3507,$$
$$c_{opt}^* = \sqrt{3} - \sqrt{2} + \tfrac{1}{\sqrt{2}} + \sqrt{\sqrt{2} - 1} \approx 1.6685,$$
$$e_{opt}^* = 1 + \tfrac{1}{\sqrt{2}} + \sqrt{\sqrt{2} - 1} \qquad \approx 2.3507,$$
with maxdiff$^* = \left(\tfrac{1}{\sqrt{2}} - \sqrt{\sqrt{2} - 1}\right) M \approx 0.0635M.$

Both solutions, $\langle a_{opt}, b_{opt}, c_{opt}, e_{opt}\rangle$ and $\langle 1, b_{opt}^*, c_{opt}^*, e_{opt}^*\rangle$, fulfil the inequalities in (2), i.e., the regularity criteria.

4   Integer Approximations, Case I

Working with real-valued local distances is generally not desirable. To find good integer local distances, we use the optimal solution for $a \equiv 1$. The candidates for integer approximations of the optimal values, denoted A, B, C, and E, are found by multiplying the real-valued solutions b, c, e by an integer scale factor A and rounding to the nearest integer. Note that it is necessary to check that the resulting distance transform, $\langle 1, \frac{B}{A}, \frac{C}{A}, \frac{E}{A}\rangle$ is within the allowed hyper-volume for

Case I! All eleven differences $E(1, \frac{B}{A}, \frac{C}{A}, \frac{E}{A})$ are computed, to find the maximum for the approximation. Good integer weighted DTs are listed in Table 1, together with the optimal solutions. The maximal differences for the integer DTs are converging towards maxdiff* as expected.

Table 1. Maximal differences to the Euclidean distance for different $\langle a, b, c, e \rangle$ weighted distance transforms in an $M \times M \times M$ image.

Case	a	b	c	e	maxdiff
$3 \times 3 \times 3$	a_{opt}	b_{opt}	c_{opt}	∞	0.0736M
$3 \times 3 \times 3$	1	b_{opt}^*	c_{opt}^*	∞	0.1024M
$3 \times 3 \times 3$	3	4	5	∞	0.1181M
I	a_{opt}	b_{opt}	c_{opt}	e_{opt}	0.0455M
I	1	b_{opt}^*	c_{opt}^*	e_{opt}^*	0.0635M
I	3	4	5	7	0.0809M
I	11	15	19	27	0.0729M
I	14	19	24	34	0.0687M
I	17	23	29	41	0.0662M
I	20	27	34	48	0.0646M

The ball for $\langle 3, 4, 5, 7 \rangle$ can be seen in Fig. 5. This ball represents the sensible choice of a weighted integer distance transform in the extended neighbourhood.

Fig. 5. A $\langle 3, 4, 5, 7 \rangle$ ball with radius 83 voxels.

5   Discussion

We have presented results when optimizing local distances for an extended neighbourhood in 3D, where we use the three weights in the $3 \times 3 \times 3$ neighbourhood together with the $(2, 1, 1)$ weight from the $5 \times 5 \times 5$ neighbourhood. For the optimal, real-valued, solution, the maximum difference compared to the Euclidean distance is 4.55% of the maximum distance occurring in the image. This can be compared to 7.37% when a $3 \times 3 \times 3$ neighbourhood is used. In the $3 \times 3 \times 3$ case the best integer approximation we can have has difference 10.24%, and the sensible $\langle 3, 4, 5 \rangle$ has 11.81%. Compare this with the best integer approximation in the $5 \times 5 \times 5$ case, 6.35%, and the sensible choice $\langle 3, 4, 5, 7 \rangle$ with 8.09% difference.

For the other cases of semi-regular $\langle a, b, c, e \rangle$ DTs, Cases II-VIII, we have seen that the expressions for the distances $D(x, y, z)$ become quite complicated. This is due to the fact that $D(x, y, z)$ is dependent often on whether an even or an odd number of steps has been taken.

We will continue our investigation of finding optimal local distances for a $5 \times 5 \times 5$ neighbourhood by adding also the steps d and f in the minimal paths.

References

1. G. Borgefors. Distance transformations in digital images. *Computer Vision, Graphics and Image Processing*, 34:344–371, 1986.
2. G. Borgefors. Discretization – problems and solutions. In H. Maître and J. Zinn-Justin, editors, *Les Houches session LVIII: Progress in Picture Processing*, pages 1–80, Les Houches, 1992. Elsevier Science B.V., 1996.
3. G. Borgefors. On digital distance transforms in three dimensions. *Computer Vision and Image Understanding*, 64(3):368–376, 1996.
4. C. Kiselman. Regularity properties of distance transformations in image analysis. *Computer Vision and Image Understanding*, 64(3):390–398, Nov. 1996.
5. E. Remy and E. Thiel. Optimizing 3D chamfer masks with norm constraints. In *Proceedings of International Workshop on Combinatorial Image Analysis*, pages 39–56, Caen, France, July 2000.
6. A. Rosenfeld and J. L. Pfaltz. Sequential operations in digital picture processing. *Journal of the Association for Computing Machinery*, 13(4):471–494, Oct. 1966.
7. A. Rosenfeld and J. L. Pfaltz. Distance functions on digital pictures. *Pattern Recognition*, 1:33–61, 1968.
8. E. Thiel. *Les Distance de chanfrein en analyse d'images: Fondements et applications*. PhD thesis, Université Joseph Fourier Grenoble I, Sept. 1994.
9. B. J. H. Verwer. Local distances for distance transformations in two and three dimensions. *Pattern Recognition Letters*, 12(11):671–682, Nov. 1991.
10. M. Yamashita and T. Ibaraki. Distance defined by neighbourhood sequences. *Pattern Recognition*, 19:237–246, 1986.

Independent Modes of Variation in Point Distribution Models

Marco Bressan and Jordi Vitrià*

Centre de Visió per Computador, Dept. Informàtica,
Universitat Autònoma de Barcelona, 08193 Bellaterra, Barcelona, Spain.
Tel. +34 93 581 30 73 Fax. +34 93 581 16 70
[marco, jordi]@cvc.uab.es

Abstract. A Point Distribution Model requires first the choice of an appropriate representation for the data and then the estimation of the density within this representation. Independent Component Analysis is a linear transform that represents the data in a space where statistical dependencies between the components are minimized. In this paper, we propose Independent Component Analysis as a representation for point distributions. We observe that within this representation, the density estimation is greatly simplified and propose solutions to the most common problems concerning shapes. Mainly, testing shape feasibility and finding the nearest feasible shape. We also observe how the description of shape deformations in terms of statistically independent modes provides a more intuitive and manageable framework. We perform experiments to illustrate the results and compare them with existing approaches.

Keywords: Point distribution model; Shape Representation; Shape Description; Independent Component Analysis; Independent Modes of Variation.

1 Introduction

The Point Distribution Model (PDM) [9] is a shape description technique based on the vectorized representation of shapes to estimate a statistical model for non-rigid shape variations. This model can be used for the generating or testing new examples. The statistical modeling for shape variation, and its combination with several image processing techniques has generated an important number of applications in the last years. These applications include tracking, recognition, biomedical imaging, special effects for film and television and registration among others [12,22,1].

The construction of an appropriate PDM for a certain type of shape we wish to learn, requires both the selection of a good representation and of an

---

* This work is supported by CICYT and EU grants TAP98-0631 and 2FD97-0220 and the Secretaría de Estado de Educación, Universidades, Investigación y Desarollo from the Ministerio de Educación y Cultura de España.

C. Arcelli et al. (Eds.): IWVF4, LNCS 2059, pp. 123–134, 2001.

appropriate density estimation method for the distribution of the shapes within this representation. For the representation we can use linear or nonlinear models. Though several nonlinear models have been proposed [20,21,5,13], a linear representation is still a common choice for their speed and straightforward interpretability. As a matter of fact, most of the nonlinear representations are applied over a linear representantion which previously performs the dimensionality reduction. On the other side, even when the training set generates complex distributions, a linear representation can be used and complexity charged to the statistical model [8]. The most successful linear representation so far, for its simplicity and straightforward interpretation, is the one obtained through Principal Component Analysis (PCA). By projecting a shape in a previously learnt PCA space, we have a set of coefficients or parameters (the principal components) which control the variation along maximum variance directions. This is why the parameters of the distribution model in the PCA space are known as (principal) modes of variation.

In this paper we propose an alternative linear representation of the PDMs using Independent Component Analysis (ICA). This linear transform represents our data in a space in which the statistical dependence between the components is minimized [2]. This can be particularly interesting for those non rigid shapes whose modes of variation are supposed statistically independent. Because of the theory underlying PCA, the modes of variation are then given by uncorrelated projections of maximum variance. The assumption that these projections are optimal for modeling shape variation is not necessarily correct. In certain cases (the fingers of the hand can prove to be a good example), higher order relationships such as independence can be important for better modeling. The most important advantage of independence is that it simplifies density estimation by transforming an N-dimensional density estimation problem in N 1-dimensional estimations. This provides simplicity and a robust framework. The direct relationship between the independent components and the shape deformations, also allows robust tagging (classification) and tracking. Both of these features are of great importance in specific applications such as Active Shape Models [9].

Before exposing the main results we introduce the basics of both Independent Component Analysis and Point Distribution Models. Specially in the first case, the theory here exposed results incomplete for the interested reader. For an extended and up to date exposition of definitions and results involving ICA we recommend Aapo Hyvärinen's survey [14].

2   Point Distribution Models

If we use n points to describe a certain shape in d dimensions, we can represent this shape by a $N = nd$ dimensional vector by concatenating the point position values. Given K samples of a certain shape, we choose n locations as key points or "landmarks points", and obtain K vectors representing each shape of the training set. In order to be able to compare these points, a certain alignment in an approximate sense is necessary. Procrustes method [11] or modifications are frequently used in this stage. The selection of a correct criteria for alignment should not be underestimated since these operations will greatly affect the final distribution by introducing or avoiding nonlinearities [9]. For the rest of this paper we will assume that our aligned training set is a sample of the random vector $\mathbf{x}$.

The next step is to find a proper representation for $\mathbf{x}$. In the choice of the representation simplicity, dimensionality reduction, statistical properties and interpretability should be considered.

2.1 The PCA Representation

From the training set, we can estimate both the mean of the data $\bar{\mathbf{x}}$ and its covariance $\mathbf{C}$. Intuitively, the covariance matrix tells us the way each landmark tends to move as the others move. Let $\phi_{\mathbf{i}}$ be the eigenvectors of $\mathbf{C}$ and λ_i their corresponding eigenvalues in decreasing order. If ϕ is the matrix built by placing the first M eigenvectors as columns, a set of parameters for the shape can be defined then by

$$\mathbf{b} = \phi^t(\mathbf{x} - \bar{\mathbf{x}}) \tag{1}$$

This is the PCA or Karhunen-Loeve transform and the parameters $\mathbf{b}$ represent the projection of the shapes in the subspace spanned by the eigenvectors of the covariance matrix. It can be seen that these projections result uncorrelated and this subspace is the best linear subspace of dimension M fit to the data. Thus, PCA besides decorrelating, also provides a way for reducing dimensionality. By projecting a shape in the PCA space, we have a set of coefficients or parameters (the principal components) which control the variation along these maximum variance directions. So we can naturally associate each principal component $\mathbf{b}$ to a mode of variation of the shape. The choice of an appropriate value for M can be done in several ways, the most frequent is based on the proportion of variance we wish to capture in the subspace.

2.2 Statistical Density Models for the PCA Representation

If we have estimated, from the training set, the distribution of the parameters $\mathbf{b} \sim p(\mathbf{b})$, reasonable way to decide over the feasibility of shape with parameters $\mathbf{b}$ is

$$p(\mathbf{b}) > p_t \tag{2}$$

where p_t is a certain threshold we consider appropriate. Usually, the threshold is chosen so that some proportion of the training set passes the threshold. If the parameters $\mathbf{b}$ are assumed Gaussian and independent, we have that

$$\log p(\mathbf{b}) = -\frac{1}{2} \sum_{i=1}^{M} \frac{b_i^2}{\lambda_i} + \kappa \tag{3}$$

where κ is constant for any parameter. In this case, the threshold represents a likelihood which constrains feasible shapes to a hyperellipsoid. The size of the hyperellipsoid can be obtained considering that the sum of the square of gaussian variables has a chi-squared distribution. Using this fact and given a certain probability, we can obtain the desired threshold p_t. Given a shape, if its likelihood is lower than our threshold, the nearest feasible shape is that shape belonging to the intersection of the hyperellipsoid and the line passing through our current shape and the origin.

Another approach, is to choose hard limits on each direction [9]. This is related with the idea of statistical independence of the components of the parameters. It is equivalent to constraining feasible shapes to a hypercube. A good heuristical value for the threshold on each direction is 3 times the standard deviation on that direction. If we assume a gaussian distribution on each direction, this choice of limit values means that a shape is plausible if it belongs to the symmetrical mean-centered interval which has a marginal probability of 0.997. In this case, for each $i = 1, \ldots, M$, the feasibility of a shape is checked by $b'_i < 3\sqrt{\lambda_i}$ and the nearest feasible shape is obtained by

$$b_i^F = \text{sign}(b'_i) * \min\left(3\sqrt{\lambda_i}, |b'_i|\right) \tag{4}$$

If a simple gaussian estimation is not enough we can use more complex models. A useful approach is to model $p(\mathbf{b})$ using Gaussian Mixture Models (GMM) [4]. The parameters for the GMM can be estimated with parameter estimation algorithm such as Expectation Maximization (EM) [10]. In this case, the plausibility of a shape is a more complex problem. Even though more precise solutions can be developed, a simple one consists on deciding that a shape is plausible if its likelihood is above the likelihood of a certain percentage of shapes in the training set. The percentage value is generally above 80%. When using a GMM, a general solution for the problem of finding the nearest feasible shape is not available so Monte Carlo and gradient descent methods are employed. Moreover, estimating the GMM parameters on high dimensions is a highly unstable problem.

3   An ICA Representation for the PDM

The ICA of an N dimensional random vector is the linear transform which minimizes the statistical dependence between its components. This representation in terms of independence proves useful in an important number of applications such as data analysis and compression, blind source separation, blind deconvolution, denoising, etc.

3.1   The ICA Model

The noise-free ICA Model can be expressed as

$$\mathbf{x} - \bar{\mathbf{x}} = \mathbf{As} \tag{5}$$

where $\mathbf{x}$ corresponds to the random vector representing our data, $\bar{\mathbf{x}}$ its mean, $\mathbf{s}$ the random vector of *independent components* with dimension $M \leq N$, and

A is called the *mixture matrix*. The pseudoinverse of **A** which we will represent as **W**, is called the filter or projection matrix, and provides an alternative respresentation of the ICA Model,

$$\mathbf{W}(\mathbf{x} - \bar{\mathbf{x}}) = \mathbf{s} \tag{6}$$

If the components of vector **s** are independent, at most one is Gaussian and its densities are not reduced to a point-like mass, it can be seen that **W** is completely determined [7].

In practice, the independent components are unknown in advance so ICA is performed by estimating **W**. For the estimation of the filter matrix several objective functions such as likelihood, network entropy, mutual information and approximations of these, have been proposed [17,3,6,16]. Though several algorithms have been tested, the method employed in this article for estimation of the independent components is the one introduced by [15] well known as FastICA. This method introduces a progressive minimization of mutual information by finding maximum negentropy[1] directions, and proves fast and efficient.

Assuming we have learnt the mixing and filter matrix for the ICA Models (5) and (6), we will call **s** the *independent modes of variation*, and assume that its components are statistically independent. The independent components have zero mean and we can assume, without loss of generality, that they have unit variance [7]. The choice of dimension is not as straightforward as in PCA but there exist several approaches [18]. In the shape domain it is conceivable to assign small variances to errors in the labelling process so, when required, we will reduce dimensionality by first performing PCA and then ICA, considering noise the discarded components.

3.2   Statistical Density Models for the ICA Representation

Due to the assumption of independence, we need only to model the one-dimensional densities corresponding to the M components of **s**. The complexity of the method employed for density estimation is not relevant since we are working with a single dimension and the calculations need only be performed while training our PDM. Depending on the problem, we can use non-parametric methods [4,19] such as histograms, kernel-based methods with Parzen Windows, Radial Basis Functions or semi-parametric methods such as Gaussian or Laplacian Mixture Models.

Let s be a random variable which, within the PDM problem, would correspond to one of the independent modes of variation. Because of the nature of PDMs, we will assume that the density of s is likely to be one of a few particular densities. Unimodal densities can be classified in subgaussian, gaussian and supergaussian, according to their kurtosis (or fourth order cumulant). Kurtosis is zero for gaussian densities, less the zero for subgaussian densities such as the

---

[1] Negentropy is a non-gaussianity measure based on differential entropy. Mutual information can be expressed in terms of negentropy.

uniform distribution, and larger than zero for supergaussian densities such as the Laplacian distribution, or the delta distribution in the extreme case. A zero-centered supergaussian density is the result of the component being mostly close to zero and only seldom significantly non-zero.

In the shape domain, the frequency of a certain position for a particular shape will affect the sub or supergaussianity of the modes of variation. A mode of variation of a shape which has a preferred position and seldom deforms will have a sparse or supergaussian distribution. On the other side, a mode of variation with almost equiprobable states along its variation range is clearly subgaussian. When preparing a training set for the generation of a PDM there is a tendency towards generating uniform distributions of the modes of variation. This also favours subgaussian distributions. Symmetrical and gradual deformations of a part of the shape correspond to continuous symmetrical distribution. This is a frequent situation. In the shape problem skewness is more related with the sampling than with the real deformations of a shape. When a shape can be found in different states but doesn´t deform continuously from one state to the other, we find clusters in the distribution of the mode of variation. Clusters are frequently introduced by the incorrect identification of landmark points. We conclude that our density model should be open to both sub and supergaussian distributions, but particularly the first. Symmetry is very frequent, and it should also include multimodal densities, unless we have some other prior knowledge.

The broadest approach can be simply to use a histogram approximation. The problem is that its discrete nature introduces complexity in the equations. Immediately related is to use a kernel method with Radial Basis Functions. This kernel method positions Gaussians at all the samples in the distribution. In our case, and if we have K samples $s_1, \ldots, s_K$, the corresponding density can be expressed as

$$P_{KER}(s) = \sum_{i=1}^{K} \frac{1}{K} G(s; s_i; \sigma) \tag{7}$$

where a good choice for σ if K is sufficiently high is $(\frac{12}{K})^{\frac{1}{5}}$ [19].

Even though the kernel method can model satisfactorily the situations presented, it is non-parametric. In the development of algorithms which make use of PDMs, parametric methods can be more useful since these are faster and provide analytic solutions to several problems such as the inverse likelihood problem. A possible simplification is to use mixture models such as GMMs or other more specific models. For instance, in a problem in which sparsity is known in advance, Laplacian (or Double-exponential) Mixture Models can be used on the marginal densities,

$$p_{LMM}(s) = \sum_{l=1}^{L} \frac{w_l}{\sqrt{2}\alpha_l} e^{-\frac{\sqrt{2}|s-\mu_l|}{\alpha_l}} \tag{8}$$

Where w_l are the weights, μ_l the means and α_l measures of variance. As mentioned, the parameters in this case can be estimated through an EM algorithm. This semi-parametric methods are good for modeling both unimodal and multimodal densities but don´t succeed when we wish to model highly subgaussian densities such as a uniform density. Experiments were performed using GMMs when possible, and RBF kernel-based methods otherwise.

3.3   Shape Plausibility with ICA

Suppose we have estimated the density for each of the independent modes of variation with one of the methods suggested above so that $s_m \sim p^m(s)$.

Given a certain probability value P_t between 0 and 1, let $p_t = P_t^{\frac{1}{M}}$. For each component there exists a union of disjoint intervals

$$I^m = [a_1^m, a_2^m] \bigcup [a_3^m, a_4^m] \bigcup \ldots \bigcup [a_{2t_m-1}^m, a_{2t_m}^m]$$

such that for all $m = 1 \ldots M$ I^m satisfies

$$\int_{I^m} p^m(s)ds = p_t$$

and

$$p(a_i^m) = l^m, \qquad\qquad \forall i = 1, \ldots, 2t_m$$

Once we have the set of intervals for each component, using the assumption of independence, it can be seen that

$$\int_{I^1 \otimes \ldots \otimes I^M} p(\mathbf{s})d\mathbf{s} = \prod_{m=1}^{M} \int_{I^m} p_m(s_m)ds_m = \prod_{m=1}^{M} p_t = P_t \qquad (9)$$

A constructive method which shows the existence of these intervals in a bimodal density obtained from experiments is exposed in Fig. 1. We first assume the probability density is continuous in $\mathbb{R}$. Given the likelihood value L, it can be seen that if the line $y = L$ intersects the function $y = p^m(s)$ it has to be in an even number of points. These points determine the interval borders. If the intersection is empty, we define I^m as the empty set. The method consists in starting at a likelihood above the maximum and decreasing the likelihood value, thus increasing the probability, until the threshold is reached.

In practice, implementation depends on the density model. For certain parametric and semi-parametric models, both the likelihood and the interval borders can be obtained analytically. This is performed in the training stage. Any algorithm working on new shapes will need only the interval information for plausibility tests. Dividing each direction in intervals divides the whole space into hyperboxes which have a geometric distribution reminiscent of that which arises from separable functions. In Fig. 1(b) the joint distribution of two independent directions obtained from real experiments are plotted. Each marginal density (clearly bimodal) was estimated with a kernel-based method. The product of the marginal densities is plotted with gray levels and contour lines. The rectangular boxes represent the cartesian product of the intervals estimated for each direction for $P_t = 0.95$.

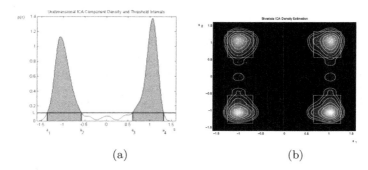

(a)   (b)

Fig. 1. (a) For a certain bimodal independent mode of variation, two intervals and a certain probability value corresponds to the likelihood value l_m. (b) The same but in the bivariate case. The curves represent contour lines of the density estimation (done with a kernel method), and the rectangles represent the cartesian product of the intervals, capturing 95% of the probability.

Since the calculation of the intervals is performed in the training stage, all complex algebraic operations are removed from the working algorithms. This is because there is no need for the calculation of likelihoods once we have the interval limits. This interval structure also provides precision. It can be seen in Fig. 1(b) that, if we decided to use a GMM, the only way to improve the estimation would have been using more than four components in the mixture. This can be really bad as dimensions increase and we have no prior knowledge of the structure.

In the interval context, plausibility is easily checked by first projecting the shape in the parameter space and then by verifying if $s^m \in I^m$ for all $m = 1, \ldots, M$.

Given the intervals I^m, and a shape with independent modes of variation s_T^m, the nearest feasible shape $\mathbf{s}_F$, with components s_F^m is

$$
s_F^m = \left\{ \begin{array}{ll} s_T^m & \text{if } s_T^m \in I^m; \\ \arg\min_{1 \leq i \leq 2t_m} |s_T^m - a_i^m| & \text{otherwise.} \end{array} \right. \tag{10}
$$

4   Experiments

An artificial set of shapes was created. In each shape we use 19 points to describe a fixed base and three deformable extensions of fixed length (see Fig. 2). Each extension can be found rotated in an angle between $-\frac{\pi}{4}$ and $\frac{\pi}{4}$. We created a training set of 400 shapes, choosing randomly the angle corresponding to each extension. We have then, three independent degrees of freedom for each shape. Since the shapes were already aligned when created, only centering was necessary. Fig. 2(a) shows the three principal modes of variation when given values between the mean and approximately three standard deviations. PCA decorrelates each of the movements but does not take in account statistics of higher order. The decorrelated movements have no relationship with the degrees of freedom chosen in the creation of the shapes. Fig. 2(b) shows the three independent modes of variation. We observe how ICA separated the deformations corresponding to each one of the extensions. This allows unidimensional density estimation for the statistical model, where each dimension clearly represents a degree of freedom. It also gives a straightforward and easily interpretable solution for the problem of classifying a shape according to its deformations. Experiments were

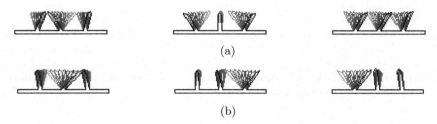

(a)

(b)

Fig. 2. (a) The three modes of variation using PCA for an artificial set of shapes. (b) The three independent modes of variation using ICA, for the same artificial set.

also performed on a dataset of shapes representing 180 hands with the five fingers extended. The limited number of samples has strong influence in the estimation of the ICA model, so dimension of the Point Distribution Model should be taken into account. We finally decided to describe the hands by 11 points each, so the resulting dimension is 22. This results in a naive shape descriptor for a hand, but a more complete set of landmark points would result in a higher dimension and the ICA Model would no longer be trustable. This, of course, can be solved by increasing the number of samples. The PCA space of parameters captures 95% of the data variation with a dimension of 5. In Fig. 3(a) we observe the five first principal modes of variation. The first two modes capture practically all the movement in the hands, mixing the movement of all fingers (except the index in the first mode). The remaining three components capture the uncorrelated variation of groups of one or two fingers. In all, except maybe for the second mode, the variations here presented do not resemble any kind of realistic hand movement.The first five independent modes of variation are shown in Fig. 3(b). These five modes were obtained by performing ICA on the principal modes of variation of dimension five. It has been observed [18] that this can bring up corrupted independent components. Even though this seems to be the case, we observe some interesting differences with the principal modes of variation. First of all, ICA isolated the thumb, which clearly is the only really single finger we can independently displace in our hand. The rest of the movements resemble much more realistic "independent" hand movements. For instance the opening of the hand moves the thumb, ring and little finger much more than the index and middle fingers. This is illustrated in the second mode. The fourth mode

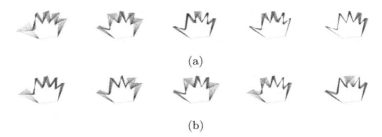

(a)

(b)

Fig. 3. (a) The five principal modes of variation for set of hand shapes. (b) The five independent modes of variation, for the same set.

shows the close dependence between index and thumb. Third and fifth modes correspond to second and third principal modes respectively.

Even though this analysis of the modes gives us some idea of what the ICA representation can achieve, the most important advantage of ICA is the way it simplifies the density estimation. Fig. 4(a) shows the distribution of the first two principal modes and what results from deciding plausibility assuming a normal bivariate distribution or hard limits (ellipse or rectangle). It can be seen that none of these assumptions hold. Fig.4(b) shows the corresponding independent modes, where sources are clearly more separated and the interval method after a kernel density estimation was employed. In all cases, the limits were chosen so that the intervals encompass for 98% of the probability.

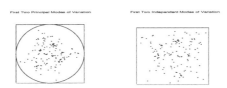

Fig. 4. Two first modes of variation for the PCA and the ICA representation respectively. The limits shown are the limits for plausible shapes. In the PCA case assuming independence (square) and normality (ellipse). In the ICA case assuming independence. The higher precision of the latter is observed (for testing shape feasibility, for instance).

5   Conclusions

In this paper, we expose ICA as an alternative representation for data corresponding to n-dimensional shapes. This representation, is based on higher order statistics, has important advantages for PDMs. The assumption of independence makes density estimation a one-dimensional problem and thus, the application of complex and accurate methods is permitted. We propose non-parametric and semi-parametric density estimation methods for each of the independent modes of variation. The density estimation is used when considering the plausibility of new shapes or when approximating a certain shape with the nearest feasible shape. A simple method based on feasibility intervals in every direction is exposed. It addresses both of these problems and is robust to multimodality. These intervals are obtained in the training stage and all the posterior operations for testing plausibility are reduced to logical order operations. When similar methods are used for PCA representations practical results can be unsatisfactory since they assume independence and only ensure decorrelation (see Fig. 4).

Moreover, in an ICA framework, modes of variation no longer represent uniquely the deviations within the shape and can now be thought of as independent deformations. This allows higher control when modeling, due to the natural association between independent deformations and each independent mode of variation. In the experiments, the independent component that captures the movement of the thumb illustrates this idea. This could be applied to shape understanding and classification. This last can be done by learning a separate ICA model for each of the classes to be considered. The independence assumption allows the fast implementation of a Bayesian decision scheme.

Still, validation on an extended dataset of real shapes showing independent deformations is necessary. Testing on Active Shape Models would test the performance of the algorithms and the shape plausibility considerations. Classification with the ICA Model and comparisons with previous approaches can be of great use in tagging and tracking of shapes.

From the results we conclude that the ICA representation can be successfully used when there are reasons to think that different shape deformations correspond to independent factors, and the shapes we observe are linear mixtures of these deformations. In this case, ICA can not only separate the deformations allowing control and classification, but can also provide a robust and simple density estimation framework.

References

1. A. Baumberg and D. Hogg. An efficient method for contour tracking using active shape models. Technical report, University of Leeds, School of Computer Studies, November 1994.
2. A. Bell and T. Sejnowski. An information-maximization approach for blind signal separation. *Neural Computation*, 7:1129–1159, 1995.
3. A. Bell and T. Sejnowski. The 'independent components' of natural scenes are edge filters. *Neural Computation*, 11:1739–1768, 1999.
4. C. Bishop. *Neural Networks for Pattern Recognition*. Oxford University Press, 1997.
5. R. Bowden, T. Mitchell, and M. Sahardi. Cluster based non-linear principle component analysis. *IEE Electronic Letters*, 33(22):1858–1858, 1997.
6. J. Cardoso. Infomax and maximum likelihood for source separation. *IEEE Letters on Signal Processing*, 4:112–114, 1997.
7. P. Comon. Independent component analysis - a new concept? *Signal Processing*, 36:287–314, 1994.

8. T. Cootes and C. Taylor. A mixture model for representing shape variation. In *Clark A.F., ed. British Machine Vision Conference 1997, BMVC'97*, volume 1, pages 110–119. University of Essex, UK:BMVA, 1997.

9. T. Cootes, C. Taylor, D. Cooper, and J. Graham. Active shape models - their training and application. *Computer Vision and Image Understanding*, 61:38–59, 1995.

10. A. Dempster, N. Laird, and D. Rubin. Maximum likelihood for incomplete data via the em algorithm. *Journal of the Royal Statistical Society*, 39:1–38, 1977.

11. J. Gower. Generalized procrustes analysis. *Psychometrika*, 40:33–51, 1975.

12. A. Heap and D. Hogg. 3d deformable hand models. In *Gesture Workshop, York, UK*, March 1996.

13. A. Heap and D. Hogg. Improving specificity in pdms using a hierarchical approach. In *Clark A.F., ed. British Machine Vision Conference 1997, BMVC'97*, volume 1, pages 80–89. University of Essex, UK:BMVA, 1997.

14. A. Hyvärinen. Survey on independent component analysis. *Neural Computing Surveys*, 2:94–128, 1999.

15. A. Hyvärinen and E. Oja. A fast fixed-point algorithm for independent component analysis. *Neural Computation*, 9:1483–1492, 1999.

16. T. Lee, M. Girolami, and T. Sejnowski. Independent component analysis using an extended infomax algorithm for mixed sub-gaussian and super-gaussian sources. *Neural Computation*, 11:609–633, 1998.

17. D. Pham, P. Garrat, and C. Jutten. Separation of a mixture of independent sources through a maximum likelihood approach. In *EUSIPCO*, pages 771–774, 1992.

18. J. Porrill and J. Stone. Undercomplete independent component analysis for signal separation and dimension reduction. Technical report, The University of Sheffield, Department of Psychology, 1998.

19. B. Silverman. *Density Estimation*. Chapman and Hall, 1986.

20. P. Sozou, T. Cootes, C. Taylor, and E. Di-Mauro. A non-linear generalization of pdms using polynomial regression. In *Hancock E., ed. British Machine Vision Conference 1994, BMVC'94*, volume 1, pages 397–406. University of York, UK:BMVA, 1994.

21. P. Sozou, T. Cootes, C. Taylor, and E. Di-Mauro. Non-linear point distribution modeling using a multi-layer perceptron. In *Pycock D., ed. British Machine Vision Conference 1995, BMVC'95*, volume 1, pages 107–116. University of Birmingham, UK:BMVA, 1995.

22. M. Stegmann. On properties of active shape models. Technical report, Technical University of Denmark, Department of Mathematical Modeling, March 2000.

Qualitative Estimation of Depth in Monocular Vision

Virginio Cantoni, Luca Lombardi, Marco Porta, and Ugo Vallone

Dipartimento di Informatica e Sistemistica, Università di Pavia
Via Ferrata,1 – 27100 – Pavia – Italy
{cantoni, luca, porta, vallone}@vision.unipv.it

Abstract. In this paper we propose two techniques to qualitatively estimate distance in monocular vision. Two kinds of approaches are described, the former based on texture analysis and the latter on histogram inspection. Although both the methods allow only to determine whether a point within an image is nearer or farther than another with respect to the observer, they can be usefully exploited in all those cases where precision is not critical or single images are the only source of information available. Moreover, combined with previously studied techniques, they could be used to provide more accurate results. Step by step algorithms will be presented, along with examples illustrating their application to real images.

1   Introduction

Monocular vision concerns the analysis of data obtainable from single images. While in stereo or trinocular vision spatial information can be drawn by comparing different images of the same scene, in monocular vision the analysis can be performed only by studying intrinsic characteristics of the representation. For example, comparisons and statistical investigations can be carried out on the distribution of gray levels (in the monochromatic case) or of the red, green and blue channels (in the RGB case) of the pixels composing the image. Therefore, results which can be obtained are influenced by the acquisition systems employed and by the spatial and tonality resolutions adopted.

The two techniques we propose in this paper are based on texture and histogram analysis. Although only qualitative information can be obtained through them, we think they can be anyway useful when the precision of the results is not critical or when there is no other source of data available. Monocular vision may be notably more advantageous than techniques exploiting couples or sequences of images, since it requires simpler acquisition systems and is computationally more efficient in terms of execution times. Moreover, qualitative estimations could be used to confirm evaluations obtained by means of other more precise techniques (such as binocular vision, infrared sensors, etc.).

The paper is structured as follows. Section 2 will describe the approach based on texture analysis. After a brief discussion about previous works regarding the topic and an introduction to the theory behind it, two practical algorithms will be presented, of which the second can be used to distinguish between nearer and farther zones within images in perspective projection. Section 3 will describe the technique based on his-

C. Arcelli et al. (Eds.): IWVF4, LNCS 2059, pp. 135–144, 2001.
© Springer-Verlag Berlin Heidelberg 2001

togram analysis. A brief introduction will precede the presentation of the implemented algorithm, which, like the texture-based one, allows relative distances within images in perspective projection to be estimated. Section 4, lastly, will draw some conclusions and suggest directions for future research. All the algorithms are followed by examples illustrating their use.

2   Texture Analysis as a Source of Spatial Information

In general, a surface can be considered as being characterized by some form of *texture* if the relevant shapes composing it are uniformly distributed, i.e. if they do not differ very much in appearance, size and density throughout the extension of the surface itself [1].

Essentially, two different approaches have been proposed for texture analysis. Some researchers (such as [2]) follow a *structural* approach, which requires the real structure of texture to be determined (periodicity, uniformity, symmetry, etc.). Although this is probably the method used by the human vision system to infer 3D structure of the environment, it is difficult to automate. Other more feasible techniques limit themselves to making assumptions about the texture arrangement. For instance, if the texture is isotropically distributed, i.e. line segments composing the real surface have not a prevalent direction, three-dimensional information can be obtained from the "preferred" direction observed. This approach was first proposed by Witkin [3] and subsequently improved by Davis et al. [4]. Gårding [5] and Blake et al. [6] perfected the method under the hypothesis of orthographic projection. As regards perspective projection, Kanatani [7] provided a rigorous mathematical description of the problem. Under the hypothesis that texture is composed of points and straight lines only, he developed a technique based on texture homogeneity and density. However, this algorithm is not able to produce good results applied to real scenes. In fact, Kanatani analyzes how the density of image points varies as distance increases, asserting that for greater distances density along a surface grows. Unfortunately, in most real cases the number of relevant points decreases, because of lenses' blur.

We will now concentrate on Witkin's algorithm, in the form corrected by Davis et al. [4]. In Section 2.2 it will be used to determine, given two points on an image in perspective projection, which one corresponds to a farther region in the real scene.

2.1   Texture Analysis in Orthographic Projection

When observing shapes on a plane surface, two different kinds of geometric distortions can be noted: (1) as the surface departs from the observer, shapes appear to be smaller and smaller; (2) the more a surface is inclined with respect to the image plane, the more shapes on the image appear to be flattened towards the tilt direction (*foreshortening* effect). As will be shown, such distortions can be usefully exploited to get spatial information about the real scene.

In image processing, when the *orthographic projection* hypothesis is satisfied, which means that every point in the three-dimensional space is orthogonally projected on the image plane (see Figure 1), effect (1) can be ignored and it becomes easier to estimate the orientation of the plane on which the real scene lies.

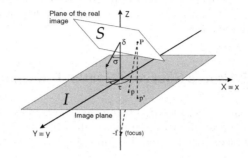

Fig. 1. Reference cartesian system. Orthographic (point p') and perspective (point p) projections of the spatial point P on the image plane

As already stated, for orthographic projection Witkin [3] was the first to face the problem of drawing the spatial arrangement of a plane from image texture analysis, under the hypotheses of *independence* and *isotropy*, which can be summarized as follows:

1. Given an image on the image plane, the possible orientations of the planes on which the original scene may lie are equiprobable.
2. On the plane on which the original image lies, the possible orientations of the tangents to the curves composing the image itself are equiprobable.
3. Orientations of the elements in the real image are statistically independent.

Witkin's studies led to a practical algorithm by which it is possible to estimate the σ and τ polar coordinates (slant and tilt) of the plane on which a scene lies (Figure 1). Let's suppose to have a plane S, on which there are curves and shapes satisfying the isotropy and independence hypotheses. Also, suppose the orthographic projection of S on the image plane I is satisfied. Then, directions β of the tangents to the curves on S will have $p.d.f.(\beta) = 1/\pi$ as probability density function, where $\beta \in [0, \pi[$. As regards parameters σ and τ, the $p.d.f.$ function can also be expressed as $p.d.f.(\sigma, \tau) = (sin\sigma)/\pi$. In fact, consider the *gaussian sphere*, which is the unit radius sphere formed by all the possible normals to all the possible plane arrangements in space. For each value of σ, the possible values of τ define a circumference on the gaussian sphere, whose radius approaches 1 as σ approaches $\pi/2$. The probability that the orientation of a normal corresponds to a certain point on a certain circumference is the same for every point on it and is proportional to the length of the circumference itself ($2\pi sin\sigma$). Therefore, the $p.d.f.$ function can be simply expressed in function of $sin\sigma$, without the need to introduce parameter τ. Since, in general, it is not possible to measure β angles directly, it is necessary to find their transformation into the corresponding angles on the image plane I (let's call them α). If the orientation of plane S is (σ, τ), the image on I can be translated into the one on S by rotating it by the same angles. Then, supposing $\tau = 0$, point $p(x, y, 0)$ on I will have $X = x cos\sigma$, $Y = y$ and $Z = x sin\sigma$ as coordinates on S. Since the orthographic projection of a spatial point (X, Y, Z) on I is (X, Y), a simple way to get the projection of a curve which lies on plane S on plane I consists in "placing" the curve on I, rotating it by (σ, τ), and projecting it again on I. Now, let's consider versor $t = [cos\beta, sin\beta]$, tangent to a curve on S in a certain point, according to an angle β. After the rotation on I, it will become $t^* = [cos\beta cos\sigma, sin\beta]$. Therefore, we have that $tan\alpha = sin\beta/(cos\beta \cdot cos\sigma) = tan\beta/cos\sigma$, where α is the orientation of versor t^* on plane I. If $\tau \neq 0$, it is sufficient to add it to α, which means that $\alpha = atan(tan\beta/cos\sigma) + \tau$ and $\beta = atan(cos\sigma \cdot tan(\alpha - \tau))$.

Considering that, in general, $p.d.f(\varphi(x)) = p.d.f.(x) \cdot dx / d\varphi(y)$, we find:

$$p.d.f.(\alpha(\beta)|\sigma,\tau) = p.d.f.(\beta|\sigma,\tau)\frac{\partial\beta}{\partial\alpha} = \frac{1}{\pi}\frac{cos\sigma}{cos^2(\alpha - \tau) + cos^2\sigma \cdot sin^2(\alpha - \tau)}$$

However, an image will be composed of many curves and hence it is also necessary to find the compound probability density function $p.d.f.(A|\sigma,\tau)$, where A is a set $\{\alpha_i\}$ containing the various α angles measured on the image plane for all the curves on it. If all α_i are independent, the following can be obtained:

$$p.d.f.(A = \{\alpha_1,...,\alpha_n\}|\sigma,\tau) = \prod_{i=1}^{n} p.d.f.(\alpha_i|\sigma,\tau) = \qquad (1)$$

$$= \prod_{i=1}^{n}\frac{\pi^{-1}cos\sigma}{cos^2(\alpha_i - \tau) + cos^2\sigma \cdot sin^2(\alpha_i - \tau)}$$

where n is the number of α angles. Substantially, the preceding expression allows us to calculate how much probable a set of values $\{\alpha_i\}$ is, given σ and τ. However, we are interested in the opposite problem, i.e. in finding the probability to have a particular couple (σ, τ) given a set of values $\{\alpha_i\}$. Applying the Bayes rule and normalizing so that the integral of the probability density function is equal to one, we have:

$$p.d.f.(\sigma,\tau|A) = \frac{p.d.f.(\sigma,\tau)p.d.f.(A|\sigma,\tau)}{\int p.d.f.(\sigma,\tau)p.d.f.(A|\sigma,\tau)d\sigma d\tau} \qquad (2)$$

That couple (σ^*, τ^*) which maximizes $p.d.f.(\sigma,\tau|A)$ is the more probable one and therefore it can be assumed as the orientation of plane S (*maximum likelihood estimate* technique). From expressions (1), and remembering that $p.d.f.(\sigma, \tau) = sin\sigma/\pi$, the value of the numerator of expression (2) can be calculated as follows:

$$p.d.f.(\sigma,\tau)p.d.f.(A|\sigma,\tau) = \frac{sin\sigma}{\pi}\prod_{i=1}^{n}\frac{\pi^{-1}cos\sigma}{cos^2(\alpha_i - \tau) + cos^2\sigma \cdot sin^2(\alpha_i - \tau)} = \qquad (3)$$

$$= exp\left(log\left(\frac{sin\sigma}{\pi}\right) + \sum_{i=1}^{n} a_i \, log \frac{\pi^{-1}cos\sigma}{cos^2(\alpha_i - \tau) + cos^2\sigma \cdot sin^2(\alpha_i - \tau)}\right)$$

where a_i is the number of measures whose value is α_i.

The previous considerations lead to the following algorithm to estimate slant and tilt of the plane on which a scene lies.

Algorithm 1. The practical algorithm we have implemented is derived from that of

Witkin and is composed of the following steps:

1. If the original image on the image plane is known to be affected by noise, it is filtered through a low-pass filter.
2. After having been normalized, the image undergoes an edge-detection process, through the Sobel operator. As a result, two different "images" are obtained from the gradient, namely the module image and the phase image.
3. The module image is thresholded against a certain value, to get a binary image (mask) identifying only the most relevant edges. We are studying for an automatic evaluation of this value.
4. For each pixel in the original image whose value in the mask is one, the direction of its tangent is calculated. This is done by simply rotating by 90 degrees the corresponding value in the phase image.
5. Interval $[0, \pi[$ is subdivided into n sub-intervals. An array $A = \{a_1,...,a_n\}$ is built in which a_i is the number of values obtained in step 4 falling into the ith sub-interval (α_i). Interval $[0, \pi/2]$ (σ values) is subdivided into m sub-intervals and interval $[0, \pi[$ (τ values) is subdivided into p sub-intervals.
6. For each possible couple (σ_j, τ_k) ($j \in [0, m-1]$, $k \in [0, p-1]$), expression (3) is calculated. That couple (σ_{j*}, τ_{k*}) for which the result is maximum is taken as the orientation of the plane containing the real scene in the three-dimensional space.

Experimental Results. We tested algorithm 1 with many images and the results obtained confirm its effectiveness in estimating slant and tilt of real scenes. Figure 2 shows three example images, for which the computed σ and τ are reported. Of course, the algorithm considers them as if they were in orthographic projection. To give an intuitive insight of the results obtained, circles oriented according to the calculated values are superimposed on the original images, and, at the center of the circles, segments normal to the estimated planes are placed. For the algorithm, the following parameter values have been assumed: $n = 64$, $m = 90$, $p = 180$, $\alpha_i = \pi(1/2+i)/n$, $\sigma_j = \pi(1/2+j)/2m$, $\tau_k = \pi(1/2+k)/p$, where $i \in [0, n-1]$, $j \in [0, m-1]$ and $k \in [0, p-1]$.

| $\sigma = 56.5°$ | $\sigma = 69.5°$ | $\sigma = 58.5°$ |
| $\tau = 90.5°$ | $\tau = 147.5°$ | $\tau = 87.5°$ |

Fig. 2. Three examples of application of Algorithm 1. Beneath each image, which is considered as if it was in orthographic projection, the calculated σ and τ parameters are reported

2.2 Texture Analysis in Perspective Projection

In perspective projection, each point in the three-dimensional space is projected on the image plane through the focus of the optical system (see again Figure 1). When, as often occurs, the orthographic projection hypothesis is not satisfied, the algorithm presented in the previous section can only produce approximated results. In fact, shape size lessening and the foreshortening effect cause the image on the image plane to be less "uniform", thus weakening the concept of texture itself. However, intuitively, if from an image in perspective projection sufficiently small regions are extracted, it will be reasonable to consider them as if they were in orthographic projection. If algorithm 1 is then applied to each of them, we will find that those regions which are farther with respect to the tilt direction produce greater values for parameter σ, because shapes turn out to be more flattened. In fact, referring to Figure 1, given a point $P(X,Y,Z)$ on plane S, the corresponding coordinates on the image plane will be $x = fX/(f+Z)$ and $y = fY/(f+Z)$. Let's suppose $\tau = 90°$ and consider the perspective projection of a circle placed on S on the image plane. The resulting image, unless $\sigma = 90°$, will be an ellipse, as shown in Figure 3.

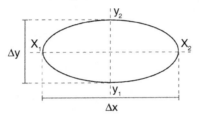

Fig. 3. Projection of a circle on the image plane

If the distance of point y_1 from the image plane is Z and the circle's diameter is D, the ellipse's Δx width can be calculated as $fD/[f+Z+(D/2)\sin\sigma]$. If y_1 has Y as the corresponding coordinate in space, the ellipse's Δy height is $\Delta y = y_2 - y_1$, where $y_1 = fY/(f+Z)$ and $y_2 = f(Y+D\cos\sigma)/(f+Z+D\sin\sigma)$. Since the equation of plane S is $Z = \tan\sigma\cdot Y + \delta$ (where δ is the coordinate of the intersection of plane S with the Z axis), we obtain: $\Delta y = y_2 - y_1 = fD\cos\sigma(f+\delta)/[(f+Z)(f+Z+D\sin\sigma)]$. If the projection was orthographic instead of perspective, relation $\cos\sigma_o = \Delta y/\Delta x$ would hold (where σ_o indicates just the slant calculated in the orthographic case). In fact, Δx would be equal to D and Δy would be the projection of D on the image plane, i.e. $\Delta y = D\cos\sigma_o$. However, if a sufficiently small area is selected on the image to be processed, algorithm 1 can provide a good approximation of the image slant. Then, supposing this is the case, from the previous expressions we have:

$$\cos\sigma_o = \frac{\Delta y}{\Delta x} = \frac{\cos\sigma_p (f+\delta)\left(f+Z+\dfrac{D}{2}\sin\sigma_p\right)}{(f+Z)(f+Z+D\sin\sigma_p)}$$

where σ_p indicates the slant calculated in the perspective projection case. It is evident that when Z increases (i.e. the distance from the observer grows) also σ_o must increase.

The previous considerations suggest a simple but effective way to take into account perspective when trying to estimate distance in images for which the orthographic projection hypothesis is not valid.

Algorithm 2. Given an image on the image plane, to determine whether a certain point is farther or nearer than another (with respect to the observer), it is sufficient to consider two areas around them and apply algorithm 1. The farther point is that for which σ is greater.

Experimental Results. Figure 4 shows some examples of application of algorithm 2. Rectangles on the images identify the regions analyzed through algorithm 1, whose estimated σ and τ are reported. Values assumed for the parameters are the same as for the examples in Section 2.1 ($n = 64$, $m = 90$, $p = 180$).

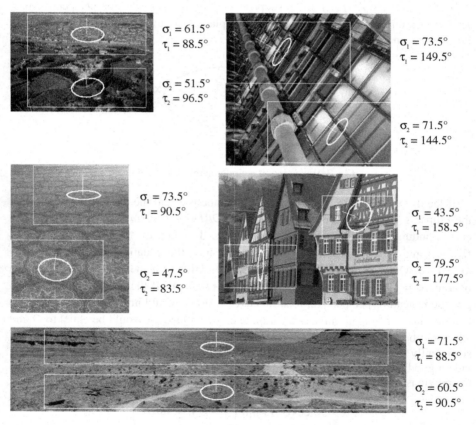

Fig. 4. Examples of application of Algorithm 2. At the right of each image, in correspondence with rectangles identifying the examined regions, the estimated σ and τ parameters are reported

3   Histogram Analysis for Distance Estimation

In this section, we will present another algorithm for qualitative estimation of distance in monocular vision: histogram analysis. Instead of using geometric parameters, it exploits physical properties of the medium in which the objects in the images to be analyzed are contained. The technique, which is primarily intended for being used in combination with other methods, is currently in a development stage and we are still working on it to automate the choice of the values for some key parameters.

Everyone can note that, for instance in images representing landscapes, edges of distant elements (such as mountains) are not very clear: colors of different objects tend to blend each other and generate zones with nearly uniform colors. Such an effect is more marked if haze or fog is present, or when underwater images are examined. In general, it is more perceivable in images with very long depth of field compared with the medium opacity.

Rays of light in an opaque medium are diffused by its molecules. For example, air contains a great number of water particles, which refract light. In a monochromatic image, the background's gray levels tend to be the weighed average of all the gray levels present in the image itself. As a result, the image becomes more and more uniform as the distance from the observer increases, thus producing a sort of blur. This blur, however, is not to be mistook for that produced by camera lenses, which must be absolutely avoided to get valid results from the technique we are now going to describe.

3.1   The Algorithm

As already stated, the zones of an image which are perceived as more blurred (i.e. which are farther with respect to the observer) have gray levels gathered around an average value. On the contrary, in those areas where edges are sharper (i.e. which are nearer with respect to the observer) gray levels are more scattered. Information about the distance of a zone with respect to another can thus be obtained by analyzing the gray level spectrum in these regions.

The algorithm we present here is based on variance applied to the image histogram.

Algorithm 3. Consider two points P_1 and P_2 on an image. To determine which one is nearer (or farther) with respect to the observer, we can proceed according to the following steps:

1. Two areas A_1 and A_2 around P_1 and P_2 are extracted from the image and their histograms are obtained.
2. Both for A_1 and A_2, the average gray levels x_{a1} and x_{a2} of their pixels are calculated. In general, indicating with $f(x)$ the number of pixels whose gray level is equal to x, the average is obtained as follows:

$$x_a = \frac{\sum_i x_i f(x_i)}{\sum_i f(x_i)}$$

3. Both for A_1 and A_2, variances σ_1^2 and σ_2^2 are calculated in the following way:

$$\sigma^2 = \frac{\sum_i (x_i - x_a)^2 f(x_i)}{\sum_i f(x_i)} = \frac{\sum_i x_i^2 f(x_i)}{\sum_i f(x_i)} - x_a^2$$

4. The point which is farther with respect to the observer will be that whose variance is lower.

Experimental results. As an example of application of algorithm 3 to a real image, Figure 6 displays the histograms relative to the three areas highlighted by rectangles (a, b and c) in Figure 5.

Fig. 5. An example of application of algorithm 3. White rectangles are used to highlight the areas analyzed (a, b and c)

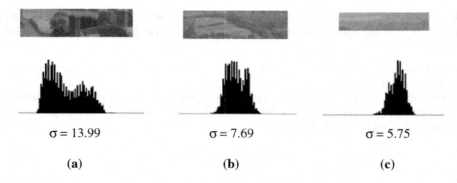

σ = 13.99	σ = 7.69	σ = 5.75
(a)	**(b)**	**(c)**

Fig. 6. Histograms relative to the areas highlighted in Figure 5, and corresponding variances

As can be noted, the histogram of the nearest zone (rectangle *a*) is more scattered than that of the farthest zone (rectangle *c*) and, accordingly, the variance of the first area (13.99) is greater than that of the second one (5.75).

Of course, dimensions of rectangles may remarkably affect the results. We are currently trying to make such a choice automatic, based on a preliminary segmentation phase to select sufficiently homogeneous areas.

4   Conclusions

In this paper we have presented two techniques which allow distances to be estimated from a relative point of view. That is, they allow to determine whether a point within an image is farther or nearer than another with respect to the observer.

Two algorithms, applicable to images in perspective projection, have been described. The first (Algorithm 2), based on texture analysis, has been tested with a great number of real images in which some form of texture was recognizable, always giving very good results. The second (Algorithm 3), which is based on histogram analysis and is still to be perfected, has shown its good applicability especially to representations with long depth of field (e.g. landscapes). Although some precautions are to be taken (there must not be any form of lens blur), the use of this method is favored by its computational lightness.

We hold that, although they cannot be exploited for actual distance estimations, the proposed algorithms, integrated with others previously studied [8], can be usefully used for qualitative evaluations. Moreover, they can be exploited as a preprocess step able to detect regions of interest for successive application of more expensive techniques, to provide more accurate results. Current work is devoted to this aspect and to the integration of the system with different sources of information.

References

1. Gibson, J.: The Perception of the Visual World. Houghton Milfin, Boston, MA (1950).
2. Blostein, D., Ahuja, N.: Shape from Texture: Integrating Texture-element Extraction and Surface Estimation. IEEE Transactions on Pattern Analysis and Machine Intelligence, 11 (1989) 325-342.
3. Witkin, A.P.: Recovering Surface Shape and Orientation from Texture. Artificial Intelligence, 17 (1981) 17-45.
4. Davis, L.S., Janos, L., Dunn, S.M.: Efficient recovery of shape from texture. IEEE Transactions on Pattern Analysis and Machine Intelligence, 5 (1983) 485-492.
5. Gårding, J.: Shape from Texture and Contour by Weak Isotropy. Artificial Intelligence, 64 (1993) 243-297.
6. Blake, A., Marinos, C.: Shape from Texture: Estimation, Isotropy and Moments. Artificial Intelligence 45 (1990) 323-380.
7. Kanatany, K., Chou, T.: Shape from Texture: a General Principle. Artificial Intelligence, 38 (1989) 1-48.
8. Matessi, A., Lombardi, L.: Vanishing Point Detection in the Hough Transform Space. Proceedings of the Fifth International Euro-Par Conference, Tolouse, France, August/September (1999) 987-994.

A New Shape Space for Second Order 3D-Variations

Per-Erik Danielsson and Qingfen Lin

Department of Electrical Engineering
Linkoping University
SE-581 83 Sweden
ped@isy.liu.se, qingfen@isy.liu.se

Abstract. A common model of second degree variation is an ellipsoid spanned by the magnitudes of the Hessian eigenvalues. We find this model incomplete and often misleading. Here, we present a more complete representation of the information embedded in second degree derivatives. Using spherical harmonics as a basis set, the rotation invariant part of this information is portrayed as an orthonormal shape-space, which is non-redundant in the sense that any local second order variation can be rotated to match one and only one unique prototype in this space. A host of truly rotation invariant and shape discriminative shape factors is readily defined.

1 Introduction

The interest in three-dimensional volume analysis is steadily increasing with the advent of more powerful imaging technologies such as helical cone beam Computer Tomography and more advanced modalities of Magnetic Resonance Imaging. As an initial step in such analysis, some authors have been measuring the local second order density variation at each grid point. If magnitude and rotation (orientation) dependency is eliminated from this variation, the remaining information is shape. The subject of this paper is how to describe the totality of this shape information, or, in other words, to define the second order *shape space* for 3D-volumes. It seems customary to assume that all of these possible shapes can be modeled as ellipsoids, something we find to be mathematically incorrect as well as a severe limitation in practical applications.

The local second order variation is measured by *convolving* the 3D-function $f(x, y, z)$ with six *derivators* $\boldsymbol{g}_2^T = (g_{xx}, g_{yy}, g_{zz}, g_{xy}, g_{xz}, g_{yz})$. The response vector consists of the derivative estimates $(f_{xx}, f_{yy}, f_{zz}, f_{xy}, f_{xz}, f_{yz})$, which also can be assembled as a symmetric 3x3 matrix, the *Hessian*. Typically, the derivators are designed as differentiated Gaussians, which calls for reasonable compromises between minimum approximation errors and computational efficiency.

In many applications the next processing step is *derotation*, which is to separate the rotation dependent information from the rotation-invariant part in the response vector. The standard procedure is to diagonalize the Hessian to obtain three eigenvalues $(\lambda_1, \lambda_2, \lambda_3)$ and their corresponding eigenvectors. After proper normalization the eigenvectors comprise the columns of the 3x3-rotator matrix R with three degrees of

C. Arcelli et al. (Eds.): IWVF4, LNCS 2059, pp. 145–154, 2001.

freedom. Let $\boldsymbol{x}^T = (x, y, z)$. The rotator R should be able to transform the local neighborhood $f(\boldsymbol{x})$ into a unique *prototype* function $p(R\boldsymbol{x})$ so that

$$(f_{xx}, f_{yy}, f_{zz}, f_{xy}, f_{xz}, f_{yz}) \xrightarrow{R} (p_{xx}, p_{yy}, p_{zz}, 0, 0, 0) \tag{1}$$

which separates *orientation* information from *magnitude* and *shape*. The orientation appears in the rotator R, magnitude and shape in the prototype vector (p_{xx}, p_{yy}, p_{zz}). It would then be tempting to just identify the three eigenvalues taken in any order with the three prototype derivatives (p_{xx}, p_{yy}, p_{zz}). Unfortunately, since there are no less than six possible permutations (indexing) of the three eigenvalues and the eigenvector columns of the rotator R, the uniqueness of such a prototype is not secured. In this paper we will show that there is a rule for selecting a permutation to secure a truly unique derotation, which at the same time delivers a two-dimensional continuous shape space residing on a part of the unit sphere.

As mentioned, several authors have been employing second derivatives and the Hessian eigenvalues to detect and discriminate for shape. Lorenz et al [1] distinguish a bright line (string) when $\lambda_1 \approx \lambda_2 \ll 0$, $|\lambda_3| \approx 0$ and derives several *stringness* factors from the eigenvalues, e.g. $1 - \dfrac{2|\lambda_3|}{|\lambda_1 + \lambda_2|}$. Sato et al [2] observe the same string detection rules (except for a different ordering among the eigenvalues) and distinguish them from sheets (planes) for which $|\lambda_3| < 0$, $|\lambda_1| \approx |\lambda_2| \approx 0$. Frangi et al [3] suggest three shape descriptors based on various combinations of the magnitudes of the eigenvalues. Similar shape descriptors are also used by Kindlmann et al [4] to span a shape space, which is visualized as a triangle with the three archetypal ellipsoidal shapes called blob-like ellipsoid, string-like ellipsoid and plane-like ellipsoid. This shape space is essentially the same as the one brought forward many years ago by Knutsson [5] although the tensor behind the latter shape space is not the Hessian but a set of quadrature filter responses. A special feature of these responses is phase-invariance with respect to the locally dominating frequency components. Similarly, Basser [6] portrays the eigenvalues of the diffusion matrix with ellipsoids, which may change in shape depending on the relative magnitudes between the eigenvalues.

For tensors like the 3D-diffusion coefficients, where all eigenvalues have the same sign, it makes some sense to model the local variation as an ellipsoid. However, the Hessian and other tensors, which are not restricted in this way, have a much richer variation. This is revealed already in the 2D-case where the two eigenvalues (λ_1, λ_2) are identified as the prototype derivatives (p_{xx}, p_{yy}) or (p_{yy}, p_{xx}). If these quantities are equal in magnitude but have different signs, the local intensity variation is shaped as a saddle surface; if the signs are equal it is a positive or negative 2D-blob. These two shapes are extremely different, in fact orthogonal. A method or a representation that fails to detect, or doesn't distinguish between a saddle and a blob is incomplete to say the least.

2 The Spherical Harmonics Orthonormal Basis Set

Let $h_0(r)$ be a rotationally symmetric 3D-function, e.g. a Gaussian. The six second degree derivators are then obtained in the signal and the Fourier domains as

$$
\begin{bmatrix} g_{xx} \\ g_{yy} \\ g_{zz} \\ g_{xy} \\ g_{xz} \\ g_{yz} \end{bmatrix} = \partial^2 \begin{bmatrix} (\partial x)^{-2} \\ (\partial y)^{-2} \\ (\partial z)^{-2} \\ (\partial x \partial y)^{-1} \\ (\partial x \partial z)^{-1} \\ (\partial y \partial z)^{-1} \end{bmatrix} h_0(r) \Leftrightarrow -4\pi^2 H_o(\rho) \begin{bmatrix} u^2 \\ v^2 \\ w^2 \\ uv \\ uw \\ vw \end{bmatrix} = -4\pi^2 \rho^2 H_o(\rho) \begin{bmatrix} \sin^2\theta\cos^2\phi \\ \sin^2\theta\sin^2\phi \\ \cos^2\theta \\ \sin^2\theta\cos\phi\sin\phi \\ \sin\theta\cos\theta\cos\phi \\ \sin\theta\cos\theta\sin\phi \end{bmatrix}.
\tag{2}
$$

$$
u = \rho\sin\theta\cos\phi \qquad v = \rho\sin\theta\sin\phi \qquad w = \rho\cos\theta
$$

The above derivators $(G_{xx}, G_{yy}, G_{zz}, G_{xy}, G_{xz}, G_{yz})$ in the Fourier domain are combined as follows in (4) into *orthonormal* spherical harmonic operators $(C_{20}, ... C_{25})$ with normalizing factors yielding

$$
\int\limits_0^{2\pi}\int\limits_0^{\pi} [C_i(\rho)]^2 \sin\theta\, d\theta\, d\phi = \tfrac{4\pi}{6}[H_2(\rho)]^2 \quad \text{for } i = 20,21,22,23,24,25.
\tag{3}
$$

$$
\begin{bmatrix} C_{20} \\ C_{21} \\ C_{22} \\ C_{23} \\ C_{24} \\ C_{25} \end{bmatrix} \overset{\Delta}{=} \begin{bmatrix} \sqrt{\tfrac{1}{6}} & \sqrt{\tfrac{1}{6}} & \sqrt{\tfrac{1}{6}} & 0 & 0 & 0 \\ -\sqrt{\tfrac{5}{24}} & -\sqrt{\tfrac{5}{24}} & \sqrt{\tfrac{5}{6}} & 0 & 0 & 0 \\ \sqrt{\tfrac{5}{8}} & -\sqrt{\tfrac{5}{8}} & 0 & 0 & 0 & 0 \\ 0 & 0 & 0 & \sqrt{\tfrac{5}{2}} & 0 & 0 \\ 0 & 0 & 0 & 0 & \sqrt{\tfrac{5}{2}} & 0 \\ 0 & 0 & 0 & 0 & 0 & \sqrt{\tfrac{5}{2}} \end{bmatrix} \begin{bmatrix} G_{xx} \\ G_{yy} \\ G_{zz} \\ G_{xy} \\ G_{yz} \\ G_{xz} \end{bmatrix} = -4\pi^2\rho^2 H_0(\rho) \begin{bmatrix} \sqrt{\tfrac{1}{6}} \\ \sqrt{\tfrac{5}{24}}(3\cos^2\theta - 1) \\ \sqrt{\tfrac{5}{8}}\sin^2\theta\cos 2\phi \\ \sqrt{\tfrac{5}{8}}\sin^2\theta\sin 2\phi \\ \sqrt{\tfrac{5}{8}}\sin 2\theta\cos\phi \\ \sqrt{\tfrac{5}{8}}\sin 2\theta\sin\phi \end{bmatrix}
\tag{4}
$$

Because orthogonality and normalization, as well as harmonic variation in the angular direction are preserved over the Fourier transform [7] we have in the signal domain

$$
\begin{bmatrix} c_{20} \\ c_{21} \\ c_{22} \\ c_{23} \\ c_{24} \\ c_{25} \end{bmatrix} = \begin{bmatrix} h_{20}(r)\sqrt{\tfrac{1}{6}} \\ \\ h_2(r)\sqrt{\tfrac{5}{8}} \begin{bmatrix} \sqrt{\tfrac{1}{3}}(3\cos^2\theta - 1) \\ \sin^2\theta\cos 2\phi \\ \sin^2\theta\sin 2\phi \\ \sin 2\theta\cos\phi \\ \sin 2\theta\sin\phi \end{bmatrix} \end{bmatrix} = \frac{1}{r^2} \begin{bmatrix} h_{20}(r)\sqrt{\tfrac{1}{6}}(x^2 + y^2 + z^2) \\ \\ h_2(r)\sqrt{\tfrac{5}{8}} \begin{bmatrix} \sqrt{\tfrac{1}{3}}(2z^2 - x^2 - y^2) \\ x^2 - y^2 \\ 2xy \\ 2xz \\ 2yz \end{bmatrix} \end{bmatrix}.
\tag{5}
$$

Existing in both domains, the polynomial forms in (2) and (5) indicate a strong relationship between moment functions and spherical harmonics. The radial variation $H_0(\rho)$, common for all six basis functions in the Fourier domain, corresponds over two Hankel transforms of different orders with the functions in the signal domain. This is because the harmonic variation is of zero order for the Laplacian operator $c_{20} - C_{20}$ and of second order for the five other basis functions.

The mapping (4) from derivators to orthonormal basis functions performed by the 6x6 matrix M carries over directly to the derivatives. Therefore, we also obtain the total response $\boldsymbol{f}_2$ and the derotated protoype response $\boldsymbol{p}_2$ as coefficients for the six orthogonal basis functions $(c_{20}c_{21}, c_{22}, c_{23}, c_{24}, c_{25})$ as follows

$$\boldsymbol{f_2} = (f_{20}, f_{21}, f_{22}, f_{23}, f_{24}, f_{25})^T = M(f_{xx}, f_{yy}, f_{zz}, f_{xy}, f_{xz}, f_{yz})^T, \tag{6}$$

$$\boldsymbol{p_2} = \begin{bmatrix} p_{20} \\ p_{21} \\ p_{22} \end{bmatrix} = \begin{bmatrix} \frac{1}{\sqrt{6}} & \frac{1}{\sqrt{6}} & \frac{1}{\sqrt{6}} \\ -\sqrt{\frac{5}{24}} & -\sqrt{\frac{5}{24}} & \sqrt{\frac{5}{6}} \\ \sqrt{\frac{5}{8}} & -\sqrt{\frac{5}{8}} & 0 \end{bmatrix} \begin{bmatrix} p_{xx} \\ p_{yy} \\ p_{zz} \end{bmatrix}. \tag{7}$$

Both vectors will now appear in this perfectly orthonormal and complete shape space. We notice that out of the six basis functions, (c_{23}, c_{24}, c_{25}) are identical to the three cross-derivators (g_{xy}, g_{xz}, g_{yz}) except for a scale factor. Hence, the above diagonalization procedure, which eliminates the responses (f_{xy}, f_{xz}, f_{yz}), will leave us with prototype responses that we, at least initially, could identify as $p_{xx} = \lambda_1$, $p_{yy} = \lambda_2$, $p_{zz} = \lambda_3$ with a randomly ordered set of eigenvalues. This preliminary response vector $\boldsymbol{p}_2'$ is shown in the (c_{20}, c_{21}, c_{22})-space in Figure 1.

Alluding to their shape in signal space indicated by the attached icons in Figure 1, the three basis functions are named Blob, Double Cone, and (four-wedge) Orange, respectively. The solid lines in these icons represent zero-crossings of the intensity, the boundaries between positive and negative density relative to the underlying DC-level and more global gradients. The smooth decay of the magnitude in the radial direction due to the functions $h_{20}(r)$ and $h_2(r)$ is indicated by a dotted contour.

3   The Non-redundant Shape Space

The (c_{20}, c_{21}, c_{22})-space in Figure 1 is redundant. Certain shapes repeat themselves along the equator of Figure 1, the subspace spanned by (c_{21}, c_{22}). Using the polynomial expressions (5), in (8), (9) we obtain two alternative sets of basis functions by rotations around the c_{20}-axis with 120° and 240°, respectively. The polynomial expressions for both these sets are identical to the ones for (c_{21}, c_{22}) except for cyclic permutations of the signal space coordinates. The shapes they represent in signal space are therefore identical except for rotation of the coordinate system (x, y, z) around the vector $(1, 1, 1)$ (the cube diagonal) with 120° and 240°, respectively.

$$\begin{bmatrix} c_{41} \\ c_{42} \end{bmatrix} = \begin{bmatrix} -\frac{1}{2} & -\frac{\sqrt{3}}{2} \\ \frac{\sqrt{3}}{2} & -\frac{1}{2} \end{bmatrix} \begin{bmatrix} c_{21} \\ c_{22} \end{bmatrix} = h_2(r)\sqrt{\frac{5}{8}} \begin{bmatrix} \sqrt{\frac{1}{3}}(2y^2 - z^2 - x^2) \\ z^2 - x^2 \end{bmatrix}. \tag{8}$$

$$\begin{bmatrix} c_{31} \\ c_{32} \end{bmatrix} = \begin{bmatrix} -\frac{1}{2} & \frac{\sqrt{3}}{2} \\ -\frac{\sqrt{3}}{2} & -\frac{1}{2} \end{bmatrix} \begin{bmatrix} c_{21} \\ c_{22} \end{bmatrix} = h_2(r)\sqrt{\frac{5}{8}} \begin{bmatrix} \sqrt{\frac{1}{3}}(2x^2 - y^2 - z^2) \\ y^2 - z^2 \end{bmatrix}. \tag{9}$$

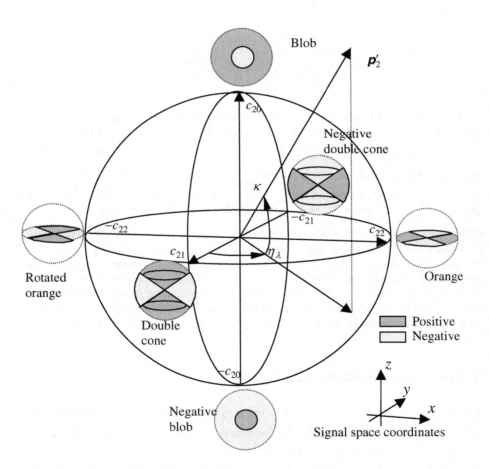

Fig. 1. The redundant shape space (c_{20}, c_{21}, c_{22})

We notice that the shapes of c_{22} and $-c_{22}$ are identical except for rotation around the z-axis with $+90^o$ or -90° in the signal space, equivalent to a shape-invariant flipping of the equatorial plane around the c_{21}-axis in the (c_{20}, c_{21}, c_{22})-space. We simply swap the roles of x^2 and y^2 in the polynomials. For symmetry reasons there are three possibilities for such a pair-wise swap. All the six possibilities to perform a mapping between the eigenvalues and the prototype derivatives are then accounted for. Only one sixth of the 360° circumference of the equator in Figure 1 is needed to describe all shapes, provided we make certain that all prototype responses end up in the same sector.

In Figure 1, let us select the 60^0-sector $\pm 30^o$ around the unit vector c_{22} as the "chosen one", knowing that any of the five other 60^0-sectors would be just as fine. The chosen sector of the non-redundant shape space is portrayed in Figure 2. The shape angle $\eta = \arg(c_{21}, c_{22})$ is then restricted by

$$\tfrac{\pi}{3} \leq \eta < \tfrac{2\pi}{3} . \tag{10}$$

We use the polynomial expressions in (5) and (7) in (10) to obtain

$$\frac{\pi}{3} \leq \arg(c_{21}, c_{22}) \quad \Rightarrow \quad \sqrt{3} \leq \frac{\sqrt{\tfrac{5}{8}}(p_{xx} - p_{yy})}{\sqrt{\tfrac{5}{24}}(2p_{zz} - p_{xx} - p_{yy})} \quad \Rightarrow \quad p_{zz} \leq p_{xx} \tag{11}$$

$$\arg(c_{21}, c_{22}) < \frac{2\pi}{3} \quad \Rightarrow \quad \sqrt{3} < \frac{\sqrt{\tfrac{5}{8}}(p_{xx} - p_{yy})}{-\sqrt{\tfrac{5}{24}}(2p_{zz} - p_{xx} - p_{yy})} \quad \Rightarrow \quad p_{yy} < p_{zz} . \tag{12}$$

Hence, we should make the eigenvalue/prototype assignment as

$$p_{xx} \Leftarrow \lambda_1, \quad p_{yy} \Leftarrow \lambda_3, \quad p_{zz} \Leftarrow \lambda_2, \qquad \text{where} \qquad \lambda_1 > \lambda_2 > \lambda_3. \tag{13}$$

In Figure 2 we find the axially symmetric shapes (*strings*, *planes*) along the boundaries of the 60° wedge. It can be shown [8] that the best match for ideal strings and planes are obtained with the prototypes called *optimal string* and *optimal plane*, which deviate significantly from the ideal shapes. Along the meridian through c_{22} we find the *orange* but also some structures that we call *ovals* which are located approximately half-way between the blobs at the poles and the orange at the equator.

In the introduction of this paper we claimed that the ellipsoidal model makes sense only when all eigenvalues (all prototype derivatives) have the same sign. In Figure 2 we find those parts of the shape space at the top and bottom, respectively. The upper "ellipsoidal" area is limited by a curve on which the smallest eigenvalue $\lambda_3 = p_{yy} = 0$, the lower one by a curve where the largest eigenvalue $\lambda_1 = p_{xx} = 0$. This in accordance with the fact that the icons, the normalized prototypes, located on these curves also lack contributions from g_{yy} and g_{xx}, respectively.

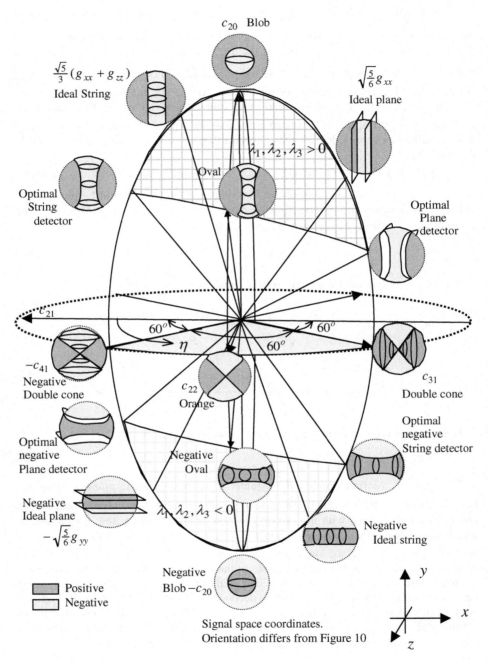

Fig. 2. The non-redundant shape space

4 Rotation Invariants. Shape Factors

Whatever pose we give a certain prototype, the derotation procedure described above will return the same coefficients (p_{20}, p_{21}, p_{22}) in the (c_{20}, c_{21}, c_{22})-space. Hence, any function using the arguments (p_{20}, p_{21}, p_{22}) or the derivatives (p_{xx}, p_{yy}, p_{zz}) is rotation invariant. One such function that have been proposed by some authors as an energy measure is

$$p_\lambda^2 = \lambda_1^2 + \lambda_2^2 + \lambda_3^2 = p_{xx}^2 + p_{yy}^2 + p_{zz}^2 = 2p_{20}^2 + \tfrac{4}{5}(p_{21}^2 + p_{22}^2), \qquad (14)$$

where we have derived the last part of this expression from the inverse of (6). p_λ^2 is rotation invariant but *not shape invariant*, which by all likelihood was the intention. It over-emphasizes the p_{20}-component and under-emphasizes the (p_{21}, p_{22})-components. Hence, for a certain given signal energy it gives a higher value for blob-like shapes, a lower value for shapes which appear at the equator in Figure 2. Let $p_2 = \|\boldsymbol{p}_2\|$, $f_2 = \|\boldsymbol{f}_2\|$. The only energy measure indifferent to both rotation and shape is the sum of the energies of the orthogonal components, which yields

$$p_2^2 = p_{20}^2 + p_{21}^2 + p_{22}^2 = p_{xx}^2 + p_{yy}^2 + p_{zz}^2 - \tfrac{1}{2}(p_{xx}p_{yy} + p_{xx}p_{zz} + p_{yy}p_{zz}) = \qquad (15)$$

$$f_2^2 = f_{20}^2 + f_{21}^2 + f_{22}^2 + f_{23}^2 + f_{24}^2 + f_{25}^2 =$$

$$f_{xx}^2 + f_{yy}^2 + f_{zz}^2 - \tfrac{1}{2}(f_{xx}f_{yy} + f_{xx}f_{zz} + f_{yy}f_{zz}) + \tfrac{5}{2}(f_{xy}^2 + f_{xz}^2 + f_{yz}^2)$$

and can be computed directly from the derivatives without derotation and serve as the first discrimination level for voxels-of-interest. In general, all functions with arguments (p_2, p_{20}) or $(p_2^2, p_2^2 - p_{20}^2) = (p_{20}^2, p_{21}^2 + p_{22}^2)$ can be computed *without derotation*, which also includes the function p_λ^2 just mentioned. To be able to discriminate for shape we may use the *shape angle* κ in Figure 1, which is defined as

$$\kappa = \arcsin\frac{f_{20}}{f_2} = \arcsin\frac{p_{20}}{p_2}. \qquad (16)$$

A rather general discriminative and energy independent *shape factor* Q_κ can be obtained, still without computing eigenvalues/eigenvectors, as

$$Q_\kappa = \sin 2\kappa = \frac{2f_{20}\sqrt{f_2^2 - f_{20}^2}}{f_2^2} = \frac{2p_{20}\sqrt{p_2^2 - p_{20}^2}}{p_2^2}, \qquad (17)$$

which is zero both for blobs, where $p_2 = |p_{20}|$, $\kappa = \pm\tfrac{\pi}{2}$, and for purely Laplacian-free shapes where $p_{20} = 0$, $\kappa = 0$. The shape factor Q_κ takes maximal and minimal values $+1$ and -1 for $\kappa = \pm\tfrac{\pi}{4}$, respectively, i.e. for shapes that are in between the poles and the equator in Figure 2. This is, approximately, where we find second-degree variations of type planes and strings to be dealt with below.

The angle η in Figure 2 is also shape specific. All shapes at the two edges of the shape space are rotationally symmetric around the x-axis and the y-axis, respectively. In general for *axially symmetric* prototypes we have

$$\sqrt{3}\,|p_{21}| = p_{22} \quad \Leftrightarrow \quad \sin\eta = \frac{\sqrt{3}}{2}. \tag{18}$$

It is interesting to note that the inner loop of the Jacobi method for diagonalization [Press88] should terminate rather quickly for axially symmetric and nearly axially symmetric shapes. The details of this argument are left out here for the sake of brevity. However, lack of fast convergence indicates non-axial shape symmetry, which could be set to trigger early termination if such shapes are considered non-interesting. We have found that the successive discrimination of voxels-of-interest by energy, by shape (e.g. with Q_κ), and by early termination of the eigenvalue computation brings about dramatic savings in computation time [8].

Assuming that we have access to the complete prototype vector we propose the following *shape factor for axial symmetry.*

$$Q_\eta = \frac{2\sqrt{3}\,p_{21}p_{22}}{3p_{21}^2 + p_{22}^2}, \tag{19}$$

which is $+1$ and -1 for shapes with a perfect symmetry axis along y and x respectively, zero for shapes along the c_{22}-meridian in Figure 2. However, it is important to be able to distinguish between strings and planes that is indicated by $Sgn(p_{20}p_{21})$. To this end we may use the product

$$Q_\kappa Q_\eta = \frac{2p_{20}\sqrt{p_{21}^2 + p_{22}^2}}{p_2}\frac{2\sqrt{3}p_{21}p_{22}}{3p_{21}^2 + p_{22}^2} \approx \left(Q_\kappa' Q_\eta'\right) = \frac{8}{\sqrt{3}}\frac{p_{20}p_{21}p_{22}}{p_2\sqrt{p_{21}^2 + p_{22}^2}}, \tag{20}$$

which takes the sign from the factor $p_{20}p_{21}$, since all other factors including p_{22} are positive. But from the properties of the two shape factors Q_κ and Q_η follows also that we obtain the maximum value $+1$ for all strings and the minimum value -1 for all planes independent of polarity. It also follows that the zero values are found on the equator and along the c_{22}-meridian. Hereby, the ovals, halfway between planes and strings, will also get the shape factor value zero.

5   Conclusions

This paper advocates that the second derivative response vector should be mapped onto an orthogonal space, with the help of spherical harmonics. Without an orthogonal space many misunderstandings tend to prevail, e.g. that the second degree derivators (g_{xx}, g_{yy}, g_{zz}) are mutually orthogonal just as the three gradient detectors

(g_x, g_y, g_z). Rotation invariance and shape are two sides of the same coin. One cannot be understood without the other. None of them can be fully grasped without portraying the complete and non-redundant shape space, the variation that is left after the object under study is derotated to match its prototype. We have presented here a method, which for the first time is able to reduce the six-dimensional measurement space into an orthogonal non-redundant three-dimensional space for shape and magnitude. All response vectors and functions thereof are rotation invariant in this space. In summary, the steps of this generic and open-ended method are as follows.

1. Convolve the object function $f(x, y, z)$ with the derivators $\boldsymbol{g}_2^T$ to retrieve H_f.
2. Compute the eigenvalues and the eigenvectors of the Hessian H_f
3. Map the eigenvalues onto the reduced prototype space (c_{20}, c_{21}, c_{22})
4. Compute orientation angles, magnitude, shape factors and other discriminators

A more complete mathematical background to the subject of this paper, as well as implementations and experiments in image enhancement and segmentation using second derivatives, are found in [8].

References

1. Lorenz, C., Carlsen, I.-C., Buzug, T.M., Fassnacht, C., Weese, J.: A Multi-Scale Line Filter with Automatic Scale Selection Based on the Hessian Matrix for Medical Image Segmentation. Lecture Notes in Computer Science, Vol. 1252, Springer (1997) 152-163
2. Sato, Y., Westin, C.F., Bhalero, A., Nakajima, S., Shiraga, N., Tamura, S., Kikinis, R.: Tissue Classification Based on 3D Local Intensity Structures. IEEE Trans. Visualization and Computer Graphics, **6** (2000) 160-180
3. Frangi, A. Niessen, W., Vincken, K., Viergever, M.: Multi-Scale Vessel Enhancement Filtering. In: Wells, W., Colchester, A., Delp, S. (eds.): Medical Image Computing and Computer-Assisted Intervention-MICCAI'98. Lecture Notes in Computer Science, Vol. 1496. Springer (1998) 130-137
4. Kindlmann, G., Weinstein, D., Hart, D.: Strategies for Direct Volume Rendering of Diffusion Tensor Fields. IEEE Transactions on Visualization and Computer Graphics, **6** (2000) 124-138
5. Knutsson, H., Representing Local Structure with Tensors. Proc. of 6th Scandinavian Conference on Image Analysis, Oulu, Finland (1989) 244-251
6. Basser, P.: Inferring Micro-Structural Features and the Physiological State of Tissues from Diffusion-Weighted Images. NMR in Biomedicine, Vol. **8** (1995) 333-344
7. Andersson, L. E.: Fourier Transforms of RIO's of Any Order. Appendix 1 in Report LiTH-ISY-1238, Dept of EE, Linköping Univ., SE-58183, Sweden (1991)
8. Danielsson, P.E., Lin, Q., Ye, Q.-Z.: Efficient Detection of Second Degree Variations in 2D and 3D Images. To appear in Journal of Visual Communications and Image Representation (2001)

Spatial Relations among Pattern Subsets as a Guide for Skeleton Pruning

Claudio De Stefano[1] and Maria Frucci[2]

[1] Facoltà di Ingegneria, Università del Sannio, Piazza Roma,
82100 Benevento Italy
cladeste@unina.it
[2] Istituto di Cibernetica, C.N.R., Via Toiano 6,
80072 Arco Felice-Napoli Italy
mfr@imagm.cib.na.cnr.it

Abstract. The skeleton is an effective tool for shape analysis if its structure can be regarded as a faithful stick-like representation of the pattern. However, contour noise may affect this structure by originating spurious skeleton branches, so that skeletonization algorithms should include a pruning phase devoted to an analysis of the peripheral skeleton branches and, possibly, to their partial or total removal. In this paper, labeled skeletons are considered and the significance of a peripheral branch is evaluated by analyzing the type of interaction between the pattern subset corresponding to the peripheral branch and the pattern subsets corresponding to the skeleton branches adjacent to the peripheral branch. The proposed criteria for skeleton pruning are expressed in terms of four parameters, which as a whole describe the role that the pattern subset corresponding to the peripheral branch plays in the characterization of the shape of the pattern.

1   Introduction

The skeleton is a useful representation to consider when, in the framework of a structural description task, a decomposition into parts of a planar pattern is desired [1]. The parts are obtained in correspondence with the elements of a partition of the skeleton, and are as much perceptually significant as the structure of the skeleton is a faithful stick-like representation of the pattern. For this reason, the skeleton should not possess branches in correspondence with details of the contour which are regarded either as due to some kind of noise, e.g., digitization noise occurring under pattern rotation, or as not relevant in the specific problem domain. Since such branches generally exist, any skeletonization algorithm is required to include a pruning phase devoted to an analysis of the peripheral skeleton branches and, possibly, to a partial or total removal of these ones [2]. As a result, a significant and manageable skeleton should be obtained.

Pruning is more effective when applied to labeled skeletons, i.e., with pixels labeled with their distance from the complement of the pattern. In fact, distance information allows one to identify the role of every skeletal pixel in the representation

C. Arcelli et al. (Eds.): IWVF4, LNCS 2059, pp. 155–164, 2001.
© Springer-Verlag Berlin Heidelberg 2001

of the pattern, in terms of the area contribution to the shape of the pattern given by the disc associated to that pixel.

In this paper, the evaluation of the significance of a peripheral skeleton branch is done by analyzing in detail the relations between the tip of the branch (the end point) and a reference pixel in the branch, in terms of their associated discs, as well as the type of interaction between the pattern subset corresponding to the peripheral branch and the pattern subsets corresponding to the skeleton branches adjacent to the peripheral branch. The analysis is carried out in the context of steady subsets of the skeleton, the presence of which is assumed to account for the significant pattern subsets characterizing any shape. These pattern subsets are regions with almost constant thickness and regions with thickness larger or smaller than the thickness of their adjacent regions. Only the parts of the skeleton which are not steady subsets can be removed, if suitable criteria are verified. The criteria are expressed in terms of four parameters, which as a whole describe the role that the pattern subset corresponding to the peripheral branch plays in the characterization of the shape of the pattern

In the next section, notations and preliminary notions are introduced, and in section 3 the four parameters are described. In section 4, the pruning method is outlined, while pointing out the two phases characterizing the procedure. Finally, in section 5, some examples are shown and the performance of the proposed algorithm is compared with the one relative to two well known jut-based pruning algorithms.

2   Preliminaries

We refer to a binary picture, rid of salt-and-pepper noise, where $F=\{1\}$ and $B=\{0\}$ are the foreground and the background, respectively. Without loss of generality, we suppose that F is a connected set, regardless of its connectivity order, and does not touch the frame of the picture. The 8-metric and the 4-metric are respectively used for F and B.

For any pair of pixels p and q in the picture, a *path* from p to q is a finite sequence of pixels starting from p and terminating in q, where each pixel belongs to the 8-neighborhood of the preceding pixel. The (d_1,d_2)-*weighted distance* from p to q (with $d_2 < 2d_1$) is the length of a shortest path from p to q, where the distance between two successive pixels in the path is either equal to d_1 if the pixels are horizontally/vertically adjacent, or equal to d_2 if the pixels are diagonally adjacent [3]. In this paper the (3,4)-weighted distance is taken into account, and $d(p,q)$ will be used to denote the distance between p and q.

For every p in F, the disc D_p centered on p is the set of pixels having distance from p not greater than the distance of p from B. A pixel p is the *center of a maximal disc* (*cmd*) if it does not belong to a shortest path between any of its neighbors and B.

The *distance transformation* of F with respect to B is the process creating a multi-valued set, replica of F, which differs from F in having each pixel labeled with its distance from B.

For a pixel p labeled k, the *reverse distance transformation* is the process building the open disc with radius $\lceil k/d_1 \rceil$ centered in p, and assigning to any of its pixels, say r, the label $(k\text{-}d(p,r))$.

The *skeletonization* of F is a process leading to the extraction of a linear subset, the skeleton, which is spatially placed along the central regions of F. The skeleton is a stick-like representation of the pattern and, depending on the problem domain, is required to account for different shape properties, such as symmetry, elongation, width, and contour curvature.

The skeletonization algorithm [4] we consider is modeled on the grassfire transformation [5] and detects the skeleton in correspondence with the regions of F where different fire fronts interact with each other. The found skeleton is union of simple digital arcs and curves, is centrally located within F and has the same topology as F. Its pixels (skeletal pixels) are either centers of maximal discs or pixels connecting pairs of such centers, and are labeled with their (3,4)-weighted distance from B. Note that, in the following, the letters used to denote skeletal pixels should be understood as denoting also their associated distance label.

For any skeletal pixel p, let N_p denote the set of skeletal pixels in the 8-neighborhood of p and let n_p be its cardinality, then p is called *end point* if $n_p=1$, *normal point* if $n_p=2$ and *branch point* if $n_p>2$. A sequence of normal points having as extremes an end point and a branch point is called *peripheral skeleton branch*.

A skeletal pixel p is called *local maximum* if all its neighbors in N_p, except one, have label not greater than p and the remaining neighbor has label less than p.

A skeletal pixel p is called *local minimum* if all its neighbors in N_p, except one, have label not smaller than p and the remaining neighbor has label greater than p.

Skeletal pixels which are local maxima, local minima, or are pixels with at least one neighbor with the same label are called *critical skeletal pixels*.

The *steady skeleton subsets* are (union of) maximal sequences of critical skeletal pixels. Only maximal sequences constituted by at least three critical pixels, or by at most two pixels which are either both local maxima or both local minima, are taken into account. Two sequences are merged if they are separated by a set, constituted by at most three consecutive pixels, which is delimited by extremes of the two sequences whose distance labels differ by no more than the amount $d_1 + d_2$.

For a peripheral skeleton branch traced from the end point, which is assumed as not belonging to a steady subset, the *hinge pixel* is either the first pixel belonging to the steady subset firstly met or the branch point, if no steady subsets are found in the branch. The sequence delimited by the end point and by the predecessor of the hinge pixel is the part of the branch to be analyzed and possibly pruned. This sequence is constituted by pixels with labels which are non decreasing, starting from the end point, and will be denoted by A.

3   Pruning Parameters

In a peripheral skeleton branch, let p and q respectively denote the end point and the hinge pixel, and let D_p and D_q be the discs created in correspondence of them by the reverse distance transformation.

Moreover, if D_p and D_q partially overlap each other, let H_p denote the subset of D_p not overlapped by D_q. H_p is the pattern subset which could not be represented by the skeleton, if the subset A would be removed. Since labels are not decreasing in A and p is a cmd not belonging to a steady subset, there results $p<q$ and D_p with size smaller than the size of D_q. The label of p and the spatial position of p with respect to q determine the shape and the size of H_p.

Four parameters are used to characterize H_p: they are IN, TH, RP, and SM. Meaning and definitions of the parameters are as follows.

- IN is used to express the size of H_p with respect to D_p.

  IN=q-d(p,q).

  When IN>d_2, p belongs to D_q and has no neighbors in the complement of D_q. H_p is at most 40% of D_p See Fig 1a and 1b.
  When 0<IN≤d_2, p belongs to D_q and has at least one of its eight neighbors belonging to the complement of D_q. H_p is at most 60% of D_p. See Fig 1c.
  When IN=0, p belongs to the complement of D_q and is adjacent to D_q. The size of H_p is at most 65% of D_q. See Fig 1d.
  When IN<0, p belongs to the complement of D_q and overlapping between D_p and D_q is not guaranteed. When overlapping occurs, H_p is at least 73% of D_p. See Fig 1e.

- TH is defined for IN≥0, and expresses the weighted distance from the background of the innermost pixel in Hp.

TH= p	if	IN=0
TH= p -d1	if	0<IN≤d1
TH= p -2d1	if	IN=(2d1-1)
TH= p -IN	if	IN>d1 and IN≠(2d1-1)

  When TH $\leq d_2$, all the pixels of H_p are adjacent to the background, see Fig. 1a. As for Figs. 1b-1e, there results TH > d_2.

- RP is defined for IN≥0 and gives the number of pixels belonging both to Hp and to the longest among the horizontal/vertical paths from the background to the innermost pixels of Hp.

RP= int (p/d_1)		if IN=0
RP= 1+maxn(p) - maxn(IN)		otherwise
with	maxn(r) = int (r/d_1)+1	if mod$(r,d_1) > 0$
	maxn(r) = int (r/d_1)	otherwise

  In Figs. 1a, 1b, 1c and 1d, RP is equal to 2, 3, 4 and 4, respectively.
- SM represents the degree of smoothness of the contour arc of the region in correspondence with the terminal part of A, and is computed when A is constituted by at least three pixels p1, p2, p3, with p1=p. For this contour arc, we define four degrees of smoothness, from 0 to 3, depending on the relations between the labels

associated to p1, p2 and p3. SM is as greater as the contour arc is smoother, see Fig. 2.

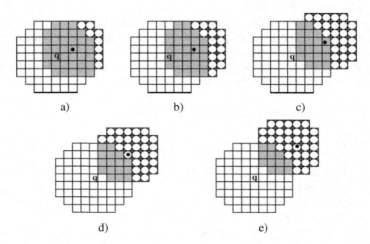

a)   b)   c)

d)   e)

Fig. 1. Different size of H_p, when IN varies. The examples are shown for $q=15$ and $p=12$. A black dot denotes p; gray squares denote pixels belonging both to D_q and to D_p; H_p is the set of circled pixels. a) IN=8; b) IN=5; c) IN=1; d) IN=0; e) IN=-1.

SM=0   if $p_1=p_2$;

SM=1   if $p_1{\neq}p_2$ and for i=1,2:

$\qquad$ $[(p_{i+1}-p_i){<}d_1$ and $d(p_{i+1},p_i)=d_1]$ or $[(p_{i+1}-p_i){<}d_2{-}1$ and $d(p_{i+1},p_i)=d_2]$

SM=2   if $p_1{\neq}p_2$ and the following condition holds either for i=1 or for i=2:

$\qquad$ $[(p_{i+1}-p_i){\geq}d_1]$ or $[(p_{i+1}-p_i)=d_1{-}1$ and $d(p_{i+1},p_i)=d_2)]$

SM=3   if $p_1{\neq}p_2$ and for i=1,2:

$\qquad$ $[(p_{i+1}-p_i){\geq}d_1]$ or $[(p_{i+1}-p_i)=d_1{-}1$ and $d(p_{i+1},p_i)=d_2)]$

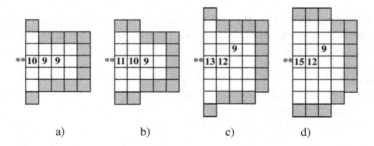

a)   b)   c)   d)

Fig. 2. SM represents the degree of smoothness of the contour arc (denoted by gray squares) of the region in correspondence with the labeled skeletal pixels. There results: a) SM=0; b) SM=1; c) SM=2; d) SM=3.

4   Pruning Process

Two phases are foreseen for pruning the skeleton subset A. In the first phase, H_p is estimated by referring only to the mutual positions of D_p and D_q. Three outputs are returned from this analysis: 1) retain A, 2) remove A, 3) more information about the context is necessary to take a decision. The second phase is accomplished only in the third case, which occurs when the hinge pixel is a branch point. This phase is concerned with the analysis of the interaction between A and the skeleton subsets adjacent to the branch point, and takes into account the parts of the pattern in correspondence with those branches, with special reference to their thickness.

4.1   First Phase

For $IN \geq 0$, A is retained when the interaction between D_p and D_q is weak, the thickness of H_p is not negligible and the contour arc delimiting H_p has high curvature. In detail, A is retained if at least one of the following conditions holds:

a) $IN < d_1$ and $TH > d_2$; b) $SM < 3$ and $RP > SM$ and $RP > 1$ and ($IN \leq d_2$ or $TH > d_2$)

If $IN \geq 0$ and the previous conditions are not verified, A is removed if the interaction between D_p and D_q is very high or if the thickness of H_p is negligible with respect to the size of D_p or if the region of the pattern associated with A is a negligible protrusion of the region associated with the steady subset including q. In detail, A is removed if at least one of the conditions following holds:

a') $IN > d_2$ and $TH \leq d_2$; b') $TH \leq d_2$ and $p \leq d_2$; c') q is not a branch point.

Note that if q is not a branch point, then it is necessarily a pixel of a steady subset and the fact that conditions a), b) ,c), d) do not hold implies that H_p is negligible.
The remaining cases occurring when $IN \geq 0$ and q is a branch point are considered in the second phase.
For $IN < 0$, A is retained if at least one of the following conditions holds:

a'') $IN < -d_2$; b'') $p > d_2$; c'') $SM < 3$

In fact, in this case the interaction between D_p and D_q is either very low or does not occur. On the contrary, if none of the previous conditions holds, A is pruned if q is not a branch point, while any decision is postponed to the second phase when q is a branch point.

4.2   Second Phase

We consider the interaction between A and the skeleton subsets branching off the hinge pixel q and adjacent to it. We distinguish two types of skeleton subsets, say C and S, and denote by NC and NS the number of times these subsets are present.

A subset C is constituted by pixels with labels which are non decreasing, starting from q, and no pixel of C is a steady pixel. See Figs. 3a, 3b, 3c.

A subset S is a steady subset and is constituted by pixels with a same label, not less than q. See Figs. 3b, 3d and 3e.

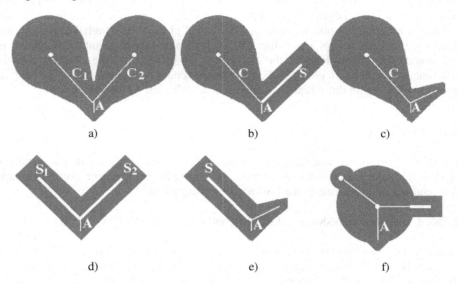

Fig. 3. Thick lines and dots represent steady subsets.

If NC>1, see Fig. 3a, A is retained.

If NC≤1, see Figs. 3b, 3c, 3d, 3e, 3f, A is pruned if H_p is negligible, i.e., at least one of the following conditions holds:

a) $p \leq d_2$ and SM=3; b) IN=d_2 and RP=1.

On the contrary, A is retained if condition a) does not hold and IN<0.

In the remaining cases occurring when NC≤1, the interaction between A and the other skeleton subsets adjacent to q is considered.

If NS>1, see Fig. 3d, or if NS=1 and NC=1, see Fig.3b, A is generally retained, but it is removed when H_p is considered a negligible protrusion of the pattern subset associated both with q and with the skeleton subsets branching off q, i.e., if at least one of the following conditions holds:

a') IN≥d_2 and SM=3; b') IN>1 and SM=3 and RP=1.

If NS≤1 and NC=0, see Figs. 3e and 3f, A is removed.

If NS=0 and NC=1, see Fig.3c, A is removed only if: IN≥d_2 and RP<SM.

5   Discussion

A pruning procedure implementing the previous criteria has been developed and tested on a data set of more than one hundred patterns, of which also a version rotated by 30 degrees has been considered. The results have been compared with the ones obtained by considering two well-known jut-based pruning algorithms (refer to [6], pp.132-133), for simplicity denoted here by *AP1* and *AP2*. For both algorithms, a peripheral skeleton branch, delimited by an end point *p*, is traced from *p* to a pixel *q*, which is the center of a maximal disc and is the firstly met pixel which satisfies a given condition. If such a pixel *q* exists, the branch is shortened up to *q*, *q* excluded, otherwise the branch is removed, the branch point excluded.

Algorithm *AP1* takes into account the sharpness of a protrusion, and in this paper *q* is the first pixel of a sequence of three pixels which are all centers of maximal discs. Pixel *q* may coincide with pixel *p*. As for *AP2*, *q* is such that $\lambda > d_2$, where $\lambda = p-q+d(p,q)$. The value λ denotes the amount by which the protrusion is shortened when the branch is trimmed down to *q*.

With reference to Figs. 4, we show in column a) the input patterns, together with their skeletons before pruning, as obtained by the algorithm described in [4]. The skeletons pruned according to the proposed method are shown in column b), while the skeletons pruned according to *AP1* and *AP2* are respectively shown in columns c) and d). As a general remark, we may note that the proposed pruning algorithm shows a better performance as regards both stability under pattern rotation and preservation of significant peripheral skeleton branches.

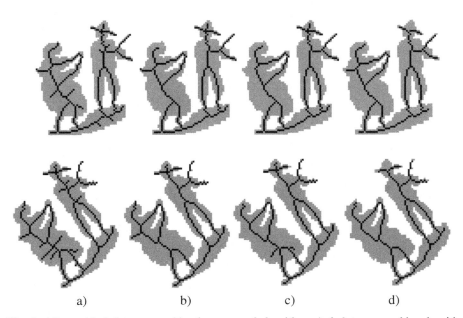

a)   b)   c)   d)

Fig. 4. a) Input; b) skeleton pruned by the proposed algorithm; c) skeleton pruned by algorithm *AP1*. d) skeleton pruned by algorithm *AP2*.

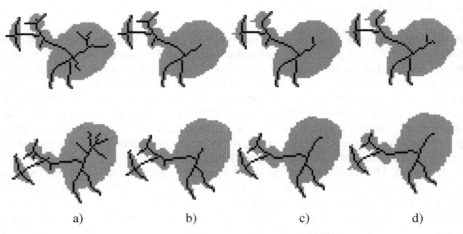

a)　　　　　　b)　　　　　　c)　　　　　　d)

Fig. 4. - Continued.

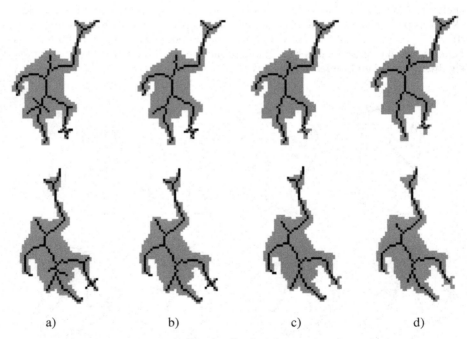

a)　　　　　　b)　　　　　　c)　　　　　　d)

Fig. 4. - Continued.

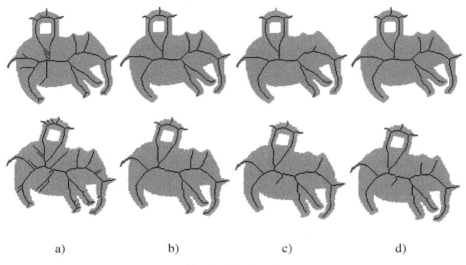

a)　　　　　　b)　　　　　　c)　　　　　　d)

Fig. 4. - Continued.

References

1. G. Sanniti di Baja and E. Thiel: (3,4)-weighted skeleton decomposition for pattern representation and description. Pattern Recognition, 27 (1994) 1039-1049.
2. D. Shaked and A. M. Bruckstein: Pruning medial axes. Computer Vision and Image Understanding, 69 (1998) 156-169.
3. G.Borgerfors: Distance transformations in digital images. CVGIP, 34 (1986) 344-371.
4. M. Frucci and A. Marcelli: Efficient skeletonization of binary figures through (d1,d2)-erosion and directional information. In: C. Arcelli, L.P. Cordella and G. Sanniti di Baja (eds.): Aspects of Visual Form Processing, World Scientific, Singapore (1994) 221-230.
5. H. Blum: Biological shape and visual science (part I). Journal of Theoretical Biology, 38 (1973) 205-287.
6. C. Arcelli and G. Sanniti di Baja: Skeletons of planar patterns. In: T.Y. Kong and A. Rosenfeld (eds.): Topological Algorithms for Digital Image Processing, North Holland, Amsterdam (1996) 99-143.

Euclidean Fitting Revisited

Petko Faber and R.B. Fisher

Division of Informatics, University of Edinburgh,
Edinburgh, EH1 2QL, SCOTLAND

Abstract. The focus of our paper is on the fitting of general curves and surfaces to 3D data. In the past researchers have used approximate distance functions rather than the Euclidean distance because of computational efficiency. We now feel that machine speeds are sufficient to ask whether it is worth considering Euclidean fitting again. Experiments with the real Euclidean distance show the limitations of suggested approximations like the Algebraic distance or Taubin's approximation. In this paper we present our results improving the known fitting methods by an (iterative) estimation of the real Euclidean distance. The performance of our method is compared with several methods proposed in the literature and we show that the Euclidean fitting guarantees a better accuracy with an acceptable computational cost.

1 Motivation

One fundamental problem in building a recognition and positioning system based on implicit 3D curves and surfaces is how to fit these curves and surfaces to 3D data. This process will be necessary for automatically constructing CAD or other object models from range or intensity data and for building intermediate representations from observations during recognition. Of great importance is the ability to represent 2D and 3D data or objects in a compact form. Implicit polynomial curves and surfaces are very useful representations. Their power appears by their ability to smooth noisy data, to interpolate through sparse or missing data, their compactness and their form being commonly used in numerous constructions. Let $f_2(\boldsymbol{x})$ be an *implicit polynomial* of degree 2 given by

$$f_2(\boldsymbol{x}) = p_0 + \boldsymbol{x}' \cdot \boldsymbol{p}_1 + \boldsymbol{x}' \cdot \boldsymbol{P}_2 \cdot \boldsymbol{x} = 0, \ \boldsymbol{x} \in \mathbb{R}^2 \text{ or } \boldsymbol{x} \in \mathbb{R}^3 \ . \tag{1}$$

Then, we only have to determine the set of parameters which describes the data best. The parameter estimation problem is usually formulated as an optimization problem. Thereby, a given estimation problem can be solved in many ways because of different optimization criteria and several possible parameterizations. Generally, the literature on fitting can be divided into two general techniques: clustering (e.g. [4,6]) and least-squares fitting (e.g. [2,5,7]). While the clustering methods are based on mapping data points to the parameter space, such as the Hough transform and the accumulation methods, the least-squares methods are centered on finding the sets of parameters that minimize some distance measures

C. Arcelli et al. (Eds.): IWVF4, LNCS 2059, pp. 165–175, 2001.
© Springer-Verlag Berlin Heidelberg 2001

between the data points and the curve or surface. Unfortunately, the minimization of the Euclidean distances from the data points to a *general* curve or surface has been computationally impractical, because there is no closed form expression for the Euclidean distance from a point to a general algebraic curve or surface, and iterative methods are required to compute it. Thus, the Euclidean distance has been approximated. Often, the result of evaluating the characteristic polynomial $f_2(\boldsymbol{x})$ is taken, or the first order approximation, suggested by Taubin [12] is used. However, experiments with the Euclidean distance show the limitations of approximations regarding quality and accuracy of the fitting results.

The quality of the fitting results has a substantive impact on the recognition performance especially in the reverse engineering where we work with a constrained reconstruction of 3D geometric models of objects from range data. Thus it is important to get good fits to the data.

2   Fitting of Algebraic Curves and Surfaces

An implicit curve or surface is the set of zeros of a smooth function $f : \mathbb{R}^n \to \mathbb{R}^k$ of the n variables: $\mathcal{Z}(f) = \{\boldsymbol{x} \, : \, f(\boldsymbol{x}) = 0\}$. In our applications we are interested in three special cases for their applications in computer vision and especially range image analysis: $\mathcal{Z}(f)$ is a *planar curve* if $n = 2$ and $k = 1$, it is a *surface* if $n = 3$ and $k = 1$ and it is a *space curve* if $n = 3$ and $k = 2$.

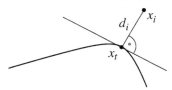

Given a finite set of data points $\mathcal{D} = \{\boldsymbol{x}_i\}$, $i \in [1, m]$, the problem of fitting an algebraic curve or surface $\mathcal{Z}(f)$ to the data set $\mathcal{D}$ is usually cast as minimizing the mean square distance

Fig. 1.   Euclidean   distance
$\mathrm{dist}(\boldsymbol{x}_i, \mathcal{Z}(f))$ of a point $\boldsymbol{x}_i$ to a
zero set $\mathcal{Z}(f)$

$$\frac{1}{m} \sum_{i=1}^{m} \mathrm{dist}\,(\boldsymbol{x}_i, \mathcal{Z}(f))^2 \to \text{Minimum} \quad (2)$$

from the data points to the curve or surface $\mathcal{Z}(f)$, a function of the set of parameters of the polynomial. The problem that we have to deal with is how to answer whether the distance from a certain point $\boldsymbol{x}_i$ to a set $\mathcal{Z}(f)$ of zeros of $f : \mathbb{R}^n \to \mathbb{R}^k$ is the (global) minimum or not. The distance from the point $\boldsymbol{x}_i$ to the zero set $\mathcal{Z}(f)$ is defined as the minimum of the distances from $\boldsymbol{x}_i$ to points $\boldsymbol{x}_t$ in the zero set $\mathcal{Z}(f)$

$$\mathrm{dist}(\boldsymbol{x}_i, \mathcal{Z}(f)) = \min \{\| \, \boldsymbol{x}_i - \boldsymbol{x}_t \, \| : f(\boldsymbol{x}_t) = 0\}. \quad (3)$$

Thus, the Euclidean distance $\mathrm{dist}(\boldsymbol{x}_i, \mathcal{Z}(f))$ between a point $\boldsymbol{x}_i$ and the zero set $\mathcal{Z}(f)$ is the minimal distance between $\boldsymbol{x}_i$ and the point $\boldsymbol{x}_t$ in the zero set whose tangent is orthogonal to the line joining $\boldsymbol{x}_i$ and $\boldsymbol{x}_t$ (see Fig.1). As mentioned above there is no closed form expression for the Euclidean distance from a point to a *general* algebraic curve or surface and iterative methods are required to compute it. In the past researchers have often replaced the Euclidean distance by an approximation. But it is well known that a different performance function can produce a very biased result. In the following we will summarize the methods used to approximate the real Euclidean distance by the algebraic distance and an approximation suggested by Taubin ([12], [13]).

Algebraic fitting. The algebraic fitting is based on the approximation of the Euclidean distance between a point and the curve or surface by the algebraic distance

$$\text{dist}_A\left(\boldsymbol{x}_i, \mathcal{Z}(f)\right) = f_2(\boldsymbol{x}_i) \ . \tag{4}$$

To avoid the trivial solution, where all parameters are zero, and any multiple of a solution, the parameter vector may be constrained in some way (e.g. [1, 5,7] and [10]). The pros and cons of using algebraic distances are a) the gain in computational efficiency, because closed form solutions can usually be obtained, on the one hand and b) the often unsatisfactory results on the other hand.

Taubin's fitting. An alternative to approximately solve the minimization problem is to replace the Euclidean distance from a point to an implicit curve or surface by the first order approximation [13]. There, the Taylor series is expanded up to first order in a defined neighborhood, truncated after the linear term and then the triangular and the Cauchy-Schwartz inequality were applied.

$$\text{dist}_T\left(\boldsymbol{x}_i, \mathcal{Z}(f)\right) = \frac{|f_2(\boldsymbol{x}_i)|}{\|\nabla f_2(\boldsymbol{x}_i)\|} \tag{5}$$

Besides the fact that no iterative procedures are required, the fundamental property is that it is a first order approximation to the exact distance. But, it is important to note that the approximate distance is also biased in some sense. If, for instance, a data point $\boldsymbol{x}_i$ is close to a critical point of the polynomial, i.e., $\|\nabla f_2(\boldsymbol{x}_i)\| \approx 0$, but $f_2(\boldsymbol{x}_i) \neq 0$, the distance becomes large. This is certainly a limitation.

Note, neither the Algebraic distance nor Taubin's approximation are invariant with respect to Euclidean transformations.

2.1   Euclidean Distance

To overcome the problems with the approximated distances, it is natural to replace them again by the real geometric distances, that means the Euclidean distances, which are invariant to transformations in Euclidean space and are not biased. For primitive curves and surfaces like straight lines, ellipses, planes, cylinders, cones, and ellipsoids, a closed form expression exists for the Euclidean distance from a point to the zero set and we use these. However, as the expression of the Euclidean distance to other 2nd order curve and surfaces is more complicated and there exists no known closed form expression, an iterative optimization procedure must be carried out. For more general curves and surfaces the following simple iterative algorithm will be used (see also Fig.2):

1. *Select* the initial point $x_t^{[0]}$. In the first step we determine the initial solution by intersecting the curve or surface with the straight line defined by the center point x_m and the point x_i. By the initial solution, the upper bound for the distance is estimated.
2. *Update* the actual estimation $x_t^{[k+1]} = F(x_t^{[k]})$, $k = 0, 1, 2, \ldots$. In the second step a new solution is determined. The search direction will be determined by the gradient of the curve $\nabla(f(x_t^{[k]}))$.

$$x_t^{[k+1]} = x_t^{[k]} + \alpha^{[k]} \nabla f \left(x_t^{[k]} \right). \tag{6}$$

The method is an adaptation of the steepest descent method. As the result we get two possible solutions, $x_t^{[k]}$ and $x_t^{[k+1]}$ (cf. Fig 2), and we have to decide by an objective function $\mathcal{F}$, if $x_t^{[k+1]}$ will be accepted as new solution.

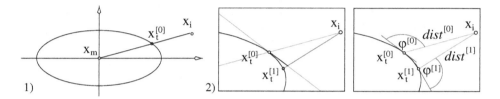

Fig. 2. Steps to estimate the Euclidean distance $\mathrm{dist}_E(x_i, \mathcal{Z}(f))$ of a point x_i to the zero set $\mathcal{Z}(f)$ of an ellipse

3. *Evaluate* the new estimation $x_t^{[k+1]}$. The set of solutions is evaluated by the objective function $\mathcal{F}(x_i, x_t^{[k+1]}, \mathcal{Z}(f)) = \min(\mathrm{dist}_E(x_i, x_t^{[k]}), \mathrm{dist}_E(x_i, x_t^{[k+1]}))$. If the distance from the new estimation $x_t^{[k+1]}$ is smaller, we accept this as the new local solution. Otherwise $x_t^{[k+1]} = x_t^{[k]}$ and $\alpha^{[k+1]} = -\tau\alpha^{[k]}$, $\tau > 0$. Then, the algorithm will be continued with step 2 until the difference between the distances of the old and the new estimation is smaller then a given threshold. To speed up the estimation a criterion to terminate the updating may be used like e.g. $\|x_t^{[k+1]} - x_t^{[k]}\| \leq \tau_d$, or $k \geq \tau_k$.

2.2   Estimation Error of Surface Fit

Given the Euclidean distance error for each point, we then compute the curve or surface fitting error as $\mathrm{dist}_E(x_i, \mathcal{Z}(f))$. The standard least-squares method tries to minimize $\sum_i \mathrm{dist}_E^2(x_i, \mathcal{Z}(f))$, which is unstable if there are outliers in the data. Outlying data can give so strong an effect in the minimizing that the parameters are distorted. Replacing the squared residuals by another function can reduce the effect of outliers. Appropriate minimization criteria including functions were discussed in for instance [3] and [14]. It seems difficult to select a function which is generally suitable. Following the results given in [11] the best choice may be the so-called L_p (*least power*) function: $L_p := |\text{dist}_E(\boldsymbol{x}_i, \mathcal{Z}(f))|^\nu / \nu$. This function represents a family of functions including the two commonly used functions L_1 (*absolute power*) with $\nu = 1$ and L_2 (*least squares*) with $\nu = 2$. Note, the smaller ν, the smaller is the influence of large errors. For values $\nu \approx 1.2$, a good error estimation may be expected [11].

2.3 Optimization

Given a method of computing the fitting error for the curves and surfaces, we now show how to minimize the error. Many techniques are readily available, including Gauss-Newton algorithm, Steepest Gradient Descent, and Levenberg-Marquardt algorithm. Our implementation is based on the Levenberg-Marquardt (LM) algorithm [8,9] which has become the standard of nonlinear optimization routines. The LM method combines the inherent stability of the Steepest Gradient Descent with the quadratic convergence rate of the Gauss-Newton method. The (iterative) fitting approach consists of three major steps:

1. *Select* the initial fitting $\mathcal{P}^{[0]}$. The initial solution $\mathcal{P}^{[0]}$ is determined by Taubin's fitting method.
2. *Update* the estimation $\mathcal{P}^{[k+1]} = F_{\text{LM}}(\mathcal{P}^{[k]})$ using the Levenberg-Marquardt (LM) algorithm.
3. *Evaluate* the new estimation $\mathcal{P}^{[k+1]}$. The updated parameter vector is evaluated using the L_p function on the basis of the $\text{dist}_E(\boldsymbol{x}_i, \mathcal{Z}(f))$. $\mathcal{P}^{[k+1]}$ will be accepted if $L_p(\mathcal{P}^{[k+1]}) < L_p(\mathcal{P}^{[k]})$ and the fitting will be continued with step 2. Otherwise the fitting is terminated and $\mathcal{P}^{[k]}$ is the desired solution.

3 Experimental Results

We present experimental results comparing Euclidean fitting (*EF*) with Algebraic fitting (*AF*), and Taubin's fitting (*TF*) in terms of quality, robustness and speed.

3.1 Robustness

To test the robustness of the proposed *EF* method, we used three different surface types: cylinders, cones, and general quadrics. Note that plane estimation is the same for all three methods. To enforce the fitting of a special surface type we include in all three fitting methods the same constraints which describe the expected surface type. The 3D data were generated by adding isotropic Gaussian noise $\sigma = \{1\%, 5\%, 10\%, 20\%\}$. Additionally the surfaces were partially occluded. The visible surfaces were varied between $1/2$ (maximal case), $5/12$, $1/3$, $1/4$, and $1/6$ of the full 3D cylinder (see Fig.4). In all our experiments the number of 3D points was 5000. And finally, each experiment runs 100 times to measure the

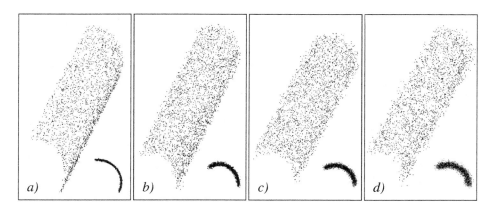

Fig. 3. View of the 3D data points for a cylinder (maximal case) with added isotropic Gaussian noise a) $\sigma = 1\%$, b) $\sigma = 5\%$, c) $\sigma = 10\%$, and d) $\sigma = 20\%$.

average fitting error. The mean least power errors ($MLPE$'s) of the different fittings are in Tab.1. We determined the real geometric distance between the 3D data points and the estimated surfaces using the method described in Sec.2.1. That means we calculated the $MLPE$ for all fitting results on the basis of the estimated Euclidean distance. Otherwise, a comparison of the results will be useless. Based on this table we evaluate the three fitting methods with respect to quality and robustness. The EF requires an initial estimate for the parameters,

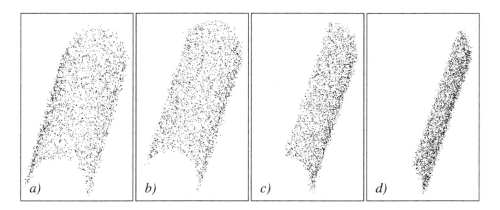

Fig. 4. View of the 3D data points of partially occluded cylinders, a) 1/2 (maximal case), b) 5/12, 1/3 (see Fig.3b)), c) 1/4, and d) 1/6. Added isotropic Gaussian noise $\sigma = 5\%$.

and we have found that the results depend on the initial choice. A quick review of the values in Tab.1 shows that the results of TF are better for initializing than the results of *AF*. Maybe another fitting method can give a better initialization, but here we use *TF* because of its advantages.

As expected, the *TF* and *EF* yield the best results respect with to the mean and standard deviation, and the mean for *EF* is always lower than for the other two algorithms. The results of *AF* are not acceptable because of their high values for mean and standard deviation. The results of *TF* are much better, compared with the *AF*. But, in the direct comparison with the *EF* these results are also unacceptable. Furthermore, note that *AF* give sometimes wrong results which means that the fitted curve or surface typo does not come up with our expectations. We removed all failed fittings out of the considerations. The percentage of failures is given as footnote in Tab.1. For *TF* and *EF* we had no failures in our experiments.

3.2   Noise Sensitivity

The second experiment is perhaps more important and assesses the stability of the fitting with respect to different realizations of noise with the same variance. The noise has been set to a relatively high level because the limits of the three methods are more visible then. It is very desirable that the performance is affected only by the noise level, and not by a particular realization of the noise. In Tab.1 the μ's and σ's are shown for four different noise levels. If we analyze the table regarding noise sensitivity, we observe:

– The stability of all fittings, reflected in the standard deviation, is influenced by the noise level of the data. The degree of occlusion has an additional influence on stability. Particularly serious is the combination of both high noise level($\sigma \geq 20\%$) and strong occlusion (visible surface < 1/4).
– *AF* is very unstable, even with a noise level of $\sigma = 1\%$. In some experiments with *AF* the fitting failed and the estimated *mean least power error* between the estimated surface and the real 3D data was greater than a given threshold. We removed all failed fittings, sometimes up to 23 percent (see Tab.1: fitting cylinder 1, 1/4 visible and $\sigma = 10\%$). Thus, the performance of the Algebraic fitting is strongly affected by the particular realization of the noise, which is absolutely undesirable.
– *TF* is also affected by particular instances of the noise, but on a significantly lower level.
– The noise sensitivity of *EF* has a similar good performance. The cause for the instability of the *EF* is the initialization.

3.3   Sample Density

In the third experiment we examined the influence of the sample density. The cardinality of the 3D point set was varied accordingly. On the basis of the *MLPE* for the several fittings (see Tab.2) it can be seen that, with increasing the number of points, the fitting becomes a) more robust and b) less noise sensitive. Note,

Table 1. Least power error fitting cylinder 1 and 2. The visible surfaces were varied between 1/2 (maximal case), 5/12, 1/3, 1/4, and 1/6 of the full 3D cylinder. Gaussian noise σ was 1%, 5%, 10%, and 20%. For AF the percentage of failed fittings is given in brackets. The number of trials was 100.

			AF $[\ \mu\ \pm\ \sigma\ ]\cdot 10^{-2}$	TF $[\ \mu\ \pm\ \sigma\ ]\cdot 10^{-2}$	EF $[\ \mu\ \pm\ \sigma\ ]\cdot 10^{-2}$
cylinder_1 (rad=50, length=500)	$\sigma = 1\%$	6/12	$[\ 9.06 \pm 1.55]_{(0.08)}$	$[\ 1.14 \pm 0.06]$	$[\ 0.71 \pm 0.03]$
		5/12	$[33.19 \pm 3.90]_{(0.22)}$	$[\ 1.30 \pm 0.13]$	$[\ 0.65 \pm 0.04]$
		4/12	$[44.91 \pm 3.72]_{(0.02)}$	$[\ 2.75 \pm 0.43]$	$[\ 0.72 \pm 0.05]$
		3/12	$[55.06 \pm 2.04]_{(0.08)}$	$[\ 3.82 \pm 0.80]$	$[\ 0.94 \pm 0.11]$
		2/12	$[58.05 \pm 3.07]_{(0.13)}$	$[\ 3.94 \pm 0.49]$	$[\ 1.30 \pm 0.10]$
	$\sigma = 5\%$	6/12	$[14.80 \pm 1.88]_{(0.03)}$	$[\ 1.32 \pm 0.17]$	$[\ 0.62 \pm 0.03]$
		5/12	$[36.11 \pm 2.21]_{(0.14)}$	$[\ 2.92 \pm 0.13]$	$[\ 0.93 \pm 0.12]$
		4/12	$[25.56 \pm 2.76]_{(0.08)}$	$[\ 5.27 \pm 0.32]$	$[\ 1.35 \pm 0.32]$
		3/12	$[55.13 \pm 3.50]_{(0.02)}$	$[\ 5.07 \pm 0.75]$	$[\ 1.97 \pm 0.45]$
		2/12	$[55.93 \pm 3.08]_{(0.15)}$	$[\ 4.13 \pm 1.03]$	$[\ 2.34 \pm 0.81]$
	$\sigma = 10\%$	6/12	$[10.44 \pm 1.09]_{(0.10)}$	$[\ 2.05 \pm 0.29]$	$[\ 0.97 \pm 0.12]$
		5/12	$[23.10 \pm 3.47]_{(0.18)}$	$[\ 3.48 \pm 0.74]$	$[\ 1.71 \pm 0.63]$
		4/12	$[37.76 \pm 3.45]_{(0.23)}$	$[\ 3.90 \pm 0.79]$	$[\ 1.78 \pm 0.49]$
		3/12	$[56.37 \pm 2.73]_{(0.09)}$	$[\ 4.28 \pm 1.04]$	$[\ 1.83 \pm 0.34]$
		2/12	$[58.46 \pm 3.33]_{(0.03)}$	$[\ 8.98 \pm 3.46]$	$[\ 3.02 \pm 1.13]$
	$\sigma = 20\%$	6/12	$[15.34 \pm 1.93]_{(0.03)}$	$[\ 2.38 \pm 0.35]$	$[\ 1.09 \pm 0.10]$
		5/12	$[49.40 \pm 3.18]_{(0.13)}$	$[\ 2.90 \pm 0.56]$	$[\ 1.07 \pm 0.07]$
		4/12	$[55.61 \pm 3.24]_{(0.11)}$	$[\ 3.69 \pm 0.64]$	$[\ 1.41 \pm 0.11]$
		3/12	$[22.49 \pm 2.66]_{(0.10)}$	$[\ 4.05 \pm 0.95]$	$[\ 1.72 \pm 0.38]$
		2/12	$[41.37 \pm 3.22]_{(0.12)}$	$[\ 9.20 \pm 3.55]$	$[\ 3.00 \pm 1.13]$
cylinder_2 (rad=250, length=500)	$\sigma = 1\%$	6/12	$[26.68 \pm 1.37]_{(0.02)}$	$[\ 6.40 \pm 0.05]$	$[\ 4.26 \pm 0.02]$
		5/12	$[21.82 \pm 0.96]$	$[\ 6.83 \pm 0.13]$	$[\ 3.68 \pm 0.21]$
		4/12	$[25.39 \pm 0.88]$	$[\ 7.28 \pm 0.25]$	$[\ 4.12 \pm 0.17]$
		3/12	$[21.93 \pm 1.76]_{(0.01)}$	$[10.67 \pm 0.74]$	$[\ 3.18 \pm 0.48]$
		2/12	$[25.80 \pm 2.26]$	$[23.24 \pm 1.67]$	$[\ 5.43 \pm 0.70]$
	$\sigma = 5\%$	6/12	$[25.31 \pm 1.26]_{(0.03)}$	$[\ 6.67 \pm 0.21]$	$[\ 4.73 \pm 0.33]$
		5/12	$[18.54 \pm 0.83]$	$[\ 8.06 \pm 0.71]$	$[\ 3.22 \pm 0.18]$
		4/12	$[25.32 \pm 1.05]$	$[\ 8.38 \pm 0.72]$	$[\ 5.26 \pm 0.70]$
		3/12	$[18.29 \pm 1.10]_{(0.07)}$	$[15.91 \pm 1.60]$	$[\ 6.47 \pm 1.08]$
		2/12	$[40.08 \pm 1.90]_{(0.02)}$	$[25.38 \pm 1.57]$	$[\ 8.88 \pm 1.36]$
	$\sigma = 10\%$	6/12	$[26.27 \pm 1.45]_{(0.09)}$	$[\ 6.71 \pm 0.23]$	$[\ 3.99 \pm 0.30]$
		5/12	$[19.31 \pm 0.84]$	$[\ 7.46 \pm 0.48]$	$[\ 3.79 \pm 0.45]$
		4/12	$[27.33 \pm 0.84]$	$[\ 8.19 \pm 0.90]$	$[\ 4.11 \pm 0.52]$
		3/12	$[23.42 \pm 1.92]$	$[15.87 \pm 1.70]$	$[\ 5.32 \pm 0.85]$
		2/12	$[31.89 \pm 1.85]_{(0.02)}$	$[25.68 \pm 2.04]$	$[\ 7.15 \pm 0.92]$
	$\sigma = 20\%$	6/12	$[24.74 \pm 1.33]_{(0.02)}$	$[\ 6.80 \pm 0.27]$	$[\ 3.49 \pm 0.13]$
		5/12	$[18.55 \pm 0.87]_{(0.07)}$	$[\ 7.11 \pm 0.56]$	$[\ 3.95 \pm 0.33]$
		4/12	$[27.08 \pm 0.90]$	$[\ 7.43 \pm 0.35]$	$[\ 5.24 \pm 0.35]$
		3/12	$[22.32 \pm 1.36]_{(0.04)}$	$[15.17 \pm 1.79]$	$[\ 6.60 \pm 0.91]$
		2/12	$[35.18 \pm 2.28]_{(0.02)}$	$[38.71 \pm 8.81]$	$[11.30 \pm 2.22]$

Table 2. Mean squares error for cylinder fitting by varied sample density. The density was 500, 1000, and 2000 3D points. Gaussian noise was $\sigma = 5\%$. For AF the percentage of failed fittings is given as a footnote.

			AF $[\ \mu\quad \sigma\ ]\cdot 10^{-2}$	TF $[\ \mu\quad \sigma\ ]\cdot 10^{-2}$	EF $[\ \mu\quad \sigma\ ]\cdot 10^{-2}$
cylinder_1	500	6/12	[15.88 ± 2.54]$_{(0.03)}$	[2.94 ± 0.80]	[1.17 ± 0.30]
		5/12	[29.73 ± 3.31]$_{(0.03)}$	[1.55 ± 0.11]	[0.86 ± 0.06]
		4/12	[32.96 ± 2.55]$_{(0.05)}$	[4.39 ± 0.90]	[2.30 ± 0.68]
		3/12	[23.67 ± 3.01]$_{(0.04)}$	[3.81 ± 0.51]	[1.55 ± 0.12]
		2/12	[24.36 ± 1.51]$_{(0.06)}$	[6.86 ± 2.45]	[4.37 ± 1.63]
	1000	6/12	[16.61 ± 2.42]$_{(0.08)}$	[1.57 ± 0.17]	[0.85 ± 0.11]
		5/12	[36.17 ± 3.52]$_{(0.17)}$	[3.52 ± 1.40]	[1.75 ± 0.59]
		4/12	[35.06 ± 2.70]$_{(0.06)}$	[2.79 ± 0.33]	[1.24 ± 0.15]
		3/12	[22.01 ± 3.03]$_{(0.02)}$	[4.51 ± 0.71]	[1.68 ± 0.16]
		2/12	[25.43 ± 1.91]	[4.10 ± 1.17]	[1.47 ± 0.12]
	2000	6/12	[16.85 ± 3.00]$_{(0.08)}$	[1.61 ± 0.26]	[0.72 ± 0.03]
		5/12	[45.99 ± 2.94]$_{(0.07)}$	[1.93 ± 0.43]	[0.73 ± 0.04]
		4/12	[38.38 ± 2.81]$_{(0.15)}$	[3.30 ± 0.76]	[1.22 ± 0.29]
		3/12	[19.79 ± 2.52]$_{(0.06)}$	[4.60 ± 1.16]	[2.05 ± 0.56]
		2/12	[24.28 ± 1.64]$_{(0.07)}$	[2.22 ± 0.28]	[1.30 ± 0.08]
cylinder_2	500	6/12	[23.27 ± 0.98]	[7.08 ± 0.24]	[4.10 ± 0.32]
		5/12	[20.05 ± 1.94]	[7.82 ± 0.56]	[3.72 ± 0.22]
		4/12	[25.19 ± 1.30]	[10.55 ± 0.62]	[4.66 ± 0.28]
		3/12	[21.12 ± 1.85]$_{(0.04)}$	[17.10 ± 1.68]	[7.18 ± 0.87]
		2/12	[38.13 ± 2.77]$_{(0.17)}$	[24.74 ± 2.86]	[7.90 ± 0.81]
	1000	6/12	[25.84 ± 1.28]$_{(0.05)}$	[7.11 ± 0.31]	[3.97 ± 0.39]
		5/12	[20.19 ± 1.59]	[7.60 ± 0.37]	[3.33 ± 0.19]
		4/12	[25.11 ± 0.89]	[8.37 ± 0.40]	[4.23 ± 0.17]
		3/12	[20.61 ± 1.33]$_{(0.06)}$	[14.82 ± 1.21]	[8.27 ± 1.03]
		2/12	[38.37 ± 2.73]$_{(0.10)}$	[21.61 ± 1.76]	[6.66 ± 0.84]
	2000	6/12	[24.30 ± 1.11]$_{(0.02)}$	[6.39 ± 0.17]	[3.74 ± 0.29]
		5/12	[18.95 ± 0.82]$_{(0.02)}$	[7.03 ± 0.32]	[3.08 ± 0.17]
		4/12	[27.31 ± 0.87]	[7.98 ± 0.34]	[4.36 ± 0.16]
		3/12	[20.31 ± 1.02]$_{(0.06)}$	[14.92 ± 1.50]	[5.36 ± 0.79]
		2/12	[35.16 ± 2.15]$_{(0.05)}$	[24.90 ± 1.76]	[8.82 ± 1.21]

not only is the absolute number of points important, but the point density is crucial.

However noise sensitivity increases with increasing occlusion for both TF and EF, so that the fitting becomes altogether more unstable. Similar conclusions about AF as in Sec.3.1 and Sec.3.2 also apply here.

3.4   Computational Cost

The algorithms have been implemented in C and the computation was performed on a SUN Sparc ULTRA 5 workstation. The average computational costs in milliseconds per 1000 points for the three algorithms are in Tab.3.

As expected. the AF and TF supply the best performance, because the EF algorithm requires a repeated search for the point x_t closest to x_i and the calculation of the Euclidean distance. A quick review of the values in Tab.3 shows that the computational costs increase if we fit an elliptical cylinder, a circular or an elliptical cone respectively a general quadric. The algorithm to estimate the distance by the closed form solution respectively the iterative algorithm is more complicated in these cases (cf. Sec.2.1).

The number of necessary iterations is also influenced by the required precision of the LM algorithm to terminate the updating process.

Table 3. Average computational costs in milliseconds per 1000 points.

	Average computational costs [msec.]		
	AF	TF	EF
Plane	0.958	1.042	2.417
Sphere	1.208	1.250	3.208
Circular cylinder	3.583	3.625	12.375
Elliptical cylinder	13.292	13.958	241.667
Circular cone	15.667	15.833	288.375
Elliptical cone	15.042	15.375	291.958
General quadric	18.208	18.458	351.083

4   Conclusion

We revisited the Euclidean fitting of curves and surfaces to 3D data to investigate if it is worth considering Euclidean fitting again. The focus was on the quality and robustness of Euclidean fitting compared with the commonly used Algebraic fitting and Taubin's fitting. Now, we can conclude that robustness and accuracy increases sufficiently compared to both other methods and Euclidean fitting is more stable with increased noise.

The main disadvantage of the Euclidean fitting, computational cost, has become less important due to rising computing speed. In our experiments the computational costs of Euclidean fitting were only about 2-19 times worse than Taubin's fitting. This relation probably cannot be improved substantially in favor of Euclidean fitting, but the absolute computational costs are becoming an insignificant deterrent to usage, especially if high accuracy is required.

Acknowledgements. The work was funded by the CAMERA (CAd Modelling of Built Environments from Range Analysis) project, an EC TMR network (ERB FMRX-CT97-0127).

References

1. A. Albano. Representation of digitized contours in terms of conic arcs and straight-line segments. *CGIP*, 3:23, 1974.
2. F. E. Allan. The general form of the orthogonal polynomial for simple series with proofs of their simple properties. In *Proc. Royal Soc. Edinburgh*, pp. 310–320, 1935.
3. P. J. Besl. *Analysis and Interpretation of Range Images*, chapter Geometric Signal Processing. Springer, Berlin-Heidelberg-New York, 1990.
4. P. J. Besl and R. C. Jain. Three-dimensional object recognition. *Computing Survey*, 17(1):75–145, März 1985.
5. F. L. Bookstein. Fitting conic sections to scattered data. *CGIP*, 9:56–71, 1979.
6. R. O. Duda and P. E. Hart. The use of Hough transform to detect lines and curves in pictures. *Comm. Assoc. Comp. Machine*, 15:11–15, 1972.
7. A. W. Fitzgibbon and R. B. Fisher. A buyer's guide to conic fitting. In *6th BMVC*. IEE, BMVA Press, 1995.
8. K. Levenberg. A method for the solution of certain nonlinear problems in least squares. *Quarterly of Applied Mathematics*, 2:164–168, 1944.
9. D. W. Marquardt. An algorithm for least squares estimation of nonlinear parameters. *J. of the Soc. of Industrial and Applied Mathematics*, 11:431–441, 1963.
10. K. A. Paton. Conic sections in chromosome analysis. *Pattern Recognition*, 2:39, 1970.
11. W. J. J. Ray. *Introduction to Robust and Quasi-Robust Statistical Methods*. Springer, Berlin-Heidelberg-New York, 1983.
12. G. Taubin. Estimation of planar curves, surfaces and non-planar space curves defined by implicit equations, with applications to edge and range image segmentation. *IEEE Trans. on PAMI*, 13(11):1115–1138, 1991.
13. G. Taubin. An improved algorithm for algebraic curve and surface fitting. In *4th Int. Conf. on Computer Vision*, pages 658–665, 1993.
14. Z. Zhang. Parameter estimation techniques: a tutorial with application to conic fitting. *Image and Vision Computing*, 15:59–76, 1997.

On the Representation of Visual Information

Mario Ferraro, Giuseppe Boccignone, and Terry Caelli

Dipartimento di Fisica Sperimentale, Università di Torino and INFM
E-mail: ferraro@ph.unito.it
Dipartimento di Ingegneria dell'Informazione
e Ingegneria Elettrica, Università di Salerno and INFM
E-mail: boccig@unisa.it
Department of Computing Science
The University of Alberta, General Services Building
Edmonton AB T6G 2E9 Canada
E-mail: tcaelli@ualberta.ca

Abstract. Loss of information in images undergoing fine-to-coarse image transformations is analized by using an approach based on the theory of irreversible transformations. It is shown that entropy variation along scales can be used to characterize basic, low-level information and to gauge essential perceptual components of the image, such as shape and texture. The use of isotropic and anisotropic fine-to-coarse transformations of grey level images is discussed, and an extension of the approach to multi-valued images is proposed, where cross-interactions between the different colour channels are allowed.

1   Introduction

The sensing process in vision can be seen as a mapping $\mathcal{T}$ from the state of the world into a set of images. Suppose that some detail of the state of the world we are interested in has a representation g; the map $\mathcal{T}$ applied to g produces an image I: $\mathcal{T}(g) = I$. The representation g can be for instance the shape of some surface, a description of the meaningful part of the image, or a function indicating if an object is present in the image. This raises the issue of how visual information, the information contained in an image, can be encoded in a way that is suited for the specific problem the visual system must solve. The most basic image representation, and indeed the one closest of the process of image formation can be derived as follows: suppose the image domain D to be lattice of N pixels and let the intensity at each pixel s be given by the number n of photons reaching s, thus the image is determined by their distribution. To each pixel, then, it is possible to associate a number of photons n_s whose distribution can be seen as the realization of a random field $\mathcal{F}$. The information, in the sense of Shannon, can then be computed by considering photons distribution on the retina, and different measures of information, can be obtained, depending on the constraints imposed on the distributions [1]. However, the information obtained this way is defined on the set of all possible images, given the constraints; hence,

C. Arcelli et al. (Eds.): IWVF4, LNCS 2059, pp. 176–185, 2001.
© Springer-Verlag Berlin Heidelberg 2001

to all images is assigned the same information and, in particular, it is not possible to discriminate among region with different information content.

Here we shall present a different approach, based on the idea that entropy, defined in the sense of statistical mechanics, measures lack of information [2]. Images are considered as an isolated thermodynamical system by identifying the image intensity with some thermodynamical variable, e.g. temperature or concentration of particles, evolving in time. It will be shown that entropy production in fine to coarse transformations of image representations implicitly provides a measure of the information contained in the original image; since, a local measure of entropy can be defined by means of the theory of irreversible transformations [3], parts of the image with different information content will be identified with those giving rise to different amount of local entropy production. Some examples will be given, illustrating how the identification of such different regions can provide a preliminary step towards the inference of shape and texture. The generalization of the approach to vector-valued images is also discussed.

2   Information from Irreversible Transformations

Let Ω be a subset of $\mathbf{R}^2$, (x, y) denote a point in Ω, and the scalar field $I : (x, y) \times t \to I(x, y, t)$ represent a gray-level image. The non-negative parameter t defines the scale of resolution at which the image is observed; small values of t correspond to fine scales, while large values correspond to coarse scales. A scale transformation is given by an operator T that takes the original image $I(\cdot, 0)$ to an image at a scale t, namely, $T_t : I(\cdot, 0) \to I(\cdot, t)$. We assume first T to be a semi-dynamical system, that is, a non-invertible or *irreversible* transformation which cannot be run backward across scale. Irreversibility of T ensures that the causality principle is satisfied, i.e. any feature at coarse level "controls" the possible features at a finer level of resolution - but the reverse need not be true. As a consequence of these assumptions every image point will, under the action of T, converge to a set of equilibrium points, denoted by the set $\mathcal{I}^*$, while preserving the total intensity [4].

Let I^* denote the fixed point of the transformation, that is, $I^*(\cdot, 0) = I^*(\cdot, t)$, for all t, and let I^* be stable, that is $\lim_{t \to \infty} I(\cdot, t) = I^*(\cdot)$. Relevant information contained in an image I can be measured in terms of the Kullback-Leibler distance between I and the corresponding fixed point I^* under the transformation T. In [4], the conditional entropy $H(f|f^*)$ was introduced as the negative of the Kullback-Leibler distance $H(f|f^*) = -\int \int_\Omega f(x, y, t) \ln \frac{f(x,y,t)}{f^*(x,y)} dx dy$. Here f and f^* are the normalized versions of I and I^*, respectively. The "dynamic information" of the image, under the process T, is provided by the evolution of $H(f|f^*)$ across scales:

$$\frac{\partial}{\partial t} H(f|f^*) = \frac{\partial}{\partial t} H(f) + \frac{\partial}{\partial t} \int \int_\Omega f(x, y, t) \ln f^*(x, y) dx dy, \qquad (1)$$

where $H(f) = -\int \int_\Omega f(x, y, t) \ln f(x, y, t) dx dy$ is the Boltzmann-Gibbs entropy measure. It has been proven [4] that $\frac{\partial}{\partial t} H(f) =$

$-\int\int_\Omega \ln f(x,y,t)\frac{\partial}{\partial t}f(x,y,t)dxdy.$ If f^* is constant over Ω, then $\frac{\partial}{\partial t}H(f|f^*) = \frac{\partial}{\partial t}H(f)$, and the entropy production as defined by the thermodynamics of irreversible transformations [3], that is $\mathcal{P} = \frac{\partial}{\partial t}H(f)$, can be used to quantify the information loss, during a transformation from fine to coarse image representations [4]. For simplicity we have dropped the scale variable t in H and $\mathcal{P}$. Fine-to-coarse transformations can be modelled, by a diffusion or heat equation[5]:

$$\frac{\partial f(x,y,t)}{\partial t} = \nabla^2 f(x,y,t). \tag{2}$$

In this case, it has been shown [4] that

$$\mathcal{P} = \int\int_\Omega f(x,y,t)\sigma(x,y,t)dxdy, \tag{3}$$

where

$$\sigma(x,y,t) = \frac{\nabla f(x,y,t) \cdot \nabla f(x,y,t)}{f(x,y,t)^2} \tag{4}$$

is the *density of entropy production* in the thermodynamical sense [3]. Since $\mathcal{P} \geq 0$, $H(f)$ is an increasing function of t. Note, also, that $\lim_{t\to\infty} \sigma = \lim_{t\to\infty} \mathcal{P} = 0$. By using $\mathcal{P}$ it is possible to gauge, at any t, the information loss of the process through its entropy increase. Furthermore, this can be done locally, at each point of the pattern. Fig. 1 shows an example of local entropy production maps computed at different scales for the image "Aerial 1". For representation purposes, $\sigma(x,y,t)$ has been linearly quantized in the interval $[0, 255]$ and the whole map rendered as a grey level image, where brighter pixels represent higher entropy production. It is easy to see that, during the transformation, σ is large along borders defining the shape of objects (e.g., buildings), smaller in textured regions (e.g, vegetation), and almost zero in regions of almost constant intensity.

In case of anisotropic transformations, e.g. anisotropic diffusion, the derivation of the formula for the evolution of $H(f|f^*)$ is obtained in a way similar to the one outlined above. Consider the well-known Perona-Malik equation for anisotropic diffusion

$$\frac{\partial}{\partial t}f(x,y,t) = div(\chi(x,y)\nabla f(x,y,t)), \tag{5}$$

where the neighborhood cliques which represent similarities between pixels can be defined by a conductance function, χ, that depends on the local structure of the image. For instance, Perona and Malik [6] assume χ to be a nonnegative, monotonically decreasing function of the magnitude of local image gradient. In this way the diffusion mainly takes place in areas where intensity is constant or varies slowly, whereas it does not affect areas with large intensity transitions. However, it has been widely noted that the choice of the conductance function, χ is critical since an ill-posed diffusion may occur, in the sense that images close

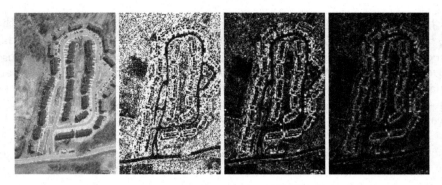

Fig. 1. The image "Aerial 1" and three local entropy production maps computed at different scales. Each map, from left to right, is obtained at 1, 5, 20 iterations of the diffusion process. Brighter points indicate high density of entropy production.

to each other are likely to diverge during the diffusion process [7]. Here, we assume Nordstrom's conjecture, namely we assume that, under a suitable choice of $\chi(x, y)$, and for $t \to \infty$ the stationary point f^* of the transformation corresponds to a piecewise constant function with sharp boundaries between regions of constant intensity [8]. In practice, such ideal fixed point can be obtained by defining $f^*(\cdot) = f(\cdot, t^*)$, with $t^* \gg 0$ (as it is actually done in the simulations). Note that, in contrast with the isotropic case, the fixed point $f^*(\cdot)$ of the transformation is constant with respect to time t, but not with respect to the direction of the spatial derivatives. It has been shown [8] that, in this case, a local measure σ_{an} of the variation rate of conditional entropy $H(f|f^*)$ can be defined by setting

$$\frac{\partial}{\partial t} H(f|f^*) = \mathcal{P} - \mathcal{S} = \int \int_\Omega f(x, y, t)\sigma_{an}(x, y, t)dxdy, \qquad (6)$$

where $\mathcal{P} = \int \int_\Omega f(x, y, t)\sigma'_{an}(x, y, t)dxdy$ and $\mathcal{S} = \int \int_\Omega f(x, y, t)\sigma''_{an}(x, y, t)dxdy$, and σ_{an} can be written as the sum of two terms, denoted by σ'_{an} and σ''_{an} respectively,

$$\sigma_{an}(x, y, t) = \sigma'_{an}(x, y, t) - \sigma''_{an}(x, y, t)$$
$$= \chi(x, y)\frac{\nabla f(x, y, t) \cdot \nabla f(x, y, t)}{f(x, y, t)^2} - \chi(x, y)\frac{\nabla f(x, y, t) \cdot \nabla f^*(x, y)}{f(x, y, t)f^*(x, y, t)}(7)$$

Numerical computations of $\partial H(f|f^*)/\partial t$ and $\mathcal{P}, \mathcal{S}$, performed on a data set of 120 natural images provide evidence that if χ is a non-negative decreasing function of $|\nabla f|$, $0 < \mathcal{S} < \mathcal{P}$, so that $\mathcal{P} - \mathcal{S} > 0$. Further, the evolution of $H(f|f^*)$, at least for t small, is determined by entropy production alone.

The term $\sigma''_{an} = \chi(x, y)\frac{\nabla f(x, y, t) \cdot \nabla f^*(x, y)}{f(x, y, t)f^*(x, y, t)}$ is calculated by using an approximated fixed point image $\tilde{f}^*(x, y) \simeq f^*(x, y)$ obtained by letting the anisotropic diffusion run for large t. The image in Fig. 2 shows the fixed point for the

"Aerial1" image obtained for $t = 2000$ iterations and using a "balanced" backward sharpening diffusion as detailed in [7]. In the same figure, plots of $\partial H(f|f^*)/\partial t$ and $\mathcal{P}, \mathcal{S}$, as functions of t are also depicted.

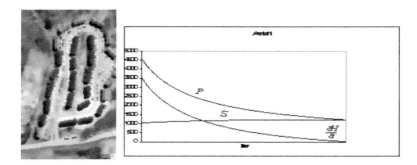

Fig. 2. The fixed point of "Aerial 1" obtained through 2000 iterations of anisotropic diffusion. The graph on the right of the image plots $\partial H(f|f^*)/\partial t$, $\mathcal{P}$, and $\mathcal{S}$ as a function of scale, represented by iterations of the anisotropic diffusion process. Units are $nats/t$.

In [4] it has been argued that $\mathcal{P}dt = dS$ is the loss of information in the transition from scale t to $t + dt$, that is, $\mathcal{P}$ is the loss of information for unit scale. Intuitively, $\mathcal{P}$ is a global measure of the rate at which the image, under the action of T, looses structure. Analogously, the density of entropy production, being a function of x, y, measures local loss of information [4], that is, the loss of information at a given pixel for unit scale. In other terms, σ depends on the local structure of the image and thus enables us to define features which can exist at different scales (in different image regions). The density of entropy production implicitly defines regions of different information content, or saliency; furthermore it is clear, by definition, that σ is large along edges, smaller in textured regions, and almost zero in regions of almost constant intensity. This property has been the basis for a method of region identification [4] [8].

Let the activity across scales, in the isotropic case, be $a_\sigma(x, y) = \int_0^\infty f(x, y, t)\sigma(x, y, t)dt$; similarly, for anisotropic diffusion, can be defined the activities $a_{\sigma_{an}}(x, y)$, $a_{\sigma'_{an}}$ and $a_{\sigma''_{an}}$, with $a_{\sigma_{an}} = a_{\sigma'_{an}} - a_{\sigma''_{an}}$. The activity is then a function defined on the image plane. It is possible, by a thresholding procedure [4], [8], to distinguish between two adjacent regions of different activity. Namely, every pixel in the image is classified as belonging to one of three classes: low information regions or l-type regions (characterized by low activity), medium information regions or m-type (medium activity), and high information regions or h-type (high activity). Such basic features represent a preliminary step, both in the isotropic and anisotropic case, towards the labelling of image regions in terms of basic perceptual components: smooth regions, textures and edges. An example, which compares different results obtained in the anisotropic case vs. the isotropic case is provided in Fig. 3 .

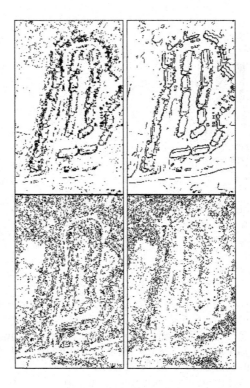

Fig. 3. Results of region identification by activity for image "Aerial1". The left column displays results obtained experimenting with the isotropic process: the top image presents h-type regions, whereas the bottom image presents m-type regions. The two images in the right column are the corresponding results obtained via anisotropic process. In both experiments, the activity has been computed by integrating over 100 iterations.

Note that the separation of regions of different activity entails the separation of well defined objects, the buildings, for instance, characterized by high activity, from a cluttered/textured background (vegetation).Further, the use of anisotropic diffusion makes localization of different features in the image more precise. In particular, textural parts are more neatly encoded by m-type regions, whereas localization of shapes via edges is improved, since anisotropic diffusion avoids edge blurring and displacement, and confined to h-type regions.

3   Deriving Spatio-Chromatic Information

In order to provide an extension of this framework to color images it is necessary to address the issue of multi-valued isotropic diffusion. A color image can be considered a vector-valued image, which we denote $\boldsymbol{f}(x,y,t) = (f_i(x,y,t))^T$, where $i = 1, 2, 3$, then a fine-to-coarse transformation must be defined for the vector field $\boldsymbol{f}(x,y,t)$. Little work has been devoted to extending Eq. 2 to color images, while some do address anisotropic diffusion of vector-valued images[9, 10]. Here the problem is tackled by representing a color image as a system of independent single-valued diffusion processes evolving simultaneously:

$$\frac{\partial f_i(x,y,t)}{\partial t} = \nabla^2 f_i(x,y,t), \tag{8}$$

where $i = 1, 2, 3$ labels the color components, or channels, in the image. Unfortunately Eq. 8 does not allow interactions among different color channel to take place, whereas there is a general agreement that the processing and interpretation of color images cannot be reduced to the separate processing of three independent channels, but must account to some extent for cross-effects between channels, whatever the channels employed, RGB, HSI, etc. Indeed, exploiting cross-effects is an important point when complex images are taken into account, where complex interactions occur among color, shape and texture (cfr. Fig. 4).

Fig. 4. The "Waterworld" image (color, in the original) presents complex interaction between colour, shape and texture

It is well known [3] that diffusion (Eq.2) can be derived from a more general equation, namely $\frac{\partial f}{\partial t} = -div\boldsymbol{J}$, where $\boldsymbol{J}$ is the flux density or flow [3]. Also, it is known that, in a large class of transformations, irreversible fluxes are linear functions of the thermodynamical forces expressed by the phenomenological laws of irreversible processes. For instance, Fourier's law states that the component of the heat flow are linearly related to the gradient of the temperature. In this linear region then $\boldsymbol{J} = L\boldsymbol{X}$ [3], where $\boldsymbol{X}$ is called the generalized force and L is a matrix of coefficients. In this case the density of entropy production is given by [3] $\sigma = \boldsymbol{J} \cdot \boldsymbol{X} = \sum_{i,j} L_{ij} X_i X_j$ where $i, j = 1, 2, 3$ label the three color channels. Hence, we define, for each color channel, the transition from fine to coarse scale through the equation $\frac{\partial f_i}{\partial t} = -div \boldsymbol{J}_i$ and, in turn, for each i, the flow density is given by $\boldsymbol{J}_i = \sum_{j=i}^{n} L_{ij}\boldsymbol{X}_j$. We have chosen to model interactions among color components setting $\boldsymbol{X}_i = \nabla\left(\frac{1}{f_i(x,y,t)}\right)$ and $L_{ij} = \kappa_{ij}f_if_j$, where $\kappa_{ij} = \kappa_{ji}$ are symmetric coefficients weighting the strength of the interactions between channels i and j and whose maximum value is $\kappa_{ii} = 1$. We then obtain the following system of coupled evolution equations:

$$\frac{\partial f_i}{\partial t} = -div\left(\sum_j L_{ij}\boldsymbol{X}_j\right) = \sum_j \nabla \cdot \left(\kappa_{ij}f_if_j\frac{\nabla f_j}{f_j^2}\right). \tag{9}$$

Eq. 9 can be developed as:

$$\frac{\partial f_i}{\partial t} = \nabla^2 f_i + \sum_{j\neq i}\kappa_{ij}(\frac{f_i}{f_j}\nabla^2 f_j +$$

$$+\frac{1}{f_j^2}(f_j(\frac{\partial f_i}{\partial x}\frac{\partial f_j}{\partial x} + \frac{\partial f_i}{\partial y}\frac{\partial f_j}{\partial y}) - f_i(\frac{\partial f_j}{\partial x}\frac{\partial f_j}{\partial x} + \frac{\partial f_j}{\partial y}\frac{\partial f_j}{\partial y}))) \tag{10}$$

Then, color evolution across scales, in the different channels, comprises a purely diffusive term and a nonlinear term that depends on the interactions among channels; if $\kappa_{ij} = \delta_{ij}$, that is if the channels are considered as isolated systems Eq. 10 reduces to Eq. 8. The local entropy production $\Sigma(x,y,t)$ can then be computed as [3]

$$\Sigma = \sum_{i,j} L_{ij}\boldsymbol{X}_i \cdot \boldsymbol{X}_j = \sum_{i,j} L_{ij}\nabla\left(\frac{1}{f_i}\right) \cdot \nabla\left(\frac{1}{f_j}\right), \tag{11}$$

and in case of grey-level images Eq. 11 reduces to Eq. 4. Again, by considering $L_{ij} = \kappa_{ij}f_if_j$, using the symmetry property of the coefficients, Eq. 11 can be written explicitly for color images:

$$\Sigma(x,y,t) = \sum_i \frac{\nabla f_i \cdot \nabla f_i}{f_i^2} + \sum_{i\neq j}\kappa_{ij}\frac{\nabla f_i \cdot \nabla f_j}{f_if_j}. \tag{12}$$

The density of entropy production Σ is, not surprisingly, made up by two terms: $\sigma_i = \frac{\nabla f_i \cdot \nabla f_i}{f_i^2}$, the density of entropy production for every channel i, considered in isolation, and the cross terms $\kappa_{ij}\frac{\nabla f_i \cdot \nabla f_j}{f_if_j}$ that accounts for the dependence of entropy production on interactions among channels. Obviously if color channels are considered isolated $\Sigma = \sum_i \sigma_i$. By definition, it is clear that Σ, depends on the local spatio-chromatic properties of the image at a given scale t. Then, Σ measures the local loss of spatio-chromatic information, that is, the loss of information at a given pixel for unit scale. More formally, given a finite number $k = 1, 2, .., K$ of iterations, it is possible, to sample the spatio-chromatic density of entropy production as $\Sigma(x,y,t_k)$. Then, at each point (x,y)

of the image, a feature vector $\boldsymbol{v} = [\Sigma(x, y, t_1), \Sigma(x, y, t_2), ..., \Sigma(x, y, t_K)]^T$ can be defined, which captures the evolution of Σ at (x, y) across different scales. A clustering procedure is then applied to the vector $\boldsymbol{v}$, a cluster being a natural group of points with similar features of interest. From a perceptual point of view, as previously discussed, we are interested in the partitioning of l-type regions, as opposed to h-type and m-type regions; thus, we partition the cluster space in three regions through the well known C-means clustering algorithm [11]. The results presented in Fig 5 show that, as expected, most of the h and m-type points are encoded in the luminance channel; chromatic information contributes mostly to m-type points and little to h-type points.

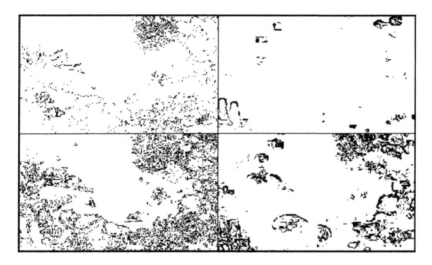

Fig. 5. Region identification for image "Waterworld". The left column displays results on the luminance channel: the top image presents h-type regions, whereas the bottom image presents m-type regions. The two images in the right column are the corresponding results on the opponent channels.

4   Discussion and Conclusion

In this note it has been presented a method to derive measure local and global information contained in a single image. It must be noted that the results depends not just on the image but also on the specific type of irreversible transformation the image undergoes to. Thus, in anisotropic diffusion the rate of change of information across scales does not depend, as in the isotropic case, solely on entropy production; due to the characteristics of the process, the loss of information is, at least, partially prevented by a term that depends on the degree of parallelism between the gradient of the image at scale t and that of the image representing the fixed point of the anisotropic diffusion process. An extension of the framework to color or vector-valued images has been presented. We have proposed an evolution equation based on the assumption that a multi-valued image is a complex isolated system, whose components, namely, the color components, interact with each others through a generalized thermodynamical force.

Note that the model proposed here for measuring pattern information is different from the classical model (*a là* Shannon) which assumes the pattern itself as a message source. In our case, the message source is represented by the whole dynamical system, namely the pattern together with the chosen transformation. Eventually, the proposed framework provides a way for representing and computing image features encapsulated within different regions of scale-space. More precisey, the method derives features as those corresponding to "entropy rich" image components. These constitute a preliminary representation towards the computation of shape and texture.

References

1. G. Boccignone and M. Ferraro, "An information-theoretic approach to interactions in images," *Spatial Vision*, vol. 12, no. 3, pp. 345–362, 1999.
2. L. Brillouin, *Science and Information Theory*. New York, N.Y.: Academic Press, 1962.
3. S. de Groot and P. Mazur, *Non-Equilibrium Thermodinamics*. Amsterdam: North-Holland, 1962.
4. M. Ferraro, G. Boccignone, and T. Caelli, "On the representation of image structures via scale-space entropy condition," *IEEE Transactions on Pattern Analysis and Machine Intelligence*, vol. 21, pp. 1199–1203, November 1999.
5. T. Lindeberg, *Scale-space theory in computer vision*. Erewhon, NC: His Publisher, 1999.
6. P. Perona and J. Malik, "Scale-space and edge detection using anisotropic diffusion," *IEEE Transactions on Pattern Analysis and Machine Intelligence*, vol. 12, pp. 629–639, 1990.
7. Y. L. You, X. Wenyuan, A. Tannenbaum, and M. Kaveh, "Behavioral analysis of anisotropic diffusion in image processing," *IEEE Transactions on Image Processing*, vol. 5, no. 11, pp. 1539–1552, 1996.
8. G. Boccignone, M. Ferraro, and T. Caelli, "Encoding visual information using anisotropic transformations," *IEEE Transactions on Pattern Analysis and Machine Intelligence*, to appear.
9. R. Whitaker and G. Gerig, "Vector-valued diffusion," in *Geometry-Driven Diffusion in Computer Vision* (B. ter Haar Romeny, ed.), (Dordrecht, The Netherlands), Kluwer Academic Publishers, 1994.
10. G. Sapiro and D. Ringach, "Anisotropic diffusion of multivalued images with application to color filtering," *IEEE Transactions on Image Processing*, vol. 5, pp. 1582–1586, November 1996.
11. R. Duda and P. Hart, *Pattern classification and scene analysis*. New York, N.Y.: Wiley-Interscience, 1973.

Skeletons in the Framework of Graph Pyramids*

Roland Glantz and Walter G. Kropatsch

Pattern Recognition and Image Processing Group 183/2
Institute for Computer Aided Automation
Vienna University of Technology
Favoritenstr. 9
A-1040 Vienna
Austria
email{glz,krw}@prip.tuwien.ac.at
phone ++43-(0)1-58801-18358
fax ++43-(0)-1-58801-18392

Abstract. Graph pyramids allow to combine pruning of skeletons with a concept known from the representation of line images, i.e. generalization of paths without branchings by single edges. Pruning will enable further generalization of paths and the latter speeds up the former. Within the unified framework of graph pyramids a new hierarchical representation of shape is proposed that comprises the skeleton pyramid, as proposed by Ogniewicz. In particular, the skeleton pyramid can be computed in parallel from any distance map.

1 Introduction

A major goal of skeletonization consists in bridging the gap between low level raster-oriented shape analysis and a semantic object description [Ogn94]. In order to create a basis for the semantic description, the medial axis [Blu62,Ser82] is often transformed into a plane graph [Ogn94]. This task has been solved using the Voronoi diagram defined by the boundary points of a shape [Ogn93,Ogn94, OK95] or by the use of special metrics on derived grids [Ber84]. In this paper we will propose a method that is not confined to a special metric (distance map) on a special grid nor on a special irregular structure like the Voronoi diagram.

The new method starts with a regular or irregular *neighborhood graph*. The neighborhood graph reflects the arrangement of the sample points in the plane. The vertices of the neighborhood graph represent the sample points and the distances from the sample points to the boundary of the shape are stored in the vertex attributes. The edges of the neighborhood graph represent the neighborhood relations of the sample points. All illustrations in this paper refer to the regular neighborhood graph, in which the sample points represent pixel centers and the edges indicate the 4-connectivity of the pixels (Fig. 1a). The vertex attributes reflect the Euclidean distance map (EDM) on a 4-connected set S of

---

* This work has been supported by the Austrian Science Fund (FWF) under grant
  P14445-MAT.

C. Arcelli et al. (Eds.): IWVF4, LNCS 2059, pp. 186–195, 2001.
© Springer-Verlag Berlin Heidelberg 2001

pixels: each pixel p of S is equipped with the Euclidean distance between the center of p and the *closest pixel center outside of S* [SL98] (Fig. 1b).

The dual of the neighborhood graph is referred to as *crack graph*. The edges of the crack graph describe the borders of the pixels. Each edge e of the neighborhood graph perpendicularly intersects exactly one edge $\bar{e}$ of the crack graph. The edge e is called the *dual* of $\bar{e}$ and vice versa. Each vertex of the crack graph stands for a pixel corner (Fig. 2b).

Dual graph contraction (DGC) [Kro95] is used to successively generalize the neighborhood graph by the removal and the contraction of edges. One level of the resulting *graph pyramid* will be called *skeleton graph* (Fig. 5a). This term is justified by the fact that all centers of maximal disks (with respect to the distance map) are represented by vertices of the skeleton graph. Furthermore, the skeleton graph is always connected.

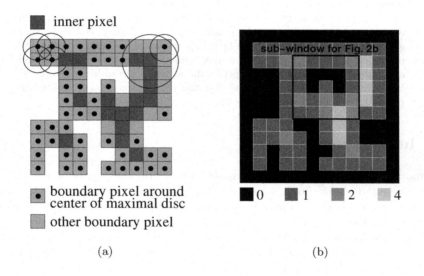

Fig. 1. (a) 4-connected pixel set. (b) Squared distances of (a).

This paper is organized as follows: Section 2 is devoted to the initialization of the attributes in the neighborhood graph and in the crack graph. In Section 3 the crack graph is contracted. Regarding the neighborhood graph this amounts to the deletion of edges that are dual to the ones contracted in the crack graph. The reduced neighborhood graph is called *extended skeleton graph* (since it contains the skeleton graph). In Section 4 the extended skeleton graph is contracted to the skeleton graph. An overview of the different graphs and their relations is given in Fig.2a.

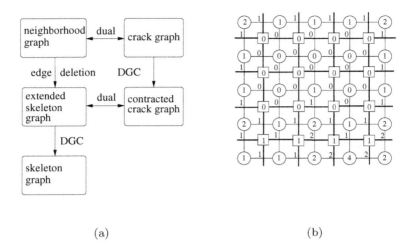

(a)   (b)

Fig. 2. (a) Overview of the graphs and their relations. (b) Neighborhood graph ($\bigcirc$) and crack graph ($\square$) restricted to the sub-window in Fig. 1b. The numbers indicate the attribute values of the vertices and the edges.

Like the skeleton, the skeleton graph is not robust. In Section 5 we propose a pruning and generalization method for the skeleton graph. It is based on DGC and yields the new shape representation by means of a graph pyramid. The pyramid proposed in [OK95] can be obtained from the new representation by threshold operations. We conclude in Section 6.

2   Initialization of the Neighborhood Graph and the Crack Graph

The neighborhood graph may be interpreted as digital elevation model (DEM), if the vertex attributes, i.e. the distances of the corresponding sampling points to the border of the shape, are interpreted as altitudes. Intuitively, the plan for the construction of the skeleton graph is to reduce the neighborhood graph such that the remaining edges describe the connections of the summits in the DEM via the crest lines of the DEM. In contrast to [KD94,NGC92] our concept is a dual one: the neighborhood relations of the basins are described by the dual of the skeleton graph. In the next two sections it will turn out that the reduction of the skeleton graph depends only on the order of the values from the distance transform. Hence, we may use *squared* distances and thus avoid non-integer numbers. The idea for the reduction of the neighborhood graph is to remove edges that do not belong to ridges - thus forming the basins represented by the dual graph. The following initialization (Fig. 2b) will allow to control this process. The first part refers to the neighborhood graph, the second to the crack graph (Fig. 2b):

- Let $dist^2(v)$ denote the squared distance of the pixel that corresponds to vertex v. The attribute value of v is set to $dist^2(v)$. The attribute value of edge $e = (u, v)$ is set to the minimum of $dist^2(u)$ and $dist^2(v)$.

- The attribute value of edge $\bar{e}$ is set to the attribute value of edge e, where e denotes the edge in the neighborhood graph that is dual to $\bar{e}$. The attribute value of vertex $\bar{v}$ is set to the minimum of the attribute values of all edges incident to $\bar{v}$.

3   Contracting the Crack Graph

Recall, that the contraction of an edge in the crack graph is associated with the removal of the corresponding dual edge in the neighborhood graph [BK99a]. The neighborhood graph can never become disconnected: The removal of an edge e would disrupt the neighborhood graph, only if the corresponding dual edge $\bar{e}$ in the crack graph was a self-loop. DGC, however, forbids the contraction of self-loops.

In order to get an intuitive understanding of the duality between contraction and deletion, we focus on the embedding of graphs on the plane (only planar graphs can be embedded on the plane) [TS92]. An embedding of the neighborhood graph on the plane divides the plane into *regions* (Fig.3). Note that the

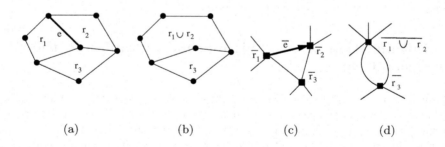

(a)   (b)   (c)   (d)

Fig. 3. Duality of edge deletion in a plane graph (a)→(b) and edge contraction in the dual of the plane graph (c)→(d). The regions $r_1, \ldots, r_4$ in (a) are represented by the vertices $\bar{r_1}, \ldots, \bar{r_4}$ in (c).

removal of an edge in the neighborhood graph is equivalent to the fusion of the regions on both sides of the edge. In terms of watersheds [MR98] it is intuitive to fuse the regions of the neighborhood graph until each of the resulting regions corresponds to exactly one basin of the landscape. **Two neighboring regions may be fused, if there is no separating ridge between the regions**. Due to the initialization of the attribute values in the crack graph, we may formulate a criterion for the fusion of two regions as follows [GEK99]: Let r_1 and r_2 denote two regions of the neighborhood graph and let $\overline{r_1}$ and $\overline{r_2}$ denote the corresponding vertices in the crack graph. The regions r_1 and r_2 ($r_1 \neq r_2$) may be fused, if there exists an edge $\overline{e}$ between $\overline{r_1}$ and $\overline{r_2}$, whose attribute value equals the attribute value of $\overline{r_1}$ or the attribute value of $\overline{r_2}$. Assume that the attribute value of $\overline{r_2}$ is smaller or equal to the attribute value of $\overline{r_2}$. Then the fusion of r_1 and r_2 is achieved by the contraction of $\overline{r_1}$ into $\overline{r_2}$ (Fig. 3c). Thus, during the whole contraction process the attribute values of a vertex in the crack graph indicates the altitude of the deepest point in region represented by the vertex.

Multiple fusions can be done by iterating the following *parallel* steps:

1. For each edge of the crack graph that meets the above criterion for contraction, mark the end vertex with the minimal attribute value. In case of equality choose one of the end vertices by a random process.
2. Form a maximal independent set (MIS) of the marked vertices as explained in [Mee89]. The MIS is a maximal subset of the marked vertices, no two elements of which are connected by an edge.
3. Contract all edges that are incident to a vertex v of the MIS and that meet the above criterion for contraction (v being the end vertex with the minimal attribute).

The iteration stops, when none of the edges in the crack graph meets the above criterion. The resulting graph is called *extended skeleton graph* (Fig. 5a). It is connected and it still contains all vertices of the neighborhood graph.

4   Contracting the Neighborhood Graph

In this section the extended skeleton graph is further reduced to the so called *skeleton graph*. The skeleton graph

– still must contain all vertices which represent maximal discs and
– still must be connected.

We focus on edges $e = (u, v)$ such that v has degree 1 in the extended skeleton graph. The idea is to contract v into u, if we can tell by a local criterion that v does not represent a center of a maximal disc. All edges that have a degree-one end vertex to fulfill this criterion may then be contracted in parallel.

Consider an edge $e = (u, v)$ such that v is the end vertex with degree one. Using the notation of Section 2, i.e. $dist(v)$ [$dist^2(v)$] for the [squared] distance of a vertex v, we formulate the following criterion: If

$$dist(u) - dist(v) = 1, \tag{1}$$

the vertex v does not represent a center of a maximal disc and v may be contracted into u [San94].

The distances $dist(u)$ and $dist(v)$ in condition(1) are integers [1]. This follows from equation

$$dist^2(u) = (dist(v) + 1)^2 = dist^2(v) + 2dist(v) + 1 \qquad (2)$$

and the fact that the squared distances are integers.

In case of grids other than the square grid or in case of irregular samplings, Equation 1 generalizes to

$$dist(u) - dist(v) = \| u - v \|_2, \qquad (3)$$

where $\| \cdot \|_2$ denotes the Euclidean length of $u - v$. In terms of [Ser82], u is *upstream* of v. Repeated contraction of edges in the extended skeleton graph yields the skeleton graph (Fig. 5a).

5   A New Hierarchical Representation for Shapes

The new hierarchy is build on top of the skeleton graph. Besides pruning we also apply generalization of paths between branchings by single edges, as proposed in [BK99b].

In order to asses the prominence of an edge e in the skeleton of a shape S, Ogniewicz [OK95]

1. defines a measure m for boundary parts b of S , i.e. the length of b,
2. for each edge e determines the boundary part b_e of S associated with e (this is formulated within the concept of Voronoi Skeletons),
3. sets the *prominence* of e to $m(b_e)$.

In our approach we also measure boundary parts by their lengths. However, we associate the boundary parts with vertices and thus define prominence measures for vertices. The initial prominence measure $prom(v)$ of v indicates the number of boundary vertices (vertices representing boundary pixels) contracted into v including v itself, if v is a boundary vertex. (Fig. 5a). This can already be accomplished during the contraction of the neighborhood graph (Section 4) by

1. setting $prom(v)$ to 1, if v is a boundary vertex, 0 otherwise (before the contraction),
2. incrementing the prominence measure of w by the prominence measure of v, if v is contracted into w.

Prominence measures will only be calculated for vertices that do not belong to a cycle of the skeleton graph. In the following, the calculation of the prominence measures from the initial prominence measures is combined with the calculation of the skeleton pyramid.

Let the degree of a vertex v in a graph be written as $deg(v)$ and let P denote a maximal path without branchings in the skeleton graph, i.e. $P = (v_1, v_2, \ldots, v_n)$ such that

---

[1] Thus condition(1) may be checked using only the squared distances and a look-up table.

- v_i is connected to v_{i+1} by an edge e_i for all $1 \leq i < n$ and
- $deg(v_i) = 2$ for all $1 < i < n$ and
- $deg(v_1) \neq 2$, $deg(v_n) \neq 2$.

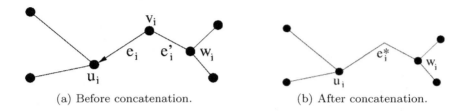

(a) Before concatenation.   (b) After concatenation.

Fig. 4. Concatenation of edges.

Let $e_i = (u_i, v_i)$ be an edge of P with $u_i \neq v_i$ and $deg(v_i) = 2$ (Fig. 4a). Since $deg(v_i) = 2$, there is a unique edge $e'_i \neq e_i$ in P with $e'_i = (v_i, w_i)$, $w_i \neq v_i$. We assume that e_i does not belong to a cycle, i.e. $w_i \neq u_i$. If $deg(w_i) \neq 1$, we allow that v_i may be contracted into u_i. The contraction of e_i can be described by the replacement of the two edges e_i and e'_i by a new edge e^*_i (Fig. 4b).

The prominence measure of u_i is updated by

$$prom(u_i) := prom(u_i) + prom(v_i). \tag{4}$$

This contraction process is referred to as *concatenation*. Due to the requirement $deg(w_i) \neq 1$ the prominence measures are successively collected at the vertices with degree 1. The result of concatenation on the skeleton graph in Fig. 5a is depicted in Fig. 5b.

After concatenation there are no vertices with degree 2. We focus on the set of *junctions with dead ends*, i.e. the set U of all vertices u with

- $deg(u) > 2$ and
- there exists an edge e and a vertex v with $e = (u, v)$ and $deg(v) = 1$.

The set of all edges that connect a vertex $u \in U$ with a vertex v of degree 1 is denoted by $Ends(u)$. Note, that for each edge e there is at most one $u \in U$ with $e \in Ends(u)$. For each $u \in U$ let $e_{min}(u)$ denote an edge in $Ends(u)$, whose end vertex with degree 1 has a minimal prominence measure.

The prominence measures of the vertices with degree 1 induce an order in $Ends(u)$, $e_{min}(u)$ being the least element. In case of $|\ Ends(u)\ | > 1$ we allow $e_{min}(u)$ to be contracted. Analogous to concatenation the prominence measure of u is updated to $prom(u) := prom(u) + prom(v)$.

In Fig. 6 the contraction is indicated by arrows: a white vertex at the source of an arrow is contracted into the black vertex at the head of the arrow. The new prominence measures are emphasized.

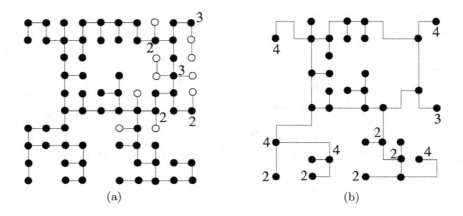

Fig. 5. (a) Extended skeleton graph. The vertices of the skeleton graph are given by the filled circles. The numbers indicate the initial prominence measures > 1. (b) After concatenation of the skeleton graph in Fig. 5a. The numbers indicate the prominence measures > 1.

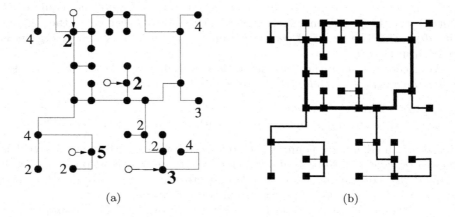

Fig. 6. (a) Contraction of vertices with degree 1. The numbers indicate the prominence measures > 1. (b) Ranks: bold 1, medium 2, thin 3.

For each set $Ends(u)$ the operation of contraction followed by updating the prominence measure takes constant time. These operations can be performed in parallel, since the sets $Ends(u)$, $u \in U$ are disjoint. *Generalization* of the skeleton graph consists of iterating concatenation followed by contraction (both include updating).

In [OK95] a hierarchy of skeleton branches is established by a skeleton traversal algorithm. The traversal starts at the most prominent edge and follows the two least steep descents (starting at each end vertex of the most prominent edge). The highest rank skeleton branch consists of all edges that have been traversed by this procedure. Skeleton branches of second highest rank originate from the highest rank branch and are also least steep descent (ignoring the edges of the highest rank branch). Skeleton branches of third highest rank and so on are defined analogously. The edges of the skeleton are labelled according to the rank of the skeleton branch they belong to.

Analogous ranks can be determined by imposing a restriction on the generalization described above: for each vertex $u \in U$ only one edge in $Ends(u)$ may be contracted. This is achieved by initializing all vertices as *vacant*. Once an edge from $Ends(u)$ was contracted, u is marked as *occupied*. The generalization up to the state, in which no further generalization can be done is summarized as *first step*. Thereafter, all vertices are marked as *vacant* again. The second step is finished, when occupation again forbids further generalization and so on. Thus, each edge of the concatenated skeleton graph that does not belong to a cycle is contracted. If n denotes the number of the last step, the *rank* of an edge contracted in step k, $(1 \leq k \leq n)$ is set to $2 + n - k$. Edges that belong to at least one cycle of the extended skeleton graph receive rank 1.

The set of edges with rank smaller or equal to k, $(1 \leq k \leq n + 1)$ always forms a connected graph. As in [OK95] these graphs can be derived by a simple threshold operation on the concatenated skeleton graph according to the ranks of the edges. The ranks of the extended skeleton graph in Fig. 5b are shown in Fig. 6c.

6   Conclusion

In this paper we have introduced a graph based hierarchical representation of shapes that comprises the skeleton pyramid as proposed by Ogniewicz. The new representation relies on the concept of graph pyramids by dual graph contraction. It allows to represent paths without branchings by single edges. This additional hierarchical feature is suggested for the hierarchical matching of shapes by means of their skeletons.

References

[Ber84]   Gilles Bertrand. Skeletons in derived grids. In *Proc. 7th International Conference on Pattern Recognition*, pages 326–329, Montreal, Canada, July 1984.

[BK99a]   Luc Brun and Walter.G. Kropatsch. Pyramids with Combinatorial Maps. Technical Report PRIP-TR-57, Institute for Computer-Aided Automation 183/2, Pattern Recognition and Image Processing Group, TU Wien, Austria, 1999.

[BK99b]   Mark J. Burge and Walter G. Kropatsch. A minimal line property preserving representation of line images. *Computing*, 62:355 – 368, 1999.

[Blu62]   H. Blum. An associative Machine for dealing with the Visual Fields and some of its biological Implications. *Biol. Prot. and Synth. Syst.*, Vol. 1:pp.244–260, 1962.

[GEK99]   Roland Glantz, Roman Englert, and Walter G. Kropatsch. Representation of Image Structure by a Pair of Dual Graphs. In Walter G. Kropatsch and Jean-Michel Jolion, editors, *2nd IAPR-TC-15 Workshop on Graph-based Representation*, pages 155–163. OCG-Schriftenreihe, Österreichische Computer Gesellschaft, 1999. Band 126.

[KD94]   J. Koenderink and A. van Doorn. Image structure. In E. Paulus and F. Wahl, editors, *Mustererkennung 1997*, pages 401–408. Springer, 1994.

[Kro95]   Walter G. Kropatsch. Building Irregular Pyramids by Dual Graph Contraction. *IEE-Proc. Vision, Image and Signal Processing*, 142(6):366 – 374, 1995.

[Mee89]   Peter Meer. Stochastic image pyramids. *CVGIP*, 45:269 – 294, 1989.

[MR98]   A. Meijsler and J. Roerdink. A Disjoint Set Algorithm for the Watershed Transform. In *Proc. of EUSIPCO'98, IX European Signal Processing Conference*, pages 1665 – 1668, Rhodes, Greece, 1998.

[NGC92]   C.W. Niblack, P.B. Gibbons, and D.W. Capson. Fast homotopy-preserving skeletons using mathematical morphology. *CVGIP*, 54-5:420 – 437, 1992.

[Ogn93]   Robert L. Ogniewicz. *Discrete Voronoi Skeletons*. PhD thesis, ETH Zurich, Switzerland, 1993. Hartung-Gorre Verlag Konstanz.

[Ogn94]   R. L. Ogniewicz. A Multiscale MAT from Voronoi Diagrams: The Skeleton-Space and its Application to Shape Description and Decomposition. In Carlo Arcelli, Luigi P. Cordella, and Gabriella Sanniti di Baja, editors, *2nd Intl. Workshop on Visual Form*, pages 430–439, Capri, Italy, June 1994.

[OK95]   Robert L. Ogniewicz and Olaf Kübler. Hierarchic voronoi skeletons. *Pattern Recognition*, 28(3):343–359, March 1995.

[San94]   Gabriella Sanniti di Baja. Well-Shaped Stable and Reversible Skeletons from the 3-4 Distance Transform. *Journal of Visual Communication and Image Representation*, pages 107–115, 1994.

[Ser82]   Jean Serra. *Image and Mathematical Morphology*, volume 1. Academic Press, London, G.B., 1982.

[SL98]   Frank Y. Shih and Jenny J. Liu. Size Invariant Four-Scan Euclidean Distance Transformation. *Pattern Recognition*, 31(11):1761–1766, 1998.

[TS92]   K. Thulasiraman and M.N.S. Swamy. *Graphs: Theory and Algorithms*. J. Wiley & Sons, New York, USA, 1992.

Computational Surface Flattening: A Voxel-Based Approach

Ruth Grossmann[1], Nahum Kiryati[*1], and Ron Kimmel[2]

[1] Dept. of Electrical Engineering – Systems, Tel Aviv University
Ramat Aviv 69978, Israel
`nk@eng.tau.ac.il`
[2] Dept. of Computer Science, Technion
Haifa 32000, Israel
`ron@cs.technion.ac.il`

Abstract. A voxel-based method for flattening a surface while best preserving the distances is presented. Triangulation or polyhedral approximation of the voxel data are not required. The problem is divided into two main subproblems: Voxel-based calculation of the minimal geodesic distances between the points on the surface, and finding a configuration of points in 2-D that has Euclidean distances as close as possible to the minimal geodesic distances. The method suggested combines an efficient voxel-based hybrid distance estimation method, that takes the continuity of the underlying surface into account, with classical multi-dimensional scaling (MDS) for finding the 2-D point configuration. The proposed algorithm is efficient, simple, and can be applied to surfaces that are not functions. Experimental results are shown.

1 Introduction

Surface flattening is the problem of mapping a surface in 3-D space into 2-D. Given a digital representation of a surface in 3-D, the goal is to map each 3-D point into 2-D such that the distance between each pair of points in 2-D is about the same as the corresponding geodesic distance between the points on the surface in 3-D space.

It is known that mapping a surface in 3-D space into a 2-D plane introduces metric distortions unless they have the same Gaussian curvature [4]. For example, a plane and a cylinder both have zero Gaussian curvature, since for the plane both principal curvatures vanish and for the cylinder one principal curvature vanishes. Therefore a plane bent into a cylindrical shape can obviously be flattened with no distortion. On the other hand, distortion-free flattening of a sphere onto the plane is impossible, because their Gaussian curvatures are different. General surfaces are likely to have nonzero Gaussian curvatures, hence their flattening introduces some distortion.

The need for surface flattening arises primarily in medical imaging (cortical surface visualization and analysis) and in computer graphics. One of the first

---

[*] Corresponding author.

C. Arcelli et al. (Eds.): IWVF4, LNCS 2059, pp. 196–204, 2001.
© Springer-Verlag Berlin Heidelberg 2001

flattening algorithms was proposed in [21]. More efficient methods were later developed: In [8] local metric properties were preserved in the flattening process but large global distortions sometimes occurred; In [6,7] a global energy functional was minimized; In [1,11,13] angle preserving mappings were used. In computer graphics, surface flattening can facilitate feature-preserving texture mapping (e.g. [2,19]). The 3-D surface is flattened, the 2-D texture image is mapped onto the flattened surface and reverse mapping from 2-D to 3-D is applied.

In principle, the flattening process can be divided into two steps. First, the minimal geodesic distances between points on the surface are estimated. Then, a planar configuration of points that has Euclidean distances as close as possible to the corresponding minimal geodesic distances is determined. This paper combines a voxel-based geodesic distance estimator with an efficient dimensionality reduction algorithm to obtain a fast, practical method for surface flattening. The method presented is suitable for general surfaces in 3-D: the surface to be flattened does not have to be a function. A unique feature of our approach is that the algorithm operates directly on voxel data, hence an intermediate triangulated representation of the surface is not necessary.

2   Voxel-Based Geodesic Distance Estimation

Finding minimal geodesic distances between points on a continuous surface is a classical problem in differential geometry. However, in the context of digital 3-D data, purely continuous differential methods are impractical due to the need for interpolation, their vast computational cost and the risk of convergence to a locally minimal solution rather than to the global solution.

The common practice is to transform the voxel-based surface representation into a triangulated surface representation (see [18] for an efficient triangulation method) prior to distance calculation. In [20] the distances between a given source vertex on the surface and all other surface vertices are computed in $O(n^2 \log n)$ time, where n is the number of edges on the triangulated surface. In [22] an algorithm that is simple to implement is given, but the algorithm runs in exponential time. In [14] and [15] the fast marching method on triangulated domains is introduced. The method computes the minimal distances between a given source vertex on the surface and all other surface vertices in $O(n \log n)$ time, where n is the number of triangles that represent the surface. Texture mapping via surface flattening with the fast marching method on triangulated domains is presented in [23].

An algorithm for geodesic distance estimation on surfaces in 3-D, that uses the voxel representation without the need to first triangulate or interpolate the surface, was presented in [17]. The method is based on a high precision length estimator for continuous 3-D curves that have been digitized and are given as a 3-D chain code [16]. The shortest path between two points on the surface is associated with the path that has the shortest length estimate and that length estimate corresponds to the geodesic distance between the two points. Therefore an algorithm that calculates length estimates between points on a path can be combined with an algorithm to find shortest paths in graphs in order to find minimal distances on the surface. The computational complexity of the method is the same as that for finding shortest paths on *sparse* graphs, i.e. $O(n \log n)$, where n is the number of surface voxels.

Voxel-based, triangulation-free geodesic distance estimation [17] is a central component in the suggested surface flattening method. The digital surface is viewed as a sparse graph. The vertices and edges of the graph respectively correspond to voxels and to 3-D digital neighborhood relations between voxels. One of three weights (derived in [16]) is assigned to each edge, depending on the digital connection between the neighboring voxels: direct, minor diagonal or major diagonal. The sparsity of the resulting weighted surface graph allows very efficient search for shortest paths based on priority queues.

3 Flattening by Multidimensional Scaling

A *dissimilarity matrix* is a matrix that stores measurements of dissimilarity among pairs of objects. Multidimensional scaling (MDS) [3] is a common name for a collection of data analytic techniques for finding a configuration of points in low-dimensional Euclidean space that will represent the given dissimilarity information by the interpoint Euclidean distances.

Since usually no configuration of points can precisely preserve the information, an objective function is defined and the task is formulated as a minimization problem. So given a set of items $\{\bar{\mathbf{x}}_k\}$, $\bar{\mathbf{x}}_k \in \Re^r$ with dissimilarities $\delta(k, l)$ between items $\bar{\mathbf{x}}_k$ and $\bar{\mathbf{x}}_l$, the goal is to find p-dimensional data vectors $\{\hat{\bar{\mathbf{x}}}_k\}$, $\hat{\bar{\mathbf{x}}}_k \in \Re^p$, $p < r$ that have Euclidean distances $\{d(k, l)\}$ that approximate $\{\delta(k, l)\}$ well.

Many variants of MDS exist, differing in the objective function and optimization algorithms. These can be divided into two basic classes: metric and non-metric MDS. In metric MDS the original distance (or dissimilarity) matrix is approximated. Non-metric MDS deals with ordinal data or data in which only the order of the distances (dissimilarities) needs to be preserved.

MDS is used as the second step in our two-step surface flattening approach. Once the minimal geodesic distances between points on the surface are estimated, they are represented as a matrix that serves as a dissimilarity matrix. The flattened 2-D point configuration that best preserves these distances is then obtained via a direct and simple metric MDS method, known as *Classical Scaling*. It provides an analytic solution, requires no iterations and is fast to compute. Classical scaling minimizes an objective function known as *Strain*.

Let $\star$ denote the pointwise product of two matrices, i.e. $\mathbf{A} \star \mathbf{B} = (a_{ij}b_{ij})$ if $\mathbf{A} = (a_{ij})$ and $\mathbf{B} = (b_{ij})$. Given a dissimilarity matrix $\boldsymbol{\Delta}$ (a symmetric matrix with nonnegative elements and zeroes on the diagonal that, in our case, contains the estimated geodesic distances between the surface points), the *Strain* minimization problem is

$$\min_{\mathbf{X}} \|\mathbf{D}(\mathbf{X}) \star \mathbf{D}(\mathbf{X}) - \boldsymbol{\Delta} \star \boldsymbol{\Delta}\|_F^2$$

where $\mathbf{X}$ is the $n \times 2$ matrix of point coordinates in $\Re^2$, $\mathbf{D}(\mathbf{X})$ is a $n \times n$ matrix of Euclidean distances between the points in $\Re^2$ and $\|\cdot\|_F$ denotes the Frobenius norm, i.e. $\|\mathbf{A}\|_F^2 = \text{trace}(\mathbf{A}'\mathbf{A})$.

The *double centering operator* [5] is defined as:

$$\mathbf{T}(\mathbf{A}) = -\frac{1}{2}\mathbf{J}\mathbf{A}\mathbf{J} \ .$$

Here $\mathbf{A}$ is a square $n \times n$ matrix, $\mathbf{J} = \mathbf{I} - \frac{1}{n}\bar{\mathbf{e}}\bar{\mathbf{e}}'$, $\bar{\mathbf{e}} = (1, \ldots, 1)'$ is an n-vector of ones and $\mathbf{I}$ is the $n \times n$ identity matrix. The double centering operator is a linear operator on square matrices. Note that $\mathbf{T}(\mathbf{A})$ is symmetric if $\mathbf{A}$ is symmetric.

The classical scaling algorithm consists of the following steps:

- Apply double centering to $\boldsymbol{\Delta} \star \boldsymbol{\Delta}$: $\quad \mathbf{B} = \mathbf{T}(\boldsymbol{\Delta} \star \boldsymbol{\Delta})$

- Compute the two ($p = 2$) largest eigenvalues (λ_1, λ_2) and their corresponding eigenvectors $[\bar{\mathbf{v}}_1|\bar{\mathbf{v}}_2]$ of $\mathbf{B}$. Create $\boldsymbol{\Lambda} = \text{diag}(\lambda_1, \lambda_2)$ and $\mathbf{Q} = [\bar{\mathbf{v}}_1|\bar{\mathbf{v}}_2]$.

- The output coordinate matrix is given by $\mathbf{X} = \mathbf{Q}\boldsymbol{\Lambda}^{1/2}$.

In this method, *partial* eigendecomposition of an $n \times n$ symmetric matrix is required. For complete eigendecomposition, the symmetric QR algorithm is usually used, requiring $O(n^3)$ time. Since only the two largest eigenvalues and their corresponding eigenvectors are required, much faster algorithms can be applied, such as bisection, the power method and Rayleigh quotient iteration and orthogonalization with Ritz acceleration. See [3] (section 7.8) and [12] for details.

4   Interpolation

Surfaces in 3-D space are often represented by tens of thousands of voxels. Although the interpoint geodesic distances between the surface voxels can be computed quite efficiently, applying the MDS algorithm to such a large amount of data may be inconvenient. We therefore first select a sample of surface voxels (say 1000) and apply the flattening procedure to this subset. Note that the minimal distance between each pair of points in this sample is still calculated using the complete surface model. After flattening the sample, the flattened position of the remaining surface voxels is obtained by interpolation. Many different interpolation algorithms can be employed; radial function interpolation [9,10] was used in this research.

The quality of the interpolation depends on the curvature of the surface and on the sampling method. Sampling adapted to the curvature of the surface, or uniform sampling of the surface, will produce the best results, but producing such samples may not be easy. However, if the curvature of the surface is not too high, other sampling methods can still lead to good results. For example, when dealing with depth images $z(x, y)$ of smooth surfaces, uniform sampling in the (x, y) plane is usually adequate.

5 Experimental Results

Fig. 1 shows a synthetically created surface in 3-D space. The surface can be regarded as a rolled rectangular planar sheet of paper, with a uniform chessboard pattern. Note that, due to the rolling, this surface is not a function.

Since the surface is a rolled plane it can, ideally, be flattened without any distortion. Applying the algorithm suggested in this paper to the surface yielded the flattened result shown in Fig. 2. The slight distortion is due to the small errors in the estimation of geodesic distances that are used as input to the MDS procedure.

Fig. 3 shows Euclidean distances on the flattened surface as a function of the corresponding estimated geodesic distances on the rolled surface in 3-D space. Since the original surface can ideally be flattened without distortion, had error-free geodesic distance estimates been available, all points in the graph would have been on the diagonal line. However, due to the slight errors in geodesic distance estimation, distances on the flattened surface cannot be identical to the geodesic distance estimates, leading to the small spread of points near the line.

One application of the suggested surface flattening algorithm is texture mapping. Fig. 4 is a depth image of a human face, in which depth is represented as brightness. Following flattening, Fig. 5 shows Euclidean distances on the flattened surface as a function of the corresponding estimated geodesic distances on the surface of the 3-D model. The larger scattering in this case around the diagonal line is due to the fact that this surface cannot be flattened without some distortion. Yet, the distortion is quite small.

Fig. 6 (left) shows a brick wall texture. Overlaying this texture onto the flattened face surface and mapping the surface back to 3-D with its texture (using the known transformation) yields the texture-mapped face shown in Fig. 6 (right).

Acknowledgments. We thank Haim Shvaytser, Gabriele Lohmann and Nira Dyn for the interesting discussions. The friendly and helpful advice of Gil Zigelman and Sharon Ganot is appreciated. We are grateful to the National Research Council of Canada for the 3-D model visualized in Fig. 4. This research was supported in part by the Adams Super-Center for Brain Studies at Tel Aviv University and by the German-Israeli Foundation for Scientific Research and Development (GIF).

Fig. 1. A synthetic surface in 3-D space obtained by rolling a planar rectangular sheet. Note that this surface is not a function.

Fig. 2. The flattened surface obtained using the method presented in this paper.

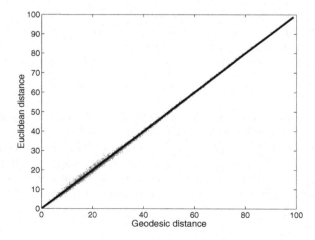

Fig. 3. Euclidean distances on the flattened surface vs. the corresponding estimated geodesic distances on the 3-D surface.

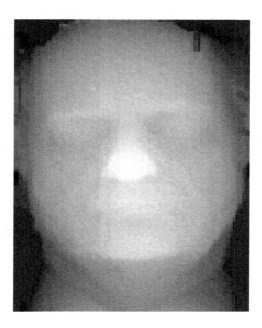

Fig. 4. A depth image of a human face. Underlying 3-D model courtesy of NRC–National Research Council of Canada, Institute for Information Technology, Ottawa, Canada K1AOR6, 1997.

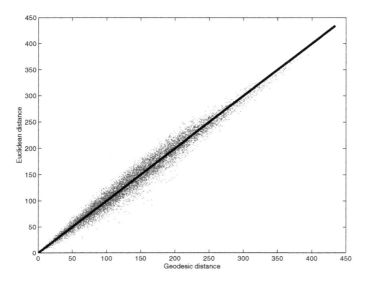

Fig. 5. Euclidean distance on flattened surface vs. estimated geodesic distance on 3D surface of the face.

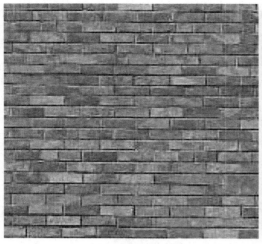 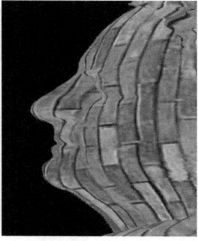

Fig. 6. *Left:* A brick-wall texture. *Right:* The brick texture mapped onto the face.

References

1. S. Angenent, S. Haker, A. Tanenbuam and R. Kikinis, "Conformal Geometry and Brain Flattening", *Proc. 2nd Intern. Conf. on Medical Image Computing and Computer-Assisted Intervention (MICCAI'99)*, Cambridge, England, pp. 269-278, 1999.
2. C. Bennis, J.M. Vezien and G. Iglesias, "Piecewise Surface Flattening for Non-Distorted Texture Mapping", *Computer Graphics*, Vol. 25, pp. 237-247, 1991.
3. I. Borg and P. Groenen, *Modern Multidimensional Scaling: Theory and Applications*, Springer, 1997.
4. M.P. Do Carmo, *Differentail Geometry of Curves and Surfaces*, Prentice-Hall, 1976.
5. F. Critchley, "On Certain Linear Mappings Between Inner-Product and Squared-Distance Matrices", *Linear Algebra and its Applications*, Vol. 105, pp. 91-107, 1988.
6. A.M. Dale, B. Fischl and M.I. Sereno, "Cortical Surface-Based Analysis: 1. Segmentation and Surface Reconstruction", *NeuroImage*, Vol. 9, pp. 179-194, 1999.
7. A.M. Dale, B. Fischl and M.I. Sereno, "Cortical Surface-Based Analysis: 2. Cortical Surface Based Analysis", *NeuroImage*, Vol. 9, pp. 195-207, 1999.
8. H.A. Drury, D.C. Van Essen, C.A. Anderson, C.W. Lee, T.A. Coogan and J.W. Lewis, "Computerized Mappings of the Cerebral Cortex: A Multiresolution Flattening Method and a Surface-Based Coordinate System", *J. of Cognitive Neuroscience*, Vol. 8, pp. 1-28, 1996.
9. N. Dyn, "Interpolation of Scattered Data by Radial Functions", in *Topics in Multivariate Approximations* (C.K. Chui, L. L. Schumaker, and F.I. Utreras, eds.), pp. 47-62, Academic Press, 1987.
10. N. Dyn, "Interpolation and Approximation by Radial and Related Functions", *Approximation Theory VI*, Vol. 1, pp. 211-234, 1989.
11. J. Gomes and O. Faugeras, "Segmentation of the Inner and Outer Surfaces of the Cortex in Man and Monkey: an Approach Based on Partial Differential Equations", *Proc. Human Brain Mapping Conf.*, Dusseldorf, Germany, June, 1999.

12. G.H. Golub and C.F. Van Loan, *Matrix Computations*, The Johns Hopkins University Press, Baltimore, 1989.
13. S. Haker, S. Angenent, A. Tannenbaum, R. Kikinis, G. Sapiro and M. Halle, "Conformal Surface Parameterization for Texture Mapping", *IEEE Trans. on Visualization and Computer Graphics*, Vol. 6, No. 2, 2000.
14. R. Kimmel and J.A. Sethian, "Computing Geodesic Paths on Manifolds", *Proceedings of the National Academy of Sciences*, Vol. 95, pp. 8431-8435, 1998.
15. R. Kimmel and J.A. Sethian, "Fast Voronoi Diagrams and Offsets on Triangulated Surfaces", in *Curve and Surface Design: Saint-Malo 1999*, Vanderbilt University Press, 1999.
16. N. Kiryati and O. Kübler, "Chain Code Probailities and Optimal Length Estimators for Digitized Three-Dimensional Curves", *Pattern Recognition*, Vol. 28, pp. 361-372, 1995.
17. N. Kiryati and G. Székely, "Estimating Shortest Paths and Minimal Distances on Digitized Three-Dimensional Surfaces", *Pattern Recognition*, Vol. 26, pp. 1623-1637, 1993.
18. W.E. Lorensen and H.E. Cline, "Marching Cubes: A High Resolution 3D Surface Construction Algortihm", *SIGGRAPH'87, ACM Computer Graphics*, Vol. 21, pp. 163-169, 1987.
19. S.D. Ma and H. Lin, "Optimal Texture Mapping", *Eurographics*, pp. 421-428, 1988.
20. J.S.B. Mitchell, D.M. Mount and C.H. Papadimitriou, "The Discrete Geodesic Problem", *SIAM J. Comput.*, Vol. 16, pp. 647-668, 1987.
21. E.L. Schwartz, A. Shaw and E. Wolfson, "A Numerical Solution to the Generalized Mapmaker's Problem: Flattening Nonconvex Polyhedral Surfaces", *IEEE Trans. Pattern Anal. Mach. Intell.*, Vol. 11, pp. 1005-1008, 1989.
22. E. Wolfson and E.L. Schwartz, "Computing Minimal Distances on Polyhedral Surfaces", *IEEE Trans. Pattern Anal. Mach. Intell.*, Vol. 11, pp. 1001-1005, 1989.
23. G. Zigelman, R. Kimmel and N. Kiryati, "Texture Mapping using Surface Flattening via Multi-Dimensional Scaling", submitted.

An Adaptive Image Interpolation Using the Quadratic Spline Interpolator

Hyo-Ju Kim and Chang-Sung Jeong

Department of Electronics Engineering, Korea University
1-5ka, Anam-dong, Sungbuk-ku, Seoul 136-701, Korea
E-mail: hyojukim@snoopy.korea.ac.kr

Abstract. In this paper we propose a novel image interpolation algorithm based on the quadratic B-spline basis function. Our interpolation algorithm preserves the original edges while not destroying the smoothness in flat area using the adaptive interpolation method according to the directional edge pattern of input image, significantly improving the overall performance of the interpolation. Our experimental result shows that it can produce higher quality and resolution than the currently existing image interpolation methods.

1 Introduction

intermediate value of continuous function from the given discrete samples. It is a conversion technique from some sampling rate to the other one [1]. Image interpolation can be used in correlation of spatial image distortion and image zooming system [2]. Image interpolation is often divided into two sub-processes: signal reconstruction and sampling. The former creates a continuous function from the discrete image data, and the latter samples this to create a new, re-sampled image. If sampling frequency of input signal satisfies Nyquist sampling condition and frequency limitation, input signal perfectly reconstructs interpolated signal using an ideal interpolation function, sinc function. However, since sinc function is not time limited, it can not be implemented in hardware. Therefore, we need to design a proper interpolator which can be implemented in hardware, but close to sinc function, that is, ideal low pass filter. The simple image interpolation methods comprise zero-order interpolation (or nearest neighborhood interpolation) and first order interpolation (bilinear interpolation) method. Both methods can be cost-effectively realized in consumer electronics devices. These days they are not used, since they may degrade the quality of the interpolated image by blocky artifacts and excessive smoothness and more advanced interpolation schemes can be implemented due to the rapid advancement in VLSI. Hou and Andrews [5] proposed a more refined interpolation method *Cubic B-spline interpolation*. However, it is hardly used due to its large computational complexity. Keys proposed another interpolation technique *cubic convolution* which reduces computational complexity of Cubic B-spline interpolation [6]. Unser presented a theoretical analysis of B-spline signal representation from a signal processing point of view, which is called *B-spline transform* [3].

C. Arcelli et al. (Eds.): IWVF4, LNCS 2059, pp. 205–215, 2001.
© Springer-Verlag Berlin Heidelberg 2001

In this paper we propose a novel image interpolation algorithm based on the quadratic B-spline basis function. Our interpolation algorithm preserves the original edges while not destroying the smoothness in flat area using the adaptive interpolation method according to the directional edge pattern of input image, significantly improving the overall performance of the interpolation. Our experimental result shows that it can produce higher quality and resolution than the currently existing image interpolation methods.

In section 2, the ideal interpolator and the conventional B-spline function are reviewed. In section 3, our interpolation algorithm is presented. In section 4, our experimental results are given in order to show that our method is superior to the other ones including Unser's *cardinal cubic spline interpolation* method. Finally, in section 5, a conclusion is given.

2   B-Spline Interpolation Methods

2.1   Ideal Interpolation

Ideal interpolator is theoretical concept for explaining interpolator. Some fundamental properties of any interpolator can be derived from this ideal interpolation function. The ideal interpolator exchanges from positive value to negative value at the unit knot points known as zero-crossing points. If interpolators satisfy the following condition, they avoid smoothing and preserve high frequency component called edge.

$$h(0) = 1, \ h(x) = 0, \ |x| = 1, 2, \dots \tag{1}$$

Ideal interpolator is spatially unlimited. There are two common approaches for spatially limited interpolators: truncation and windowing methods. Truncation of ideal interpolator produces ringing effects referred to as the Gibbs's Phenomenon in the frequency domain because a mount of energy is discarded. Truncation technique is not sufficient in the image interpolation. Another approach, windowing is equivalent to the multiplication of ideal interpolator with less severe window than the rectangular function. With respect to the flat frequency response of pass-band, windowing is preferable to truncation method.

2.2   B-Spline Interpolations

commonly used family of spline function. B-splines of order n are piecewise polynomial functions of degree n. These functions are differentiable $n - 1$ times, i.e. they have derivations up to order $n - 1$. Any continuous $n - th$ degree polynomial piecewise function which is also differentiable $n - 1$ times can be represented using B-spline functions of the same order. In the case of uniform spacing between knot points, such a function can be represented in the form [3]:

$$\phi^n(x) = \sum_{k=-\infty}^{+\infty} c_n(k) \bullet \beta^n(x - k) \tag{2}$$

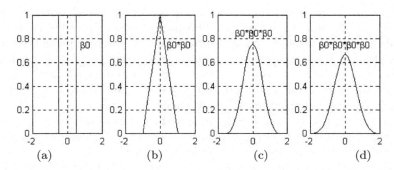

Fig. 1. Convolution property of B-spline weight functions: (a) Zero order (b) first order (c) quadratic (d) cubic

where $\beta^n(x)$ is $n-th$ order B-spline weight function; n is a degree of the piecewise polynomials that are connected at the knot points. The function $\phi^n(x)$ is uniquely determined by its B-spline coefficient $c_n(k)$.

The property of B-spline can be derived by several self-convolutions of a so-called basis function. If we define $\beta^0(x)$ with

$$\beta^0(x) = \begin{cases} 1, & -\frac{1}{2} \leq \text{x} \leq \frac{1}{2} \\ 0, & \text{otherwise} \end{cases} \tag{3}$$

then B-spline weight functions of any order satisfy the convolution property:

$$\beta^n(x) = \beta^{n-1}(x) * \beta^0(x) \tag{4}$$

as illustrated by figure 1(a)~(d).

Actually the first order B-spline can be considered as the result of convolving the rectangular zero order B-spline. For $n = 2$, we obtain the quadratic B-spline function defined by the following equation:

$$\beta^2(x) = \begin{cases} \frac{3}{4} - |x|^2, & |x| \leq \frac{1}{2} \\ \frac{1}{2}(|x| - \frac{3}{2})^2, & \frac{1}{2} \leq |x| \leq \frac{3}{2} \\ 0, & \text{otherwise} \end{cases} \tag{5}$$

Quadratic B-spline functions have been disregarded because they are known to be space variant and to introduce phase distortion in the output signal. But Dodgson showed this is not a general case, and recently derived a family of quadratic interpolator that is better behaved [7]. Also, Toraichi proposed another family of quadratic interpolator that has linear phase [4]. For $n = 3$, we obtain the cubic B-spline function:

$$\beta^3(x) = \begin{cases} \frac{1}{2}|x|^3 - |x|^2 + \frac{2}{3}, & |x| \leq 1 \\ \frac{1}{6}(2 - |x|)^3, & 1 \leq |x| \leq 2 \\ 0, & \text{otherwise} \end{cases} \tag{6}$$

Cubic B-spline function is hardly used as interpolation function, since it fails to satisfy the interpolation condition (1).

Table 1. Transfer function of B-spline weight function of order 0 to 3

n	0	1	2	3
$\beta^n(w)$	$sinc(w/2\pi)$	$sinc^2(w/2\pi)$	$\frac{4sinc^3(w/2\pi)}{3+cos(w)}$	$\frac{3sinc^4(w/2\pi)}{2+cos(w)}$

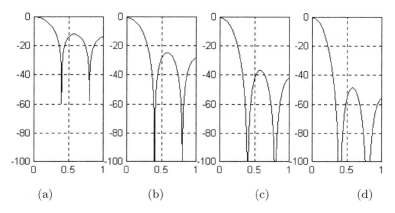

(a)   (b)   (c)   (d)

Fig. 2. Log-magnitude plot of B-spline weight functions: (a) Zero order (b) first order (c) quadratic (d) cubic

Table 1 shows the transfer function of B-spline weight functions of order 0 to 3. Figure 2 shows Log-magnitude response plot of B-spline weight functions. The transfer functions decrease too early, indicating that B-spline interpolation performs too much averaging. Increasing an order of spline not only improves the quality of interpolation but also increases the smoothing effects.

3   Adaptive Image Interpolation Algorithm

steps: Edge estimation and interpolation. In the edge estimation step, five types of edge are classified by investigating the direction of the edge. In the interpolation step, quadratic spline interpolation is performed by manipulating interpolation coefficients adaptively according to the types of edges.

3.1   Edge Estimation

We shall describe how to classify edges according to their directions. Using the four neighboring pixels around the center in the interpolation area as shown in figure 3, d1~d4 are obtained by calculating the absolute difference between them as follows:

$$d1 = |x1 - x2|, \; d2 = |x1 - x3|, \; d3 = |x3 - x4|, \; d4 = |x2 - x4| \qquad (7)$$

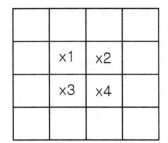

Fig. 3. Adjacent 4 pixels for determining the edge

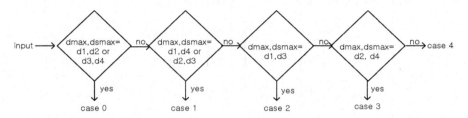

Fig. 4. Flow diagram of determined directional edges

Let d_{max} and d_{smax} be the first and second largest values among them respectively. Then, five types of edge patterns are determined according to d_{max} and d_{smax} by the edge detection algorithm shown in figure 4. We can easily figure out the direction of edge for each type as illustrated in figure 5. In the next subsection, we shall show how to exploit the edge directional pattern in the quadratic spline interpolation.

3.2 Adaptive Quadratic Spline Interpolation Algorithm

In this subsection we shall show that a quadratic spline interpolator is derived from a quadratic B-spline weight function by calculating the weighted sum of quadratic B-spline weight functions, and it is a smooth piecewise polynomial of degree two with linear phase characteristics in the frequency domain. In interpolation step, a continuous signal is reconstructed from its discrete samples as follows:

$$f(x) = \sum_{k=-\infty}^{+\infty} c(k) \bullet q(x - k) \tag{8}$$

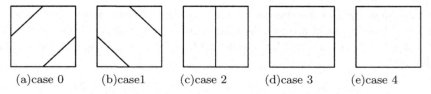

(a)case 0   (b)case1   (c)case 2   (d)case 3   (e)case 4

Fig. 5. 5 types of edge for proposed adaptive interpolation

where $c(k)$ is a discrete-signal sample, and $q(x)$ is a continuous impulse response of quadratic spline interpolator. Let us consider a discrete signal $f(k)$ on $k = -\infty \sim +\infty$ defined as

$$f(k) = \sum_{m=-\infty}^{+\infty} c(m) \bullet q(k-m) \tag{9}$$

If we assume that $q(x)$ is a quadratic B-spline weight function $\beta^2(x)$ given by equation (5), $\beta^2(-1) = 1/8$, $\beta^2(0) = 6/8$, and $\beta^2(+1) = 1/8$, and we obtain

$$f(k) = \frac{1}{8}\{c(k-1) + 6 * c(k) + c(k+1)\} \tag{10}$$

which can be described in convolution form

$$f(k) = c(k) * q(k) \tag{11}$$

From equations (10) and (11), we can derive the following equation

$$f(k) = c(k) * \frac{1}{8}\{\delta(k-1) + 6 * \delta(k) + \delta(k+1)\} \tag{12}$$

In the frequency domain, Fourier transform of $f(k)$ is given by

$$F(w) = C(w) \bullet \frac{1}{4}\{3 + \cos(w)\} \tag{13}$$

and hence Fourier transform of $c(k)$ yields

$$C(w) = F(w) \bullet \frac{4}{\{3 + \cos(w)\}} \tag{14}$$

Then, the inverse Fourier transform of $C(w)$ is obtained by

$$c(x) = f(x) * \sqrt{2} \sum_{x=-\infty}^{+\infty} (2\sqrt{2} - 3)^{|x|} = \sqrt{2} \sum_{k=-\infty}^{+\infty} (2\sqrt{2} - 3)^{|x-k|} \cdot f(k) \tag{15}$$

Therefore, the continous function $f(x)$ can be written by the weighted sum of the initial sample values $f(i)$ as follows:

$$f(x) = \sum_{k=-\infty}^{+\infty} c(k) \bullet q(x-k) = \sum_{i=-\infty}^{+\infty} f(i) \bullet h_{quad}(x-i) \tag{16}$$

where the quadratic spline interpolator $h_{quad}(x)$ corresponding to the interpolation coefficient is determined by

$$h_{quad}(x) = \sqrt{2} \sum_{k=-\infty}^{+\infty} (2\sqrt{2} - 3)^{|k|} \beta^2(x-k) \tag{17}$$

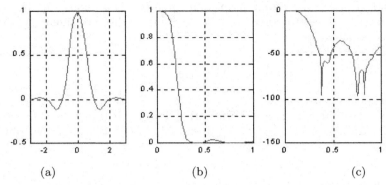

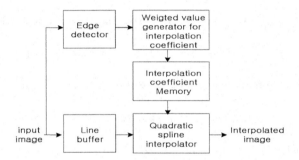

Fig. 6. Quadratic spline interpolator and its magnitude response: (a) interpolator, (b) its linear magnitude response, (c) its Log-magnitude response

Fig. 7. Block diagram of proposed algorithm

Figure 6 illustrates quadratic spline interpolator, its magnitude frequency and log magnitude responses. Comparison with B-spline weight functions in figure 2 shows that our quadratic spline interpolator has the excellent pass-band characteristic while satisfying the zero-crossing condition. Figure 7 shows the block diagram of our algorithm which adaptively manipulate the interpolation coefficients according to the edge directional patterns. Our method consists of following steps: First, edge detection is performed by using the proposed edge detection algorithm, which classifies five types of edge. At the same time, original image is stored into line buffer for calculating quadratic spline interpolator. In order to provide adaptive interpolation, weighted value generator produces parameters which controls interpolation coefficient according to the five types of edge. If edge is not detected in the support area of interpolation, interpolation is performed using the original coefficients; otherwise using the new coefficients which is converted from the original one by using the control parameters.

4   Experimental Results

The performance of the proposed interpolation is tested in PSNR and subjective visual quality. Two types of images have been tested using the conventional interpolation methods and the proposed interpolation algorithm: natural image of 256x256x8 bit Lena image and graphical text image of 256x256x8 bit text image with more high contrast.

4.1   PSNR Measure

In order to have an objective evaluation of the proposed scheme, the test images are 4 pixels (2x2 mask) average low-pass filtered, and decimated to the half-row and half-column size. Then, several interpolation methods with a scale factor 2 are applied to produce the original image including B-splines, cubic convolution, quadratic spline interpolation, cubic spline interpolation, and our proposed interpolation method. Table.2 illustrates PSNR for each of the interpolation methods, and show that our proposed interpolation method is superior to the other interpolation methods.

Table 2. PSNR of the each interpolation methods(dB)

Intp. Method	Zero order B-spline	First order B-spline	Quad. B-spline	cubic B-spline	Cubic conv.	Quad. intp.	Cubic intp.	Proposed intp.
Lena image	52.23	43.40	59.29	58.9	59.46	62.99	63.01	63.03
Text image	28.83	26.08	32.27	32.09	32.33	33.54	33.58	33.63

4.2   Image Interpolation Results

Figure 8 shows the images obtained by applying several interpolation techniques for the magnification of a Lena image with a zooming factor 8. The first two images are the ones obtained by simple zero order B-spline(figure 8 (a)) and first order B-spline interpolation(figure 8 (b)) respectively. They show mosaic blocks (figure 8 (a)) that make the image visually degraded and image blurring (figure 8 (b)). Figure 8 (c) and figure 8 (d) show the images obtained by high order B-spline, that is, second order and third order B-spline respectively. They show severely blurred image. Therefore, we can see that increasing the order of the spline not only improves the quality of interpolated image but also increases the smoothing effect. Figure 8(e) and (f) show the images by Unser's and our algorithms respectively. In both of them, the blocky effect is significantly reduced, while maintaining the fidelity of the original image compared with the other interpolation methods.

Figure 9 shows scale up images with scale factor 2 for text image by using the various interpolation methods: first order B-spline, second order B-spline,

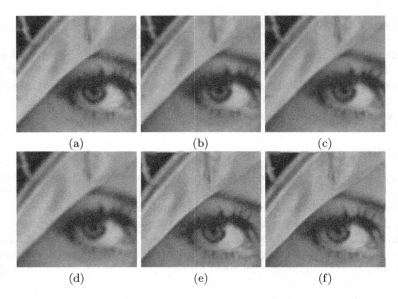

Fig. 8. Magnification images of several interpolation with zoom factor 8: (a) Zero order, (b) First order (c) Second order, (d) Third order (e) Cardinal cubic spline (f) Proposed interpolation

cardinal cubic spline interpolation and proposed interpolation algorithm. Figure 9 (a) is the original text image that has 256 gray level and size of 256 x 256 pixels. Figure 9(b) is low-pass filtered, and decimated to the half-row and half-column size of original text image. Then, several interpolation methods with a scale factor 2 are applied to produce the original image. In order to calculate the PSNR, decimation and interpolation processes are performed. Images obtained by applying first order B-spline, second order B-spline, cardinal cubic spline interpolation, and our interpolation method to the image in figure 9(b) are shown in figure 9 (c) through (f). The objective comparison of our method with the other ones are shown in Table.2 using PSNR. It can be observed that the better result is achieved by our method than the other conventional methods. Compared to Unser's method, our method shows better performance for text images rather than natural image.

5   Conclusion

In this paper we have presented a new adaptive image interpolation algorithm based on the quadratic B-spline basis function. Our interpolation algorithm consists of 2 major steps: Edge estimation and interpolation. In the edge estimation step, five types of edge are classified by investigating the direction of the edge. In the interpolation step, interpolation is performed by manipulating interpolation

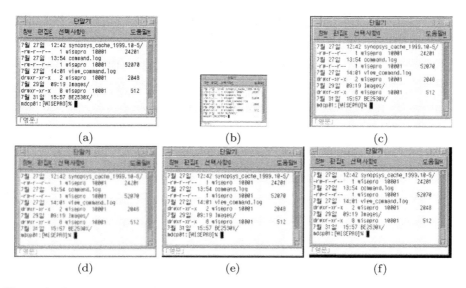

Fig. 9. Scale up images with scale factor 2 using the various interpolation methods: (a) Original text image (b) Decimated image with scale down factor 2 from original image (c) First order B-spline (d) Second order B-spline (e) Cardinal cubic spline (f) Proposed interpolation

coefficients corresponding to quadratic spline interpolator adaptively according to the types of edges obtained in the edge estimation step. We have shown that a quadratic spline interpolator is derived from a quadratic B-spline weight function by calculating the weighted sum of quadratic B-spline weight functions, and its adaptation according to the edge patterns enables us to preserve the original edges while not destroying the smoothness in flat area, greatly enhancing the overall performance of the interpolation. In our experiments our method is compared with the previous ones for graphic and text images, and we have shown that it can produce higher quality and resolution than the currently existing image interpolation methods. Moreover, our method, which is close to ideal low pass filter characteristics, is being implemented in hardware now, since it can be done with less hardware than the Unser's cardinal cubic spline interpolator.

References

1. J. S. Lim, Two-Dimensional Signal and Image Processing, Prentice-hall, pp.495-497,1990
2. J. K. Paik and S. W. Park, "A Real Time Image Interpolation For Digital Zooming." Proc. 1st Korea-Japan Joint Conf. Computer Vision, pp. -322, Seoul, Korea, October 1991.

3. M. Unser, A. Aldroubi and M. Eden,"Fast B-spline Transforms for Continuous Image Representation and Interpolation",IEEE Trans. Patt. Anal. Machine Intell., vol.13, no.3, pp.277-285, March 1991.
4. K. Toraichi, "A Quadratic Spline Function Generator", IEEE Trans. On ASSP, vol.37, no.4, April 1989
5. H.S. Hou and H.C.Andrews, "Cubic splines for image interpolation and digital filtering." IEEE Trans. Acoust., Speech, Signal Processing, vol. ASSP-26, no. 6, pp.508-517, 1978.
6. R. G. Keys, "Cubic convolution interpolation for digital image processing", IEEE Trans. Acoust., Speech, Signal Processing, vol. ASSP-29, no. 6, pp.1153-1160, 1981.
7. N. A. Dodgson, "Quadratic interpolation for image resampling", IEEE Trans. Image Processing, vol.6, pp.1322-1326, 1997.

The Shock Scaffold for Representing 3D Shape

Frederic F. Leymarie and Benjamin B. Kimia

Brown University, Division of Engineering, Providence RI, USA,
{leymarie,kimia}@lems.brown.edu

Abstract. The usefulness of the 3D Medial Axis (*MA*) is dependent on both the availability of accurate and stable methods for computing individual MA points and on schemes for deriving the local structure and connectivity among these points. We propose a framework which achieves both by combining the advantages of *exact bisector* computations used in computational geometry, on the one hand, and the local nature of propagation-based algorithms, on the other, but without the computational complexity, connectivity, added dimensionality, and post processing issues commonly found in these approaches. Specifically, the notion of *flow of shocks* along the MA manifold is used to identify flow along special points and curves which define a *shock scaffold*. This 1D scaffold is of lower dimensional complexity than the typical geometric locus of medial points which are represented as 2D sheets. The scaffold not only organizes shape information in a hierarchical manner, but is a tool for the efficient recovery of the scaffold itself and can lead to exact reconstruction. We present examples of this approach for synthetic data, as well as for sherd data from the domain of digital archaeology.
Keywords: 3D Medial Axis, 3D Skeletons, 3D Symmetry Sets, shock hypergraph, shape representation.

1   Introduction

The Medial Axis (*MA*) or skeleton representation [3] has shown great potential in object recognition, in solid modeling for designing and manipulating shapes, in organizing a cloud of points into surfaces, for mesh generation, path planning, numerical tool machining, animation, *etc.* However, for the MA to be useful in these applications, it must often first be organized in a *graph* structure which embeds not only the qualitative aspects of shapes, *e.g.*, parts, in a hierarchy of scales, but also the more detailed quantitative features. Traditionally, algorithms for 3D skeleton computation have typically focused on deriving the geometric locus of skeletal surfaces, thus leaving unclear the local connectivity in the interior of each MA sheet as well as in the joints, where three or more sheets come together. Also, while the MA as a transformation from object to symmetry coordinates is useful in itself, it does not address the issue of data reduction, since it is not a priori clear how to summarize the 2D MA sheets into a lower dimensional structure. While the interesting notion of curve skeletons has been proposed earlier [4,29], this is not for free form shapes of arbitrary complexity. A key goal of this paper is to address both problems by proposing the notion of a *shock scaffold*, upon which the remaining parts of the MA can be constructed

C. Arcelli et al. (Eds.): IWVF4, LNCS 2059, pp. 216–228, 2001.

in a robust manner, and by developing an efficient computational scheme for obtaining this scaffold.

Techniques developed to extract MA symmetries in 3D, can be roughly organized into six main classes: *(i)* Thinning [18], *(ii)* Boundary modeling [30], *(iii)* Voronoi diagram [23,2,19,1], *(iv)* Distance Transform [9,21,31,4,5], *(v)* Surface evolution [22,24,10], *(vi)* Bisectors computations and trimming in Computational Geometry [6,20]. We cannot review these here due to space limitations, but in some sense the ideal algorithm for the recovery of the 3D MA should combine the advantages of these approaches. Specifically, we seek a method which on the one hand features the exactness of bisector computations and Voronoi diagrams, but when stripped of their tremendous computational burden, and on the other hand features the flow-based nature of Blum's grassfire [3,14], which underly thinning, distance transforms, and surface evolutions, but without their connectivity, added dimension and post-processing issues.

A key insight which unifies these approaches in this work and the earlier 2D version [28,25,27,26] is that the full bisector need not be considered if a flow-based approach is adopted. Specifically, if the initial sources of flow are completely classified, one may only compute bisectors which initiate from these and ignore all others completely, leading to substantial savings. This also immediately leads to a graph structure which captures local connectivity and exact results.

A second key feature which is specific to 3D is the need for *dimension reduction* in computing 3D MA, which is accomplished by employing the notion of a shock scaffold. The complete classification of the 3D MA points and the 3D *shock* points, *i.e.*, MA points augmented with a sense of flow, was reported in [8]. Specifically, the MA points are formally classified into five types: one type corresponding to the interior of MA sheets, two types for MA curves, at the boundary of skeletal sheets, and two types for MA nodes, at the intersection of these boundaries. This classification, together with the notion of flow along the curves, is the basis of constructing the *shock scaffold*, a *reduced-dimension summary of the MA*. The proposed algorithm identifies points of propagation from initial *shocks sources*, propagates along the shock scaffold, computes intersections among the junctions of this scaffold structure, until this propagation computation terminates at *shock sinks*.

The resulting shock scaffold is a graph structure consisting of nodes (isolated points) and links (curve segments). Together with hyperlinks (surface patches), the shock scaffold gives rise to the *shock hypergraph* which is a complete representation of shape. The scaffold essentially allows us to ignore or approximate the medial surface patch geometry, or even the medial curve geometry, while retaining the connectivity among nodes and links which proportionally contain the most significant aspects of the MA. The algorithm is generic for any initial shape geometry, whether described by a cloud of points as described in this paper, or as a collection of surface patches as will be described in future work. The advantages of this framework are the reduced dimensionality, the exactness of the results, the efficiency of the algorithm, its applicability to unsegmented and unorganized data, and the immediate availability of a graph structure which can be used in recognition and other applications. We illustrate the results for a set of synthetic examples as well as for sherds to be used for grouping fragments into reconstructed pots and other complex shapes found, in particular, for a project involving the archaeological site of Petra, Jordan [13].

2   The Shock Scaffold

The classification of shock points described in [8] is based on the notion of contact with spheres, *i.e.*, the loci of spheres osculating sources. Let A_k^n denote a circle (in 2D) or a sphere (in 3D) osculating a boundary element at n distinct points, each with $k + 1$ degree of contact, Figure 1: $k = 1$ denotes regular tangency; $k = 2$ denotes a sphere of curvature for a surface patch; $k = 3$ denotes a sphere of curvature at a *ridge* point; $k = 4$ denotes a sphere of curvature at a turning point of a ridge, etc. [11].

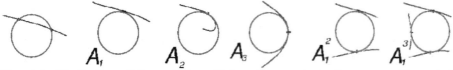

Figure 1. Illustration of the notation A_k^n based on contact of a curve with a circle (from [8]). $k + 1$ counts order or degree of contact: A_1 is regular tangent contact, A_2 is regular "curvature" contact, A_3 is a curvature maximum contact. The superscript n counts the number of contact points, so that A_1^2 means two A_1 contacts. A similar definition holds for the contact of surfaces with spheres.

Only odd orders of contact (*i.e.*, $k = 1, 3$) can contribute to a MA type of shock, that is, as being the center of a *maximal* sphere, S. A classification based on the number and order of contact [8] leads to *five* principal types of shock points: A_1^2, A_1^3, A_3, A_1^4 and $A_1 A_3$: *(i)* A_1^2 *contact*: a sphere with ordinary A_1 contact at two source points generates a shock *sheet* point. The centers of such spheres sweep out a piece of the Symmetry Set (SS) which is locally smooth. *(ii)* A_3 *contact*: this is the limiting case of two A_1^2 points, which corresponds in 2D to curvature extrema and in 3D to *ridges* on the boundary. *(iii)* A_1^3 *contact*: the sphere, S, has triple tangency on the bounding surface elements, M. Choose any 2 of these 3 tangency points and move the sphere so that it remains bitangent to M at points close to these two. This results in a smooth sheet of the SS or MA for each pair, leading to a total of three such smooth sheets passing through the center of S. *(iv)* $A_1 A_3$ *contact*: it contains the centers of spheres which have contact with the surface in two places, one near the original A_1 point (*i.e.*, ordinary tangency) and one near the ridge point A_3. *(v)* A_1^4 *contact*: the sphere is tangent to 4 source points - this is generic. At the center of the sphere passes 6 smooth sheets of the SS (*i.e.*, 6 pairs from 4 source points), two of which are not manifested in the MA, leading to four intersection of MA sheets. An alternative view of this event, is as the combination/intersection of four axial A_1^3 curves.

Two observations are significant here. First, the topology of each of these types is as follows: A_1^2 points are interior points of a medial surface; A_3 points organize into curves representing *ridges* on surfaces and are the "exterior" boundary of medial surface sheets; A_1^3 points organize into curves which are the intersection of three A_1^2 medial surfaces sheets; these curves often correspond to "generalized axis" as well as to "interior" boundary of MA sheets; A_1^4 and A_1A_3 are isolated points where four A_1^3 or a pair of A_1^3 and A_3 curves intersect, respectively.

(a)

(b)

Figure 2. Flows at A_1^3 and A_3 shock curves (a) and at A_1^4 shock points (b).

Shock flow	1	2	3	4
Sheet	A_1^2-1	A_1^2-2	A_1^2-3	A_1^2-4
Ridge	A_3-1	A_3-2	A_3-3	A_3-4
Axis	A_1^3-1	A_1^3-2	A_1^3-3	A_1^3-4
Ridge end	-	A_1A_3-2	A_1A_3-3	A_1A_3-4
Axis end	-	A_1^4-2	A_1^4-3	A_1^4-4

Table 1. Final classification of 18 possible shock points based on contact with spheres, A_k^n, and flow type (see text).

Second, one can construct a notion of flow for each MA point in the direction of increasing radius, which leads to a further subclassification of points as was done in [12] in 2D. Shocks, *i.e.*, MA points endowed with a sense of flow, can flow along sheets (A_1^2) or curves (A_3 and A_1^3) in various ways: they can flow monotonically (1st order), can act as a source and initiate flow (2nd order), or can act as a sink and terminate flow (4th order), Figure 2a. Third-order shocks represent infinitely "fast" flows, which are not generic, but must be considered, especially for man-made objects. For nodes (A_1A_3 and A_1^4), the classification is based on the number of incoming branches, Figure 2b. Table 1 summarizes the notation. This classification of the medial axis and shock points leads to an intriguing data structure for representing them.

Definition 1. *Let the A_1^4 and A_1A_3 points of a 3D MA denote* **nodes,** *and the A_1^3 and A_3 curves which connect these nodes denote* **links,** *which together with the radius function attribute form the* **shock scaffold** *graph. In addition, let the A_1^2 surface patches, whose boundary is described by an ordered, closed sequence of nodes and links, act as* **hyperlinks** *on the graph to form the* **shock hypergraph.**

We illustrate the scaffold concept in Figure 3. Clearly, the scaffold representation arises from a recognition of special types of points, *i.e.*, an understanding of the local topology of each type of points as isolated point, curve, or sheet, and the connectivity among the five types. When this graph structure is ignored, a trace of MA points remains which is the classical view of the MA found in the literature. The advantage of the graph structure is that it organizes the MA information into groups and specifies their connectivity. It is precisely the connectivity among groups which contains the qualitative information, while the remaining information allows for an exact reconstruction or approximation of the shape from the shock hypergraph [7]. From the shock scaffold alone, we are still able to get a fairly good idea of the shape of the object due to the remaining connectivity, in the same way that a generalized axis (curve) represents a

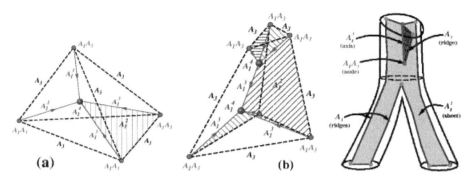

Figure 3. The shock scaffold is illustrated for a few simple shapes. The dark broken lines are surface ridges (A_3), the smaller dots are surface vertices (A_1A_3), the larger nodes are A_1^4 shocks, the interior links have arrows to indicate flow (all A_1^3's here), the hashed sheets are hyperlinks (A_1^2; not all shown). The shock scaffold of a tetrahedron consists of 5 nodes, 10 links and 6 hyperlinks; for a truncated tetrahedron we have 8 nodes, 7 links and 9 hyperlinks. (c) Sketch of the shock scaffold for a branching structure which at the top is a cylinder whose base grows from a triangle to an ellipse, and which splits into two cylindrical structures with elliptic bases.

cylinder well. The MA can be approximated by interpolating the missing MA sheets, stretching smooth elastic surfaces over the bounding curves, much as is done when a "tent" is constructed. Similar arguments hold for the geometry of the curves such that at the very coarsest level only nodes need be retained.

3   3D MA recovery by flow along scaffold curves

A key idea of this paper is that the shock scaffold is not merely a post-processing tool for organizing traced MA geometry. Rather, it is an essential element in the recovery process itself. The argument is based on substantial savings in 2D if a flow-based recovery of 2D shocks from boundary sources is adopted [25,27], where the flow permits the consideration of only the *relevant* bisectors.

Consider the problem of deriving the MA of M surface patches G_i, $i = 1. \ldots, M$. In computational geometry, the pairwise bisector B_{ij}, *i.e.*, the equidistant surface between a pair of models G_i and G_j, is computed for all such pairs and the results "trimmed" by removing those portions of B_{ij} which are closer to a third source G_k, Figure 4a. This results in the *exact* MA, but is computationally *prohibitive*. However, note that shocks can be considered as flowing along bisectors. The shock flow begins at certain initial sources and terminates at sinks. If the initial sources are identified and traced, only the viable bisectors are considered, thus tremendously reducing the computational effort, Figure 4b. While in 2D the flow along shocks is a 1D process, in 3D, shocks flow along a vector field on a sheet. A key insight, which reduces the underlying dimension of the computational effort is that flow along shock curves (A_1^3 and A_3) is sufficient to recover the MA *exactly*. This is because all MA sheets are bounded by curves, and it is clear from each curve which two sources generate each sheet bisector,

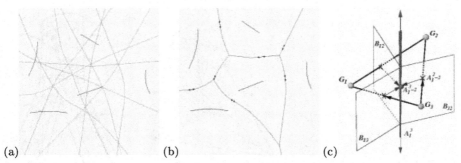

(a)   (b)   (c)

Figure 4. In computational geometry, pairs of sources, represented as dark segments in (a), are used to compute the set of bisectors, shown in (a) as the remaining curves. (b) These bisectors are then "trimmed" to obtain the MA [27]. In our approach, we use a notion of propagation along bisectors which are initiated from valid sources (double arrows), thus avoiding the need to consider the numerous irrelevant bisectors. (c) The situation is similar in 3D: three point sources G_1, G_2 and G_3 give rise to 3 bisector sheets and 1 trisector curve. The computation of only valid sources and flows along curves leads to tremendous improvements in efficiency.

thus leading to an exact identification of sheets, Figure 4c, if the shock curves are available.

Initial shock sources for curves are identified as follows. Along A_1^3 and A_3 curves, A_1^3-2 and A_3-2 are the only initial sources, respectively. Thus, it is sufficient to identify A_1^3-2 and A_3-2 MA points and propagate along A_1^3 and A_3 curves from these points until the curves come to an end, which can only happen at A_1A_3 and A_3-4 points for A_3 curves, and at either A_1A_3, A_1^3-4 or A_1^4 points for A_1^3 curves. The recipe for continuing the propagation at junctions is also straightforward: Shock curves entering an A_1^4 can either terminate (A_1^4-4), leave in a single (A_1^4-3) or two (A_1^4-2) outcoming branches; similarly, A_1A_3 shocks can be A_1A_3-3 (with A_3 flowing in and A_1^3 flowing out), A_1A_3-2, or A_1A_3-4. The general algorithm for M sources is thus described at an abstract level as summarized below.

1. Identify all MA initial shock sources.
2. Identify the next junction by considering intersections of the A_3 and A_1^3 curves with neighboring shock waves, and propagate these for each source.
3. If an outcoming shock is available, use the junction as a source and go to step 2 (iteration).
4. Output the shock scaffold graph.

The details of this algorithm for extracting the shock scaffold of unorganized clouds of points can be found in [15].

4 Unorganized clouds of points

In this paper we only consider shapes specified by a collection of unorganized point sources, *e.g.*, as they arise from a laser scanner. The bisector for a pair of point sources, $G_i = (x_i, y_i, z_i)$ and $G_j = (x_j, y_j, z_j)$ is a planar sheet orthogonal to the line joining the sources and passing through their midpoint, $M_{ij} = \frac{1}{2}(x_i + x_j, y_i + y_j, z_i + z_j)$. The point M_{ij} is the initial shock source A_1^2-2 on the sheet B_{ij}, Figure 4c, which is described by the implicit polynomial:

$$2\left[(x_i - x_j)x + (y_i - y_j)y + (z_i - z_j)z\right] + \left[(x_j^2 + y_j^2 + z_j^2) - (x_i^2 + y_i^2 + z_i^2)\right] = 0.$$

Similarly, for three point sources, G_i, G_j and G_k, the A_1^3 shock curve is computed as the intersection of two shock sheets. The initial point of flow along this curve, the A_1^3-2 point, is the *circumcenter*, O_3, of the *triangle* defined by the three sources, which can be efficiently and robustly computed in terms of the edge lengths of the triangle as,

$$O_3 = G_i + \left(\frac{b^2(\overrightarrow{a} \wedge \overrightarrow{b})}{8\,\triangle^2}\right) \wedge \overrightarrow{a} + \left(\frac{a^2(\overrightarrow{b} \wedge \overrightarrow{a})}{8\,\triangle^2}\right) \wedge \overrightarrow{b},$$

where $\overrightarrow{a} = \overrightarrow{G_iG_j}$, $\overrightarrow{b} = \overrightarrow{G_iG_k}$, $a = \|\overrightarrow{a}\|$, $b = \|\overrightarrow{b}\|$ and $\triangle$ denotes the *area* of the triangle. Next, for four point sources, G_i, G_j, G_k and G_l, the A_1^4 node is computed as the *circumcenter*, O_4, of the *tetrahedron* defined by the four sources as,

$$O_4 = G_i + \frac{1}{12V}\left[c^2(\overrightarrow{a} \wedge \overrightarrow{b}) + b^2(\overrightarrow{c} \wedge \overrightarrow{a}) + a^2(\overrightarrow{b} \wedge \overrightarrow{c})\right],$$

where $\overrightarrow{c} = \overrightarrow{G_iG_l}$, $c = \|\overrightarrow{c}\|$, V denotes the *volume* of the tetrahedron, and $\overrightarrow{a}$, $\overrightarrow{b}$, a and b are defined as for the triangle case.

Finally, no A_3 curves or A_1A_3 points are possible for point sources. Detailed computation analysis, including numerical complexity and pseudocode listings are provided in [15]. We note that the computations for unorganized polygonal patches are much more intricate but nevertheless fully computable [16].

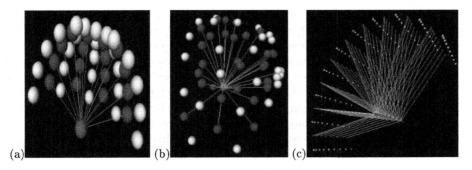

(a)   (b)   (c)

Figure 5. The shock scaffold is depicted for collections of (white) dot samplings. (a) Regular and (b) irregular samplings of a spherical cap. (c) Regular sampling on a cylindrical segment.

5   Results and Discussion

We now present examples for dot samples of geometrically simple shapes, *i.e.*, spherical caps, cylinders and a parallelepiped, as well as for dots sampling of the surface of an aorta section, a pot sherd and a full pot. An intrinsic challenge in presenting 3D MA results is the visualization of the results, which can be best seen interactively in 3D, and we invite the reader to visit `www.lems.brown.edu/vision/researchAreas/Shocks3D/`, but which must be conveyed with 2D snapshots here.

Figure 6. (a) Input samples randomly distributed along half-cylindrical sections (shown as white spheres). Darker spheres indicate A_1^3-2 and A_1^4 points. (b) Pruning: cutting away initial curve shock sources A_1^3-2 and associated branches leads to the central axis.

The first set of examples illustrates the effect of grid sampling along spherical caps and cylindrical segments, Figure 5. Note the correct placement of A_1^3-2 points (grey) which identify initial shock curve sources. We only keep and show those curve segments which are connected to valid A_1^4 nodes. These curves also propagate to infinity (not shown), but this is easily and explicitly detected via the circumcenters of associated shock nodes. In Figure 5a the regular sampling of a spherical cap results in a single A_1^4 at the sphere center, as expected. In Figure 5b perturbations along the tangent space keep this geometry intact. The shock scaffold for a cylindrical segment is shown in Figure 5c, where we note that the "generalized axis" is readily visible. In Figure 6, randomly distributed points (white spheres) along sections of a half-cylinder are used as input. Application of a structural pruning strategy [26,17] lead us to directly retrieve the main axis of the original cylinder. Note that a significant goal in object representation is the approximation of data by generalized cylinder descriptions which are highly symmetric.

While the first set of examples examines the correctness of the algorithm for simple situations, the next example examines the more complex geometry of a parallelepiped, Figure 7. Observe that A_1^4 points are placed as expected. This

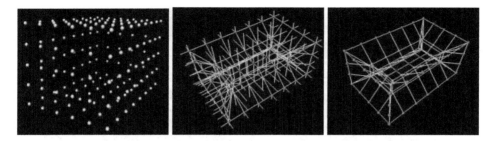

Figure 7. Left to right: Initial sample points on the surface of a parallelepiped; full scaffold; scaffold with initial shock curves pruned away.

parallelepiped has a large number of degeneracies, *i.e.*, overlaps of A_1^3-2 and A_1^4 shock points. A large portion of the scaffold is due to initial shock curves (Figure 7, middle). Removing the initial curve shocks and their associated links reveals the internal structure of the scaffold (Figure 7, right). Shock links with infinite flow velocity are also identified. Furthermore, the direction of flow of regular links is also available explicitly (not illustrated here).

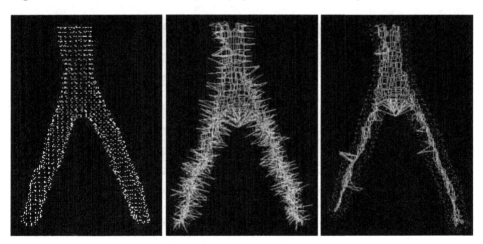

Figure 8. Left: bottom part of an aorta scan (data from [29]). Middle: scaffold without the initial curve shocks. Right: simplified scaffold after using the combined structural-saliency pruning.

In Figure 8 we show the result of applying the combined structural-saliency pruning [17] on the bottom branching part of the aorta data. Note that here, we did not display the medial structure in between the two "legs", by using a maximum distance criterion, therefore displaying the scaffold structure only in the vicinity of the input data.

In Figure 9, we show the effect of this pruning on a close-up of the top-section for the aorta data of Figure 8 (transversal section with respect to the aorta's

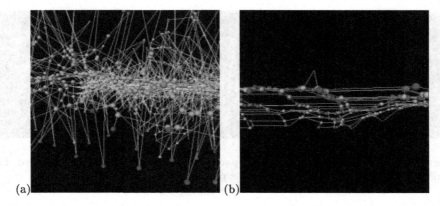

Figure 9. Section through the scaffold of the aorta: (a) before pruning; (b) after.

main vertical orientation, seen from the side). Note that each loop (in Figures 8 and 9) in the scaffold corresponds to a shock sheet (hyperlink).

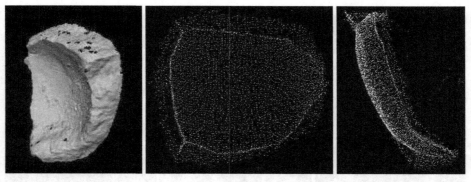

Figure 10. Left: original pot sherd, 5000 samples, obtained from a laser scanner. The frontal (middle) and side (right) views of the shock scaffold nodes.

Our last examples are shown for samples taken from pottery excavated at the archaeological site of Petra, Jordan. The samples are obtained via laser scanning. The pruned shock scaffold (nodes only) is depicted in frontal (b) and side (c) views. Results for a full pot is illustrated in Figure 11 where we overlap input samples and scaffold nodes; this pot has a complex symmetry structure due to a neck, two handles, a hole at the bottom and input samples on part of the internal surfaces. Our use of skeletons in this project aims to single out the curves along the shock scaffold as a representation of the pot sherd. The latter can then be matched to other sherds, *i.e.*, used in the stitching of sherds to ultimately achieve automatic reconstruction of the full pot [13].

In conclusion, we have presented an approach to the recovery and representation of the 3D Medial Axis based on the notion of a hierarchically organized shock scaffold and presented a specific method for extracting the shock scaf-

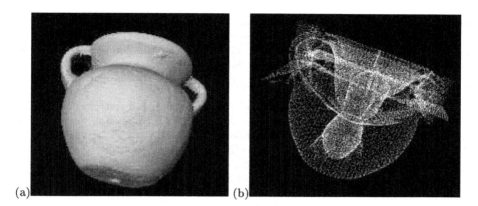

(a)   (b)

Figure 11. Point samples (51 000) of a full pot (a) give rise to the shock scaffold nodes in (b). Source samples are also shown.

fold of a cloud of points. The approach combines computational geometry and propagation-based methods, is exact, efficient, applicable to unorganized points [15] and surface patches [16], and results in a graph structure which can be used in recognition and matching applications [17].

Acknowledgements

Major funding for this project comes from the USA NSF/KDI grant #BCS-9980091. F. Leymarie was partially supported by a PhD fellowship from IBM. We thank A.Verroust and F.Lazarus for providing us the aorta data (cf. [29]).

References

1. N. Amenta et al. A new Voronoi-based surface reconstruction algorithm. *Computer Graphics (SIGGRAPH '98)*, pages 415–421, July 1998.
2. D. Attali and A. Montanvert. Computing and simplifying 2D and 3D continuous skeletons. *CVIU*, 67(3):261–273, 1997.
3. H. Blum. Biological shape and visual science. *J. Theo. Bio.*, 38:205–287, 1973.
4. G. Borgefors, I. Nyström, and G. Sanniti di Baja. Computing skeletons in three dimensions. *Pattern Recognition*, 32(7):1225–1236, July 1999.
5. S. Bouix and K. Siddiqi. Divergence-based medial surfaces. In *Sixth European Conference on Computer Vision*, Trinity College, Dublin, Ireland, June 2000.
6. G. Elber and M.-S. Kim. Computing rational bisectors. *IEEE Computer Graphics and Applications,*, 19(6), Nov./Dec. 1999.
7. P. Giblin and B. Kimia. On the intrinsic reconstruction of shape from its symmetries. In *Proc. of CVPR*, pages 79–84. IEEE Computer Society, 1998.
8. P. Giblin and B. Kimia. On the local form of symmetry sets, and medial axes, and shocks in 3D. In *Proc. of CVPR*, pages 566–573, June 2000.
9. G.Malandain and S.Fernandez-Vidal. Euclidean skeletons. *Image and Vision Computing*, 16(5):317–327, 1998.

10. J. Gomes and O. Faugeras. Reconciling distance functions and level sets. *Journal of Visual Communication and Image Representation*, 11:209–223, 2000.
11. P. Halliman, G. Gordon, A. Yuille, P. Giblin, and D. Mumford. *Two- and Three-Dimensional Patterns of the Face*. A. K. Peters, 1999.
12. B. Kimia, A. Tannenbaum, and S. Zucker. Shapes, shocks, and deformations. *IJCV*, 15:189–224, 1995.
13. F. Leymarie et al. The SHAPE Lab: New technology and software for archaeologists. In *CAA 2000: Computing Archaeology for Understanding the Past*, BAR International Series. Archaeopress, Oxford, UK, 2001.
14. F. Leymarie and B. Kimia. Discrete 3D wave propagation for computing morphological operations from surface patches and unorganized points. volume 18 of *Computational Imaging and Vision*, pages 351–360. Kluwer Academic, 2000.
15. F. Leymarie and B. Kimia. Computation of the shock scaffold for unorganized point samples. Technical Report LEMS-186, Brown University, February 2001.
16. F. Leymarie and B. Kimia. Computation of the shock scaffold for unorganized polygonal patches. Technical Report LEMS-187, Brown University, 2001.
17. F. F. Leymarie. *3D Shape Representation via Shock Flows*. PhD thesis, Brown University, 2001.
18. C. Ma and M. Sonka. A fully parallel 3D thinning algorithm and its applications. *CVIU*, 64(3):420–433, 1996.
19. M. Näf et al. 3D Voronoi skeletons and their usage for the characterization and recognition of 3D organ shape. *CVIU*, 66(2):147–161, 1997.
20. M. Peternell. Geometric properties of bisector surfaces. *Graphical Models*, 62(3):202–236, May 2000.
21. C. Pudney. Distance-ordered homotopic thinning: A skeletonization algorithm for 3D digital images. *CVIU*, 72(3):404–413, 1998.
22. J. Sethian. *Level Set Methods*. Cambridge University Press, 1996.
23. E. Sherbrooke et al. An algorithm for the medial axis transform of 3D polyhedral solids. *IEEE Trans. Visu. & Comp. Graphics*, 2(1):44–61, 1996.
24. H. Tek and B. Kimia. Volumetric segmentation of medical images by three-dimensional bubbles. *CVIU*, 65(2):246–258, 1997.
25. H. Tek and B. Kimia. Curve evolution, wave propagation, & mathematical morphology. volume 12 of *Comp. Ima. & Vision*, pages 115–126. Kluwer Acad., 1998.
26. H. Tek and B. Kimia. Perceptual organization via symmetry map and symmetry transforms. In *Proc. of CVPR*, Forth Collins, Colorado, June 1999.
27. H. Tek and B. Kimia. Symmetry maps of free-form curve segments via wave propagation. In *Proc. 7th ICCV*, pages 362–369, Kerkrya, Greece, Sept. 1999.
28. H. Tek, F. Leymarie, and B. Kimia. Interpenetrating waves and multiple generation shocks via the CEDT. In *Advances in Visual Form Analysis*, pages 582–593. World Scientific, 1997.
29. A. Verroust and F. Lazarus. Extracting skeletal curves from 3D scattered data. In *IEEE Proc. of Shape Modeling International*, Aizu, Japan, March 1999.
30. M. Zerroug and R. Nevatia. Three-dimensional descriptions based on the analysis of the invariant and quasi-invariant properties of some curved-axis generalized cylinders. *IEEE Trans. on PAMI*, 18(3):237–253, 1996.
31. Y. Zhou, A. Kaufman, and A. Toga. Three-dimensional skeleton and centerline generation based on an approximate minimum distance field. *The Visual Computer*, 14(7):303–314, 1998.

Curve Skeletonization by Junction Detection in Surface Skeletons

Ingela Nyström[1], Gabriella Sanniti di Baja[2], and Stina Svensson[3]

[1] Centre for Image Analysis, Uppsala University
Lägerhyddvägen 17, SE-75237 Uppsala, SWEDEN
ingela@cb.uu.se
[2] Istituto di Cibernetica, National Research Council of Italy
Via Toiano 6, IT-80072 Arco Felice (Naples), ITALY
gsdb@imagm.cib.na.cnr.it
[3] Centre for Image Analysis, Swedish University of Agricultural Sciences
Lägerhyddvägen 17, SE-75237 Uppsala, SWEDEN
stina@cb.uu.se

Abstract. We present an algorithm that, starting from the surface skeleton of a 3D solid object, computes the curve skeleton. The algorithm is based on the detection of curves and junctions in the surface skeleton. It can be applied to any surface skeleton, including the case in which the surface skeleton is two-voxel thick.

1 Introduction

Reducing discrete structures to lower dimensions is desirable when dealing with volume images. This can be done by skeletonization. The result of skeletonization of a 3D object is either a set of surfaces and curves, or, if even more compression is desired and the starting object is a solid object, a set of only curves. In the latter case, the curve skeleton of the object is obtained. The curve skeleton is a 1D set centred within the object and with the same topological properties. Although the original object cannot be recovered starting from its curve skeleton, this is useful to achieve a qualitative shape representation of the object with reduced dimensionality.

There are two different approaches to compute the curve skeleton of a 3D object. One approach is to directly reduce the 3D object to its curve skeleton. See for example [1,5,7]. Another approach is to first obtain a surface skeleton from the 3D object. Thereafter, the curve skeleton can be computed from the surface skeleton. See for example [2,10,12]. For both approaches, maintenance of the topology is not too hard to fulfil as topology preserving removal operations are available. The more crucial point is the detection of the end-points, i.e., the voxels delimiting peripheral branches in the curve skeleton. These voxels could in fact be removed without altering topology, but their removal would cause unwanted shortening and thereby important shape information would be lost. Different end-point detection criteria can be used. Criteria based on the number (and possibly position) of neighbouring voxels are blind in the sense that it is not

C. Arcelli et al. (Eds.): IWVF4, LNCS 2059, pp. 229–238, 2001.

known a priori which end-points (and, hence, which branches) the curve skeleton will have. In fact, local configurations of object voxels, initially identical to each other, may evolve differently, due to the order in which voxels are checked for removal. Thus, the end-point detection criterion is sometimes fulfilled and sometimes not for identical configurations. Preferable end-point detection criteria are based on geometrical properties, i.e., end-points are detected in correspondence with convexities on the border of the object, or with centres of maximal balls. As far as we know, only blind criteria were used, when the first approach (object $\rightarrow$ curve skeleton) was followed. Therefore, we regard the second approach (object $\rightarrow$ surface skeleton $\rightarrow$ curve skeleton) as preferable, especially when a distance transform based algorithm is used in the object $\rightarrow$ surface skeleton phase. In fact, in this case the surface skeleton can include all the centres of maximal balls, [9]. This guarantees that end-points delimiting curves in the surface skeleton are automatically kept. The problem of not removing those end-points and correctly identifying other end-points during the surface skeleton $\rightarrow$ curve skeleton phase still needs to be carefully handled.

We present an algorithm to compute the curve skeleton from the surface skeleton and use geometrical information to ascribe to the curve skeleton voxels placed in curves and in (some of the) junctions, including voxels that will play the role of end-points in the curve skeleton.

A surface skeleton consists, in most cases, of surfaces and curves crossing each other. The basic idea behind our algorithm is to detect the curves and the junctions between different surfaces and prevent their removal. This would automatically prevent unwanted shortening of curves and junctions without need of any end-point detection criterion. Indeed, we start from an initial classification of voxels in the surface skeleton, [8]. We distinguish *junction, inner, edge,* and *curve* voxels, see Fig.1. The border of the surface skeleton is the set including both edge and curve voxels. All curve voxels should be ascribed to the curve skeleton. Junctions shown in Fig.1 are placed in the innermost regions of the surface and should also be ascribed to the curve skeleton. However, junctions are not always in the innermost part of the surface. In fact, junctions may group in such a way that they delimit a surface in the surface skeleton. See Fig.2, where all junctions in the surface skeleton are shown to the right. Only junctions that could be interpreted as *peripheral* branches in the set of junctions should be kept in the skeleton, while junctions grouped into *loops* should not, as this would prevent the curve skeleton to be obtained. (Note also that loops in the curve skeleton correspond to tunnels in the object, and no tunnels exist in the surface skeleton shown in Fig.2.)

Our algorithm computes the curve skeleton from the surface skeleton in two steps, both based on iterated edge voxel removal. During the first step, all curve and junction voxels found in the original surface skeleton are always prevented from being removed. During the second step, voxels initially classified as junction voxels are prevented from removal, only if they are now classified as curve voxels; all voxels detected as edge voxels during this step are possibly removed. (Note

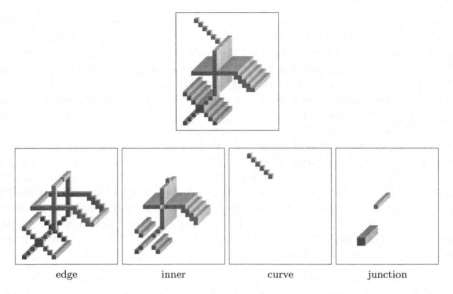

Fig. 1. A surface, top, with its classification, bottom.

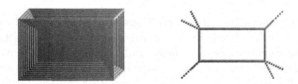

Fig. 2. A surface skeleton, left, and its junction voxels, right.

that also voxels that were classified as junction voxels during the classification done on the original surface skeleton are now possibly removed.)

The algorithm outlined above can be applied after any surface skeletonization algorithm and results in a curve skeleton. Moreover, the classification that we use can deal also with two-voxel thick surfaces so that our curve skeletonization algorithm can also be applied after algorithms resulting in two-voxel thick surface skeletons.

2   Notions

We refer to bi-level images consisting of object and background. In particular, in this paper the object is a surface, e.g., the one resulting after a 3D solid object has been reduced to its surface skeleton. The 26-connectedness is chosen for the object and the 6-connectedness for the background. Any voxel v has three types of neighbours: face, edge, and point neighbours.

We will use two different surface skeletonization algorithms for the examples in this paper. One is based on the D^6 metric, i.e., the 3D equivalent of the city-block metric, and was introduced in [9]. We will call the resulting set D^6 *surface skeleton*. The other algorithm is based on the D^{26} metric, i.e., the 3D equivalent of the chess board metric, and was introduced in [11]. We will call the resulting set D^{26} *surface skeleton*.

Classification of the voxels in a surface was suggested in [3,6]. A voxel is classified after investigating its $3 \times 3 \times 3$ neighbourhood. Of course, that classification works for an "ideal" surface, i.e., a surface which is one-voxel thick everywhere. Complex cases consisting of surfaces crossing each other would not produce a consistent classification at junctions, whenever these are more than one-voxel thick, see Fig.3. In Fig.3, left, voxels where the two surfaces cross each other, shown in dark grey, are classified as junction voxels, while in Fig.3, right, voxels where the two surfaces cross, marked by •, are classified as inner voxels. That classification also fails when applied to surface skeletons of 3D objects having regions whose thickness is an even number of voxels. These surface skeletons are in fact likely to be two-voxel thick, [9,11].

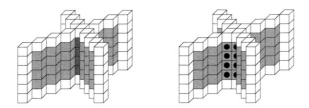

Fig. 3. Simple examples of junctions between surfaces. Edge voxels are shown in white, inner voxels in grey, and junction voxels in dark grey for the classification introduced in [3,6]. Voxels marked by • should be classified as junction voxels to be consistent.

In this paper, we use the classification introduced and thoroughly described in [8]. There some criteria suggested in [3,6] are used in combination with other criteria, where a slightly larger neighbourhood of each voxel is taken into account. Two-voxel thick regions are singled out with a linear four-voxel configuration $(4 \times 1 \times 1, 1 \times 4 \times 1, 1 \times 1 \times 4)$, which identifies portions of the surface skeletons being exactly two-voxel thick in any of the x, y, z-directions. The classification requires a number of different criteria and the same voxels are likely to be checked against many criteria before they are eventually classified. It is then not possible to summarize it here. A more detailed description is given in a recently submitted paper [4].

3   Curve Skeletonization by Junction Detection

The different classes of voxels in the surface skeleton we are interested in are *junction*, *inner*, *edge*, and *curve* voxels, see Fig.1. The curve skeleton is obtained in two steps by an iterative algorithm. Each iteration of both steps includes two subiterations dealing with i) detection of edge voxels and ii) voxel removal by means of topology preserving removal operations, respectively. The two steps differ from each other for the selection of the voxels that, at each iteration, are checked to identify the edge voxels, i.e., the set of voxels candidate for removal.

In the first step, only voxels *initially* (i.e., on the original surface skeleton) classified as inner (or edge) voxels are checked during the identification of the edge voxels. Voxels are actually interpreted as edge voxels, if their neighbourhood has been suitably modified due to removal of some neighbouring voxels. Note that we do not need any end-point detection criterion because (curve and) junction voxels are never checked to establish whether they have become edge voxels, iteration after iteration. An undesirable branch shortening would be obtained if also the voxels initially classified as junction voxels were checked. In fact, voxels placed on the tips of junctions, i.e., the junction voxels that should play the role of end-points, could be classified as edge voxels and, as such, could be removed.

In the second step, also voxels initially classified as junction voxels are checked during the identification of the edge voxels. Edge voxels that have been transformed into curves during the first step are not interpreted as edge voxels and, hence, are automatically preserved from removal. The remaining voxels initially classified as junction voxels can be now interpreted as edge voxels, if their neighbourhood has been suitably modified.

In both steps, on the current set of edge voxels, removal is done unless voxels are necessary for topology preservation. Standard topology preserving removal operations, e.g., those described in [11], are sequentially applied. After each subiteration of removal of edge voxels, a new iteration starts and a new set of edge voxels is determined. Removal operations are then applied on the new set of edge voxels. Edge detection and voxel removal are iterated until no more edge voxels are identified and possibly removed.

If the set of junctions of the surface skeleton has only peripheral junctions, i.e., no junctions are grouped into loops, the curve skeleton is obtained directly after the first step. It consists of the initial curve and junction voxels, as well as voxels necessary for connectedness. Otherwise, also the second step is necessary. In this case, the effect of the first step is to cause some junctions initially forming loops (see Fig.2) to become edge voxels. This allows skeletonization to continue towards voxels in the innermost part of the surface skeleton.

Our algorithm is first illustrated on the D^6 surface skeleton of a simple object, a cube, for which the first step is enough to compute the curve skeleton, Fig.4. The D^6 surface skeleton of the cube is shown in the middle. The resulting curve skeleton, shown to the right, coincides with the set of junction voxels.

A slightly more complex case, the D^6 surface skeleton of a box, is shown in Fig.5(b). The set resulting at completion of the first step of our algorithm is shown in Fig.5(c). Voxels detected as junction voxels during the initial clas-

Fig. 4. A cube with its D^6 surface skeleton and the curve skeleton computed by our algorithm.

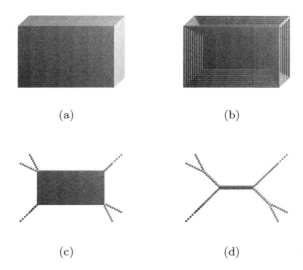

(a) (b)

(c) (d)

Fig. 5. A box with its D^6 surface skeleton, top. The intermediate result and the final result of the curve skeletonization algorithm, bottom.

sification are partly transformed into curve voxels, and partly into edge voxels surrounding the rectangular surface found in the middle of the box. The curve skeleton is the set resulting after the second step of our algorithm, see Fig.5(d).

The box in Fig.5(a) is of size $60 \times 40 \times 20$ voxels, i.e., it has an even number of voxels in every direction. The rectangular surface in the middle of the surface skeleton is hence two-voxel thick. Therefore, also the obtained curve skeleton is two-voxel thick in the central part. Final reduction to a one-voxel thick curve skeleton could be achieved by identifying tip of protrusions and iteratively removing voxels not necessary for topology preservation as was shown in [2].

For the sake of completeness, we point out that the surface skeleton could have been reduced to one-voxel thickness before extracting the curve skeleton. One reason why we prefer not to do so, is that the resulting curve skeleton could be more than one-voxel thick anyway (the rectangular surface can be one-voxel thick in depth, but an even number of voxels in other directions, as in the case above). Also, we have found that if reduction to one-voxel thickness is postponed until the curve skeleton has been obtained, the risk of creating spurious branches in the curve skeleton is significantly reduced.

4 Some Examples

Projections of thin complex structures are hard to visualize in a descriptive way. We are showing the results of our algorithm on rather small synthetic objects. In Figs. 6 and 7, a pyramid rotated 45° with its D^6 and D^{26} surface skeletons and the curve skeletons computed by our algorithm are shown. In Figs. 8 and 9, a cylinder with its D^6 and D^{26} surface skeletons and the curve skeletons computed by our algorithm are shown.

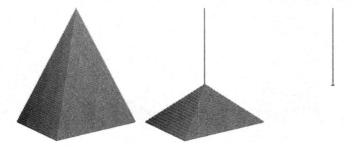

Fig. 6. A pyramid rotated 45° with its D^6 surface skeleton and the curve skeleton computed by our algorithm.

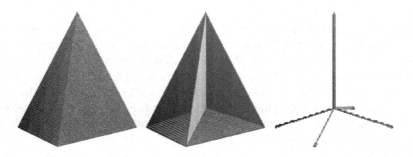

Fig. 7. The same pyramid as in Fig.6 with its D^{26} surface skeleton and the curve skeleton computed by our algorithm.

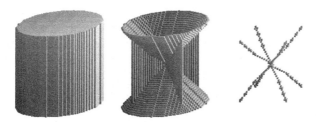

Fig. 8. A cylinder with its D^6 surface skeleton and the curve skeleton computed by our algorithm.

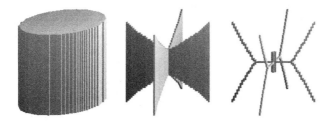

Fig. 9. The same cylinder as in Fig.8 with its D^{26} surface skeleton and the curve skeleton computed by our algorithm.

We remark that the curve skeletonization algorithm can be applied regardless of which algorithm has been used to compute the surface skeleton. In any case, the curve skeleton is a satisfactory shape descriptor.

The curve skeleton of the D^{26} surface skeleton of the cylinder, Fig.9, right, has a number of peripheral branches besides those including voxels initially classified as junction voxels. This is due to the fact that the edges of the surfaces are characterized by convexities (angle less than $90°$), which during iterated voxel removal are shrunk to curves. These curves are very short as they consist of one or two voxels only. However, once voxels have been classified as curve voxels they are ascribed to the skeleton and this causes further voxels to be prevented from removal during curve skeletonization.

5   Conclusion

In this paper, we have presented an algorithm that computes the curve skeleton of a 3D solid object starting from its surface skeleton. The algorithm is based on the detection of curve and junction voxels in the surface skeleton. One of the advantages of this approach is its independence of the choice of surface skeletonization algorithm. This is not always the case with other algorithms. For example, the surface skeleton $\to$ curve skeleton part of the algorithm presented in [2] can only be computed when starting from a D^6 surface skeleton, to obtain a reasonable result. In fact, it includes a blind end-point detection criterion tailored specifically to the D^6 case, which would not work nicely in other cases, e.g., for a D^{26} surface skeleton.

The computational cost of the curve skeletonization algorithm is quite high (a couple of minutes for complex real objects in images of size $128 \times 128 \times 128$ voxels). This is due to the non-optimized classification process that has to be repeatedly used during curve skeletonization.

We have tested our algorithm on a large number of surface skeletons, one-voxel and two-voxel thick, and have in all cases obtained satisfactory results.

Acknowledgement. We are thankful to Prof. Gunilla Borgefors, Centre for Image Analysis, Uppsala, Sweden, for useful discussions on skeletonization methods.

References

1. G. Bertrand and Z. Aktouf. A three-dimensional thinning algorithm using sub-fields. In R. A. Melter and A. Y. Wu, editors, *Vision Geometry III*, pages 113–124. Proc. SPIE 2356, 1994.
2. G. Borgefors, I. Nyström, and G. Sanniti di Baja. Computing skeletons in three dimensions. *Pattern Recognition*, 32(7):1225–1236, 1999.
3. G. Malandain, G. Bertrand, and N. Ayache. Topological segmentation of discrete surfaces. *International Journal of Computer Vision*, 10(2):183–197, 1993.
4. I. Nyström, G. Sanniti di Baja, and S. Svensson. Curve skeletonization guided by surface voxel classification. Submitted to *Pattern Recognition Letters*, 2001.
5. K. Palágyi and A. Kuba. A parallel 3D 12-subiteration thinning algorithm. *Graphical Models and Image Processing*, 61:199–221, 1999.
6. P. K. Saha and B. B. Chaudhuri. Detection of 3-D simple points for topology preserving transformations with application to thinning. *IEEE Transactions on Pattern Analysis and Machine Intelligence*, 16(10):1028–1032, Oct. 1994.
7. P. K. Saha, B. B. Chaudhuri, and D. D. Majumder. A new shape preserving parallel thinning algorithm for 3D digital images. *Pattern Recognition*, 30(12):1939–1955, Dec. 1997.
8. G. Sanniti di Baja and S. Svensson. Classification of two-voxel thick surfaces: a first approach. Internal Report 19, Centre for Image Analysis, 2000. Available from the authors.
9. G. Sanniti di Baja and S. Svensson. Surface skeletons detected on the D^6 distance transform. In F. J. Ferri, J. M. Iñetsa, A. Amin, and P. Pudil, editors, *Proceedings of S+SSPR 2000: Advances in Pattern Recognition*, volume 1876 of *Lecture Notes in Computer Science*, pages 387–396, Alicante, Spain, 2000. Springer-Verlag, Berlin Heidelberg.
10. S. N. Srihari, J. K. Udupa, and M.-M. Yau. Understanding the bin of parts. In *Proceedings of International Conference on Cybernetics and Society, Denver, Colorado*, pages 44–49, Oct. 1979.

11. S. Svensson, I. Nyström, and G. Borgefors. Fully reversible skeletonization for volume images based on anchor-points from the D^{26} distance transform. In B. K. Ersbøll and P. Johansen, editors, *Proceedings of The 11th Scandinavian Conference on Image Analysis (SCIA'99)*, pages 601–608, Kangerlussuaq, Greenland, 1999. The Pattern Recognition Society of Denmark.
12. Y.-F. Tsao and K.-S. Fu. Parallel thinning algorithm for 3-D pictures. *Computer Graphics and Image Processing*, 17(4):315–331, Dec. 1981.

Representation of Fuzzy Shapes

Binh Pham

Faculty of Information Technology
Queensland University of Technology,
GPO Box 2434 Brisbane Q4001 AUSTRALIA
b.pham@qut.edu.au

Abstract. Exact mathematical representations of objects are not suitable for applications where object descriptions are vague or object data is imprecise or inadequate. This paper presents representation schemes for basic inexact geometric entities and their relationships based on fuzzy logic. The aim is to provide a foundation framework for the development of fuzzy geometric modelling which will be useful for both creative design and computer vision applications.

1 Introduction

The success of object recognition depends very much on how an object is represented and processed. In many cases, an exact low-level geometric representation for the object such as edges and vertices, or control vertices for parametric surfaces, or CSG (Constructive Solid Geometry) primitives might not be possible to obtain. There might not be sufficient information about the objects because they were not previously known, or because the image data is too noisy. A representation scheme for fuzzy shapes, if exists, would be able to reflect more faithfully the characteristics of data and provide more accurate object recognition. Similarly, commercial CAD packages require designers to specify object shapes in exact low-level geometric representations. The current practice is for designers to manually sketch many alternative designs before a final design is chosen and a detailed model is constructed from it, using a CAD package. The necessity to work with exact object representations is counter-intuitive to the way designers work. What needed is an intuitive and flexible way to provide designers with an initial rough model by specifying some fuzzy criteria which are more in tune with the fuzziness in human thought process and subjective perception. This need for fuzziness also arises from our inability to acquire and process adequate information about a complex system. For example, it is difficult to extract exact relationships between what humans have in mind for objects' shape and what geometric techniques can offer due to the complexity of rules and underlying principles, viewed from both perceptual and technical perspectives. In addition, designers often start a new design by modifying an existing one, hence it would also be advantageous to have a library of fuzzy objects which can be specified and retrieved based on fuzzy specifications.

Another problem that would benefit from a fuzzy representation of shapes is how to overcome the lack of robustness in solid modeling systems. Although objects in

C. Arcelli et al. (Eds.): IWVF4, LNCS 2059, pp. 239–248, 2001.

these systems are represented by ideal mathematical representations, their coordinates are represented approximately in a computer in floating point arithmetic which only has finite precision. The inaccuracy arisen from rounding off errors causes ill-conditioned geometric problems. For example, gaps or inappropriate intersections may occur and result in topological violations.

The fuzziness in shape may therefore be categorised into two main types: ambiguity or vagueness arising from the uncertainty in descriptive language; and imprecision or inaccuracy arising from the uncertainty in measurement or calculation. Although fuzzy logic has been used extensively in many areas, especially in social sciences and engineering (e.g. [9,12,13]), fewer attempts have been made to apply fuzzy logic to shape representation, modelling and recognition. Rosenfeld and Pal [8, 13] discussed how to represent a digital image region by a fuzzy set and how to compute some geometric properties and measurements for a region which are commonly used in computer vision (e.g. connectedness, convexity, area, compactness). Various fuzzy membership functions and index of fuzziness were also introduced to deal with uncertainty in image enhancement, edge detection, segmentation and shape matching (e.g. Pal and Majumder [7], Huntsberger [5]). To address the problem of lack of robustness in solid modeling, a few methods have been presented. For example, Barker [2] defined a fuzzy discrimination function to classify points in space. Hu et al. [4] introduced a comprehensive scheme to use interval arithmetic as a basis for an alternative geometric representation to allow some notion of fuzziness.

In previous papers, we analysed the needs for fuzziness in computer-aided design [9,10] and presented a scheme for shape specification using fuzzy logic in order to realise some aesthetic intents of designers using fuzzy logic [11]. The intention is to bridge the gap between the impreciseness of artistic interpretation and the preciseness of mathematical representations of shapes. We also constructed a database of fuzzy shapes based on superquadric representations and investigated appropriate retrieval schemes for these fuzzy shapes [14]. For some applications, fuzzy systems often perform better than traditional systems because of their capability to deal with non-linearity and uncertainty. One reason is that while traditional systems make precise decisions at every stage, fuzzy systems retain the information about uncertainty as long as possible and only draw a crisp decision at the last stage. Another advantage is that linguistic rules, when used in fuzzy systems, would not only make tools more intuitive, but also provide better understanding and appreciation of the outcomes.

This paper deals with theoretical aspects of fuzzy geometry, in particular, how to represent basic fuzzy geometric entities and their relationships using fuzzy logic, and how to perform operations on such entities. These entities cover fuzzy points, lines, curves, polygons, regions and their 3D counterparts. The aim is to provide a unified foundation framework for the development of fuzzy geometric modelling which will benefit both creative design and computer vision applications.

2   Fuzzy Geometric Representations

This section explores some fundamental concepts of fuzzy geometric entities, how they can be defined in terms of fuzzy sets and how they are related to exact geometry. Although there are a number of different ways to introduce fuzziness into exact geometry, the representations chosen here follow a unified approach which allows a fuzzy entity to be intuitively visualized, constructed and extended to other entities in the shape hierarchy. It is assumed that readers have some basic knowledge of fuzzy set and fuzzy system design which can be found in many text books (e.g. [3,6]).

2.1   Fuzzy Points

In classical geometry, the coordinates of a point defines its position in space and is expressed as an ordered set of real numbers. This definition implies the uniqueness of the point.   However, if the position of the point is uncertain (e.g. vagueness in position specification), or imprecise (e.g. due to round off error in the calculation of its coordinates), then this definition is no longer valid and cannot be used for computational purposes.   A fuzzy point is introduced to capture the notion of vagueness and impreciseness, and at the same time, overcome the inflexibility and inaccuracy arisen if a crisp point is used.

A fuzzy point is defined as a fuzzy set P such that $P = \{(p, \mu_P(p))\}$, where $\mu_P(p)$ is the membership function which can take value between 0 and 1. A crisp point is a special case where $\mu_P(p)$ has the value 1 if $p \in P$ and 0 if $p \notin P$. Figure 1 displays an example of the membership function of a fuzzy point with respect to x-coordinate. To simplify the explanation and illustration of these new fuzzy geometrical elements, we assume a symmetrical membership function around the region of plausibility. However, all these concepts are also applicable to the cases where membership functions are unsymmetrical.

A fuzzy point therefore is represented as a set of points defined within the support of this membership function, where each of these points has a different degree of possibility of belonging to this set. The corresponding support is a circle for a 2D point and a sphere for a 3D point respectively (Figure 2).   Thus, a fuzzy point may be viewed as an extension of an interval point defined by Hu et al. [4]. In the latter case, there is an equal possibility that any point within an interval be the point of interest while in the former case, such possibility may be different for each point within the support of the membership function.

2.2   Incidence of Fuzzy Points

Two fuzzy points P and Q are incident if there exists a point which is common to both of them. In other words, if there exists a point lying within both the supports of these two fuzzy membership functions. The membership value for this incident point is the minimum of its membership values in these two fuzzy sets. Thus, the notion of *weak* and *strong* incidence may be introduced based on the membership value. This notion could be useful for decision making tasks in computer vision and computer-aided design. The incidence is symmetric but not transitive because the existence of a common point in fuzzy sets P and Q , and a common point in fuzzy sets Q and

R does not necessarily imply that there exists a common point in fuzzy sets P and R.

Fig. 1. The fuzzy membership function of a fuzzy point   **Fig. 2.** A fuzzy point

2.3   Fuzzy Connectedness of Two Points

To identify points that belong to the same object (e.g. fuzzy line segment, curve, region or volume), it is necessary to define the concept of fuzzy connectedness. Given a fuzzy set of points F , Rosenfeld [10] defined the degree of fuzzy connectedness of two points p and q within F as

$C_F(p,q) = \max[\min \mu_F(r)]$, where the maximum is taken over all paths

connecting these two points and the minimum is taken over all points r on each path.

Two points p and q are said to be fuzzily connected in F if $C_F(p,p) \geq \min[\mu_F(p), \mu_F(q)]$. In other words, two points are connected in a fuzzy set of points if there exists a path between them which is composed of only points which also belong to this fuzzy set. This definition is also consistent with the concept of connectedness of two points within a crisp set whose membership values are all 1.

2.4   Fuzzy Lines and Planes

A fuzzy line PQ which connects two fuzzy points P and Q with membership functions $\mu_P(p)$ and $\mu_Q(q)$ is defined as a fuzzy set each of whose members is a linear combination of a pair of points p and q with a membership function defined as $\mu_{PQ}(pq) = \min(\mu_P(p), \mu_Q(q))$.

A fuzzy line may be visualised as a centre line with variable thickness (Figure 3). This thin area of space (or thin volume of space for 3D case) bounds a family of crisp lines which are formed by pairs of endpoints belonging to the two fuzzy sets of endpoints.

A fuzzy plane which is an extension of a fuzzy line is a thin planar shell with variable thickness. This shell encloses a family of crisp planes which is an extension of the family of crisp lines representing the fuzzy line. These concepts of fuzzy lines and planes encapsulate exact lines and exact planes as special cases.

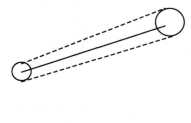

Fig. 3. A fuzzy point

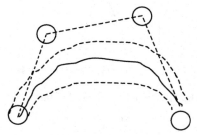

Fig. 4. A fuzzy Bezier curve

2.5   Fuzzy Polygons and Polyhedra

A fuzzy polygon is composed of fuzzy vertices and fuzzy edges. Thus, it may be visualised as having edges of variable thickness, as described for fuzzy lines. This concept is readily extended to a fuzzy polyhedron for 3D case.

2.6   Fuzzy Regions and Volumes

Image segmentation or volume segmentation often result in fuzzy regions or volumes when the discrimination between texture characteristics is low or the quality of images is not good. In exact geometry, a 2D region is defined by the space bounded by a closed boundary. A fuzzy region is defined as the space bounded by a fuzzy boundary, where a fuzzy boundary is a fuzzy set of connected points in space, with the notion of connectedness being defined as above.   Thus, a fuzzy region consists of two types of points: inner points and boundary points. In exact geometry, a point is an inner point if all points in its local 4 or 8-neighbourhood belong to the region while for a boundary point, there exists at least one point in its local neighbourhood that lies outside the region. In order to determine if a point is a boundary point or an inner point of a fuzzy region, we therefore need to consider the characteristics of the local neighbourhood of a point.

A point is an inner point of a fuzzy region if all points in its local neighbourhood have membership values of 1. If at least one point in its neighbourhood has a membership value between $(0,1)$, then the point is a boundary point. On the other hand, if points in its local neighbourhood either have membership values equal 0 or are boundary points, then the point is outside the region.   These definitions cover the notion of inner, boundary and outer points of an exact region as special cases.

For 3D volumes, pixels are replaced by voxels, and a fuzzy boundary curve is replaced by a fuzzy boundary surface. The local 4 and 8 neighbourhood of a point are extended to local 6 and 14- neighbourhood and a scheme for classification of points can be readily defined in a similar way.

2.7   Fuzzy Bezier and B-Splines Curves and Surfaces

Fuzzy free-form curves and surfaces can be represented by fuzzy polynomial splines such as Bezier and B-splines which have fuzzy control points. Each of these control points may be visualized as a fuzzy set of points located within a circle (in 2D case) or a sphere (in 3D case). These points may have different membership values. A fuzzy spline curve is therefore represented by a family of curves lying within a thin tube and a fuzzy spline surface is represented by a family of surfaces lying within a thin shell. Figure 4 shows an example of a fuzzy Bezier curve.

In solid modeling and CAD / CAM applications, fuzzy geometry serves the same purpose as interval geometry. Both representation schemes are useful for overcoming the problems of topological violation or gaps and inappropriate intersections caused by floating point arithmetic. However, one advantage of fuzzy geometry over interval geometry is that a notion of degree of possibility or plausibility that a curve or a surface in these families of curves and surfaces is the element of interest can be represented. This fact is also relevant in image processing where many tasks such as edge detection and segmentation often result in edges and contours whose degree of fuzziness depends on the variation of contrast in intensity or colour.

2.8   Minimum Bounding Rectangle for a Fuzzy Shape

The minimum bounding rectangle for a shape has been used as an approximation to the shape to speed up many tasks such as searching for an area of interest and checking the intersection or occlusion of objects.   The true minimum bounding rectangle is defined as the smallest rectangle with its sides being parallel to the major axes of the shape. However, for many applications, it is more efficient and sufficient to use the minimum bounding rectangle whose sides are parallel to the coordinate axes.   In the exact case, this rectangle is usually determined by the minimum and maximum coordinates of points belonging to the shape.   For a fuzzy shape, we obtain a thin rectangular tube whose coordinates correspond to the extreme coordinates (leftmost, rightmost, topmost, bottommost) of the boundary points of the fuzzy shape.

3   Operations on Fuzzy Geometry Entities

In this section, we discuss how basic operations such as intersection, union, decomposition and de-fuzzification are performed on fuzzy geometry entities.

3.1   Intersection of Fuzzy Lines, Curves, Planes, and Surfaces

The intersection of two fuzzy lines PQ and RS is a fuzzy point I which is represented by a fuzzy set $I = (i, \mu_I(i))$, where

$$\mu_I(i) = \min(\mu_{PQ}(i), \mu_{RS}(i)) \tag{1}$$

Figure 5 shows the intersection point of two 2D fuzzy lines represented by a fuzzy set of points which are located within a quadrilateral bounded by the limiting lines defining fuzzy lines.   For 3D case, the domain of the fuzzy set that represents the fuzzy point of intersection is the intersection volume of two thin tubes representing two fuzzy lines.

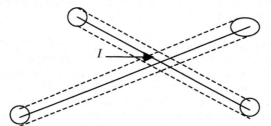

Fig. 5. Intersection of two fuzzy lines

The intersection of two fuzzy curves in 2D or 3D is a fuzzy set defined in the same fashion. Similarly, we can extend these concepts to cover the intersection of a fuzzy line and a crisp plane, or of a fuzzy line and a fuzzy plane, or of two fuzzy planes, or of two fuzzy surfaces. Thus, the intersection of these geometry entities can be performed as two separate tasks. The first task is to compute the intersection of pairs of crisp geometry entities (which belong to the two families of fuzzy entities) in the same way as in exact geometry. The second task is to compute the membership value for each resulting entity.

3.2   Union, Intersection, and Adjacency of Fuzzy Regions and Volumes

The union of two fuzzy regions R_1 and R_2 is defined as a fuzzy subset U of common points u whose membership value for either of these regions is non-zero, where $U = (u, \mu_U(u))$ and $\mu_U(u) = \max(\mu_{R_1}(u), \mu_{R_2}(u))$. This formula for $\mu_U(u)$ is generally accepted for the union of two fuzzy sets. However, if we wish to model more accurately the probability of u being a member of the new combined set, then it is more appropriate to use the following formula:

$$\mu_U(u) = 1 - (1 - \mu_{R_1})(1 - \mu_{R_2}) = \mu_{R_1} + \mu_{R_2} - \mu_{R_1}\mu_{R_2} \tag{2}$$

The intersection of two fuzzy regions R_1 and R_2 is defined as a fuzzy subset J of common points j whose membership values for both regions are non-zero, where

$$J = (j, \mu_J(j)) \text{ and } \mu_J(j) = \min(\mu_{R_1}(j), \mu_{R_2}(j)) \tag{3}$$

A common boundary of two fuzzy regions R_1 and R_2 is a special case of the intersection where the fuzzy subset of intersection do not include the inner points of either region.

In exact geometry, two regions are adjacent if they have a common border. In the fuzzy case, two fuzzy regions are adjacent if they have a fuzzy common border. That means if there exists a fuzzy subset $B = (b, \mu_B(b))$ of common boundary points b whose membership values for these two regions are non-zero, where the membership function is calculated as $\mu_B(b) = \min(\mu_{R_1}(b), \mu_{R_2}(b))$.

These concepts of intersection and adjacency can be readily extended to the case of fuzzy volumes if we use a voxel instead of a pixel to represent a point and replace boundary curves by boundary surfaces.

3.3   Decomposition of a Fuzzy Shape

A fuzzy shape may need to be decomposed into smaller shapes according to certain criteria and for specific purposes. The question is how to re-define the membership value for each point in these new shapes? To facilitate the processing or understanding of a shape, a common way is to reduce its complexity by subdividing it into subcomponents which have more simple shape (e.g. convex). This case is simple because it is reasonable to retain the membership value of a point in the original shape as its membership value for the new subcomponent it belongs to. The main reason is that the criterion used for splitting up the shape does not have any direct effects on the probability of membership. However, if the criterion for decomposition is more complex, for example, to obtain subregions with more homogeneous texture, then the membership value of each point needs to be recomputed based on how closely its textural characteristics resemble that of the new subregion. Thus, in general, after a decomposition of a fuzzy shape, it is necessary to examine if the decomposition has affected the membership and if so, how can it be re-computed. Methods for re-calculation of the membership values will depend on the context of each specific problem.

3.4   Defuzzification of a Fuzzy Shape

It is advantageous to retain the notion of fuzziness in shapes as long as possible during a computational or decision-making process in order to avoid early commitments of using the approximation of shapes at each stage because errors would accumulate as a result. However, in real life applications, there would come to a stage where a crisp representation of a shape is required, e.g. an exact description of an object is required for manufacturing purposes. The question is how to derive a viable exact geometry entity from a fuzzy one?

The most common defuzzification method is the centroid (or centre of gravity) of a fuzzy set A which is calculated as its weighted mean, where

$$x^{'} = (\sum_i x_i \mu_A(x_i)) / (\sum_i \mu_i(x_i))$$

(4)

This definition is appropriate for a fuzzy point because the 'balance' point is obtained. For a fuzzy line, there are two ways to interpret this weighted mean. The first way is to defuzzify each fuzzy endpoint first, before joining them to obtain an exact line. The second way is to apply the weighted mean directly to the family of crisp lines which make up the fuzzy lines. Each of these crisp lines is represented by a tuple (m_i, c_i, μ_i) where m_i, c_i are the slope and intercept of the line respectively and μ_i is the membership value of the line. Hence the weighted mean for the slope and intercept may be computed using the same formula to obtain a defuzzified line.

The first method can be applied to defuzzify a polynomial spline if its control points are known. However, if the control points are not known, the defuzzified curve can be computed as the weighted medial axis, where the membership value of each point is used as a weight for the distance calculation in the medial axis transform. Due to page limit, details on this transform are not covered here, but they may be found in [1, page 252].

For the case of a fuzzy region, once its fuzzy boundary is identified (by excluding the inner points and outer points of the region as explained in the previous main section), the boundary can be defuzzified in a similar way to that for a general curve.

4   Conclusion

A set of complete representation schemes for fuzzy geometry entities and basic operations has been presented which is based on fuzzy logic. These schemes may be seen as the extensions of exact geometry representations. They have potential uses in many important applications in order to cater for the ambiguity of human reasoning and linguistic descriptions as well as the impreciseness of numerical computation. It is also envisaged that these representations will provide a foundation framework for fuzzy geometric modelling, where fuzziness may be found not only in geometry entities, but in their spatial relationships and other attributes. They would provide a systematic way to construct fuzzy object models for design, matching and recognition. However, an important issue that needs to be investigated is the trade-off between efficiency and the retention of fuzzy information. What type of fuzzy information should be incorporated in the model and at what stage should such information be defuzzified? Our related work on similarity measures and evolution of fuzzy shapes are currently in progress.

Acknowledgements. This work is sponsored by the Australian Research Council.

References

1. Ballard D.H. and Brown C.M.: Computer Vision, Prentice-Hall (1982).
2. Barker S.M.: Towards a topology for computational geometry, Computer-Aided Design 27 (4) (1995) 311-318.
3. Berkan R.C. and Trubatch S.L.: Fuzzy System Design Principles, IEEE Press,NY (1997).
4. Hu C.Y., Patrikalakis N.M. and Ye X.: Robust interval solid modeling Part I: representations, Computer-Aided Design 28 (10) (1996) 807-817.
5. Huntsberger T.L., Rangarajan C. and Jayaramamurthy S.N.: Representattion of uncertainty in computer vision using fuzzy sets, IEEE Trans. Comput. C-35, (2) (1986) 145-156.
6. McNeill D. and Freioberger P.: Fuzzy Logic, Simon & Scuster, NY (1993).
7. Pal S.K. and Majumder D.D.: Fuzzy Mathematical Approach to Pattern Recognition, Wiley (Halsted Press), NY (1986).
8. Pal S.K and Rosenfeld A.: Image enhancement and thresholding by optimization of fuzzy compactness, Pattern Recog. Letters 7 (1988) 77-86.
9. Pham B.: A hybrid representation for aesthetic factors in design, International Jour.on Machine Graphics & Vision 6 (2) (1997) 237-246.
10. Pham B., Fuzzy Logic Applications in CAD, in Reznik L., Dimitrov V., Kacprzyk J. (eds.): Fuzzy System Design: Social and Engineering (1998) 73-85.
11. Pham B. and Zhang J.: A fuzzy shape specification system to support design for aesthetics, in Reznik L. (ed.): Soft Computing in Measurement and Information Acquisition , Physica-Verlag, Heidelberg, in print.
12. Reznik L., Dimitrov V. and Kacprzyk J. (Eds): Fuzzy System Design: Social and Engineering, Physica-Verlag, Heidelberg (1998).
13. Rosenfeld A.: The fuzzy geometry of image subsets, in Dubois D., Prade H. and Yager R.R. (eds.): Readings in Fuzzy Sets for Intelligent Systems, Morgan Kaufmann (1993).
14. Zhang J, Pham B. and Chen P.: Construction of a Fuzzy Shape Database, Second International Discourse on Fuzzy Logic in the New Millennium, Great Barrier Reef, Australia September 2000, in print.
15. Zadeh L.A. and Kacprzyk J. (Eds.): Computing with Words in Information / Intelligent Systems 1and 2 – Foundations, Physica-Verlag, Heidelberg (1999).

Skeleton-Based Shape Models with Pressure Forces: Application to Segmentation of Overlapping Leaves

Gilles Rabatel[1], Anne-Gaëlle Manh[1], Marie-José Aldon[2], and
Bernard Bonicelli[1]

[1] Cemagref Montpellier GEAF, BP 5095, 34033 Montpellier cedex 1, France,
{rabatel, manh, bonicelli}@montpellier.cemagref.fr
[2] LIRMM, 161 rue Ada, 34392 Montpellier Cedex 5, France
Marie-Jose.Aldon@lirmm.fr

Abstract. Deformables templates, because they contain *a priori* knowledge information on searched shapes, can be useful in the segmentation of complex images including partially occluded objects. In this paper, we propose a generic deformable template for shapes that are built around a flexible symmetry axis. The shape model is based on a parametrical skeleton, the distance between this skeleton and the contour points of the shape being determined by a parametrical envelope function. Various examples of skeleton and envelope models are given, and a general scheme for identification and matching suitable for a reusable code implementation is proposed. An application to image segmentation of partially overlapping leaves in natural outdoor conditions is then presented.

1 Introduction

Because the vision process makes a projection of a 3D world into 2D data, image interpretation is basically an underconstrained problem: extrinsic information about the shape of the objects we are looking for will be often necessary to overcome ambiguities in image interpretation of complex scenes. Such extrinsic information about object shape is refered as a shape model. Many types of shape models have been proposed in the litterature [1], and can be classified in two main groups:

- free-form models, or active contours: this type of model has first been introduced by Kass et al.[2]. Free-form models just integrate local constraints such as smoothness (curvature limitation) and elasticity (length limitation). They can be represented by a set of contour sampling points [2], or under an analytical form, such as B-Splines [3].
- deformable templates: these models aim to introduce more strong and specific shape constraints, based on an a priori knowledge on the objects of interest. They can be defined either analytically ([4,5]), or by a prototype shape with associated deformation modes issued from a training set [6]. Because they include strong information on shape, deformable templates have a

C. Arcelli et al. (Eds.): IWVF4, LNCS 2059, pp. 249–259, 2001.
© Springer-Verlag Berlin Heidelberg 2001

better ability to deal with specific vision problems such as partially occluded objects recovering.

Recovering objects in an image with shape models is usually seen as an optimisation problem, either using a Bayesian approach ([7,8]), or using an energy minimisation approach ([2,4]). Though some authors solve this optimisation problem by stochastic methods, such as simulated annealing [9], methods based on gradient descent are often preferred, because they are much less time consuming.

A major problem when fitting shape models in images is that the initial state of the model must be close enough to the expected solution to get a correct convergence. In [10], Cohen proposed a solution to this problem for free-form models, by introducing a "balloon force" that inflate the active contour until the image attracting features are reached. This method was proposed for cavity edges detection in medical imagery, and requires that no intermediary local minimum can be encountered during the inflating process which starts from inside the cavity.

However, this kind of solution can be generalized to more complex images, under two conditions:

- sufficiently specific shape constraints are necessary to avoid intermediary minima
- adapted forces must be defined to make the model evolve in the image.

This is the approach that we propose here, in the framework of leaf recognition in natural outdoor scenes. The shapes that we are looking for are characterized by their natural variability, and by partial overlapping situations. In counterpart, generic shape properties can be highlighted, and the high colour contrast between leaves and background allows to design efficient model evolution forces.

The paper is organized as following: in parts 2 and 3, we propose a generic type of deformable template, called skeleton-based model, which adresses shapes presenting a deformable symmetry axis. We also propose associated principles for identification and matching which allows reusable code implementation for various detailed shape definitions. The application to leaf segmentation is presented in part 4.

2   Skeleton-Based Shape Models: Definition and Identification

2.1   About Parametric Shape Model Identification

We assume here that a shape in a 2D image can be described by a single closed contour. Under this restrictive condition, a parametric shape model can be considered as a function:

$$g_P : \begin{cases} [0,1[\to I\!\!R^2 \\ s \mapsto (x,y) \end{cases} \qquad \text{or in the discrete case:} \qquad f_P : \begin{cases} \{0,\ldots,N-1\} \to I\!\!R^2 \\ i \mapsto (x_i,y_i) \end{cases}$$

where $P = \{p_0, \ldots, p_{K-1}\}$ is a set of K parameters, and N is the number of sampling points representing the contour.

Newton-Based Identification Procedure. The objective of the identification procedure is to determine the set of parameters P fitting the model to a given set of image points. First, it will allow us to verify if a given parametrical model is pertinent for the type of shape we are interested in. Then, as we will see later, it can be used in the iterative process of model matching in the image.

Let us call $X = \{M_0, M_1, \ldots, M_{N-1}\}$ a set of N image points on which we want to fit a parametrical model f_P, and $F_P = \{f_P(0), \ldots, f_P(N-1)\}$ the current state of the model. Identification can be obtained using the Newton algorithm to solve the equation:

$$X - F_P = 0. \tag{1}$$

The Newton method to solve a multivariable equation $Y(P) = 0$ consists in applying to the variable P the iterative correction $dP = -J^I.Y$, where J^I is the pseudo-inverse of the Jacobian matrix of $Y(P)$. In our case, it comes:

$$dP = H^I.(X - F_P). \tag{2}$$

where H is the $N \times K$ Jacobian matrix:

$$H = \begin{bmatrix} \cdots & \cdots & \cdots \\ \frac{\partial f_p(i)}{\partial p_0} & \cdots & \frac{\partial f_p(i)}{\partial p_{K-1}} \\ \cdots & \cdots & \cdots \end{bmatrix}. \tag{3}$$

The correction dP given by (2) will be applied to the set of parameters P until it is nearly null. Notice that if H is a constant, it can be easily shown that the solution is obtained in one iteration.

2.2 Skeleton-Based Shape Models

Definition. The kind of model that we introduce here adresses 2D shapes that present a native axial symmetry, but are subject to deformations of their symmetry axis. Those kind of shapes can be encountered in manufactured objects (e.g. pipes). They are also particularly important in agricultural objects such as leaves, fruits, etc., where the symmetry axis is induced by the biological mechanism of growth, but subject to random deformations. Therefore, we defined a generic skeleton-based model, illustrated in Fig. 1, which comprises:

- a parametric skeleton (open curve), representing the symmetry axis of the shape
- a parametric envelope, which is a scalar function giving the shape contour distance on each side of the skeleton.

In these conditions, each point i of the model is defined by:

$$\begin{pmatrix} x(i) \\ y(i) \end{pmatrix} = \begin{pmatrix} x_s(i_s) \\ y_s(i_s) \end{pmatrix} \pm f_E(i) \begin{pmatrix} \boldsymbol{x}_N(i_s) \\ \boldsymbol{y}_N(i_s) \end{pmatrix}. \tag{4}$$

where:

- the number of sample points of the skeleton is: $Ns = N/2 + 1$
- i_s is the skeleton corresponding point index ($i_s = i$ for $i \leq N/2$, and $i_s = N - i$ otherwise)
- (x, y) are the contour point coordinates
- (x_s, y_s) are the skeleton point coordinates
- f_E is the envelope scalar function
- $(x_N, y_N) = \overrightarrow{N}(i_s)$ is the vector normal to the skeleton at the point (x_s, y_s)

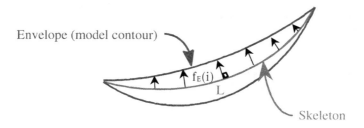

Fig. 1. Generic skeleton-based model

As an example, a shape model made of a parabolic skeleton and a parabolic envelope could be defined by:

$$x_s(i_s) = a_x i_s^2 + b_x i_s + c_x. \tag{5}$$
$$y_s(i_s) = a_y i_s^2 + b_y i_s + c_y. \tag{6}$$
$$f_E(i_s) = b.l_s.(1 - l_s). \tag{7}$$

where $l_s = i_s/(N_s - 1)$, and N_s is the number of skeleton samples.

Identification. As shown in equation (2), the identification procedure requires the Jacobian matrix H of the parametrical model. For a skeleton-based shape model as defined by (4), it comes:

$$\frac{\partial f_P(i)}{\partial p_j} = H_s \pm \left(\frac{\partial f_E(i_s)}{\partial p_j} \overrightarrow{N}(i_s) + f_E(i_s) \frac{\partial \overrightarrow{N}(i_s)}{\partial p_j} \right). \tag{8}$$

where H_s is the skeleton model Jacobian matrix. Therefore, the identification just requires to know the derivatives of the envelope function and of the skeleton normal vector, and the Jacobian matrix H_s. This result allows us to define, from a software design point of view, a generic skeleton-based model class, on which identification process will be applied regardless of the particular skeleton and envelope model chosen: required derivatives just have to be available as virtual member functions.

Fig. 2 shows the object hierarchy that we have developed, with various derivated classes of skeletons and envelopes.

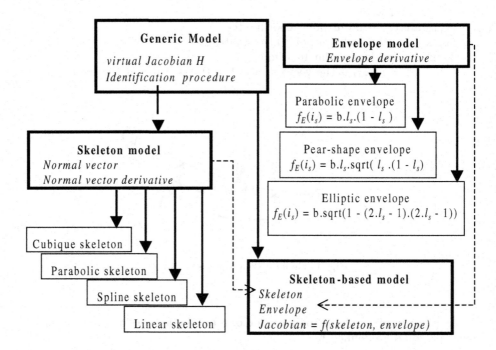

Fig. 2. Object-oriented hierarchy of skeleton-based models (bold arrows show inheritance)

3   Shape Model Matching

3.1   Pressure Forces

The more a shape model is strongly constrained, the more it will have the ability to bypass undesired local minima, and thus to start the matching process from an initial state far from the final target position, and even to manage object overlapping situations. However, for this purpose, attraction forces directly derived from an image energy function are not sufficient: evolution forces allowing to control the model modification process, and thus depending on the current

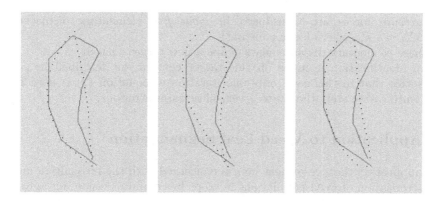

Fig. 3. Identification examples with a parabolic skeleton and various envelope models (*from left to right*: parabolic, pear-shaped and elliptic envelopes)

shape of the model itself are required. We will call them "pressure forces", by extension of the previously ones defined in [10]. As an example, in order to match the model on a binary shape (white on black), we can define a pressure force which aims to make the model expand or retract itself according to the value of the pixels under the model. Thus, for each point $f_P(i)$ of the model contour, this force will have:

− the direction of the vector normal to the contour at this point
− a positive intensity if the pixel at this point is white, negative otherwise (Fig. 4).

As we will see in part 4, such pressure forces allow to start the model from a very small portion of the object candidate, as soon as the principal axis of the initial model is correctly chosen.

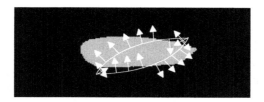

Fig. 4. Pressure forces for a binary image

3.2 Iterative Procedure

Once the evolution forces are defined, they have to be applied iteratively in order to modify the shape model parameters until the final position is reached. This is made by the following procedure:

- pressure forces are transduced in *model-free* elementary displacements $\Delta X_i = k.F_i$ (where k is a constant)
- these elementary displacements define a new target contour.
- an identification is made on this target contour, as described in part 2. Notice that in this case, only one iteration of equation (2) is used in the identification step, thanks to a very close target contour.

4   Application to Weed Leaf Segmentation

The application that we present here is concerned with the recognition and the characterisation of weed populations in crops by computer vision, for agronomical purposes (optimisation of herbicide usage). In this framework, shape models have been considered in order to overcome the image segmentation problems, which are mainly due to the biological variability of plants and overlapping leaves situations.

We present here the first results that have been obtained with skeleton-based models for a very common weed variety in maize crop, called green foxtail (*Setaria viridis*). The model that has been chosen for this variety is based on a parabolic skeleton and a parabolic envelope, as defined in 2.2.

4.1   Definition of the Pressure Forces

As described in 3.1, pressure forces are normal to the model contour. Pressure force intensities are computed from color information in the images. As a first step, the mean value μ and the covariance matrix C of the RGB values of leaf pixels have been determined. It allows us to calculate, for each pixel of the image with a color value $x = (R, G, B)$, the Mahalanobis distance d from the plant color class by:

$$d^2 = (x - \mu)^T C^{-1}(x - \mu). \tag{9}$$

Then a threshold value s is applied on the square distance d^2 to perform a binary segmentation, that will be used for the model initialisation. However, the intensities of the pressure forces are not simple binary values as suggested in 3.1. They are given in the neighbourhood of s by the relation: $Fi = s - d^2$ (see Fig. 5).

4.2   Model Initialisation

Models are initialised by searching for the leaf tips in the binary image issued from color segmentation. Whenever a leaf tip is found (i.e. a small leaf-type window surrounded by a majority of background pixels), the best model orientation is determined by rotating a small elliptic model around the centre of the window, and keeping the most included position in vegetation (see Fig. 6).

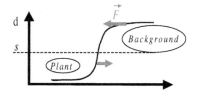

Fig. 5. Intensity of color pressure forces

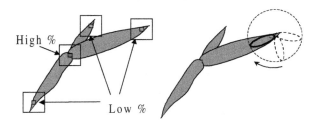

Fig. 6. Model initialisation procedure

4.3   Model Evolution

The evolution process has been described in 3.2. However, experimentations have shown that there still exists some undesirable deformations: the sample points of the model can cluster on some parts of the contour, creating irregular behaviours which have to be controlled. In [11], Berger proposed to resolve this problem for snakes by adding a term in their internal energy, which attempts to maintain equidistance between the points along the curve. In our case, we applied this supplementary constraint to the skeleton of the model, by adding corresponding forces on each contour point. An example of model evolution is given in Fig. 7.

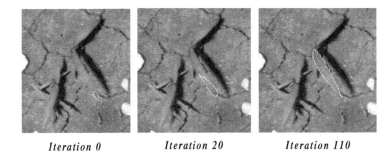

Iteration 0   *Iteration 20*   *Iteration 110*

Fig. 7. Model evolution on a weed leaf

The termination criterion is based on the skeleton speed of evolution, which decreases when the model approaches its final position.

4.4   Results

Forty images of green foxtail have been processed, representing about 600 leaves. These images were acquired using a photographic flash, in order to minimise the effect of the variability of natural outdoor lighting conditions. For each image, models have been initialised on every leaf tips found and then been expanded as described above. Because in some cases, several tips can be detected for a same leaf, redundant resulting models have then been searched and removed.

Under these conditions, 83.6% of the leaves were detected. Leaves that are not detected (16.4%) correspond to occluded tips. In order to measure the adjustment quality of models in their final position, we defined a criterion based on the similarity, for each point of the model, between the direction of the gradient of colorimetric distance d and the normal to the model: this allows to check if the model contour is actually aligned on a color edge. The quality criterion is then the percentage of points of the model verifying this similarity, defined by a maximum angle θ. A value of 40° has been set as the best value.

Figure 8 shows an example of detected leaves and the corresponding quality criterion values. We can see that well-adapted models have a high criterion value. Much lower criterion values (less than 80%) correspond not only to erroneous models, but also to well-adapted models, that are placed on partially occluded leaves. As a consequence, if the quality criterion is sufficiently high, models are assumed to be well-adapted. Otherwise, the quality criterion is not reliable enough to draw any conclusion on the adjustment.

In our case, 34.3% of the resulting models have a quality criterion up to 80%. For those models, we can assume that they are well-adjusted on leaves. Concerning the others, as the quality criterion is not reliable enough, the next step will consist in carrying on with the analysis of the model. In particular, the study of the spatial relative position between each model should help to get information on the adjustment of the models, as it would be possible to interpret models not individually but at a plant scale.

5   Conclusion

We have introduced here a particular type of deformable shape model, based on a parametrical symmetry axis and a parametrical envelope, and shown that this concept was flexible enough to easily support various specific implementations for various shape modelisations, with the same basic procedures. Because this kind of models contain strong internal constraints, they can be fitted in images starting from far initial positions, using non-gradient evolution forces. An application to the segmentation of partially occluded vegetation leaves detection has been presented. Results show that a satisfactory rate of leaves can be correctly detected. However, in this particular case, the quality criterion that has been

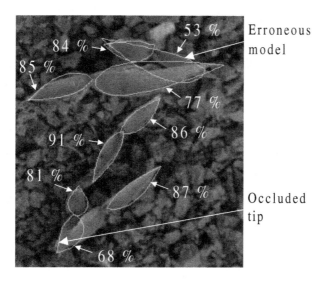

Fig. 8. Example of results

defined is not reliable enough to isolate erroneous detections. Further studies are necessary to overcome this problem by analysing relative position of leaves with respect to the possible plant structures.

References

1. A.K. Jain, Y.Zhong, M.-P. Dubuisson-Jolly. Deformable Template Models: a Review. Signal Processing **71** (1998) pp 109-129.
2. M. Kass, A.Witkin, D. Terzopoulos. Snakes: Active Contour Models. (1988) pp 321-331.
3. S. Menet, P. Saint-Marc, G. Medioni. B-Snakes: Implementation and Application to Stereo. DARPA Image Understanding Workshop (1990) pp 720-726.
4. A. Yuille, P. Hallinan, D. Cohen. Feature Extraction from Faces Using Deformable Templates. International Journal of Computer Vision **8** (2) (1992) pp 99-110.
5. M. Dubuisson-Jolly, S. Lakshmanan, A.K. Jain. Vehicle Segmentation and Classification Using Deformable Templates. IEEE Transactions on Pattern Analysis and Machine Intelligence **18** (3) (1996) pp 293-308.
6. U. Grenander. General Pattern Theory: a Mathematical Study of Regular Structures. Oxford University Press, Oxford (1993).
7. D. Mumford. The Bayesian Rationale for Energy Functionals. Kluwer Academic, Dordrecht (1994) pp 141-153.
8. M. Figueiredo, J. Leitao. Bayesian Estimation of Ventricular Contours in Angiographic Images. IEEE Transactions on Medical Imaging **11** (3)(1992) pp 416-429.
9. N. Friedland, D. Adam. Automatic Ventricular Cavity Boundary Detection from Sequential Ultrasound Images Using Simulated Annealing. IEEE Transactions on Medical Imaging **8** (4)(1989) pp 344-353.

10. L. Cohen. Note on Active Contours Models and Balloons. CVGIP: Image Understanding **53** (2)(1991) pp 211-218.
11. M.-O. Berger. Les Contours Actifs: Modélisation, Comportement et Convergence (*Active Contours: Modeling, Study and Convergence*). INRIA, Institut Polytechnique de Lorraine. 116 p. PhD Thesis (1991).

A Skeletal Measure of 2D Shape Similarity

Andrea Torsello and Edwin R. Hancock

Dept. of Computer Science, University of York
Heslington, York, YO10 5DD, UK
`atorsell@cs.york.ac.uk`

Abstract. This paper presents a geometric measure that can be used to gauge the similarity of 2D shapes by comparing their skeletons. The measure is defined to be the rate of change of boundary length with distance along the skeleton. We demonstrate that this measure varies continuously when the shape undergoes deformations. Moreover, we show that ligatures are associated with low values of the shape-measure. The measure provides a natural way of overcoming a number of problems associated with the structural representation of skeletons. The first of these is that it allows us to distinguish between perceptually distinct shapes whose skeletons are ambiguous. Second, it allows us to distinguish between the main skeletal structure and its ligatures, which may be the result of local shape irregularities or noise.

1 Introduction

The skeletal abstraction of 2D and 3D objects has proved to be an alluring yet highly elusive goal for over 30 years in shape analysis. The topic is not only important in image analysis, where it has stimulated a number of important developments including the medial axis transform and iterative morphological thinning operators, but is also an important field of investigation in differential geometry and biometrics where it has lead to the study of the so-called Blum skeleton [4]. Because of this, the quest for reliable and efficient ways of computing skeletal shape descriptors has been a topic of sustained activity. Recently, there has been a renewed research interest in the topic which as been aimed at deriving a richer description of the differential structure of the object boundary. This literature has focused on the so-called shock-structure of the reaction-diffusion equation for object boundaries.

The idea of characterising boundary shape using the differential singularities of the reaction equation was first introduced into the computer vision literature by Kimia Tannenbaum and Zucker [9]. The idea is to evolve the boundary of an object to a canonical skeletal form using the reaction-diffusion equation. The skeleton represents the singularities in the curve evolution, where inward moving boundaries collide. The reaction component of the boundary motion corresponds to morphological erosion of the boundary, while the diffusion component introduces curvature dependent boundary smoothing. In practice, the skeleton can be computed in a number of ways [1,11]. Recently, Siddiqi, Tannenbaum and Zucker have shown how the eikonal equation which underpins the

C. Arcelli et al. (Eds.): IWVF4, LNCS 2059, pp. 260–271, 2001.

reaction-diffusion analysis can be solved using the Hamilton-Jacobi formalism of classical mechanics [6,17].

One of the criticisms that can be levelled at existing skeletonisation methods is their sensitivity to small boundary deformations or ligatures. Although these can be reduced via curvature dependent smoothing, they may have a significant effect on the topology of the extracted skeleton.

Once the skeletal representation is to hand then shapes may be matched by comparing their skeletons. Most of the work reported in the literature adopts a structural approach to the matching problem. For instance, Pelillo, Siddiqi and Zucker use a sub-tree matching method [14] This method is potentially vulnerable to structural variations or errors due to local deformations, ligature instabilities or other boundary noise. Tithapura, Kimia and Klein have a potentially more robust method which matches by minimising graph-edit distance [10,20].

One of the criticisms of these structural matching methods is that perceptually distinct shapes may have topologically identical skeletons which can not be distinguished from one-another. Moreover, small boundary deformations may significantly distort the topology of the skeleton.

We draw two observations from this review of the related literature. The first is that the existing methods for matching are based on largely structural representations of the skeleton. As a result, shapes which are perceptually different but which give rise to the same skeleton topology are ambiguous with one-another. For this reason we would like to develop a metrical representation which can be used to assess the differences in shape for objects which have topologically identical skeletons. Secondly, we would also like to be able to make comparisons between shapes that are perceptually close, but whose skeletons exhibit topological differences due to small but critical local shape deformations.

To meet these dual goals, our shape-measure must have three properties. First, it must be continuous over local regions in shape-space in which there are no topological transitions. If this is the case then it can be used to differentiate shapes with topologically identical skeletons. Secondly, it must vary smoothly across topological transitions. This is perhaps the most important property since it allows us to define distances across transitions in skeleton topology. In other words, we can traverse the skeleton without encountering singularities. Thirdly, it must distinguish between the principal component of the skeleton and its ligatures [2]. This will allow us to suppress instabilities due to local shape deformations.

Commencing from these observations, we opt to use a shape-measure based on the rate of change of boundary length with distance along the skeleton. To compute the measure we construct the osculating circle to the two nearest boundary points at each location on the skeleton. The rate of change of boundary length with distance along the skeleton is computed by taking neighbouring points on the skeleton. The corresponding change in boundary length is computed by determining distance along the boundary between the corresponding points of contact for the two osculating circles. The boundary distances are averaged for the boundary segments either side of the skeleton.

This measurement has previously been used in the literature to express *relevance* of a branch when extracting or pruning the skeleton [11,12]. We show that rate of change of boundary length with distance along the skeleton has a number of interesting properties. The consequence of these properties is that the descriptive content of the measure extend beyond simple feature saliency, and can be used to attribute the relational structure of the skeleton to achieve a richer description of shape. Furthermore, we demonstrate that there is an intimate relationship between the shape measure and the divergence of the distance map. This is an important observation since the divergence plays an central role when the skeleton is computed using the Hamilton-Jacobi formalism to solve the eikonal equation.

2   Skeleton Detection

A great number of papers have been written on the subject of skeleton detection. The problem is a tricky one because it is based on the detection of singularities on the evolution of the eikonal equation on the boundary of the shape.

The eikonal equation is a partial differential equation that governs the motion of a wave-front through a medium. In the case of a uniform medium the equation is $\frac{\partial}{\partial t} C(t) = \alpha N(t)$, where $C(t) : [0, s] \to \mathbb{R}^2$ is the equation of the front at time t and $N(t) : [0, s] \to \mathbb{R}^2$ is the equation of the normal to the wave front in the direction of motion and α is the propagation speed. As the wave front evolves, opposing segments of the wave-front collide, generating a singularity. This singularity is called a shock and the set of all such shocks is the skeleton of the boundary defined by the original curve. This realisation of the eikonal equation is also referred to as the reaction equation.

To detect the singularities in the eikonal equation we use the Hamilton-Jacobi approach presented by Siddiqi, Tannenbaum, and Zucker [6,17]. Here we review this approach.

We commence by defining a distance-map that assigns to each point on the interior of an object the closest distance D from the point to the boundary (i.e. the distance to the closest point on the object boundary). The gradient of this distance-map defines a field F whose domain is the interior of the shape. The field is defined to be $F = \nabla D$, where $\nabla = (\frac{\partial}{\partial x}, \frac{\partial}{\partial y})^T$ is the gradient operator. The trajectory followed by each boundary point under the eikonal equation can be described by the ordinary differential equation $\dot{x} = F(x)$, where x is the coordinate vector of the point. This is a Hamiltonian system, i.e. wherever the trajectory is defined the divergence of the field is zero [13]. However, the total inward flux through the whole shape is non zero. In fact, the flux is proportional to the length of the boundary.

The divergence theorem states that the integral of the divergence of a vector-field over an area is equal to the flux of the vector field over the enclosing boundary of that area. In our case, $\int_A \nabla \cdot F \, d\sigma = \int_L F \cdot n \, dl = \Phi_A(F)$, where A

is any area, F is a field defined in A, $d\sigma$ is the area differential in A, dl is the length differential on the border L of A, and $\Phi_A(F)$ is the outward flux of F through the border L.

By virtue of the divergence theorem we have that, within the interior, there are points where the system is not conservative. The non-conservative points are those where the boundary trajectory is not well defined, i.e. where there are singularities in the evolution of the boundary. These points are the so-called shocks or skeleton of the shape-boundary. Shocks are thus characterised by locations where $\nabla \cdot F < 0$. Unfortunately, skeletal points are, also, ridges of the distance map D, that is $F = \nabla D$ is not uniquely defined in those points, but have different values on opposite sides of the watershed. This means that the calculation the derivatives of F gives raise to numerical instabilities. To avoid this problem we can use the divergence theorem again. We approximate the divergence with the outward flux through a small area surrounding the point. That is $\nabla \cdot F(x) \approx \Phi_{\mathcal{U}}(F)(x)$, where $\mathcal{U}$ is a *small* area containing x. Thus, calculating the flux through the immediate neighbors of each pixel we obtain a suitable approximation of $\nabla \cdot F(x)$.

2.1 Locating the Skeleton

The thinning of the points enclosed within the boundary to extract the skeleton is an iterative process which involves eliminating points with low inward flux. The steps in the thinning and localisation of the skeleton are as follows

- At each iteration of the thinning process we have a set of points that are candidates for elimination. We remove from this set the point with the lowest inward flux.
- Next and we check whether the point is topologically simple, i.e. whether it can be eliminated without splitting the remaining point-set.
- If the point is not simple, then it must be part of the skeleton. Thus we retain it.
- If the point is simple, then we check whether it is an endpoint. If the point is simple and not an endpoint, then we eliminate it from the image. If this is the case then we add to the candidate set the points in its 8-neighborhood that are still part of the thinned shape (i.e. points that were not previously eliminated).
- If a simple point is also an endpoint, then the decision of whether or not it will be eliminated is based on the inward flux value. If the flux value is below a certain threshold we eliminate the point in the manner described above. Otherwise we retain the point as part of the skeleton.

We initialise this iterative process by placing every boundary point in the candidate set. We iterate the process until we have no more candidates for removal. The residual points will all belong to the skeleton.

3   The Shape-Measure and Its Properties

When the skeleton is computed in this way, then the eikonal equation induces a map from a point in the skeleton to a set of points on the boundary of the shape. That is, there is a correspondence between a point on the skeleton and the set of points on the boundary whose trajectories intercept it under the motion induced by the eikonal equation. The cardinality of this set of corresponding points on the boundary can be used to classify the local topology of the skeleton in the following manner

- the cardinality is greater than or equal to 3 for junctions.
- for endpoints the cardinality is number from 1 to a continuum.
- for the general case of points on branches of the skeleton, the cardinality is exactly 2.

As a result of this final property, any segment of a skeleton branch s is in correspondence with two boundary segments l_1 and l_2. This allows us to assign to a portion of the skeleton the portion of the boundary from which it arose. For each internal point in a skeleton branch, we can thus define the local ratio between the length of the generating boundary segment and the

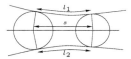

Fig. 1. Geometric quantities

length of the generated skeleton segment The rate of change of boundary length with skeleton length is defined to be $dl/ds = dl_1/ds + dl_2/ds$. This ratio is our measure of the relevance of a skeleton segment in the representation of the 2D shape-boundary.

Our proposal in this paper is to use this ratio as a measure of the local relevance of the skeleton to the boundary-shape description. In particular we are interested in using the measure to identify ligatures [2]. Ligatures are skeleton segments that link the logically separate components of a shape. They are characterised by a high negative curvature on the generating boundary segment. The observation which motivates this proposal is that we can identify ligature by attaching to each infinitesimal segment of skeleton the length of the boundary that generated it. Under the eikonal equation, a boundary segment with high negative curvature produces a rarefaction front. This front will cause small segments to grow in length throughout their evolution, until they collide with another

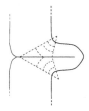

Fig. 2. Ligature points are generated by short boundary segments

front and give rise to a so-called shock. This means that very short boundary segments generate very long skeleton branches. Consequently, when a skeleton branch is a ligature, then there is an associated decrease in the boundary-length to shock-length ratio. As a result our proposed skeletal shape measure 'weights" ligature less than other points in the same skeleton branch.

To better understand the rate of decrease of the boundary length with skeletal length, we investigate its relationship to the local geometry of the osculating circle to the object boundary. We have

$$dl_1/ds = \frac{\cos\theta}{1 - rk_1} \quad \text{and, similarly,} \quad dl_2/ds = \frac{\cos\theta}{1 - rk_2} \tag{1}$$

where r is the radius of the osculating circle, k_i is the curvature of the mapped segment on the boundary, oriented so that positive curvatures imply the osculating circle is in the interior of the shape, and, finally, θ is the angle between the tangent to the skeleton and the tangent to the corresponding point on the boundary. These formulae show that the metric is inversely proportional to negative curvature and radius. That is, if we fix a negative curvature k_1, the measure decreases as the skeleton gets further away from the border. Furthermore, the measure decreases faster when the curvature becomes more negative.

Another important property of the shape-measure is that its value varies smoothly across shape deformations, even when these deformations impose topological transitions to the skeleton. To demonstrate this property we make use of the taxonomy of topological transition of the skeleton compiled by Giblin and Kimia [7]. According to this taxonomy, a smooth deformation of the shape induces only two types of transition on the skeleton (plus their time reversals). The transitions are *branch contraction* and *branch splicing*. A deformation *contracts* a branch joining two junctions when it moves the junctions together. Conversely, it *splices* a branch when it reduces in size, smoothes out, or otherwise eliminates the protrusion or sub-part of a shape that generates the branch.

A deformation that contracts or splices a skeleton branch, causes the global value of the shape-measure along the branch to go to zero as the deformation approaches the topological transition. This means that a deceasing length of boundary generates the branch, until the branch disappears altogether.

When a deformation causes a contraction transition, both the length of the skeleton branch and the length of the boundary segments that generate the branch go to zero. A more elusive case is that of splicing. Through a splicing deformation, a decreasing length of boundary maps to the skeleton branch. This is because either the skeleton length and its associated boundary length are both reduced, or because the deformation allows boundary points to be mapped to adjacent skeleton branches. For this reduction in the length of the generating boundary, we do not have a corresponding reduction of the length of the skeleton branch. In fact, in a splice operation the length of the skeleton branch is a lower bound imposed by the presence of the ligature. This is the major cause of the perceived instability of the skeletal representation. Weighting each point on the boundary which gave rise to a particular skeleton branch allows us to eliminate the contributions from ligatures, thus smoothing the instability. Since a smooth shape deformation induces a smooth change in the boundary, the total shape-measure along the branch has to vary smoothly through any deformation.

Just like the radius of the osculating circle, key shape elements such as necks and seeds are associated with local variations of the length ratio. For instance, a neck is a point of high rarefaction and, thus, a minimum of the shape-measure along the branch. A seed is a point where the front of the evolution of the eikonal equation concentrates, and so is characterised by a maximum of the ratio.

Another important property of the shape-measure is its invariance to bending of the shape. Bending invariance derives from the fact that, if we bend the shape, we loose from one side the same amount of boundary-length that we gain on the opposite side. To see this we let k be the curvature on the skeleton, and k_1 and k_2 be the inward curvature on the corresponding boundary points. Further, suppose that θ is the angle between the border tangent and the skeleton tangent. Let $p = 2k + \frac{k_1 \cos \theta}{1 - rk_1} = \frac{k_2 \cos \theta}{1 - rk_2}$, thus we have $p = k_2(\cos \theta + pr)$ and

$$k_2 = \frac{p}{\cos \theta + pr} = \frac{2k + \frac{\cos \theta}{1 - rk_1}k_1}{\cos \theta + 2rk + \frac{r \cos \theta}{1 - rk_1}k_1} = \frac{2k(1 - rk_1) + k_1 \cos \theta}{2rk(1 - rk_1) + \cos \theta}$$

Substituting the above in (1), we have

$$dl2/ds = \frac{\cos \theta}{1 - r\frac{2k(1 - rk_1) + k_1 \cos \theta}{2rk(1 - rk_1) + \cos \theta}} = \frac{2rk(1 - rk_1) + \cos \theta}{1 - rk_1}$$

Thus we find that $dl_2/ds = 2rk + dl_1/ds$, or $dl_2/ds - rk = dl_1/ds + rk$. That is, if we bend the image enough to cause a curvature k in the skeleton, what we lose on one side we get back on the other.

4   Measure Extraction

The extraction of the skeletal shape measure is a natural by-product which comes for free when we use the Hamilton-Jacobi approach for skeleton extraction. This is a very important property of this shape-measure. Using the divergence theorem we can transport a quantity linked to a potentially distant border to a quantity local to the skeleton. Using this property, we can prove that the border length to shock length ratio is proportional to the divergence of the gradient of the distance map.

As we have already mentioned, the eikonal equation induces a system that is conservative everywhere except on the skeleton. That is, given the field F defined as the gradient of the distance map, the divergence of F is everywhere zero, except on the skeleton.

To show how the shape-measure can be computed in the Hamilton-Jacobi setting, we consider a skeleton segment s and its ϵ-envelope. The segment s maps to two segment borders l_1 and l_2. The evolution of the points in these border segments define two areas A_1^ϵ and A_2^ϵ enclosed within the ϵ-envelope of s, the segments of boundary l_1 and l_2, and the trajectories b_1^1 and b_1^2, and b_2^1 and b_2^2 of the endpoints of l_1 and l_2. The geometry of these areas is illustrated in figure 3.

Since $\nabla \cdot F = 0$ everywhere in A_1^ϵ and A_2^ϵ, by virtue of the divergence theorem we can state that the flux from the two areas are both zero, i.e. $\Phi_{A_1^\epsilon}(F) = 0$ and $\Phi_{A_2^\epsilon}(F) = 0$. The trajectories of the endpoints of the border are, by construction, parallel to the field, so the normal is everywhere normal to the field and thus there is no flux through the segments b_1^1, b_1^2, b_2^1 and b_2^2. On the other hand the field on the shape-boundary is always normal to the boundary. Hence, the flux through the border segments l_1 and l_2 is equal to the length $\ell(l_1)$ and $\ell(l_2)$ of the segments l_1 and l_2 respectively.

Since $\Phi_{A_1^\epsilon}(\boldsymbol{F}) = 0$ and $\Phi_{A_2^\epsilon}(\boldsymbol{F}) = 0$ the flux that enters through the border segments l_1 and l_2 has to exit through the ϵ-envelope of s. That is, if ϵ_1 and ϵ_2 are the sides of A_1^ϵ and A_2^ϵ on the ϵ-envelope of s, we have $\Phi_{\epsilon_1}(\boldsymbol{F}) = \Phi_{l_1}(\boldsymbol{F})$ and $\Phi_{\epsilon_2}(\boldsymbol{F}) = \Phi_{l_2}(\boldsymbol{F})$. This, in turn, implies that the flux through the whole ϵ-envelope of s is $\Phi_\epsilon(\boldsymbol{F}) = \ell(l_1) + \ell(l_1)$.

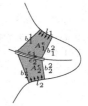

Since $lim_{\epsilon \to 0} \int_\epsilon \nabla \cdot \boldsymbol{F} \, d\epsilon = \int_s \nabla \cdot \boldsymbol{F} \, ds$, and the value of the flux through the ϵ-envelope of s is independent of ϵ, we have $\int_s \nabla \cdot \boldsymbol{F} \, ds = \ell(l_1) + \ell(l_2)$.

Taking the first derivative with respect to ds we have, for each non-singular point in the skeleton, $\nabla \cdot \boldsymbol{F} = dl_1/ds + dl_2/ds$.

Fig. 3. The flux through the border and through ϵ are equal

4.1   Computing the Distance between Skeletons

This result allows us to calculate a global shape-measure for each skeleton branch during the branch extraction process. For our matching experiments we have used a simple graph representation where the nodes are junctions or endpoints, and the edges are branches of the skeleton. When we have completed the thinning of the shape boundary and we are left only with the skeleton, we pick an endpoint and start summing the values of the length ratio for each skeleton points until we reach a junction. This sum $\sum_{i \in s} \nabla \cdot \boldsymbol{F}(\boldsymbol{x}_i)$ over every pixel x_i of our extracted skeleton branch is an approximation of $\int_s \nabla \cdot \boldsymbol{F} \, ds = \int_s (dl_1/ds + dl_2/ds) = \ell(l_1) + \ell(l_2)$ the length of the border that generates the skeleton branch.

At this point we have have identified a branch and we have calculated the total value of the length-ratio along that branch, or, in other words, we have computed the total length of the border that generated the branch. We continue this process until we have spanned each branch in the entire skeleton. Thus we obtain a weighted graph representation of the skeleton. In the case of a simple shape, i.e. a shape with no holes, the graph has no cycles and thus is an (unrooted) tree.

Given this representation we can cast the problem of computing distances between different shapes as that of finding the tree edit distance between the weighted graphs for their skeletons.

Tree edit distance is a generalization to trees of *String edit distance*. The edit distance is based on the existence of a set S of basic edit operation on a tree and a set C of costs, where $c_s \in C$ is the cost of performing the edit operation $s \in S$. The choice of the basic edit operations, as well as their cost, can be tailored to the problem, but common operations include leaf pruning, path merging, and, in case of an attributed tree, change of attribute. Given two trees T_1 and T_2, the set S of basic edit operations, and the cost of such operation $C = c_s, s \in S$, we call an *edit path* from T_1 to T_2 a sequence $s_1, \ldots, s_n$ of basic edit operations that transform T_1 into T_2. The length of such path is $l = c_{s_1} + \cdots + c_{s_n}$; the *minimum length edit path* from T_1 to T_2 is the path form T_1 to T_2 with minimum length. The length of the minimum length path is the tree edit distance.

With our measure assigned to each edge of the tree, we define the cost of matching two edges as the difference of the total length ratio measure along the branches. The cost of eliminating an edge is equivalent to the cost of matching it to an edge with zero weight, i.e. one along which the total length ratio is zero.

5   Experimental Results

In this section we asses the ability of the proposed measure to discriminate between different shapes that give rise to skeletons with the same topology. We will also asses how smoothly the overall measure goes through transitions.

As demonstrated earlier in the paper, we know that the length ratio measure should be stable to any local shape deformation, including those that exhibit an instability in shock length. This kind of behaviour at local deformations is what has led to the idea that the skeleton is an unstable representation of shape.

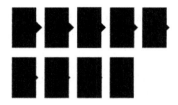

Fig. 4. A "disappearing" protrusion which causes instability in shock-length, but not in our measure

To demonstrate the stability of the skeletal representation when augmented with the length ratio measurement, we have generated a sequence of images of a rectangle with a protrusion on one side (Figure 4). The size of the protrusion is gradually reduced throughout the sequence, until it is completely eliminated in the final image. In figure 5 we plot the global value of the length ratio measure for the shock branch generated by the protrusion. It is clear that the value of the length ratio measure decreases monotonically and quite smoothly until it becomes zero when the protrusion disappears.

In a second set of experiments we have aimed to assess the ability of the length ratio measure to distinguish between structurally similar shapes. To do this we selected two shapes that were perceptually different, but which possessed skeletons with a very similar topology. We, then, generated an image sequence in which the two shapes were morphed into one-another. Here the original shapes are the start and end frames of the sequence. At each frame in the sequence we calculated the distance between the start and end shapes.

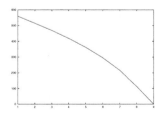

Fig. 5. The measure of the skeleton segment generated by a protrusion

We repeated this experiment with two morphing sequences. The first sequence morphed a sand shark into a swordfish, while the second morphed a donkey into a hare.

Fig. 6. Morphing sequences and their corresponding skeletons: sand shark to swordfish on the left, and donkey to hare on the right

To determine the distance between two shapes we used is the Euclidean distance between the normalized weights of matched edges. In other words, the distance is $D(A, B) = \sqrt{\sum_i (e_i^A - e_i^B)^2}$ where e_i^A and e_i^B are the normalised weights on the corresponding edges indexed by i on the shapes denoted by A and B. The normalised weights are computed by dividing the raw weights by the sum of the weights of each tree.

We apply this normalized length ratio measure to ensure scale invariance: two identical shapes scaled to different proportion would have different ratios due to the scale difference, but measure along equivalent branches of the two shapes would vary by a constant scale factor: the ratio of the lengths of the borders. Since the the sum of the weights of the edges of a tree is equal to the total length of the border, dividing the weights in each branch by this quantity we have reduced the two measurements to the same scale. In this way the relevant quantity is not the absolute magnitude for a branch, but the magnitude ratio with other branches.

There is clearly an underlying correspondence problem involved in calculating the distance in this way. In other words, we need to know which edge matches with which. To fully perform a shape recognition task we should solve the correspondence problem. However, the aim of the work reported here was to analyze the properties of our length ratio measure and not to solve the full recognition problem. Thus for the experiments reported here we have located the edge correspondences by hand.

For each morphing sequence, in figure 7 we plot the distance between each frame in the sequence and the start and end frames. The monotonicity of the distance is evident throughout the sequences.

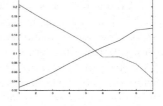

(a) Distances in fish morphing sequence

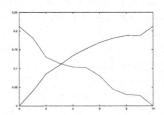

(b) Distances in donkey to hare morphing sequence

Fig. 7. Distances from first and last frame of the morphing sequences

This is a proof of capacity of our length ratio measure to disambiguate between shapes with topologically similar skeletons.

To further asses the ability to discriminate between similar shapes, we selected a set of topologically similar shapes from a database of images of tools. In the first column of figure 8 we show the selected shapes. To their right are the remaining shapes sorted by increasing normalized distance. Each shape is annotated by the value of the normalized distance.

It is clear that similar shapes are usually closest to one-another. However, there are problems due to a high sensitivity to occlusion. This can be seen in the high relative importance given to the articulation angle. This is due to the fact that, in the pliers images, articulation occludes part of nose of pliers. While sensitivity to occlusion is, without a doubt, a drawback of the measure, we have to take into account that skeletal representation in general are highly sensitive to occlusion.

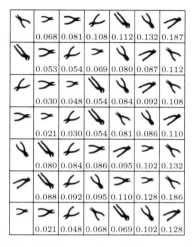

Fig. 8. Some tools and the normalized distance between them

6   Conclusions

In this paper we presented a shape measure defined on the skeleton. This measure has been used in the literature as a branch relevance measure during skeleton extraction and pruning. We state that the informational content of the measure goes beyond this use, and can be used to augment the purely structural information residing in a skeleton in order to perform shape indexation and matching tasks.

We show that the shape measure has a number of interesting properties that allow it to distinguish between structurally similar shapes. In particular, the measure a) changes smoothly through topological transitions of the skeleton, b) is able to distinguish between ligature and non-ligature points and to weight them accordingly, and c) it exhibits invariance under "bending". What makes the use of this measure particularly appealing is the fact that it can be calculated with no added effort when the skeleton is computed using the Hamilton-Jacobi method of Siddiqi, Tannenbaum and Zucker.

References

[1] C. Arcelli and G. Sanniti di Baja. A width-independent fast thinning algorithm. *IEEE Trans. on PAMI*, 7(4):463-474, 1985.

[2] J. August, K. Siddiqi, and S. W. Zucker. Ligature instabilities in the perceptual organization of shape. *Comp. Vision and Image Und.*, 76(3):231-243, 1999.

[3] J. August, A. Tannenbaum, and S. W. Zucker. On the evolution of the skeleton. In *ICCV*, pages 315–322, 1999.

[4] H. Blum. Biological shape and visual science (part I). *Journal of theoretical Biology*, 38:205-287, 1973.

[5] G. Borgefors, G. Ramella, and G. Sanniti di Baja. Multi-scale skeletons via permanence ranking. In *Advances in Visual Form Analysis*, 31-42, 1997.

[6] S. Bouix and Kaleem Siddiqi. Divergence-based medial surfaces. In *Computer Vision ECCV 2000*, 603-618. Springer, 2000. LNCS 1842.

[7] P. J. Giblin and B. B. Kimia. On the local form and transitions of symmetry sets, medial axes, and shocks. In *ICCV*, 385-391, 1999.

[8] B. B. Kimia and K. Siddiqi. Geometric heat equation and nonlinear diffusion of shapes and images. *Comp. Vision and Image Understanding*, 64(3):305-322, 1996.

[9] B. B. Kimia, A. R. Tannenbaum, and S. W. Zucker. Shapes, shocks, and deformations I. *International Journal of Computer Vision*, 15:189-224, 1995.

[10] P. Klein et al. A tree-edit-distance algorithm for comparing simple, closed shapes. In *ACM-SIAM Symp. on Disc.e Alg.*, 1999.

[11] R. L. Ogniewicz. A multiscale mat from voronoi diagrams: the skeleton-space and its application to shape description and decomposition. In *Aspects of Visual Form Processing*, 430-439, 1994.

[12] R. L. Ogniewicz and O. Kübler. Hierarchic voronoi skeletons. *Pattern Recognition*, 28(3):343-359, 1995.

[13] S. J. Osher and J. A. Sethian. Fronts propagating with curvature dependent speed: Algorithms based on hamilton-jacobi formulations. *J. of Comp. Physics*, 79:12-49, 1988.

[14] M. Pelillo, K. Siddiqi, and S. W. Zucker. Matching hierarchical structures using association graphs. *PAMI*, 21(11):1105-1120, 1999.

[15] D. Sharvit, J. Chan, H. Tek, and B. B. Kimia. Symmetry-based indexing of image database. *J. of Visual Comm. and Image Rep.*, 9(4):366–380, 1998.

[16] A. Shokoufandeh, S. J. Dickinson, K. Siddiqi, and S. W. Zucker. Indexing using a spectral encoding of topological structure. In *CVPR*, 1999.

[17] K. Siddiqi, S. Bouix, A. Tannenbaum, and S. W. Zucker. The hamilton-jacobi skeleton. In *ICCV*, 828-834, 1999.

[18] K. Siddiqi and B. B. Kimia. A shock grammar for recognition. In *CVPR*, 507-513, 1996.

[19] K. Siddiqi, A. Shokoufandeh, S. J. Dickinson, and S. W. Zucker. Shock graphs and shape matching. *International Journal of Computer Vision*, 35(1):13–32, 1999.

[20] S. Tirthapura et al. Indexing based on edit-distance matching of shape graphs. In *SPIE Int. Symp. on Voice, Video, and Data Comm.*, 25-36, 1998.

Perception-Based 2D Shape Modeling by Curvature Shaping

Liangyin Yu* and Charles R. Dyer**

Computer Sciences Department
University of Wisconsin
Madison, Wisconsin 53706 USA
{yuly,dyer}@cs.wisc.edu

Abstract. 2D curve representations usually take algebraic forms in ways not related to visual perception. This poses great difficulties in connecting curve representation with object recognition where information computed from raw images must be manipulated in a perceptually meaningful way and compared to the representation. In this paper we show that 2D curves can be represented compactly by imposing shaping constraints in curvature space, which can be readily computed directly from input images. The inverse problem of reconstructing a 2D curve from the shaping constraints is solved by a method using curvature shaping, in which the 2D image space is used in conjunction with its curvature space to generate the curve dynamically. The solution allows curve length to be determined and used subsequently for curve modeling using polynomial basis functions. Polynomial basis functions of high orders are shown to be necessary to incorporate perceptual information commonly available at the biological visual front-end.

1 Introduction

The first goal of visual perception is to make the structure of the contrast variation in the image explicit. For stationary images, the structure is organized through the curvilinear image contours. From the point of view of information theory, the probability that an image contour is formed by some random distribution of contrast is extremely small and thus is highly informative. For the contour itself, it is also more meaningful to identify the part of the image contour that is more informative than other parts of the same contour. Though this principle is important from either the view of information theory or data compression, it is nonetheless essential to inquire about the inverse problem, i.e., how can the less informative part be recovered from the more informative part? This paper is about both problems in the 2D curve domain with main emphasis on the inverse problem.

---

* Current address: Canon Research Centre Europe, 1 Occam Court, Occam Road, Guildford, Surrey GU2 7YJ, United Kingdom
** The support of the National Science Foundation under Grant No. IIS-9988426 is gratefully acknowledged.

C. Arcelli et al. (Eds.): IWVF4, LNCS 2059, pp. 272–282, 2001.

Methods for representing 2D curves are usually segment-based with each segment defined by either a straight line (polygon) or a parameterized curve (spline). The segmentation points that separate segments are determined from properties computed along the curve, among them curvature is the most commonly used [4,5]. However, the properties for curve segmentation are generally highly sensitive to scale changes because the computation is commonly conducted after the contours are identified, which is notoriously dependent on the scale used. This problem can be avoided by using methods that compute the curvature along the contour directly from the image and a carefully designed selection scheme for the scales used in the computation [11].

Curvature has been considered to be one of the major perceptual properties of 2D shapes [2,5,10]. It is invariant to rigid transformations and can be computed by our physiological system [3,7]. It has been used extensively in shape matching [8] and object recognition [6] as well as for shape modeling in both 2D [9] and 3D [1].

From these observations regarding a 2D curve and its perception, the problem of 2D curve modeling can be formulated as a two-stage process: first, the perception-based selection of the local parts on the curve to be modeled, and second, the measurement of relevant modeling parameters regarding the shape. In this paper we also formulate the inverse problem of constructing a curve from a given set of parameters that has been selected previously as modeling parameters. The combination of the significance of curvature in visual perception and the importance of geometrical modeling of image contours is motivation for developing a new framework for 2D curve representation and reconstruction using curvature.

2 Background

2.1 Direct Curvature Computation from an Image

Curvature computation on image contours is generally sensitive to noise due to the way the computation is conducted, i.e., compute curvature from contour detected. This problem can be remedied greatly by computing curvature directly from an image at a tentative contour position [11]. The method can be extended to the computation of higher-order differential invariants such as the derivative of curvature, which will be used extensively in this paper.

Let $I(x, y)$ be the input image. A 2D Gaussian kernel is separable and defined by $\psi_{00}(x, y; \sigma) = \psi_0(x; \sigma)\psi_0(y; \sigma)$, with σ being a scale parameter and $\psi_0(x; \sigma) = (1/\sqrt{2\pi}\sigma)\exp(-x^2/2\sigma^2)$ an 1D Gaussian kernel. The ith-order and jth-order differentiations of ψ with respect to x and y are given by $\psi_{ij}(x, y; \sigma) = \psi_i(x; \sigma)\psi_j(y; \sigma)$. It can be shown that the curvature at location (x_0, y_0) of an image contour is given by

$$\kappa(x_0, y_0) = \frac{\psi_{20}(x^r, y^r) * I(x, y)}{\psi_{01}(x^r, y^r) * I(x, y)} \tag{1}$$

where $*$ is the convolution operator and $(x^r, y^r) = (x\cos\theta + y\sin\theta, -x\sin\theta + y\cos\theta)$ with θ being the orientation of the contour. The derivative of curvature is then given by $d\kappa/ds = \kappa\lambda - (\psi_{30}(x^r, y^r) * I(x, y))/\Phi_\theta$, where $\Phi(x, y, \theta; \sigma) \stackrel{\triangle}{=} -(\partial\psi_{01}/\partial\theta) * I(x, y)$, $\lambda = -\nabla\Phi \cdot \boldsymbol{n}/\Phi_\theta$, and $\Phi_\theta = \partial\Phi/\partial\theta$ with $\boldsymbol{n}$ being the unit normal vector to the contour. An example of the curvature computed in this way is shown in Figure 1.

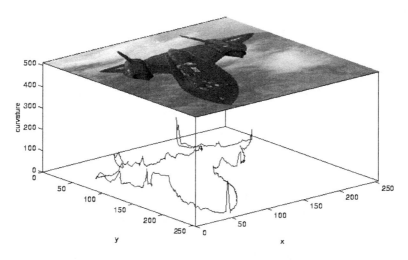

Fig. 1. The curvature along the contour of an airplane image.

2.2   Geometry of 2D Curves

Given a curve $\boldsymbol{c}(s) = (x(s), y(s))$ parameterized by its curve length s, the fundamental theorem of differential geometry for planar curves enables us to describe the curve uniquely (up to a rotation θ_0 and a translation (a, b)) using its curvature $\kappa(s)$. This is explicitly formulated by the intrinsic equations: $x(s) = \int \cos(\theta(s))\, ds + a$, $y(s) = \int \sin(\theta(s))\, ds + b$, $\theta(s) = \int \kappa(s)\, ds + \theta_0$ with three boundary conditions given to specify (a, b) and θ_0. The curve $\kappa(s)$ is the *curvature space* for $\boldsymbol{c}(s)$.

Hence the problem of shape representation in 2D image space is equivalent to the representation of the function $\kappa(s)$ in curvature space. The primary difficulty in using the intrinsic equations directly in curve reconstruction from curvature space is that there is no well-defined computational procedure for constraining the shape in either the 2D image space or the curvature space from the changing shape in the other space. In other words, when taking into account noise, both spaces are unstable by themselves. However, by incorporating both spaces into a reconstruction algorithm, satisfactory results can be achieved.

We will subsequently consider the following geometrical parameters of a curve. Given a smooth curve $c(s)$ that is C^2 continuous, two points P_0, P_1 on $c(s)$ at s_0, s_1 are such that the respective curvatures $\kappa(s_0), \kappa(s_1)$ are curvature extrema, i.e., $\kappa'(s_0) = \kappa'(s_1) = 0$. The points P_0, P_1 are called *feature points* for $c(s)$ (Figure 2).

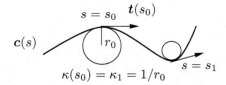

Fig. 2. The geometrical factors that determine a 2D curve.

Given this background, the problem of *2D shape representation using feature points* will be to locate the curvature extrema along a curve and construct the curve using these extremal points. Traditionally this goal is achieved through piecewise interpolation using cubic splines and matching boundary conditions at the knots. However, this approach is unable to incorporate the higher-order constraints of κ and κ'. The other problem is the relatively straight segments provided by this model, requiring more knots for more curved regions. This fact may not be favorable when the scale-space factor is taken into account, which requires a more or less even distribution of knot positions along the curve at a given scale. These problems can be alleviated by using higher order splines. Another problem is the extra feature points inserted by the basis functions. To solve this problem, a different approach that works on both image space and curvature space is required.

The parameter used for the interpolation is also problematic, especially when curve length parameterization is required. For a given image, the curve length along an image contour can generally be estimated quite accurately, and the relevant geometrical and modeling parameters can be computed. However, the inverse problem of finding the curve from a given set of boundary conditions does not provide information on curve length. The method presented in the next section provides the curve length information as one of its results. This information can then be used for modeling the curve using the high-order polynomial basis functions presented in Section 4.

3   Curve Representation by Curvature Space Shaping

Given two feature points, P_0, P_1, on an unknown curve $c(s)$ in an image and their associated tangent orientations, θ_0, θ_1, as well as their signed curvatures, κ_0, κ_1, we now present a method to solve the problem of finding the curve that satisfies the given boundary conditions with the property that there is no computable curvature extremum in between P_0 and P_1 other than those at P_0 and P_1.

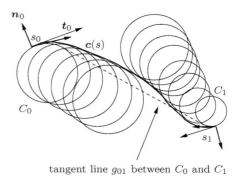

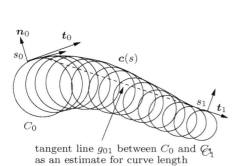

tangent line g_{01} between C_0 and C_1
as an estimate for curve length

tangent line g_{01} between C_0 and C_1
as an estimate for curve length

Fig. 3. Dynamical moves for a curve segment between two convex points.

Fig. 4. Dynamical moves for a curve segment between one convex and one concave point.

Let the osculating circle at P_0, P_1 be C_0, C_1 respectively, and the tangent line between C_0, C_1 in the direction of $t_0 = (\cos\theta_0, \sin\theta_0)$ be g_{01} (the unit vector g along g_{01} at the C_0 end has the property $g \cdot t_0 > 0$). Among the four tangent lines, g_{01} is chosen to be one of the two non-crossing ones closest to P_0 if $\kappa_0\kappa_1 > 0$, and one of the two crossing ones if $\kappa_0\kappa_1 < 0$. The curve $c(s)$ is constructed by dynamically moving stepwise from P_0 along a direction that will gradually changed into the direction of $t_1 = (\cos\theta_1, \sin\theta_1)$ while gradually changing the curvature of corresponding osculating circle in the process (Figures 3 and 4). Since $c(s)$ is unknown, the curve length between P_0 and P_1 cannot be determined in advance. Rather, we use the length of the tangent line g_{01} as an initial estimate for the curve length.

Let the desired steps for reaching P_1 from P_0 be n, and let the length of tangent line g_i at each step be s_i^g for $i = 1, \ldots n$. The direction of movement at each step i is determined by the corresponding θ_i and g_i by $(t_i + g_i)/2$, i.e., move half way in between t_i and g_i. The curvature κ_i of the osculating circle C_i is given by $\kappa_0 + (\kappa_1 - \kappa_0)/n$. The distance d_i to be moved consists of a movement along g_i followed by a movement along C_i and is given by $d_i = s_i^g/(n - i + 1)$. The orientation change caused by this movement is then given by $\Delta\theta = \kappa_i d_i = (\kappa_1 - \kappa_0)d_i/n$. This is formulated in such a way that if the estimated curve length s_i^g is indeed the curve length, then at each step we will move precisely $1/(n - i + 1)$ of the curve length and will complete our journey in n steps. The curve length is thus given by

$$s = \sum_{i=1}^{n} \frac{s_i^g}{n - i + 1}$$

The corresponding curvature spaces of the curves in Figures 3 and 4 are given in Figures 5 and 6, respectively. Under the solvability conditions explained in the next section, it can be shown that the movement will approach P_1 in n steps with

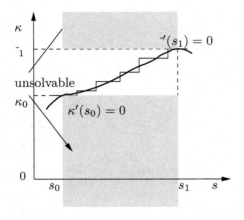

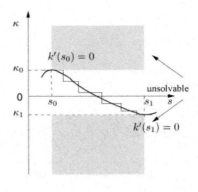

Fig. 5. The curvature space for a curve segment with two convex points.

Fig. 6. The curvature space for a curve segment with one convex and one concave point.

the desired boundary conditions, and the following limits can be established:

$$\lim_{n\to\infty} P_n = P_1, \qquad \lim_{n\to\infty} \theta_n = \theta_1, \qquad \lim_{n\to\infty} k_n = k_1$$

Each of the n segments of curve $c(s)$, according to the curvature space, is a partial arc on the osculating circle C_i with constant curvature κ_i (Figure 7), which can be approximated by a piecewise straight segment. Even though the curves have great similarity, their curvature spaces have completely different shapes. This illustrates the difficulty in working from only one of the spaces. In comparison, a constant curvature segment can better track the tangent line and converge faster to the destination than the straight segment counterpart because the osculating circle at each point bends toward rather than away from the line. This implies fewer steps are required and better precision.

3.1   Solvability Conditions

There are two conditions governing whether this problem has a solution. One corresponds to the "sidedness" of the object, since the sign of curvature is defined according to which side of the object the normal vector lies by the Frenet equation $t' = \kappa n$. The tangent line g_i gives an estimate of the curve length of $c(s)$ and at the same time defines on which side the object lies. It is unsolvable when the boundary conditions create an impossible object. The other condition of solvability is whether during the process the curvature enters an area in which extra extrema have to be created. These areas are denoted *unsolvable* in Figures 5 and 6. These two unsolvability conditions are illustrated in Figure 8.

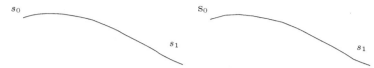

curve with piecewise constant curvature   curve with straight line segments

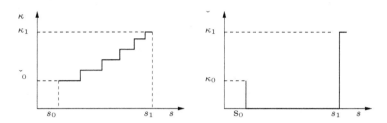

Fig. 7. The two segment models of a curve.

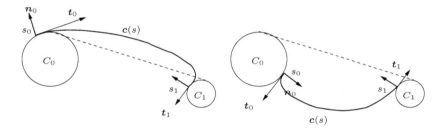

Fig. 8. The two unsolvability conditions between two convex points.

4   Curve Representation by Polynomial Basis

The curve length of an arbitrarily parameterized curve $c(t)$ is $s = \int |c'(t)|\, dt$. From this formulation it is clear that if $c(t)$ is represented by polynomial basis functions, its curve length will not be polynomial, and vice versa. Hence, to use a polynomial basis we can either work in the image space of $c(t) = (x(t), y(t))$ by fitting the boundary conditions $c_0, c_1, \theta_0, \theta_1, \kappa_0, \kappa_1, \kappa_0', \kappa_1'$, or work in curvature space through the intrinsic equations on the same set of boundary conditions. Both methods result in a set of highly nonlinear equations with the existence of a solution questionable. In this section we introduce a compromise method using polynomial basis functions in image space that satisfy all the boundary conditions but do not guarantee that new curvature extrema will not be inserted in the models.

4.1   Curves from Hermite Splines

The Hermite polynomials $H_{i,j}(t)$ of order L with $i = 0, \ldots, L$ and $j = 0, 1$ satisfy the *cardinal property*: $H_{i,0}^k(0) = \delta_{ki}, H_{i,0}^k(1) = 0, H_{i,1}^k(0) = 0, H_{i,0}^k(1) =$

δ_{ki}, where the superscript k indicates the order of differentiation. This set of equations defines polynomials of order 3 for $L = 1$, of order 5 for $L = 2$, and of order 7 for $L = 3$.

The cardinal property is useful for fitting boundary conditions of various differential orders. We will consider here the case of first, second and third order differentials. Let P_j^i be the ith derivative of the curve at $\boldsymbol{P}_j = \boldsymbol{c}(t_j)$. The curve segment connecting the two points $(t = t_0, t = t_1)$ using Hermite splines is

$$\boldsymbol{c}(t) = \sum_{j=0}^{1} \sum_{i=0}^{l} \boldsymbol{P}_j^i H_{i,j}\left(\frac{t - t_0}{t_1 - t_0}\right), \qquad l = 1, \dots, 3$$

This formulation is given in terms of the differentials at two boundary points, which are not readily available since the problem is given the conditions of location $(\boldsymbol{P}_0, \boldsymbol{P}_1)$, orientation (θ_0, θ_1), curvature (κ_0, κ_1), and differential of curvature (κ_0', κ_1'). Since the estimation of $\boldsymbol{P}_j^i$ is generally noisy, it is necessary to use curve length as a parameter. Hence, the curvature $\kappa = x'y'' - x''y'$ and $\kappa' = x'y''' - x'''y'$ since $(x')^2 + (y')^2 = 1$. Given κ and κ' allows us freedom to choose two additional conditions to fix x'', y'' and x''', y'''. This can be done arbitrary since the problem itself does not dictate these conditions.

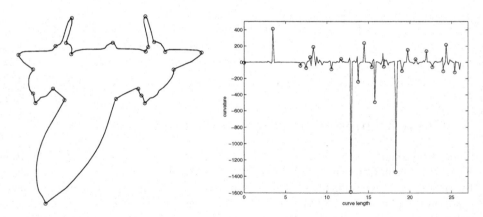

Fig. 9. The image contour of the air-plane in Figure 1.

Fig. 10. The curvature space and extrema for the airplane contour.

5   Examples

For the airplane contour in Figure 9, the curvature space is given in Figure 10, in which the curvature extrema are identified. The corresponding feature points are also marked in Figure 9. These points are the component partition points in which the concave points separate components while the convex points mark the partition of different segments within the same component. One of these components with feature points within the segment identified is shown in Figure 11. The curve length is estimated from the curve computed using the method of curvature shaping, and subsequently used in the representation by Hermite basis functions. Three different orders of the basis functions were used. Bases of order 3 used the location and tangent orientation information only. Bases of order 5 also matched boundary conditions for the curvature, while Hermite bases of order 7 satisfied the additional condition that these points are actually feature points with extremal curvature.

Fig. 11. Parts of the airplane represented by Hermite basis of order 3(···), 5 (−·),7 (−−), compared to the original (—).

6   Discussion

6.1   Scale Space

Scale space manifests its effect mostly in computation. From the formulation in Section 2.1, it can be observed that after the image contours are computed, the effect of the 2D scale-space kernel parameterized by rectangular Cartesian coordinates is equivalent to the effect of an 1D scale-space kernel parameterized by curve length. This is because the image contour is computed by orienting the kernel ψ_{01} in the direction of the contour [11]. This essentially creates a "curvature scale space" [9], in which variations within a fixed scale are gradually lost when the scale getting coarser. This results in a separation of feature points on the curve with a distance proportional to the scale used for the computation. Hence, even though the curvature $\kappa(s)$ is a highly nonlinear function of $x(s)$

and $y(s)$ and its shape cannot be exactly predicted for a given scale, the feature points with extremal curvature can nonetheless be located by searching the curvature space of finest scale and partitioning the curve with segments of length proportional to the scale without actually computing the curvature scale space for coarser scales.

6.2   Perceptual Boundary Conditions

The measure of distance in biological visual perception is provided by comparison between a reference length and what is to be measured, i.e., there is no intrinsic metric. This renders computations using distance measure (such as optical flow and curvature) imprecise. On the other hand, an orientation measure has built-in mechanisms with a certain degree of precision. From these observations, the primary measurement of the local shape of a curve will be the position and orientation relative to a 2D Cartesian coordinate system, while curve length and curvature are much less precise in terms of measurement. Hence the primary boundary conditions related to perception are locations and orientations. However, we do show that when secondary boundary conditions such as curvature and its derivative are available, the representation is much more compact and precise. For example, by using cubic or quintic Hermite bases for the component in Figure 11, precision can be augmented by adding more knot points. However, it is not clear which ones to choose since there is no special attribute in curvature space to facilitate this choice.

6.3   Component Partitions Using Curvature Space

There are two different kinds of information presented through the curvature space when the whole contour is considered. The prominent ones are actually component partition points (negative extrema) of the shape or segment partition points (positive extrema) within a component. This can be seen in Figures 1 and 10. Feature points for each segment have to be identified within the segment rather than compared to every segment in the contour, and their extremal property is essential to the modeling. The inadequacy of lower-order Hermite bases to represent a curve segment is clearly seen in Figure 11, since these do not take into account the extremal property at these points.

7   Conclusions

A compact description of a smooth curve was presented in this paper, based on curve features that are perceptually-important. The geometrical model of these features was defined by location, orientation and curvature, with the additional property that the curvature reaches extremal values at feature points. Being able to describe compactly an object contour is of great importance in object recognition, especially when the description is independent of the viewpoint. We also develop a method to identify the curve from a given set of perception-based boundary conditions at prescribed feature points. One of the results is a good estimation of curve length that can subsequently be used by polynomial basis functions for curve modeling. However, to satisfy the given boundary conditions, much higher-order polynomials are needed than what are commonly used.

References

[1] H. Asada and M. Brady. The curvature primal sketch. *IEEE Trans. Patt. Anal. Machine Intell.*, 8(1):2–14, 1986.

[2] F. Attneave. Some informational aspects of visual perception. *Psychology Review*, 61:183–193, 1954.

[3] A. Dobbins, S. Zucker, and M. Cynader. Endstopping and curvature. *Vision Research*, 29(10):1371–1387, 1989.

[4] M. Fischler and R. Bolles. Perceptual organization and curve partitioning. *IEEE Trans. Patt. Anal. Machine Intell.*, 8(1):100–105, 1986.

[5] M. Fischler and H. Wolf. Locating perceptually salient points on planar curves. *IEEE Trans. Patt. Anal. Machine Intell.*, 16(2):113–129, 1994.

[6] D. Hoffman and W. Richards. Parts of recognition. *Cognition*, 18:65–96, 1985.

[7] J. J. Koenderink and A. J. van Doorn. Receptive field families. *Biological Cybernetics*, 63:291–297, 1990.

[8] E. Milios. Shape matching using curvature processes. *Computer Vision, Graphics, and Image Processing*, 47:203–226, 1989.

[9] F. Mokhtarian and A. Mackworth. A theory of multiscale, curvature-based shape representation for planar curves. *IEEE Trans. Patt. Anal. Machine Intell.*, 14(8):789–805, 1992.

[10] W. Richards, B. Dawson, and D. Whittington. Encoding contour shape by curvature extrema. *J. Opt. Soc. Amer. A*, 2:1483–1491, 1986.

[11] L. Yu and C. Dyer. Direct computation of differential invariants of image contours from shading. In *Proc. Int. Conf. Image Processing*, 1998.

Global Topological Properties of Images Derived from Local Curvature Features

Erhardt Barth[1], Mario Ferraro[2], and Christoph Zetzsche[3]

[1] Institute for Signal Processing
University of Lübeck, Ratzeburger Allee 160, 23538 Lübeck, Germany
barth@isip.mu-luebeck.de, http://www.isip.mu-luebeck.de
[2] Dipartimento di Fisica Sperimentale and INFM
via Giuria 1, Torino, Italy
ferraro@to.infn.it
[3] Institut für Medizinische Psychologie, Goethestr. 31, 80336 München, Germany
chris@imp.med.uni-muenchen.de

Abstract. In this paper we show that all images are topologically equivalent. Nevertheless, one can define useful pseudo-topological properties that are related to what is usually referred to as topological perception. The computation of such properties involves low-level structures, which correspond to end-stopped and dot-responsive visual neurons. Our results contradict the common belief that the ability for perceiving topological properties must involve higher-order, cognitive processes.

Keywords: Topology, Euler number, closure, curvature, visual perception

1 Introduction

A fundamental issue of image analysis and understanding is the mathematical description, or representation, of the input data. The basic representation of an image and indeed the closest to the physical process of vision is that of a graph of a function $I : (x, y) \rightarrow I(x, y)$ from a subset U of R^2 to R; here U is the retinal plane and $I(x, y)$ is the light intensity at the point $(x, y) \in U$. From a geometrical point of view, the graph of I is a Monge patch, that is a surface L of the form $L = \{x, y, I(x, y)\}$. Alternatively, L can be considered as the image of the mapping $\phi : (x, y) \rightarrow (x, y, I(x, y))$. Then, the process of image formation can be seen as a mapping from the surface M of a physical object to the Monge patch L representing the image. Let S be the surface of an object in R^3. Consider a coordinate system $\{x, y, z\}$, whose origin coincides with the position of the viewer and let $\{x, y\}$ be the image plane. The visible part M of S can be given a Monge-patch representation $M = (x, y, f(x, y))$, where $z = f(x, y)$ is the distance of the point $r = (x, y)$ on the image plane of the observer to $s = (x, y, z)$ in M. Note that, if orthographic projection is assumed, f and I share the same domain U and there is a one-to-one correspondence

C. Arcelli et al. (Eds.): IWVF4, LNCS 2059, pp. 285–294, 2001.

between $p = (x_0, y_0, f(x_0, y_0)) \in M$ and $q = (x_0, y_0, I(x_0, y_0)) \in L$ [8]. The extension to the case of perspective projection is immediate. If a surface M is transformed by a continuous transformation into a surface M', a new image L' will be generated by M'; the relation between the transformation of the image and the transformation of the underlying object has been made precise in [15].

The representation of images as surfaces has two advantages: first it is close to the original data structure and does not require any high level process to be generated, second it allows one to use the technical machinery of geometry to investigate it. In particular the condition of differentiability allows to make use of a very powerful theorem of differential geometry, the Gauss-Bonnet theorem, which provides a link between global and local properties of surfaces. Surfaces can be given a global classification based on the notion of topological invariants. A property of a surface is called a topological invariant if it is invariant under homeomorphism, that is under an one-to-one continuous transformation that has a continuous inverse. Two surfaces are said to be topologically equivalent if they can be transformed one in the other by a homeomorphism. For instance, the number of holes in a surface is a topological invariant. In particular here we want to investigate, how the topological properties of object surfaces are reflected in the images. It will be shown in the next section that all images are topologically equivalent, i.e. any two images can be transformed one into the other by means of an homeomorphism. From this result it follows that the topological properties of an object's surface are not intrinsic properties of its image.

We shall show how topological properties of the object underlying the images can be found by using $2D$-operators, i.e. operators whose output is different from zero only in case of $2D$-features such as corners, line-ends, curved lines and edges. These operators can be associated to the activity of nonlinear end-stopped neurons. Mechanisms possibly underlying the activity of such neurons have been investigated by a few authors, e.g. [7,9,14,16]. A general theory for end-stopping and $2D$-operators, however, is still missing, but attempts to identify the basic ingredients for such a theory have been made [16,17].

Finally, we will show how $2D$-operators can be used to provide an alternative explanation for some experimental findings [4,5], which have suggested that the human visual system might be quite sensitive to global, "topological" characteristics of images.

2   The Gauss-Bonnet Theorem

The Gauss-Bonnet theorem is one of the most important theorems in differential geometry, in that it provides a remarkable relation between the topology of a surface and the integral of its Gaussian curvature.

Let R be a compact region (e.g. a Monge patch) whose boundary ∂R is the finite union of simple, closed, piecewise regular curves C_i. Consider a polygonal decomposition of R, that is a collection of polygonal patches that cover R in a way such that if any two overlap, they do so in either a single common vertex or a single common edge [12].

Thus a polygonal decomposition $\mathcal{D}$ carries with it not only the polygonal patches, called faces, but also the vertices and edges of these patches. Suppose $\mathcal{D}$ has f faces, e edges and v vertices, then $\chi = f - e + v$ is the Euler-Poincare' characteristic of R and it is the same for all polygonal decompositions of R [12]. The Euler-Poincare characteristic can be extended, in a natural way, to regular, compact surfaces. In Fig. 1 some examples of closed surfaces with different values of χ are shown. Two compact surfaces are topologically equivalent if and

Fig. 1. Surfaces with different Euler-Poincare' characteristics ($\chi = 2, 0, -2$ from left to right).

only if they have the same Euler-Poincare' characteristic χ [12,6]. Here, we are interested in Monge patches, that represent both the visible part of surfaces and their images. If the Monge patch M is a simple region, that is is homeomorphic to an hemisphere, $\chi = 1$ [6] , if it has an hole $\chi = 0$, and in general

$$\chi = (1 - n_{holes}).   \qquad (1)$$

Note that if, instead of regions, we consider regular surfaces, equation 1 becomes $\chi = 2(1 - n_{holes})$. The definition of χ holds for connected surfaces; we shall now extend the definition of χ to the case of not connected surfaces. Suppose there are n object in the scene, the global Euler-Poincare' characteristic χ_T is then simply $\chi_T = \sum_j^n \chi_j$.

Consider a region R. The Euler-Poincare' characteristic is related to the curvature of R by the celebrated Gauss-Bonnet formula

$$\iint_R K dA + \sum_i \int_{C_i} k_g ds + \sum_i \theta_i = 2\pi\chi,   \qquad (2)$$

here dA is the infinitesimal area element, C_i are regular curves forming the boundary ∂R of R, k_g is the geodetic curvature computed along the curve C_i and θ_i are the external angles at the vertices of ∂R.

If S is a compact orientable surface then $\int\int_S K dA = 2\pi\chi(S)$ [12,6].

This is a striking result: how is it possible that, when integrating a local property of a surface, we obtain global topological invariants of that surface? Let us consider the surface of a sphere S^2 in R^3. For a given radius r, the Gaussian curvature is $K = 1/r^2$ and $\int\int_{S^2} K dA = 4\pi$. Note that this result does not depend on r, since as r increases K decreases but the area of the surface increases. More importantly, the result is the same for all surfaces topologically equivalent to a sphere. Suppose S^2 is transformed into a surface S by deforming the sphere with a dent. In this case the area increases, and the elliptic part of the dent gives additional positive values. At the same time, however, new negative values of K are produced at the hyperbolic regions of the dent, and, as a consequence, positive and negative curvatures cancel out and the total curvature remains constant.

3   All Images Are Topologically Equivalent

In this section we give a formal proof that all images are topologically equivalent. Topological equivalence of two images implies that, given any pair of images with representation $L = (x, y, I(x, y))$ and $N = (x, y, J(x, y))$, there exists an homeomorphism taking one into the other. First we prove the following

Lemma 1 *Let I be a function from U to R and let $\phi : (x, y) \to (x, y, I(x, y))$ from U to R^3. If I is continuous then ϕ is an homeomorphism.*

Proof. The map ϕ is one-to-one and can be written as $\phi = i \times I$ where $\times$ is the Cartesian product and i is the identity map of U onto itself, which is obviously continuous. Since I is continuous by hypothesis, then ϕ is continuous [6]. The inverse of ϕ, is nothing else than the orthographic projection from M to U, which is continuous [6] and the assertion follows.
   From the lemma it follows that

Proposition 1 *Let L and N be images defined as the graph of continuous functions I and J respectively; then there exists an homeomorphism $h : L \to N$.*

   Proof. Consider the maps $\phi : (x, y) \to (x, y, I(x, y))$ from U to L and $\psi : (x, y) \to (x, y, J(x, y))$ from U to N and define $h = \psi \circ \phi^{-1}$, which is a map from L to N. The map h is an homeomorphism being the composition of two homeomorphisms ϕ and ψ.
   If f is supposed to be smooth, that is to have continuous partial derivatives of any order in a open set U in R^2, then the surface is regular [6], and then it is easy to prove that all surfaces, which are the graph of differentiable functions, are diffeomorphically equivalent; that is to say that for any pair of surfaces there exists a bijective map, smooth with its inverse, taking one surface onto the other.
   From proposition 1 it follows that all images must have the same Euler-Poincare' characteristic χ, that can now be computed by making use of the Gauss-Bonnet theorem.
   The Gaussian curvature of a Monge patch $L = \{x, y, I(x, y)\}$ is given by [6]
   $K = \mid H_I \mid (1 + (\partial I / \partial x)^2 + (\partial I / \partial y)^2)^{-2}$ where $\mid H_I \mid$ is the determinant of the Hessian matrix H_I of the function I.
   Let now $V \subset R^2$ be a disk of radius r with boundary C; V is a flat surface, hence $\int \int_V K dA = 0$ and $\sum_i \theta_i = 0$ because there are no vertices. In this case the geodetic curvature along C is equal to the curvature $k = 1/r$ of C and it follows

that $\int_C k_g ds = 2\pi$. Therefore $\chi = 1$, and, since χ is a topological invariant, it must be the same for all images.

It must be pointed out that the result applies only to Monge patches, and hence not to any regular surface in R^3; however we are interested in images, which indeed are graphs of functions, and hence Monge patches.

In our proof we assumed image intensity I to be a continuous function. This is a common assumption justified by the observation that most images are band limited due to the imaging system. In human vision there is a good match between the band limitation and the sampling density.

We have seen here that all images are topologically equivalent. Of course this is is not true for the objects that generate different images. The surfaces of these objects may have different topological properties, e.g. for a sphere and a torus $\chi = 2$ and $\chi = 1$ respectively (see Fig. 1), and their visible parts have $\chi = 1$ and $\chi = 0$ respectively (see Eq. 1); however, their images have $\chi = 1$. Thus the topological properties of an object's surface cannot be determined as topological properties of the corresponding image surfaces. In other words, characteristics such as holes or discontinuities do not exist in the images *per se*, indeed there is no a priori reason to interpret dark or light blobs as holes. Nevertheless, one can do so and successful methods for estimating the Euler characteristic of binary images have been presented [13,10].

4   Pseudo-Topological Properties of Images

It has been mentioned before that there is some experimental evidence suggesting that the human visual system can discriminate on the basis of what appear to be different topological properties of the objects underlying the image. Then it is of interest to search for image properties, which reflect topological properties of the associated objects. We call such properties pseudo-topological properties of images, and we shall investigate, which kind of operators are appropriate to detect pseudo-topological image properties. To compute pseudo-topological properties of images by integrating local features as the outputs of some operator, then only 2D-operators, that capture the local curvature of the image, seem appropriate, even though not all $2D$ global operators will work in detecting pseudo-topological properties of images. Indeed let L be the geometrical representation of an image and suppose that its boundary C is a regular closed curve contained in a planar region such that, in C, $|\nabla I| = 0$. Then we have, as seen before,

$$\sum_i \theta_i + \sum_i \oint_{C_i} k_g ds = 2\pi.$$ (3)

From the Gauss-Bonnet formula $\int\int_V KdA + 2\pi = 2\pi\chi$, and, since $\chi = 1$, $\int\int_V KdA = 0$. If we now extend an image by a planar frame we can find a curve C such that Eq. (3) holds. Then, for all "framed" images the total curvature is equal to zero. For any practical purpose this also implies that any deviation from zero will be a boundary effect. Therefore, a straightforward application of the Gauss-Bonnet theorem to image analysis cannot lead to useful image properties. We will make no attempt to develop a general theory about the invariance properties of integral 2D-operators. Instead, we will show how a specific 2D-operator, namely the clipped-eigenvalues (CEV) operator, can be used to compute pseudo-topological properties of images. This operator has been introduced as a model for dot-responsive cells in [16], and described in more detail in [1]. Nevertheless, it seems useful to understand in the present context how the operator can be derived from the expression for the Gaussian curvature K. The determinant of the Hessian of I can be written as

$$|H_I| = \frac{1}{4}\left(\frac{\partial^2 I}{\partial x^2} + \frac{\partial^2 I}{\partial y^2}\right)^2 - \frac{1}{4}\left(\frac{\partial^2 I}{\partial x^2} - \frac{\partial^2 I}{\partial y^2}\right)^2 - \left(\frac{\partial^2 I}{\partial x \partial y}\right)^2, \tag{4}$$

that is,

$$|H_I| = \left(\nabla^2 I\right)^2 - \epsilon^2, \tag{5}$$

where $\nabla^2 I$ is the Laplacian on the intensity I and ϵ is the eccentricity; they determine the eigenvalues of H_I through the formula $\lambda_{1,2} = \nabla^2 I \pm \epsilon^2$. The operation of clipping is defined as $\lambda^+ = Max(0, \lambda)$ and $\lambda^- = Min(0, \lambda)$. The CEV operator is then $CEV(I) = \lambda_2^+(I) - \lambda_1^-(I)$. Note that in case of isotropic patches, where $\epsilon = 0$, $CEV(I) = \nabla^2 I$. But when a pattern becomes elongated, CEV will be less than $\nabla^2 I$ and will become zero for straight patterns. The local operator CEV yields a global measure $< CEV >$ defined as the average of CEV on the whole image.

The main difference between the Gaussian curvature and the operator CEV is that the latter is zero for hyperbolic surface patches. Using the clipping operation one obtains different signs for positive- and negative-elliptic patches, and only patches with absolute CEV values above a small threshold contribute to $< CEV >$. Fig. 2 presents some examples of the action of CEV on different images and the corresponding values of $< CEV >$ are also shown. Note that $|< CEV >|$ is, within a very small error due to the numerical approximation, equal to $|\chi_T|$, the total Euler-Poincare' characteristic of the visible parts of surfaces in the scene.

The $< CEV >$ measure exhibits the pseudo-topological invariance illustrated above as long as the patterns have the same contrast. A contrast independent version of the CEV operator can be defined as $CEV_N = \frac{(\lambda_2^+(I) - \lambda_1^-(I))}{(c + (\partial I/\partial x)^2 + (\partial I/\partial y)^2)^{1/2}}$, where c is a small constant. With this measure we have obtained results where $< CEV_N >$ varied with less than 1% for patterns with different contrasts.

Pseudo-topological invariance is also limited by the scale of the operators and the size and shape of the patterns that are involved. We should mention here, that before computing the partial derivatives, the images are low-pass filtered with a Gaussian filter, which defines the scale of the CEV operator (that can be evaluated on multiple scales). However, any image can be zoomed (by nearest-neighbor interpolation) such that patterns are approximated by saw-tooth contours (like the two tilted polygons in the third row of Fig. 2). In this

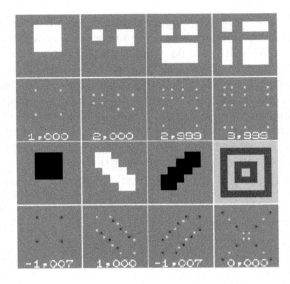

Fig. 2. Responses of the CEV operator (second and fourth row) to 8 different patterns (first and third row). In addition, the mean values of $< CEV >$ (normalized to the first image) are given.

case the scale on which the CEV operator is computed can be finer than the highest frequencies of the original patterns. On such zoomed images, we can obtain the above pseudo-topological invariance independent of the shape of the patterns.

As concerns simulations of visual functions, however, the issue of strong invariance is less relevant. More important is whether we can predict the experimental results with reasonable model assumptions.

5   Simulations of Experimental Data

Due to Minsky and Papert [11] it is a widespread belief that topological properties of images can neither be easily computed by simple neural networks, nor easily perceived by human observers. In a series of experiments, Chen has shown that subjects are sensitive to "topological" properties of images, and he has interpreted his results as being a challenge to computational approaches to vision. For example in [4] he has shown that if two images are presented briefly (5 ms), subjects discriminate better between images of a disk and a torus, than they do in case of images of a disk and a square or a disk and a triangle, respectively. Note that the objects disk, square, and triangle are topologically equivalent, $\chi = 2$, whereas is case of a torus $\chi = 0$, compare Eq. (1) and Fig. 1.

Further indications of this kind of performance, that Chen attributed to some topological process, have been found in reaction-time experiments [5] where subjects were as fast (750 ms on average) in finding the quadrant containing a "closed" pattern (a triangle), with the 3 other quadrants containing an "open" pattern (an arrow), as they were in finding an empty quadrant (with the other 3 quadrants containing a square) - see Fig. 4.

Proposition 1, on the other hand, demonstrates that the image does not directly exhibit the topological properties of the underlying surfaces and that some type of further processing is needed, which we have attributed to the action of the CEV operator. To simulate the results obtained by Chen, the image shown at the left of Fig. 3 was used as an input, to which the CEV operator was applied with a result shown in the middle of Fig. 3. The final results are displayed on the right of Fig. 3. Here we have not computed the global mean values $< CEV >$ but have integrated the local CEV values by low-pass filtering.

Therefore the intensity map shown on the right of Fig. 3 can be interpreted as a local estimate of $< CEV >$, denoted by CEV_{LP}, that varies with (x, y) and depends on the chosen scale. Obviously, if CEV_{LP} were the representation, which subjects use for discrimination, the difference between the ring and the disc would be larger than the differences between the disc and the rectangle and triangle (as found in the experiment). The existence of an end-stopped CEV-like representation is well motivated by neurophysiological and some psychophysical results [17]. The spatial filtering is common in many other models, e.g, of texture perception [3]. What it assumes is that the similarity metric involves some spatial integration. Of course, the CEV_{LP} representation depends on a few parameters, mainly the scale of CEV itself and of the low-pass filter. However, the point here is that it is easy to predict the experimental results with reasonable values of the parameters, and we found the predictions to be stable with respect to variations of the parameters. A more comprehensive analysis of how the pseudo-topological properties depend on the spatial scales of the underlying operations is beyond the scope of this paper. The results shown in Fig. 4 have been obtained

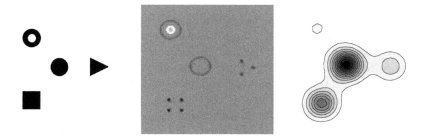

Fig. 3. Input image (left), output of the CEV operator (middle) and, on the right, low-pass filtered CEV operator (CEV_{LP}). The result on the right predicts the experimental findings that humans are more sensitive to the difference between a circle and a ring, than between a circle and a square or a triangle.

in a similar way. Here the CEV_{LP} result shown on the right illustrates the large difference between the triangle and the arrow in this representation.

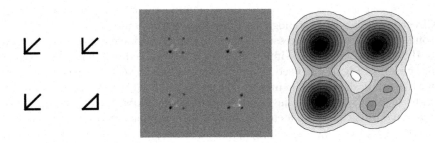

Fig. 4. Simulations as in Fig. 3 but for a different input. In this example results predict the large perceptual difference between the open and the closed shapes arrow and triangle.

6   Conclusions

We have shown here that all images are topologically equivalent. From this we can conclude that any "topological" properties of images depend on, and are restricted to, some additional abstractions or computational rules.

Chen's experiments reveal that the human visual system is sensitive to "topological" properties of the input patterns. Our simulations show that the results can be explained by assuming that the visual system evaluates integral values of specific curvature measures. These integral values can be seen as corresponding to activities of end-stopped, dot-responsive cells that are averaged over space. A possible interpretation is that in certain cases, e.g. at short-time presentations, the system evaluates integral values of an underlying, end-stopped representation. End-stopped neurons in cortical areas V1 and V2 of monkeys are oriented and more complex than dot-responsive cells. For simplicity, we have restricted our simulations to the CEV operator and have argued that the basic requirement for pseudo-topological sensitivity is that straight features are not represented. However, we have shown before that even the retinal output could be endstopped, depending on the dynamics of the input, and that the quasi-topological sensitivity is not limited to the use of the CEV operator [2].

The evaluation of integral values is assumed to be relevant to texture perception also. Indeed, we have been able to show that certain human performances in texture segmentation can be predicted by integrating the output of $2D$-operators in general and of the CEV operator in particular [3]. Thus, we are confident to be dealing with a rather general principle of visual processing.

Acknowledgment. This work is based on an earlier manuscript, which had been supported by a grant from the Deutsche Forschungsgemeinschaft (DFG-Re 337/7) to I. Rentschler and C. Z. We thank C. Mota and the reviewers for valuable comments.

References

1. E. Barth, T. Caelli, and C. Zetzsche. Image encoding, labelling and reconstruction from differential geometry. *CVGIP:GRAPHICAL MODELS AND IMAGE PROCESSING*, 55(6):428–446, 1993.
2. E. Barth and C. Zetzsche. Endstopped operators based on iterated nonlinear center-surround inhibition. In B. Rogowitz and T. Papathomas, editors, *Human Vision and Electronic Image Processing*, volume 3299 of *Proc. SPIE*, pages 67–78, Bellingham, WA, 1998.
3. E. Barth, C. Zetzsche, and I. Rentschler. Intrinsic two-dimensional features as textons. *J. Opt. Soc. Am. A*, 15(7):1723–1732, July 1998.
4. L. Chen. Topological structure in visual perception. *Science*, 218(12):699–700, 1982.
5. L. Chen. Topological perception: a challenge to computational approaches to vision. In e. a. P Pfeifer, editor, *Connectionism in perspective*, pages 317–329. Elsevier Science Publishers, North-Holland, 1989.
6. M. P. Do Carmo. *Differential Geometry of Curves and Surfaces*. Prentice-Hall, Inc., Englewood Cliffs, NJ, 1976.
7. A. Dobbins, S. W. Zucker, and M. S. Cynader. Endstopped neurons in the visual cortex as a substrate for calculating curvature. *Nature*, 329:438–41, 1987.
8. M. Ferraro. Local geometry of surfaces from shading analysis. *Journal Optical Society of America*, A 11:1575–1579, 1994.
9. J. J. Koenderink and W. Richards. Two-dimensional curvature operators. *J. Opt. Soc. Am. A*, 5(7):1136–1141, 1988.
10. C.-N. Lee and A. Rosenfeld. Computing the Euler number of a 3D image. Technical Report CAR-TR-205, CS-TR-1667, AFOSR-86-0092, Center for Automation Research, University of Maryland, 1986.
11. M. Minsky and S. Papert. *Perceptrons*. MIT Press, Cambridge MA, 1969.
12. B. O'Neill. *Elementary Differential Geometry*. Academic Press, San Diego CA, 1966.
13. A. Rosenfeld and A. C. Kak. *Digital Picture processing*. Academic Press, Orlando, FL, 1982.
14. H. R. Wilson and W. A. Richards. Mechanisms of contour curvature discrimination. *J Opt Soc Am A*, 6(1):106–15, 1989.
15. A. Yuille, M. Ferraro, and T. Zhang. Image warping for shape recovery and recognition. *Computer Vision and Image Understanding*, 72:351–359, 1998.
16. C. Zetzsche and E. Barth. Fundamental limits of linear filters in the visual processing of two-dimensional signals. *Vision Research*, 30:1111–1117, 1990.
17. C. Zetzsche, E. Barth, and B. Wegmann. The importance of intrinsically two-dimensional image features in biological vision and picture coding. In A. Watson, editor, *Digital images and human vision*, pages 109–138. MIT Press, Cambridge, MA, 1993.

Adaptive Segmentation of MR Axial Brain Images Using Connected Components[*]

Alberto Biancardi[1] and Manuel Segovia-Martínez[2]

[1] DIS — Università di Pavia, Via Ferrata,1 Pavia, Italy,
alberto@vision.unipv.it,
[2] CVSSP — University of Surrey, Guildford, UK,
m.segovia@ee.surrey.ac.uk

Abstract. The role of connected components and connected filters is well established. In this paper a new segmentation procedure is presented based on connected components and connected filters. The use of connected components simplified the development of the algorithm. Moreover, if connected components are available as a basic data type, implementation is achievable without resorting to pixel level processing. Using parallel platforms with hardware support for connected components, the algorithm can fully exploit its data parallel implementation. We apply our segmentation procedure to axially oriented magnetic resonance brain images. Novel ideas are presented of how connected components operations (e.g. moments and bounding boxes) and connected filtering (e.g. area close-opening) can be effectively used together.

Keywords: Connected components, connected filters, medical image analysis, magnetic resonance imaging (MRI), brain segmentation.

1 Introduction

Magnetic Resonance (MR) imaging has become a widespread and relied upon source of information for medical diagnosis and evaluation of treatments, especially as far as the brain is concerned. The first step in any computer-based procedure has to identify the brain region of interest from the surrounding non-brain areas in order to prevent any mismatch in the other stages of computation [2,7].

If we look at the degree of autonomy, proposed solutions range from fully interactive tools [10,8], where the program supplies special tools that help human operators limit their fatigue and improve their performance, to semiautomatic tools [16], where a limited amount of human intervention is required (e.g. to set initial condition, to validate and possibly correct the final result), to fully automatic methods with no human action. Given the specific goal for this segmentation task, the ideal solution calls for a method of the last class, because it is a repetitive operation, with the quality of the first group, because it is very difficult to do it well and right.

[*] This work was in part supported by the Erasmus Intensive Programme IP2000

C. Arcelli et al. (Eds.): IWVF4, LNCS 2059, pp. 295–302, 2001.

In the fully automatic class existing methods rely on multiple passes to achieve their result; they may be roughly grouped into two classes and summarised as follows:

1. extract close-enough regions (mainly by filtering and thresholding each image) and then refine their contours [1,5,20];
2. classify pixels or voxels and then interpolate or extract the boundaries [11, 2];

In both cases the actual segmentation stage requires a refinement because boundaries cannot are not reliable enough to use them directly.

Thanks to the use of boundary-preserving filters, our method avoids the need for further processing. Another novelty of our approach is the extensive use of operators based on connected components that allows it to be implemented in a data parallel way.

After an overview of the mathematical and processing tools used by the method, each step of the procedure will be presented. A brief analysis and conclusions will complete the paper.

2   Connected Components and Connected Filters

The development of image analysis applications is a deeply involved operation where multiple stages take place before reaching satisfactory results. While moving through these stages the process is faced with variations and options. Being able to describe and implement all the algorithms involved in the application without loosing focus inside low-level details is fundamental.

The data parallel paradigm [9] has proved highly valuable as far as the early vision stage is concerned. However, when dealing with intermediate-level vision, the processing of pixel aggregates or the performing of irregular data movements has limited its appeal. It is also true, on the other hand, that most of these image-analysis transformations may be expressed in terms of operations on connected sets of pixels, which can represent a region, a contour, or other connected parts with respect to the chosen image topology. Connected components may, thus, represent a kind of structural information that is easily computed even if, by definition, it is image-data dependent. By using the shape of connected components as a modifier of data-parallel operations, an extended data-parallel paradigm can be defined [3] that preserves the benefits of the plain one and is able to tackle the needs of image analysis tasks.

An additional benefit is that algorithms can be implemented using only data parallel commands, making them suitable to a porting on a massively parallel architecture with hardware support for connected components [14,15].

2.1   Basic Definitions

As connected components play such a key role, some basic definitions are given to support the description of the proposed algorithm.

Given an image (of $\mathcal{N} \times \mathcal{M}$ pixels), the concept of neighbouring pixels can be formally described using the image planar graph representation [18] where vertices represent pixels and edges represent the belonging of pixels to each other neighbourhood.

Binary image: A binary image **B** is a subset of $\mathbb{N} \times \mathbb{N}$

Relationship: If two pixels $U, V \in \mathbf{B}$ are related by the binary relation $\mathcal{C} \subseteq \mathbf{B} \times \mathbf{B}$, they will be noted $U.\mathcal{C}.V$

Connectivity: For each image pixel a (symmetrical) connectivity relationship $\mathcal{C}$ is defined so that $U.\mathcal{C}.V$ iff pixels U and V are connected. If $V \in \mathbf{B}$ the subset $\mathbf{N} = \{U \in \mathbf{B} : V.\mathcal{C}.Q\}$ is the (nearest) neighbourhood of V

Connected components: The partitions of **B** by $\mathcal{C}$ are called connected components. Equivalently, two pixels $U, V \in \mathbf{B}$ belong to the same connected component iff there exist a set $\mathbf{P} = \{P_0, P_1, \ldots, P_n\}$ such that $P_0 = U, P_n = V$ and $\forall i \in \mathbb{N}, 0 \leq i < n : P_i.\mathcal{C}.P_{i+1}$

2.2   Filters by Reconstruction

Filters by reconstruction [17] belong to the family of connected operators and collect openings by reconstruction and closings by reconstruction.

In particular we define *opening by reconstruction* any operation that is the composition of any pixel-removing operation composed with a trivial connected opening, which actually reconstructs any connected component that has not been completely removed; on the other hand *closing by reconstruction* is the dual operation in that it is the composition of a pixel-adding operation composed with a trivial connected closing, which completely removes any component which is not entirely preserved. Connected openings and connected closings are also known under the names of geodesic dilations and geodesic erosions [13] or propagations [6] depending on the different points of view they were first introduced.

Filters by reconstruction for grey-scale images are computed by stacking (i.e. adding) the result of their binary counterparts applied to each of the (grey-scale) image cross sections [19].

Area filters, which are used by this method, belong to the class of filters by reconstruction; in particular area openings and area closings use a size criterium for the pixel-removing or pixel-adding operations: any component whose size is less then the required amount is removed.

3   An Overview of the Method

MR brain images are characterised by being heavily textured. But, while texture may be essential for the discrimination of sickness types, it makes the segmentation more difficult. Segmentation tries to extract one or more shapes that operators readily perceive notwithstanding the many variations of pixel intensities.

Our method is modelled after the following conjecture: operators are able to elicit some homogeneity criteria for image regions and then they choose only those regions that match a higher level goal. Filters by reconstruction were seen as the key operation for driving the two stages, as they have the property of simplifying the images while preserving contours, while thresholding is used as the core of the region selection step. In more detail, the sequence of operations is as follows:

- an area filtering regularises the brain image;
- a thresholding level is determined by processing an image that evaluates the correlation between distance from the image centre and intensity of all the image pixels;
- a thresholding operation that uses the previously computed level highlights completely the brain area and, possibly, some other extraneous pieces (e.g. ocular bulbs) lying outwardly with respect to the image centre;
- a final filtering by reconstruction removes all the unwanted regions.

Since thresholds can be found by means of an automatic procedure in the upper frames only, an adaptive procedure extrapolates the missing values in following frames going downward.

3.1   Area Filtering

Aiming at preserving the integrity of the brain region, and therefore of its boundary, the need of keeping the thin areas that may be present in the border pixels was seen as crucial for a successful filtering. Hence, area filters [4] were chosen among the filters by reconstruction because of their shape-preserving ability; at the same time these filters reduce variation among pixel values, which again plays favourably in limiting the bad effects textures may cause.

One problem with their application to gray-scale images is that they may cause an intensity drift in the resulting image. This is why after applying the filter, our method replaces the value of each connected component with its average gray level, re-computed on the original (un-filtered) frame.

3.2   Threshold Level

In the upper frames, the brain area is made of a single, elliptical area. This area has no dark parts so our algorithm uses this a-priori knowledge to select an optimal threshold.

Figure 1 shows the correlation between a pixel distance from the image centre and its intensity, where the distance is placed along the abscissas and the intensity is placed along the ordinates (The origin, i.e. the upper left corner, shows the number of pixels having zero distance and zero intensity). Since there are no dark points at distance 0 (close to the image centre) the threshold can be found by looking for the topmost pixel in the low distance range; this means finding the maximum ordinate of the connected component whom the pixel at co-ordinates $(0, 0)$ belongs to.

In order to detect when the learning phase must stop, the minimum distance of any point, whose intensity is within a pre-defined range for dark pixels, is

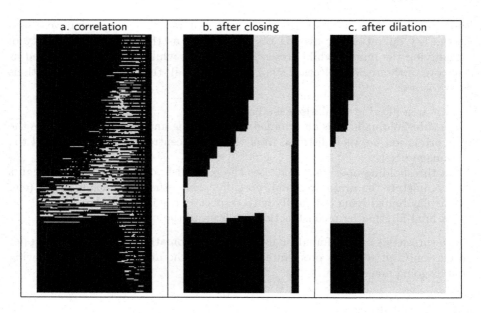

Fig. 1. The three stages for finding the threshold level.

determined by finding the maximum abscissa of the same component containing the pixel at $(0,0)$. The range is expressed thanks to a closing of the correlation image by a vertical line whose length is equal to the maximum value of the range for dark pixels; a dilation, on the other hand, by a horizontal line is required to limit the connected component at the topmost pixel within the range of allowed distances for bright pixels.

When the learning stage is completed, the average of the threshold levels found up to that point is used as threshold level for all the following frames.

3.3 Filtering Out Unrelated Components

The thresholding operation alone is not able to segment the brain region. By looking at the correlation image (Figure 1.a) it is clearly visible that the brain region, being the closest one to the image centre, contributes the pixels the are on the left side; selecting any threshold level that keeps those points will preserve some of the surrounding components as well. A new filtering stage is therefore mandatory.

As long as the threshold selection is in the *learning* stage, it is possible to select the brain region simply by reconstructing the only connected component that covers the image centre (which is not be the visual centre of the brain region but it is close enough to allow a recovery of the correct region to the program). Figure 2 shows the boundary of the extracted region superimposed to the size-filtered image and to the original MR image.

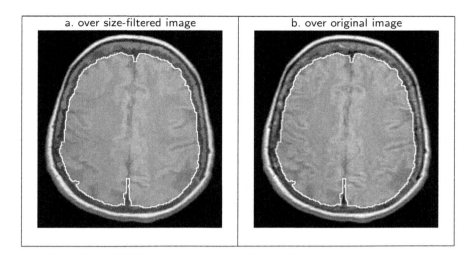

Fig. 2. Boundary of the final segmented region.

When the threshold selection switches to the extrapolating stage, then some extra care must be taken because of the possible presence of brain regions not connected to the main central area. The brain area generated by each frame, therefore, is used as a validation mask for the following frame (always going downwards): if the overlap between each of the current-frame regions and the validation mask is below a certain percentage threshold, regions are removed from the final brain area. Figure 3.a shows in dark grey the outline of the brain area at the previous frame together with the current frame result in white; Figure 3.b shows only the outline of the selected area for the current frame superimposed to the original image.

4   Results and Discussion

The method was developed using a processing environment that supplies an augmented data-parallel environment with direct support for connected components [3], and it was tested on several PD brain scans obtained from the Harvard Brain Atlas (http://www.med.harvard.edu/AANLIB/home.html).

The main problem arises when the brain region is kept connected to the bright surrounding border that survives thresholding. This case is easy to detect and may be solved by an apt number of successive erosions (that will be mirrored by an equal number of dilations after filtering by reconstruction). Even if this procedure reaches its goal, it has the unfortunate effect of removing some of fine details of the region. We are currently studying a data-parallel approach at finding the optimal *cut* for these extraneous paths.

The results achieved by our method seem on a par with those reported in literature. Many methods take advantage of the "Snakes" algorithm [12] to refine

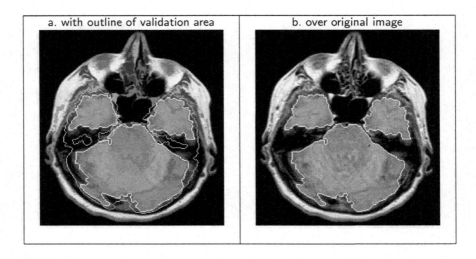

| a. with outline of validation area | b. over original image |

Fig. 3. Example of segmentation with multiple regions.

the brain region border; by so doing they loose the ability to track thin features that go toward region centres. Our method tries to avoid any border correction, actually preserving thin details.

5   Conclusion

An automatic method for brain extraction from axial MR images was presented. An adaptive method for threshold selection and widespread use of connected filtering are the main novelties of this method. Results are on a par with other methods, future work will attempt at improving quality and robustness.

References

1. Atkins, M.S., Mackiewich, B.T.: Fully Automatic Segmentation of the Brain in MRI. IEEE Trans. Med. Imag. **17** (1998) 98–107
2. Bezdek, J.C., Hall, L.O., Clarke, L.P.: Review of MR Image Segmentation Techniques Using Patter Recognition. Medical Physics **20** (1993) 1033-1048
3. Biancardi, A., Mérigot, A.: Connected component support for image analysis programs. Proc. 13th Inter. Conf. on Pattern Recognition **4** (1996) 620–624
4. Crespo, J., Maojo, V.: Shape preservation in morphological filtering and segmentation. Proc. XII Brazilian Symposium on Computer Graphics and Image Processing (1999) 247–256
5. Davatzikos, C.A., Prince, J.L.: An Active Contour Model for Mapping the Cortex. IEEE Trans. Med. Imag. **14** (1995) 65–80
6. M. J. B. Duff, *Propagation in Cellular Logic Arrays,* Proc. Workshop on Picture Data Description and Management (1980) 259–262

7. Freeborough, P.A., Fox N.C.: MR Image Texture Analysis Applied to the Diagnosis and Tracking of Alzheimer's Disease. IEEE Trans. Med. Imag. **17** (1998) 475–479
8. Freeborough, P.A., Fox N.C., Kitney, R.I.: Interactive Algorithms for the Segmentation and Quantitation of 3D MRI Brain Scans Comp. Meth. Programs in Biomed. **53** (1997) 15–25
9. Hillis, W. D., Steele Jr., G. L.: Data parallel algorithms. Communications of the ACM **29** 12 (1986) 1170-1183
10. Hohne, K.H., Hanson, W.: Interactive 3D Segmentation of MRI and CT Volumes using Morphological Operations. Journal of Computer Assisted Tomography **16** (1994) 285–294
11. Kapur, T., Grimson, W.E.L., Wells, W.M., Kikinis, R.: Segmentation of Brain Tissue from Magnetic Resonance Images Med. Imag. Anal. **1** (1996)
12. Kass, M., Witkin, A., Terzopulos, D.: Snakes: Active Contour Models. Intern. Jour. Comput. Vision (1998) 321–331
13. Lantuéjoul, C.: Geodesic Segmentation in Preston Jr., K., Uhr, L. (eds.): Multi-computers and Image Processing. Academic Press, New York (1982)
14. Mèrigot, A.: Associative nets: A graph-based parallel computing model. IEEE Trans. on Computer **46** (1997) 558–571
15. Mèrigot, A., Dulac, D., Mohammadi, S.: A new scheme for massively parallel image analysis. Proc. 12th Intern. Conf. on Pattern Recognition **3** (1994) 352–356
16. Niessen, W.J., Vincken, K.L., Weickert J., Viergrever, M.A.: Three-dimensional MR Brain Segmentation. 6TH Intern. Conf. on Computer Vision (1998) 53–58
17. Salembier, P., Serra, J.: Flat Zones Filtering, Connected Operators, and Filters by Reconstruction. IEEE Trans. on Image Process. **4** (1995) 1153–1160
18. Serra, J.: Image Analysis and Mathematical Morphology. Academic Press, New York (1982)
19. Soille, P.: Morphological Image Analysis. Springer-Verlag, Berlin Heidelberg New York (1999)
20. Zijdenbos, A.P., Dawant, B.M., Margolin, R.A., Palmer, A.C.: Morphometric Analysis of White Matter Lesions in MR Images: Method and Validation. IEEE Trans. Med. Imag. **13** (1994) 716–724

Discrete Curvature Based on Osculating Circle Estimation

David Coeurjolly, Serge Miguet, and Laure Tougne

Laboratoire E.R.I.C.
Université Lyon 2
5 av. Pierre Mendès-France
69676 Bron cedex, France
{dcoeurjo,miguet,ltougne}@eric.univ-lyon2.fr

Abstract. In this paper, we make an overview of the existing algorithms concerning the discrete curvature estimation. We extend the Worring and Smeulders [WS93] classification to new algorithms and we present a new and purely discrete algorithm based on discrete osculating circle estimation.

Keywords. Discrete Geometry, Curvature Calculus, Discrete Circles.

Introduction

Boundary analysis is an important step in many computer vision paradigms in which contours are used to extract conceptual information from objects. In such a contour analysis, the curvature is a commonly used geometrical invariant to extract characteristic points. In 3D medical imaging or in 3D snow sample analysis, the curvature measurement on surfaces is also used as a registration tool [MLD94,TG92] or as a physical or mechanical characteristics extraction tool [Pie99].

In classical mathematics, the curvature calculus is clearly defined and its properties are well known but when we want to apply this calculus on discrete data (2D or 3D discrete images), two different approaches are possible: we can first change the model of the data and put them into the classical continuous space by using interpolations or parameterizations of mathematical objects (B-splines, quadratic surfaces) on which the continuous curvature can be easily computed [Cha00,HHTS98]. Otherwise, we can try to express discrete curvature definitions and properties, and make sure that these new definitions are coherent with the continuous ones.

In the first approach, we have two main problems: the first one is that there exists a great number of parameterization algorithms in which some parameters have to be set according to the inputs. In order to provide a given accuracy, we have to reduce the input area and thus to limit our method. The second problem is that these algorithms have got a prohibitif computational time when we use large input data such as in medical imaging.

C. Arcelli et al. (Eds.): IWVF4, LNCS 2059, pp. 303–312, 2001.
© Springer-Verlag Berlin Heidelberg 2001

In a discrete approach, there are three classical ways to define this differential operator : we have the tangent orientation based algorithms, the second discrete derivative estimation based methods and finally the osculating circle based methods. In the continuous space, these definitions are equivalent but in the digital space, they lead to specific classes of algorithm.

In this paper, we propose an optimal algorithm to compute the curvature of a discrete curve in 2 or 3 dimension based on circle estimation. This algorithm holds two important properties: it's a purely low level algorithm that does not need neither preprocessing tool nor data linked parameters, and it only lies on the discrete model.

1   Framework and Related Methods

In this section, we present the different definitions of the continuous curvature we can find in the literature. We explain what they become on the discrete space and we extend the Worring and Smeulder 's [WS93] classification to new discrete algorithms.

1.1   Continuous Definitions

Given a continuous objet $\mathcal{X}$ with boundary $\partial\mathcal{X}$, we consider a curvilinear abscissa parametrization $x(s)$ of the boundary. We have three classical ways to define the curvature of a curve or a path x. The first one is based on the norm of the second derivative of the curve.

Definition 1 (Second derivative based curvature)

$$k(s) = sign\|x''(s)\|$$

where $sign$ is either -1 or 1 according to the local convexity of the curve.

We can also define the curvature using the directional changes of the tangent. The curvature is obtained by computing the angle variations between the tangent t of the curve and a given axis.

Definition 2 (Tangent Orientation based curvature)

$$k(s) = \theta'(s) \quad where \quad \theta(s) = \angle(t(s), axis)$$

Finally, we have a geometrical approach of the curvature definition given by the osculating circle of radius $r(s)$.

Definition 3 (Osculating circle based curvature)

$$k(s) = sign(\frac{1}{r(s)})$$

In a continuous space, all these definitions are obviously equivalent whereas in discrete space, they lead to specific algorithms.

1.2   Discrete Curvature Computation

Now, we consider X as a digitization of the object $\mathcal{X}$ on a discrete grid of resolution h using a digitization process D_h (*i.e.* $X = D_h(\mathcal{X})$). The goal of this paper is not to study the best digitization process. Based on the discrete object X, we can define the boundary dX of X as the set of eight-connected points such that each of them have a four-connected neighbor in the complement of X [KR96].

We can now formulate the discrete curvature computation: *given a discrete boundary dX how can we compute the curvature at each point?*

First of all, we detail existing algorithms that can be classified according to the previous definitions. Just remark that due to the difficulty to provide an accurate first derivative calculus of a discrete curve, the second derivative based methods are not really tractable. Indeed, Worring and al. have presented a definition of the curvature using second derivative at a point p_i using derivative Gaussian kernels [WS93].

Tangent-Orientation based methods. A first definition of the tangent orientation based discrete curvature is computed using a local angle [BD75,CM91]: given a neighborhood m, the discrete curvature at a point p_i of dX is:

$$k_h(p_i) = \frac{\angle(p_{i-m}p_i, p_ip_{i+m})}{|p_{i-m}p_i| + |p_ip_{i+m}|}$$

In [RJ73], Rosenfeld and al. proposed a new curvature calculus that adapts the window size m to local characteristics of the discrete curve.

In such a way, Worring and Smeulders [WS93] proposed curvature definitions based on the variation of the angle between the best straight line fitting to the data and an axis in a given neighborhood of p_i:

$$k_h(p_i) = \frac{\theta(p_i) * \mathcal{G}'_\sigma}{1.107} \quad \text{where} \quad \theta(p_i) = \angle(Linefitting(p_i, m), x - axis)$$

Worring and al. computed the line fitting using an optimization in a window of size m. $\mathcal{G}'_\sigma$ corresponds to the derivative of a Gaussian kernel with parameter σ to estimate the variation of angle.

In [Via96], Vialard proposed a purely discrete definition of the line fitting process. Let us remember the definition of a discrete straight line given by Réveillès [Rev91]. A set of discrete points belong to the arithmetical discrete straight line $D(a, b, \mu, \omega)$ (with $a, b, c, \mu, \omega \in \mathbb{Z}$) if and only if each point $p(x, y)$ of the set satisfies:

$$\mu \leq ax - by < \mu + \omega$$

a/b denotes the slope of the line, μ the lower bound on the grid and ω the thickness (in this paper, we only consider *naive straight lines* with $\omega = sup(|a|, |b|)$).

Based on this definition, Vialard defines the discrete tangent at a point p of a curve as the longest discrete straight line centered in p that belongs to the curve. This calculus is driven by the Debeld's straight line recognition algorithm [DR95]. Vialard defines the curvature at a point p of a discrete curve with the equation:

$$k_h(p_i) = \frac{\theta(p_i) * \mathcal{G}'_\sigma}{1.107} \quad \text{where} \quad \theta(p_i) = \angle(\mathcal{T}(p_i), x - axis)$$

where $\mathcal{T}(p_i)$ denotes the discrete tangent centered at p_i.

In [FT99], Feschet and Tougne proposed an optimal algorithm to compute the tangent at each point of a discrete curve.

Osculating Circle based Methods. In this approach, the estimation of the best fitting circle often leads to statistical or optimization processes. Worring [WS93] proposed to optimize a distance between the smoothed data in a given window and Euclidean circles. In [Kov90], Kovalevsky proposed a geometrical approach to compute the set of possible arcs (maybe empty) separating two discrete sets of points using a *Voronoï-like* algorithm. We can use this algorithm to solve our arc fitting problem with computing the separating circle between the complement of the discrete boundary which has two connected components. Obviously we can reduce the size of the two sets considering the points connected to the boundary (see figure 1). The problem of this algorithm is that the computational cost of the calculus of the Voronoï cell in $O(n^2)$ (n denotes the size of the curve) makes this approach not be tractable as compared with the Vialard's algorithm computational cost in $O(n)$.

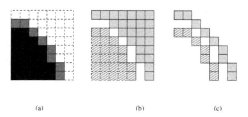

(a)   (b)   (c)

Fig. 1. Preprocessing to the Kovalevsky's algorithm : (a) the discrete object X with its discrete boundary dX, (b) the two connected components of the complement of the boundary, (c) the resulting two sets which are the input of the Kovalevsky's algorithm.

Discussion. In the tangent orientation based approach, Vialard's algorithm shows how the discrete model and discrete operators solve the statistical problem of the line fitting process and so the curvature calculus in an optimal time.

However, such an algorithm needs a derivative filter and this leads to two problems: we have to estimate a smoothing parameter σ according to the data, and we use a non discrete operator. If we want to define in a purely discrete way the curvature on a discrete curve, the only possible approach is the geometrical one if we are able to recognize or estimate osculating discrete circles.

2   A Purely Discrete Optimal Time Algorithm

2.1   Discrete Circle Analysis

First, we have to define and to characterize discrete circles. Andres [And94] proposed a discrete arithmetical circle definition based on a double diophantian inequation: $p(x, y)$ belongs to the discrete arithmetical circle $C(x_o, y_o, r)$ (with $x_o, y_o, r \in \mathbb{Z}$) if and only if it statisfies:

$$(r - \frac{1}{2})^2 \leq (x - x_o)^2 + (y - y_o)^2 < (r + \frac{1}{2})^2 \tag{1}$$

In the discrete straight line recognition algorithms, two main approaches can be extracted: the first one is a geometrical and a arithmetical approach [DR95,Kov90] roughly based on computing the narrowest strip that encloses the points. The second one is an algebraic approach [FST96] that solves the inequation system in $\mathbb{Z}^n$ given by the double diophantian inequation at each point. Due to the non-linearity of the equation 1, none of these two approaches can be easily extended to the discrete circle recognition problem in a similar computational time as the Vialard's algorithm. Therefore the idea is to develop an approximation algorithm.

In [DMT99] Tougne and al. present a new approach to define discrete circles: whatever the digitization process is, a discrete circle is the result of the intersections between the continuous circle $x^2 + y^2 = r^2$ and the discrete grid. In the first octant, if we analyze the intersections between a continuous circle and the vertical discrete lines $x = r - k$ ($0 \leq k \leq r$), we obtain a bundle of parabolas $\mathcal{H}_k\colon y = \sqrt{2kx + k^2}$. Hence, each kind of discrete circle (arithmetical circles, Pitteway's circles or Breshenam's circles) is the result of a specific digitization of these parabolas (see figure 2 for an example of the discrete bundle of parabolas associated to the Pitteway's circles).

Using this bundle of parabolas, a discrete circle can be built in the first octant only with vertical patches. For example, if we consider the case of Pitteway's circles, the height of the first vertical patch is given by the formula[1]:

$$h = [\sqrt{2x + 1}]$$

Just note that any kind of discrete circle can be entirely characterized with its bundle of discrete parabolas.

In the following, since Tougne shown the equivalence of discrete circles given by their parabolas, we choose to estimate arithmetical osculating circle and base

---

[1] $[x]$ denotes the rounding to the closest integer.

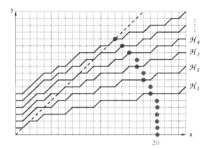

Fig. 2. The Pitteway's circle of radius 20 and its discrete parabolas given by the equation: $y = \left[\sqrt{2kx + k^2}\right]$.

our calculus on the discrete bundle of parabolas that generates the arithmetical circles. This bundle of parabolas can be computed considering the intersection between the Euclidean circles $x^2 + y^2 = (r - 1/2)^2$ and the grid, this leads to the bundle:

$$y = \left[\sqrt{r(1 + 2k) - k^2 + 1/4}\right] \quad \text{with } k \text{ a positive integer}$$

2.2   The Algorithm

We only use arithmetical circles due to their analogy to Euclidean rings: a discrete point belongs to an arithmetical circle if and only if it belongs to the Euclidean ring of radii $r + 1/2$ and $r - 1/2$. Furthermore, the discrete tangent can be viewed as a digitization of the longest tangent of the circle of radius r that lies in the ring (see figure 3).

In the continuous space, it's clear that the length of the inner tangent uniquely characterize the radius of the ring. In the discrete space, since we are

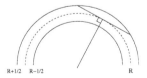

Fig. 3. Continuous ring and the inner tangent.

not able to recognize discrete circles, we have to make a link between discrete circles and well known discrete objects such as discrete straight lines. At this point of the discussion, we make the assumption that the length of the discrete tangent given by the Vialard's algorithm is constant at each point of an arithmetical circle. In concrete case, we have a small variation of the tangent length on a discrete circle but we will see that this variation does not interfere the results in the curvature calculus. Based on this assumption, we can propose a discrete osculating circle estimation: given a discrete curve, let l be the half length of the discrete tangent at a given point of the curve. Since the length of the discrete tangent on an arithmetical circle is constant, we can link the half discrete tangent to the first vertical patch given by the discrete parabolas associated to the arithmetical circles. Hence, by inverting the discrete equation of the first discrete parabola linked to arithmetical circle, we can compute the set $\mathcal{S}_l$ of radii of discrete circles which have a first vertical patch of half length l:

$$\mathcal{S}_l = \{r_{inf}, \dots, r_{sup}\}$$
$$r_{inf} = \left\lceil (l - 1/2)^2 - 1/4 \right\rceil$$
$$r_{sup} = \left\lfloor (l + 1/2)^2 - 1/4 \right\rfloor$$

Now, given this set of possible discrete circles, we can compute the possible curvatures by inverting the radii. In practise, we return as an estimation of curvature the invert of the mean radius of this set:

$$k_h(\mathcal{S}_l) = \frac{2}{r_{inf} + r_{sup}}$$

Since the discrete tangent can be computed in an optimal time, we present, in algorithm 1, an optimal in time algorithm for the curvature computation of a discrete curve.

Algorithm 1 Curvature Estimation

COMPUTE-CURVATURE(CURVE c)
 1: **for all** pixel p in c **do**
 2:    Compute the tangent τ in p
 3:    Estimate the radii r_{sup} and r_{inf} using $l = length(\tau)/2$
 4:    **return** $\left(\frac{2}{r_{inf} + r_{sup}} \right)$
 5: **end for**

2.3   Discussion

In this section, we will discuss about the link between the grid of resolution h and our curvature calculus. Let us denote by k the curvature at the point $x(s)$ of $\partial \mathcal{X}$. We consider a digitization of $\partial \mathcal{X}$ on a grid of step h denoted by dX_h.

Our algorithm estimates the curvature $\tilde{k}_h$ at the discrete point associated to s. A classical way in discrete geometry to justify a new discrete definition of an Euclidean tool is to prove the asymptotic convergence of the measure. Thus, if $\tilde{k}_h$ is the estimation on a grid of size h and k the expected Euclidean curvature, we must proof the following equality :

$$lim_{h \to 0} \frac{\tilde{k}_h}{h} = k$$

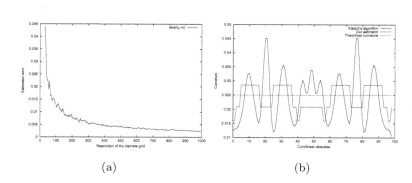

(a)   (b)

Fig. 4. Accuracy of the estimator: (a) Error in the curvature estimation ($|1/r_h - \tilde{k}_h|$) when the digitization grid is refined (b) comparison between the Vialard's algorithm with $\sigma = 1$ and our estimator on a discrete circle of radius 40.

This equality means that the error of the estimation converges to 0 when we refine the discrete grid.

Since there is no formal proof of the convergence of the discrete tangent, we can only show an experimentally asymptotic convergence considering Euclidean disks and their discretization. Let R denotes the radius of an Euclidean disk $\mathcal{X}$. The radius of the discrete approximating circle of $\partial \mathcal{X}$ is given by $r_h = [R/h]$. Decreasing h to 0, the estimated curvature $\tilde{k}_h$ should converge to $1/r_h$. Instead of making h decreasing, we can compute the mean curvature of increasing discrete circles and check if the estimation error $|\tilde{k}_h - 1/r_h|$ converges to 0. This experimental result can be found on figure 4-a.

2.4   Results

First of all, we have to check the accuracy of our estimation for a given step grid h. Hence, we compare our algorithm to the Vialard's one on a discrete circle (see figure 4-b). The results are quite similar but note that in our case, there is no smoothing process and our computation is purely discrete. We have also tested the accuracy in sense of localization of high curvature points (see figure 5).

Since Figueiredo and Réveillès [FR95] have proposed an arithmetical definition of 3D discrete lines based on the two dimensional ones: a 3D set of point is a 3D discrete straight line if and only if two of its three canonical projections on the grid planes are 2D discrete straight lines. Based on this definition Debled [DR95] has proposed an optimal time recognition algorithm of 3D lines. In the same way we can define a 3D discrete tangent using the recognition algorithm and thus we can use our algorithm to compute in an optimal time the curvature of a 3D discrete curve (see figure 6).

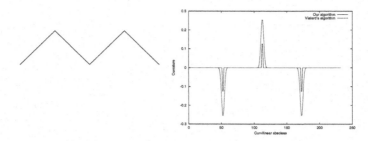

Fig. 5. Curvature of an unfolded square.

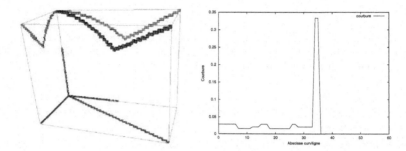

Fig. 6. An example of a 3D discrete curve (drak gray) composed of a circular arc and a straight lines, and its unsigned curvature graph.

Conclusion

In this article, we have extended the Worring and Smeulders's [WS93] classification to new discrete algorithms and have presented a purely discrete and optimal curvature algorithm based on circle estimation that solves the arc fitting problem. We have shown the good results of this algorithm compared to the Vialard's algorithm which needs a smoothing parameter. Furthermore, we do not provide any filtering process to make our results stable. As future work, we will try to extend our method to 3D discrete surface curvature and thus give an useful tool for snow sample analysis [Pie99] or medical imaging [MLD94].

References

[And94]   Eric Andres. *Cercles Discrets et Rotations discretes*. PhD thesis, Université Louis Pasteur - Strasbourg, 1994.

[BD75]   J.R. Bennett and J.S. Mac Donald. On the measurement of curvature in a quantized environnement. *IEEE Transactions on Computers*, 24(8):803–820, August 1975.

[Cha00]   Raphaelle Chaine. *Segmentation d'ensembles non organisés de points 3D d'une surface : propagation anisotropique d'etiquettes basée sur les graphes.* PhD thesis, Université Claude Bernard - Lyon 1, January 2000.

[CM91]   J-M. Chassery and A. Montanvert. *Géométrie discrète en analyse d'images.* Éditions Hermes, 1991.

[DMT99]   M. Delorme, J. Mazoyer, and L. Tougne. Discrete parabolas and circles on 2d cellular automata. *Theoretical Computer Science*, 218:347–417, 1999.

[DR95]   Isabelle Debled-Rennesson. *Etude et reconnaissance des droites et plans discrets.* PhD thesis, Université Louis Pasteur, 1995.

[FR95]   Oscar Figueiredo and Jean-Pierre Reveillès. A contribution to 3D digital lines. *Discrete Geometry for Computer Imagery*, 1995.

[FST96]   J. Françon, J.M. Schramm, and M. Tajine. Recognizing arithmetic straight lines and planes. *Discrete Geometry for Computer Imagery*, 1996.

[FT99]   Fabien Feschet and Laure Tougne. Optimal time computation of the tangent of a discrete curve : Application to the curvature. In *Discrete Geometry for Computer Imagery*, 1999.

[HHTS98]   H. Hagen, S. Heinz, M. Thesing, and T. Schreiber. Simulation based modeling. In *International Journal of Shape Modeling*, volume 4, pages 143–164, 1998.

[Kov90]   V.A. Kovalevsky. New definition and fast recognition of digital straight segments and arcs. *Proceedings of the tenth international conference on Pattern Analysis and Machine Intelligence*, June 1990.

[KR96]   Kong and Rosenfeld. Digital topology – A brief introduction and bibliography. In Kong and Rosenfeld, editors, *Topological Algorithms for Digital Image Processing (Machine Intelligence and Pattern Recognition, Volume 19), Elsevier.* 1996.

[MLD94]   O. Monga, R. Lengagne, and R. Deriche. Crest lines in volume 3D medical images : a multi-scale approach. Technical report, INRIA, Projet SYNTIM, 1994.

[Pie99]   J.B. Brzoska B. Lesaffre C. Coléou K. Xu R.A. Pieritz. Computation of 3D curvature on a wet snow sample. *The European Physical Journal Applied Physics*, 1999.

[Rev91]   J.P. Reveilles. *Géométrie discrète, calcul en nombres entiers et algorithmique.* PhD thesis, Université Louis Pasteur - Strasbourg, 1991.

[RJ73]   A. Rosenfeld and E. Johnston. Angle detection on digital curves. *IEEE Transactions on Computers*, pages 875–878, September 1973.

[TG92]   J.P. Thirion and A. Gourdon. The 3D Marching Lines Algorithm and its Application to Crest Lines Extraction. Technical Report 1672, INRIA, April 1992.

[Via96]   Anne Vialard. Geometrical parameters extraction from discrete paths. *Discrete Geometry for Computer Imagery*, 1996.

[WS93]   Marcel Worring and Arnold W. M. Smeulders. Digital curvature estimation. *Computer Vision, Graphics, and Image Processing. Image Understanding*, 58(3):366–382, November 1993.

Detection and Enhancement of Line Structures in an Image by Anisotropic Diffusion

Koichiro Deguchi[1], Tadahiro Izumitani[2], and Hidekata Hontani[3]

[1] Graduate School of Information Sciences, Tohoku University, Sendai, Japan
[2] Faculty of Engineering, University of Tokyo, Tokyo, Japan
[3] Faculty of Engineering, Yamagata University, Yonezawa, Japan

Abstract. This paper describes a method to enhance line structures in a gray level image. For this purpose, we blur the image using anisotropic gaussian filters along the directions of each line structures. In a line structure region the gradients of image gray levels have a uniform direction. To find such line structures, we evaluate the uniformity of the directions of the local gradients. Before this evaluation, we need to smooth out small structures to obtain line directions. We, first, blur the given image by a set of gaussian filters. The variance of the gaussian filter which maximizes the uniformity of the local gradient directions is detected position by position. Then, the line directions in the image are obtained from this blurred image. Finally, we blur the image using anisotropic filter again along the directions, and enhance every line structure.

Keywords: Line structure enhancement, multi-resolution analysis, anisotropic filter, structure tensor

1 Introduction

Generally a figure in an image has local structure and global structure simultaneously [1]. For example, the figure in the image shown in the Fig.1(a) has characters locally and character strings globally. By a local operation, the global line structures of the character strings cannot be caught from this image. The purpose of the method proposed in this paper is detecting and emphasizing the global line structures to recognize the global image structure. For example, when the figure (a) of Fig.1 is given, we intend to obtain the figure (b) before applying the OCR technology to read out the string. For this purpose, we need to shape out an in-line distribution of local small gray-level profiles into a global simple line structure, as shown in (c).

In order to enhance the global line structure, we must devise a method to disregard local small structures. Blurring off them by a gaussian filter is a common technique for this purpose [2]. However, as shown in Fig.2, for the case where line structures are not isolated in a image, we cannot achieve this by simple applications of a gaussian filters even though we must be careful to choose the proper size of the filter.

C. Arcelli et al. (Eds.): IWVF4, LNCS 2059, pp. 313–322, 2001.

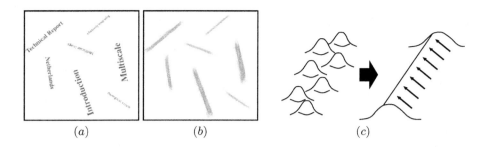

Fig. 1. (a) Image including local structures and global structures. (b) Global structures in the left image. (c) An in-line distribution of local gray-level profiles(left) is shaped out to be a global linear profile(right) which has gradients of the same direction

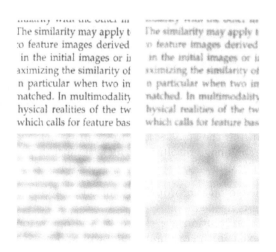

Fig. 2. Upper left: the original image (151×133 pixels). The interval of character lines is about 15 pixels. Upper right, lower left, and lower right images are the results of blurring with gaussian filters with the sizes $\sigma = 1.0, 16.0$, and 49.0, respectively.

As shown in this figure, increasing the size of the gaussian filter turns out that small local structures disappear next to next. But, the gaussian filter of too large size will smooth out also the line structures themselves (see lower right figure of Fig.2). They will be blurred out under the influence of the neighbor structures. When the size is chosen well so as not to be influenced by the next neighbor line structures, it in turn could not enhance the line structure of character sequences (see lower left figure of Fig.2).

Thus, line structure cannot necessarily be enhanced by the straightforward application of the gaussian filters. We must employ the technique of diffusing the local gray-level only along the direction of the line structure.

Some techniques to apply the gaussian filter having different sizes in its directions, that is, the shape of the gaussian filter is anisotropic, have been proposed[3][4][5]. These methods smooth out only in the direction of line structure. But, these methods need knowledge about the direction of line structures in the image and also about the upper amount of the size of local structures to be smoothed out, in advance of the processing. In this paper, we propose a method to determine the proper parameters of the gaussian filter to smooth out only local small structures and enhance global line structures adaptively to a given image, and to each positions in the image. In this method, first, we apply blurring to the given image with some various size gaussian filters. Next, we evaluate the "line-likeness", which expresses the similarity of the directions of the local gray-level gradients, position by position in the blurred image. And we determine the proper sizes and directions of the anisotropies of the gaussian filters for each position of the image.

2   Multi-resolution Image Analysis

For catching the global structure of an image, first, a device which disregards local small structures is needed. The most important consideration is to determine how small local structures must be disregarded for the ability to detect global structures. Disregarding the small structures can be achieved by employing an operation reducing the image resolution. But, we must impose the following two conditions to the operation.

- No prior knowledge on the proper resolution of a given image to catch the global structure is needed.
- When the image resolution is reduced by the operation, no new structure which did not exist in the original image will appear.

The only operation which satisfies these requirements is the diffusion of the image $f(x, y)$ according to the next differential equation.

$$\begin{cases} \partial_t u(x, y, t) = \text{div}(\nabla u(x, y, t)) \\ u(x, y, 0) \quad = f(x, y) \end{cases} \tag{1}$$

The solution of this diffusion equation just agrees with the result of the blurring of $f(x, y)$ by the gaussian filter with the variance t [2][7], as

$$u(x, y, t) = f(x, y) * G_t \tag{2}$$

where

$$G_t = \frac{1}{2\pi t} \exp(\frac{-(x^2 + y^2)}{2t}) \tag{3}$$

and $*$ means the convolution.

Fig. 3. (Left) Original image. (Center) and (right) Blurred images by the gaussian filters with $t = 16$ and 200, respectively.

Resolution of $u(x, y, t)$ becomes low gradually as t grows large. For example, the left figure in Fig.3 becomes the center figure by blurring with a gaussian filter at $t = 16$, then the right figure at $t = 200$. With t being large, the ring structure which is the global line structure in the original image will appear, and then disappear. It is necessary to catch the proper moment when this global line structure appears, if we intend to recognize such the global shape of the figure.

In the next section, then, we introduce a criterion of "line-likeness" to find the proper value of t.

3   Evaluation of Line-Likeness

In the neighborhood of a line structure on an image, the gradients $(\frac{\partial f}{\partial x}, \frac{\partial f}{\partial y})^\top$ of the image gray-level $f(x, y)$ have the same direction toward the center of the line, that is normal to the line structure. When the gradients of gray-levels have equal direction in a small neighbor region, the image is defined to have a line structure at the point. Then, we define line-likeness by the amount how similar the directions of the gradients of $u(x, y, t)$ are in the neighbor of the image point.

Three examples of figures having different line-likeliness are shown in Fig.4. The right-hand one has the structure which is more likely to line.

To show this line-likeliness qualitatively, we introduce the gradient space which is spanned by $\frac{\partial f}{\partial x}$ and $\frac{\partial f}{\partial y}$. The gray-level gradients at each pixel in the image of Fig.4 (left), (center) and (right) distribute as shown in Fig.5 (left), (center) and (right), respectively. These figures show that, when an image has line structure, its gray-level gradients distribute linearly in the direction normal to the image line structure.

The deviation of this distribution and its direction are evaluated by the eigen vectors and the eigen values of the covariance matrix of the gray-level gradients of

$$J(f(x, y)) = \begin{pmatrix} \iint (f_x)^2 dxdy & \iint f_x f_y dxdy \\ \iint f_x f_y dxdy & \iint (f_y)^2 dxdy \end{pmatrix} \tag{4}$$

Fig. 4. Examples of three figures with different line-likeliness

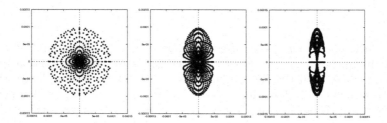

Fig. 5. Distributions of gradients of gray-levels of images shown in Fig.4

This covariance matrix J represent the total line-likeness of whole image. Then, in order to evaluate a local distribution of gray-level gradients, we introduce a structural-analysis tensor as

$$J_\rho(f(x,y)) = \begin{pmatrix} G(x,y,\rho^2) * (f_x)^2 & G(x,y,\rho^2) * f_x f_y \\ G(x,y,\rho^2) * f_x f_y & G(x,y,\rho^2) * (f_y)^2 \end{pmatrix} \qquad (5)$$

where $G(x,y,\rho^2)$ is the gaussian function with the variance ρ^2.

The eigen vectors of this structural-analysis tensor at a position (x,y) show the two principal directions of the gradient vectors in the neighborhood of (x,y), and the eigen values show the deviation of the distribution in those directions.

The value of ρ determines the area of the neighborhood in which we evaluate the distribution of the gray-level gradients. When $\rho \to \infty$, the structural-analysis tensor becomes equal to the covariance matrix of (4).

Then, we evaluate the line-likeness at a position (x,y) by the following $S(x,y)$ in (6) which is defined using the eigen values λ_1 and λ_2 ($\lambda_1 > \lambda_2$) of the structural-analysis tensor.

$$S(x,y) = \frac{\lambda_1 - \lambda_2}{\lambda_1 + \lambda_2} \qquad (6)$$

The value of $S(x,y)$ spans in $[0,1]$. When $S(x,y) \approx 1$, the gray levels around (x,y) has a line-like structure, and when $S(x,y) \approx 0$, they have not. When

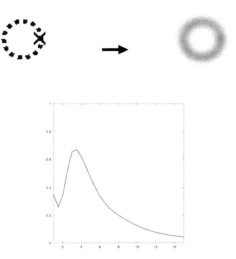

Fig. 6. Plottings of the value of the line-likeliness $S(x, y)$ at the position indicated with × in the upper left image with respect to the change of the value of σ.

$S(x, y) \approx 1$, the direction of the eigen vector corresponding to λ_1 is normal to the direction of the line structure at the point.

4   Multi-scale Evaluation

For the evaluation of the line likeliness using the $S(x, y)$, we must be careful to choose the value of ρ. To determine proper ρ, we employ a series of next two-step evaluations of the line-likeliness. As the first step, we blur the original image with a gaussian filter with variance σ^2. Then, we evaluate $S(x, y)$ for all pixels. We apply this two-step evaluation next by next by changing the blurring parameter σ^2. In every evaluation, we set $\rho = 2\sigma$. With larger value of σ, we evaluate more global line-likeliness.

Fig.6 shows the change of the line-likeness at the position indicated with × in the upper left image with respect to the value of σ. The upper right image is the blurred image with the value of σ with which the line-likeliness $S(x, y)$ becomes the maximal. This shows that we just detect the global line structure of the original image by blurring it with the σ which makes $S(x, y)$ become maximal. At the same time, the directions of the line structure at every position can be also detected from the direction of the eigen vector of structural-analysis tensor corresponding to the eigen value of λ_1.

Of course, we have the case where $S(x, y)$ has multiple maximals. It should be noted that, usually, an image has several sizes of structure hierarchically. In our method, such a multi-level hierarchy of structures can be detected as a set of maximal of the line-likeliness $S(x, y)$. In the next section, we show example images having the multi-level structures. They also shows, our method works to detect such the structures.

5   Anisotropic Diffusion to Enhance Line Structure

We just have shown the the global line structures could be detected by evaluating the line-likeliness $S(x, y)$. However, the obtained result image was, as shown in upper right of Fig.6, a blurred one and the detected line structure was faded.

Then, by using the eigen values and the eigen vectors which are corresponding to the detected line structure, we apply the blurring only along with the direction of the line structure. This process results in smoothing out gray level changes only within the line structure and enhancing it with clear contour edges. To blur an image only within a specific direction, so called the anisotropic diffusion of (7) has been proposed [4][5][7]. It diffuses an image according to a dynamical process expressed in (7).

$$\begin{cases} \partial_t L(x, y, t) = \text{div}(D(x, y)\nabla L(x, y, t)) \\ L(x, y, 0) \quad = f(x, y) \end{cases} \tag{7}$$

where $D(x, y)$ is a 2×2 matrix defined at every position in the image, so that its eigen values determine the degrees of the diffusion in the directions of its respective eigen vectors. Here after, we call $D(x, y)$ as a diffusion tensor.

When all the elements of the diffusion tensor $D(x, y)$ belong to C^∞ with respect to x and y, and $D(x, y)$ is positive definite, the solution of (7) corresponding to t always exists uniquely, and the solutions $L(x, y, t)$ do not have new line structure which the original image has not.

In this paper, we propose the determination of the suitable diffusion tensor to enhance line structures by using the evaluation of the line likeliness $S(x, y)$. According to the results of the previous sections, those diffusion tensors will be obtained by defining the eigenvalues Λ_1 and Λ_2, and the corresponding eigen vectors V_1 and V_2 as follows.

Let us denote the value σ at (x, y) with which the line likeliness $S(x, y)$ becomes maximal with $\sigma_0(x, y)$. The directions of line structures at each positions are given as the directions of the eigen vectors of the structural-analysis tensor $J_\rho(f(x, y) : \sigma_0^2(x, y))$. Therefore, letting the eigen vectors be $v_1(x, y)$ and $v_2(x, y)$ which are corresponding to λ_1 and λ_2, respectively, we define

$$\begin{aligned} V_1 &\equiv v_2, & \Lambda_1 &\approx 0, \\ V_2 &\equiv v_1, & \Lambda_2 &\approx 1. \end{aligned} \tag{8}$$

Figure 7 shows an example of the emphasizing proposed here. The global line structure in the left image was enhanced without blurring its contour edge as shown in the right image.

Fig. 7. (Left) Original image. (Right) Result image of the emphasizing of the global line structure

Fig. 8. Anisotropic blurring of the image of Fig.2(upper left) with the selected directions and variances for each positions in the image. The line structures of the original image were enhanced

6   Experimental Results

We have apply the proposed method to many types of images.

We showed in Fig.2 a document image having rows of character strings. We also showed the original line structures of the strings could not be enhanced by an isotropic and uniform gaussian filters. By applying the proposed method and selecting proper directions and variances of the filters for each positions, we obtained the image of Fig.8. The line structures were enhanced along with the strings using anisotropic and non-uniform blurring with the parameters of the maximal point of $S(x, y)$ for every image positions.

Next example shown in Fig.9 has multi-level line structures. Every elements of the image consist of dots in wave curved lines. More globally, those elements composes rows of line structures. Figure 10 shows $S(x, y)$ at a point in this image. Almost all points in the image has such the two maximals as this point. Figure 11 shows the results of the line-structure enhancements using the first and the second maximal points of the $S(x, y)$, respectively. Two-level line structures were enhanced separately.

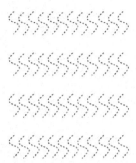

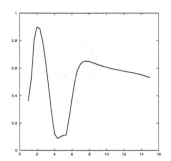

Fig. 9. An image having multi-level line structures. Every elements consist of dots in wave curved lines. More globally, those elements composes rows of line structures. (216 × 256) pixels

Fig. 10. The line likeliness $S(x, y)$ at (185,97) with respect to σ. $S(x, y)$ has two maximal points of σ

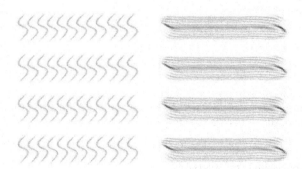

Fig. 11. (Left) Small line structures were enhanced with the parameters based on the first maximal point of $S(x, y)$. (Right) Global line structures were enhanced based on the second maximal of $S(x, y)$

The final example shows an application of line structure enhancement. Figure 12 is an image of a finger print corrupted with heavy noise. Figure 13 is the result of line structure enhancement by the proposed method using anisotropic and non-uniform blurring. Finger print pattern were clearly detected.

7   Conclusions

The technique of emphasizing the global line structure of a gray-level image was proposed. First, by carrying out multiple resolution analysis of the image, we obtained proper resolution to make global line structures appear for every

Fig. 12. A finger print image corrupted with noise

Fig. 13. Enhancement of the line structures of Fig.12 by the proposed method with anisotropic and non-uniform blurring

positions in the image. Then, we obtained the direction of the line structures from the image of this resolution, applied the diffusion process to the image with the blurring along with the line structures. The original gray levels are only smoothed out in the direction of line structure, global line structures have been enhanced, without obscuring the outline of line structure.

References

1. J.B. Antoine Maintz, Petra A. vanden Elsen, and Max A. Viergever, Evaluation of Ridge Seeking Operators for Multi-modality Medical Image Matching, IEEE Trans. on Pattern Analysis and Machine Intelligence, vol. 18, no. 3, 353-365, 1996.
2. T. Lindeberg, Scale-space theory in computer vision, Kluwer, Boston, 1994.
3. J. Bigun, G.H. Granlund and J. Wiklund, Multidimensional Orientation Estimation With Applications to Texture Analysis and Optical Flow, IEEE Trans. on Pattern Analysis and Machine Intelligence, vol. 13, no. 8, 775-790, 1991.
4. J. Weickert, Multiscale texture enhancement, Lecture Notes in Comp. Science, vol.970, Springer, Berlin, 230-237, 1995.
5. J. Weickert, A Review of Nonlinear Diffusion Filtering, Lecture Notes in Comp. Science, vol.972, Springer, Berlin, 3-28, 1997.
6. Bart M. ter Haar Romeny(Ed.), Geometry-Driven Diffusion in Computer Vision, Kluwer Academic Publishers, 1-38, 1994.
7. Bart M. ter Haar Romeny, Introduction to Scale-Space Theory: Multiscale Geometric Image Analysis, Technical Report No. ICU-96-21, Utrecht University, 1996.

How Folds Cut a Scene

Patrick S. Huggins and Steven W. Zucker*

Yale University, New Haven CT 06520, USA
{huggins,zucker}@cs.yale.edu

Abstract. We consider the interactions between edges and intensity distributions in semi-open image neighborhoods surrounding them. Locally this amounts to a kind of figure-ground problem, and we analyze the case of smooth figures occluding arbitrary backgrounds. Techniques from differential topology permit a classification into what we call folds (the side of an edge from a smooth object) and cuts (the background). Intuitively, cuts arise when an arbitrary scene is "cut" from view by an occluder. The condition takes the form of transversality between an edge tangent map and a shading flow field, and examples are included.

1 Introduction

On which side of an edge is figure; and on which ground? This classical Gestalt question is thought to be locally undecidable, and ambiguous globally (Fig. 1(a)). Even perfect line drawing interpretation is combinatorially difficult (NP-complete for the simple blocks world) [13], and various heuristics, such as closure or convexity, have been suggested [7]. Nevertheless, an examination of natural images suggests that the intensity distribution in the neighborhood of edges does contain relevant information, and our goal in this paper is to show one basic way to exploit it.

The intuition is provided in Fig. 1(b). From a viewer's perspective, edges arise when the tangent plane to the object "folds" out of sight; this naturally suggests a type of "figure", which we show is both natural and commonplace. In particular, it enjoys a stable pattern of shading (with respect to the edge). But more importantly, the fold side of the edge "cuts" the background scene, which implies that the background cannot exhibit this regularity in general.

Our main contribution in this paper is to develop the difference between *folds* and *cuts* in a technical sense. We employ the techniques of differential topology to capture qualitative aspects of shape, and propose a specific mechanism for classifying folds and cuts based on the interaction between edges and the shading flow field. The result is further applicable to formalizing an earlier classification of shadow edges [1].

---

* Supported by AFOSR

C. Arcelli et al. (Eds.): IWVF4, LNCS 2059, pp. 323–332, 2001.
© Springer-Verlag Berlin Heidelberg 2001

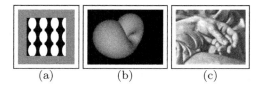

| (a) | (b) | (c) |

Fig. 1. (a) An ambiguous image. The edges lack the information present in (b), a Klein bottle. The shading illustrates the difference between the "fold", where the normal varies smoothly to the edge until it is orthogonal to the viewer, and the "cut". (c) An image with pronounced folds and cuts.

2   Folds and Cuts

Figure-ground relationships are determined by the positions of surfaces in the image relative to the viewer, so we are specifically interested in edges resulting from surface geometry and viewing, which we now consider.

Consider an image $(I : Z \subset \mathbb{R}^2 \to \mathbb{R}^+)$ of a smooth (C^2) surface $\Sigma : X \subset \mathbb{R}^2 \to Y \subset \mathbb{R}^3$; here X is the surface parameter space and Y is 'the world'. For a given viewing direction $\mathbf{V} \in \mathbb{S}^2$ (the unit sphere), the surface is projected onto the image plane by $\Pi_{\mathbf{V}} : Y \to Z \subset \mathbb{R}^2$. For simplicity, we assume that Π is orthographic projection, although this particular choice is not crucial to our reasoning. Thus the mapping from the surface domain to the image domain takes $\mathbb{R}^2$ to $\mathbb{R}^2$. See Fig. 2.

Fig. 2. The mappings referred to in the paper, from the parameter of a curve (U), to the coordinates of a surface (X), to Euclidean space (Y), to the image domain (Z).

Points in the resulting image are either regular or singular, depending on whether the Jacobian of the surface to image mapping, $d(\Pi_{\mathbf{V}} \circ \Sigma)$ is full rank or not. An important result in differential topology is the Whitney Theorem for mappings from $\mathbb{R}^2$ to $\mathbb{R}^2$ [5][10], which states that such mappings generically have only two types of singularities, folds and cusps. (By generic we mean that the singularities persist under perturbations of the mapping.)

Let $T_x[A]$ denote the tangent space of the manifold A at the point x.

Definition 1. *The* FOLD *is the singularity locus of the surface to image mapping,* $\Pi_{\mathbf{V}} \circ \Sigma$*, where* Σ *is smooth. In the case of orthographic projection the fold is the image of those points on the surface whose tangent plane contains the view direction.*

$$\gamma_{fold} = \{z_p \in Z \mid \mathbf{V} \in T_{y_p}[\Sigma(X)], \; y_p = \Sigma(x_p), \; z_p = \Pi_{\mathbf{V}}(y_p)\}$$

We denote the *fold generator*, i.e. the pre-image of γ_{fold} on Σ, by

$$\Gamma_{fold} = \{y_p \in Y \mid x_p \in X, \mathbf{V} \in T_{y_p}[\Sigma(X)], \, y_p = \Sigma(x_p)\}$$

Since the singularities of $\Pi_{\mathbf{V}} \circ \Sigma$ lead to discontinuities if we take Z as the domain, they naturally translate into edges in the image corresponding to the occluding contour and its end points.

Note that due to occlusion and opacity, not all of the singularities present in a given image mapping will give rise to edges in the image. For example, the edge in an image corresponding to a fold actually corresponds to two curves on the surface: the fold generator and another curve, the locus of points occluded by the fold. We call this the *contour shadow*,

$$\Gamma_{\Gamma-shadow} = \{y_p \in Y \mid \exists t \in \mathbb{R}^+, \, y_p = y_q + t\mathbf{V}, \, y_q \in \Gamma_{fold}\}$$

Now suppose Σ is piecewise smooth. We now have two additional sources of discontinuity in the image mapping: points where the surface itself is discontinuous,

$$\Gamma_{boundary} = \{y_p \in Y \mid \exists \delta \in \mathbb{S}^1, \, \lim_{\varepsilon \to 0} \Sigma(x_p + \varepsilon\delta) \neq \Sigma(x_p), \, y_p = \Sigma(x_p)\}$$

and points where the surface normal is discontinuous,

$$\Gamma_{crease} = \{y_p \in Y \mid \exists \delta \in \mathbb{S}^1, \, \lim_{\varepsilon \to 0} N(x_p + \varepsilon\delta) \neq N(x_p), \, y_p = \Sigma(x_p)\}$$

Fig. 2 summarizes the points we've defined.

Definition 2. *The* CUT *is the set of points in the image where the image is discontinuous due to occlusion, surface discontinuities, or surface normal discontinuities.*

$$\gamma_{cut} = \{z_p \in Z \mid z_p \in \Pi_{\mathbf{V}}(\Gamma_{\Gamma-shadow} \cup \Gamma_{boundary} \cup \Gamma_{crease})\}$$

Note that $\gamma_{fold} \subset \gamma_{cut}$, while their respective pre-images are disjoint, except at special points such as T-junctions.

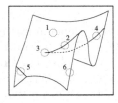

Fig. 3. Categories of points of a mapping from $\mathbb{R}^2$ to $\mathbb{R}^2$: (1) a regular point, (2) a fold point, (3) a cusp, (4) a Γ-shadow point, (5) a crease point, (6) a boundary point. The viewpoint is taken to be at the upper left. From this position the fold (solid line) and the fold shadow (dashed line) appear aligned.

If a surface has a pattern on it, such as shading, the geometry of folds gives rise to a distinct pattern in the image. Identifying the fold structure is naturally useful as a prerequisite for geometrical analysis [9][14]. It is the contrast of this structure with that of cuts which is intriguing in the context of figure-ground. Our contribution develops this as a basis for distinguishing between γ_{fold} and γ_{cut}.

2.1   Curves and Flows at Folds and Cuts

Consider a surface viewed such that its image has a fold, with a curve on the surface which runs through the fold. In general, the curve in the image osculates the fold (Fig. 4).

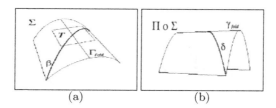

(a)   (b)

Fig. 4. A curve, $\Sigma \circ \alpha$, passing through a point on the fold generator, Γ_{fold}. (a) The tangent to the curve $T[\Sigma \circ \alpha]$, lies in the tangent plane to the surface, $\mathbf{T}$, as does the tangent to the fold generator, $T[\Gamma_{fold}]$. (b) In the image, the tangent plane to the surface at the fold projects to a line, and so the curve is tangent to the fold.

Let α be a smooth (C^2) curve on Σ; $\alpha : U \subset R \to X$. If α passes through point $y_p = \Sigma \circ \alpha(u_p)$ on the surface then $T_{y_p}[\Sigma \circ \alpha(U)] \subset T_{y_p}[\Sigma(X)]$. An immediate consequence of this for images is that, if we choose $\mathbf{V}$ such that $z_p = \Pi_{\mathbf{V}}(y_p) \in \gamma_{fold}$, then the image of α is tangent to the fold, i.e. $T_{z_p}[\Pi \circ \Sigma\alpha(U)] = T_{z_p}[\gamma_{fold}(Y)]$.

There is one specific choice of $\mathbf{V}$ for which this does not hold: $\mathbf{V} \in T_{y_p}[\Sigma \circ \alpha(U)]$. At such a point $\Pi \circ \Sigma \circ \alpha(U)$ has a cusp and is transverse (non-tangent) to γ_{fold}.

Intuitively, it seems that the image of α should be tangent to γ_{fold} "most of the time". Situations in which the image of α is not tangent to γ_{fold} result from the "accidental" alignment of the viewer with the curve. The notion of "generic viewpoint" is often used in computer vision to discount such accidents [4][17]. We use the analogous concept of *general position*, or *transversality*, from differential topology, to distinguish between typical and atypical situations.

Definition 3. *[6]: Let M be a manifold. Two submanifolds $A, B \subset M$ are* IN GENERAL POSITION, *or* TRANSVERSAL, *if* $\forall p \in A \cap B$, $T_p[A] + T_p[B] = T_p[M]$.

We call a situation typical if its configuration is transversal, atypical (accidental) otherwise. See Fig. 5.

We show that if we view an arbitrary smooth curve, on an arbitrary smooth surface, from an arbitrary viewpoint, then typically at the point where the curve crosses the fold in the image, the curve is tangent to the fold. We do so by showing that in the space of variations, the set of configurations for which this holds is transversal, while the non-tangent configurations are not transversal.

For the image of α to appear transverse to the fold, we need $T_{y_p}[\Sigma \circ \alpha(U)] = \mathbf{V}$ at some point $y_p \in \Gamma_{fold}$. $T[\Sigma \circ \alpha(U)]$ traces a curve in $\mathbb{S}^2$, possibly with self

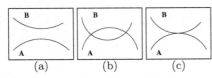

Fig. 5. Transversality. (a) A and B do not intersect, thus they are transversal. (b) A and B intersect transversally. A small motion of either curve leaves the intersection intact. (c) A non-transverse intersection: a small motion of either curve transforms (c) into (a) or (b).

intersections. $\mathbf{V}$ however is a single point in $\mathbb{S}^2$. At $T[\varSigma \circ \alpha(U)] = \mathbf{V}$ we note that $T_{\mathbf{V}}[T[\varSigma \circ \alpha(U)]] \cup T_{\mathbf{V}}[\mathbf{V}] = T_{\mathbf{V}}[T[\varSigma \circ \alpha(U)]] \cup \emptyset \neq T_{\mathbf{V}}[\mathbb{S}^2]$, thus this situation is not transversal. If $T[\varSigma \circ \alpha(U)] \neq \mathbf{V}$ then $T[\varSigma \circ \alpha(U)] \cap \mathbf{V} = \emptyset$. See Fig. 6. This is our first result:

Result 1 *If, in an image of a surface with a curve lying on the surface, the curve on the surface crosses the fold generator, then the curve in the image will typically appear tangent to the fold at the corresponding point in the image.*

Examples of single curves on surfaces where this result can be exploited include occluding contours [12][16] and shadows [2][8].

Fig. 6. The tangent field of α, $C = T[\varSigma \circ \alpha(U)]$, traces a curve in $\mathbb{S}^2$. When $\mathbf{V}$ intersects C, the curve α is tangent to the fold in the image. This situation ($\mathbf{V}_1$) is not transversal, and thus only occurs accidentally. The typical situation ($\mathbf{V}_2$) is α tangent to the fold when it crosses.

For a family of curves on a surface, the situation is similar: along a fold, the curves are typically tangent to the fold. However, along the fold the tangents to the curves vary, and may at some point coincide with the view direction. The typical situation is that the curves are tangent to the fold, except at isolated points on the fold, where they are transverse.

Let $A : (U, V) \subset \mathbb{R}^2 \to X$ define a family of curves on a surface. As before, a curve appears transverse to the fold if its tangent is the same as the view direction: $T_{y_p}[\varSigma \circ A(U, V)] = \mathbf{V}$, and $\mathbf{V}$ is a point in $\mathbb{S}^2$. Now $T_U[\varSigma \circ A(U, V)]$ is a surface in $\mathbb{S}^2$. The singularities of such a field are generically folds and cusps (again applying the Whitney Theorem), and so $\mathbf{V}$ does not intersect the singular points transversally. However, $\mathbf{V}$ will intersect the regular portion of $T_U[\varSigma \circ A(U, V)]$, and such an intersection is transversal: $T_{\mathbf{V}}[T_U[\varSigma \circ A(U, V)]] = T_{\mathbf{V}}[\mathbb{S}^2]$. The dimensionality of this intersection is zero: and so non-tangency occurs at isolated points along γ_{fold}. The number of such points depends on the singular stucture of the vector field [15]. This gives us:

Result 2 *In an image of a surface with a family of smooth curves on the surface, the curves crossing the fold generator typically are everywhere tangent to the fold in the image, except at isolated points.*

Similar arguments can be made for more general projective mappings. Dufour [3] has classified the possible diffeomorphic forms families of curves under mappings from $\mathbb{R}^2$ to $\mathbb{R}^2$ can take; one feature of this classification is that the tangency condition just described is satisfied by those forms describing folds.

For a discontinuity in the image not due to a fold, the situation is reversed: for a curve to be tangent to the edge locus, it must have the exact same tangent as the edge (Fig. 7).

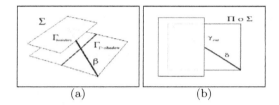

Fig. 7. The appearance of a curve intersecting a cut. (a) At a cut, the tangent plane to the surface does not contain the view direction. As a result there is no degeneracy in the projection, and so the curve will appear transverse to the cut in the image (b).

As before, we consider the behaviour of a curve α on a surface, now in the vicinity of a cut, γ_{cut}. For $\Pi_{\mathbf{V}} \circ \Sigma \circ \alpha$ to be tangent to γ_{cut}, we need $T_{z_p}[\Pi \circ \Sigma \circ \alpha(U)] = T_{z_p}[\gamma_{cut}]$, which only occurs when $T_{y_p}[\Sigma \circ \alpha(U)] = T_{x_p}[\Gamma_{cut}]$, or equivalently $T_{x_p}[alpha(U)] = T_{x_p}[\Sigma^{-1} \circ \Gamma_{cut}]$. Consider the space $\mathbb{R}^2 \times \mathbf{S}^1$. $\alpha \times T[\alpha]$ traces a curve in this space, as does $\Sigma^{-1} \circ \Gamma_{cut} \times T[\Sigma^{-1} \circ \Gamma_{cut}]$. We would not expect these two curves to intersect transversally in this space, and indeed: $p \in \alpha \times T[\alpha] \cap \Sigma^{-1} \circ \Gamma_{cut} \times T[\Sigma^{-1} \circ \Gamma_{cut}] \neq T_p[\mathbb{R}^2 \times \mathbf{S}^1]$.

Result 3 *If, in an image of a surface with a curve lying on the surface, the curve on the surface crosses the cut generator, then the curve in the image will typically appear transverse to the cut at the corresponding point in the image.*

We now derive the analogous result for a family of curves A. For $\Pi_{\mathbf{V}} \circ \Sigma \circ A(U,V)$ to be tangent to γ_{cut}, we need $T_{z_p}[\Pi \circ \Sigma \circ A(U,V)] = T_{z_p}[\gamma_{cut}]$, which only occurs when $T_{y_p}[\Sigma \circ A(U,V)] = T_{y_p}[\Gamma_{cut}]$. In $\mathbb{R}^2 \times \mathbf{S}^1$, $A \times T[A]$ is a surface, and $\Sigma^{-1} \circ \Gamma_{cut} \times T[\Sigma^{-1} \circ \Gamma]$ is a curve. The intersection of these two objects is transverse: $p \in A \times T[A] \cap \Sigma^{-1} \circ \Gamma_{cut} \times T[\Sigma^{-1} \circ \Gamma_{cut}] = T_p[\mathbb{R}^2 \times \mathbf{S}^1]$. See Fig. 8.

Result 4 *In an image of a surface with a family of smooth curves on the surface, the curves crossing the cut generator typically are everywhere transverse to the cut in the image, except at isolated points.*

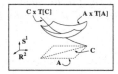

Fig. 8. $A \times T[A(U,V)]$, traces a surface in $\mathbb{R}^2 \times \mathbb{S}^1$, while, letting $C = \Sigma^{-1} \circ \Gamma_{cut}$, $C \times T[C]$ traces a curve. When the two intersect, the curves of A are tangent to the cut in the image. This situation is transversal, but has dimension zero.

Thus, in an image of a surface with a family of curves on the surface, there are two situations: (FOLD) the curves are typically tangent to the fold, with isolated exceptional points or (CUT) the curves are typically transverse to the cut, with isolated exceptional points.

3   The Shading Flow Field at an Edge

Now consider a surface Σ under illumination from a point source at infinity in the direction L. If the surface is Lambertian then the shading at a point p is $s(p) = N \cdot L$ where N is the normal to the surface at p; this is the standard model assumed by most shape-from-shading algorithms. We define the *shading flow field* to be the unit vector field tangent to the level sets of the shading field: $\mathbf{S} = \frac{1}{\|\nabla s\|}(-\frac{\partial s}{\partial y}, \frac{\partial s}{\partial x})$. The structure of the shading flow field can be used to distinguish between several types of edges, e.g. cast shadows and albedo changes [1]. Applying the results of the previous section, the shading flow field can be used to categorize edge neighborhoods as *fold* or *cut*.

Since Σ is smooth (except possibly at Γ_{cut}), N varies smoothly, and as a result so does s. Thus $\mathbf{S}$ is the tangent field to a family of smooth curves. Consider $\mathbf{S}$ at an edge point p. If p is a fold point, then in the image $\mathbf{S}(p) = T_p[\gamma_{fold}]$. If p is a cut point, then $\mathbf{S}(p) \neq T_p[\gamma_{cut}]$. (Fig. 9)

Proposition 1. *At an edge point $p \in \gamma$ in an image we can define two semi-open neighborhoods, N_p^A and N_p^B, where the surface to image mapping is continuous in each neighborhood. We can then classify p as follows:*

1. FOLD-FOLD: *The shading flow is tangent to γ in N_p^A and in N_p^B, with exception at isolated points.*
2. FOLD-CUT: *The shading flow is tangent to γ at p in N_p^A and the shading flow is transverse to Γ at p in N_p^B, with exception at isolated points.*
3. CUT-CUT: *The shading flow is transverse to γ at p in N_p^A and in N_p^B, with exception at isolated points.*

Figs. 10, 11, and 12 illustrate the applicability of our categorization.

3.1   Computing Folds and Cuts

To apply our categorization requires the computation of the shading flow in the neighborhood of edges. This presents a problem, as both edge detection and

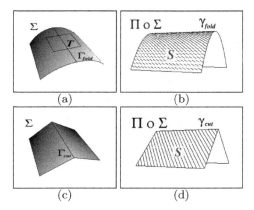

Fig. 9. The shading flow field at an edge. Near a fold (a) the shading flow field becomes tangent to the edge in the image (b). At a cut (c), the flow is transverse (d).

shading flow computation typically involve smoothing, and at discontinuities the shading flow will be inaccurate.[1] The effects of smoothing across an edge must either be minimized (e.g., by adaptively smoothing based on edge detection), or otherwise accounted for (e.g., relaxation labeling [1]). For simplicity only uniform smoothing is applied; note that this immediately places an upper limit on the curvature of folds we can discern (consider how occluding polyhedral edges may appear as folds at high resolution). This raises the question as to what the appropriate measurement should be in making the categorization. In the examples shown, the folds are of greater extent than the applied smoothing filter (see Fig. 10(d)), and so our classification is applicable by observing the shading flow orientation as compared to edge orientation (a filter based on orientation averaging suffices); in cases where high surface curvature is present, higher order calculations may be appropriate (i.e. shading flow curvature).

4   Conclusions

The categorizations we have presented are computable locally, and are intimately related to figure-ground discrimination (see Fig. 12). Furthermore, the advantage of introducing the differential topological analysis for this problem is that it is readily generalized to more realistic, or even arbitrary, shading distributions. For example, shading that results from diffuse lighting can be expressed in terms of an aperture function that smoothly varies over the surface [11], meeting the conditions we described in Section 2, thus enabling us to make the fold-cut distinction. The same analysis can be applied to texture or range data.

---

[1] This also affects the gradient magnitude which can be used to aid our classification.

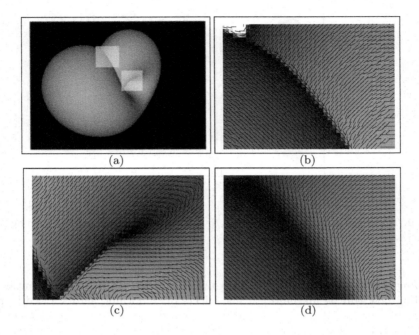

(a)   (b)

(c)   (d)

Fig. 10. The Klein bottle (a) and its shading flow field at a fold (b) and a cusp (c). The blur window-size is 3 pixels. On the fold side of the edge, the shading flow field is tangent to edge, while on the cut side it is transverse. In the vicinity of a cusp, the transition is evident as the shading flow field swings around the cusp point and becomes discontinuous. (d) shows the shading flow of (a) computed with a larger blur window (7 pixels).

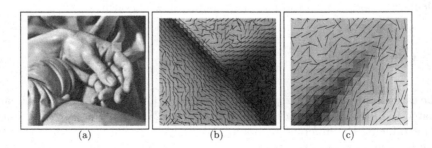

(a)   (b)   (c)

Fig. 11. (a) A scene with folds and cuts (a close-up of Michelangelo's *Pieta*). (b) The shading flow field in the vicinity of an occlusion edge (where the finger obscures the shroud). Observe how the shading flow is clearly parallel to the edge on the fold side of the edge. (c) A fold-fold configuration of the shading flow at a crease in the finger.

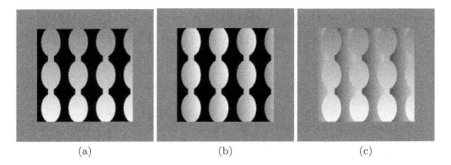

(a)   (b)   (c)

Fig. 12. Shaded versions of the ambiguous figure from Fig.1, after Kanizsa[7]. In (a) the shading is transverse to the edges. In (b) the shading is approximately tangent to the edges. Notice how flat the convex blobs in (a) look compared to in (b). In (c) both types of shading are present.

References

1. Breton, P. and Zucker, S.W.: Shadows and shading flow fields. CVPR (1996)
2. Donati, L., Stolfi, N.: Singularities of illuminated surfaces. International Journal Computer Vision **23** (1997) 207–216
3. Dufour, J.P.: Familles de courbes planes differentiables. Topology **4** (1983) 449–474
4. Freeman, W.T.: The generic viewpoint assumption in a framework for visual perception. Nature **368** (1994) 542–545
5. Golubitsky M., Guillemin, M.: Stable Mappings and Their Singularities. (1973) Springer-Verlag
6. Hirsch, M.: Differential Topology. (1976) Springer-Verlag
7. Kanizsa, G.: Organization in Vision. (1979) Praeger
8. Knill, D.C., Mamassian, P., Kersten D.: The geometry of shadows. J Opt Soc Am A **14** (1997)
9. Koenderink, J.J.: What does the occluding contour tell us about solid shape? Perception **13** (1976) 321–330
10. Koenderink, J.J., van Doorn, A.J.: Singularities of the visual mapping. Biological Cybernetics **24** (1976) 51–59
11. Langer, M.S., Zucker, S.W.: Shape from shading on a cloudy day. J Opt Soc Am A **11** (1994) 467–478
12. Nalwa, V.S.: Line-drawing interpretation: a mathematical framework. International Journal of Computer Vision **2** (1988) 103–124
13. Parodi, P.: The complexity of understanding line-drawings of origami scenes, International Journal of Computer Vision **18** (1996) 139–170
14. Rieger, J.H.: The geometry of view space of opaque objects bounded by smooth surfaces. Artificial Intelligence **44** (1990) 1–40
15. Thorndike, A.S., Cooley, C.R., Nye, J.F.: The structure and evolution of flow fields and other vector fields. J. Phys. A. **11** (1978) 1455–1490
16. Tse, P.U., Albert, M.K.: Amodal completion in the absence of image tangent discontinuities. Perception **27** (1998) 455–464
17. Weinshall, D., Werman, M.: On view likelihood and stability. IEEE-PAMI **19** (1997) 97–108

Extraction of Topological Features from Sequential Volume Data

Yukiko Kenmochi[1]*, Atsushi Imiya[2], Toshiaki Nomura[1], and Kazunori Kotani[1]

[1] School of Information Science, JAIST, Japan
[2] Department of Information and Image Sciences, Chiba Univ., Japan
{kenmochi,tnomura,ikko}@jaist.ac.jp, imiya@icsd7.tj.chiba-u.ac.jp

Abstract. In this paper, we present an incremental algorithm to efficiently extract topological features of deformable objects from a given sequential volume data. Even though these features are global shape information, we show that there are possibilities that we can extract them by local operations avoiding global operations.

1 Introduction

In this paper, we consider a sequence of 3D digital images is given as volume data of objects. From such a sequential volume data, we present an algorithm to efficiently extract the shape information of the objects of interest. Shape information is mainly divided into geometric and topological features and we focus on the latter ones such as a number of objects, numbers of cavities and tunnels for each object. Topological features are global shape information and some algorithms for their calculation using the entire volume data at a moment are already presented [1,2,3]. In this paper, since we deal with a sequence of volume data, we use the topological features at a previous time slot which are already calculated for those at a current time slot. In other words, we obtain their difference equations with respect to time. If the shape difference caused by deformation is small, we expect that global calculation for topological features can be local calculation at the part where small deformation occurs.

For such topological feature extraction, we need a deformable object model and its deformation model. Several models with consideration on topological changes are already studied for segmentation of 3D digital images [4,5]. Those models use geometric constraints and image density constraints to change topology of object surfaces. Therefore, it is possible to change topologies of objects by using their geometric and image density information though it is impossible to change geometries of objects by using their topological information. Since one of our final goals is 3D shape analysis by integration of geometric and topological information, we look for models such that topological features can be treated more explicitly. In thinning algorithms, topology-based deformable object and

---

* The current address of the first author is Laboratoire A2SI, ESIEE, Cité Descartes, B.P. 99, 93162 Noise-Le-Grand Cedex, France thanks to JSPS Postdoctoral Fellowships for Research Abroad from 2000.

C. Arcelli et al. (Eds.): IWVF4, LNCS 2059, pp. 333–345, 2001.

deformation models are actually considered [6,7]. However, their deformation reduces the dimensions of objects of interest such as from three to one dimension, while our deformation preserves the dimensions of objects since we consider any 3D object keep being a 3D object in a sequence of volume data.

In order to consider the topologies of objects including their dimensions, we take the approach of combinatorial topology [8] for our deformable object model and its deformation model. We first introduce a polyhedral representation based on discrete combinatorial topology [9] and show that this representation enables us to describe deformation of 3D objects by set operations of polyhedra. We then present our difference algorithms for calculation of topological features from a sequential volume data by using set operations.

2 Polyhedral Representation of 3D Discrete Objects

Setting $\mathbf{Z}$ to be the set of integers, $\mathbf{Z}^3$ becomes the set of points whose coordinates are all integers in the three-dimensional Euclidean space $\mathbf{R}^3$. Those points are called lattice points and a set of volume data is given as a finite subset of $\mathbf{Z}^3$. Our polyhedral representation is based on the approach of discrete combinatorial topology [9]. We first define primitives of polyhedra in $\mathbf{Z}^3$ and then define general polyhedra as combination of those primitives. To combine the primitives, we define set operations since our polyhedron is described by a set of polygons.

2.1 Primitives of Discrete Polyhedra

This subsection is devoted to defining sets of primitives of n-dimensional polyhedra whose vertices are all lattice points. They are called n-dimensional unit discrete polyhedra. Setting a unit cubic region

$$\mathbf{D}(i, j, k) = \{(i + \epsilon_1, j + \epsilon_2, k + \epsilon_3) \ : \ \epsilon_l = 0 \text{ or } 1, \ l = 1, 2, 3\},$$

we consider that each point in $\mathbf{D}(i, j, k)$ has a value of either one or zero. Such a point whose value is one or zero is called a 1- or 0-point, respectively. For every arrangement of 1- and 0-points in $\mathbf{D}(i, j, k)$, a convex hull of all 1-points is obtained. The dimension of a convex hull depends on the number denoted by p and arrangement of 1-points in $\mathbf{D}(i, j, k)$. For instance, a convex hull becomes a 0-dimensional isolated point when $p = 1$, a 1-dimensional line segment when $p = 2$, and a 2-dimensional triangle when $p = 3$. When $p = 4$, it becomes a 2-dimensional rectangle if all 1-points lie on a plane, and otherwise a 3-dimensional tetrahedron. When $p \geq 5$, it becomes a 3-dimensional polyhedron with p vertices.

Definition 1. *If the convex hull of p 1-points $\boldsymbol{x}_1$, $\boldsymbol{x}_2$, ..., $\boldsymbol{x}_p$ in $\mathbf{D}(i, j, k)$ has n dimensions where $n = 0, 1, 2, 3$, then $\{\boldsymbol{x}_1, \boldsymbol{x}_2, \ldots, \boldsymbol{x}_p\}$ is called an n-dimensional unit discrete polyhedron.*

In $\mathbf{Z}^3$, a set of the points neighboring a point $\boldsymbol{x}$ is defined as

$$\mathbf{N}_m(\boldsymbol{x}) = \{\boldsymbol{y} \in \mathbf{Z}^3 \ : \ \|\boldsymbol{x} - \boldsymbol{y}\| \leq \sqrt{t}\}$$

Table 1. All 2- and 3-dimensional unit discrete polyhedra for the 6-, 18- and 26-neighborhood systems. We omit unit discrete polyhedra which differ from those in the table by rotations.

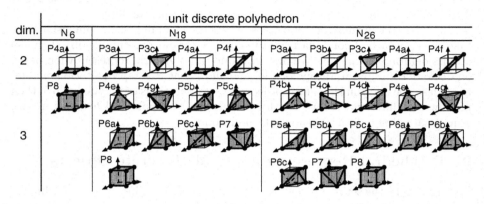

where $t = 1, 2, 3$ for $m = 6, 18, 26$. Those neighborhood systems are called 6-, 18-, 26-neighborhood systems, respectively. If any pair of adjacent vertices which are connected via an edge of a unit discrete polyhedron are m-neighboring, such a unit discrete polyhedron is classified into the set of unit discrete polyhedra for the m-neighborhood system. In this paper, we focus on the dimensions of $n = 2$ and 3. Hereafter, we refer to 2- and 3-dimensional unit discrete polyhedra as unit discrete polygons and polyhedra, respectively. Table 1 shows all unit discrete polygons and polyhedra defined for each neighborhood system. A unit discrete polygon consisting of p points $\boldsymbol{x}_1, \boldsymbol{x}_2, \ldots, \boldsymbol{x}_p$ is denoted by

$$\mathbf{S} = \{\boldsymbol{x}_1, \boldsymbol{x}_2, \ldots, \boldsymbol{x}_p\} .$$

Because a unit discrete polyhedron is bounded by a set of unit discrete polygons, it is denoted by

$$\mathbf{P} = \{\mathbf{S}_1, \mathbf{S}_2, \ldots, \mathbf{S}_q\} . \tag{1}$$

Each $\mathbf{S}_i = \{\boldsymbol{x}_{i1}, \boldsymbol{x}_{i2}, \ldots, \boldsymbol{x}_{ip}\}$ for $i = 1, 2, \ldots, q$ is oriented such that the sequence of $\boldsymbol{x}_{i1}, \boldsymbol{x}_{i2}, \ldots, \boldsymbol{x}_{ip}$ has a counterclockwise order from a viewpoint exterior to $\mathbf{P}$.

2.2   Recursive Definition of Discrete Polyhedra and Set Operations

Let $\mathcal{P}$ be the family of sets of oriented unit discrete polygons. For any pair of finite sets $\mathbf{A}$ and $\mathbf{B}$ in $\mathcal{P}$, we consider

$$\mathbf{X}_{\mathbf{A}}(\mathbf{B}) = \{\mathbf{S} \in \mathbf{A} \; : \; \mathbf{S} = \mathbf{T}^{-1} \text{ for any } \mathbf{T} \in \mathbf{B}\} ,$$

where the notation $\mathbf{S}^{-1}$ represents a unit discrete polygon whose orientation is opposite to that of $\mathbf{S}$ and the relation $\mathbf{S} = \mathbf{T}$ means that $\mathbf{S}$ and $\mathbf{T}$ are equivalent oriented unit discrete polygons. Using $\mathbf{X}_{\mathbf{A}}(\mathbf{B}) \subseteq \mathbf{A}$ and $\mathbf{X}_{\mathbf{B}}(\mathbf{A}) \subseteq \mathbf{B}$, we define

Fig. 1. An example of addition between two unit discrete polyhedra **P** and **Q** for the 26-neighborhood system.

the addition operation with the notations of $\cup$ and $\setminus$ which are the union and difference sets respectively, such that

$$\mathbf{A} + \mathbf{B} = (\mathbf{A} \setminus \mathbf{X_A}(\mathbf{B})) \cup (\mathbf{B} \setminus \mathbf{X_B}(\mathbf{A})) \,. \tag{2}$$

Since the empty set is regarded as the neutral element for any $\mathbf{A} \in \mathcal{P}$ such that

$$\mathbf{A} + \emptyset = \emptyset + \mathbf{A} = \mathbf{A} \,,$$

the inverse element for any $\mathbf{A} \in \mathcal{P}$ is defined as

$$-\mathbf{A} = \{\mathbf{S}^{-1} \, : \, \mathbf{S} \in \mathbf{A}\}$$

so that

$$\mathbf{A} + (-\mathbf{A}) = (-\mathbf{A}) + \mathbf{A} = \emptyset \,.$$

We then define the subtraction for $\mathbf{A}$ and $\mathbf{B}$ in $\mathcal{P}$ such that

$$\mathbf{A} - \mathbf{B} = \mathbf{A} + (-\mathbf{B}) = (\mathbf{A} \setminus \mathbf{B}) \cup (-(\mathbf{B} \setminus \mathbf{A})) \,.$$

Let us consider the addition of two unit discrete polyhedra **P** and **Q** in $\mathcal{P}$. Figure 1 illustrates an example of $\mathbf{P} + \mathbf{Q}$ in the 26-neighborhood system. After combining **P** and **Q** at their common faces of $\mathbf{S} \in \mathbf{X_P}(\mathbf{Q})$ and $\mathbf{T} \in \mathbf{X_Q}(\mathbf{P})$, we exclude all **S** and **T** from a union of **P** and **Q** and obtain $\mathbf{P} + \mathbf{Q}$. If we set $\mathbf{P'} = \mathbf{P} + \mathbf{Q}$ in Figure 1, then we also see $\mathbf{P'} - \mathbf{Q} = \mathbf{P}$. By using the addition operation given in (2), we define the set of discrete polyhedra from the finite set of unit discrete polyhedra for each m-neighborhood system.

Definition 2. *Discrete polyhedra $\mathbf{P}_m$ are recursively constructed for each $m = 6, 18, 26$ as follows:*

1. *a unit discrete polyhedron for an m-neighborhood system is considered to be a discrete polyhedron $\mathbf{P}_m$, and we set*

$$\mathbf{W}(\mathbf{P}_m) = \{\mathbf{D}(\boldsymbol{x}) \, : \, \underset{\mathbf{S} \in \mathbf{P}_m}{\cup} \mathbf{S} \subseteq \mathbf{D}(\boldsymbol{x})\} \,;$$

2. *if $\mathbf{P}_m$ and $\mathbf{A}_m$ are, respectively, a discrete polyhedron and a unit discrete polyhedron such that they satisfy the following conditions:*
   a) $\mathbf{W}(\mathbf{P}_m) \cap \mathbf{W}(\mathbf{A}_m) = \emptyset$;

b) if there exist a pair of $\mathbf{S} \in \mathbf{P}_m$ *and* $\mathbf{T} \in \mathbf{A}_m$ *in* $\mathbf{D}(\boldsymbol{x}) \cap \mathbf{D}(\boldsymbol{y}) \neq \emptyset$ *so that*
$\mathbf{D}(\boldsymbol{x}) \in \mathbf{W}(\mathbf{P}_m)$ *and* $\mathbf{D}(\boldsymbol{y}) \in \mathbf{W}(\mathbf{A}_m)$, *then* $\mathbf{S} = \mathbf{T}^{-1}$,
then $\mathbf{P}'_m = \mathbf{P}_m + \mathbf{A}_m$ *becomes a discrete polyhedron, and we set*

$$\mathbf{W}(\mathbf{P}'_m) = \mathbf{W}(\mathbf{P}_m) \cup \mathbf{W}(\mathbf{A}_m) \,.$$

From Definition 2, we see that a discrete polyhedron $\mathbf{P}_m$ for $m = 6, 18, 26$ is constructed by combining unit discrete polyhedra $\mathbf{A}_m$s in all $\mathbf{D}(\boldsymbol{x}) \in \mathbf{W}(\mathbf{P}_m)$ one by one.

2.3   Discrete Polyhedron Construction from Volume Data

Let $\mathbf{V}$ be a set of volume data which is a finite subset of $\mathbf{Z}^3$. Setting all points in $\mathbf{V}$ to be 1-points, $\mathbf{Z}^3 \setminus \mathbf{V}$ becomes the set of all 0-points in $\mathbf{Z}^3$. Considering a discrete polyhedron $\mathbf{P}_m(\mathbf{V})$ constructed from any $\mathbf{V}$, we obtain

$$\mathbf{P}_m(\mathbf{V}) = \underset{\mathbf{D}(\boldsymbol{x}) \in \mathbf{W}}{+} \mathbf{P}_m(\mathbf{V} \cap \mathbf{D}(\boldsymbol{x})) \tag{3}$$

where
$$\mathbf{W} = \{\mathbf{D}(\boldsymbol{x}) \; : \; \mathbf{D}(\boldsymbol{x}) \cap \mathbf{V} \neq \emptyset, \; \boldsymbol{x} \in \mathbf{Z}^3\} \,.$$

Each $\mathbf{P}_m(\mathbf{V} \cap \mathbf{D}(\boldsymbol{x}))$ for $\mathbf{D}(\boldsymbol{x}) \in \mathbf{W}$ represents a 3-dimensional unit discrete polyhedron which is constructed with respect to 1-points of $\mathbf{V}$ in $\mathbf{D}(\boldsymbol{x})$ as shown in Table 1. The following theorem is proved in [9].

Theorem 1. *For any given* $\mathbf{V}$, *we can uniquely construct* $\mathbf{P}_m(\mathbf{V})$ *for each* $m = 6, 18, 26$ *by* (3).

In [9], the efficient algorithm which directly constructs $\mathbf{P}_m(\mathbf{V})$ from $\mathbf{V}$ is also presented. From theorem 1, we have the proof of unique construction of $\mathbf{P}_m(\mathbf{V})$ for each $\mathbf{V}$ in a sequence. In other words, we obtain a unique sequence of $\mathbf{P}_m(\mathbf{V})$s corresponding to a sequence of $\mathbf{V}$s. In the following sections, we omit m of $\mathbf{P}_m(\mathbf{V})$ because the following discussion is common for any m.

3   Deformation of Polyhedral Objects

3.1   Deformation Description by Set Operations

Deformation of a discrete polyhedron $\mathbf{P}$ is mainly classified into two types of simple deformation. If $\mathbf{P}_t$ and $\mathbf{P}_{t+1}$ are discrete polyhedra before and after deformation, respectively, then two types of deformation from $\mathbf{P}_t$ to $\mathbf{P}_{t+1}$ are described using the addition and subtraction operations such that

$$\mathbf{P}_{t+1} = \mathbf{P}_t + \Delta\mathbf{P} \,, \tag{4}$$

$$\mathbf{P}_{t+1} = \mathbf{P}_t - \Delta\mathbf{P} \,, \tag{5}$$

where $\Delta\mathbf{P}$ is a discrete polyhedron which is a difference between $\mathbf{P}_t$ and $\mathbf{P}_{t+1}$. Equations (4) and (5) are called expanding and shrinking, respectively.

3.2   Polyhedral Deformation from Sequential Volume Data

Any deformation of discrete polyhedra $\mathbf{P}(\mathbf{V}_t)$s constructed from a sequence of volume data $\mathbf{V}_t$s is described by a combination of expanding and shrinking operations (4) and (5). First, we consider two kinds of such combinatorial deformation for the cases of adding a point $\boldsymbol{x}$ to $\mathbf{V}_t$ and removing $\boldsymbol{x}$ from $\mathbf{V}_t$, respectively.

For the deformation such as adding (resp. removing) $\boldsymbol{x}$ to (resp. from) $\mathbf{V}_t$, we should consider only the region such as the union of

$$\mathbf{E}_{\boldsymbol{x}} = \{\mathbf{D}(\boldsymbol{y}) \ : \ \boldsymbol{x} \in \mathbf{D}(\boldsymbol{y})\}$$

which consists of eight unit cubes $\mathbf{D}(\boldsymbol{y})$s around $\boldsymbol{x}$ because such deformation affects only the polyhedral shape at the union of $\mathbf{E}_{\boldsymbol{x}}$. By adding $\boldsymbol{x} \in \mathbf{Z}^3 \setminus \mathbf{V}_t$ to $\mathbf{V}_t$, $\mathbf{P}(\mathbf{V}_t)$ is expanded to $\mathbf{P}(\mathbf{V}_t \cup \{\boldsymbol{x}\})$ which is also uniquely determined by (3), such that

$$\mathbf{P}(\mathbf{V}_t \cup \{\boldsymbol{x}\}) = \mathbf{P}(\mathbf{V}_t) - \Delta\mathbf{P}_1 + \Delta\mathbf{P}_2 \tag{6}$$

where

$$\Delta\mathbf{P}_1 = \mathop{+}_{\mathbf{D}(\boldsymbol{y})\in\mathbf{E}_{\boldsymbol{x}}} \mathbf{P}(\mathbf{V}_t \cap \mathbf{D}(\boldsymbol{y}))$$
$$\Delta\mathbf{P}_2 = \mathop{+}_{\mathbf{D}(\boldsymbol{y})\in\mathbf{E}_{\boldsymbol{x}}} \mathbf{P}((\mathbf{V}_t \cup \{\boldsymbol{x}\}) \cap \mathbf{D}(\boldsymbol{y})) \ .$$

In (6), we first subtract a unit discrete polyhedron $\Delta\mathbf{P}_1$ in $\mathbf{E}_{\boldsymbol{x}}$ from $\mathbf{P}(\mathbf{V}_t)$ and then add the replacement $\Delta\mathbf{P}_2$ reconstructed in $\mathbf{E}_{\boldsymbol{x}}$ after adding $\boldsymbol{x}$ to $\mathbf{V}_t$.

Similarly, an operation for shrinking $\mathbf{P}(\mathbf{V}_t)$ to $\mathbf{P}(\mathbf{V}_t \setminus \{\boldsymbol{x}\})$ by removing $\boldsymbol{x}$ from $\mathbf{V}_t$ is described by

$$\mathbf{P}(\mathbf{V}_t \setminus \{\boldsymbol{x}\}) = \mathbf{P}(\mathbf{V}_t) - \Delta\mathbf{P}_1 + \Delta\mathbf{P}_3 \tag{7}$$

where

$$\Delta\mathbf{P}_3 = \mathop{+}_{\mathbf{D}(\boldsymbol{y})\in\mathbf{E}_{\boldsymbol{x}}} \mathbf{P}((\mathbf{V}_t \setminus \{\boldsymbol{x}\}) \cap \mathbf{D}(\boldsymbol{y})) \ .$$

We execute each deformation of (6) and (7) in two stages such as subtraction and then addition. We do not calculate the differences such as $-\Delta\mathbf{P}_1 + \Delta\mathbf{P}_2$ and $-\Delta\mathbf{P}_1 + \Delta\mathbf{P}_3$ first because there is no guarantee that the differences can be calculated. In order to obtain the guarantee such that the calculation result becomes a discrete polyhedron, $-\Delta\mathbf{P}_1$ and $\Delta\mathbf{P}_2$ (resp. $\Delta\mathbf{P}_3$) need to satisfy the similar conditions to those in Definition 2 and it is easy to find some counter examples that do not satisfy the conditions. To maintain the uniqueness property of $\mathbf{P}(\mathbf{V}_t \cup \boldsymbol{x})$ and $\mathbf{P}(\mathbf{V}_t \setminus \boldsymbol{x})$, we need the two-stage procedures for (6) and (7).

Given a sequence of volume data $\mathbf{V}_1, \mathbf{V}_2, \ldots$, we set $\Delta\mathbf{V}_t$ to be the difference between $\mathbf{V}_t$ and $\mathbf{V}_{t+1}$. If either (6) or (7) is manipulated for every point in $\Delta\mathbf{V}_t$ in order, consequently we obtain $\mathbf{P}(\mathbf{V}_{t+1})$ from $\mathbf{P}(\mathbf{V}_t)$. Note that $\mathbf{P}(\mathbf{V}_{t+1})$ is independent of the order of points chosen from $\Delta\mathbf{V}_t$ because of the uniqueness property in Theorem 1. Thus, given a sequence of $\mathbf{V}_1, \mathbf{V}_2, \ldots$, we obtain a unique sequence of $\mathbf{P}(\mathbf{V}_1), \mathbf{P}(\mathbf{V}_2), \ldots$ incrementally by (6) and (7) and each $\mathbf{P}(\mathbf{V}_t)$ is equivalent to the unit polyhedron directly constructed from $\mathbf{V}_t$ by (3).

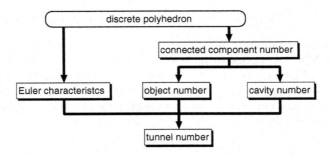

Fig. 2. The calculation procedure for topological features of a polyhedral object.

4   Topological Feature Extraction during Deformation

In this paper, we consider topological features such as the number of objects in volume data $\mathbf{V}_t$ and the numbers of cavities and tunnels in objects. To calculate these numbers from our polyhedral representation $\mathbf{P}(\mathbf{V}_t)$, we make use of the number of connected components [2,9] and the Euler characteristics [1,2,3] of $\mathbf{P}(\mathbf{V}_t)$. The calculation procedure of them is derived from their relations and illustrated in Figure 2. In this section, assuming that we already have a sequence of $\mathbf{P}(\mathbf{V}_1)$, $\mathbf{P}(\mathbf{V}_2)$, ... which are obtained from a sequence of $\mathbf{V}_1$, $\mathbf{V}_2$, ... as shown in the previous section and also the values of topological features for a $\mathbf{P}(\mathbf{V}_t)$, we calculate the values of topological features for $\mathbf{P}(\mathbf{V}_{t+1})$. Consequently, in this section we try to obtain the deference equations for calculation of the values of topological features for $\mathbf{P}(\mathbf{V}_{t+1})$.

4.1   Connected Component Number

The number of connected components for any set in $\mathcal{P}$ which is the family of sets of oriented unit discrete polygons is defined.

Definition 3. *For any pair of* $\mathbf{S}$ *and* $\mathbf{T}$ *in* $\mathbf{A} \in \mathcal{P}$, *if there exists a path such that*

$$\mathbf{S}_1 = \mathbf{S}, \mathbf{S}_2, \ldots, \mathbf{S}_a = \mathbf{T}$$

where every $\mathbf{S}_i \in \mathbf{A}$ *and* $\mathbf{S}_i \cap \mathbf{S}_{i+1} \neq \emptyset$, *then* $\mathbf{A}$ *is connected.*

The definition is different from that in [2]. While the connectivity is considered between points in [2], we consider that between unit discrete polygons and it is based on the definition of connectivity in combinatorial topology [8,9]. To decompose $\mathbf{P}(\mathbf{V}_1)$ into connected components, we apply an algorithm in [9] and set $b_c(\mathbf{P}(\mathbf{V}_1))$ to be the initial number of connected components in $\mathbf{P}(\mathbf{V}_1)$ for the sequence $\mathbf{V}_t$.

Let us assume that $b_c(\mathbf{P}(\mathbf{V}_t))$ is given and we calculate $b_c(\mathbf{P}(\mathbf{V}_{t+1}))$ from $b_c(\mathbf{P}(\mathbf{V}_t))$. During the deformation from $\mathbf{P}(\mathbf{V}_t)$ to $\mathbf{P}(\mathbf{V}_{t+1})$, there may be connectivity changes in a small local area around the points in $\mathbf{V}_{t+1} \setminus \mathbf{V}_t$, such as

Fig. 3. The changes of global connectivity caused by connectivity changes in a small local area, such as joining((a), (b)) and separating ((c), (d)).

joining and separating as shown in Figure 3. The similar local changes are considered in [5] and they are called melting and constriction from the topological sense. For joining, we have two possible cases for $b_c(\mathbf{P}(\mathbf{V}_{t+1}))$ such that

$$b_c(\mathbf{P}(\mathbf{V}_{t+1})) = b_c(\mathbf{P}(\mathbf{V}_t)) - 1 \ , \tag{8}$$

$$b_c(\mathbf{P}(\mathbf{V}_{t+1})) = b_c(\mathbf{P}(\mathbf{V}_t)) \ . \tag{9}$$

Equation (8) describes two different components in $\mathbf{P}(\mathbf{V}_t)$ are joined into one as shown in Figure 3 (a), and (9) describes two different parts in one component are joined as shown in Figure 3 (b). To distinguish between these cases (8) and (9), we need to know if two joining parts are in the same component or not and such information is automatically obtained if we have information on decomposition of $\mathbf{P}(\mathbf{V}_t)$ by connectivity.

For separating, we also have two possible cases such that

$$b_c(\mathbf{P}(\mathbf{V}_{t+1})) = b_c(\mathbf{P}(\mathbf{V}_t)) + 1 \ , \tag{10}$$

$$b_c(\mathbf{P}(\mathbf{V}_{t+1})) = b_c(\mathbf{P}(\mathbf{V}_t)) \ . \tag{11}$$

Equation (10) describes that one component is separated into two as shown in Figure 3 (c). In (11), even though one component is separated locally, the component is still joined at some other part as shown in Figure 3 (d). In any case of local separation, therefore, we need to distinguish between (10) and (11) by a global operation, which is similar to that for decomposition by connectivity [9], to check if two separated parts are still in the same component or not.

If there is no local connectivity change before and after the deformation from $\mathbf{P}(\mathbf{V}_t)$ to $\mathbf{P}(\mathbf{V}_{t+1})$, we keep the connected component number such as (9) and (11). The local connectivity change is checked by using local connectivity conditions which is similar to simple points for the thinning algorithms [6,7] rather than by using geometric conditions [5]. We currently study to clarify the relations between our conditions for the local connectivity change and the simple point conditions. In this paper, however, we do not give the further discussion and leave this topic for our future work. Note that the local connectivity change is checked only in the union of $\mathbf{E}_x$s for all x in $\mathbf{V}_{t+1} \setminus \mathbf{V}_t$ and the algorithm complexity will be $O(k)$ where k is the number of points in $\mathbf{V}_{t+1} \setminus \mathbf{V}_t$.

4.2   Object and Cavity Numbers

From (1), $\mathbf{P}(\mathbf{V}_{t+1})$ is given as a set of unit discrete polygons and it represents object surfaces. Because $\mathbf{P}(\mathbf{V}_{t+1})$ does not contain the inside structure of objects, connected components in $\mathbf{P}(\mathbf{V}_{t+1})$ are classified into two types of object surfaces: surfaces which face the exterior space and surfaces which face the cavity space. These surfaces are called exterior and interior surfaces, respectively. We see that numbers of exterior and interior surfaces correspond to numbers of objects and cavities, respectively. Let $b_0(\mathbf{P}(\mathbf{V}_{t+1}))$ and $b_2(\mathbf{P}(\mathbf{V}_{t+1}))$ be numbers of objects and cavities in $\mathbf{V}_{t+1}$. Then, we have the following relation

$$b_c(\mathbf{P}(\mathbf{V}_{t+1})) = b_0(\mathbf{P}(\mathbf{V}_{t+1})) + b_2(\mathbf{P}(\mathbf{V}_{t+1})) \ .$$

For each component $\mathbf{C}$ in $\mathbf{P}(\mathbf{V}_{t+1})$, we distinguish between exterior and interior of surfaces by Algorithm 1. After classifying all $\mathbf{C}$ in $\mathbf{P}(\mathbf{V}_{t+1})$ into two sets of interior and exterior surfaces, we obtain $b_0(\mathbf{P}(\mathbf{V}_{t+1}))$ and $b_2(\mathbf{P}(\mathbf{V}_{t+1}))$. Note that $b_0(\mathbf{P}(\mathbf{V}_{t+1}))$ and $b_2(\mathbf{P}(\mathbf{V}_{t+1}))$ are obtained only for $\mathbf{P}(\mathbf{V}_{t+1})$ which have had a new connectivity structure in the previous subsection.

Algorithm 1

input: C.
output: 0 *(interior)* or 1 *(exterior).*
begin
  1. *Setting $\boldsymbol{x}$ which is a point further away from all points in $\mathbf{P}(\mathbf{V}_{t+1})$ and $\boldsymbol{y}$ in $\mathbf{C}$, draw a line l from $\boldsymbol{x}$ to $\boldsymbol{y}$;*
  2. *choose a point $\boldsymbol{z}$ in $\mathbf{C}$ which is nearest to $\boldsymbol{x}$ lying on l;*
  3. *find a unit discrete polygon $\mathbf{S}$ which includes $\boldsymbol{z}$ and set the normal vector $\boldsymbol{n}$ which is oriented to the exterior of $\mathbf{P}(\mathbf{V}_{t+1})$ for $\mathbf{S}$;*
  4. *if $(\boldsymbol{y} - \boldsymbol{x}) \cdot \boldsymbol{n} > 0$, then return 0;*
  5. *else if $(\boldsymbol{y} - \boldsymbol{x}) \cdot \boldsymbol{n} < 0$, then return 1;*
  6. *else if $(\boldsymbol{y} - \boldsymbol{x}) \cdot \boldsymbol{n} = 0$, then go back to step 1 and choose another $\boldsymbol{x}$.*
end

4.3 Euler Characteristics

There are two Euler characteristics for polyhedral objects in $\mathbf{R}^3$: E_s which is based on triangulation of polyhedral surfaces and E_v which is based on tetrahedrization of polyhedral objects [2,3]. Since $\mathbf{P}(\mathbf{V}_t)$ is a surface representation, only E_s is calculated from $\mathbf{P}(\mathbf{V}_t)$ for any t such that

$$E_s(\mathbf{P}(\mathbf{V}_t)) = s_0(\mathbf{P}(\mathbf{V}_t)) - s_1(\mathbf{P}(\mathbf{V}_t)) + s_2(\mathbf{P}(\mathbf{V}_t)) \tag{12}$$

where $s_0(\mathbf{P}(\mathbf{V}_t))$ and $s_1(\mathbf{P}(\mathbf{V}_t))$ denote the numbers of 0- and 1-dimensional unit discrete polyhedra in the union of discrete polygons of $\mathbf{P}(\mathbf{V}_t)$ and $s_2(\mathbf{P}(\mathbf{V}_t))$ denotes the number of all unit discrete polygons in $\mathbf{P}(\mathbf{V}_t)$, respectively. For calculation of the number $b_1(\mathbf{P}(\mathbf{V}_t))$ of tunnels in the next subsection, however, we use the relation [2]

$$E_v(\mathbf{P}(\mathbf{V}_t)) = b_0(\mathbf{P}(\mathbf{V}_t)) - b_1(\mathbf{P}(\mathbf{V}_t)) + b_2(\mathbf{P}(\mathbf{V}_t)) \ . \tag{13}$$

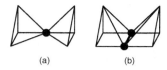

Fig. 4. Two kinds of connection in a discrete polyhedron for which the relation (14) is not established.

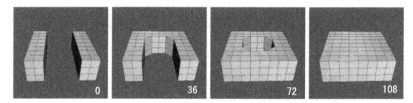

Fig. 5. Parts of a sequence of discrete polyhedra constructed from a sequence of 109 sets of volume data which is used for an experiment. The numbers in the figures correspond to the time slots in a sequence.

Since we already have $b_0(\mathbf{P}(\mathbf{V}_t))$ and $b_2(\mathbf{P}(\mathbf{V}_t))$ in the previous subsection, $E_v(\mathbf{P}(\mathbf{V}_t))$ is only required for $b_1(\mathbf{P}(\mathbf{V}_t))$ from (13).

There is an important relation between $E_s(\mathbf{P}(\mathbf{V}_t))$ and $E_v(\mathbf{P}(\mathbf{V}_t))$ which is introduced in [3];

$$E_s(\mathbf{P}(\mathbf{V}_t)) = 2E_v(\mathbf{P}(\mathbf{V}_t)) . \tag{14}$$

However, this relation is not satisfied when $\mathbf{P}(\mathbf{V}_t)$ contains connection of unit discrete polyhedra as illustrated in Figure 4. In this paper, therefore, we consider only simple polyhedra, namely 2-manifolds, which do not have such connection in $\mathbf{P}(\mathbf{V}_t)$ to make use of (14).

From (12) and (14), the difference equation for $E_v(\mathbf{P}(\mathbf{V}_t))$ is obtained by

$$E_v(\mathbf{P}(\mathbf{V}_{t+1})) = E_v(\mathbf{P}(\mathbf{V}_t)) + \frac{1}{2}(\Delta s_0 - \Delta s_1 + \Delta s_2) \tag{15}$$

where for $n = 0, 1, 2$

$$\Delta s_n = s_n(\mathbf{P}(\mathbf{V}_{t+1})) - s_n(\mathbf{P}(\mathbf{V}_t)) .$$

4.4   Tunnel Number

From (13), we obtain

$$b_1(\mathbf{P}(\mathbf{V}_{t+1})) = b_0(\mathbf{P}(\mathbf{V}_{t+1})) + b_2(\mathbf{P}(\mathbf{V}_{t+1})) - E_v(\mathbf{P}(\mathbf{V}_{t+1}))$$
$$= b_1(\mathbf{P}(\mathbf{V}_t)) + \Delta b_0 + \Delta b_2 - \Delta E_v \tag{16}$$

where for $i = 0, 2$

$$\Delta b_i = b_i(\mathbf{P}(\mathbf{V}_{t+1})) - b_i(\mathbf{P}(\mathbf{V}_t)) ,$$
$$\Delta E_v = E_v(\mathbf{P}(\mathbf{V}_{t+1})) - E_v(\mathbf{P}(\mathbf{V}_t)) .$$

4.5   Algorithm

Summarizing the section, we present an incremental algorithm for topological feature extractions of $\mathbf{P}(\mathbf{V}_{t+1})$ given those of $\mathbf{P}(\mathbf{V}_t)$ following the procedure as illustrated in Figure 2.

Algorithm 2

input: $b_c(\mathbf{P}(\mathbf{V}_t))$, $b_i(\mathbf{P}(\mathbf{V}_t))$ *for* $i = 0, 1, 2$, $E_v(\mathbf{P}(\mathbf{V}_t))$, $\mathbf{P}(\mathbf{V}_t)$, $\mathbf{P}(\mathbf{V}_{t+1})$.
output: $b_c(\mathbf{P}(\mathbf{V}_{t+1}))$, $b_i(\mathbf{P}(\mathbf{V}_{t+1}))$ *for* $i = 0, 1, 2$, $E_v(\mathbf{P}(\mathbf{V}_{t+1}))$.
begin

1. *Around the points in $\mathbf{V}_{t+1} \setminus \mathbf{V}_t$, check the local connectivity change (see subsection 4.1 for more information);*
2. *if joining is seen at the local regin, then*
   2.1 *if joining is the type of Figure 3 (a), apply (8) for $b_c(\mathbf{P}(\mathbf{V}_{t+1}))$; otherwise, apply (9);*
   2.2 *obtain $b_0(\mathbf{P}(\mathbf{V}_{t+1}))$ and $b_2(\mathbf{P}(\mathbf{V}_{t+1}))$ by Algorithm 1;*
3. *else if separating is seen at the local regin, then*
   3.1 *if two separated parts are in the same component (global check; see subsection 4.1), apply (10) for $b_c(\mathbf{P}(\mathbf{V}_{t+1}))$; otherwise, apply (11);*
   3.2 *obtain $b_0(\mathbf{P}(\mathbf{V}_{t+1}))$ and $b_2(\mathbf{P}(\mathbf{V}_{t+1}))$ by Algorithm 1;*
4. *else if no local change is seen, $b_c(\mathbf{P}(\mathbf{V}_{t+1})) = b_c(\mathbf{P}(\mathbf{V}_t))$, $b_0(\mathbf{P}(\mathbf{V}_{t+1})) = b_0(\mathbf{P}(\mathbf{V}_t))$ and $b_2(\mathbf{P}(\mathbf{V}_{t+1})) = b_2(\mathbf{P}(\mathbf{V}_t))$;*
5. *obtain $E_v(\mathbf{P}(\mathbf{V}_{t+1}))$ from (15);*
6. *obtain $b_1(\mathbf{P}(\mathbf{V}_{t+1}))$ from (16).*

end

We see that only step 3.1 requires the global operation such as checking the whole $\mathbf{P}(\mathbf{V}_{t+1})$ and the worst complexity will be $O(l^2)$ where l is the number of discrete polygons in $\mathbf{P}(\mathbf{V}_{t+1})$. If we will have an efficient data structure for discrete polyhedra by using the spatial location information of discrete polygons in a discrete polyhedron, it may be possible to reduce the complexity; the further discussion will be left for our future work. Since the complexity of every step in Algorithm 2 except for step 3.1 will be $O(k)$ where k is the number of elements in $\mathbf{V}_{t+1} \setminus \mathbf{V}_t$, if no local separation occurs during the deformation, the algorithm works very efficiently by using only local operations.

5   Experimental Results

We apply Algorithm 2 to extract the topological features from a sequence of 109 sets of volume data. Note that we also use (6) to obtain a sequence of expanding discrete polyhedra from the sequence of volume data. Expansion of discrete polyhedra in the sequence is shown in Figure 5. The calculation results of topological features are shown in Figure 6. In this experiment, we succeeded to avoid step 3.1 in Algorithm 2, namely global operations.

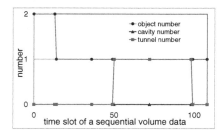

Fig. 6. The topological feature changes of a sequential volume data given in Figure 5.

6   Conclusions

In this paper, we have presented a procedure for extraction of topological features from a sequential volume data using a polyhedral object representation and its set operations. In the steps of the procedure, we have obtained the difference equations for calculation of the values of topological features based on the relations between the features. We have also shown the experimental results for a sequential volume data in which deformation of objects cause the changes of their topological features.

The authors thank to the reviewers for their helpful comments to improve the paper.The fist author also thanks to Prof. Gilles Bertrand and Prof. Michel Couprie at ESIEE for the fruitful discussions on the topic. A part of this work was supported by JSPS Grant-in-Aid for Encouragement of Young Scientists (12780207).

References

1. Imiya, A., Eckhardt, U.: The Euler characteristics of discrete object and discrete quasi-objects. Computer Vision and Image Understanding **75 3** (1999) 307–318
2. Yonekura, T., Yokoi, S., Toriwaki, J., Fukumura, T.: Connectivity and Euler number of figures in the digitized three-dimensional space (in Japanese). IEICE Transactions **J65-D 1** (1982) 80–87
3. Toriwaki, J., Yokoi, S.: Basics of algorithms for processing three-dimensional digitized pictures (in Japanese). IEICE Transactions **J68-D 4** (1985) 426–432
4. Caselles, V., Kimmel, R., Sapiro, G.: Geodesic active contours. International Journal of Computer Vision **22 1** (1997) 61–79
5. Lachaud, J. O., Montanvert, A.: Deformable meshes with automated topology changes for coarse-to-fine three-dimensional surface extraction. Medical Image Analysis **3 2** (1998) 187–207
6. Kong, T. Y., Rosenfeld, A.: Digital topology: introduction and survey. Computer Vision Graphics Image Processing **48** (1989) 357–393
7. Bertrand, G.: Simple points, topological numbers and geodesic neighborhoods in cubic grids. Pattern Recognition Letters **15** (1994) 1003–1011

8. Aleksandrov, P. S.: Combinatorial topology I. Graylock Press, Rochester, N.Y. (1956)
9. Kenmochi, Y.: Discrete combinatorial polyhedra: theory and application. PhD thesis, Chiba University (1998)

Using Beltrami Framework for Orientation Diffusion in Image Processing

Ron Kimmel[1] and Nir Sochen[2]

[1] Computer Science Dept. Technion-Israel Institute of Technology, Haifa 32000, Israel
[2] Dept. of Applied Mathematics, University of Tel-Aviv, Tel-Aviv 69978, Israel

Abstract. This paper addresses the problem of enhancement of noisy scalar and vector fields, when they are known to be constrained to a manifold. As an example, we address selective smoothing of orientation using the geometric Beltrami framework. The orientation vector field is represented accordingly as the embedding of a two dimensional surface in the spatial-feature manifold. Orientation diffusion is treated as a canonical example where the feature (orientation in this case) space is the unit circle S^1. Applications to color analysis are discussed and numerical experiments demonstrate again the power of this framework for non-trivial geometries in image processing.

1 Introduction

Feature enhancement is part of algorithms that play a major role in many image analysis and processing applications. These applications include texture processing in a transform space, disparity and depth estimation in stereo vision, robust computation of derivatives, optical flow and orientation vector fields in color processing. We are witnessing the emergence of a variety of methods for feature denoising that generalize the traditional image denoising. Here, we show how harmonic map methods, defined via the Beltrami framework, can be used to perform adaptive feature and shape denoising.

We concentrate on the example of direction diffusion that gives us information on the preferred direction at every pixel. This example is useful in applications like texture and color analysis and can be generalized to motion analysis. It incorporates all the problems and characteristics of more involved feature spaces.

The input to the denoising process is a vector field on the image. The values of this vector field are in the unit circle S^1. The given vector field is noisy and we wish to denoise it under the assumption that, for the class of images we are interested in, this vector field is piecewise smooth, see [2] for our analysis of higher dimensional spheres S^n.

Two approaches for this problem are known: Perona was the first to addresses directly this issue [6], he uses a single parameter θ as an internal coordinate in S^1. Next, Tang, Sapiro and Casseles [14,15] embedded the unit circle S^1 in $\mathbb{R}^2$ (the sphere S^2 in $\mathbb{R}^3$) and work with the external coordinates, see also [16] for a related effort. The first approach is problematic because of the periodicity of

C. Arcelli et al. (Eds.): IWVF4, LNCS 2059, pp. 346–355, 2001.
© Springer-Verlag Berlin Heidelberg 2001

S^1. Averaging small angles around zero such as $\theta = \epsilon$ and $\theta = 2\pi - \epsilon$ leads to the erroneous conclusion that the average angle is $\theta = \pi$. Perona solved this problem by exponentiating the angle such that $V = e^{i\theta}$. This is actually the embedding of S^1 in $\mathbb{C}$ which is isometric to $\mathbb{R}^2$. This method is specific to two-dimensional embedding space where complex numbers can be used. The problem in using only one internal coordinate manifests itself In the numerical implementation of the PDE through the braking of rotation invariance. In the second approach we have to make sure that we stay always in S^1 along the flow. This problem is known as the projection problem. It is solved in the continuum by adding a projection term. Chan and Shen [1] also use external coordinates with a projection term but suggest to use a Total Variation (TV) measure [8] in order to better preserve discontinuities in the vector field. This works well for the case where the codimension is one, like a circle. Yet it is difficult to generalize to higher codimensions like the sphere. Moreover, the flow of the external coordinates is difficult to control numerically since errors should be projected on S^1 and no well-defined projection exist.

We propose a solution to these problems and introduce an adaptive smoothing process, that preserves orientation discontinuities. The proposed solutions work for all dimensions and codimensions, and overcome possible parameterization singularities by introducing several internal coordinates on different patches (charts) such that the union of the patches is S^n. Adaptive smoothness is achieved by the description of the vector field as a two-dimensional surface embedded in three- and four-dimensional spatial-feature manifold for the S^1 and S^2 cases respectively. We treat here the S^1 case only due to space limitations.

The problem is formulated, in the Beltrami framework [12,3] in terms of the embedding map

$$Y : (\Sigma, g) \to (M, h)$$

where Σ is the two-dimensional image manifold, and M, in the following examples is $\mathbb{R}^n \times S^1$ with $n = 2$ ($n = 4$) for gray-level (color) images. The key point is the choice of **local coordinate systems** for **both** manifolds. Note the difference w.r.t. [14,15,1] where the image metric is flat. At the same time we should verify that the geometric filter does not depend on the specific choice of coordinates we make.

Once a local coordinate system is chosen for the embedding space and the optimization is done directly in these coordinates the updated quantities lie always in M! Other examples of enhancement by the Beltrami framework of non-flat feature spaces, like the color perceptual space and the derivatives vector field, can be found in [13,10].

2   The Beltrami Framework

Let us briefly review the Beltrami geometric framework for non-linear diffusion in computer vision [12].

2.1   Representation and Riemannian Structure

We represent an image and other local features as an embedding map of a Riemannian manifold in a higher dimensional space. The simplest example is the image itself which is represented as a 2D surface embedded in $\mathbb{R}^3$. We denote the map by $Y : \Sigma \to \mathbb{R}^3$. Where Σ is a two-dimensional surface, and we denote the local coordinates on it by (x^1, x^2). The map Y is given in general by $(Y^1(x^1, x^2), Y^2(x^1, x^2), Y^3(x^1, x^2))$. We choose on this surface a Riemannian structure, namely, a metric. The metric is a positive definite and a symmetric 2-tensor that may be defined through the local distance measurements

$$ds^2 = g_{11}(dx^1)^2 + 2g_{12}dx^1 dx^2 + g_{22}(dx^2)^2. \tag{1}$$

We use below the Einstein summation convention in which the above equation reads $ds^2 = g_{\mu\nu}dx^\mu dx^\nu$ where repeated indices are summed over. We denote the inverse of the metric by $g^{\mu\nu}$.

2.2   Image Metric Selection: The Induced Metric

A reasonable assumption is that distances we measure in the embedding spatial-feature space, such as distances between pixels and difference between grey-levels, correspond directly to distances measured on the image manifold. This is the assumption of isometric embedding under which we can calculate the image metric in terms of the embedding maps Y^i and the embedding space metric h_{ij}. It follows directly from the fact that the length of infinitesimal distances on the manifold can be calculated in the manifold and in the embedding space with the same result. Formally, $ds^2 = g_{\mu\nu}dx^\mu dx^\nu = h_{ij}dY^i dY^j$. By the chain rule, $dY^i = \partial_\mu Y^i dx^\mu$, we get $ds^2 = g_{\mu\nu}dx^\mu dx^\nu = h_{ij}\partial_\mu Y^i \partial_\nu Y^i dx^\mu dx^\nu$. From which we have

$$g_{\mu\nu} = h_{ij}\partial_\mu Y^i \partial_\nu Y^j. \tag{2}$$

Intuitively, we would like our filters to use the support of the image surface rather than the image domain. The reason is that edges can be considered as 'high cliffs' in the image surface, and a Gaussian filter defined over the image domain would smooth uniformly everywhere and will not be sensitive to the edges. While, a Gaussian filter defined over the image manifold, would smooth along the walls of the edges and preserve these high cliffs of the image surface.

As an example let us take the gray-level image as a two-dimensional image manifold embedded in the three dimensional Euclidean space $\mathbb{R}^3$. The embedding maps are

$$(Y^1(x^1, x^2) = x^1, Y^2(x^1, x^2) = x^2, Y^3(x^1, x^2) = I(x^1, x^2)). \tag{3}$$

We choose to parameterize the image manifold by the canonical coordinate system $x^1 = x$ and $x^2 = y$. The embedding, by abuse of notation, is $(x, y, I(x, y))$. The induced metric g_{11} element is calculated as follows

$$g_{11} = h_{ij}\partial_{x^1}Y^i \partial_{x^1}Y^j = \delta_{ij}\partial_x Y^i \partial_x Y^j = \partial_x x \partial_x x + \partial_x y \partial_x y + \partial_x I \partial_x I = 1 + I_x^2. \tag{4}$$

Other elements are calculated in the same manner.

2.3   A Measure on the Space of Embedding Maps

Denote by (Σ, g) the image manifold, and its metric and by (M, h) the space-feature manifold and its metric. Then the functional $S[\cdot, \cdot, \cdot]$ attaches a real number to a map $Y : \Sigma \rightarrow M$

$$S[Y^i, g_{\mu\nu}, h_{ij}] = \int dV \langle \nabla Y^i, \nabla Y^j \rangle_g h_{ij}, \tag{5}$$

where $dV = dx^1 dx^2 \cdots dx^m \sqrt{g}$ is a volume element and the scalar product $\langle, \rangle_g$ is defined with respect to the image metric i.e. $\langle \nabla Y^1, \nabla Y^2 \rangle_g = g^{\mu\nu} \partial_\mu Y^1 \partial_\nu Y^2$. This functional is known in high-energy physics as the Polyakov action [5]. Note that the image metric and the feature coordinates i.e. intensity, color, orientation etc. are independent variables. The minimization of the functional with respect to the image metric can be solved analytically in the two-dimensional case (see for example [11]). The minimizer is the induced metric. If we choose, a-priory, the image metric induced from the metric of the embedding spatial-feature space M, then the Polyakov action is reduced to an area (volume) of the image manifold.

Using standard methods in the calculus of variations (see [11]), the Euler-Lagrange equations with respect to the embedding are

$$-\frac{1}{2\sqrt{g}} h^{il} \frac{\delta S}{\delta Y^l} = \frac{1}{\sqrt{g}} \partial_\mu (\sqrt{g} g^{\mu\nu} \partial_\nu Y^i) + \Gamma^i_{jk} \langle \nabla Y^j, \nabla Y^k \rangle_g. \tag{6}$$

Since $(g_{\mu\nu})$ is positive definite, $g \equiv \det(g_{\mu\nu}) > 0$ for all x^μ. This factor is the simplest one that does not change the minimization solution while giving a geometric (reparameterization invariant) expression. The operator that is acting on Y^i in the first term is the natural generalization of the Laplacian from flat spaces to manifolds and is called *the second order differential parameter of Beltrami* [4], or for short *Beltrami operator*, and is denoted by Δ_g. The second term involves the Levi-Civita connection whose coefficients are given in terms of the metric of the embedding space

$$\Gamma^i_{jk} = \frac{1}{2} h^{il} \left(\partial_j h_{lk} + \partial_k h_{jl} - \partial_l h_{jk} \right). \tag{7}$$

This is the term that guarantees that the image surface flows in a non-Euclidean manifold and not in $\mathbb{R}^n$.

A map that satisfies the Euler-Lagrange equations $-\frac{1}{2\sqrt{g}} h^{il} \frac{\delta S}{\delta Y^l} = 0$ is a **harmonic map**. The one- and two-dimensional examples are a geodesic curve on a manifold and a minimal surface.

The non-linear diffusion or scale-space equation emerges as the gradient descent minimization flow

$$Y^i_t = \frac{\partial}{\partial t} Y^i = -\frac{1}{2\sqrt{g}} h^{il} \frac{\delta S}{\delta Y^l} = \Delta_g Y^i + \Gamma^i_{jk} \langle \nabla Y^j, \nabla Y^k \rangle_g. \tag{8}$$

This flow evolves a given surface towards a minimal surface, and in general it changes continuously a map towards a harmonic map.

There are few major differences between this flow and those suggested in [6, 14,1]. Notably, the metric that is used in those papers is flat while we use the induced metric that combines the information about the geometry of the signal and that of the feature manifold.

3 Beltrami S^1 Direction Diffusion

We are interested in attaching a unit vector field to an image. More precisely we would like to construct a non-linear diffusion process that will enhance a given noisy vector field of this form while preserving the unit magnitude of each vector.

3.1 The Embedding Space Geometry

Hemispheric coordinate system. Denote the vector field by two components $(U, V)^T$ such that $U^2 + V^2 = 1$. This description is actually an **extrinsic** description. The unit circle S^1 is a one-dimensional curve and one parameter should suffice as an internal description. Since S^1 is a simply connected and compact manifold without boundaries, we need at least two coordinate systems to cover all the points of S^1 such that the transition function between the two patches is infinitely differentiable at all points that belong to the intersection of the two patches. We define the two patches as follows: The coordinate system on $S^1 - \{(\pm 1, 0)\}$ is U, with the induced metric

$$ds_{S^1}^2 = dU^2 + dV^2 = dU^2 + (d(\sqrt{1 - U^2}))^2 = \frac{1}{1 - U^2} dU^2. \tag{9}$$

The coordinate system on $S^1 - \{(0, \pm 1)\}$ is V with the induced metric

$$ds_{S^1}^2 = dU^2 + dV^2 = (d(\sqrt{1 - V^2}))^2 + dV^2 = \frac{1}{1 - V^2} dV^2. \tag{10}$$

It is clear the the transformations $V(U) = \sqrt{1 - U^2}$ and $U(V)$ are differentiable anywhere on the intersection $S^1 - \{(\pm 1, 0), (0, \pm 1)\}$.

The embedding is of the two-dimensional image manifold in the three-dimensional space $\mathbb{R}^2 \times S^1$. The canonical embedding for the first patch is $(Y^1(x, y) = x, Y^2(x, y) = y, Y^3(x, y) = U(x, y))$, and for the second patch is

$$(Y^1(x, y) = x, Y^2(x, y) = y, Y^3(x, y) = V(x, y)). \tag{11}$$

Stereographic coordinate system. Another possibility is to use the stereographic coordinate system. The stereographic transformation gives the values of Y^i as functions of the points on the north (south) hemispheres of the hypersphere. Explicitly, for S^1 it is given (after shifting the indices by two for a coherent notation with the next sections) as $Y^3 = \frac{U^3}{1 - U^4}$. Inverting these relations we find

$$U^3 = \frac{2Y^3}{1 + (Y^3)^2} \quad , \quad U^4 = \frac{-1 + (Y^3)^2}{1 + (Y^3)^2}, \tag{12}$$

and the induced metric is

$$h_{ij} = \frac{4}{(1 + (Y^3)^2)^2} \delta_{ij}$$

Due to space limitations we defer further analysis on the stereographic coordinate system. Below we analyze the hemispheric coordinate system.

3.2   The S^1 Beltrami Operator

The line element on the image manifold is

$$ds^2 = ds_{\mathbb{R}^2}^2 + ds_{S^1}^2 = dx^2 + dy^2 + \frac{1}{1 - U^2} dU^2, \tag{13}$$

and by the chain rule

$$ds^2 = (1 + A(U)U_x^2)dx^2 + 2A(U)U_xU_y dxdy + (1 + A(U)U_y^2)dy^2, \tag{14}$$

where $A(U) = \frac{1}{1-U^2}$, and similarly for V.

The induced metric is therefore

$$(g_{\mu\nu}) = \begin{pmatrix} 1 + A(U)U_x^2 & A(U)U_xU_y \\ A(U)U_xU_y & 1 + A(U)U_y^2 \end{pmatrix}, \tag{15}$$

and the Beltrami operator acting on U is $\Delta_g U = \frac{1}{\sqrt{g}}\partial_\mu(\sqrt{g}g^{\mu\nu}\partial_\nu U)$, where $g = 1 + A(U)(U_x^2 + U_y^2)$ is the determinant of $(g_{\mu\nu})$, and $(g^{\mu\nu})$ is the inverse matrix of $(g_{\mu\nu})$.

3.3   The Levi-Civita Connection

Since the embedding space is non-Euclidean we have to calculate the Levi-Civita connection. Remember that the metric of the embedding space is

$$(h_{ij}) = \begin{pmatrix} 1 & 0 & 0 \\ 0 & 1 & 0 \\ 0 & 0 & A(U) \end{pmatrix}. \tag{16}$$

The Levi-Civita connection coefficients are given by the fundamental theorem of Riemannian geometry in the following formula $\Gamma^i_{jk} = \frac{1}{2}h^{il}\left(\partial_j h_{lk} + \partial_k h_{jl} - \partial_l h_{jk}\right)$, where the derivatives are taken with respect to Y^i for $i = 1, 2, 3$.

The only non-vanishing term is Γ^3_{33} that reads $\Gamma^3_{33} = Uh_{33}$.

The second term in the EL equations in this case reads $Uh_{33}||\nabla U||_g^2$. We can rewrite this expression as

$$h_{33}||\nabla U||_g^2 = 2 - \frac{1}{g}(1 + g), \tag{17}$$

where we used the induced metric identity Eq. (2).

3.4   The Flow and the Switches

The Beltrami flow is

$$Y_t^i = \Delta_g Y^i + \Gamma_{jk}^i(Y^1, Y^2, Y^3)\langle \nabla Y^j, \nabla Y^k \rangle_g, \tag{18}$$

for $i = 3$, and similarly for the other coordinate charts. Gathering together all the pieces we finally get the Beltrami flow

$$U_t = \Delta_g U + U\frac{g-1}{g} \quad , \quad V_t = \Delta_g V + V\frac{g-1}{g}. \tag{19}$$

In the implementation we compute the diffusion for U and V simultaneously and take the values $(U, \text{sign}(V)\sqrt{1-U^2})$ for the range $U^2 \le V^2$, and the values $(\text{sign}(U)\sqrt{1-V^2}, V)$ for the range $V^2 \le U^2$.

4   Color Diffusion

There are many coordinate systems and models of color space which try to be closer to human color perception. One of the popular coordinate systems is the HSV system [9]. In this system, color is characterized by the Hue, Saturation and Value. The Saturation and Value take their value in $\mathbb{R}_+$, while the Hue is an angle that parameterizes S^1.

In order to denoise and enhance color images by a non-linear diffusion process which is adapted to human perception we use here the HSV system. We need to have special treatment of the Hue coordinate in the lines of Section 3.

Let us represent the image as a mapping $\mathbf{Y} : \Sigma \to \mathbb{R}^4 \times S^1$ where Σ is the two-dimensional image surface and $\mathbb{R}^4 \times S^1$ is parameterized by the coordinates (x, y, H, S, V). As mentioned above, a diffusion process in this coordinate system is problematic. We define therefore two coordinates

$$U = \cos H \qquad ; \qquad W = \sin H$$

and continue in a similar way to Section 3. The metric of $\mathbb{R}^4 \times S^1$ on the patch where U parameterizes S^1 and $W(U)$ is non-singular is

$$h_{ij} = \begin{pmatrix} 1 & 0 & 0 & 0 & 0 \\ 0 & 1 & 0 & 0 & 0 \\ 0 & 0 & A(U) & 0 & 0 \\ 0 & 0 & 0 & 1 & 0 \\ 0 & 0 & 0 & 0 & 1 \end{pmatrix}, \tag{20}$$

where $A(U) = 1/(1 - U^2)$.

The induced metric is therefore

$$\begin{aligned} ds^2 &= dx^2 + dy^2 + A(U)dU^2 + dS^2 + dV^2 \\ &= dx^2 + dy^2 + A(U)(U_x dx + U_y dy)^2 + (S_x dx + S_y dy)^2 + (V_x dx + V_y dy)^2 \\ &= (1 + A(U)U_x^2 + S_x^2 + V_x^2)dx^2 + \end{aligned}$$

$$2(A(U)U_xU_y + S_xS_y + V_xV_y)dxdy + (1 + A(U)U_y^2 + S_y^2 + V_y^2)dy^2. \quad (21)$$

Similar expressions are obtained on the other dual patch.

The only non-vanishing Levi-Civita connection's coefficient is $\Gamma_{33}^3 = Uh_{33}$. The resulting flow is

$$
\begin{aligned}
U_t &= \Delta_g U + 2U - U(g^{11} + g^{22}) \\
W_t &= \Delta_g W + 2W - W(g^{11} + g^{22}) \\
S_t &= \Delta_g S \\
V_t &= \Delta_g V.
\end{aligned}
\quad (22)
$$

Note that the switch between U and W should be applied not only to the U and W equations but also to the S and V evolution equations where, at each point, one needs to work with the metric that is defined on one of the patches.

5   Experimental Results

Our first example deals with the gradient direction flow via the Beltrami framework. Figure 1 shows a vector field before and after the application of the flow for a given evolution time. The normalized gradient vector field extracted from an image is presented before and after the flow and show the way the field flows into a new smooth orientation transactions field.

Next, we explore a popular model that captures some of our color perception. The HSV (hue, saturation, value) model proposed in [9] is often used as a 'user oriented' color model, rather than the RGB 'machine-oriented' model.

Figure 2 shows the classical representation of the HSV color space, in which the hue is measured as an angle, while the value (sometimes referred to as brightness) and the color saturation are mapped onto finite non-periodic intervals. This model lands itself into a filter that operates on the spatial x, y coordinates, the value and saturation coordinates, and the hue periodic variable. Our image is now embedded in $\mathbb{R}^4 \times S^1$.

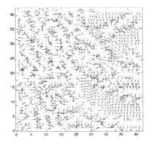 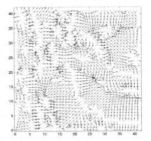

Fig. 1. Two vector fields before (left) and after (right) the flow on S^1.

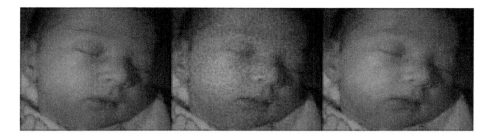

Fig. 2. The HSV color model captures human color perception better than the RGB model which is the common way our machines represent colors. The original image (left), the noisy image (middle) and the filtered image (right) demonstrate the effect of the flow as a denoising filter in the HSV color space. For further examples follow the links at $http://www.cs.technion.ac.il/ \sim ron/pub.html$

6   Concluding Remarks

There are two important issues in the process of denoising a constrained feature field. The first is to make the process compatible with the constraint in such a way that the latter is never violated along the flow. The second is the type of regularization which is applied in order to preserve significant discontinuities of the feature field while removing noise.

These two issues are treated in this paper via the Beltrami framework. First a Riemannian structure, i.e. a metric, is introduced on the feature manifold and several local coordinate systems are chosen to describe *intrinsically* the constrained feature manifold. The diffusion process acts on these coordinates and the compatibility with the constraint is achieved through the intrinsic nature of the coordinate system. The difficulty in working on a non-Euclidean space transforms itself to the need to locally choose the best coordinate system to work with.

A preservation of significant discontinuities is dealt with by using the induced metric and the corresponding Laplace-Beltrami operator acting on feature coordinates only. This operation is in fact a projection of the mean curvature normal vector on the feature direction(s). This projection slows down the diffusion process along significant (supported) discontinuities, i.e. edges.

The result of this algorithm is an adaptive smoothing process for a constrained feature space in every dimension and codimension. As examples we showed how our geometric model coupled with a proper choice of charts handles the orientation diffusion problem. This is a new application of the Beltrami framework proposed in [11]. We tested our model on vector fields restricted to the unit circle S^1, and hybrid spaces like the HSV color space. The integration of the spatial coordinates with the color coordinates yield a selective smoothing filter for images in which some of the coordinates are restricted to a circle.

Acknowledgments. We thank Alfred Bruckstein from the Technion Israel for stimulating discussions on diffusion and averaging, and color analysis. We also thank Guillermo Sapiro from University of Minnesota for sharing his ideas and results on direction diffusion with us.

References

1. T. Chan and J. Shen, "Variational restoration of non-flat image features: Models and algorithms", SIAM J. Appl. Math., to appear.
2. R. Kimmel and N. Sochen, "How to Comb a Porcupine?", ISL Technical Report, Technion Tel-Aviv University report, Israel, June 2000. Accepted to special issue on PDEs in Image Processing, Computer Vision, and Computer Graphics, Journal of Visual Communication and Image Representation.
3. R. Kimmel and R. Malladi and N. Sochen, "Images as Embedding Maps and Minimal Surfaces: Movies, Color, Texture, and Volumetric Medical Images", International Journal of Computer Vision, 39(2) (2000) 111-129.
4. E. Kreyszing, "Differential Geometry", Dover Publications, Inc., New York, 1991.
5. A. M. Polyakov, "Quantum geometry of bosonic strings", *Physics Letters*, **103B** (1981) 207-210.
6. P. Perona, "Orientation Diffusion" *IEEE Trans. on Image Processing*, 7 (1998) 457-467.
7. "Geometry Driven Diffusion in Computer Vision", Ed. B. M. ter Haar Romeny, Kluwer Academic Publishers, 1994.
8. L. Rudin and S. Osher and E. Fatemi, "Nonlinear total variation based noise removal algorithms", *Physica D*, 60 (1991) 259-268.
9. A. R. Smith, "Color gamut transform pairs", SIGGRAPH'79, 12-19,
10. N. Sochen and R. M. Haralick and Y. Y. Zeevi, "A Geometric functional for Derivatives Approximation" EE-Technion Report, April 1999.
11. N. Sochen and R. Kimmel and R. Malladi, "From high energy physics to low level vision", Report, LBNL, UC Berkeley, LBNL 39243, August, Presented in ONR workshop, UCLA, Sept. 5 1996.
12. N. Sochen and R. Kimmel and R. Malladi, "A general framework for low level vision", *IEEE Trans. on Image Processing*, 7 (1998) 310-318.
13. N. Sochen and Y. Y. Zeevi, "Representation of colored images by manifolds embedded in higher dimensional non-Euclidean space", IEEE ICIP'98, Chicago, 1998.
14. B. Tang and G. Sapiro and V. Caselles, "Direction diffusion", International Conference on Computer Vision, 1999.
15. B. Tang and G. Sapiro and V. Caselles, "Color image enhancement via chromaticity diffusion" Technical report, ECE-University of Minnesota, 1999.
16. J. Weickert, "Coherence-enhancing diffusion in colour images", Proc. of VII National Symposium on Pattern Rec. and Image Analysis, Barcelona, Vol. 1, pp. 239-244, 1997.

Digital Planar Segment Based Polyhedrization for Surface Area Estimation

Reinhard Klette and Hao Jie Sun

CITR Tamaki, University of Auckland
Tamaki Campus, Building 731, Auckland, New Zealand
{rklette, hsun009}@cs.auckland.ac.nz

Abstract. Techniques to estimate the surface area of regular solids based on polyhedrization are classified to be either local or global. Surface area calculated by local techniques generally fails to be multigrid convergent. One of the global techniques which is based on calculating the convex hull shows a tendency to be multigrid convergent. However this algorithm only deals with convex sets. The paper estimates the surface area using another global technique called DPS (Digital Planar Segment) algorithm. The projection of these DPSes into Euclidean planes is used to estimate the surface area. Multigrid convergence experiments of the estimated surface area value are used to evaluate the performance of this new method for surface area measurement.

1 Introduction

Gridding techniques are widely used to represent volume data sets in three-dimensional space in the field of computer-based image analysis. The problem of multigrid convergent surface area measurement had been discussed for more than one hundred years [8,16] and not yet reached a satisfactory result. In [9], *regular solids* are defined as the models of '3D objects' and as sets in $\mathcal{R}^3$ being candidates for multigrid surface area studies. The surface areas of such sets are well defined.

C. Jordan [8] studied the problem of volume estimation based on gridding techniques in 1892, and C. F. Gauss (1777-1855) studied the area problem for planar regions also based on this technique. Gridding is in today's point of view also considered as digitization which maps a 'real object' into a grid point set with certain resolution r defined as being the number of grid points per unit. *Jordan digitization* is characterized by *inclusion digitization*, being the union of all cubes completely contained in the topological interior of the given regular solid, and *intersection digitization*, being the union of all cubes having a non-empty intersection with the given set. *Gauss center point digitization* is the union of all cubes having a center point in the given set. The Gauss center point digitization scheme is used in this paper. It maps given regular solids into orthogonal grid sets, in which each grid edge (i.e. an edge of a grid square or grid cube) is of uniform length (grid constant).

C. Arcelli et al. (Eds.): IWVF4, LNCS 2059, pp. 356–366, 2001.
© Springer-Verlag Berlin Heidelberg 2001

The effect of studying objects enlarged r times considering the grid constant as 1 is the same as studying the object of original size having resolution r (or grid constant $1/r$). The advantages of the former (preferred by Jordan and Minkowski) are that calculations of surface areas are integer arithmetic operations and it is chosen in our implementation later illustrated. This is also called the general *duality principle for multigrid studies*.

The paper [9] proposed a way to classify polyhedrization schemes for surfaces of digitized regular solids based on the notion of *balls of influence* $B(p)$. Let $D_r(\theta)$ be the digitized set of a regular solid θ with a grid resolution r. For each polygon **G** of the constructed polyhedron there is a subset of $D_r(\theta)$ such that only grid points in the subset have influence on the specification of polygon **G**. This subset is the ball of influence. Due to the finite number of calculated polygons, there is a maximum value $R(r, \Theta)$ of all radii of the ball of influence. The polyhedrization techniques are then classified based on the following criterion:

Definition 1. *A polyhedrization technique is* local *if there exists a constant κ such that $R(r, \Theta) \leq \kappa/r$, for any regular solid Θ and any grid resolution r. If a polyhedrization method is not local then it is* global.

For the surface detection algorithm of [2], the marching cubes algorithm and marching tetrahedra algorithm, the constant κ is not more than $\frac{\sqrt{3}}{2}$ [11].

The soundness of a grid technique as pointed out by [10] should meet in short the following properties with respect to surface area estimation:

1. higher resolutions should lead to convergence for the calculated value of surface area, and
2. convergence should be towards the true value.

Obviously, surface area calculation by counting the surface faces (grid faces of the digital surface) of the digitized regular solid is not meeting the criteria. Since the result of doing this may converge to the value from 1 to $\sqrt{3}$ times the true value depending on the shape and position of the given object when the resolution r goes to infinity [13]. Other local polyhedrization techniques as investigated in [9] such as marching cube and dividing cube algorithms, although all converge, fail to converge to the true value.

On the other hand, a proposed global technique, relative-convex hull polyhedrization [17] which is implemented for convex sets by the calculation of the convex hull, converges to the true value when dealing with these convex sets [9]. Our method is also a global polyhedrization technique. In this method the given digital object is first polyhedrized by agglomerating surface faces into maximum digital planar segments (DPSes). Then the surface area of polyhedra is calculated by summing the area of the those digital planar segments and is used to estimate the surface area of the digitized object. Our global polyhedrization technique is called *DPS method*.

Section 2 gives our definition of a digital plane segment. Section 3 explains the algorithm to recognize a digital plane segment, and to perform the surface area calculation. The experimental results are presented and discussed in Section 4.

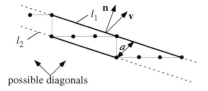

possible diagonals

Fig. 1. Definition of a DSS based on main diagonals: l_1 is the line containing the maximum length tangential line segment of the 4-curve, l_2 is the second tangential line parallel to l_1, and $a < \sqrt{2}$ is the main diagonal distance between both. Vector n is the normal of l_1, and v is the vector of length $\sqrt{2}$ of the main diagonal

We conclude in Section 5. In this paper, we restrict our interest on implementation and experimental results. The theoretical convergence analysis is left to another paper.

2   Digital Planar Segments

We generalize from a definition of a digital straight segment (DSS) in the plane as introduced in arithmetic geometry [7,12,14]. Two parallel lines having a given 4-path in between are called *bounding lines*. There are two possible diagonals in 2D grid space, see Fig. 1. The *main diagonal* is the one which maximizes the dot product with the normal of these lines. The *main diagonal distance* between these parallel lines is measured in direction of the main diagonal. A 4-path is a DSS iff there is a pair of bounding lines having a main diagonal distance less than $\sqrt{2}$.

We will define a digital planar segment (DPS) in 3D space by a main diagonal distance between two parallel planes. We distinguish them by defining one to be a supporting plane and the other as tangential plane. For a finite set of surface faces we consider a slice defined by these two planes. The supporting plane is defined by at least three non-collinear vertices of this face set, and all faces of this set are in only one of the (closed) halfspaces defined by this plane. Note that any non-empty finite set of faces has at least one supporting plane. Any supporting plane defines a tangential plane which is touching the vertices of this set 'from the other side'. Figure 2 gives a rough illustration of a DPS. Again, n denotes the normal, and v is the vector of length $\sqrt{3}$ in direction of the main diagonal.

A grid cube has eight directed diagonals. The main diagonal of a pair of such two planes is that diagonal direction out of these eight directed diagonals which has the maximum dot product (inner product) with the normal of the planes. Note that there may be more than one main diagonal direction for the planes, and we can choose any of these as our main diagonal. The distance between supporting plane and tangential in the main diagonal direction is called *main diagonal distance*.

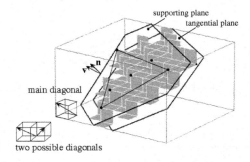

Fig. 2. A 3D view of a DPS: the main diagonal distance between the two bounding planes is less than $\sqrt{3}$

Definition 2. *A finite set of grid cubes in 3D space, being edge-connected, is a DPS iff there is a pair of parallel planes (a supporting plane and a tangential plane) containing this set in-between, and having a main diagonal distance less than $\sqrt{3}$. A supporting plane of a DPS is called effective supporting plane.*

Therefore the main diagonal distance between the effective plane and its corresponding tangential plane is less than $\sqrt{3}$. For each DPS, there is possible to have more than one effective supporting plane.

Let v be a vector in a main diagonal direction with length of $\sqrt{3}$, n be the normal vector of the pair of planes and $d = n \cdot p$ be the equation for one of these two planes. According to Def. 2, all the vertices p of the grid cubes of a DPS must satisfy the following inequality,

$$0 \leq n \cdot p - d < n \cdot v \tag{1}$$

Let $n = (a, b, c)$. Since v is a vector of which all three elements have an absolute value 1 and the angle between n and v is always less than or equals to $90°$, Equ. 1 becomes

$$0 \leq ax + by + cz - d < |a| + |b| + |c| \tag{2}$$

This equation is equivalent to that of the standard plane defined originally by Reveillés [14]. Let $w = |a| + |b| + |c|$, where w is called the thickness of a discrete plane. This thickness guarantees that the DPS remains tunnel free and without simple points [1]. This is the optimal arithmetical thickness for discrete planes. It follows that the normals of the effective supporting planes of a DPS are uniquely defined. Given the boundary of a DPS and one of the effective supporting planes, it is possible to reconstruct the original digital plane. Further digital plane models and algorithms may be found in [4,5,19].

3   The Algorithm

The digital surface of a regular solid consists of grid faces, and these may be traced using the surface tracking algorithm [2] to build a surface graph. Each node of the surface graph contains a grid face as well as four pointers to all of its four edge-adjacent faces. By this representation, we can implement a breadth-first search of all the faces to agglomerate faces into segments such that all faces in one segment belong to one DPS. We are interested in calculating maximum sets of faces satisfying the DPS definition.

The approach to agglomerate faces according to Equ. 1 is a classical problem which is also known as recognition of a digital plane in computer imagery. Of course, the definition of a digital plane may vary. The problem can be represented as follows: given n points $\{p_1, p_2, \ldots, p_n\}$, does there exist a DPS such that each point satisfies the inequality Equ. 1, i.e. is it possible to solve the following inequality system:

$$0 \leq \boldsymbol{n} \cdot \boldsymbol{p}_i - d < \boldsymbol{n} \cdot \boldsymbol{v} \quad i = 1, \ldots, n \tag{3}$$

This inequality system contains four scalar unknowns, d and $\boldsymbol{n} = (a, b, c)$. Since $\boldsymbol{n}$ has to be a normal vector in this case (i.e. of unit length), Equ. 3) is not just a linear inequality system and hence not a trivial problem to find out whether a feasible solution exists or not. However, by eliminating d, we get a new inequality system Equ. 4 which has n^2 inequalities:

$$\boldsymbol{n} \cdot \boldsymbol{p}_i - \boldsymbol{n} \cdot \boldsymbol{p}_j < \boldsymbol{n} \cdot \boldsymbol{v}, \quad i, j = 1, \ldots, n \tag{4}$$

This inequality system no longer requires $\boldsymbol{n}$ to be a normal vector, thus it is turned into a linear homogeneous inequality system with three unknowns a, b, c and it can be solved in various ways. For example, the paper [7] presents a Fourier elimination approach. Computationally this turned out to be not time-efficient. More efficient algorithms include operation research's linear programming, which is to find an optimal solution given an object function and a set of linear constraints. If such a solution exists, the inequality system is solved. Quick Hull [20] is another method which constructs a convex hull by cutting the 3D space given a finite set of half-spaces. By finding out whether a non-empty convex hull exists, the inequality system is then solved. The CDD [21] (C implementation of Double Description Method of Motzkin) algorithm generates all vertices, i.e. extreme points and extreme rays of a general convex polyhedron in R^d given by a system of linear inequalities. Our experiments show that the CDD works fastest in our case. However when increasing the grid resolutions, the number of inequalities becomes very large, solve the inequality system over and over is proved very slow and not sufficient for our multigrid convergence study. It is critical to have an elegant and efficient algorithm to solve the inequality system.

Using the idea of effective supporting planes, an incremental algorithm can be achieved by keeping a list of the effective supporting planes. Given a DPS (the effective supporting planes are known and kept in a list) and to test a new vertex, we first need to find out whether the existing effective supporting planes are still effective to the new vertex. Those not are deleted from the list. New supporting planes then will be constructed with the new vertex. Those new supporting planes are added to the list if they are effective. If after all the list is empty, then the vertex failed to add to the current DPS. Constructing new supporting planes with a new vertex is done by incrementally constructing the convex hull of a DPS ([3]).

We start the DPS recognition process from a face graph which is obtained by applying the surface tracking algorithm introduced in [2]. After the face graph is constructed, we start breadth-first search in the face graph to agglomerate the faces into DPSes. This breadth-first search algorithm is implemented using two queues. One is called a seeds queue containing all searched faces not yet belonging to any recognized DPS. Whenever a face cannot be added to the current DPS, it is enqueued to the seeds queue. The next DPS will start from one face chosen from the seeds queue. The second queue is used to maintain the breadth-first search so that the growing of the DPS looks like the propagation of a circular wave. Searching in this way the shape of the DPS is expected more close to a circle. Applying different search strategy will result in different DPS segmentation. For example, depth-first search will generate DPSes with narrow shape. We will show the impacts of these factors on our surface area measurement in the experimental results section.

When an adjacent face is found, we try to add it to the current DPS. Only when all four vertices of a face belongs to the current DPS (those vertices already on the DPS need not test), it can be added to the current DPS and deleted from the seeds queue if it is also there. Otherwise we enqueue this face to the seeds queue and try another adjacent face. If no further adjacent face can be added, we start a new DPS from a face in the seeds queue.

There is a few consideration to further speed up the process. For example the set of surface faces are all in one direction, these faces must belong to a DPS. If all the surface faces are in one direction and the next face has a different direction, this next face must belong to the same DPS too. At most three directions of the surface faces are allowed in one DPS.

The algorithm is summarized in Fig. 3. The computational complexity of our surface tracking algorithm is $\mathbf{O}(n^2)$. Incrementally compute the convex hull costs $\mathbf{O}(n^2)$ too. They are outside each other's loop. Therefore the overall cost of our DPS recognition algorithm is $\mathbf{O}(n^2)$.

The constructed polyhedron is composed of DPSes, which are in turn composed of faces. Those faces are not coplanar in the sense of Euclidean geometry. To evaluate the surface area of a DPS, we must first project the surface faces on one of the bounding planes of the DPS and then sum up the projected area. The surface area of the original regular solid is then estimated by the sum of areas of all DPSes.

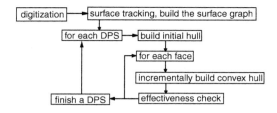

Fig. 3. Steps involve in our incremental DPS recognition algorithm

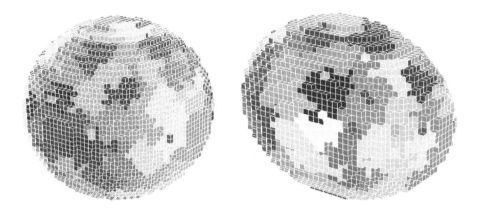

Fig. 4. A sphere and a 20*16*12 ellipsoid digitized at $r = 40$ and polyhedrized by DPSes with a face search depth limited to 7.

4   Experimental Results

The calculation of the surface area of an ellipsoid with all three semi-axes a, b, c being allowed to be different is known to be a complicate task. If two radii coincide, i.e. in case of an ellipsoid of revolution, the surface area can by analytically specified in terms of standard functions. The surface area formula in the general case may be based on elliptic integrals. Example 2 in [11] specifies an analytical method to estimate the surface of such ellipsoids. The value calculated by this method for surface area is used in this paper as 'ground truth' to evaluate the performance of our DPS algorithm.

Fig. 4 shows a digitized sphere and ellipsoid where faces are grouped into DPSes with the breadth-first search strategy. The search depth is restricted to 7. The total numbers of faces of digitized sphere and ellipsoid are 7584 and 4744, respectively, at resolution $r=80$. The number of DPSes of the sphere and ellipsoid are 247 and 160, respectively. The average size of one DPS is approximately 30 faces in both cases.

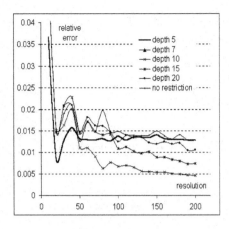
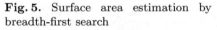

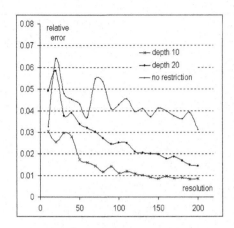

Fig. 5. Surface area estimation by breadth-first search

Fig. 6. Surface area estimation by depth-first search

With depth restriction on search depth, the sizes of the DPSes are more evenly distributed, and the DPSes have more regular shapes. Histograms for various search depth values can be found in [18]. More evenly distributed sizes and a more regular shape of the DPSes leads to more accurate surface area estimations for general ellipsoids. Figure 5 illustrates surface area measurement for a 20 by 16 by 12 ellipsoid in standard position (i.e. its axes are parallel to the coordinate axes). It shows that the algorithm performs best when the depth limitation is chosen appropriately. This is because if the search depth is limited at a small value such as 5 the global technique is transformed into a local one.

Results using a depth-first search strategy are shown in Fig. 6. Since the shapes of the DPSes are more irregular in this case, without search depth limitation, the result is less accurate than the breadth-first search. When search depth applies, the accuracy is closer to the breadth-first search.

Both search strategies of the DPS technique show a convergence tendency to the true value, but not homogeneously. This is due to the fact that with different resolution and the choice of initial faces for each DPS, the resulting polyhedron may differ.

The marching cubes and the convex hull techniques are compared with our DPS algorithm, and and results are shown in Fig. 7. Compared to the convex hull or marching cubes algorithm at the same resolutions, the accuracy of our DPS method is far better than the other two. The convex hull method and marching cubes have a relative error of 3.2% and 8.7% respectively at $r = 100$, while the DPS method is only 0.67%. Note that this result is somewhat dependent on the initial face and search strategy.

Compared to algorithms based on solving the inequality system, our incremental algorithm dramatically improves the computational efficiency. Figure 8 shows the computational time for polyhedrizing a $20 * 16 * 12$ ellipsoid with res-

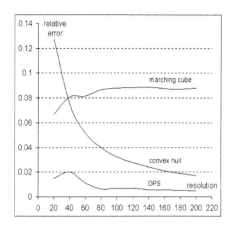

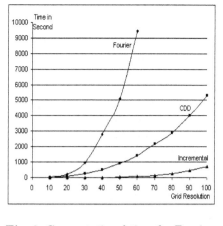

Fig. 7. Surface area estimation by marching cubes, convex hull and DPS.

Fig. 8. Computational time for Fourier elimination, CDD and DPS algorithm.

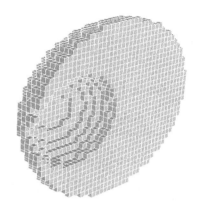

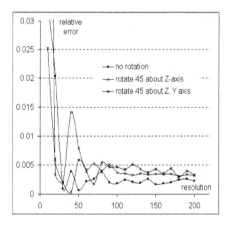

Fig. 9. Cut through a non-convex solid

Fig. 10. Surface area estimation of a non-convex solid

olutions from 10 to 100 and with search depth restriction of 7. The comparison was done on a Pentium II 350 running Linux.

An impossible situation for the convex hull method is given when non-convex solids have to be considered. Hence we also test our algorithm using non-convex regular solids defined by an ellipsoid having an 'inner tangential ellipsoidal hole'. Half of such a solid is visualized in Fig. 9. Results for three positions of such a solid are shown in Fig. 10.

5   Conclusion

We designed and implemented a global polyhedrization algorithm and estimated the surface area of regular solids. Compared to other polyhedrization techniques such as the marching cubes and convex hull technique, our DPS method converges fast and is converging towards the true value. Our algorithm applies to non-convex objects as well. Although the relative error is already less than 0.5% with resolution 200, we can see that the convergence speed slows down and we can hardly tell whether it does finally converge to the true value. This is a general limitation of experimental studies. Note that the intersection of the Euclidean projections of the DPSes does not form a complete polyhedron in general, i.e. this might be a reason for a possible (not verified!) difficulty to ensure absolutely perfect convergence to the true value. Further investigations are needed.

Acknowledgements. Thanks are due to Garry Tee (Auckland University) who provided the program to calculate the surface area of an arbitrary ellipsoid which makes the comparison to the true value possible. We also thank Laurent Papier (Université Louis Pasteur de Strasbourg) for a valuable discussion on [6]. Our program is built on top of a source written by Feng Wu (Auckland University).

References

1. E. Andres, R. Acharya and C. Sibata: Discrete analytical hyperplanes. *Graphical Models and Image Processing*, **59** (1997) 302–309.
2. E. Artzy, G. Frieder, and G. T. Herman: The theory, design, implementation and evaluation of a three-dimensional surface detection algorithm. *CVGIP*, **15** (1981) 1–24.
3. M. de Berg: *Computational Geometry: Algorithms and Applications.* Springer, New York (1997).
4. I. Debled-Renesson and J.P. Reveillés: An incremental algorithm for digital plane recognition. *DGCI'94*, Grenoble, France (1994) 207–222.
5. J. Françon: Discrete combinatorial surfaces. *Graphical Models Image Processing*, **57** (1995) 20–26.
6. J. Françon and L. Papier: Polyhedrization of the boundary of a voxel object. *DGCI'99*, **LNCS 1568**, Springer, Berlin (1999) 425–434.
7. J. Françon, J.-M. Schramm, and M. Tajine: Recognizing arithmetic straight lines and planes. *DGCI'96*, **LNCS 1176**, Springer, Berlin (1996) 141–150.
8. C. Jordan: Remarques sur les intégrales deéfinies. *Journal de Mathématiques 4e série* (1892) T. 8, 69–99.
9. Y. Kenmochi and R. Klette: Surface area calculation for digitized regular solids. *Vision Geometry IX*, **SPIE-4117** (2000) 100–111.
10. R. Klette: Approximation and representation of 3D objects. in: *Advances in Digital and Computational Geometry*, Springer, Singapore (1998) 161–194.
11. R. Klette and B. Yip: The length of digital curves. *Machine GRAPHICS & VISION*, **9** (2000) 673–703.
12. V. A. Kovalevsky: New definition and fast recognition of digital straight segments and arcs. *Proc. IEEE Conf. ICPR-10*, Los Almitos, vol. II (1994) 31–34.

13. A. Lenoir, R. Malgouyres, and R. Revenu: Fast Computation of the normal vector field of the surface of a 3-d discrete object. *DGCI'99* **LNCS 1176**, Berlin, Springer (1999) 101–112.
14. J. P. Reveillés: Géométrie Discrète, Calcul en nombres entiers et Algorithmique. Thése d'état soutenue à l'Université Louis Pasteur, (1991).
15. A. Rosenfeld. Digital straight line segments. *IEEE Trans. Computers*, **23** (1974) 1264–1269.
16. H. A. Schwarz: Sur une définition erronée de l'aire d'une surface courbe. *Ges. math. Abhandl.*, **2** (1890) 309–311.
17. F. Sloboda and B. Zaťko: On polyhedral form for surface representation. Technical report, Institute of Control Theory and Robotics, Slovak Academy of Sciences, Bratislava (2000).
18. H. Sun: Surface Area Measurement Based on Polyhedrization. *MSc Thesis*, CITR Tamaki, Auckland (2001).
19. J. Vittone and J.-M. Chassery: Recognition of digital naive planes and polyhedrization. *DGCI'2000*, **LNCS 1953**, Springer, Berlin (2000) 296–307.
20. http://www.geom.umn.edu/software/qhull/
21. http://www.ifor.math.ethz.ch/~fukuda/cdd_home/cdd.html

Shape-Guided Split and Merge of Image Regions

Lifeng Liu[1] and Stan Sclaroff[2]

[1] MD Online, 99 Hayden Ave, Lexington, MA 02421, USA
[2] Computer Science, Boston Univ., 111 Cummington St, Boston, MA 02215, USA
lliu@mdol.com, sclaroff@cs.bu.edu

Abstract. A method for deformable shape-based image segmentation is described. Regions in an image are merged together and/or split apart, based on their agreement with an a priori distribution on the global deformation parameters for a shape template. Perceptually-motivated criteria are used to determine where and how to split regions, based on the local shape properties of the region group's bounding contour. In general, finding the globally optimal region partition for an image is an NP hard problem; therefore, two approximation strategies are employed: the highest confidence first algorithm and shape indexing trees. Experiments show that the speedup obtained through use of the approximation strategies is significant, while accuracy of segmentation remains high. Once trained, the system autonomously segments shapes from the background, while not merging them with adjacent objects or shadows.

1   Introduction

Segmentation using traditional low-level image processing methods, such as region growing or region split/merge, requires a considerable amount of interactive guidance in order to get satisfactory results. One solution is to exploit prior knowledge to sufficiently constrain the segmentation problem. For instance, a model based segmentation scheme can be used in concert with image preprocessing to guide and constrain region grouping [30,40]. However, due to shape deformation and variation within object classes, a simple rigid model-based approach will break down in general. This realization has led to the use of deformable shape models in image segmentation [8,24,37,42].

Unfortunately, the above mentioned techniques are going to make mistakes in merging regions, even in constrained contexts. This is because local constraints are in general insufficient; to gain a more reliable segmentation, global consistency must be enforced. This idea is embodied in the *principle of global coherence* [39]: the best partitioning is the one that globally and consistently explains the greatest portion of the sensed data [39]. Unfortunately, finding *the* globally optimal image partition is an NP hard problem; therefore, approximation strategies were proposed to achieve a practical system; e.g., simulated annealing [15], highest confidence first (HCF) [7], or agglomerative clustering methods [34].

In this paper, a method for deformable shape-based image segmentation is described. Regions in an image are merged together and/or split apart, based

C. Arcelli et al. (Eds.): IWVF4, LNCS 2059, pp. 367–377, 2001.

on their agreement with an a priori distribution on the global deformation parameters for a particular shape class. The likelihood of a region merge or split is evaluated using a cost measure that includes region compatibility, region/model area overlap, and a deformation likelihood term. Perceptually-motivated criteria are used to determine where/how to split regions, based on the local shape properties of the region group's bounding contour. An initial version of our system used HCF in region merging [26]. This paper describes the basic segmentation formulation, plus two substantial improvements: region splitting and shape indexing trees. Experimental results show that these extensions improve both the speed and accuracy of shape-based segmentation. Once trained, the system autonomously segments deformed shapes from the background, while not merging them with adjacent objects or shadows.

2   Related Work

Previous work in deformable shape segmentation stems from the active contours formulation [8,12,22,24,37]. The active contours formulation can be extended to include a term that enforces homogeneous properties over the region contained within the contour during region growing [6,20,23,32]. This hybrid approach offers the advantages of both region-based and deformable modeling techniques, and tends to be more robust with respect to model initialization and noisy data. However, it requires hand-placement of the initial model, or a user-specified seed point on the region's interior.

Other approaches use special-purpose deformable templates [22,27,41]. The inclusion of object-specific knowledge in the model is used to further constrain segmentation, resulting in enhanced robustness to occlusion and noise. Furthermore, the recovered template parameters can be used for shape recognition. These methods require the careful construction and parameterization of templates. Deformable templates can be derived semi-automatically, via statistical analysis of shape training data [10,29]. The estimated probability density function (PDF) for the shape deformation parameters can be used in Bayesian segmentation methods.

From another view, image segmentation is a labeling problem; the ideal segmentation should be consistent or nearest to the one with maximum likelihood. This has led to various relaxation labeling or stochastic labeling methods that are related to general optimization algorithms [2,14,19]. These techniques require prior information, such as the number of labels needed and the probability distribution of these labels in the image. However, such prior information is not always available or is inaccurate for general imagery. One common solution is to use the Minimum Description Length (MDL) principle in segmentation [11,17, 25,42]. MDL has a strong fundamental grounding, being based on information-theoretic arguments: the simplest model explaining the observations is the best, and it can result in an objective function with no arbitrary thresholds.

After defining the criterion function for labeling, the next problem is computing the solution to the optimization problem. Generally speaking, finding the globally-consistent image labeling is an NP hard problem; therefore, approximation algorithms are needed in solving any segmentation problems of realistic size. Simulating annealing methods are frequently used [5,7,15,16,38]. In simulated annealing, the choice of the temperature sequence involves a tradeoff between the convergence speed and the correctness of the result. In general, the temperature must be lowered at a very slow (logarithmic) rate; however, it has been shown that the temperature can be lowered more rapidly (e.g., exponentially) if moves are selected from a size distribution proportional to the Cauchy distribution [16]. Despite these insights, the simulated annealing approach to segmentation remains prohibitively slow.

Chou and Brown [7] used highest confidence first (HCF) to infer a unique labeling from the *a posteriori* distribution that is consistent with both the prior knowledge and evidence. In the HCF algorithm, the computational complexity is generally less than that needed to obtain similar quality segmentation results via the simulated annealing algorithm [26]. In HCF, the number of different merging configurations tested is $O(\mathbf{n}^2)$, where $\mathbf{n}$ is the number of regions in the initial over-segmentation of the image [26].

The above mentioned work leads to the development of our approach. Deformable shape templates are used to partition the image into a globally consistent interpretation, determined in part by the MDL principle. The highest confidence first algorithm will be used to obtain an approximation to the globally consistent labeling of the image. As will be described in the next section, two extensions of this basic approach are needed to make segmentation efficient and accurate: 1.) shape indexing trees for faster model fitting, and 2.) perceptually motivated region splitting methods for (generally) more accurate segmentation.

3   Basic Approach

In [26] we proposed a method for region merging that uses a deformable model to guide grouping of image regions. We would like to review it briefly here. As show in Fig. 1, the deformable model-based segmentation system includes a pre-processing (over-segmentation and edge detection) stage, and a model-based region grouping stage.

In the pre-processing stage, the input image is over-segmented via standard region merging algorithms [1,9]. An example input image, and the resulting over-segmented output are shown in Fig. 1. The output of this module includes a standard region adjacency graph. An edge map is also computed; notable edges and their strengths are detected via standard image processing methods. The resulting edge map will be used to constrain consideration of possible grouping hypotheses later in region merging.

The system then tests various combinations of candidate region groupings to obtain an optimal labeling of the image. The shape model is deformed to match each grouping hypothesis $\mathbf{g}_i$ in such a way as to minimize a cost function:

$$E(\mathbf{g}_i) = \alpha E_{color} + (1 - \alpha)((1 - \beta)E_{area} + \beta E_{deform}), \tag{1}$$

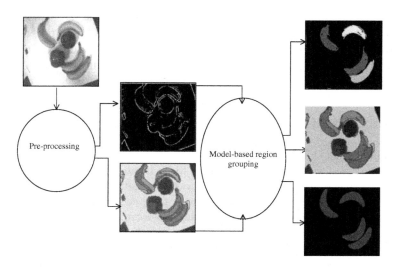

Fig. 1. Basic system overview. The color image (image of bananas) undergoes pre-processing, which results in an over-segmentation and an edge map. These are inputs to the model-based region grouping stage (using a banana template). The final output includes region groupings for detected objects (four bananas),and recovered models for the objects.

where α and β are scalar constants with values in the range $[0, 1]$ that control the relative importance of the three terms: E_{color} is a region color compatibility term for the region grouping, E_{area} is a region/model area overlap term, and E_{deform} is a deformation energy for the shape model. A model fitting procedure is used to compute the cost in Eq. 1 via the downhill simplex method.

The deformation term enforces *a priori* constraints on the amounts and types of deformations allowed for the template; *i.e.:*

$$E_{deform} \propto -\log P(\mathbf{a}|\Omega), \tag{2}$$

where $P(\mathbf{a}|\Omega)$ gives the prior distribution on global deformation parameters, $\mathbf{a}$, for a particular shape class Ω. In our experience, the use of a Gaussian model for the prior distribution on global deformation leads to reliable shape-based image segmentation. An estimate of the prior distribution is computed in a supervised fashion, for a given set of training examples for that shape class. In our implementation, linear and quadratic polynomials are used to model deformation due to stretching, shearing, bending, and tapering.

Further, to test the quality of a possible partitioning, a global cost function for partitioning the whole image is defined:

$$\mathcal{E} = (1 - \gamma) \sum_{i=1}^{n} \mathbf{r}_i \mathbf{E}(\mathbf{g}_i) + \gamma \mathbf{n}, \tag{3}$$

where γ is a scalar constant, $\mathbf{n}$ is the number of the groupings in the current image partitioning, $\mathbf{r}_i$ is the ratio of $\mathbf{i}^{th}$ group area to the total area, and $\mathbf{E}(\mathbf{g}_i)$ is the cost function for group $\mathbf{g}_i$ (Eq. 1). The highest confidence first (HCF) algorithm is used to find an approximately optimal value for Eq. 3. Region groupings obtained via HCF and recovered shape models are shown in Fig. 1.

4   Index Trees

One problem with the system proposed in Section 3 is that segmentation can be slow for images of moderate complexity. This is because the shape model fitting procedure must be invoked many times in order to get the cost values of different configurations. Although we utilize methods to speed up the fitting procedure, such as multi-resolution fitting, and caching deformation parameters, most of the CPU time (over 90%) is still used in model fitting. We therefore propose to use an *index tree* method to accelerate the model fitting procedure.

The basic idea behind index trees can be explained as follows. We first generate many deformed instances of the object class by sampling in the deformation space according to the prior distribution of the deformation parameters. We then compute a shape feature vector for each generated instance. In our implementation, the features employed are the seven normalized central moments. The shape feature vector and the deformation parameters are stored with the instance. Then, in the fitting process, we compute the shape feature vector for a potential region group. Via comparing the shape feature vectors, the most similar one for the region group is fetched from the set of generated instances (called an *instance set*). Its associated deformation parameters are used as the parameters for the region group, or as a starting point to invoke a refining process.

To speed up search, we organize the instances in a tree such that the retrieval time can be logarithmic to the number of instances. We use a hierarchical clustering method (minimum variance) to process the shape features of the instances, and get the tree structure [21]. In our experiments, we have used the cophenetic correlation coefficient (CPCC) [21] to validate clustering. Although we uniformly sample in the deformation space, there is indeed a hierarchical structure in the corresponding shape feature space.

By searching for the best match in the index tree, the searching time is reduced but it does not guarantee that the nearest match is always found. We tried to use the mean feature of instances in each non-leaf node to select a branch and go to the next level. However, the covariance and distribution for the instances in each node are not the same; furthermore, their distributions are not Gaussian in general. In order to overcome this problem, we use linear discriminant functions[13] at the non-leaf nodes. Our experiments verified that this method can increase the success rate in finding the nearest neighbors.

Another problem with index tree search is that the retrieved result is the nearest neighbor in the shape feature space. However, the distance metric in shape feature space is not the same as the fitting cost (Eq. 1) nor is it monotonic to the fitting cost. A neural network (NN) can be used to map from the difference in the shape feature space to the fitting cost measure. We use a three layer back-propagation network with bias terms and momentum [33]. The NN is only used for mapping in the leaf nodes of the index tree to reduce the on-line computation. In summary, a linear discriminant based on the shape feature vector is used in the first level (coarse level) of the index tree, and the NN mapping from shape feature difference to fitting cost is used in the second level (fine level).

The index tree approach was tested on over one hundred cluttered images of objects taken from a number of different shape classes. It was observered that the CPU time needed for segmentation was decreased by one order of magnitude, while the number of errors in segmentation did not increase appreciably over HCF without index trees.

5   Shape-Based Region Splitting

The shape-based region merging may not yield accurate segmentation in all cases. A common problem is that in some images there is no distinguishable boundary between touching or overlapping objects; therefore, it is not possible to separate them in the over-segmentation stage. If we only use model-based merging operations in the second stage, then the correctness of the final result will depend on the correctness of the initial over-segmentation. Therefore, a model-based region splitting operation is needed.

A natural question is: *How does the human vision system parse regions into parts?* Perhaps such strategies can be adapted to solve our problem. Some theorists postulate that there is a set of basic shape primitives [3,4,28,31] that are useful in finding parts and describing them. Others postulate that there are rules, based on geometric properties alone, by which humans perceive boundaries between parts for a given shape [18,35]. In our system, we make use of both shape constraints (our deformable template and statistical priors) and geometric properties of the region contour (curvature minima) in guiding splits.

The first problem is determining if a region should be split at all. Candidate regions for splitting can be detected based on the model fitting cost value (Eq. 1) and a specified threshold. The threshold can be obtained through statistical analysis over a example set. If the threshold is lower than required, it does not degrade the accuracy of segmentation; instead, more possible splits have to be tested, but in our experience the resulting segmentation is similar.

Given a candidate region for splitting, possible cuts need to be determined and their feasibility tested. Following [18], cut points can be chosen as concave cusps and negative minima of curvature of the region's boundary. In human perception, it has been found that there is a preference for certain cuts over others [18,35]. For instance, there is the *short cut rule* [36]: if boundary points can be joined in more than one way to parse a silhouette, we prefer the parsing which uses the shortest cuts. Thus, a cut is defined to be (1) a straight line which (2) crosses an axis of local symmetry, (3) joins two points on the outline of a silhouette, such that(4) at least one of the two points has negative curvature.

To select the best cut, the direct strategy is to test all possible cuts, and choose the split that yields the greatest decrease in the global cost (Eq. 3). The drawback of this exhaustive search is the computational requirement. If the number of candidate cuts is small, then the strategy is suitable. In general, however, we employ modified short-cut rule: select the shortest cut, if this cut is a feasible cut, stop; otherwise, delete the candidate point with least significance from this cut, and repeat to test the short cuts in the remaining candidate points until a feasible cut is found or there is no cut to test. The significance of a candidate point can be defined as the distance between the point and its corresponding point on the model boundary.

6   Experiments

The system has been implemented and tested in a color image segmentation application that uses 2D shape models and global deformations. In the implementation, the prior distribution on global deformations for each shape is assumed Gaussian, and estimated using region segmentations provided in a training set. The system was tested on cluttered images of objects taken from a number of different shape classes (*e.g.*, fish, blood cells, leaves, fruit, and vegetables), and results are encouraging. Example input images and segmentation results obtained with the system are shown in Fig. 2.

To quantitatively evaluate performance of the system, we also employed synthesized imagery. Fig. 2 shows an example taken from our database of synthetic fish images. The shape deformation parameters for generating fish were randomly sampled based on the distribution information obtained from the fish model training stage. We also added 5% white noise to the synthesized images, and triangulation methods were used to assign different color values to the different parts of the generated fish. As a result, in the over-segmentation results shown in the figure, every fish object breaks into many small regions. From the figure, we can see that there are some incorrect mergings during the HCF region merge stage, and that the boundaries of some fish are imprecise. The results are improved after model-based splitting and re-merging.

To evaluate the improvement gained by using HCF split/merge over HCF with merging only, we conducted experiments using a database of 22 leaf images (210 leaf objects), 20 synthesized fish images (160 fish objects), and 21 blood cell micrographs (about 700 cell objects). The success rate of object detection and mean fitting cost for objects before and after the splitting stage are shown in Table 1. The mean fitting cost can be regarded as a measure of the boundary accuracy in object detection. As is evident in the table, segmentation improves when model-based splitting is used, followed by a merge step. Combined split/merge yielded at least a 7% improvement in the object detection rate, as well as a uniform reduction in model fitting cost.

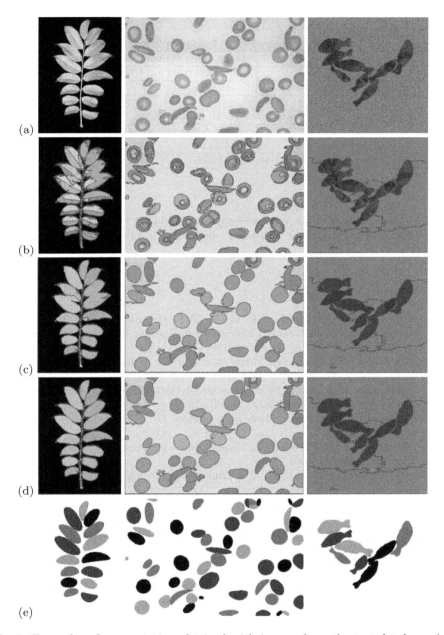

Fig. 2. Examples of segmentation obtained with images from the test database: (a) original images, (b) initial over-segmentation (note some regions overlap more than one object), (c) shape-based region merging result obtained via HCF, (d) result obtained via split followed by another merge step, (e) recovered shape models.

Table 1. Statistical analysis of region splitting algorithm for the method based on distance from model boundary and short-cut rule. See Sec. 6 for discussion.

Accuracy Before Region Splitting

	images of leaves	images of cells	images of fish
mean fitting cost	1.1338	1.1863	1.0626
success rate of object detection	85.24%	85.79%	76.25%

Accuracy After Region Splitting

	images of leaves	images of cells	images of fish
mean fitting cost	1.1118	1.0531	1.0528
success rate of object detection	92.86%	93.90%	85.00%

Acknowledgments. This work was supported in part through US ONR Young Investigator Award N00014-96-1-0661, and NSF grants IIS-9624168 and EIA-9623865.

References

1. J.R. Beveridge, J.S. Griffith, R.R. Kohler, A.R. Hanson, and E.M. Riseman. Segmenting images using localized histograms and region merging. *IJCV*, 2(3):311–352, 1989.
2. B. Bhanu and O.D. Faugeras. Shape matching of two-dimensional objects. *PAMI*, 6(2):137–156, 1984.
3. I. Biederman. Recognition by components: A theory of human image understanding. *Psychological Review*, 94(2):115–147, 1987.
4. T.O. Binford. Visual perception by computer. In *IEEE Conf. on Systems and Control*, 1971.
5. G. Bongiovanni and P. Crescenzi. Parallel simulated annealing for shape detection. *CVIU*, 61(1):60–69, 1995.
6. A. Chakraborty, L.H. Staib, and J. Duncan. Deformable boundary finding influenced by region homogeneity. In *Proc. CVPR*, pages 624–627, 1994.
7. P. B. Chou and C. M. Brown. The theory and practice of Bayesian image labeling. *IJCV*, 4:185–210, 1990.
8. L.D. Cohen. On active contour models and balloons. *CVGIP:IU*, 53(2):211–218, 1991.
9. D. Comaniciu and P. Meer. Robust analysis of feature spaces: Color image segmentation. In *Proc. CVPR*, pages 750–755, 1997.
10. T.F. Cootes, A. Hill, C.J. Taylor, and J. Haslam. Use of active shape models for locating structure in medical images. *Image and Vision Comp.*, 12(6):355–365, 1994.
11. T. Darrell and A.P. Pentland. Cooperative robust estimation using layers of support. *PAMI*, 17(5):474–487, 1995.
12. A. Del Bimbo and P. Pala. Visual image retrieval by elastic matching of user sketches. *PAMI*, 19(2):121–132, 1997.
13. R.O. Duda and P. E. Hart. *Pattern Classification and Scene Analysis*. Wiley, NY, 1973.

14. O. D. Faugeras and M. Berthod. Improving consistency and reducing ambiguity in stochastic labeling: An optimization approach. *PAMI*, 3(4):412–424, 1981.
15. S. Geman and D. Geman. Stochastic relaxation, Gibbs distribution, and Bayesian restoration of images. *PAMI*, 6(11), 1984.
16. R. P. Grzeszczuk and D. N. Levin. Brownian strings: segmentating images with stochastically deformable contours. *PAMI*, 19(10):1100–1114, 1997.
17. H. Gu, Y. Shirai, and M. Asada. Mdl-based spatiotemporal segmentation from motion in a long image sequence. *Proc. CVPR*, 448–453, 1994.
18. D.D. Hoffman and W.A. Richards. Parts of recognition. *Cognition*, 18:65–96, 1985.
19. R. A. Hummel and S. W. Zucker. On the foundations of relaxation labeling processes. *PAMI*, 5(3):267–287, 1983.
20. J. Ivins and J. Porrill. Active-region models for segmenting textures and colors. *Image and Vision Comp.*, 13(5):431–438, June 1995.
21. A. K. Jain and R. C. Dubes. *Algorithms for clustering data*. Prentice Hall, 1988.
22. A.K. Jain, Y. Zhong, and S. Lakshmanan. Object matching using deformable templates. *PAMI*, 18(3):267–278, 1996.
23. T.N. Jones and D. Metaxas. Segmentation using deformable models with affinity-based localization. *CVRMed*, 53–62, 1997.
24. M. Kass, A.P. Witkin, and D. Terzopoulos. Snakes: Active contour models. *IJCV*, 1(4):321–331, 1988.
25. Y. G. Leclerc. Constructing simple and stable descriptions for image partitioning. *IJCV*, 3:73–102, 1989.
26. L. Liu and S. Sclaroff. Deformable shape detection and description via model-based region grouping. *CVPR*, 2:21–27, 1999.
27. K.V. Mardia, W. Qian, D. Shah, and K.M.A. Desouza. Deformable template recognition of multiple occluded objects. *PAMI*, 19(9):1035–1042, 1997.
28. D. Marr. Analysis of occluding contour. *Proc. Royal Society*, B-197:441–475, 1977.
29. J. Martin, A. Pentland, and R. Kikinis. Shape analysis of brain structures using physical and experimental modes. In *Proc. CVPR*, 1994.
30. M. Nagao, T. Matsuyama, and Y. Ikeda. Region extraction and shape analysis in aerial photographs. *Computer Graphics and Image Processing*, 10:195–223, 1979.
31. A. Pentland. Automatic extraction of deformable part models. *IJCV*, 4(2):107–126, 1990.
32. R. Ronfard. Region-based strategies for active contour models. *IJCV*, 13(2):229–251, 1994.
33. D. Rumelhart, G. Hinton, and R.J. Williams. *Learning Internal Representations by Error Propagation*. Parallel Distributed Processing 1, MIT Press, 1986.
34. J. Silverman and D. Cooper. Bayesian clustering for unsupervised estimation of surface and texture models. *PAMI*, 10(4):482–495, 1988.
35. M. Singh and B. Landau. Parts of visual shape as primitives for categorization. *Behavioral and Brain Sciences*, 21(1):36–37, 1998.
36. M. Singh, G. Seyranian, and D. Hoffman. Parsing silhouettes: The short-cut rule. *Perception and Psychophysics*, 61(4):636–660, 1999.
37. L. Staib and J.S. Duncan. Boundary finding with parametrically deformable models. *PAMI*, 14(11):1061–1075, 1992.
38. G. Storvik. Bayesian approach to dynamic contours through stochastic sampling and simulated annealing. *PAMI*, 16(10):976–986, 1994.
39. T. M. Strat. *Natural Object Recognition*. Springer-Verlag, 1992.
40. J.M. Tenenbaum and H.G. Barrow. Experiments in interpretation guided segmentation. *Artificial Intelligence*, 8:241–274, 1977.

41. A.L. Yuille, D.S. Cohen, and P.W. Hallinan. Feature extraction from faces using deformable templates. *IJCV*, 8(2):99–111, 1992.
42. S.C. Zhu and A. Yuille. Region competition: Unifying snakes, region growing, and Bayes/MDL for multiband image segmentation. *PAMI*, 18(9):884–900, 1996.

A Rotation-Invariant Morphology for Shape Analysis of Anisotropic Objects and Structures

Cris L. Luengo Hendriks and Lucas J. van Vliet

Pattern Recognition Group, Delft University of Technology, the Netherlands.
`cris@ph.tn.tudelft.nl`

Abstract. In this paper we propose a series of novel morphological operators that are anisotropic, and adapt themselves to the local orientation in the image. This new morphology is therefore rotation invariant; i.e. rotation of the image before or after the operation yields the same result. We present relevant properties required by morphology, as well as other properties shared with common morphological operators. Two of these new operators are increasing, idempotent and absorbing, which are required properties for a morphological operator to be used as a sieve. A sieve is a sequence of filters of increasing size parameter, that can be used to construct size distributions. As an example of the usefulness of these new operators, we show how a sieve can be build to estimate a particle or pore length distribution, as well as the elongation of those features.

1 Introduction

When analyzing images without a preferred orientation, or images with an unknown orientation (as is the case, for example, of an image acquired after placing a sample randomly under a microscope), it is desirable to use rotation invariant operations. A rotation invariant operation yields an output that is independent of the orientation of the sample with respect to the sampling grid. There are three different ways of constructing rotation invariant operators:

– using a single isotropic operator (the kernel itself is rotation invariant),
– using a data-driven anisotropic operator (the kernel is anisotropic, but is oriented to the local gradient in the image), or
– by combining a set of anisotropic operators.

Non-rotationally invariant filters will almost certainly produce incorrect results if they are not aligned with the image under study, and an isotropic filter is often limited in its capabilities. Therefore, it is worthwhile to study rotation invariant operators based on anisotropic kernels.

For example, consider an isotropic morphological closing, which has a disk as the structuring element (we regard 2D images for now). If we apply such a filter to an image with dark objects, such as the microscopical image in Fig. 1, all dark objects smaller than the structuring element will be removed from the image. If we see the image as a landscape where the dark features are the valleys and the light ones the hills, as in Fig. 1, we can imagine the closing as filling up the valleys such that no valleys remain in which the structuring element cannot fit (see Fig. 2).

C. Arcelli et al. (Eds.): IWVF4, LNCS 2059, pp. 378–387, 2001.

This closing operation can be used as a sieve to detect features larger than a certain size. The problem is that this size is only determined by the smallest diameter of the features. To measure length, an anisotropic structuring element is required.

In this paper we will introduce new morphological operators based on isotropic structuring elements with a *lower* dimensionality than the image under study (and thus *anisotropic* in the space of the image). By dropping one or more dimensions, the structuring element gets some degrees of rotational freedom that allows it to align itself with the features in the image. By selecting the orientation that causes minimum or maximum response (pixel by pixel), we create a rotation-invariant operator. In the two-dimensional case, the structuring element would be one-dimensional, with one degree of rotational freedom. A closing in this new morphological framework would remove an object only if the line element could not fit. This would mean that its largest diameter (supposing convex objects) is smaller than the structuring element (see Fig. 3).

We will call this new morphological framework Rotation-Invariant Anisotropic (RIA) morphology. We can call it morphology because it satisfies the four principles of morphology [1]:

- Translation invariance,
- Compatibility under change of scale,
- Local knowledge, and
- Semi-continuity.

The first three principles are expressed as properties of the operators in Sect. 3, and proven elsewhere [2]. The principle of semi-continuity requires that the theory in the continuous world has an approximate counterpart in the discrete world, and is responsible for this theory to be applicable in practice [3]. Although the discretization of the operators presented here is beyond the scope of this article, it certainly is possible to apply these operators to discrete images.

In Sect. 4 we will apply the new closing and opening introduced here to do segmentation-free measurements using morphological sieves. Sieves are used to build multi-parameter (length, width, depth) size distributions that characterize the shapes of objects, structures or textures in grey-value images. These measured distributions can be used for image recognition or characterization, and are applicable in a wide variety of situations.

2   Definitions

In this paper we use the notation as specified in Table 1. We will use Greek characters (especially φ and θ) for rotation angles, and Latin characters (especially x and y) for translation vectors and image coordinates. f and g denote continuos functions $\mathbb{R}^N \to \mathbb{R}$ (the image being processed). Vectors are not distinguished typographically because it is obvious from the context which variables are vectors and which ones are scalars.

2.1   Dilation

A flat, isotropic structuring element D of radius r can be decomposed into (an infinite amount of) rotated line segments L_φ of length $\ell = 2r$. The dilation then becomes, with $\varphi \in [0, \pi)$,

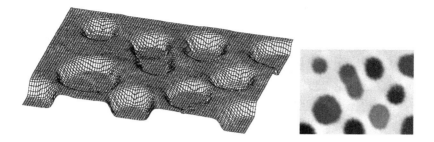

Fig. 1. A portion of the image 'cermet', after some processing.

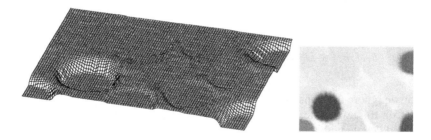

Fig. 2. The image from Fig. 1, after closing with a circular structuring element.

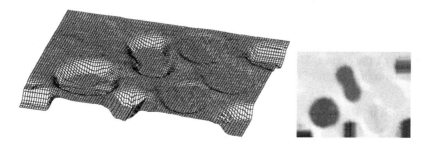

Fig. 3. Result of the new, rotation invariant, anisotropic morphological closing applied to the image in Fig. 1. Compare with the result of the isotropic closing in Fig. 2.

Table 1. Notation used in this paper

f_φ	rotation of f over an angle φ
f_x	translation of f over x: $\qquad\qquad f_x(t) = f(t - x)$
$f_{\varphi,x}$	rotation of f over an angle φ and then a translation over x
$S_b f$	scaling of f with a factor b: $\qquad\quad S_b f(x) = f(\frac{x}{b})$
$\oplus$	Minkowski addition
$\check{S}$	transpose of S: $\qquad\qquad\qquad\quad \check{S}(x) = S(-x)$
$\delta_S f$	dilation of f with S, any flat structuring element: $$\delta_S f = f \oplus \check{S} = \bigvee_{x \in \check{S}} f_x$$
$\varepsilon_S f$	erosion of f with structuring element S
$\gamma_S f$	opening of f with structuring element S
$\phi_S f$	closing of f with structuring element S
$A \triangleq B$	definition: "Let A be defined as B"
$A \doteq B$	equality by definition: "A is equal to B by definition"

$$\delta_D f \doteq f \oplus D = f \oplus \bigcup_\varphi L_\varphi = \bigvee_\varphi (f \oplus L_\varphi) \doteq \bigvee_\varphi \bigvee_{x \in L_\varphi} f_x \ . \tag{1}$$

Note that we ignore the transpose operation since $\check{D} = D$ and $\check{L}_\varphi = L_\varphi$.

Based on this, we define a new morphological operator, which we will call RIA dilation, and denote with the symbol $\delta^\triangleleft$,

$$\delta_L^\triangleleft f \triangleq \bigwedge_\varphi \bigvee_{x \in L_\varphi} f_x \doteq \bigwedge_\varphi \delta_{L_\varphi} f \ . \tag{2}$$

This operator takes the maximum of the image over a line segment rotated in such a way as to minimize this maximum. Figure 4 gives an example of the effect that the operator has on an object boundary. Note that a convex object boundary is not changed, but a concave one is.

We like to compare this dilation operator with a train running along a track. The train wagons (which are joined at both ends to the track) require some extra space at the inside of the curves. This dilation, applied to a train track, and using a structuring element with the length of the wagons, reproduces the area required by them.

2.2 Erosion

RIA erosion is defined as the dual of the RIA dilation, and will be denoted with the symbol $\varepsilon^\triangleleft$.

$$\varepsilon_L^\triangleleft f \triangleq -\delta_L^\triangleleft(-f) \doteq -\bigwedge_\varphi \bigvee_{x \in L_\varphi} (-f_x) = \bigvee_\varphi \bigwedge_{x \in L_\varphi} f_x \doteq \bigvee_\varphi \varepsilon_{L_\varphi} f \ . \tag{3}$$

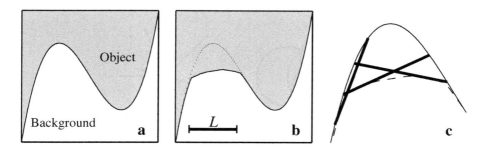

Fig. 4. Effect of the RIA dilation on an object boundary. **a:** The original boundary. **b:** The boundary after the dilation, together with the line segment used as a structuring element. **c:** Construction of the dilated object boundary.

2.3 Closing

The closing is usually defined as a dilation followed by an erosion,

$$\phi_D f \doteq \varepsilon_D \delta_D f \ . \tag{4}$$

However, it is easier to understand (and modify) if we see it as the maximum of the image over the support of the structuring element D, after shifting it in such a way that it minimizes this maximum, but still hits the point t at which the operation is being evaluated, (see Fig. 5a). Or, in other words, the 'lowest' position we can give D by shifting it over the 'landscape' defined by the function f:

$$\phi_D f = \bigwedge_{x \in D} \bigvee_{y \in D_x} f_y \left[= \bigwedge_{x \in D} \left(\bigvee_{y \in D} f_y \right)_x = \varepsilon_D \delta_D f \right] \ . \tag{5}$$

In accordance to this, we define a new morphological operation, RIA closing, as the 'lowest' position we can give the linear structuring element L, by shifting and rotating it over the 'landscape' f, such that it still hits the point x being evaluated (see Fig. 5b). It will be denoted by $\phi^{\triangleleft}$, and defined by

$$\phi_L^{\triangleleft} f \triangleq \bigwedge_{\varphi} \bigwedge_{x \in L_\varphi} \bigvee_{y \in L_{\varphi,x}} f_y \ , \tag{6}$$

which is analogous to the definition of the RIA dilation, where we also changed the disk for a line, and added a minimum over the orientation of that line. As it turns out, this is the same as the minimum of the closings, at all orientations, with a line segment as structuring element,

$$\phi_L^{\triangleleft} f \doteq \bigwedge_{\varphi} \bigwedge_{x \in L_\varphi} \left(\bigvee_{y \in L_\varphi} f_y \right)_x = \bigwedge_{\varphi} \varepsilon_{L_\varphi} \delta_{L_\varphi} f \doteq \bigwedge_{\varphi} \phi_{L_\varphi} f \ , \tag{7}$$

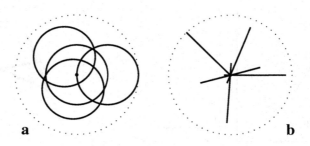

Fig. 5. a: The closing with an isotropic structuring element (disk) is determined by shifting the disk in such a way that it still hits the point being evaluated, and minimizes the supremum of the image over its support. **b:** The RIA closing is determined by shifting and rotating the line segment in such a way that it still hits the point being evaluated, and minimizes the supremum of the image over its support.

but not equal to a RIA dilation followed by a RIA erosion.

We will show elsewhere [2] that this transformation is increasing, idempotent and extensive, and therefore we can call it an algebraic closing [4]. Moreover, Matheron has shown that any intersection of morphological closings is an algebraic closing [5]. We can interpret $\bigwedge_\varphi \phi_{L_\varphi} f$ as the intersection of an infinite series of closings, in which case the increasingness, idempotence and extensivity are proven by Matheron. For previous work using rotated line segments see Soille [6].

2.4 Opening

The RIA opening is defined as the dual of the RIA closing, and denoted by the symbol $\gamma^\triangleleft$.

$$\gamma_L^\triangleleft f \triangleq -\phi_L^\triangleleft(-f) \doteq -\bigwedge_\varphi \bigwedge_{x \in L_\varphi} \bigvee_{y \in L_{\varphi,x}} (-f_y) = \bigvee_\varphi \bigvee_{x \in L_\varphi} \bigwedge_{y \in L_{\varphi,x}} f_y \doteq \bigvee_\varphi \gamma_{L_\varphi} f \ . \quad (8)$$

2.5 Extension to Higher Dimensionalities

Until now we have only talked about operations on two-dimensional images. However, it is very easy to extend the RIA morphology to higher dimensionalities. For example, in the 3D case, it would be possible to have structuring elements with either one or two dimensions (i.e. a disk or a line segment); both have two degrees of rotational freedom. A closing with these two structuring elements can be used to measure the first and second largest diameters of the (convex) object: the line segment can not fit if it is longer than the largest diameter; the disk can not fit it is wider than the second largest diameter. To measure the smallest diameter, the isotropic closing would be used.

3   Properties

Properties that are valid for all operators are only specified for the RIA dilation. Properties mentioned only for the RIA closing are by duality also true for the RIA opening but not for the dilation or erosion.

Property 1. *Translation invariance:*

$$\delta_L^{\triangleleft} f_x = \left(\delta_L^{\triangleleft} f\right)_x$$

Property 2. *Compatibility under change of scale:*

$$S_b \delta_L^{\triangleleft} f = \delta_{b \cdot L}^{\triangleleft} S_b f$$

The result of the operation is scaled by b if both the image and the structuring element are scaled by b.

Property 3. *Local knowledge:*

$$W_1 \cdot \delta_L^{\triangleleft}(W_2 \cdot f) = W_1 \cdot \delta_L^{\triangleleft} f$$

This property simply states that the result of the operator inside some window W_1 is independent of the image outside some other window W_2. This implies that $W_1 \subset W_2$.

These first three properties are the cornerstones of morphology, without which it is not possible to define shape. Together with the principle of semi-continuity, they are the requirements for operators to belong to morphology.

Property 4. *Rotation invariance:*

$$\delta_L^{\triangleleft} f_\theta = \left(\delta_L^{\triangleleft} f\right)_\theta$$

Rotation invariance of the RIA morphology is a key property, necessary for the correct analysis of images with an unknown orientation, or images without a single dominant orientation.

Property 5. *Contrast invariance:*

$$\delta_L^{\triangleleft}(c \cdot f) = c \cdot \delta_L^{\triangleleft} f$$

This property can be taken further, by stating that both the RIA dilation and the RIA closing commute with any anamorphoses (which is defined as an increasing and continuous mapping $\mathbb{R} \to \mathbb{R}$) [1].

Property 6. *Increasingness:*

$$f \leq g \quad \Longrightarrow \quad \delta_L^{\triangleleft} f \leq \delta_L^{\triangleleft} g$$

Property 7. *Extensivity / anti-extensivity:*

$$\varepsilon_L^\triangleleft f \leq f \leq \delta_L^\triangleleft f$$
$$\gamma_L^\triangleleft f \leq f \leq \phi_L^\triangleleft f$$

Property 8. *Extended extensivity:*

$$\varepsilon_L^\triangleleft f \leq \gamma_L^\triangleleft f \leq f \leq \phi_L^\triangleleft f \leq \delta_L^\triangleleft f$$

Property 9. *Idempotence:*

$$\phi_L^\triangleleft \phi_L^\triangleleft f = \phi_L^\triangleleft f$$

Property 10. *Absorption:*

$$\ell_1 \geq \ell_2 \quad \Longrightarrow \quad \begin{cases} \phi_{L^{(2)}}^\triangleleft \phi_{L^{(1)}}^\triangleleft f = \phi_{L^{(1)}}^\triangleleft f \\ \phi_{L^{(1)}}^\triangleleft \phi_{L^{(2)}}^\triangleleft f = \phi_{L^{(1)}}^\triangleleft f \end{cases}$$

Where $L^{(i)}$ is a linear structuring element with length ℓ_i.

This property states that applying a RIA closing at a large scale to the result of the RIA closing at a smaller scale yields the same results as applying it to the original image. Furthermore, applying other RIA closings at smaller scales after that has no effect.

Note that idempotence is a special case of absorption, where $\ell_1 = \ell_2$. Also, the comutativity of the RIA closing follows from the absorption property, since only the largest-scale operator influences the result, independently from the order in which they are applied.

Property 11. *Sieving:*

$$\ell_1 \leq \ell_2 \quad \Longrightarrow \quad \phi_{L^{(1)}}^\triangleleft f \leq \phi_{L^{(2)}}^\triangleleft f$$

The sieving property is a requirement for granulometric applications, and is implied by the increasing, extensivity and absorption properties [5]. Basically, it states that all features removed at a smaller scale will also be removed at a larger scale. This allows a sequence of operators of increasing size to 'sieve' the features in an image and classify them according to size (see Sect. 4).

Property 12. *Commutativity:*

$$\phi_{L^{(1)}}^\triangleleft \phi_{L^{(2)}}^\triangleleft f = \phi_{L^{(2)}}^\triangleleft \phi_{L^{(1)}}^\triangleleft f$$

This property follows from Property 10, and does not hold for the RIA dilation and RIA erosion.

Property 13. *Non-distributivity: Unlike the common dilation and erosion, the RIA dilation and erosion do not distribute with the extremum operators.*

Property 14. *Comparison with regular morphology:*

$$f \leq \delta_L^{\triangleleft} f \leq \delta_D f$$
$$f \geq \varepsilon_L^{\triangleleft} f \geq \varepsilon_D f$$
$$f \leq \phi_L^{\triangleleft} f \leq \phi_D f$$
$$f \geq \gamma_L^{\triangleleft} f \geq \gamma_D f$$

4 Granulometry

Since the RIA closing and opening comply with the sieving property (Property 11), it is possible to use them as sieving functions in a granulometric application. A sieve is composed of a sequence of morphological filters with increasing size parameter [4]. The filters are applied either in series or in parallel (which produces the same result due to Property 10, absorption), each one removing a group of image features of certain size. This size is directly proportional to the filter parameter, and the measure that determines this size depends on the filter construction. Because of the sieving property, each filter removes all image features also removed by the smaller filters, and never adds new ones.

The difference between the result of subsequent filters is called a *granule image* [7], and contains only image features in a known size range. These granule images can be used to construct a size distribution. As said before, the measure used to determine the size of an image feature depends on the filter used. A closing with an isotropic structuring element (disk) measures the width of dark features. A RIA closing measures the length of dark features. Openings do the same with light features.

The set of granule images form a scale-space, which allows to measure the size of the feature that each pixel belongs to. The 'trace' of a pixel through the scales is some sort of local size distribution, which gives (for example through a mean or median) a scale parameter for that pixel. By going through this process with different filter types, we can assign different scale parameters to each pixel; for example the length and width of the pore that it belongs to. Knowing these values, it is easy to construct a distribution for the elongation.

5 Conclusions

We have defined some new morphological operators, based on the premise that, by dropping one or more dimensions, an isotopic structuring element in a subspace becomes anisotropic in the full image space, but also gains some degrees of rotational freedom. This freedom can be used to have the structuring element align itself to the features in the image, and thus become rotation invariant.

We have shown that the dilation with such a structuring element, giving it the orientation that causes the result to be maximal, is in fact an isotropic dilation. This comes from the fact that the isotropic structuring element is the same as the union of (an infinite amount of) lower-dimensional isotropic structuring elements with all possible different orientations. In contrast, if we give the structuring element the orientation that causes the result of the operation to be minimal, we get the dilation operator proposed here.

In the same manner, we have defined a new erosion, closing and opening operators. We have stated that all of these operators are rotation, translation, scaling and contrast invariant, as well as increasing and extensive. We have also mentioned that the closing and opening defined in this article are idempotent, commutative and absorbing. These properties are important if we want to use the new operators in the same way we use other morphological operators.

The morphological framework proposed in this article has been defined in two dimensions, but it has been shown that it is easy to extend to higher dimensional spaces. In two dimensions, the closing and opening as defined here can be used to measure the length of image features. In three dimensions, different versions of the same operator can measure both the first and second largest diameters. The smallest diameter is measured in all cases using an isotropic structuring element.

Finally, we have explored an example application for the new operators, that shows that they are useful in granulometric applications.

References

[1] J. Serra. *Image Analysis and Mathematical Morphology*. Academic Press, London, 1982.

[2] C.L. Luengo Hendriks and L.J. van Vliet. A rotation-invariant morphology using anisotropic structuring elements. *in preparation*.

[3] R. van den Boomgaard. *Mathematical Morphology: Extensions Towards Computer Vision*. Ph.d. thesis, University of Amsterdam, Amsterdam, 1992.

[4] P. Soille. *Morphological Image Analysis*. Springer-Verlag, Berlin, 1999.

[5] G. Matheron. *Random Sets and Integral Geometry*. John Wiley and Sons, New York, 1975.

[6] P. Soille. Morphological operators with discrete line segments. In G. Borgefors, I. Nyström, and G. Sanniti di Baja, editors, *DGCI 2000, 9th Discrete Geometry for Computer Imagery*, volume 1953 of *LNCS*, pages 78–98, Uppsala, Sweden, 2000.

[7] J.A. Bangham, P. Chardaire, C.J. Pye, and P.D. Ling. Multiscale nonlinear decomposition: The sieve decomposition theorem. *IEEE Transactions on Pattern Analysis and Machine Intelligence*, 18(5):529–539, 1996.

Multiscale Feature Extraction from the Visual Environment in an Active Vision System

Youssef Machrouh[1], Jean-Sylvail Liénard[1], and Philippe Tarroux[1,2]

[1] Situated Perception Group LIMSI-CNRS BP 133 F-91403 Orsay Cedex, France
[2] ENS 45 rue d'Ulm F-75230 Paris cedex 05, France
{Youssef.Machrouh,
Jean-Sylvain.Lienard, Philippe.Tarroux}@limsi.fr

Abstract. This paper presents a visual architecture able to identify salient regions in a visual scene and to use them to focus on interesting locations. It is inspired by the ability of natural vision systems to perform a differential processing of spatial frequencies in both time and space and to focus their attention on a local part of the visual scene. The present paper analyzes how this differential processing of spatial frequencies is able to provide an artificial system with the information required to perform an exploration of its visual world based on a center-surround distinction of the external scene. It shows how the salient locations can be gathered on the basis of their similarities to form a high level representation of the visual scene.

1 Introduction

The use of active mechanisms seems to be a way to improve the abilities of machine vision systems. Active systems search salient features in the visual scene through a dynamic exploration. They can direct their search toward the most meaningful stimuli using attentional mechanisms leading to a reduction of the computational load [1, 2]. Thus, natural vision is a behavioral task, not a passive filtering process. An exploration of the visual world that relates perception and action allows for a labeling of the external space with natural landmarks associated with the exploratory behavior. In this respect, the relationships between agents and natural systems suggest that certain aspects of natural perception can be successfully incorporated in artificial agents.

Otherwise, during the past few years, several studies have been devoted to the understanding of the essence of vision considered as an information processing mechanism [3]. This approach is grounded on Barlow's proposal [4], which stated that the main organizational principle in visual systems is the reduction of the redundancy of the incoming stimuli.

These considerations, issued from information theory, led several authors to analyze the statistical organization of natural images. They demonstrated that natural images (those which do not exhibit any specific bias in their pixel distribution) have a

C. Arcelli et al. (Eds.): IWVF4, LNCS 2059, pp. 388–397, 2001.
© Springer-Verlag Berlin Heidelberg 2001

stationary statistics and an auto-similar structure. As a consequence of these characteristics, their power spectra fall off as $1/f^2$ [5].

In this context, different authors [6, 7] demonstrated that a way to transform the initial redundancy was to improve the statistical independence of the image descriptors. According to this hypothesis, an image can be viewed as a linear superposition of several underlying independent sources.

The filters that provide this statistical independence can be computed through the application of the source separation adequate algorithms (Infomax, BSS, ICA).

One can show [6, 7] that the optimal filters computed according to these principles are multiscale local orientation detectors similar to a Gabor wavelet basis [8].

However, although a lot of work has been devoted to the understanding of these theoretical bases of information processing in natural visual system, few attempts have been made thus far to use these principles in artificial vision systems. Practical implementations impose some limitations that require analyzing what is really obtained with simplified models based on these general principles. On the other hand, no artificial vision system has been designed to include both multiscale wavelet analysis and differential spatial and temporal processing of spatial frequencies. A prerequisite to the design of such a system is to be able to characterize the information obtained from a bank of wavelet filters in different frequency channels.

We thus analyzed here the information issued from various combinations of high and low frequencies of statistically uncorrelated signals. Our aim was to determine how to build a multivariate representation of the scene that allows a dynamic grouping of image points on the basis of their similarities in a given context and for a given task.

2 System Overview

2.1 Image Data

A set of 11 natural images selected from a larger database was used in the present study. Pictures that include too many traces of human activity (buildings, roads...) were avoided. Only images with similar initial resolution (around 256x512 pixels) were retained.

Fig. 1. Sample image from the set of natural images used in the present work. (Original size 512x256)

The images were discarded when their power spectrum did not fit the $1/f^2$ characteristics [5]. Figure 1 shows one typical example of an image used in the present study.

2.2 Initial Filters

A guideline for this work was to retain among the filtering characteristics of the primate visual system those that can be useful for the elaboration of an artificial system of situated and active vision.

Two characteristics have attracted our attention: the elimination of image redundancy in the processing steps designed to maximize the statistical independence of the scene descriptors and the differences in the processing of spatial frequencies between the center and the surround of the visual field.

The visual scene was filtered by a first bank of Gabor wavelets in four spatial orientations and four spatial frequencies (1/8, 1/16, 1/32, 1/64). For each initial image, we got 32 resulting images (two for each quadrature pair of each of the 16 Gabor filters). This multiscale processing was implemented using a Burt pyramid according to the method proposed by Guérin-Dugué [9].

For the purpose of this study and in order to obtain a complete view of what information is obtained from a detector during a systematic exploration of the visual scene, the whole scene was filtered by the entire bank of filters. In an operational system with a focal vision only a small part of these computations are needed.

2.3 Simple Cells – Complex Cells

An important distinction between the use of wavelets in image processing and the filtering steps in the visual system is the presence of strong non-linearities in the latter. Primary visual cortex shows several cell types according to the non-linearities they implement. Simple cells (SC) perform an additive combination of their inputs. They respond to an oriented stimulus localized at the center of their receptive field. The so-called complex cells (CC), on the contrary, exhibit a kind of translational invariance and respond to a stimulus whatever its position in the receptive field of the cell.

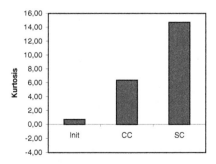

Fig. 2. Effects of filtering of the statistical independence criterion. Init: Initial image, SC: Simple Cells, CC: Complex cells

Other cell types (mainly in extrastriate cortex) combine these outputs in order to be sensitive to curvature and terminations (end-stop cells).

To model simple cells we used additive units with a zero threshold ramp transfer function which amounts to take into account only the positive part of Gabor filters. These cells do indeed not transmit the inhibitory part.

According to Field [5], we modeled complex cells output as the norm of quadrature pair Gabor filters. We verified that this implementation effectively leads to a reduction of the redundancy for both cell types by a comparison of the kurtosis before and after filtering (Figure 2). Kurtosis is indeed a good measurement of the statistical independence of a set of detectors [10].

A third type of detector with large receptive fields and designed to provide a contextual information will be considered in the following section.

In order to build a set of higher-level detectors suitable for the extraction of complex features we performed a Karhunen-Loeve transform of the outputs. A set of 1744 image patches (5x5) extracted randomly from the initial 11 natural images was used to build these spaces. We thus obtained 8 eigen-vectors at the output of simple cells and 4 eigen-vectors at the output of complex cells for each frequency band. These computations amount to a non-linear principal component projection of the initial image performed with two different types of non-linearities.

2.4 Global Energy – Local Context

As stated above, we assumed the existence of detectors sensitive to the global energy in the different orientations. In an image region corresponding to the fovea, the system computes a global energy vector for each of the four orientations. This vector is used to build a signature that can be used to classify the region. Such an analysis provides us with contextual information [11, 12]. We consider the identification of these contexts as a prerequisite for the recognition of objects. The importance of contextual information in natural systems can be deduced from the experimental observation that object recognition is effectively facilitated if the objects are viewed in congruent contexts [12].

Thus, the system computes three output sets on each image: (i) an output directly issued from the Gabor filters filtered by a ramp function (SC), (ii) an output giving the local energy at the output of these filters analogous to the output of complex cells (CC) and (iii) a large field output providing contextual information.

3 Results

3.1 Simple Cells

For each image point the system provides a high dimensional vector made of 32 orientation components spread over 4 frequency bands for SC detectors and 16 orientation components in 4 frequency bands for CC detectors.

Although Gabor detectors maximize the statistical independence of their outputs, in practice they are not strictly independent. The analysis of these outputs through a Karhunen-Loeve transform leads to a data representation basis that sorts the representations according to their greatest statistical significance.

The first axis corresponding to the highest eigen-value shows highly variable details from one scene to another (figure 3 left). It emphasizes details related to the structures present in the scene. This probably results from the fact that these structures are correlated in a given scene due to the correlation induced by the presence of objects. They are uncorrelated from one scene to another because each scene has a different organization.

Fig. 3. Output of SC filters: projection of the output along the first (left) and the last (right) eigen-vector of the output space

On the contrary, details filtered by the axes corresponding to the lowest eigen-values (figure 3 right) are expected to weakly contribute to the total variance. They correspond to features most frequently observed from one image to another.

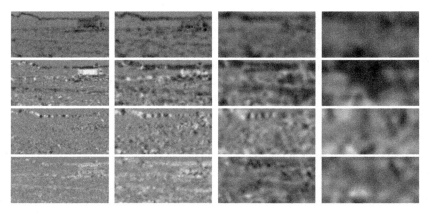

Fig. 4. Eigen-images from CC filters. The images are computed as the projection of the CC outputs on the eigen-vectors defining the output space of these filters. Columns range from high to low frequencies (from left to right: 1/8 to 1/64). Lines show the filter outputs along the principal components (top: highest variance, bottom: lowest variance).

The same region revealed by the first projection axis (Figure 3 left)(% initial variance: 29.4%) of the KL transform and the last projection axis (Figure 3 right)(% initial variance: 2.47%) shows that, while the first axis tends to reveal long edges that contribute significantly to the general structure of the objects, the last axis tends to reveal termination and curvature points that are not characteristic of the image structure.

We obtain a complex set of features along the different axes. The most representative of the presence of objects correspond to the first axes. On the others, features representing complex combinations of stimuli frequently observed in natural images seem to be sorted according to their level of abstractness.

3.2 Complex Cells

The same transform can be applied to the output of complex cells. Figure 4 shows the main axes of the KL transform following the computation of the Gabor norm for different spatial frequency bands.

The projection axes (rows in the figure) extract distinct features from the initial image as well within the same frequency band (rows) as between different frequency bands (columns)(note that for instance the building vanishes in axis 3 projection. Figure 4 3rd row). These features are entirely different from those extracted by the output transform of SC.

One can observe that high frequency details disappear in low frequency channels except for objects that exhibit frequency similarities (high frequency details repeated over a large area like the building).

Objects in the foreground, which are apparently characterized by low frequencies, appear in low frequency channels while they are not represented in high frequency band. Low frequency channels are able to distinguish features that have some spatial extension (the building or the foreground bushes).

A comparison of the lowest frequency channels (Figure 4 right column) shows that the locations revealed on the different axes are largely uncorrelated, thus corresponding to different points of view on the scene.

The lesser number of low frequency features (figure 4 right column) defines a small set of landmarks able to characterize the visual space and to guide exploratory saccades. This low-frequency information is the only one available in the periphery of the visual field.

3.3 Correlation between Channels

One of the important questions raised by this analysis is how different are the indices obtained from the different frequency channels. If two channels correspond to the same combination of basic features, the corresponding eigen-vectors should be similar. Thus, a measure of the similarity between the eigen-vectors in different frequency bands is given by the product of the eigen-matrices in these frequency bands. Using this method we compared the output spaces of respectively simple and complex cells for different frequency bands. We obtained strongly different results for the comparison of output spaces in SC channels and in CC channels.

For simple cells, the correlation between the axes of the spaces corresponding to different frequencies are low and distributed over the different axes (data not shown) while in complex cells the respectively high and low frequency bands exhibit similarities (table 1).

Table 1. Analysis of the output space for CC detectors. The eigen-vectors corresponding to the same axes show a very high correlation between respectively high and low frequency channels. The cross-correlation between eigen vectors corresponding to different axes is usually low (not reprinted here)

	Frequencies					
Axes	f0/f1	f0/f2	f0/f3	f1/f2	f1/f3	f2/f3
F1	0.990	0.442	0.410	0,365	0,330	0,996
F2	0.997	0.517	0.501	0.507	0.486	0.997
F3	0.991	0.363	0.370	0.425	0.424	0.995
F4	0.994	0.656	0,653	0.641	0.630	0.996

These results lead to the conclusion that the combination of simple cells outputs across the frequency bands underline uncorrelated details, whereas the outputs in high (resp. low) frequency bands correspond most frequently to similar stimuli.

A pyramidal decomposition of the scene allows combining these characteristics to identify spatial positions characterized by spectral compositions as diverse as possible.

This diversity seems to lead to a greater separability of these spatial positions and seems to be able to facilitate object discrimination.

3.4 Identification of Global Contexts

Cells sensitive to low frequencies have large receptive fields. However, in higher layers of the visual system cell types that encode intermediate representations also exhibit larger receptive fields. They combine the output of the cells in the preceding layers and gather the information coming from brighter regions of the visual field.

A vector that combines the global energy components associated with each frequency channel provides a suitable code for representing the whole fovea. It has been shown that such vectors can be used to classify visual scenes according to the context they belong to [11, 12]. In the present study, we build such detectors in computing the mean energy provided by the output of CC cells in the four frequency bands already mentioned.

To determine how spatial indices provided by the channels previously described can be used for the identification of interesting locations in the scene, we performed the following experiment:

A set of salient locations is computed from the eigen-images defined previously. Points in the image are selected at random or on the basis of these salient locations. At each point the mean energies of the CC outputs in an image window corresponding to the fovea were computed for each frequency. We thus obtained an energy vector for each of the selected point. A PCA analysis was performed on this set of vectors. One should keep in mind that this use of PCA differs from its use in the previous sections. The Karhunen-Loeve transform was previously used as a self-organization tool leading to a set of linear combination defining complex features frequently occurring in natural images. In this section, PCA should be considered as a means to analyze the structure of the space at the output of the SC and CC filters.

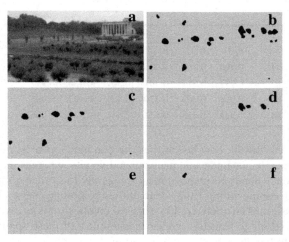

Fig. 5. Clustering of fixation points corresponding to different regions of the visual scene. Clusters were identified on the first three principal components and the fixation points corresponding to each cluster plotted on the diagrams at their position in the initial image (a). (b) Fixation points obtained from the second eigen-image and the second frequency channel shown Fig. 4. The other diagrams show the location of some clusters gathering salient points on the basis of their spatial frequencies and orientation properties: (c) trees and bushes, (d) building, (e) strong curvature at the border between hill and sky (f) another region of interest at the same limit

When the locations in the image are selected at random no obvious structure were observed in the PCA space. On the contrary, when they are selected on the basis of their saliencies, clusters were identified in the PCA space. Figure 5 shows the locations of some of these clusters on the original image. Points corresponding to a similar context are grouped into the same cluster. The example shows for instance the ability of the method to separate fixation points on the basis of their natural or artificial nature (Figure 5 c and d).

It should be noted that Figure 5 shows only a small sample of the structures that can be identified. Only 1/16 of the available dimensions is presented here. Thus, the method transforms the initial image into a huge set of clusters each characterized by similar spectral signatures.

4 Discussion and Conclusion

The visual filter system proposed in the present work produces a set of features that can be used to guide the exploration of the external scene. The features extracted by the non linear combination of SC channels seem rather suitable for object recognition. Features obtained from the computation of local energy (CC channels) allow a partition of the image into salient regions arranged according to their frequency composition. The computation of the global energy provides local context information and can be used to segment the scene on the basis of its spectral characteristics.

Thus, the output of this filtering system provides on one-hand locations of interest able to guide an attentional system and on the other hand clusters of locations arranged according to their spectral signature.

This approach can be considered as an extension of textures segmentation methods [13] to the question of the identification of contexts and an extension of the method proposed by Hérault [11] to the analysis of local contexts. However, it emphasizes the relativity of the context notion; the segmentation of the visual scene in (i) a global context and (ii) objects is an oversimplification

The visual scene is thus scattered into a set of projections on several disjoint subspaces. In each of these subspaces, salient points form clusters according to their similarities. These salient points are projected into disjoint sets of clusters and the corresponding objects can thus be grouped according to different points of view.

An object class is not characterized by a unique high-level representation, but by the transient association of a subset of properties. This association can thus dynamically depend on the current task. Objects are not considered as similar and grouped on the basis of their intrinsic properties but according to those of their properties linked to a given goal.

A further step in this work will be to demonstrate how such coding abilities could indeed facilitate object classification. This requires incorporating the present algorithms in the control architecture of a perceptive agent such that it can build a hierarchy of perception-action links based on the dynamic grouping of the perceived features.

Acknowledgments. This work was supported by a grant from the "GIS Sciences de la cognition" CNRS.

References

1. Allport, A.: Visual attention. In: Foundations of cognitive science. M.I. Posner (ed). The MIT Press. (1989)
2. Aloimonos, Y. (ed) Active Perception. Lawrence Erlbaum, Hillsdale,NJ (1993)
3. Atick, J.J. and A.N. Redlich: Towards a Theory of Early Visual Processing. Neural Computation 2 (1990) 308-320

4. Barlow, H.B.: Possible principles underlying the transformation of sensory messages. In: Sensory Communication. W. Rosenblith (ed). The MIT Press: Cambridge, MA. (1961) 217-234
5. Field, D.J.: Relations between the statistics of natural images and the response properties of cortical cells. Journal of the Optical Society of America A 4 (1987) 2379-2394
6. Bell, A.J. and T.J. Sejnowski: The "independent components" of natural scenes are edge filters. Vision Research 37 (1997) 3327-3338
7. Olshausen, B.A. and D.J. Field: Emergence of simple-cell receptive field properties by learning a sparse code for natural images. Nature 381 (1996) 607-609
8. Daugman, J. and C. Downing: Gabor wavelets for statistical pattern recognition. In: The Handbook of Brain Theory and Neural Networks. M.A. Arbib (ed). The MIT Press: Cambridge, MA. (1995) 414-420
9. Guérin-Dugué, A. and P.M. Palagi: Implantations de filtres de Gabor par pyramide d'images passe-bas. Traitement du signal 13 (1996) 1-11
10. Field, D.J.: What is the goal of sensory coding? Neural Computation 6 (1994) 559-601
11. Hérault, J., A. Oliva, and A. Guérin-Dugué: Scene categorisation by curvilinear component analysis of low frequency spectra. In: ESANN'97. Bruges (1997) 91-96
12. Oliva, A. and P.G. Schyns: Coarse blobs or fine edges? Evidence that information diagnosticity changes the perception of complex visual stimuli. Cognitive Psychology 34 (1997) 72-107
13. Andrey, P. and P. Tarroux: Unsupervised segmentation of Markov Random Field modeled textured images using selectionist relaxation. IEEE Transactions on Pattern Analysis and Machine Intelligence 20 (1998) 252-263

2-D Shape Decomposition into Overlapping Parts

Amin Massad[1] and Gerard Medioni[2]

[1] University of Hamburg, Dept. of CS/IMA, Hamburg, Germany
massad@informatik.uni-hamburg.de
[2] University of Southern California, Dept. of CS/IRIS, Los Angeles, CA, USA
medioni@iris.usc.edu

Abstract. We propose a method to generate component-based shape descriptions by the application of a perceptual grouping approach known as *tensor voting*. Based on previously described results on the generation of region, curve and junction saliencies and motivated by psychological findings about shape perception, we introduce extensions by a voting between junctions to create amodal completions, by a labeling of the junctions according to a catalog of junction types, and by a traversal algorithm to collect the local information into globally consistent part decompositions. In contrast to commonly used partitioning schemes, our method is able to create layered representations of overlapping parts. We consider this a major advantage together with the use of local operations and low computational costs whereas other approaches are based on highly iterative processes.

1 Introduction

Several classic publications – e. g. [MN78], [Bie87], [HR84], and [Pen87] – emphasize that the search for appropriate shape descriptions could be regarded as the key problem of computer vision. As shown by [RN88] and [Rom93], a representation scheme with expressiveness, discriminating power, stability, invariance against viewing conditions and occlusion necessarily has to be a component-based description. However, existing algorithms for segmentation and the generation of shape descriptions have several drawbacks which will be discussed in the following review of fundamental approaches.

Many methods can be traced back to a symmetric axis transform (SAT) introduced by [Blu67] and [BN78] leading towards a skeletal form representation. Main disadvantages are, besides the difficulty of computation, sensitivity to noise and problems with the handling of overlapping forms (Fig. 1a-c). Furthermore the aspect of symmetry is overemphasized leading to disagreements with recent psychophysical results on shape perception which see the relation to the Gestalt law of *good form* as follows: "... simplicity and regularity of form are outcomes, not causes, of the unit formation process" [KS91].

Other techniques are based on the transversality principle described by [HR84] which postulates part decompositions at points of maximal negative

C. Arcelli et al. (Eds.): IWVF4, LNCS 2059, pp. 398–409, 2001.
© Springer-Verlag Berlin Heidelberg 2001

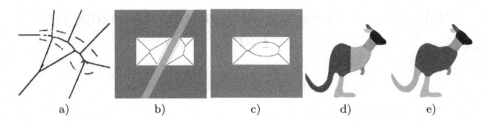

Fig. 1. Skeletal form representations have problems with a) the presence of gaps, b) overlapping parts, and c) noise (after [TSK97]). Curve evolution approaches can yield intuitively implausible parts – body and tail in (d) – as opposed to the desired segmentation (e), after [KTZ95].

curvature along a contour. Even though computation of extremal curvature is considered to be based on ecological vision theory, negative curvature can only give preliminary indications where to split segments without being a sufficient condition. Hence, it commonly yields too many decompositions due to the fact that all possible locations for segmentations along the contour are examined independently from each other without the consideration of combinations between potentially cooperating positions.

According to [Ley88] and [Ley89] the duality of extremal curvature and symmetric axes can be expressed by a so-called process-grammar describing a form as a history of deformations from an initial shape. This idea of evolving contours has been further explored by [KTZ95] with analogies to physical reaction-diffusion models and by [CKS97]. An improved curve evolution scheme which prevents the diffusion from blurring corners and dislocating feature positions can be found in [LL99]. While curve evolution models provide an elegant formalism and nice results for additive parts, they still tend to yield undesired segmentations (Fig. 1d-e) and have deficiencies in handling overlapping parts and negative parts (i.e. bridging indentation gaps) because interactions between related segmentation positions along the contour are neglected. Additionally, the input usually is required to be a closed contour. Furthermore, the edge polarity is not taken into consideration despite of its importance observed in human shape perception (e. g. a Kanizsa-Triangle with outlined circles does not yield the perception of virtual contours due to the absence of polarity information). Above all, the inherently iterative process seems to be inconsistent with the phenomenon of spontaneous shape perception known since [Kof35] and [Pet56]. The function of the iterations is rather a contribution to different abstraction levels of a multi-scale representation.

Serious drawbacks of all mentioned segmentation schemes arise from the strategy to decompose the input image into disjunct partitions, which prevents them from an adequate handling of occlusions. For that reason, we intend to find a method which facilitates the generation of overlapping forms.

Several aspects discussed here can be solved by employing methods of perceptual grouping. Firstly, there is strong support from the theory of visual interpolation theory established by Kellman and Shipley [KS91] on the basis of an extensive collection of psychological findings about shape perception (see also [SR94]). Using the relatability of discontinuities, they achieve a unified explanation for form completion in the case of partly occluded objects, virtual contour illusions and transparent occlusions. Accordingly, these phenomena, where connections of edge elements occur even across areas without physically existing edges, are based on a common grouping process for unit formation. Secondly, there is an adequate means for the realization of perceptual grouping provided by the *tensor voting* technique which has been developed by [MLT00] and [GM96] and belongs together with [SU98], [SB93], and [TW96] to the field of perceptual saliency theories. Tensor voting has the following advantages over similar methods and over techniques from other fields, e. g. regularization, consistent labeling as in [Sau99], clustering, robust methods, and connectionist models: The method is not iterative, does not require manual setting of starting parameters, handles the presence of multiple curves, regions, and surfaces at the same time, it is highly stable against noise and still preserves discontinuities. The only parameter is scale which is in agreement with the scale-dependency of human shape perception. The universality of tensor voting has been demonstrated in [MLT00] by application to a bunch of early vision problems. In contrast to commonly applied vector fields, this method profits from the definition of appropriate tensor fields which represent the information propagated from sparse input locations into their neighborhoods. All these fields are combined by means of a tensor addition (in contrast to a vector addition) which simultaneously yields the saliency for junctions, curves, and surfaces. The result of the perceptual grouping is derived by extracting the maxima in the saliency maps, for curves and surfaces a marching algorithm is employed to trace these structures along their highest saliencies.

2   Shape Descriptions from Perceptual Grouping

2.1   Overview

We will present a method to derive shape descriptions by decomposition of forms into multiple, possibly overlapping layers. Such a representation is supposed to be a natural solution for the handling of occlusions. It is not only capable to extract the overlapping parts but even allows estimations of their depth placement. The application to complex objects and object recognition tasks clearly will profit from the higher level of the representation compared to the matching of uncombined features. However, this has to be regarded as a long term aim. Here, we will discuss the foundations necessary for the achievement of layered shape representation. Therefore, we have chosen as the most basic case input images consisting of binary polygonal shapes. It is important to note that this reduction does not bound the proposed method to this domain but rather shows its basic properties; see section 3 for possible extensions.

As mentioned before, this kind of silhouette images has been studied in psychological experiments under the term *spontaneously splitting figures*. According to different studies reviewed in [KS91], the segmentation is initiated by the discontinuities, i. e. junctions, along the contour where interpolated edges start to connect so-called *relatable* edges to form part groupings. The relatability of two edges depends on the alignment of their orientations and expresses to what extent they could be connected in the sense of a good continuation. This aspect together with the consideration of spatial neighborhood is implicitly encoded in the layout of tensor voting fields applied to each input token. Thus, we have an efficient means to implement processes which have mainly been described qualitatively.

The outline of our implementation is as follows (Fig. 2): First, we have to find the contour in the given input image in order to compute the junctions contained in this form. Both tasks can be solved by tensor voting, too, as shown in [MLT00]. The next step is a newly introduced voting between the junctions: Each junction sends out voting fields in the opposite directions of its two incoming edges, hence looking for continuations of the "abruptly" ending line segments. The overlapping fields of interacting junctions will create high curve saliencies between these junctions. By extracting the curves along the locally maximal saliencies, we get candidates for the amodal completions of part boundaries which we will call briefly virtual contours.

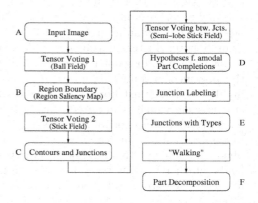

Fig. 2. Flowchart of the part decomposition algorithm. The right half shows new stages introduced in this paper. The letters A-F refer to outputs for a running example depicted in Figs. 3, 4, and 8.

In addition to the interactions of junctions, there can be interactions between a junction and a nearby contour creating the important class of T-junctions. A T-junction will in turn vote backwards to the junction by which it has been created, thus giving rise to a virtual contour between them. Finally, we have to collect the information about real and virtual contours to connect them to closed part boundaries. This step is non-trivial because it requires the integration of local operations with global constraints. We will define a process which uses some kind of cursor, called walker, to find local interpretations based on the conservation of direction and region polarity at the different junction types to get a globally consistent part boundary. The method facilitates the generation overlapping layers for which we will discuss in chapter 3 strategies to achieve meaningful depth orderings.

2.2   Computation of Real Contours and Junctions

Given an input image as shown in Fig. 3A, we first have to find the boundaries of the contained regions. For that purpose we use the region inference method already introduced in [MLT00]. In summary, a radial isotropic voting field is applied to each input location and information from different sites is collected by a first order moment computation which aggregates the incoming votes as a vector sum. Points along a region boundary will mostly receive votes from one side, in contrast to isolated points and points inside the region for which the incoming votes are equally distributed over all directions. Hence, the norm of the vector sum can be regarded as a measure for the boundary saliency (Fig. 3B) while its direction represents the direction of the region polarity. Having found the boundary points, we can get the curves and junctions by another voting step which applies a so-called stick field perpendicularly to each polarity vector and uses tensor additions for the collection of the votes. The tensors of the resulting map are decomposed into undirected components, called ball tensors, and directional components, called stick tensors. The norm of these components represents a measure for the junction saliency or the curve saliency, respectively. By extracting the locally maximal saliencies, we simultaneously obtain the junctions and the contour of the input form (Fig. 3C).

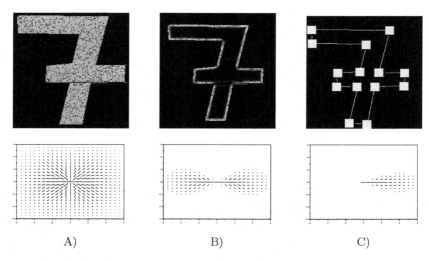

A)   B)   C)

Fig. 3. Computation of real contours and junctions: A) input image, B) region saliency map, C) extracted contours and junctions. The graphs illustrate the tensor voting field applied at each stage. See Fig. 4 for the result of the rightmost field.

2.3   Computation of Virtual Contours

In order to create interactions between the detected junctions, we first have to determine for each junction the directions of the incoming real contours. Naturally, the curve saliency in the vicinity of a junction is very unreliable due to the simultaneous influence of two different lines. Therefore, we discard this weak curve information within a predefined radius around a junction and look for the edges crossing a circle with similar radius. In addition, we store for each incoming edge its region polarity which has been computed previously. As the input consists of binary forms, the junctions found so far always consist of exactly two crossing edges of the real contour. Hence, we call them L_0-junctions where the index denotes the number of incoming virtual contours.

The first step of junction voting is necessary to detect T-junctions on the real contour. For that purpose, at each L-junction semi-lobes of a stick voting field are applied in opposite directions of the two incoming edges. Note that the stick fields used for both steps of junction voting have been adapted to cover a smaller opening angle in order to give higher preference to a straight continuation and to reduce position uncertainty which is not needed to the same extent as for finding structures in general input domains.

The interaction of these votes with the real contour transformed to its tensor representation yields high junction saliencies at locations of T-junctions which are then extracted by non-maxima suppression. The stem of each T-junction is computed to point in the direction of the L_0-junction by which it has been created, and the region polarity of this L_0-junction is used as an estimation for the region polarity at the T-stem.

Then, the second step creates all possible virtual contours as high curve saliencies by applying junction voting to the set of L_0- and T-junctions (Fig. 4pre-D.). While the voting fields for L_0-junctions are defined as described before, T-junctions send out semi-lobes of a stick field in the direction of their stem. The resulting curve saliency map is fed to a marching process which traces the locally maximal curve saliencies in order to yield the virtual contours (Fig. 4D.).

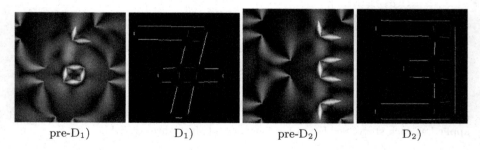

pre-D_1) D_1)   pre-D_2) D_2)

Fig. 4. Virtual contours: pre-D_1) Curve saliency map resulting from the second junction voting step applied to the shape of a seven as input. D_1) Extracted virtual contours (gray) overlaid over the previously extracted real contours (white). pre-D_2), D_2) Results for another example with the shape of a three as input.

2.4   Deriving a Shape Description

In order to introduce a unit formation process yielding multi-layered shape descriptions, we first have to take a closer look at the different kinds of junctions and the information they contain with respect to the indicated shape decompositions. A classification of the junction types is made by inspecting the L_0-junctions after the computation of virtual contours and relabeling them to L_1-junctions in case of one incoming virtual contour in addition to the two real contours or to L_2-junctions in case of two incoming virtual contours. Subsequently, a T-junction will be relabeled to T_1 if it has been induced by an L_1-junction or to T_2 if induced by an L_2-junction, respectively.

Among these types an L_0-junction does not contain any information about a decomposition. It merely represents a discontinuity on the contour.

An L_1-junction indicates a shape decomposition along the virtual contour yielding two disjunct parts[1]. The virtual contour is included in the outline of both parts. However, its polarity vector in the decomposition points into two opposite directions in dependency of the part to which it has been assigned.

Similarly, each virtual contour of an L_2-junction potentially belongs to the outline of two disjunct parts, i. e. creates decompositions with polarities in both directions (Fig. 5a). However, it is also possible that both virtual contours represent continuations of two overlapping parts (Fig. 5b). In that case only one direction of each polarity vector pair will be used in the decomposition. The discrimination of both cases is achieved by the walking algorithm described below.

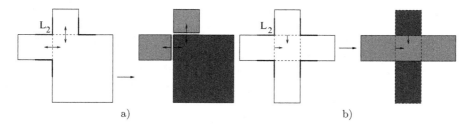

Fig. 5. L_2-junction: a) Each virtual contour belongs to two disjunct parts (both potential polarity vectors occur in the decomposition). b) Virtual contours belong to different overlapping parts (only one of each polarity vector is used).

As depicted in Fig. 6, the two types of T-junctions carry different information for the suggested decompositions: While the stem of a T_1-junctions belongs to disjunct parts and thus bears opposite polarity assignments, the stem of an T_2-junction represents an occluded contour created by two overlapping parts and possesses in the decomposition only one polarity direction which corresponds to the contour polarity at the adjacent L_2-junction. However, the corresponding half of the T_2-bar belongs to two parts, both times with the same region polarity.

---

[1] Note that the interacting junction has not necessarily to be an L_1-junction. It could also be an L_2- or a T-junction.

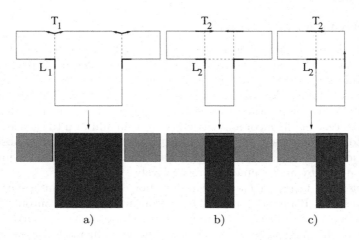

Fig. 6. T-junctions: a) T_1-junction, b), c) two examples of T_2-junctions.

To complete this junction catalog, we finally mention the possibility of an L_2'-junction which is merely created by two virtual contours. However, we will postpone the discussion of this junction type because it does not initiate any decompositions. It rather ensures correct continuations of straight lines by preserving the discontinuity at the occluded corner.

So far all computations have been local operations. As we intend to derive decompositions of the input form, we need to introduce a mechanism which collects this information to achieve a globally consistent description. For that purpose, we first build an adjacency graph for the junctions, label the junctions by the types mentioned above and assign the potential polarity vectors to the contours meeting at these junctions. It is sufficient to consider the L_1- and L_2-junctions as seeds for possible decompositions (other types don't induce decompositions or cannot occur independently).

Then, for each seed we subsequently start outgoing "walks" in one of the unvisited directions. This process is based on a cursor, called walker, which stores the current position on the contour together with the walking direction and the region polarity. The walker is advanced from the current junction to the adjacent junction in walking direction and the chosen polarity vectors along the traversal are marked as visited. At this junction the outgoing continuation of the walk has to be inferred from the incoming direction and polarity. The decision is based on a set of predefined rules of which examples are depicted in Fig. 7. In general, a continuation tries to preserve direction and polarity. If such a straight continuation is not possible, characteristics of the current junction type together with the information about already visited contour parts determine a change in direction and an adaptation of the polarity vector.

A walk is stopped on arriving back at the starting junction or in case of no possible continuation. However, a closed loop is only regarded as a successful solution if the continuation of the walker returning to the start yields the same

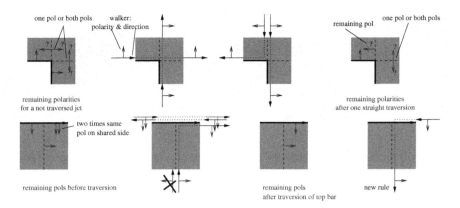

Fig. 7. Rules how to continue walks at junctions: The examples show the rules for L_2-junctions (first row) and T_2-junctions (second row).

polarity and direction as the initially outgoing walk, i. e. the walk would be cyclic. Such a walk indicates a successfully decomposed part. Otherwise an undo of the unsuccessful walk is a created by restoring all visited polarity vectors.

The walking process is repeated for all remaining polarities along virtual contours until no more decompositions are found. It terminates because the number of potential polarity vectors is reduced for each part decomposed and there is always one outgoing walk among the seeds which yields a successful part decomposition.

Some results are illustrated by the examples of Fig. 8. It shows for each shape that the junctions, including the virtual T-junctions, are extracted correctly and labeled according to junction types introduced here. Finally, the walking algorithm yields a decomposition into overlapping parts as proposed for the generation of an adequate shape description of the input images.

3   Conclusions and Perspectives

We have introduced a framework for a unit formation process based on the local computation of perceptual groupings by tensor voting and a decomposition scheme which collects polarity and directional information to extract overlapping parts. Although this last step seems to be iterative and serial, it actually could be implemented as a highly parallel process where successful decompositions are highlighted by closed walks creating positive feedback loops or synchronized cell activities. This approach would be similar to the particle system model of [UKD+99] but with a considerably lower number of iterations and the ability to create layered representations.

In agreement with psychophysical findings, we explicitly don't make use of depth information for the inference of units. We rather intend to achieve depth information from the unit formation process. Depth arrangements for sponta-

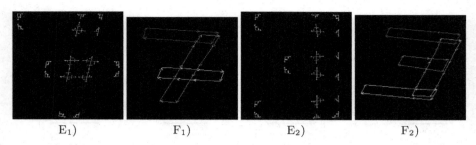

E₁)   F₁)   E₂)   F₂)

Fig. 8. Resulting part decompositions (F.) based on the detected junctions (E.) depicted for two input images, the shape of seven and of a three. Junctions are illustrated by the incoming directions (long arrows) and the initially possible polarity vectors (small arrows). Hence, the junction type L_0 is formed by two, L_2 by four, and T_2 by three long arrows. Parts of the decomposition are depicted in a 3-D display as overlapping layers of different heights where the arrows along the contour trace the positions of the walker.

neously splitting figures from binary input images can be deduced from the length of the virtual contours. Parts with short virtual contours - reflected by higher saliencies - are assumed to lie on top of parts with longer virtual contours. However, further research is needed in order to handle ambiguous orderings where the virtual contours of one part have different lengths or of two overlapping parts have similar lengths. This suggests the introduction of bistable depth orderings as known to occur in psychophysical experiments, too.

While the unit formation process should be universally applicable to other input domains with additional attributes like color and texture, the influence of this information on different depth orderings requires further investigations. Nevertheless, the units created solely depend on the form, hence should be the same for all domains as for binary inputs.

Furthermore, we intend to generalize the implementation from polygonal shapes to rounded forms which is already facilitated by the fact that the tensor voting scheme detects a saliency for "cornerness" and thus allows to handle junctions on rounded shapes as corners with lower saliency.

Currently, only one fixed scale parameter is used by the tensor voting between junctions. As long as the size of the voting fields is large enough to bridge the biggest gaps, the scale parameter is not critical for the outcome. In contrast to other approaches with continuous scale spaces, we think a small number of scales will be sufficient to get descriptions on different levels of abstraction. In addition, we plan to extend the voting scheme to an adaptive scale parameter proportional to the length of the line ending at a junction. This can also be realized by the tensor voting approach which allows to define a degree of "endpointness" at the line endings. Thus long lines are enabled to bridge large gaps while short lines are restricted to a smaller radius to find a continuation.

Finally, the application of the unit formation to real input images seems to be a very challenging goal. However, the restriction to such a simple domain as presented here helps to focus on the fundamental properties of the unit formation process which are still open to a wide range of investigations.

References

[Bie87]   I. Biederman. Recognition by components: A theory of human image understanding. *Psychol. Rev.*, 94:115–147, 1987.

[Blu67]   H. Blum. A transformation for extracting new descriptors of shape. In W. Wathen-Dunn, editor, *Models for the Perception of Speech and Visual Form*, pages 362–380, Cambride, MA, 1967. MIT Press.

[BN78]   H. Blum and R. Nagel. Shape description using weighted symmetric axis features. *Pattern Recognition*, 10(3):167–180, 1978.

[CKS97]   V. Caselles, R. Kimmel, and G. Sapiro. Geodesic active contours. *Int. J. of Computer Vision*, 22:61–79, 1997.

[GM96]   G. Guy and G. Medioni. Inferring global perceptual contours from local features. *Int. J. Computer Vision*, 20:113–133, 1996.

[HR84]   D. Hoffman and W. Richards. Parts of recognition. *Cogn.*, 18:65–96, 1984.

[Kof35]   K. Koffka. *Principles of Gestalt psychology.* Harcourt Brace, New York, 1935.

[KS91]   P. Kellman and T. Shipley. A theory of visual interpolation in object perception. *Cognitive Psychology*, 23:141–221, 1991.

[KTZ95]   B. Kimia, R. Tannenbaum, and S. Zucker. Shapes, shocks, and deformations I: The components of shape and the reaction-diffusion space. *Int. J. Computer Vision*, 15:189–224, 1995.

[Ley88]   M. Leyton. A process-grammar for shape. *A.I.*, 34:213–247, 1988.

[Ley89]   M. Leyton. Inferring causal history from shape. *Cog. Sci.*, 13:357–387, 1989.

[LL99]   L. Latecki and R. Lakmper. Conexity rule for shape decomposition based on discrete contour evolution. *CVIU*, 73:441–454, 1999.

[MLT00]   G. Medioni, M.-S. Lee, and C.-K. Tang. *A computational framework for segmentation and grouping.* Elsevier, Amsterdam, 2000.

[MN78]   D. Marr and H.K. Nishihara. Representation and recognition of the spatial organization of three-dimensional shapes. In *Proc. R. Soc. Lond.*, volume 200, pages 269–294, 1978.

[Pen87]   A. Pentland. Recognition by parts. In *Proc. IEEE Int. Conf. on Computer Vision*, pages 612–620, London, 1987.

[Pet56]   G. Petter. Nuove ricerche sperimentali sulla totalizzazione percettiva [New experimental research on perceptual totalization]. *Rivista di Psicologia*, 50:213–227, 1956.

[RN88]   K. Rao and R. Nevatia. Computing volume descriptions from sparse 3-d data. *Int. J. of Computer Vision*, 2:33–50, 1988.

[Rom93]   H. Rom. Part decomposition and shape description. Technical Report IRIS-93-319, Univ. of Southern California, 1993.

[Sau99]   E. Saund. Perceptual organization of occluding contours of opaque surfaces. *CVIU*, 76:70–82, 1999.

[SB93]   S. Sarkar and K. Boyer. Integration, inference, and managment of spatial information using bayesian networks: Perceptual organization. *IEEE Trans. Pattern Anal. and Machine Intel.*, 15:256–274, 1993.

[SR94]   R. Shapley and D. Ringach. Similar mechanisms for illusory contours and amodal completion. *Invest. Ophtalm. & Visual Sci.*, 35:1089, 1994.

[SU98]   A. Sha'ashua and S. Ullman. Structural saliency: the detection of globally salient structures using a locally connected network. In *Proc. Int. Conf. on Computer Vision*, pages 312–327, 1998.

[TSK97]   H. Tek, P. Stoll, and B. Kimia. Shocks from images: Propagation of orientation elements. In *CVPR*, pages 839–845, 1997.

[TW96]   K. Thornber and L. Williams. Analytic solution of stochastic completion fields. *Biol. Cybern.*, 75:141–151, 1996.

[UKD⁺99]   W. Uttal, R. Kakarala, S. Dayanand, T. Shepherd, J. Kalki, C. Lunskis, and N. Liu. *Computational Modeling of Vision.* Dekker, New York, 1999.

Fast Line Detection Algorithms Based on Combinatorial Optimization

Marco Mattavelli[1], Vincent Noel[1], and Edoardo Amaldi[2]

[1] Integrated Systems Laboratory - ISL - EPFL Swiss Federal
Institute of Technology,
CH 1015 Lausanne, Switzerland
{Marco.Mattavelli,Vincent.Noel}@epfl.ch
[2] DEI, Politecnico di Milano, Piazza Leonardo da Vinci 32,
20133 Milan, Italy
amaldi@elet.polimi.it

Abstract. In this paper we present a new class of algorithms for detecting lines in digital images. The approach is based on a general formulation of a combinatorial optimization problem. It aims at estimating piecewise linear models. A linear system is constructed with the coordinates of all contour points in the image as coefficients and the line parameters as unknowns. The resulting linear system is then partitioned into a close-to-minimum number of consistent subsystems using a greedy strategy based on a thermal variant of the perceptron algorithm. While the partition into consistent subsystems yields the classification of the corresponding image points into a close-to-minimum number of lines. A comparison with the standard Hough Transform and the Randomized Hough Transform shows the considerable advantages of our combinatorial optimization approach in terms of memory requirements, time complexity, robustness with respect to noise, possibility of introducing "a priori" knowledge, and quality of the solutions regardless of the algorithm parameter settings.

1 Introduction

The Hough Transform (HT) and its numerous variants are the classical approaches used to detect and recognize straight lines in digital images [1], [2]. The various HT variants have been developed to try to overcome the major drawbacks of the standard HT, namely, its high time complexity and large memory requirements. In some cases, variants such as the randomized, probabilistic and hierarchical HT [7], [8], achieve an effective complexity reduction. In others, however, they face serious difficulties and fail to provide solutions of the desired quality. This happens, for instance, when several line segments need to be simultaneously detected or when there is a relatively high level of noise [1], [2]. In general, selecting small values for the thresholds of those HT variants may yield erroneous solutions, while selecting larger values may substantially increase the computational load and therefore jeopardize their nice features of reduced time

C. Arcelli et al. (Eds.): IWVF4, LNCS 2059, pp. 410–419, 2001.

complexity and lower memory requirements. Thus, in the presence of several lines to be detected, one has to find a delicate trade-off between time/memory requirements and quality of solutions. In this paper we present a new approach for detecting lines in digital images that differs from the HT-based ones. The basic idea is to formulate the problem as that we introduced in [3] for estimating general piecewise linear models, namely as the combinatorial optimization problem of partitioning an inconsistent linear system into a close-to-minimum number of consistent subsystems. The method we devise to find approximate solutions to those problem formulations provides results of equivalent or even higher quality than the HT and it compares very favorably in terms of time and memory requirements as well as robustness. The paper is organized as follows; section 2 describes the combinatorial optimization formulation, some of its properties as well as a greedy strategy to find good approximate solutions in a short amount of time. Then some details of the algorithms as well as the convergence and projection strategy of the algorithms are presented in section 3, while some possible optimizations are presented in section 4. Some typical results are reported in section 5 and compared with those provided by the basic and randomized HT. The paper is concluded by presenting some general remarks and perspectives.

2 The MIN-PCS Based Formulation of Line Detection

The problem of classifying the points of an image into line segments can be formulated as that of partitioning an inconsistent linear system into consistent subsystems. Indeed, the coordinates of points belonging to a line segment satisfy a simple linear system whose solution corresponds to the parameters of the line. In the presence of several lines and noise distributed in the image, the linear system is inconsistent, i.e. there exists no solution satisfying the equations corresponding to all image points. In such cases, regressions and robust regression based approaches are clearly inadequate. The breakdown point of classical robust regression methods limits, for instance, their applicability to a very restricted type of situations. In particular, there must be a dominant subsystem that corresponds to at least 50% break-down limits [6], or for other approaches [10] the solution is guaranteed only for uniform or "a priori" known noise distributions. In the case of inconsistent systems corresponding to several "unknown" consistent subsystems and noise with "unknown" distributions other approaches have to be found. For these reasons, alternative approaches generally based on the HT and its numerous variants have been extensively investigated [1], [2], [7]. Although, in general, these approaches tend to provide reasonable results and to be relatively robust with respect to noise, they have high time and memory requirements and they are quite sensitive with respect to the threshold settings. In this paper we show that accurate solutions to the problem of line detection can be found by considering the following combinatorial optimization problem that we have introduced in [3].

MIN PCS: *Given an inconsistent linear system $A : x = b$ where A is an $m \cdot n$ matrix and x,b are n-dimensional vectors, find a **P**artition of the set of equations into a **MIN**imum number of **C**onsistent **S**ubsystems.*

In the case of line detection, the coefficients of each row of the inconsistent linear system correspond to the coordinates of one of the contour points at hand. Any partition into a number s of consistent subsystems is then clearly equivalent to a partition of all contour points into s line segments. Given the choice of the objective function, we look for the simplest set -for the smallest number- of line segments that account for all contour points. To cope with noise and quantization or acquisition errors, it suffices to replace each equation $a^k x = b_k$, where a^k is the kth row of A and b_k is the kth component of b, by the two complementary inequalities:

$$\begin{cases} a^k \cdot x \leq b_k + \varepsilon \\ a^k \cdot x \geq b_k - \varepsilon \end{cases} \tag{1}$$

where ε is the maximum tolerable error. See [3] for the description of a simple geometric interpretation of this version of MIN PCS. In the present setting, it amounts to finding a minimum number of hyperslabs in n-dimensions whose width is proportional to ε and such that each point corresponding to the coordinates of one contour point is contained in at least one hyperslab.

Although we proved in [3], [4] that MIN PCS is an NP-hard problem, and hence, it is unlikely that any algorithm is guaranteed to find an optimal solution in polynomial time, we have developed a heuristic which works well in practice and finds good approximate solutions in a short amount of time. The results obtained for other problems and time series modeling clearly confirm this assertion [3] [4]. Since in practical applications we are interested in finding close-to-minimum partition rapidly, we adopt a greedy strategy in which the original problem is subdivided into a sequence of smaller subproblems. Several projection schemes can be used for the solution of each subproblem depending on its nature. Here we introduce one scheme designed for line detection (i.e. MIN-PCS problems with two variables). Starting with the original inconsistent system of pairs of inequalities (1), close-to-maximum consistent subsystems are extracted iteratively. Clearly the iterative extraction of consistent subsystems yields the partition into consistent subsystems. The extraction of close-to-maximum consistent subsystems is performed by using a thermal variant of the perceptron procedure that originally comes from machine learning (see [4], [5], [6] and the references therein). The algorithm can be described as follows, see also [3]:

- **Problem:** Given any system $Ax = b$ and any maximum admissible error $\varepsilon > 0$, look for an $x_{\max} \in \mathbb{R}^n$ such that the couple of complementary inequalities $a^k x_{\max} \leq b^k + \varepsilon$ and $a^k x_{\max} \geq b^k - \varepsilon$ is satisfied for the maximum number of indices $k \in \{1, \ldots, p\}$.
- **Initialization:** Take an arbitrary $x_0 \in \mathbb{R}^n$, set $c := 0$, and initial temperature $t := t_0$, select a predefined number of cycles C as well as function $\gamma(c, C)$ decreasing for increasing c and such as $\gamma(C, C) = 0$.

begin

    $i \leftarrow 0$;

    repeat

        $c \leftarrow c + 1$; $t \leftarrow t_0 \cdot \gamma(c, C)$; $S \leftarrow \{1, \ldots, p\}$;

        while $S \neq \emptyset$ **do**

        pick $s \in S$ randomly and remove s from S

        $k_i \leftarrow s$; $E^{k_i} := b^{k_i} - a^{k_i} \cdot x_i$;

$$\delta_i := \frac{t}{t_0} \exp\left(\frac{-\left|E_i^{k_i}\right|}{t}\right);$$

        if $\left(a^{k_i} x_i \leq b^{k_i} - \varepsilon\right)$ $x_{i+1} := x_i + \delta_i a^{k_i}$;

        else if $\left(a^{k_i} x_i \geq b^{k_i} + \varepsilon\right)$ $x_{i+1} := x_i - \delta_i a^{k_i}$;

        $i \leftarrow i + 1$;

        project the current solution onto the unit cylinder

    while $c < C$

    take x_{i+1} as an estimate of $x_{\max}$

end

where t_0 is determined by the average deviation from consistency (average inequality error) for the current solution x_t at the beginning of each cycle. Intuitively, the behavior of the algorithm can be explained as follows. At high normalized temperature t/t_0, all equations with both high or low deviations from consistency lead to a significant correction of the current solution x_i. Conversely, at low temperatures, only those equations with small deviations from consistency yield relevant corrections to the current solution. Convergence of the procedure is guaranteed because when t is decreased to zero, the amplitude of the corrections tends to zero. In our experiments we have used exponentially decreasing functions for t, from an initial t_0 to 0 in a predefined maximum number of cycles C through the equation set. See [3], [4] for more details on the algorithm and the annealing schedules.

3   Convergence Analysis of MIN-PCS Based Line Detection Algorithms

In this section we study the convergence behavior of the proposed algorithms. So as to better clarify convergence issues let us consider a linear system composed by only two equations in two variables:

$$Ax = b \Leftrightarrow \begin{cases} a_{11}x_1 + a_{12}x_2 = b_1 \\ a_{21}x_1 + a_{22}x_2 = b_2 \end{cases} \tag{2}$$

These two equations represent two straight lines in $\mathbb{R}^2$. Assuming that the two lines are not parallel, the relaxation algorithm described before can be illustrated as in figure 1 (see also [3], [6]). The angle α between the two lines and

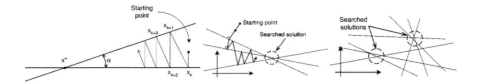

Fig. 1. Left: graphical representation of the relaxation algorithms to extract close-to-maximum consistent subsystems, center: example of convergence for the case of a consistent system, right: example for a case of an inconsistent system, i.e. two consistent sub-systems.

the convergence rate are related as follows:

$$\frac{\|X_{i+1} - X^*\|}{\|X_i - X^*\|} = \cos^2 \alpha, \frac{\|X_{i+n} - X^*\|}{\|X_i - X^*\|} = \cos^{2n} \alpha \qquad (3)$$

It is clear that when the two lines are nearly parallel, i.e. when α is close to 0, the algorithm will converge very slowly (since $\cos \alpha$ is close to 1). These situations occur very frequently in line detection problems. It can be easily shown that this is the case of relatively short image segments located far from the (conventional) coordinate origin in the binary image. Conversely, when α is close to $\pi/2$, the algorithm converges very quickly. In practice, one should try to make the row vectors in the system of linear equations mutually orthogonal by using classical techniques see for instance [9]. Unfortunately, these kind of orthogonalization procedures cannot be applied in the case of inconsistent systems (i.e. several consistent subsystems) because the "line pencil" is not unique. In fact there are several consistent subsystems corresponding to different lines as illustrated in Figure 1.

So as to devise a fast line detection algorithm, we have to guarantee a fast convergence for all possible input data (i.e. line positions in the image). The idea is to find a suitable surface on which to project the current solution x_i that is as much as possible orthogonal to all possible lines so as to speed up the convergence and to constrain the solution to a desired region of the space. With this objective the working space (parameter space) has been extended to a three-dimensional space. Each point $X = (x, y)$ in $\mathbb{R}^2$ (image space) is mapped into a plane in the three-dimensional parameter space $\mathbb{R}^3$ according to the linear equation $ax + by + c = 0$. Then each line to be extracted in $\mathbb{R}^2$ defined by $\{(x, y) \in \mathbb{R}^2 / a_i x + b_i y + c_i = 0\}$ corresponds to a line in $\mathbb{R}^3$ defined by $\{(x, y, z) \in \mathbb{R}^3 / \exists \gamma \in \mathbb{R}, (a, b, c) = \gamma(a_i, b_i, c_i)\}$. Actually, in the case of an inconsistent linear system we have several "plane pencils" corresponding respectively to each straight line in the image and each intersects to a line (all these lines contains the origin since all linear equation are homogeneous). Thus, the problem we have to solve is that in three dimensions each solution line contains the origin (i.e. $(a, b, c) = (0, 0, 0)$) so that the algorithm always converge to the trivial solution 0. To avoid this occurrence, at each solution update (correction) we perform a projection to the unit cylinder of equation $\{(a, b, c) \in \mathbb{R}^3 / \sqrt{a^2 + b^2} = 1\}$. For each image point we alternate a projection to the corresponding solution plane

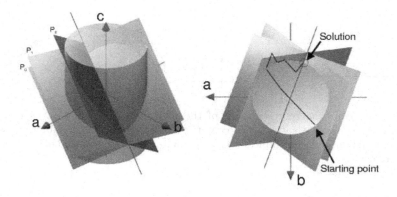

Fig. 2. 3-D representation of the projection scheme in case of a consistent system (3 aligned points in the image correspond to the pencil defined by planes P1, P2 and P3).

and one to the unit cylinder. Hence, this procedure constrains the current iterative solution to remain close to the unit cylinder and its intersecting planes. Specifically the following non-orthogonal projection is performed:

$$a \leftarrow \frac{a}{\sqrt{a^2 + b^2}}, b \leftarrow \frac{b}{\sqrt{a^2 + b^2}}, c \leftarrow \frac{c}{\sqrt{a^2 + b^2}} \qquad (4)$$

Applying this procedure the speed of convergence can now be expressed as:

$$\cos\alpha = \frac{\vec{n_1} \cdot \vec{n_2}}{\|\vec{n_1}\| \cdot \|\vec{n_2}\|} \qquad (5)$$

with $\vec{n_1}$ and $\vec{n_2}$ the normal vectors corresponding to two consecutive planes. A rigorous study of the convergence speed in the general case of an inconsistent system is obviously much more complex. This type of analysis must also take into account the statistical behavior of the consistent subsystems distributions and of the corrections versus the temperature scheme.

4   Line Detection Algorithms for Shape Detection and Tracking

In the basic scheme described in the previous paragraph, consistent subsystems (i.e. lines in an image) are extracted iteratively by the MIN-PCS greedy strategy starting from random initial solution and letting the relaxation algorithm to converge. This is the most generic approach to cope with applications for which no "a-priori" knowledge is available (number of line to be extracted, probable positions, object shape, geometrical relation between lines, etc...). In many applications, additional information such as object shape, approximated position, number of line to extract, fixed angle (or distance) between lines, etc ... might indeed be available. The nature of the approach allows the addition of various type of a-priori information that can dramatically improve robustness and performance. This "a priori" information can be easily embedded into the kernel of the MIN-PCS based algorithms while classical approaches have to consider it in the post-processing stage.

Multi Solutions Algorithms. An example of such inclusion of "a priori" information can be the following. If the number of line is known (or at least, the minimum number) several solutions can be combined in the same temperature annealing scheme, saving the computation of the temperature decrease scheme and the associated post processing stage for each line. There are two possible options to perform the correction:

- For each geometrical point randomly picked into the image (in the inner loop), each solutions is updated. This is the simplest implementation, but because of their independence, some solutions could merge and converge to the same value. This option is powerful if a good approximation of the solution is known (for instance in tracking applications).
- Only the closest solution is updated. This option avoids the possible merging of solutions.

Geometrical Constraints. The projection scheme introduced previously yield directly the Hough parameters of the line (solutions are constrained to the unit cylinder) as follow:

$$
\begin{cases}
\cos(\alpha_h) = \delta_c \cdot a \\
\sin(\alpha_h) = \delta_c \cdot b \\
d_h = \delta_c \cdot c = |c|
\end{cases}
\quad \text{with} \quad
\delta_c = \begin{cases}
1 & \text{if } c \geq 0 \\
-1 & \text{if } c < 0
\end{cases}
\tag{6}
$$

With α_h and d_h being the Hough parameters of the line. This approach allows an easy implementation of additional geometrical constrains into the inner loop. For instance, in the case of two simultaneous lines, the second line can be forced to be parallel to the first one by imposing that:

- Only the nearest solution is projected,
- The other is updated so as to be parallel (just a copy of the a & b parameter).

The same strategy can be applied for more than two parallel lines, perpendicular or with any given angle.

Although the general strategy can always be used with very good results, a suitable correction strategy for the application at hand can considerably improve the overall performance. For instance when MIN-PCS methods are used for initial features detection, it is only important that the temperature is high enough, while the choice of an initial random solution is irrelevant. The only drawback is the need of more computation to let the algorithm converge. Conversely, for tracking application, the "a priori" information available from the past images (i.e. the previous position of the features we are interested in) can be used to dramatically reduce the computation load. The initial solution is not chosen randomly, but corresponds to one of the probable solutions and the initial temperature is set to lower values compatibly with the maximum admissible distance from the past solution. If the current solution is very near to the past solution the algorithm converges very rapidly, if not the algorithm is capable of finding the new one within the admissible variation range, thus still saving computation if compared to the general detection algorithm not using any "a priori" knowledge.

5   Some Simulation Results

The generic MIN PCS algorithm has been applied to some synthetic and natural test images with different levels of noise. Fig 3 reports an example of results for images without noise and with $2\backslash\%$ of the total image or equivalently of 118% of contour points as randomly distributed noise. The same images have been processed by the classical HT and by the randomized HT (RHT) [1], [7] with a $256 \cdot 256$ accumulator arrays equal to the image resolution. Table 1 summarizes the results of a profiling analysis of HT, RHT and MIN PCS algorithms on a SUN UltraSparc WS. All results are normalized to 3.57 seconds. The lower part of the table indicates the (subjective) quality of the obtained results.

As confirmed by all our experiments, the MIN PCS approach provides the highest quality results in all noise conditions. Moreover, the time requirements are much lower than the HT for low levels of noise and comparable for higher noise levels, while yielding higher quality results. At high noise levels no results can be obtained with the HT. RHT shows its limits and fails to provide any results even for medium levels of noise. It has to be pointed out that the comparison would be much more favorable for larger image sizes. For instance, considering the same image, but at double resolution ($512 \cdot 512$ pixel instead of $256 \cdot 256$) HT memory requirements and processing time increase by a factor of 4, while the time and space complexity of our algorithm only increases with the number of contour points and number of subsystems. A more detailed analysis of memory and complexity of MIN PCS algorithms versus the HT is omitted here for brevity. Another result that demonstrates the excellent performance and robustness versus noise of the MIN PCS based algorithm is reported in Fig. 4. The image contains 500 points obtained by adding a Gaussian noise of $\sigma = 10$

Table 1. Comparison of speed (in seconds) and quality of results obtained with HT, RHT and MIN PCS algorithms fort he image of Fig. {house} at different levels of noise. (*) The noise % refers to total image points (256×256). (**) The noise % w.r.t. total contour points. [++] all information is correctly detected, [+] all information can be extracted by a simple post-processing stage, [-] information is missing, [/] no results can be obtained.

Noise % (*)	0	1	2	5	10	15
(**)	0	59	118	295	590	885
HT	1	1.16	1.29	1.77	4.79	4.76
RHT	0.08	0.13	0.14	0.46	3.30	4.01
MIN-PCS	0.06	0.49	0.93	2.19	11.2	28.0
HT	++	++	++	-	-	/
RHT	++	++	+	-	/	/
MIN-PCS	++	++	++	+	+	-

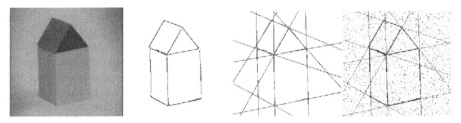

Fig. 3. From left to right: original gray scaled test image, original binary image (after basic edge detection), results of MIN PCS based line detection, results with 295% of the points as random noise (i.e. 5% of the total number of points).

Fig. 4. From left to right: MIN-PCS approach (original image: two lines whose points have been displaced by quantity with a Gaussian distribution), MIN-PCS with 250% of additional random noise (w.r.t. the points of the original image), HT with 250% of additional noise, RHT with 250% of additional noise.

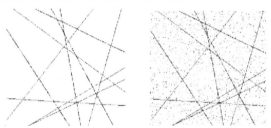

Fig. 5. Left: Synthetic image composed by 10 randomly distributed lines. Right: MIN PCS solution when 50% are the original image points and 50% are noise (randomly distributed points).

pixels to two original segments. For various noise levels, the MIN PCS approach always recovers the two original segments with 3.65 seconds of processing time. As shown in Fig. 4 (right), RHT never provides a correct result and 65% of the segments determined by the HT (in 6.25 seconds) are not correctly grouped or do not have the correct parameters.

Another example of results is reported in Fig. 5. These tests are based on images containing 10 randomly distributed segments with additional speckle noise corresponding to 50% of the line points. The MIN PCS approach always provides the correct results. Since in average each subsystem contains 3–5% of the total points, no method based on robust regression techniques can be applied.

6   Conclusion

We have presented a class of algorithms based on a general combinatorial optimization formulation able to detect lines in images. The problem is formulated as that of partitioning inconsistent linear systems into a minimum number of consistent subsystems. The linear system is obtained by the contour lines extracted by the images. A generic algorithm as well as possible variants able to include "a priori" information available in some applications have been described. A projection strategy avoiding critical convergence speed for some data distributions, short segments located far from the conventional origin of the reference system, has been also developed. The MIN PCS approach can be applied to a variety of other problems for which piecewise linear models are valuable. Since higher degree polynomials can be viewed as linear functions with respect to their coefficients, the approach can also be extended to the estimation of piecewise polynomial models with submodel of bounded degree, thus also for detecting other shapes than lines.

References

[1] P. Kultaken, L. Xu, and E. Oja, "A new curve detection method: Randomized Hough transform," Pattern Recognition Letters, vol. 11, pp. 331-338, 1995.

[2] E. Zapata, N. Guil, and J. Villalba, "A fast Hough transform for segment detection," IEEE Trans. on Image Processing, vol. 4-11, pp. 1541-1548, 1995.

[3] M. Mattavelli and E. Amaldi, "A combinatorial optimization approach to extract piecewise linear structure in nonlinear data and an application to optical flow segmentation," School of Operations Research and Industrial Engineering, Cornell University 1997.

[4] M. Mattavelli, "Motion analysis and estimation: from ill-posed discrete inverse linear problems to MPEG-2 coding," in Department of Communication Systems. Lausanne: EPFL - Swiss Federal Institute of Technology, 1997.

[5] M. Mattavelli, V. Noel, and E. Amaldi, "An efficient line detection algorithm based on a new combinatorial optimization formulation," presented at ICIP98, Chicago, 1998.

[6] P. Meer, D. Minz, and A. Rosenfeld, "Robust Regression Methods for Computer Vision: a Review," International Journal of Computer Vision - Kluwer Academic Publisher, vol. 6-1, pp. 59-70, 1991.

[7] E. Oja, H. Kalviainen, and P. Hirvonen, "Houghtool a software package for hough transform calculation," presented at 9th Scandinavian Conf. on Image Analysis, Uppsala, Sweden, 1995.

[8] P. Palmer, M. Petrou, and J. Kittler, "A Hough Transform Algorithm with a 2-D Hypothesis Testing Kernel," CVGIP, vol. 58-2, pp. 221-234, 1993.

[9] H. Stark and Y. Yang, Vector Space Projection - A numerical approach to signal and image processing, neural nets, and optics, John Wiley & Sons ed. New York, 1998.

[10] C. V. Stewart, "MINPRAN: "A New Robust Estimator for Computer Vision"," PAMI, vol. 17-10, pp. 925-938, 1995.

Koenderink Corner Points

Mads Nielsen[1], Ole Fogh Olsen[1], Michael Sig[2], and M. Sigurd[2]

[1] IT University of Copenhagen, Denmark
[2] DIKU, Copenhagen, Denmark

Abstract. Koenderink characterizes the local shape of 2D surfaces in 3D in terms of the shape index and the local curvedness. The index characterizes the local type of surface point: concave, hyperbolic, or convex. The curvedness expresses how articulated the local shape is, from flat towards very peaked. In this paper we define corner points as point on a shape of locally maximal Koenderink curvedness. These points can be detected very robustly based on integration indices. This is not the case for other natural corner points like extremal points. Umbilici can likewise be detected robustly by integral expressions, but does not correspond to intuitive corners of a shape. Furthermore, we show that Koenderink corner points do not generically coincide with other well-known shape features such as umbilici, ridges, parabolic lines, sub-parabolic lines, or extremal points. This is formalized through the co-dimension of intersection of the different structures.

Keywords: Shape features, shape evolution, differential geometry, singularity theory

1 Introduction

A shape is, in our context, a 2D surface embedded in 3D space: the boundary of a physical object. Problems like geometrical alignment (registration), interpolation, and recognition of slightly deformed shapes are far from trivial. The local properties of a shape may be used for defining local shape features such as corners, edges, cylindrical (parabolic) lines, sub-parabolic lines, etc. [10]. These may then be used for geometrical alignment [15] and interpolation [14,1] whereas their topological properties may be used for creating graph representations of shapes in turn used for recognition and indexing[12].

A visual shape is defined through measurements. That may be optical or range images of physical objects or density measurements in volumetric (medical) images. A well-founded and robust way to formalize measurements is through the concept of Gaussian scale-space theory [7,18]. In this case, the shape will generically be differentiable owned to the differentiability of the Gaussian kernel. Hence, in this paper we examine analytically parametrized shapes:

$$S : \mathbb{R}^2 \mapsto \mathbb{R}^3, \quad \begin{pmatrix} x(u,v) \\ y(u,v) \\ z(u,v) \end{pmatrix}$$

where x, y, z are analytical functions.

C. Arcelli et al. (Eds.): IWVF4, LNCS 2059, pp. 420–430, 2001.
© Springer-Verlag Berlin Heidelberg 2001

In this paper, the canonical computational example is surfaces defined as iso-intensity surfaces in a 3D volumetric (medical) image. Generically iso-surfaces of an analytical (scale-space) image are analytical surfaces. However, generically we will also see not analytical surfaces for isolated intensity values. We will in this paper also touch upon the shape properties in transitions through these isolated cases.

In next section, we will review well-known local shape features, and we will in this context define Koenderink corner points and Koenderink corner lines. In section 3 we give proofs that they are unrelated to classical shape features [10]. In section 4 we give examples on robust detection of Koenderink corner points on an artificial example and on a CT scan of a mandible.

2   Local Shape Properties

The local properties of an analytical shape is most easily accessed through its local differential structure. In the following we briefly review well-know concepts from differential geometry [4] to establish notation.

Locally (in a Monge patch) an analytical shape can be described as the orthogonal deviation from a tangent plane. Parameterizing the tangent plane (x, y) with origo in the osculation point yields the local description:

$$z(x, y) = z_{xx}\frac{x^2}{2} + z_{xy}xy + z_{yy}\frac{y^2}{2} + O((x, y)^3)$$

It is always possible to rotate the (x, y) coordinate system so that $z_{xy} = 0$ and $z_{xx} \geq z_{yy}$. Hence

$$z(x, y) = k_1 x^2 + k_2 y^2 + O((x, y)^3)$$

where k_1, k_2 are the local *principal curvatures* corresponding to the curvature in the first (x) and second (y) *principal directions*. We denote by (t_1, t_2) the two orthogonal principal directions. Moving the tangent plane along the shape, the principal directions turn. The integral curves of the principal directions are denoted the *lines of curvature*. Normally [4] the local shape is characterized through the Gaussian curvature $k_1 k_2$ and the mean curvature $\frac{k_1 + k_2}{2}$.

In the Table 1, definitions of some local shape features are given [4,3,15]. The dimension denotes the generic dimensionality of sub-manifolds having the respective properties. A negative dimension denotes the co-dimension, i.e. the number of free control parameters necessary for the property to generically appear in a point on the surface for fiducial parameter values.

Elliptical and hyperbolic points appear in areas separated by lines of parabolic points. Umbilic points generically appear as isolated points in the elliptical regions, or in co-dimension 1 in a parabolic point. These latter points are denoted planar points since the total second order structure vanishes in these points. The lines of curvature form closed curves on the surface. They start and

Table 1. Generic dimensionality of shape features.

Name	criterion	dimension
Elliptical	$k_1 k_2 > 0$	2
Hyperbolic	$k_1 k_2 < 0$	2
parabolic	$k_1 k_2 = 0$	1
Umbilic	$k_1 = k_2$	0
Planar	$k_1 = k_2 = 0$	-1
Ridge	$\frac{d}{dt_1} k_1 = 0 \text{ or } \frac{d}{dt_2} k_2 = 0$	1
sub-parabolic	$\frac{d}{dt_1} k_2 = 0 \text{ or } \frac{d}{dt_2} k_1 = 0$	1
Extremal	$\frac{d}{dt_1} k_1 = 0 \text{ and } \frac{d}{dt_2} k_2 = 0$	0

end in umbilic points. Generically three types of umbilici exist [10]: lemon, mon-star, and star. They respectively have 1, 2, and 3 lines of curvatures passing through the umbilic.

Including the 3rd order structure in the characterization, we may define ridges. At a ridge the principal curvature is extremal in its corresponding principal curvature direction. These points generically form closed curves. The curves may again be sub-divided in 8 cases according to whether it is the first or second curvature, whether it is a maximum or a minimum, and according to the sign of the curvature. The most interesting types of lines are those where the positive first principal curvature is maximal or the negative second principal curvature is minimal. These respectively form convex and concave edge (or corner-line) candidates. They have successfully been applied to non-rigid registration of medical images under the name of *crest lines* [15]. Here the ridges has also been denoted the extremal mesh. The points of intersection of ridges in the first and second principal curvature are denoted *extremal points* and form corner candidates. In the mathematical literature they have also been denoted purple fly-overs as the ridges in the first and second principal curvature have respectively been noted red and blue ridges.

The sub-parabolic lines are in some sense dual to ridges. They denote points that locally look like a surface of revolution. They are defined as the set of points where the first principal curvature is extremal in the second principal curvature direction, or vice versa. They also correspond to the points where the *focal surface* generated by an osculating circle along the lines of curvature has a parabolic point.

3   Koenderink Corner Points

The classical way of describing the local shape of a surface is through the Gaussian and mean curvature. However, the local shape is much more intuitively described in log-polar coordinates [8]:

$$\theta = \operatorname{atan} \frac{k_2}{k_1} \in [-\frac{3\pi}{4}, \frac{\pi}{4}]$$

$$c = \log(k_1^2 + k_2^2)$$

The *shape index* θ describes the local type of shape as a continuous parameter. It travels from a concave sphere through a concave cylinder, a balanced saddle, and a convex cylinder to a convex sphere as the index increases (see figure 1). The *curvedness* c describes how articulated the shape is. A large value denotes a very articulated local shape: A small sphere has a larger curvedness than a larger sphere. Unlike the mean and the Gaussian curvature, the shape index is invariant to scalings of the shape. A scaling only adds a constant value to the curvedness in all points. The shape index and curvedness, in this way forms a very natural description of the local shape [8].

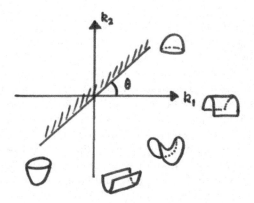

Fig. 1. The Koenderink shape index θ. Drawings must be interpreted such that the outward normal is upwards.

We make the following assertion:

Definition 1 (Koenderink corner point). *A point of locally maximal curvedness is a Koenderink corner points.*

Likewise we denote points of minimal curvedness as Koenderink flat points. In the following we will show some of the generic properties of Koenderink points and only comment on the type (flat or corner) in the cases where they have different properties. We show the generic properties by referring to the transversality theorem: it is generic that manifolds in jet-space intersect transversally [17]. This means, we can show generic properties of local shapes by showing that they intersect transversally in the space formed by the local derivatives. Since the logarithm does not influence the extremality, we analyze the simpler expression $k_1^2 + k_2^2$ in the following.

Theorem 1 (Morse property). *Koenderink corner points are generically isolated points on a surface.*

Proof. The criteria for a point being a Koenderink corner point may be written

$$z_{xx}z_{xxx} + 2z_{xxy}z_{xy} + z_{xyy}z_{yy} = 0$$
$$2z_{xy}z_{xyy} + z_{yyy}z_{yy} + z_{xxy}z_{xx} = 0$$

We can always turn the coordinate system such that z_{xy} is zero. In the case where $k_2 \neq 0$ we may rewrite the conditions as:

$$z_{xyy} = -\frac{z_{xx}z_{xxx}}{z_{yy}} \equiv \alpha z_{xxx}$$
$$z_{yyy} = -\frac{z_{xx}z_{xxy}}{z_{yy}} \equiv \alpha z_{xxy}$$

These two manifolds intersect transversally. They both have co-dimension 1 in jet-space, so generically their intersection is of co-dimension 2. Since a surface is a 2D manifold, the points are generically of dimension 0. That is isolated points.

Left is only to analyze the case of $z_{yy} = 0$. Owned to the algebraic symmetry in z_{xx} and z_{yy} we obviously may obtain the same result if instead $z_{xx} \neq 0$. Left is the case $z_{xx} = z_{yy} = z_{xy} = 0$. This happens in co-dimension 3 in jet-space. This is a planar point, and also a Koenderink flat point.

The planar points occur in co-dimension 1. Parabolic points occur where the determinant of the Hessian of z is zero. This is the case on a cone in jet-space. The planar points occur on the tip of this cone.

The singularities in Koenderink curvedness occur generically as isolated points and in co-dimension one they also occur in planar points. We have already here seen the first first example of how to prove the dimensionality of a feature. The last case of a planar point shows how to prove the coincidence of two features: We simply augment the system of equations for being a Koenderink corner point with the defining equation of the feature in mind, and solves the equations for some variables, and finally analyze the situations of any denominator being zero [9]. Following this scheme we come up with the following table of dimensionality of coincidence[13]:

	Corner	Planar
Umbilic	-2	-1
Parabolic	-1	-1
Ridge	-1	-
Extremal	-2	-
Sub-parabolic	-1	-

In all cases, for corner points not being planar points, this corresponds to what one would expect from a simple analysis of dimensionality. In the special case of a planar point, it is the matter of definition whether these points are also ridge, extremal, and/or sub-parabolic. The definition of all these types refers to the principal directions. In umbilici (of which the planar points is a special example) this coordinate system breaks down. The structure of ridges round umbilici is very elegantly analyzed by Bruce, Porteous, and Giblin [3,5,10].

The conclusion is: The Koenderink corners are unrelated to the other structures. In umbilici in general, the definition of ridge, extremal, and sub-parabolic points break down. In the following, we will analyze the structure of the corner points on a changing shape (including topology change), and look at the global constraints through the index theory. In turn, we are going to use the index theory also for robustly detecting the Koenderink corner points.

3.1 Transitions

The singularities in curvedness appear in three different types on the surface: minima, saddles, and maxima. The extrema correspond respectively to maximally flat and maximally curved points. Traversing through a generic saddle in one fiducial direction, it is a minimum in curvedness while it is a maximum in curvedness in the orthogonal direction. The curvedness is defined as a function value at any location of the surface. Like a function $\mathbb{R}^2 \mapsto \mathbb{R}$ generically will exhibit structures like ridges, iso-level lines, watersheds, etc., so will the curvedness defined on a shape.

In co-dimension 1, the Koenderink curvedness generically exhibit the same structure as a general analytical function:

$$c(x, y) = x$$
$$c(x, y) = \pm x^2 \pm y^2$$
$$c(x, y) = x^3 + tx \pm y^2$$

Where t is a general control parameter. The structure is defined up to a coordinate transformation on the shape [17,3]. This corresponds exactly to the structure of general smooth functions. Thereby is not said that the curvedness generically will reveal the same structure as an general function in any co-dimension.

One interesting case is still the planar point, that occur in co-dimension 1. However, in a planar point, the curvedness does not change structure at all. All the way through transition of a convex point becoming planar and then concave, it will be a minimum in curvedness.

Another interesting transition happening in co-dimension 1, is the change of topology of an iso-surface in an volumetric image. A surface change topology by either having a new hole in the shape or by a merge with a nearby shape. In the case of iso-surfaces of volumetric images this happens when we vary the intensity defining the iso-intensity surface through the value of a saddle. Since the curvedness is independent of a sign-change of the curvatures, the curvedness of the boundary of a solid shape and the curvedness of the boundary of the complement of the solid shape are identical. That means that the situation of creation of a hole and the merging of two surface create the same transitions. The generic transition is that two maxima of curvedness (one on each shape) meet in a non-differentiable surface points from where two saddles and two maxima appear (see Figure2).

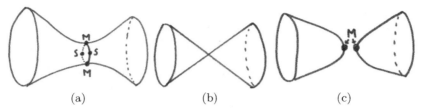

Fig. 2. The generic change of topology when the level of an iso-intensity surface in a volumetric image is varied. The shape may be either the interior or the exterior of the time glass shape. This corresponds respectively to the splitting of a shape going from (a) through (b) to (c) and to the creation of a hole going from (c) through (b) to (a). **M** indicate a point of maximal curvedness. **S** indicate of saddle point in curvedness.

3.2 Global Structure

The curvedness is a genuine function defined all over the shape, $c : S \mapsto \mathbb{R}$. This means that it must satisfy standard topological constraints formalized through the index theorem [6]. First we define the index of function to be the number of turns of the gradient field around the point in a sufficiently small loop. If the point is a regular point the index is zero. If the point is a maximum or minimum the index is plus one. If the point is a saddle the index is minus one (see Figure 3). The index theorem reads:

Theorem 2 (Surface index). *I, the point index summed over all points of surface is related to the Euler number E of the surface so that $I = 2(1 - E)$*

The Euler number is, in intuitive terms, the number of holes in the shape.

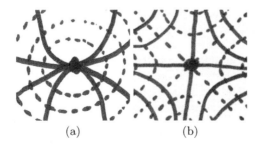

Fig. 3. Solid lines indicate lines of steepest descend and dotted lines indicate iso-level curves round an extremum (a) and a saddle (b). Following the rotation of the lines round the singularity yields one turn in positive direction (a) and one turn in negative direction (b) respectively. That is, the local surface index round an extremum is 1, while it is -1 round a saddle.

This yields a global constraint on the number of the different types of singularities in the curvedness on a shape. As long as the shape does not change topology the number of extrema minus the number of saddles is constant. Since transitions where singularities interact are local, they must also be index neutral, as the transition through a planar point mentioned above. However, the case where the topology changes also changes the shape index. When a hole is created the Euler number is increased by one, and thereby the number of saddles must increase by two compared to the number of extrema. This is the case in the above mentioned transition. Two saddles $I = 2$ is transformed into two saddles and two maxima $I = 0$ when the Euler number is increased by one (a new hole is created). Summing the index over several shapes yields the intuitive interpretation as the index being twice the number of shapes minus twice the number of holes. In the case where the change of topology corresponds to a merging of two shapes the total index is decreased by two corresponding to one less shape with the same total number of holes.

3.3   Shape Edges

So far we have looked at points of singular curvedness. Ridges have been proposed as edge lines. Especially the subset of ridge lines where the absolute largest principal curvature is absolute maximal in its principal direction have been denoted *crest lines* [15] and been used as semi landmarks [2] for non-rigid registration (geometrical alignment) of medical images [15,16].

Using the curvedness for defining edges (or ridges) opens for an alternative definition. Since the curvedness is a field defined over the surface, ridges in this measure may alternatively be defined as the watersheds. The watersheds form closed curves along steepest descend lines from Koenderink corners towards saddles of curvedness. In each saddle two ascending and two descending steepest descend lines start. A subset of the ascending lines form the watersheds. The watersheds can not be defined locally. It is a semi global property whether a steepest descend line ends in a saddle or not. The watersheds are our favorite definition of shape edges for several reasons. They form closed curves. Crest-lines may end in umbilici as the ridge through a lemon (as an example) changes from the first to the second principal curvature [3]. Secondly, the edges of a shape are not according to our intuition locally definable. Finally any number of edges may generically start from a given corner. The crest-lines generically meet one, two, or three in umbilici and two or four in extremal points. These are unnatural constraints on their topology.

An alternative local definition of shape edges is similar to the definition of crest lines: lines where the curvedness is locally maximal in a principal direction. This definition allows for local analysis of the generic coincidence with other shape features. Similar as for the corner points we may conclude that these lines do not coincide with the above mentioned shape features, the following table of dimensionality of coincidence[13] summarizes:

	Koenderink edge
Umbilic	-1
planar	-1
Parabolic	0
Ridge	0
Extremal	-1
Sub-parabolic	0

4   Detection of Shape Characteristics

In our setting, we define shapes as iso-intensity surfaces in volumetric images. We may find the local geometric structure of the surface by applying the implicit function theorem [15]. From here it, at first glance, seems straight forward to find the ridges as zero crossings of an appropriate non-linear combination of image derivative. However it turns out that the appropriate expressions are not invariants with respect to change of the coordinate system. This problem is a serious problem and not possible to avoid. In the following we give an intuitive explanation of the problem, as identified by Thirion et al.. We argue that this problem is unavoidable, and show that for the Koenderink corner points, a robust and consistent method exist.

The problem of detecting ridges origins in the fact that the lines of curvature are not oriented. An odd order derivative along the line thereby has an arbitrary sign. The extremal mesh and the crest lines are defined in terms of a first order derivative along the lines of curvature. Spurious lines will be detected where we arbitrarily change direction along the lines of curvature. A shape with topology different from a torus can not be globally parametrized. We see that immediately from the index theorem. As long as the sum of index over the shape is different from zero, at least one singularity must be present. This is normally denoted the Hairy Ball Theorem: a ball of tangential hair must have at least one whirl. Complex algorithms may overcome these parameterization problems by looking at stability of lines when coordinate systems are changed or by making parameterizations of the surface with only one singularity, in a point in advance known not to be of interest. However, for the Koenderink corner points a much easier alternative exists:

Since the Koenderink corner points appear as extrema of a field, they may easily be extracted by use of the index theorem. Similar has been done for umbilici [11] and for general singularities [6]. The index is evaluated as an integral of an analytical function, and thereby by definition a robust measure. The computation is well posed. In the Figures 4 and 5 examples of Koenderink corners and extremal points are shown. For sake of the comparison they have here both been extracted by use of zero crossings.

The algorithm for using the index computation may be outlined as follows [11]:
– compute the 3D gradient of the curvedness
– project this onto the surface of interest

– divede the surface into cells
– compute the index by adding orientation differences round every cell [6]

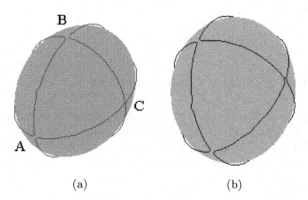

(a)   (b)

Fig. 4. (a) are the Koenderink corner lines ($\partial_{t_1} c = 0$ or $\partial_{t_2} c = 0$) and (b) are the ridges lines ($\partial_{t_1} k_1 = 0$ or $\partial_{t_2} k_2 = 0$) on the same ellipsoidal. This ellipsoidal shape is non-generic, and in this case the the Koenderink corner lines and the ridge lines coincide. Computations are based on intersection of the surface with zero-loci of the product of differential expressions for each of the two types of lines respectively [15].

Fig. 5. Koenderink edge lines (black) and ridges (gray) on the same mandible in three different projections using same algorithm as in the above figure. The lines do not generally coincide as this is a generic shape. However

5   Summary

We have seen that the Koenderink corner points are simple geometrical local features of a surface. They intuitively correspond well to our notion of shape corners. In practice, they are often very close to the extremal points [15]. The evolution of Koenderink corner points under deformation of the surface is simple, it corresponds to what we know from general functions. They can be detected using the index theorem (we have not shown this in practice here, but refer to Sander and Zucker [11]). We expect the extrema of Koenderink's curvedness to be able to play a significant role as shape landmarks in biomedical applications.

References

[1] Per R. Andresen and Mads Nielsen, *Non-rigid registration by geometry-constrined diffusion*, Medical Image Analysis (2000).

[2] F.L. Bookstein, *Morphometric tools for landmark data: Geometry and biology*, Cambridge University Press, 1991.

[3] J. W. Bruce and P. J. Giblin, *Curves and singularities*, Cambridge University Press, 1992.

[4] M. P. Do Camo, *Differential geometry of curves and surfaces*, Prentice Hall, 1976.

[5] P.J. Giblin and M.G. Soares, *On the geometry of a surface and its singular profiles*, IVC **6** (1988), 225–234.

[6] S.N. Kalitzin, B.M. ter Haar Romeny, A.H. Salden, P.F.M. Nacken, and M.A. Viergever, *Topological numbers and singularities in scalar images: Scale-space evolution properties*, JMIV **9** (1998), no. 3, 253–269.

[7] J.J. Koenderink, *The structure of images*, BioCyber **50** (1984), 363–370.

[8] J.J. Koenderink, *Solid shape*, MIT Press, 1990.

[9] Ole F Olsen, *Ph.d. thesis: Generic image structure*, DIKU, Department of Computer Science, 2000.

[10] Ian R. Porteous, *Geometric differentiation*, Cambridge University Press, 1994.

[11] P.T. Sander, *Generic curvature features from 3-d images*, SMC **19** (1989), no. 6, 1623–1636.

[12] K. Siddiqi, A. Shokoufandeh, S.J. Dickinson, and S.W. Zucker, *Shock graphs and shape matching*, IJCV **35** (1999), no. 1, 13–32.

[13] Michael Sig and Mikkel Sigurd, *Koenderink corner points*, DIKU, Depertment of Computer Science, University of Copenhagen, 1998.

[14] G. Subsol, J.P. Thirion, and N. Ayache, *Non-rigid registration for building 3d anatomical atlases*, ICPR94, 1994, pp. A:576–578.

[15] J.P. Thirion, *Extremal points: Definition and application to 3d image registration*, CVPR94, 1994, pp. 587–592.

[16] — *New feature points based on geometric invariants for 3d image registration*, IJCV **18** (1996), no. 2, 121–137.

[17] Rene Thom, *Structural stability and morphogenesis: an outline of a general theory of models*, Perseous, 1989.

[18] A.P. Witkin, *Scale-space filtering*, IJCAI83, 1983, pp. 1019–1022.

Dynamic Models for Wavelet Representations of Shape

Fernando Pérez Nava[1] and Antonio Falcón Martel[2]

[1]Dep. de Estadística, Inv. Op. y Computación.
Universidad de La Laguna. Tenerife. Spain
fdoperez@ull.es
[2]Dep. de Informática y Sistemas,
Universidad de Las Palmas de Gran Canaria.Gran Canaria.Spain
afalcon@dis.ulpgc.es

Abstract. In this paper, a dynamic model for contours using wavelets is presented. First it is shown how to construct probabilistic shape priors for modeling contour deformation using wavelets. Then a dynamic model for shape evolution in time is presented. This allows this formulation to be applied to the problem of tracking a contour using the stochastic model to predict contour location and appearance in successive image frames. Computational results for two real image problems are given for the Condensation (Conditional Density Propagation) tracking algorithm. It is shown that this formulation successfully tracks the objects in the image sequences.

1 Introduction

Tracking is a topic of considerable interest in computer vision due to its large number of applications to autonomous systems [1], object grasping [2] or augmentation of computer's user interface [3]. In this work we will assume that an object can be tracked by its shape. Therefore two important problems have to be addressed: shape representation and shape dynamics.

Recently, a new multiscale technique for shape representation has been developed based on wavelets [4],[5] and multiwavelets [6]. There are a number of salient features in wavelet transforms that make wavelet-domain statistical contour processing attractive:

☐ Locality: Each wavelet coefficient represents the signal content localized in spatial location and frequency

☐ Multiresolution: The wavelet transform analyzes the signal at a nested set of scales

☐ Energy Compaction: A wavelet coefficient is large only if singularities are present within the support of the wavelet basis. Because most of the wavelet coefficients tend to be small, we need to model only a small number of coefficients. This is of particular importance in real time applications.

☐ Decorrelation: The wavelet transform of real world signals tend to be approximately decorrelated.

Given a wavelet representation of shape it is desirable the definition of dynamic models since they greatly improve tracking algorithms by establishing a prior for possible motions. This dynamical prior distribution applies between pairs of

C. Arcelli et al. (Eds.): IWVF4, LNCS 2059, pp. 431–439, 2001.
© Springer-Verlag Berlin Heidelberg 2001

successive frames. In this work we will show how to integrate a probabilistic priors for shape deformation based on wavelets in a dynamic model that can be used for tracking shape in an image sequence. In particular we will use the Condensation [7] tracker to present several computational results of this model.

This paper is divided in five parts: in Section 2 the wavelet based probabilistic shape model is presented and its relation with Besov spaces is established. In Section 3 we introduce the dynamic model, then in Section 4 we show several tracking applications of this formulation both in an indoor and outdoor scene and finally in Section 5 we present the conclusions of this work.

2 Probabilistic Priors for Wavelet Shape Representation

In this section probabilistic priors for wavelet representations of shape will be introduced and its properties will be discused, specially a parametric specification for contour deformation.

2.1 Wavelet Shape Representation

A wavelet basis uses translations and dilations of a scaling function ϕ and a wavelet function φ. If translations and dilations of both functions are orthogonal a 1-D function f can be expressed as:

$$f(x) = \sum_{k \in Z} c_{j_0,k} 2^{\frac{j_0}{2}} \phi(2^{j_0} x - k) + \sum_{j=j_0}^{\infty} \sum_{k \in Z} d_{j,k} 2^{\frac{j}{2}} \varphi(2^j x - k) \tag{1}$$

Let then $r(s) = (x(s), y(s))$ be a discrete parametrized closed planar curve that represents the shape of an object of interest. If the wavelet transform is applied independently to each of the $x(s)$, $y(s)$ functions, we can describe the planar curve in terms of a decomposition of $r(s)$:

$$r(s) = \sum_{k \in Z} c_{j_0,k} 2^{\frac{j_0}{2}} \phi(2^{j_0} s - k) + \sum_{j=j_0}^{\infty} \sum_{k \in Z} d_{j,k} 2^{\frac{j}{2}} \varphi(2^j s - k) \tag{2}$$

$$c_{j,k} = \begin{pmatrix} c_{j,k;x} \\ c_{j,k;y} \end{pmatrix}, d_{j,k} = \begin{pmatrix} d_{j,k;x} \\ d_{j,k;y} \end{pmatrix}. \tag{3}$$

where subindex x and y represent coordinate function pertenence.

2.2   Wavelet Based Probabilistic Modeling of Curve Deformation

The simplest wavelet transform statistical models [8] are obtained by assuming that the coefficients are independent. Under the independence assumption, modelling reduces to simply specifying the marginal distribution of each wavelet coefficient.
Wavelet coefficients are generally modeled using the generalized gaussian distribution, in this work the usual gaussian distribution will be used as an approximation

For the tractability of the model, all coefficients at each scale are assumed to be independent and identically distributed. That is:

$$d_{j,k} = \begin{pmatrix} d_{j,k,x} \\ d_{j,k,y} \end{pmatrix} \sim N_2(\bar{d}_{j,k}, \sigma_{j,k}^2 I) \tag{4}$$

where I denotes the identity matrix.

Assuming a exponential decay of the variances, the final model is:

$$d_{j,k} \sim N_2(\bar{d}_{j,k}, 2^{-2j\beta} \sigma_D^2 I) \tag{5}$$

In order to complete the definition of the model we have to specify the distribution for the coefficient associated with the scaling function $c_{0,0.}$ This coefficient is associated with a rigid traslation of shape and we will assume that it is normally distributed and independent of the non-translation components $d_{j,k.}$

$$c_{0,0} = \begin{pmatrix} c_{0,0,x} \\ c_{0,0,y} \end{pmatrix} \sim N_2(\bar{c}_{0,0}, \sigma_C^2 I) \tag{6}$$

We will use this distributions to model smooth changes of shape between frames. A justification for the proposed model comes from the following theorem that is adapted from [8] to the one dimensional case:

Theorem

Let $f(x)$ a real function where x is a real variable. Let it be decomposed in wavelet coefficients and suppose each coefficient is independently and identically distributed as:

$$d_{j,k} \sim N(0, \sigma_j^2) \text{ with } ,\sigma_j = 2^{-j\beta} \sigma_0 \tag{7}$$

with $\beta > 0$ and $\sigma_0 > 0$ then, for $0 < p,q < \infty$, the realizations of the model are almost surely in the Besov Space $B_q^\alpha(L_p(I))$ if and only if $\beta > \alpha + 1/2$.

Besov spaces are smoothness spaces: roughly speaking, the parameter α represents the number of well behaved derivatives of f.

With the above assumptions we can then define a prior probabilistic shape model for curve deformation as:

$$p(X) \propto \exp\left(-\frac{1}{2}(X - \overline{X})^T \, \Sigma^{-1} \, (X - \overline{X})\right),$$

$$X = (c_{0,0,x}, d_{0,0,x}, ..., d_{j,k,x}, ... d_{J-1,2^{J-1}-1,x},$$

$$, c_{0,0,y}, d_{0,0,y}, ..., d_{j,k,y}, ... d_{J-1,2^{J-1}-1,y})$$

$$j = 0...J-1, \; k = 0..2^j - 1$$

(8)

where $n = 2^J$ is the number of points in the discretized curve, X is a vector of $2n$ wavelet coefficients and Σ is a diagonal matrix with the variances defined in (5),(6).

2.3 Parameter Estimation

We will show two methods for estimating the parameters in the model, the first one is based in the maximum likelihood equations and the second one in the mean square displacement.

2.3.1 Maximum Likelihood Estimation (MLE)

Given a set of samples for the wavelet descriptors $\{X_1, X_2, ..., X_m\}$ The equations for the MLE are:

$$\hat{\overline{X}} = \overline{X} \qquad\qquad \hat{\sigma}_T^2 = \frac{s_{00} + s_{nn}}{2}$$

(9)

$$\frac{\displaystyle\sum_{i=1}^{\log_2 n} 2^{2(i-1)\hat{\beta}}(i-1)r_i}{\displaystyle\sum_{i=1}^{\log_2 n} 2^{2(i-1)\hat{\beta}} r_i} = \frac{n(\log_2 n - 2) + 2}{(n-1)} \qquad \hat{\sigma}_D^2 = \frac{\displaystyle\sum_{i=1}^{\log_2 n} 2^{2(i-1)\hat{\beta}} r_i}{2(n-1)}$$

where :

$$\overline{X} = \frac{1}{m}\sum_{i=1}^{m} X_i, \; S = (s_{ij}) = \frac{1}{m}\sum_{i=1}^{m}(X_i - \overline{X})(X_i - \overline{X})^T, \; r_i = \sum_{j=2^{i-1}}^{2^i - 1} s_{jj} + s_{j+n\,j+n}$$

(10)

All parameters can be calculated easily except β. Its equation must be solved numerically.

2.3.2 Estimation by Mean Square Displacement

In case we have an estimate of β and the mean square deformation from the reference shape we can use the following result:

Proposition
Let a curve be described as **(8)** with arbitrary Σ then the mean square displacement along the curve is given by:

$$\overline{\rho}^2 = \frac{\text{Trace}(\Sigma)}{n} \tag{11}$$

if Σ is defined as **(8)** then:

$$\overline{\rho}^2 = \overline{\rho}_T^2 + \overline{\rho}_D^2, \quad \overline{\rho}_T^2 = \frac{2}{n}\sigma_T^2, \quad \overline{\rho}_D^2 = \begin{cases} \dfrac{2}{n}\dfrac{n^{-2\beta+1}-1}{2^{-2\beta+1}-1}\sigma_D^2 & \beta \neq 1/2 \\ \dfrac{2}{n}\log_2 n\sigma_D^2 & \beta \neq 1/2 \end{cases} \tag{12}$$

Therefore σ_T^2 and σ_D^2 can be estimated based on β and the mean square displacement due to translation and non-translation mean displacement $\overline{\rho}_T^2$ and $\overline{p}_D^2$ respectively.

In the following figure we can see a reference shape (in discontinuous line) with some realizations of the probabilistic model for various values of parameter β. As expected, when the parameter increases smoother deformations arise. Around the figure we can see in light grey a 99% confidence interval for the points in the curve.

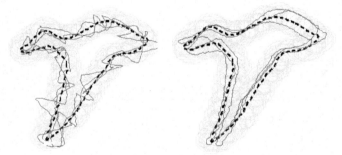

Fig. 1. Realizations of the probabilistic model. Parameter values are $\beta=0$ (no deformation smoothing) for the left image and $\beta=1.6$. in the right image.

3 Wavelet Dynamic Modeling

In order describe shape motion a second order autoregressive AR(2) process in shape space will be used

$$X(t_k) - \overline{X} = A_2(X(t_{k-1}) - \overline{X}) + A_1(X(t_{k-2}) - \overline{X}) + \Sigma^{1/2}w_k, \tag{13}$$
$$w_k \sim N_{2n}(0, I)$$

where $\overline{X}$ represents a mean shape and Σ the noise covariance.

Therefore motion is decomposed as a deterministic drift plus a diffusion process that is assured to be smooth using the above derivations.

3.1   Mean Displacement Parameter Determination

In order to make the model usable it is necessary to show how to determine the parameters A_2, A_1, Σ. We first decompose motion in several orthogonal linear subspaces P_i determined by their projection matrix P_i. Typically these subspaces are translation, rotation and deformation (euclidean similarities) or translation, affine change and deformation (planar affine motion).

Therefore we can write:

$$X(t_k) - \overline{X} = \sum_i P_i(X(t_k) - \overline{X}) \tag{14}$$

and we will model dynamics into each subspace:

$$A_1 = \sum_i a_i^1 P_i, \quad A_2 = \sum_i a_i^2 P_i, \quad \Sigma^{1/2} = \sum_i b_i P_i, \quad a_i^1, a_i^2, b_i \in \mathrm{R} \tag{15}$$

and use the following theorems:

Theorem
Let contour dynamics be given by (13) and (15) and suppose that a steady state distribution exists. Then the distribution is normal with mean $\overline{X}$ and its covariance matrix C_∞ verifies:

$$\mathrm{Trace}(C_\infty) = \sum_i \mathrm{Trace}(C_{\infty i}) \tag{16}$$

Where $C_{\infty i}$ is the covariance of the steady-state into subspace P_i.

Theorem
Let contour dynamics be given by (13) and (15) and suppose that a steady state distribution exists into subspace P_i. Then its distribution is normal with mean $P_i \overline{X}$ and its covariance matrix $C_{\infty i}$ verifies:

$$\mathrm{Trace}(C_{\infty i}) = \frac{(b_i)^2(1-a_i^2)}{(a_i^2+1)(a_i^2-1+a_i^1)(a_i^2-1-a_i^1)}\mathrm{Trace}(P_i \Sigma) \tag{17}$$

Corollary
Let contour dynamics be given by (13) and (15) and suppose that a steady state distribution exists into subspace P_i. To obtain a mean displacement $\overline{\rho}_i$ we must set b_i to:

$$b_i = \sqrt{n}\,\overline{\rho}_i \sqrt{\left(1 - \left(a_i^2\right)^2 - \left(a_i^1\right)^2 - \frac{2a_i^2\left(a_i^1\right)^2}{1 - a_i^2}\right) \frac{1}{\mathrm{Trace}(P_i\Sigma)}} \qquad (18)$$

This leads us to determine the parameters associated with the random noise if we can estimate the deterministic motion and the mean square displacement of shape.

In case no steady-state distribution exists we can use the following theorem:

Theorem
Let contour dynamics be given by (13) and (15) and suppose that no steady state distribution exists into subspace P_i. Then the mean displacement $\overline{\rho}_i\,(k)$ into subspace P_i at time t_k verifies:

$$\overline{\rho}_i(k) \approx \sqrt{\frac{n}{3}\mathrm{Trace}(P_i\Sigma)}\; b_i\; k^{3/2} \qquad (19)$$

3.2 Maximum Likelihood Parameter Determination

Parameters can also be estimated by maximum likelihood leading to a set of equations similar to (9), (10)

$$\hat{\overline{X}} = \overline{X} \qquad\qquad \hat{\sigma}_T^2 = \frac{c_{00} + c_{nn}}{2}$$

$$\frac{\displaystyle\sum_{i=1}^{\log_2 n} 2^{2(i-1)\hat{\beta}}\,(i-1)\,p_i}{\displaystyle\sum_{i=1}^{\log_2 n} 2^{2(i-1)\hat{\beta}}\,p_i} = \frac{n(\log_2 n - 2) + 2}{(n-1)} \qquad \hat{\sigma}_D^2 = \frac{\displaystyle\sum_{i=1}^{\log_2 n} 2^{2(i-1)\hat{\beta}}\,p_i}{2(n-1)} \qquad (20)$$

where:

$$R_i = \sum_{k=3}^{m} X_i, \quad R_{ij} = \sum_{k=3}^{m} X_{k-i}X_{k-j}^{\,T}, \quad R'_{ij} = R_{ij} - \frac{1}{m-2}R_i R_j^T$$

$$D = \frac{1}{m-2}\left(R_0 - \hat{A}_2 R_2 - \hat{A}_1 R_1\right)$$

$$C = \frac{1}{m-2}\left(R_{00} - \hat{A}_2 R_{20} - \hat{A}_1 R_{10} - DR_0^T\right) \quad p_i = \sum_{j=2^{i-1}}^{2^i-1} c_{jj} + c_{j+n\; j+n} \qquad (21)$$

As in (9) all equations are easily computed except equation in β that has to solved numerically.

4   Applications to Contour Tracking

A set of experiments have been carried out to test the validity of the approach both in an indoor and outdoor scene. To track shape over different frames the Condensation algorithm has been employed. It uses a set of samples S from the pdf of the contour distribution. Then the algorithm uses the dynamic model to propagate the pdf over time using the samples, obtaining a prediction of shape appearance and position. Using measurements in the image the pdf tends to peak in the vicinity of observations leading to distributions for shape at succesive time steps.

The number of wavelet coefficients used has been 16, the wavelet function used is Daubechies LA(8) and the number of elements in set S has been 250.

In the first example (Fig. 2) motion is modelled as traslation plus deformation with:

Traslation subspace $\quad P_1: \quad a_1^1 = 2, \quad a_1^2 = -1, \quad \overline{\rho} = 3$ (No steady-state)

Deformation subspace $P_2: a_2^1 = 0, \quad a_2^2 = 0, \quad \overline{\rho} = 0.5 \quad \beta = 2.25$

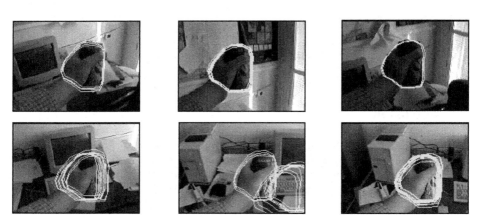

Fig. 2. An indoor scene. Frames are numbered 1-6 from top to bottom and left to right

To visualize the pdf a set of 20 curves has been sampled from the contour distribution in all frames. In frame 5 we can see how a false matching appears (the distribution becomes multimodal), however the algorithm recovers as the hands goes on moving as can be seen in frame 6.

In the second case (Fig. 3) the background is cluttered with changes between light and shadows and there are several moving elements interacting with the person being tracked. Parameters in this case have been:

Traslation subspace $\quad P_1: \quad a_1^1 = 2, \quad a_1^2 = -1, \quad \overline{\rho} = 1$ (No steady-state)

Deformation subspace $P_2: a_2^1 = 0, \quad a_2^2 = 0, \quad \overline{\rho} = 0.1$

and the smoothness parameter for frame to frame deformation has been $\beta = 2.25$

As we can see the head of the girl is in general successfully tracked. In frame 5 the dynamic model fails because the girl suddenly stops leading to the curves being more disperse around the head. However the condensation algorithm recovers as the girl goes on moving as can be seen in frame 6.

Fig. 3. An outdoor scene. Frames are numbered 1-6 from top to bottom and left to right

5 Conclusions

In this work, a dynamic autoregressive model for shape change in wavelet space has been presented. It is shown how to use this formulation to predict contour location and appearance in successive image frames. These components are integrated in the Condensation tracking algorithm and computational results show that this formulation successfully tracks the objects in the image sequences.

References

1. Reid I. and Murray D., Active tracking of foveated feature clusters using affine structure. International Journal of Computer Vision, 18(1):41--60, April 1996.
2. Taylor M. *Visually guided grasping* PhD thesis, Dept of Engeniering Science. Oxford University, 1995
3. P. Wellner. Interacting with paper on the DigitalDesk. Communications of the ACM, 36 (7):86--96, July 1993.
4. Chuang G.and Kuo C., "Wavelet descriptor of planar curves: Theory and applications", IEEE Trans. on Image Processing, 1(5), pp. 56-70, 1996
5. Chung K., "The generalized uniqueness wavelet descriptor for planar closed curves", IEEE Trans. on Image Processing 9(5) pp 834-845, 2000
6. F. Pérez and A. Falcón, "Planar shape representation based on multiwavelets", *Proceedings of the European Signal Processing Conference*, 2000
7. Blake A. and Isard M. *Active Contours*, Springer, London, 1998.
8. Choi H. and Baraniuk R., "Wavelet Statistical Models and Besov Spaces" , Proceedings of SPIE Technical Conference on Wavelet Applications in Signal Processing VII 1999.

Straightening and Partitioning Shapes

Paul L. Rosin

Department of Computer Science, Cardiff University, UK.
Paul.Rosin@cs.cf.ac.uk

Abstract. A method for partitioning shapes is described based on a global convexity measure. Its advantages are that its global nature makes it robust to noise, and apart from the number of partitioning cuts no parameters are required. In order to ensure that the method operates correctly on bent or undulating shapes a process is developed that identifies the underlying bending and removes it, straightening out the shape. Results are shown on a large range of shapes.

1 Introduction

Shape obviously plays an important role in biological vision. However, the task of shape perception is inherently complex, as demonstrated by the slow developmental process of learning undertaken by children to recognise and use shape. At first, they can only make topological discriminations. This is then followed by rectilinear versus curvilinear distinctions, then later by angle and dimension discrimination, and then continuing to more complex forms, etc. [11].

The application of shape in computer vision has been limited to date by the difficulties in its computation. For instance, in the field of content based image retrieval, simple methods based on global colour distributions have been reasonably effective [19]. However, although attempts have been made to incorporate shape, they are still relatively crude [6,14].

One approach is to simplify the problem of analysing a shape by breaking it into several simpler shapes. Of course, there is a difficulty in that the process of partitioning will require some shape analysis. However, to avoid a chicken and egg problem, low level rules based on limited aspects of shape understanding can be used for the segmentation.

There are numerous partitioning algorithms in the computer vision literature. Many are based on locating significant concavities in the boundary [16,18]. It has been shown that humans also partition forms based on, or at least incorporating, such information [4,8,10]. While this lends credence to such an approach, the computational algorithms employed to detect the concavities generally depend on measuring the curvature of the shape's boundary. Unfortunately curvature estimates are sensitive to noise. Although the noise can be reduced or eliminated by smoothing for instance, it is not straightforward to determine the appropriate degree of filtering. In addition, purely local boundary-based measures do not

C. Arcelli et al. (Eds.): IWVF4, LNCS 2059, pp. 440–450, 2001.
© Springer-Verlag Berlin Heidelberg 2001

capture the important, more global aspects, of shape. An alternative approach that can incorporate more global information operates by analysing the skeleton of the shape [1,3]. Nevertheless, it remains sensitive to local detail and tends to be error prone since even small amounts of noise can introduce substantial variations into the skeleton. Thus substantial post-processing generally needs to be applied in an attempt to correct the fragmentation of the skeleton.

2   Convexity-Based Partitioning

To overcome the unreliability of the curvature and skeleton based methods an approach to partitioning was developed based on global aspects of the shape [12]. The criterion for segmentation was convexity. Although convexity had been used in the past, previous algorithms still required various parameters to tune their performance, generally to perform the appropriate amount of noise suppression [7,15]. In contrast, the only parameter in Rosin's formulation was the number of required subparts. Moreover, an approach for automatically determining this number was also suggested.

The convexity of a shape was measured as the ratio of its area to the the area of its convex hull. The total convexity of a partitioned shape was defined as the sum of the individual convexity values of the subparts, each weighted by their area relative to the overall shape's area. Thus the convexity and combined subpart convexity values range from zero to one. Partitioning was performed by selecting the decomposition maximising convexity. As with most partitioning schemes, straight lines were used to cut the shape into subparts, and the cuts were constrained to lie within the shape.

The convexity measure is robust since small perturbations of the shape boundary only result in small variations in the convex hull. Thus noise has a minor effect on the areas of the shape and its convex hull, and therefore on the convexity measure itself. Although the measure is based on global properties of the shape it produces good partitions, often locating the cuts at significant curvature extrema even though no curvature computation is necessary.

3   Shape Straightening

Despite its general success, there are also instances in which the convexity based scheme fails [12]. In figure 1 the effect of the ideal cut (shown dotted) would be to split the crab shape into the inner convex part and the outer non-convex part. The latter would score very poorly according to convexity, and so the crab would actually receive a better score without performing any partitioning. This is counterintuitive since we would expect that partitioning should always lead to simplification. In general, we make the observation that many bent objects will be given a low convexity rating even though human perception might suggest that they are suitable representations of simple, single parts.

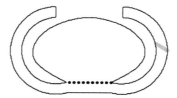

Fig. 1. Convexity is not always appropriate for partioning as demonstrated in the crab figure. Instead of the ideal cut (shown dotted) the gray cut is selected.

In this section we describe a solution to overcome the deficiency of convexity in this context. Since it is the bending of the shape that creates the problem, the bending is removed. Conceptually, if the outer portion of the crab were straightened it would receive a high convexity score since there would be no concavities. Of course, the straightening process should not eliminate all concavities *per se* since these are required to enable the convexity measure to discriminate good and bad parts. Instead, the most basic underlying bending should be removed while leaving any further boundary details unchanged.

Some work in computer vision and computer graphics has looked at multi-scale analysis and editing of shapes. For instance, Rosin and Venkatesh [13] smoothed the Fourier descriptors derived from a curve in order to find "natural" scales. The underlying shape was then completely removed by modifying the lower descriptors such that on reconstruction just the fine detail occurring at higher natural scales was retained and superimposed onto a circle. Another approach was taken by Finkelstein and Salesin [17] who performed wavelet decompositions of curves and then replaced the lower scale wavelets extracted from one curve with those of another. Although both methods enabled the high resolution detail to be kept while the underlying shape was modified there were several limitations. The Fourier based approach operates globally, and therefore assumes uniform detail spatially distributed over the curve, which is not necessarily correct. Wavelets have the advantage that they cope with spatial localisation, but Finkelstein and Salesin did not provide any means for automatically selecting which wavelets to retain such that they correspond to significant curve features.

The approach taken in this paper is to determine the appropriate straightening of a shape by first finding its medial axis. Its sensitivity to noise can be overcome since the axis is only required to describe of the shape at a very coarse level, and so heavy smoothing can be applied to eliminate all branches as shown in figure 2. More precisely, the boundary is repeatedly smoothed, and at each step the branches in the resulting medial axis are identified by checking for vertex pixels. If no vertex pixels are found the smoothing terminates and the final boundary and axis is returned. Our current implementation uses: Gaussian blurring of the boundary, Zhang and Suen's thinning algorithm to extract the medial axis [20], and vertices are identified by checking at each axis pixel for three of more black/white or white/black transitions while scanning in rotation around its eight neighbours.

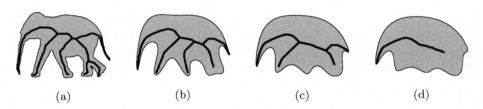

(a)   (b)   (c)   (d)

Fig. 2. Repeated boundary smoothing applied until all medial axis branches are eliminated.

Once the axis is found it is used to straighten the shape. First each boundary point needs to be assigned to a point on the axis. Conceptually this can be performed by regenerating the shape by growing the axis. In practise we just run a distance transform [2] taking the axis as the feature set. In addition to propagating the distances the originating co-ordinates of the closest feature are also propagated. These then provide the corresponding axis points for each boundary point. At this stage the smoothed boundary points are still used (after integer quantisation) rather than the original boundary set.

Second, a local co-ordinate frame is determined for each boundary point. The frame is centred at the corresponding axis point and orientated to align with the local section of axis. The orientation is calculated by fitting a straight line to the ten axis pixels on either side of the centre. The position of each point in the original shape boundary is now represented in polar co-ordinates with respect to the local co-ordinate frame determined for its corresponding smoothed point.

The third step performs the straightening of the boundary by first straightening the medial axis. The axis points $(x_i, y_i)_{i=1...n}$ are mapped to $(i, 0)$, giving the straight line $(0, 0) \rightarrow (n, 0)$. Transforming the local co-ordinate frames to be appropriately centred and oriented the transformed boundary points have now been straightened.

An example of the full process is given in figure 3. The irregular map of Africa is smoothed until its medial axis represents just the underlying bent shape. The distance transform of the medial axis is shown in figure 3c where low intensities represent small distances. The final, straightened map of africa in figure 3d clearly demonstrates that the original major bend has been removed

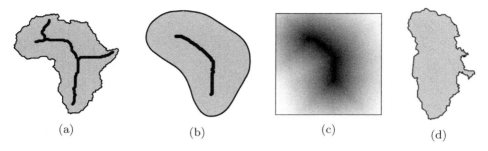

Fig. 3. The straightening process. (a) The irregular outline of the input shape produces a skeleton with several branches. (b) The shape is iteratively smoothed until its skeleton consists of a single spine. (c) The distance transform of the skeleton is generated, and the X and Y co-ordinates of the closest axis point are recorded. This enables the appropriate transformation to be applied to the original shape, resulting in (d).

while the local boundary features have been retained, although slightly distorted in some instances.

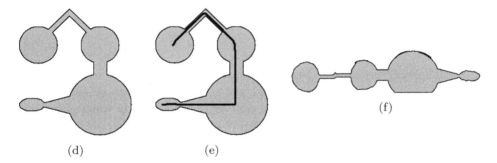

Fig. 4. Examples of shape straightening. The first column contains the original shape; the second column contains the smoothed shape with the medial axis; and the third column contains the straightened shape.

The validity of the approach is demonstrated on the synthetic examples in figure 4. Due to the nature of the data the medial axis is easily found, and reliably represents the basic form of the underlying shape. The final results show that the straightening is performed correctly.

Further examples of shape straightening are provided in figure 5. For simple elongated shapes the technique is generally successful as the medial axis is representative of the bending underlying the shape. Cases in which there are several competing axes are more problematic. For instance, in the donkey there is a dominant elongated horizontal portion as well as three vertical elongated portions (the donkey's fore feet and rear feet, and the rider). No single unbranch-

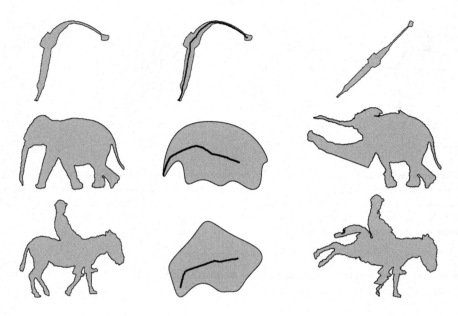

Fig. 5. Examples of performing shape straightening of natural data

ing axis can capture all this. Nevertheless, the result successfully straightens the donkey's rear feet and tail even though the rider and fore feet remain protruding. A similar partial straightening is seen on the elephant. In some cases local distortions are evident. Such errors creep in from a combination of sources such as the distance transform approximation, fitting of the local co-ordinate frame, and the mapping itself.

4   Partitioning

The partitioning algorithm is now complete. Its operation is much as before: candidate cuts are assessed and the one maximising the weighted sum of convexities is selected. However, before calculating convexity the subpart is first straightened out. Since this transformation can distort the size as well as shape of the subpart the total convexity of the set of subparts is combined using the individual subpart convexities weighted according to the relative area of the *unstraightened* subparts.

The results of the two algorithms are compared in the following figures in which the different levels of performance have been grouped. It can be seen that in many cases both methods produce the same or very similar results (figures 6 and 7). Sometimes the original algorithm is still successful despite the shape containing significant bending. For instance, although the fish's tail wiggles back and forth it is thin. Thus its low area causes its low convexity to only contribute weakly to the overall high convexity generated mainly by the highly convex fish body. The results in figure 7 verify that the incorporation of straightening does not prevent the new algorithm from performing satisfactorily.

Fig. 6. Similar partitioning using convexity alone

Fig. 7. Similar partitioning using convexity in combination with straightening

Figures 8 and 9 contain results that differ significantly between the algorithms, although it is not clear that either one is superior. For instance, the original algorithm has cut off one of the curved arms in the second shape. By incorporating straightening the new algorithm has managed to successfully combine both arms into a single part.

In some cases we find that the addition of straightening worsens the effectiveness of the method (see figures 10 and 11). The head is better partitioned by the old algorithm, although the new algorithm's result is still reasonable. By making a cut from the nose to the back of the head it has created a region that was straightened into a fairly convex shape. On the last shape the new algorithm's

Fig. 8. Different partitioning using convexity alone

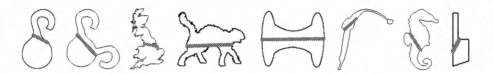

Fig. 9. Different partitioning using convexity in combination with straightening

result is poor, although a contributing factor is that it needs to be partitioned into more than two subparts.

Fig. 10. Better partitioning using convexity alone

Fig. 11. Worse partitioning using convexity in combination with straightening

Finally, examples in which straightening has provided a clear benefit are given in figures 12 and 13. In most cases the failings of using convexity alone are self-evident – sections are chopped off with no regard for their fitness as subparts.

Fig. 12. Worse partitioning using convexity alone

Fig. 13. Better partitioning using convexity in combination with straightening

5   Discussion

In this paper we have shown how shapes can be straightened, and how this can be applied to aid partitioning. Several issues remain, relating to the efficiency and effectiveness of the technique.

Currently, the straightening process is time consuming, and can take several seconds. Since it is applied repeatedly as part of the evaluation of many candidate cuts this slows down the overall analysis of a shape containing a thousand points to several hours. The actual time depends on the shape since if it contains many concavities such as the spiral then many of the trial cuts will lie outside the shape and can therefore be rejected without requiring the more time consuming straightening and convexity calculations.

Two approaches to speeding up the process are possible. The first is to improve the efficiency of the straightening process. The current implementation involves some image based operations (for the medial axis calculation and axis branch checking). A significant improvement could be made by determining the medial axis directly from the shape boundary. Efficient algorithms exist, in particular, Chin *et al.* [5] recently described an algorithm that runs in linear time (with respect to the number of polygon vertices).

Another complementary approach is to apply the convexity calculation only at selected cuts. Rather than exhaustively considering all possible pairwise combinations of boundary points as potential cuts, a two stage process can be employed. For example, the cuts can be restricted to include only a subset of boundary points such as dominant (i.e. corner) points. Although corner detectors are typically unreliable, if a low threshold is used then the significant points will probably be detected, at the cost of also including additional spurious points. Alternatively, a simpler but less reliable partioning algorithm can be used to produce a set of candidate cuts by running it over a set of parameter values. These can then be evaluated and ranked by convexity.

At the moment the only speed-up implemented is to process the data at multiple scales. First the curve is subsampled, typically at every fourth point. The best cut is determined and this initialises a second run at full resolution in which only cuts around the best low resolution cut are considered. Currently the window centred at the first cut is six times the sampling rate

Regarding the effectiveness of the straightened convexity measure some improvements could be made. As discussed previously, the measure does not explicitly take curvature extrema into account. Nevertheless these are important local features even though their reliable detection is problematic.

On the issue of the saliency of alternative partitionings, Hoffman and Singh [9] ran psychophysical experiments to determine three factors affecting part salience: relative area, amount of protrusion, and normalised curvature across the part boundary. Previously the convexity measure was demonstrated to match fairly well on a simple parameterised shape with these human saliency judgements [12]. However, there remain examples in which the basic convexity and the straighted convexity measures cannot discriminate between alternative partitions of different quality. For instance, most humans would judge the first segmentation in figure 14 as more intuitive than the second, but after straightening both receive perfect convexity scores.

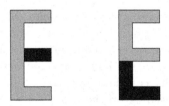

Fig. 14. Alternative partitions with identical straightened convexity ratings

Finally, the straightening process works well for elongated, bent shapes, but can run into problems with shapes containing several competing dominant axes. Simplifying the axes in order to remove all the vertices requires a large amount of smoothing leading to distortion of the shape.

References

1. C. Arcelli and G. Sanniti di Baja. An approach to figure decomposition using width information. *Computer Vision, Graphics and Image Processing*, 26(1):61–72, 1984.
2. G. Borgefors. Distance transformations in digital images. *Computer Vision, Graphics and Image Processing*, 34(3):344–371, 1986.
3. G. Borgefors, G. Ramella, and G. Sanniti di Baja. Permanence-based shape decomposition in binary pyramids. In *10th Int. Conf. on Image Analysis and Processing*, pages 38–43, 1999.

4. M.L. Braunstein, D. Hoffman, and A. Saidpour. Parts of visual objects: an experimental test of the minima rule. *Perception*, 18:817–826, 1989.
5. F. Chin, J. Snoeyink, and C. Wang. Finding the medial axis of a simple polygon in linear time. *Discrete Computational Geometry*, 21(3):405–420, 1999.
6. M. Flickner *et al.* Image and video content: The QBIC system. *IEEE Computer*, 28(9):23–32, 1995.
7. A. Held and K. Abe. On the decomposition of binary shapes into meaningful parts. *Pattern Recognition*, 27(5):637–647, 1994.
8. D. Hoffman and W. Richards. Parts of recognition. *Cognition*, 18:65–96, 1984.
9. D. Hoffman and M. Singh. Salience of visual parts. *Cognition*, 63:29–78, 1997.
10. C. Lamote and J. Wagemans. Rapid integration of contour fragments: from simple filling-in to parts-based shape description. *Visual Cognition*, 6:345–361, 1999.
11. J. Piaget and B. Inhelder. *The Child's Conception of Space*. Routledge and Kegan Paul, 1956.
12. P.L. Rosin. Shape partitioning by convexity. *IEEE Trans. on Sys., Man, and Cybernetics*, part A, 30(2):202–210, 2000.
13. P.L. Rosin and S. Venkatesh. Extracting natural scales using Fourier descriptors. *Pattern Recognition*, 26:1383–1393, 1993.
14. S. Sclaroff. Deformable prototypes for encoding shape categories in image databases. *Pattern Recognition*, 30(4):627–641, 1997.
15. L.G. Shapiro and R.M. Haralick. Decomposition of two-dimensional shapes by graph-theoretic clustering. *IEEE Trans. PAMI*, 1(1):10–20, 1979.
16. K. Siddiqi and B.B. Kimia. Parts of visual form: Computational aspects. *IEEE Trans. PAMI*, 17(3):239–251, 1995.
17. E. Stollnitz, T. DeRose, and D. Salesin. *Wavelets for Computer Graphics*. Morgan Kaufmann, 1996.
18. K. Surendro and Y. Anzai. Decomposing planar shapes into parts. *IEICE Trans. Inf. and Syst.*, E81-D(11):1232–1238, 1998.
19. M.J. Swain and D.H. Ballard. Color indexing. *Int. J. Computer Vision*, 7(1):11–32, 1991.
20. T.Y. Zhang and C.Y. Suen. A fast parallel algorithm for thinning digital patterns. *Comm. ACM*, 27(3):236–240, 1984.

Invariant Signatures from Polygonal Approximations of Smooth Curves

Doron Shaked

Hewlett Packard Laboratories Israel*

Abstract. In this paper we propose to use invariant signatures of polygonal approximations of smooth curves for projective object recognition. The proposed algorithm is not sensitive to the curve sampling scheme or density, due to a novel re-sampling scheme for arbitrary polygonal approximations of smooth curves. The proposed re-sampling provides for weak-affine invariant parameterization and signature. Curve templates characterized by a scale space of these weak-affine invariant signatures together with a metric based on a modified Dynamic Programming algorithm can accommodate projective invariant object recognition.

1 Introduction

An invariant signature of a planar curve is a unique description of that curve that is invariant under a group of viewing transformations. Namely, all curves which are transformations of each other, have the same signature, whereas the signatures of all other curves are different. Invariant signatures may be used to index [2,3,6,8,11,13,20], or to detect symmetries [5,17] of planar curves under viewing transformations.

Invariant signatures of planar curves are the subject of many research papers [1,2,3,8,12,13,19,20], to name a few. Generally speaking, in order to describe planar curves under a group of transformations, one has to employ two independent local descriptors, which are invariant under the required group of transformations. Namely, two numbers, which are well defined on small curve segments and change whenever the curve is changed, unless the change is a transformation from the required group.

When the two descriptors have a differential formulation, the curve is a solution of the differential equations and the required initial conditions. Two independent descriptors are needed since each limits the locus of the 'next curve point' to a one dimensional manifold (the 'next point' is the intersection of these manifolds).

Consider the above mentioned boundary condition. Since the description is invariant under a group of transformations, one should be able to reconstruct all the instances of that transformation from the same signature (though necessarily

---

* HP Labs Israel, Technion City, Haifa 32000, Israel. dorons@hpli.hpl.hp.com

C. Arcelli et al. (Eds.): IWVF4, LNCS 2059, pp. 451–460, 2001.
© Springer-Verlag Berlin Heidelberg 2001

from different boundary conditions). In order to accommodate the dimensionality of the transformation groups[1] we necessarily need more boundary conditions for more complicated groups.

It is indeed a known fact that differential signatures of more complicated groups of transformations have higher degrees of derivatives (see e.g. [3]). This, in turn causes numerical problems when one tries to implement the theory for complicated groups.

An accepted solution to the high derivative problem is to trade derivatives for features, which are integral in nature, and thus more stable. The integral features are local applications of global geometric invariants [1,20]. They are sometimes called semi-differential invariants [12], and in other cases they are formulated as stable numerical schemes for differential invariants [6,7]. For example, setting a grid of equally spaced points on the curve is the integral equivalent to the Euclidean arclength described differentially[2] as $ds_E = \sqrt{X_p^2 + Y_p^2}$. Alternatively, setting a grid such that the area enclosed between edges and curve segments is equal is the integral equivalent to the weak-affine invariant arclength $dS_a = \sqrt[3]{X_p Y_{pp} - Y_p X_{pp}}$.

Invariant grids imply polygonal curve sampling, though only few go all the way analyzing polygonal signatures and specifically polygonal signatures of smooth curves. In this context we address two problems:

- **Polygonal Signatures:** The simplest parametric curve description is polygons. Polygonal signatures have been addressed before [5,9], however, the following practical issues have not been addressed: Polygons are often approximations of smooth curves, rather then intrinsically polygonal shapes. One can not assume invariant curve sampling as in [7]. Furthermore, sampling is not necessarily stationary i.e. different sampling density in different, and possibly even within the same polygonal approximation.
- **Complex Transformation Groups:** Both differential an non-differential descriptors for transformation groups that are more complicated than the weak-affine group are complex and unstable. With the notable exception of the robust methods by Weiss [19], who fits canonical curves to discrete curve segments. This method can however not be implemented for sparsely sampled curves - i.e. polygons. Another effort to approximate complex invariants by simpler ones [11], is more in the spirit of this paper. Although we do not approximate complex invariants, but rather accommodate the difference between the required invariance and the available (simpler) signature by way of a tailored metric.

In this paper we propose to solve these problems. Specifically we propose a re-sampling method for polygonal shapes that is invariant under weak affine

---

[1] Degrees of freedom, for example: The weak-affine group has 5 degrees of freedom (2 for translation, 1 rotation, 1 aspect ration, and 1 skew), the affine group has 6 (5 as the weak-affine, plus scale), and the projective group has 8 (6 as the affine group, plus 2 for tilt).

[2] We use the following notations $X_p = \frac{dX}{dp}$, $X_{pp} = \frac{d^2 X}{dp^2}$.

transformations. Namely, two polygonal versions of the same curve, or its weak affine transform, with arbitrarily different sampling schemes can be re-sampled so that the new sampling density on both curves is similar, see Figure 1. Consequently the polygonal signatures of the re-sampled polygons are similar. In addition we propose a method to use our weak affine signature, to index or detect symmetries in curves under affine and projective transformations.

Fig. 1. Polygonal up-sampling of arbitrarily pre-sampled curves.

In the next section we describe the proposed polygonal re-sampling. In Section 3 we present the second invariant descriptor used for signature value. In Section 4 we propose a metric for matching two signatures. Section 5 is a brief Summary of this paper.

2   Re-sampling

Calabi et al. [6,7] have proposed a sampling method for smooth curves that converges to the weak affine arclength. In this section we present a method to sample polygonal approximations of curves in a manner that is consistent with the weak affine arclength.

It as already been suggested [1,7] to sample curves in a manner invariant to weak affine transformations by setting sampling points so that the area enclosed between the curve and line segments connecting them is constant. However, this and other area measures proposed originally for smooth curves do not generalize well to polygons, particularly not for up-sampling (i.e. adding vertices).

The solution proposed in this paper is based on the fact that area is invariant to weak affine transformations. Specifically we propose the following scheme applicable to polygons as well as smooth shapes, see Figure 2a. Given a point S on the curve, the next point S' is found by:

1. Determining an anchor point A based on an enclosing area of predefined size $\epsilon > 0$.
2. Determining S' such that the line segment based on A and sweeping the curve starting at S covers an area of $\alpha \cdot \epsilon$ for some $\alpha \in [0, 1]$.

Note that both A and S' are, by definition, invariant under weak affine transformations.

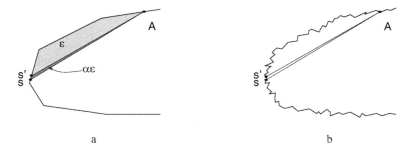

a   b

Fig. 2. Polygonal re-sampling invariant to weak affine transformations.

When $\epsilon \to 0$ the resulting sub-sample density is proportional to the weak-affine arclength, nevertheless, one can use any $\epsilon > 0$ and still keep the invariance to weak affine transformations.

The parameter ϵ is, in a certain sense, a scale parameter, filtering out small curve perturbations, so that perturbations as in Figure 2b are filtered out. Since our main interest is polygonal shapes, it should be noted that polygons have intrinsic artifacts (the vertices), whose influence we need to filter out. A rule of thumb to select ϵ is therefore to make it large enough, so that any boundary segment delimiting an area of ϵ contains at least two vertices (note that for any $\epsilon > 0$ delimiting boundary segments contain at least one vertex). Since the re-sampling needs to be invariant to a wide range of original polygonal approximations, we need to determine ϵ considering the worst expected sampling density in a given application. Let us note again that ϵ can be arbitrarily large and still be invariant to weak affine transformations. If for a large ϵ one still needs a dense sub-sampling, one can always resort to small α.

3   Signature Value

The invariant grid described in the previous section is used as the first invariant descriptor (arclength). In this section we describe the proposed second invariant, the signature value. Like the first invariant used for invariant arclength, the signature value is based on a local application of global geometric invariants. Specifically, we advance an invariant curve segment on the curve in both directions, and define the signature value to be the area of the triangle defined by the current, forward, and backward points.

It is not recommended to use the invariant arclength proposed in the previous section as a measure for the forward/backward advance unless the invariant distance advanced to either sides is larger than $1/\alpha$. Otherwise, polygon artifacts might influence the signature value (e.g. all the invariant points might be located on the same polygon edge). Therefore, the proposed signature uses the anchor point described in the previous section as the forward invariant point. The backward point is obtained symmetrically by enclosing an ϵ area in the opposite direction. The signature value is the area ratio between the triangle area and ϵ. Note that:

- The proposed signature value does not approximate the invariant curvature.
- For $\epsilon \to 0$ on smooth curves the absolute value of the signature converges to 6, however since in our implementations ϵ is constrained from below, this is not the case with the proposed signature.
- Like the invariant arclength, the proposed signature value is not full affine invariant because of the need to determine ϵ.

4   Signature Matching

In this section we discuss the metric used to compare signatures. We also show that it is possible to use the proposed metric in order to overcome, the otherwise complicated problem, of invariance to affine and projective transformations.

4.1   Weak Affine Transformations

Since the object of our study is curves that have already been sampled by arbitrary methods, two polygonal approximations of the same curve are, strictly speaking, different curves, and we cannot expect their invariant signatures to be identical. Although, if the scale parameter ϵ is chosen appropriately[3], the signature functions of two polygonal instances of the same curve should be *similar* in both the arclength and value dimensions (respectively x and y dimensions in the graphs of Figure 3).

Value perturbations are trivially dealt with by standard metrics (e.g. l^2). However, to deal with arclength perturbations we have to resort to more complicated metrics. Figure 3 is an example of the combined value/arclength deformation problem. It depicts different polygonal approximations of the same smooth curve, and the corresponding weak-affine invariant signatures. Notice that approximation c is slightly too sparse for the chosen ϵ[3], nevertheless, the proposed metric will handle this case well.

We propose to employ a composite measure based on standard metrics (e.g. l^2) both in the value and in the arclength dimensions. The proposed metric is the well known warp metric[4] used in many fields of engineering, see e.g. [10,14, 15,16,18]. It is based on the following minimization problem:

Given two signature functions $V_Q(i)$, $i \in \{1, 2, \ldots L_Q\}$ for the query curve and $V_T(i)$, $i \in \{1, 2, \ldots L_T\}$ for the template, we look for the optimal warp or reparameterization function $\Psi : \{1, 2, \ldots L_Q\} \to \{1, 2, \ldots L_T\}$ to minimize the composite distance function

$$D\left(V_Q, V_T\right) = \min_{\Psi} \left\{ \Sigma_{i=2}^{L_Q} \left(V_Q(i) - V_T(\Psi(i))\right)^2 + \Lambda \cdot \left(\nabla \Psi(i)\right)^2 \right\}$$

---

[3] See discussion in the end of Section 2.

[4] Known also as Dynamic Warping, reparameterization, and Viterbi algorithm.

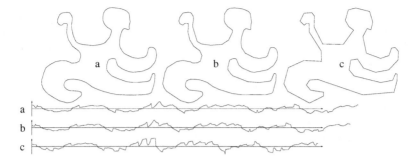

Fig. 3. Polygonal approximations of a smooth curve (a. 276, b. 125, and c. 55 vertices) and corresponding weak-affine invariant signatures.

subject to constraints on Ψ (we use: $\Psi(1) = 1$, and $0 \leq \nabla\Psi(i) \leq 2$). In the above Λ is a scalar constant, and $\nabla\Psi(i) = \Psi(i) - \Psi(i-1)$.

The optimization problem described above is solved in $O(L_Q \cdot L_T)$ time by the *Dynamic Programming algorithm.* Dynamic programming is based mainly on the following recursion[5]:

Given a series of solutions to the warping problems of a given part of the query signature $V_Q(i), \quad i \in \{1, 2, \ldots k\}$, to all the sub-parts of the template signature $V_T(i), \quad i \in \{1, 2, \ldots j\}$, with $j \in \{1, 2, \ldots L_T\}$, we can solve the series of warping problems from a longer part of the query $V_Q(i), \quad i \in \{1, 2, \ldots k+1\}$, to each of the sub-parts of $V_T(\cdot)$. For each of the problems, simply select one of the optimal paths of the given set of solutions, and extend it by a single further match to minimize the total warp error composed of: (1) The total error at the given path (2) The warp error due to the required step (3) The additional match. Figure 4 depicts the recursive completion process.

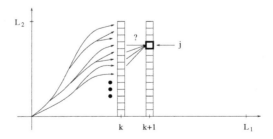

Fig. 4. Recursive path extension in Dynamic Programming.

The recursive process is initiated according to the boundary conditions. In our implementation we assume $\Psi(1) = 1$, and thus $\{V_Q(1)\}$ matches $\{V_T(1)\}$,

---

[5] Refer to [15] for a detailed description of Dynamic Programming.

and none of the other sub-parts of V_T. Thus, $D\left(\{V_Q(1)\}, \{V_T(1)\}\right) = 0$, and $D\left(\{V_Q(1)\}, \{V_T(j)\}\right) = \infty,\ \forall j \neq 1$. In real applications one can not take initial matching for granted. A more realistic initial condition for signatures on S^1 is described in [21].

The recursion is terminated similarly, according to the boundary conditions. If we insisted that $\Psi(L_Q) = L_T$, we would have chosen the resulting match of the full signature of V_Q to the full signature of V_T. However, we chose to apply a more relaxed boundary condition, as in [16], and to select the best match of the full signature V_Q to either of the longer sub parts of V_T.

4.2   Affine Transformations

The weak affine signature is invariant to affine transformations only up to the scale parameter ϵ^6. Note that the signature's arclength and value are derived from ϵ via area ratios, which are in turn invariant to affine transformations [1]. Thus, representing a plane curve by a set of signatures representing a range of ϵ values, instead of a single signature corresponding to a specific ϵ_0, makes it possible to identify the planar curve under affine transformations. A similar signature scale space has been employed in [4] for a different purpose.

The practical question is naturally, how many signatures to keep. The range of ϵ should be determined by the range of scales relevant to the application in mind. Specifically, if we expect the affine scale of queries to be in the range of $\times 0.5$ to $\times 2$ relative to the template, we will need signatures with scale parameters in the range $0.25 \times \epsilon_0$ to $4 \times \epsilon_0$. As for the number of ϵ values in the range, they have to be selected so that the signatures of intermediate scales will be similar to one of the represented signatures.

Although we have not proven the following conjecture, we found out empirically, that signatures change slowly with ϵ. Moreover, the fact that we use a warp metric reduces the influence of arclength deformations, leaving us mainly with the relatively gradual value changes. Figure 5 depicts the way signatures change with ϵ.

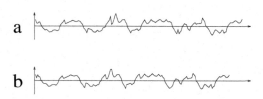

Fig. 5. Weak-affine invariant signatures for different ϵ values (a. ϵ_0, b. $1.2 \times \epsilon_0$).

---

[6] If we knew that the affine scale parameter between the query and the template is e.g. β we could use $\beta^2 \cdot \epsilon$ instead of ϵ, and thus be 'invariant' to the affine transformation.

4.3   Projective Transformations

In this section we show how a modified warp distance can accommodate object recognition under projective transformations. Let us first define the affine and projective transformations in the plane. An affine transform of a point $X = (x, y)^T$ is $X_a = AX + V$, where V is a translation vector, and A a non singular 2x2 matrix parameterizing rotations ,scale, skew, and aspect ratio. A projective transform is $X_p = \frac{1}{1+W^T \cdot X} AX + V$, where W parameterizes tilt.

Given a small neighborhood in the plane a projective transformation can be approximated by an affine transformation with a scale parameter $\frac{\sqrt{\det(A)}}{1+W^T \cdot X}$. Thus a projective transform can be approximated by an affine transformation with space varying affine scale parameter. For our purposes we have $\epsilon(X, Y) = \frac{\det(A)}{(1+W^T \cdot X)^2}$. Note that for continuous curves $\epsilon(X, Y)$ changes continuously on the curve.

Before we detail the proposed projective invariant matching, let us recall the affine matching described in Subsection 4.2, where we proposed the following procedure:

Calculate the warp distance from the signature V_Q of the query curve to a set of signatures $V_T^{\epsilon m}$ corresponding to the template curve, and a set of ϵ values. The query is considered an affine transformation of the template if at least one of the warp distances is below a predetermined threshold. Evidently, it is sufficient to consider the smallest warp distance over all m. Thus, the proposed algorithm is equivalent to an algorithm combining the dynamic programming algorithms of the different warps to a single algorithm matching $\{V_Q\}$ to a huge set of states $\{V_T^{\epsilon m}(j)\}$ $\forall j, \forall m$. Note that:

1. The initialization should facilitate equal conditions for matching $\{V_Q(1)\}$, to the initial states $\{V_T^{\epsilon m}(1)\}$ of each of the template's signatures.
2. The recursion should restrict path extension to within the same m.
3. The warp representing the template is the path corresponding to the best final state.

Now consider the case of projective invariant matching. The only difference to the above algorithm is the need to enable the recursion step to extend paths across similar scales.

Figure 6 describes the signature (b) of a projective transformation of a curve and the signature corresponding to the best warp into the template's signature scale space. Namely, values of the template signature were compiled by tracking the best-path selected by the algorithm. They were taken from weak-affine signatures corresponding to different scales. The signature of the template has been slightly lowered, otherwise it would have been difficult to distinct the two signatures. This match quality would not have been possible had we limited the algorithm to any single scale.

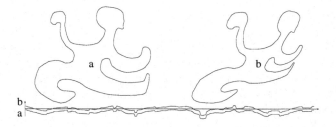

Fig. 6. A weak-affine signature of the projectively transformed curve, and the warped template signature (lowered to allow distinction).

5   Summary

In this paper we have presented a weak-affine invariant re-sampling method for polygonal approximations of smooth curves. The weak-affine signature of the resulting polygon is invariant to the original curve sampling method.

We proposed to use a signature scale space similar to the one described in [4], and argued that a metric based on a modified Dynamic Programming algorithm accommodates projective invariant object recognition.

References

1. A. M. Bruckstein, R. J. Holt, A. N. Netravali, and T. J. Richardson, "Invariant Signatures for Planar Shapes Under Partial Occlusion", *CVGIP: Image Understanding*, Vol. 58, pp. 49–65, 1993.
2. A. M. Bruckstein, N. Katzir, M. Lindenbaum, and M. Porat, "Similarity Invariant Signatures and Partially Occluded Planar Shapes", *Int. J. Computer Vision*, Vol. 7, pp. 271–285, 1992.
3. A. M. Bruckstein, and A. N. Netravali, "On Differential Invariants of Planar Curves and the Recognition of Partly Occluded Planar Shapes", AT&T Technical memo, July 1990; also in *Proc. of the Int. Workshop on Visual Form*, Capri, May 1992.
4. A. M. Bruckstein, E. Rivlin, and I. Weiss, "Scale Space Local Invariants", *CIS Report No. 9503*, Computer Science Department, Technion, Haifa, Israel, February 1995,
5. A. M. Bruckstein, and D. Shaked, "Skew Symmetry Detection via Signature Functions", *Pattern Recognition*, Vol. 31, pp. 181-192, 1998.
6. E. Calabi, P. J. Olver, C. Shakiban, A. Tannenbaum, and S. Haker, "Differential and Numerically Invariant Signature Curves Applied to Object Recognition", *Int. J. Computer Vision*, Vol. 26, pp. 107–135, 1998.
7. E. Calabi, P. J. Olver, and A. Tannenbaum, "Affine Geometry, Curve Flows, and Invariant Numerical Approximations", *Adv. in Math*, Vol. 124, pp. 154–196, 1996.
8. D. Cyganski, J.A. Orr, T.A. Cott, and R.J. Dodson, "An Affine Transform Invariant Curvature Function", *Proceedings of the First ICCV*, London, pp. 496–500, 1987.
9. T. Glauser and H. Bunke, "Edge Length Ratios: An Affine Invariant Shape Representation for Recognition with Occlusions", *Proc. of the 11th ICPR*, Hague, 1992.

10. P. S. Heckbert, "Color Image Quantization for Frame Buffer Display", http://www-cgi.cs.cmu.edu/asf/cs.cmu.edu/users/ph/www/ciq_thesis

11. R. J. Holt, and A. N. Netravali, "Using Affine Invariants on Perspective Projections of Plane Curves", *Computer Vision and Image Understanding*, Vol. 61, pp. 112–121, 1995.

12. T. Moons, E. J. Pauwels, L. Van Gool, and A. Oosterlinck, "Foundations of Semi Differential Invariants", *Int. J. of Computer Vision*, Vol. 14, 25–47, 1995.

13. E. J. Pauwels, T. Moons, L. Van Gool, and A. Oosterlinck, "Recognition of Planar shapes Under Affine Distortion", *Int. J. of Computer Vision*, Vol. 14, 49–65, 1995.

14. L. R. Rabiner, and B. H. Juang, "Introduction to Hidden Markov Models", *IEEE ASSP Magazine*, pp. 4–16, 1986.

15. L. R. Rabiner, and S. E. Levinson, "Isolated and Connected Word Recognition - Theory and Selected Applications", *IEEE COM*, Vol. 29, pp. 621–659, 1981.

16. L. R. Rabiner, A. E. Rosenberg, and S. E. Levinson, "Considerations in Dynamic Time warping Algorithms for Discrete Word Recognition", *IEEE ASSP*, Vol. 26, pp. 575–582, 1978.

17. L. Van Gool, T. Moons, D. Ungureanu, and A. Oosterlinck, "The Characterization and Detection of Skewed Symmetry" *CVIU*, Vol. 61, pp. 138–150, 1995.

18. A. Viterbi, "Error Bounds on Convolution Codes and an Asymptotically Optimal Decoding Algorithm", *IEEE Information Theory*, Vol. 13, pp. 260–269, 1967.

19. I. Weiss, "Noise Resistant Invariants of Curves", *IEEE Trans. on PAMI*, Vol. 15, pp. 943-948, 1993.

20. I. Weiss, "Geometric Invariants and Object Recognition", *Int. J. of Computer Vision*, Vol. 10, pp. 207–231, 1993.

21. N. Weissberg, S. Sagie, and D. Shaked, "Shape Indexing by Dynamic Programming", in *Proc. of the Israeli IEEE Convention*, April 2000.

On the Learning of Complex Movement Sequences

Walter F. Bischof and Terry Caelli

Department of Computing Science, University of Alberta, Edmonton, T6G 2E8,
Canada, (wfb,tcaelli)@ualberta.ca

Abstract. We introduce a rule-based approach for the learning and
recognition of complex movement sequences in terms of spatio-temporal
attributes of primitive event sequences. During learning, spatio-temporal
decision trees are generated that satisfy relational constraints of the
training data. The resulting rules are used to classify new movement
sequences, and general heuristic rules are used to combine classification
evidences of different movement fragments. We show that this approach
can successfully learn how people construct objects, and can be used to
classify and diagnose unseen movement sequences.

1   Introduction

Over the past years, we have explored new methods for the automatic learning of spatio-temporal patterns [1,2,4,3]. These methods combine advantages of
numerical learning methods (e.g. [9]) with those of relational learners (e.g. [7]),
and lead to a class of learners which induce over numerical attributes but are
constrained by relational pattern models. Our approach, Conditional Rule Generation (CRG), generates rules that take the form of numerical decision trees
that are linked together so that relational constraints of the data are satisfied.
Relational pattern information is introduced adaptively into the rules, i.e. it is
added only to the extent that is required for disambiguating classification rules.

In contrast to Conditional Rule Generation, traditional numerical learning
methods are not relational, and induce rules over unstructured sets of numerical
attributes. They thus have to assume that the correspondence between candidate
and model features is known *before* rule generation (learning) or rule evaluation
(matching) occurs. This assumption is inappropriate when complex models have
to be learned, as is the case when complex movements of multiple limb segments
have to be learned. Many symbolic relational learners (e.g. Inductive Logic Programming), on the other hand, are not designed to deal efficiently with numerical
data. Although they induce over relational structures, they typically generalize
or specialize only over symbolic variables. It is thus rare that the symbolic representations *explicitly* constrain the permissible numerical generalizations. It is
these disadvantages of numerical learning methods and inductive logic programming that CRG is trying to overcome.

Since CRG induces over a relational structure it requires general model assumptions, the most important being that the models are defined by a labeled

C. Arcelli et al. (Eds.): IWVF4, LNCS 2059, pp. 463–472, 2001.
© Springer-Verlag Berlin Heidelberg 2001

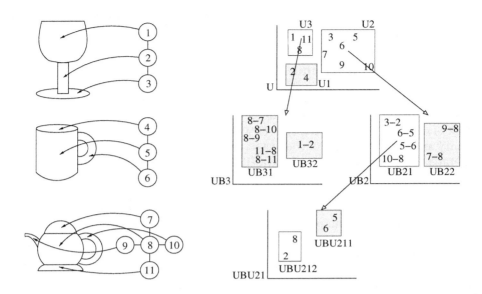

Fig. 1. Example of input data and conditional cluster tree generated by CRG method. The left panel shows input data and the attributed relational structures generated for these data, where each vertex is described by a unary feature vector $\boldsymbol{u}$ and each edge by a binary feature vector $\boldsymbol{b}$. We assume that there are two pattern classes, class 1 consisting of the drinking glass and the mug, and class 2 consisting of the teapot. The right panel shows a cluster tree generated for the data on the left. Numbers refer to the vertices in the relational structures, rectangles indicate generated clusters, grey ones are unique, white one contain elements of multiple classes. Classification rules of the form $U_i - B_{ij} - U_j \ldots$ are derived directly from this tree.

graph where relational attributes are defined only with respect to neighbouring vertices. Such assumptions constrain the types of unary and binary features which can be used to resolve uncertainties (Figure 1).

Recently, we have successively extended the CRG method for learning spatial patterns (for learning of objects and recognizing complex scenes) to the learning of spatio-temporal patterns. This method, CRG_{ST} , was successively applied to the learning and recognition of very brief movement sequences that lasted up to 1-2 seconds. In this paper, we describe how CRG_{ST} can be applied to the recognition of very long and complex movement sequences that last over much longer time periods. Specifically, we test the suitability of CRG_{ST} for the recognizing how people assemble fairly complex objects over time periods up to half a minute.

In the following, we introduce the spatial CRG method and the spatio-temporal CRG_{ST} method. We discuss representational issues, rule generation, and rule application. We then show first results on the application of CRG$_{ST}$ to the recognition of complex construction tasks.

2   Spatial Conditional Rule Generation

In Conditional Rule Generation [1], classification rules for patterns or pattern fragments are generated that include structural pattern information to the extent that is required for classifying correctly a set of training patterns. CRG analyzes unary and binary features of connected pattern components and creates a tree of hierarchically organized rules for classifying new patterns. Generation of a rule tree proceeds in the following manner (see Figure 1).

First, the unary features of all parts of all patterns are collected into a unary feature space U in which each point represents a single pattern part. The feature space U is partitioned into a number of clusters U_i. Some of these clusters may be unique with respect to class membership (e.g. cluster U_1) and provide a classification rule: If a pattern contains a part p_r whose unary features $\boldsymbol{u}(p_r)$ satisfy the bounds of a unique cluster U_i then the pattern can be assigned a unique classification. The non-unique clusters contain parts from multiple pattern classes and have to be analyzed further. For every part of a non-unique cluster we collect the binary features of this part with all adjacent parts in the pattern to form a (conditional) binary feature space UB_i. The binary feature space is clustered into a number of clusters UB_{ij}. Again, some clusters may be unique (e.g. clusters UB_{22} and UB_{31} and provide a classification rule: If a pattern contains a part p_r whose unary features satisfy the bounds of cluster U_i, and there is another part p_s, such that the binary features $\boldsymbol{b}(p_r, p_s)$ of the pair $\langle p_r, p_s \rangle$ satisfy the bounds of a unique cluster UB_{ij} then the pattern can be assigned a unique classification. For non-unique clusters, the unary features of the second part p_s are used to construct another unary feature space UBU_{ij} that is again clustered to produce clusters UBU_{ijk}. This expansion of the cluster tree continues until all classification rules are resolved or a maximum rule length has been reached.

If there remain unresolved clusters at the end of the expansion procedure (which is normally the case), the clusters and their associated classification rules are split into more discriminating rules using an entropy-based splitting procedure. The elements of an unresolved cluster (e.g. cluster UBU_{212} in Figure 1) are split along a feature dimension such that the normalized partition entropy $H_P(T)$

$$H_P(T) = (n_1 H(P_1) + n_2 H(P_2))/(n_1 + n_2). \tag{1}$$

is minimizes, where H is entropy. Rule splitting continues until all classification rules are unique or some termination criterion has been reached. This results in a tree of conditional feature spaces (Figure 1), and within each feature space, rules for cluster membership are developed in the form of a decision tree.

From the empirical class frequencies of all training patterns one can derive an expected classification (or evidence vector) $\boldsymbol{E}$ associated with each rule (e.g. $\boldsymbol{E}(UBU_{212}) = [0.5, 0.5]$), given that it contains one element of each class). Similarly, one can compute evidence vectors for partial rule instantiations, again from empirical class frequencies of non-terminal clusters (e.g. $\boldsymbol{E}(UB_{21}) = [0.75, 0.25]$). Hence an evidence vector $\boldsymbol{E}$ is available for every partial or complete rule instantiation.

3 Spatio-Temporal Conditional Rule Generation: CRG_{ST}

We now turn to CRG_{ST}, a generalization of CRG from a purely spatial domain into a spatio-temporal domain. Here, data consist typically of time-indexed pattern descriptions, where pattern parts are described by unary features, spatial part relations by (spatial) binary features, and changes of pattern parts by (temporal) binary features. In contrast to more popular temporal learners like hidden Markov models [5] and recurrent neural networks [6], the rules generated from CRG_{ST} are not limited to first-order time differences but can utilize more distant (lagged) temporal relations depending on data model and uncertainty resolution strategies. At the same time, CRG_{ST} can generate non-stationary rules, unlike e.g. multivariate time series which also accommodate correlations beyond first-order time differences but do not allow for the use of different rules at different time periods.

We now discuss the modifications that are required for CRG to deal with spatiotemporal patterns, first with respect to pattern representation and then with respect to pattern learning. This should give the reader a good idea of the representation and operation of CRG_{ST}.

Representation of Spatio-Temporal Patterns

A spatio-temporal pattern is defined by a set of labeled time-indexed attributed features. A pattern P_i is thus defined in terms of $P_i = \{p_{i1}(\boldsymbol{a} : t_{ij}), \ldots, p_{in}(\boldsymbol{a} : t_{in})\}$ where $p_{ij}(\boldsymbol{a} : t_{ij})$ corresponds to part j of pattern i with attributes $\boldsymbol{a}$ that are true at time t_{ij}. The attributes $\boldsymbol{a} = \{\boldsymbol{u}, \boldsymbol{b}_s, \boldsymbol{b}_t\}$ are defined with respect to specific labeled features, and are either unary (single feature attributes) or binary (relational feature attributes), either over space or over space-time (see Figure 2). Examples of unary attributes $\boldsymbol{u}$ include area, brightness, position; spatial binary attributes $\boldsymbol{b}_s$ include distance, relative size; and temporal binary attributes $\boldsymbol{b}_t$ include changes in unary attributes over time, such as size, orientation change, or long range position change.

Our data model, and consequently our rules, are subject to spatial and temporal adjacency (in the nearest neighbour sense) and temporal monotonicity, i.e. features are only connected in space and time if they are spatially or temporally adjacent, and the temporal indices for time must be monotonically increasing (in the "predictive" model) or decreasing (in the "causal" model). Although this limits the expressive power of our representation, it is still more general than strict first-order discrete time dynamical models such as hidden Markov models or Kalman filters.

For CRG_{ST} finding an "interpretation" involves determining sets of linked lists of attributed and labeled features, that are causally indexed (i.e. the temporal indices must be monotonic), that maximally index a given pattern.

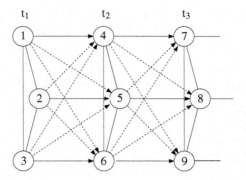

Fig. 2. A spatio-temporal pattern consisting of three parts over three time-points. Undirected arcs indicate spatial binary connections, solid directed indicate temporal binary connections between the same part at different time-points, and dashed directed arcs indicate temporal binary connections between different parts at different time-points.

Rule Learning

CRG_{ST} generates classification rules for spatio-temporal patterns involving a small number of pattern parts subject to the following constraints: First, the pattern fragments involve only pattern parts that are adjacent in space and time. Second, the pattern fragments involve only non-cyclic chains of parts. Third, temporal links are followed in the forward direction only to produce causal classification rules that can be used in classification and in prediction mode.

Rule learning proceeds in the following way: First, the unary features of all parts (of all patterns at all time points), $u(p_{it})$, $i = 1, \ldots, n$, $t = 1, \ldots, T$, are collected into a unary feature space U in which each each point represents the feature vector of one part at one time point. From this point onward, cluster tree generation proceeds exactly as described in Section 2, except that expansion into a binary space can now follow either spatial binary relations b_s or temporal binary relations b_t. Furthermore, temporal binary relations b_t can be followed only in strictly forward direction, analyzing recursively temporal changes of either the same part, $b_t(p_{it}, p_{it+1})$ (solid arrows in Figure 2), or of different pattern parts, $b_t(p_{it}, p_{jt+1})$ (dashed arrows in Figure 2) at subsequent time-points t and $t + 1$. Again, the decision about whether to follow spatial or temporal relations is simply determined by entropy-based criteria.

4   Rule Application

A set of classification rules is applied to a spatio-temporal pattern in the following way. Starting from each pattern part (at any time point), all possible sequences (chains) of parts are generated using parallel, iterative deepening, subject to the constraints the only adjacent parts are involved and no loops are generated. (Note that the same spatio-temporal adjacency constraints and temporal mono-tonicity constraints were used for rule generation.) Each chain is classified using the classification rules. Expansion of a chain $S_i = <p_{i1}, p_{i2}, \ldots, p_{in}>$ terminates if one of the following conditions occurs: 1) the chain cannot be expanded with-out creating a cycle, 2) all rules instantiated by S_i are completely resolved (i.e. have entropy 0), or 3) the binary features $b_s(p_{ij}, p_{ij+1})$ or $b_t(p_{ij}, p_{ij+1})$ do not satisfy the features bounds of any rule.

If a chain S cannot be expanded, the evidence vectors of all rules instantiated by S are averaged to obtain the evidence vector $E(S)$ of the chain S. Further, the set $\mathcal{S}_p$ of all chains that start at p is used to obtain an initial evidence vector for part p:

$$E(p) = \frac{1}{|\mathcal{S}_p|} \sum_{S \in \mathcal{S}_p} E(S). \tag{2}$$

where $|\mathcal{S}|$ denotes the cardinality of the set $\mathcal{S}$. Evidence combination based on (2) is adequate, but can be improved by noting that nearby parts (both in space and time) are likely to have the same classification. To the extent that this assumption of spatio-temporal coherence is justified, the part classification based on (2) can be improved.

We use general heuristics for implementing spatio-temporal coherence among pattern parts. one such rule is based on the following idea. For a chain $S_i = < s_{i1}, s_{i2}, \ldots, s_{in} >$, the evidence vectors $E(s_{i1})$, $E(s_{i2})$, $\ldots$, $E(s_{in})$ are likely to be similar, and dissimilarity of the evidence vectors suggests that S_i may contain fragments of different movement types. This similarity can be captured in the following way (see [10] for further details): For a chain $S_i =< p_{i1}, p_{i2}, ..., p_{in} >$,

$$w(S_i) = \frac{1}{n} \sum_{k=1}^{n} E(p_{ik}) \tag{3}$$

where $E(p_{ik})$ refers to the evidence vector of part p_{ik}. Initially, this can be found by averaging the evidence vectors of the chains which begin with part p_{ik}. Later, the compatibility measure is used for updating the part evidence vectors in an iterative relaxation scheme

$$E^{(t+1)}(p) = \Phi \left(\frac{1}{Z} \sum_{S \in S_p} w^{(t)}(S) \otimes E(S) \right), \tag{4}$$

where Φ is the logistic function $\Phi(z) = (1 + \exp[-20(z - 0.5)])^{-1}$. Z a normalizing factor, and where the binary operator $\otimes$ is defined as a component-wise vector multiplication $[a\ b]^T \otimes [c\ d]^T = [ac\ bc]^T$. Convergence of the relaxation scheme 4 is typically obtained in about 10-20 iterations.

5   Learning to Recognize Complex Construction Tasks

Learning and recognition was tested with an example where a person constructed three different objects (a "sink", a "spider" and a "U-turn") using pipes and connectors (see Figure 3). Each construction took about 20-30 s to finish, and was repeated five times. Arm and hand movements were recorded using a Polhemus system [11] with four sensors located on the forearm and hand of both arms. The sensors were recording at 120 Hz, and the system was calibrated to an accuracy of ± 5 mm. From the position data $(x(t), y(t), z(t))$ of the sensors, 3D velocity $v(t)$, acceleration $a(t)$, and curvature $k(t)$ were extracted, all w.r.t. arc length $ds(t) = (dx^2(t) + dy^2(t) + dz^2(t))^{1/2}$ [12]. Sample time-plots of these measurements are shown in Figure 4. These measurments were smoothed with a Gaussian filter with $\sigma = 0.25s$ (see Figure 4) and then sampled at intervals of $0.25s$.

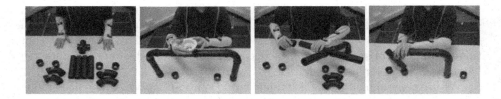

Fig. 3. Pictures of the construction tasks learned by CRG$_{ST}$. The leftmost image shows the starting position. The Polhemus movement sensors can clearly be seen on the left and right forearms and hands. The other three images show one stage of the three construction tasks used, the "sink" construction, the "spider" construction, and the "U-turn" construction. Each construction took about 20-30 seconds to finish.

The spatio-temporal patterns were defined in the following way: At every time point t, the patterns consisted of four parts, one for each sensor, each part being described by unary attributes $\boldsymbol{u} = [v, a, k]$. Binary attributes were defined by simple differences, i.e. the spatial attributes were defined as $\boldsymbol{b}_s(p_{it}, p_{jt}) = \boldsymbol{u}(p_{jt}) - \boldsymbol{u}(p_{it})$, and the temporal attributes were defined as $\boldsymbol{b}_t(p_{jt}, p_{jt+1}) = \boldsymbol{u}(p_{jt+1}) - \boldsymbol{u}(p_{jt})$.

Performance of CRG$_{ST}$ was tested with a leave-one-out paradigm, i.e. in each test run, the movement of all construction tasks were learned using all but one sample, and the resulting rule system was used to classify the remaining instance. Learning and classification proceeded exactly as described Sections 3 and 4, with rule length restricted to five levels (i.e. the most complex rules were of the form $UBUBU$). For the parameter values reported before, 73.3% of the tasks were correctly recognized on average.

An example of a classification rule generated by CRG$_{ST}$ is the following rule that has the form $U - B_t - U - B_t - U$, where $V = $ velocity, $A = $ acceleration,

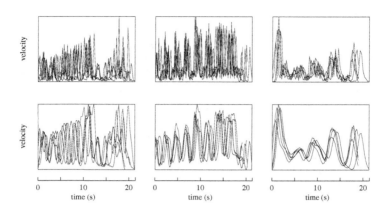

Fig. 4. Time plots for $v(t)$ of the right hand, for "sink" construction, "spider" construction and "U-turn construction" from left to right. The time range is 22 seconds and velocity has been normalized. The top row shows the raw input data, the bottom row the same data after filtering with a Gaussian filter with $\sigma = 0.25s$. Each plot shows five replications.

and ΔA = acceleration difference over time, and ΔK = curvature difference over time:

if $U_i(t)$	$-1.65 \leq V \leq 2.43$
and $B_{ij}(t, t+1)$	$-0.57 \leq \Delta K \leq 1.14$
and $U_j(t+1)$	$-1.65 \leq A \leq 0.78$
and $B_{jk}(t+1, t+2)$	$-2.49 \leq \Delta K \leq 0.73$ and $1.56 \leq \Delta A \leq 2.92$
and $U_k(t+2)$	$1.7 \leq V \leq 2.4$
then	this is part of a "spider" construction

$\mathrm{CRG_{ST}}$ makes minimal assumptions about the data. First, given that it classifies data within small temporal windows only, it can classify partial data (e.g. a short subsequence of a construction task). Second, it can easily deal with spatio-temporal mixtures of patterns. In other words, it can equally well classify sequences of different construction tasks (e.g. a person starting one task and continuing with another, or a person starting one task and then doing something completely different) or even two persons doing different constructions at the same time. Obviously, one could incorporate stronger constraints into $\mathrm{CRG_{ST}}$ (e.g. incorporating the assumption that only a single construction task is present) and thus improve classification performance further. This is, however, not our intent, as we plan to use $\mathrm{CRG_{ST}}$ to detect and diagnose tasks that are only partially correct, i.e. where some parts of the construction task are done incorrectly or differently.

6   Conclusions

Most current learners are based upon rules defined iteratively in terms of expected states and/or observations at time $t + 1$ given those at time t. Examples include hidden Markov models and recurrent neural networks. Although these methods are capable of encoding the variations which occur in signals over time and can indirectly index past events of varying lags, they do not have the explicit expressiveness of CRG_{ST} for relational time-varying structures.

In the current paper, we have extended our previous work on the recognition of brief movements to the recognition of long and very complex movement sequences, as they occur in construction tasks. We have shown that CRG_{ST} can successfully deal with such data. There remains, however, much to be done. One obvious extension is to extend CRG_{ST} to the analysis of multiple, concurrent time scales that would allow a hierarchical analysis of such movement sequences. A second extension will involve an explicit representation of temporal relations between movement subsequences, and a third extension involves introducing knowledge-based model constraints into the analysis.

In all, there remains much to be done in the area of spatio-temporal learning, and the exploration of spatio-temporal data structures which are best suited to the encoding and efficient recognition of complex spatio-temporal events.

References

1. W. F. Bischof and T. Caelli, "Learning structural descriptions of patterns: A new technique for conditional clustering and rule generation," *Pattern Recognition*, vol. 27, pp. 1231–1248, 1994.
2. W. F. Bischof and T. Caelli, "Scene understanding by rule evaluation," *IEEE Transactions on Pattern Analysis and Machine Intelligence*, vol. 19, pp. 1284–1288, 1997.
3. W. F. Bischof and T. Caelli, "Learning actions: Induction over spatio-temporal relational structures - CRGst," in *Proceedings of the Workshop on Machine Learning in Vision, European Conference on Artificial Intelligence*, pp. 11-15, 2000.
4. T. Caelli and W. F. Bischof, eds., *Machine Learning and Image Interpretation*. New York, NY: Plenum, 1997.
5. L. Rabiner and B.-H. Juang, *Fundamentals of Speech Recognition*. New York, NY: Prentice Hall, 1993.
6. T. Caelli, L. Guan, and W. Wen, "Modularity in neural computing," *Proceedings of the IEEE*, vol. 87, pp. 1497–1518, 1999.
7. S. Muggleton, *Foundations of Inductive Logic Programming*. Englewood Cliffs, NJ: Prentice-Hall, 1995.
8. J. R. Quinlan, "MDL and categorical theories (continued)," in *Proceedings of the 12th International Conference on Machine Learning*, pp. 464–470, 1995.
9. J. R. Quinlan, *C4.5: Programs for Machine Learning*. San Mateo, CA: Morgan Kaufmann, 1993.
10. B. McCane and T. Caelli, "Fuzzy conditional rule generation for the learning and recognition of 3d objects from 2d images," in *Machine Learning and Image Interpretation* (T. Caelli and W. F. Bischof, eds.), pp. 17–66, New York, NY: Plenum, 1997.

11. F. H. Raab, E. B. Blood, T. O. Steiner, and H. R. Jones, "Magnetic position and orientation tracking system," *IEEE Transactions on Aerospace and Electronic Systems*, vol. AES-15, pp. 709–, 1979.
12. F. Mokhtarian, "A theory of multiscale, torsion-based shape representation for space curves," *Computer Vision and Image Understanding*, vol. 68, pp. 1–17, 1997.

Possibility Theory and Rough Histograms for Motion Estimation in a Video Sequence

Frederic Comby, Olivier Strauss, and Marie-José Aldon

LIRMM, UMR CNRS/Universite Montpellier II, 161 rue Ada,
34392 Montpellier cedex 5, France
{comby, strauss, aldon}@lirmm.fr
http://www.lirmm.fr

Abstract. This article proposes to use both theories of possibility and rough histograms to deal with estimation of the movement between two images in a video sequence. A fuzzy modeling of data and a reasoning based on imprecise statistics allow us to partly cope with the constraints associated to classical movement estimation methods such as correlation or optical flow based-methods. The theoretical aspect of our method will be explained in details, and its properties will be shown. An illustrative example will also be presented.

1 Introduction

In a static scene, the movement of a camera entails an apparent motion on the video sequence it acquires. Phenomena such as occlusions, moving objects or variations of the global illumination can involve parasite motion. Classical methods dealing with apparent motion estimation aim at finding the main apparent motion. They can be divided in two different approaches: correlation-based and optical flow-based approaches. The validity of these approaches relies on strong hypothesis which are frequently transgressed, thus limiting their reliability.

For the matching methods it is necessary to have a model of the motion to estimate. This kind of method requires a discretization of the search area. This will limit the precision of the estimation to the sampling interval of the motion parameter's space. Besides, the correspondence between two images is said to exist. Hence, the method will return an estimation even for two totally different images. The user has then to choose a threshold under which the estimation is considered irrelevant. The estimation is less robust because of the threshold arbitrarily chosen and not set by the data.

Methods based on the optical flow computation are among the most studied for main motion estimation [1]. The optical flow links the spatio-temporal gradients of the irradiance image using the constraint equation:

$$E_x * u + E_y * v + E_t = \xi \tag{1}$$

where E_x, E_y are the spatial gradients and E_t the temporal gradient of the luminance; (u, v) is the projection of the 3D velocity field in the focal plane; ξ is the variation of the global illumination.

C. Arcelli et al. (Eds.): IWVF4, LNCS 2059, pp. 473–483, 2001.
© Springer-Verlag Berlin Heidelberg 2001

The computation of optical flow is based on the image irradiance continuity. Thus, only small movements can be estimated. Moreover this approach is based on two antagonist assumptions. On the one hand the image must be sufficiently textured so that the motion can be visible; on the other hand, the computation of local gradients E_x, E_y and E_t is made through a low-pass filter which requires a low texture of the image.

The effects of these constraints can be reduced. For example Bouthemy in [2] proposes to use robust statistics to deal with the contamination of the main motion due to the parasite motion of moving objects. He also proposes a multiresolution process to cope with the small displacement constraint.

This article presents a new method, based on possibility theory and rough histograms, to estimate the main motion (rigid transformation between two images of a video sequence). It is structured as follows: some updates about fuzzy concepts are given in section2. Section 3 briefly presents rough histograms and explains their extension in 2D. Section 4 deals with the method of motion estimation in details. Section 5 explains how a multiresolution process can improve the method. In section 6 some results are presented. Finally in section 7, a conclusion and an extension of this work are proposed.

2   Update on Fuzzy Concepts

Some of the tenets of fuzzy subset theory are now reviewed to set the stage for the discussion that follows. Further details on fuzzy subsets are given in [3].

2.1   Imprecise Quantity Representation

Fuzziness is understood as the uncertainty associated with the definition of ill-defined data or values. A fuzzy subset of a set Ω is a mapping (or membership function) from Ω to [0;1]. When $\Omega = \mathbb{R}$ (*i.e.* data are real numbers) then a fuzzy subset is called a fuzzy quantity. Crisp concepts as intervals can be easily generalized as fuzzy quantities. A fuzzy interval is a convex fuzzy quantity, that is, one whose membership function is quasiconcave:

$$\forall u, v, \forall w \in [u, v], \mu_Q(w) \geq \min(\mu_Q(u), \mu_Q(v)) \tag{2}$$

A fuzzy interval is fully characterized by its core and support (Fig. 2.1). A fuzzy interval with a compact support and a unique core value is called a fuzzy number (Fig. 2.1).

2.2   Restricted Possibility and Necessity

To compare two imprecise data (H and D), a solution given by [3], is to use two measures: the possibility of D with respect to H ($\Pi(H; D)$) and the necessity of D with respect to H ($N(H; D)$) defined by:

$$\Pi(H; D) = \sup_{\omega} \min(\mu_H(\omega); \mu_D(\omega)) \tag{3}$$

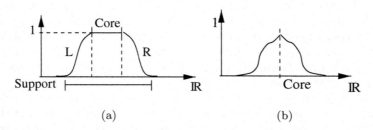

Fig. 1. Fuzzy quantities: (a) Fuzzy interval (b) Fuzzy number.

$$N(H; D) = \inf_{\omega} \max(\mu_H(\omega); 1 - \mu_D(\omega)) \tag{4}$$

The measure $\Pi(H; D)$ estimates how it is possible for H and D to refer to the same value ω. $N(H; D)$ estimates how it is certain that the value to which D refers to is among the ones compatible with H [4].

3 Rough Histograms

3.1 1-D Rough Histograms

Rough histograms are a generalization of crisp histograms [5]. Hence they can deal with imprecise data and tend to reduce the influence of the arbitrary choice of space partitioning. Using a rough histogram amounts to replacing the classical partition on which it is built by a fuzzy one, and using an imprecise accumulator. This accumulator is defined by its two boundaries called lower and upper accumulator.

Let $(Di)_N$ $(i = 1...N)$ be N imprecise data whose density of probability has to be estimated on the interval $[e_{\min}, e_{\max}]$. Rough histograms give an imprecise estimation of this density of probability on the considered interval. This density is approximated by two accumulators built on a fuzzy partition on $[e_{\min}, e_{\max}]$ of p cells H_k (Fig. 3.1).

Each cell H_k is associated with an upper accumulator $\overline{Acc_k}$ and a lower accumulator $\underline{Acc_k}$ defined by :

$$\overline{Acc_k} = \sum_{i=1}^{N} \Pi(H_k; D_i) \tag{5}$$

$$\underline{Acc_k} = \sum_{i=1}^{N} N(H_k; D_i) \tag{6}$$

where $\Pi(H_k; D_i)$ is the possibility of H_k with respect to D_i and $N(H_k; D_i)$ is the necessity of H_k with respect to D_i [3].

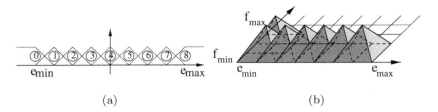

Fig. 2. Fuzzy partitions: (a) 9 cells fuzzy partition (b) 2D-Fuzzy partition.

3.2   Rough Histograms in 2D

If the data whose density has to be estimated is a two-dimensional one, the 1-D partition has to be replaced by a 2-D partition ($[e_{min}, e_{max}] * [f_{min}, f_{max}]$). If the t-norm min is used, the membership function is pyramidal (Fig. 3.1).

The formula 5 and 6 becomes:

$$\overline{Acc_{k,l}} = \sum_{i=1}^{N} \sum_{j=1}^{M} \Pi(H_{(k,l)}; D_{(i,j)}) \tag{7}$$

$$\underline{Acc_{k,l}} = \sum_{i=1}^{N} \sum_{j=1}^{M} N(H_{(k,l)}; D_{(i,j)}) \tag{8}$$

with:

$$\Pi(H_{(k,l)}; D_{(i,j)}) = \min\left(\Pi\left(H_k; D_i\right); \Pi\left(H_l; D_j\right)\right) \tag{9}$$

$$N(H_{(k,l)}; D_{(i,j)}) = \min\left(N\left(H_k; D_i\right); N\left(H_l; D_j\right)\right) \tag{10}$$

where $D_{(i,j)} = D_i \times D_j$, (D_i, D_j) are intervals and $\times$ is the Cartesian product obtained with the t-norm min.

4   Presentation of the Method

The method presented is akin to the correlation- and optical flow methods. It requires a discretization of the parameters' space. However, rough histograms are used to reduce the effects due to this discretization. Moreover, the vagueness induced on the density estimation by the data imprecision can be easily taken into account by the imprecise accumulation process. Furthermore the constraint linking spatial and temporal changes in the image irradiance, has been released: variation of gray level pixel values are observed through two dual classes. This enhances the robustness of the estimation regarding to the motion's assumptions.

The process described can be decomposed in three modules: the first one gives an estimation of the eventuality of the pixel's spatial change based on irradiance temporal change. This estimation is used by a second module to built a rough histogram in the parameter's space. Then, the last module gives an estimation of the main apparent motion.

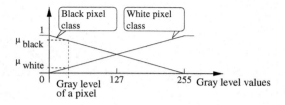

Fig. 3. Fuzzy membership function to pixels' class.

4.1 Observations Processing

This part explains how spatial change is related to irradiance change. As a matter of fact, the optical flow-based methods estimate the projection of the motion on the gradient of the image irradiance, while correlation-based methods use a statistical distance between the gray level pixel values. We propose to estimate the contingency of the displacement (u, v) of a pixel (i, j), while looking if the pixels (i, j) of the first picture and $(i + u, j + v)$ of the second one, belong to the same class. For this purpose, the gray level space is divided in 2 fuzzy dual classes : black pixels and white pixels (Fig. 3). The eventuality of displacement - or contingency - is measured in an imprecise manner with two antagonist values. These values are the displacement possibility and the non-displacement possibility.

The displacement (u, v) of the pixel (i, j) can be planned if (i, j) and $(i\prime, j\prime) = (i + u, j + v)$ belong to the same class. Likewise, the displacement cannot be planned if (i, j) and $(i\prime, j\prime)$ belong to different classes. This can be written as:

$$E_\Pi(u, v) = \max(\min(\mu_{N1}, \mu_{N2}); \min(\mu_{B1}, \mu_{B2})) \tag{11}$$

$$E_\Pi^c(u, v) = \max(\min(\mu_{N1}, \mu_{B2}); \min(\mu_{B1}, \mu_{N2})) \tag{12}$$

where μ_{N1}(resp. μ_{B1}) is the pixel (i, j) membership degree of the black pixel class (resp. white pixel class) and μ_{N2} (resp. μ_{B2}) is the pixel $(i\prime, j\prime)$ membership degree of the black pixel class (resp. white pixel class). Rather than using E_Π^c, we prefer using its complement E_N defined by:

$$E_N = 1 - E_\Pi^c \tag{13}$$

This process provides an upper and lower measure of the reliance in the displacement (u, v). This method is akin to optical flow-based ones in trying to link spatial and temporal variations of irradiance together. Meanwhile no assumption on the image's texture is claimed.

4.2 Matching between Observation and the Motion Model

The main apparent motion on image space due to the motion of a camera in 3-D environment can generally be modeled with 6 parameters (2 translations, 3

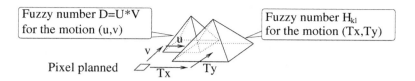

Fig. 4. Compatibility planned displacement / motion model.

rotations, and a scaling factor). However, [2] has shown that an estimation with a 4-parameter model is more robust. These parameters are: 2 translations on each axes of the image, 1 rotation around the normal to the focal plane and a scaling factor representing the variations of distance between the scene and the camera. To simplify the statement of the method, this paper will focus on a 2-parameter movement (Tx, Ty). In a further paper, a method with a 4-parameter movement will be presented.

A rough histogram built upon the parameters' space is used to estimate the main mode of the motion. For this purpose, the planned displacement of each pixel computed at the previous stage is used to build the imprecise accumulator. Then, the chief mode of the density gives an estimation of the main motion.

As mentioned in chapter 1, a discretization of the parametric space (Tx, Ty) is needed. We then create a 2-dimensional fuzzy partition built upon the displacement parameters' space. Let $H_{k,l}$ be the cell (k, l) of the fuzzy partition ($k = 1...K$ and $l = 1...L$) whose core is reduced to its mode (Tx_k, Ty_l) and whose support is defined by:

$$|H_{k,l}|_0 = [Tx_k - \Delta Tx, Tx_k + \Delta Tx] \times [Ty_l - \Delta Ty, Ty_l + \Delta Ty] \qquad (14)$$

where ΔTx, ΔTy are the spreads of the cell $H_{k,l}$. The spatial position of a pixel (i, j) is considered as an imprecise quantity of $\mathbb{R}^2$. It is usually modeled by a fuzzy number whose mode is located at the center of the pixel and whose spreading is linked to the pixel's width.

As u and v are defined by the subtraction of imprecise quantities ($u = (i - i\prime), v = (j - j\prime)$), they are also imprecise quantities (U, V) characterized by their supports $[umin, umax], [vmin, vmax]$ and by their modes. The displacement (u, v) is then modeled by a pyramidal fuzzy number $D = U \times V$. It is now necessary to find a relation between a planned movement (u, v) of a pixel (i, j) and the motion's model (Tx, Ty). The compatibility between this 2 magnitudes is estimated in a bi-modal way by: $\Pi(H_{k,l}; D)$ and $N(H_{k,l}; D)$ which are the possibility and the necessity of the cell $H_{k,l}$ - representing the motion (Tx, Ty) - with respect to the planned movement (u, v).

We now assume that the contingency of the motion (u, v) for a pixel (i, j) has been evaluated using (11) and (13). Then the pixel (i, j) will vote for the cell $H_{k,l}$ under assumption of a motion (u, v) if this displacement can be planned and if such a motion is compatible with the overall motion (Tx, Ty).

Like the displacement contingency evaluation, two votes are obtained for the cell $H_{k,l}$ which are accumulated in the upper and lower accumulators. These votes are called:

$$\text{Most favorable vote} = \min(E_\Pi(u, v); \Pi(H_{k,l}; D)) \qquad (15)$$

$$\text{Less unfavorable vote} = \min(E_N(u, v); N(H_{k,l}; D)) \qquad (16)$$

This procedure is repeated for each pixel of the image, for each planned displacement (u, v) and for each cell $H_{k,l}$ of the fuzzy partition. The two accumulators divided by the total number of votes set an upper and lower boundary of the motion probability density. One of the major changes of this bi-modal representation is that it provides an estimation of the motion but also a confidence in this estimation. That depends on the gap between the upper and lower accumulators.

4.3   Motion Estimation

Finding the main mode in a classical histogram consists in searching the bin whose accumulator is maximal. The precision of this localization is related to the spread of the histogram's bins. Using rough histograms almost comes down to the same thing. However, the precision of the mode's localization is more accurate than the one imposed by the partition. The search for the maximum involves a kind of interpolation between the discrete values of the partition [5].

That is, searching for a mode in a histogram amounts to searching for the position of a crisp or fuzzy interval Φ of precision Γ whose quantity of votes is maximal - locally or globally. The number of votes purporting to this interval Φ - given the distribution of votes on the fuzzy partition - has then to be estimated. This estimation is obtained by transposing the imprecise number of votes towards the interval Φ. This transfer uses pignistic probability [6], [7] to provide a reasonable interval $[\underline{Nb(\Phi)}, \overline{Nb(\Phi)}]$ of the sought after number of votes. This transfer can be written as:

$$\overline{Nb(\Phi)} = \sum_{\substack{i=1:m \\ j=1:n}} BetP(\Phi/H_{k,l})\overline{Nb_{(k,l)}} \qquad (17)$$

$$\underline{Nb(\Phi)} = \sum_{\substack{i=1:m \\ j=1:n}} BetP(\Phi/H_{k,l})\underline{Nb_{(k,l)}} \qquad (18)$$

with $\overline{Nb_{(k,l)}}$ (resp. $\underline{Nb_{(k,l)}}$) the value of the histogram's upper accumulator (resp. lower accumulator) associated to the cell (k, l).

The pignistic probability is defined by:

$$BetP(\Phi/H_{k,l}) = \frac{|\Phi \cap H_{k,l}|}{|H_{k,l}|} \qquad (19)$$

where $|A|$ means A cardinality.

Using the pignistic probability amounts to transferring some of the votes in favor of $H_{k,l}$ towards Φ in the overlapping proportion of $H_{k,l}$ and Φ.

The position of the maximum of $(\overline{Nb(\Phi)} + Nb(\Phi))$ is sought to find the chief motion corresponding to the maximum of an "average probability". Formulas (17) and (18) are functions of $\mathrm{Pos}(\Phi)$ - position of the interval Φ. The position of Φ fulfilling:

$$\frac{\partial(\overline{Nb(\Phi)} + Nb(\Phi))}{\partial Pos(\Phi)} = 0 \qquad (20)$$

corresponds to an extremum of the number of votes according to the imprecise accumulator. If this extremum is a maximum, its position then corresponds to the apparent motion between the two images.

The pignistic probability involves the computation of the volume of intersection between a pyramid and a volumetric shape representing the membership function of Φ. Even though this membership function is selected as the most simple ever - parallelepipedal shape - the computation of the pignistic probability is very time consuming. Increasing the number of parameters of the motion model will end up complicating or even preventing the pignistic probability computation.

Thus, using rough histograms transform non-monotonous data into monotonous ones just like classical histograms transform non-regular data into regular ones. If the data are monotonous within the interval $[T_x - \Delta T_x, T_x + \Delta T_x] \times [T_y - \Delta T_y, T_y + \Delta T_y]$, then the maximum of $(\overline{Nb(\Phi)} + Nb(\Phi))$ can be locally found on the projections of $[T_x - \Delta T_x, T_x + \Delta T_x] \times [T_y - \overline{\Delta T_y, T_y} + \Delta T_y]$. The problem then comes down to searching for two 1-D modes ; this research process is explained in [5]. This property reduces the complexity of computation brought about by the use of pignistic probability.

5   Bi-Modal Multiresolution Process

The complexity of this algorithm increases with the image size, the number of planned displacement and the number of histogram's cells. The computation time would be prohibitive when large displacement are planned. This complexity can be reduced with a multiresolution process.

Classical multiresolution methods consist in reducing the amount of data by averaging gray level values of a pixel unit. The multiresolution process we use is slightly different. To keep the imprecision information about the pixels' gray level value, a 2×2 pixels unit gray level value will be represented by two magnitudes: the pixel unit possible membership of the black-pixel and white-pixel classes. This two values are defined as the maximum of the 4 pixels membership to the black-pixel (resp. white-pixel) class.

At the finest resolution, this two boundaries are reduced to the same value. The multiresolution process aims at getting an incremental estimation of the apparent motion through a pyramidal process in a coarse to fine refinement. It uses the following property: a displacement (u, v) at a given resolution N-1

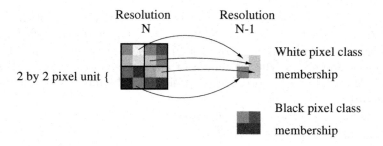

Fig. 5. Bi-modal multiresolution process.

amounts to a displacement $(2 * u, 2 * v)$ at a resolution N. Thus, large displacements can be estimated without increasing the computation time, i.e., using a reduced number of histogram bins.

6   Some Results

The mosaic presented here (Fig. 6) has been run out on a 30-images sequence shot in front of the LIRMM. The video camera operates a counter-clock-wise and then clock-wise rotation around the vertical axis. This camera being held at arm's length, the image is not steady and the sequence has many vertical variations. Finally the images recording rate being rather slow, the global brightness varies a lot between 2 successive images.

The mosaic is created by estimating the displacement (Tx, Ty) between two consecutive images and by superposing them on the resulting image. Fig. 6 presents the readjusted superposition of the 30 images. We can notice that:

- The first and the last images of the sequence are perfectly superposed while only linked through the 28 intermediate images; this estimation is then accurate and seems not to be biased.

- The motion model - two translations - is not a good match to real motion (motion includes rotations); this approximation does not bother the method. This shows its robustness as far as the model is concerned.

- The variations of brightness and the stream of people coming in the lab does not disturb the detection of the main motion mode; this shows the robustness of the method as far as the data contamination is concerned.

7   Perspectives and Conclusion

A new method of main motion estimation between two images has been presented in this paper. This method partly copes with small displacements assumptions and with texture-linked constraints. A discretization of motion parameters space is needed. However, using rough histograms involves some kind of natural interpolation, leading to a motion estimation less sensitive to discretization. This

Fig. 6. Mosaic picture of the LIRMM.

estimation method seems to be robust regarding the model and data contamination. Moreover, reliability in the detection of the main motion mode allows to qualify the given estimation.

Nonetheless, this method has some drawbacks. Defining an arbitrary research area is necessary and leads to a limitation of the motion to be estimated. Searching the main mode of the votes' distribution provides a reliable estimation of the main apparent motion if the overlapping between two consecutive images is about 80%. Under this overlapping rate, the probability of detecting the mode is bigger than the detection's own precision, so detection is not guaranteed. Finally, the computing time needed for this algorithm is rather long, which can confine its application field. This last aspect is being improved.

In a further paper the method will be extended to a 4-parameter motion model. We will also use it for a purpose of disparity estimation between two stereo images. Finally, the theoretical research on imprecise probability needs to be further looked upon.

References

1. B. K. P. Horn and B. G. Schunck. Determining optical flow. *Artificial Intelligence*, 17:185–203, 1981.
2. J. M. Odobez and P. Bouthemy. Robust multi-resolution estimation of parametric motion models applied to complex scenes. Pi 788, IRISA, 1994.
3. D. Dubois and H. Prade. *Possibility Theory An Approach to Computerized Processing of Uncertainty*. Plenum Press, 1988.
4. D. Dubois, H. Prade, and C. Testemale. Weighted fuzzy pattern matching. *Fuzzy sets and systems*, 28:313–331, May 1988.
5. O. Strauss, F. Comby, and M. J. Aldon. Rough histograms for robust statistics. In *International Conference on Pattern Recognition*, volume 2, pages 688–691. IAPR, September 2000.

6. P. Smets and R. Kennes. The transferable belief model. *Artificial Intelligence*, 66:191–243, 1994.
7. T. Denoeux. Reasoning with imprecise belief structures. Technical Report 97/44, Universite de Technologie de Compiegne, Heudiasyc Laboratory, 1997.

Prototyping Structural Shape Descriptions
by Inductive Learning

L.P. Cordella[1], P. Foggia[1], C. Sansone[1], F. Tortorella[2], and M. Vento[1]

[1]Dipartimento di Informatica e Sistemistica
Università di Napoli "Federico II"
Via Claudio, 21 I-80125 Napoli (Italy)
E-mail: {cordel,foggiapa,carlosan,vento}@unina.it

[2]Dip. di Automazione, Elettromagnetismo, Ing. dell'Informazione e Matematica Industriale
Università degli Studi di Cassino
via G. di Biasio, 43 I-03043 Cassino (Italy)
E-mail: tortorella@unicas.it

Abstract. In this paper, a novel algorithm for learning structured descriptions, ascribable to the category of symbolic techniques, is proposed. It faces the problem directly in the space of the graphs, by defining the proper inference operators, as graph generalization and graph specialization, and obtains general and coherent prototypes with a low computational cost with respect to other symbolic learning systems. The proposed algorithm is tested with reference to a problem of handwritten character recognition from a standard database.

1. Introduction

Graphs enriched with a set of attributes associated to nodes and edges (called Attributed Relational Graphs) are the most commonly used data structures for representing structural information, e.g. associating the nodes and the edges respectively to the primitives and to the relations among them. The attributes of the nodes and of the edges represent the properties of the primitives and of the relations.

Structural methods [1,2] imply complex procedures both in the recognition and in the learning process. In fact, in real applications the information is affected by distortions, and consequently the corresponding graphs result to be very different from the ideal ones. The learning problem, i.e. the task of building a set of prototypes adequately describing the objects (patterns) of each class, is complicated by the fact that the prototypes, implicitly or explicitly, should include a model of the possible distortions. The difficulty of defining effective algorithms for facing this task is so high that the problem is considered still open, and only few proposals, usable in rather peculiar hypotheses, are now available.

A first approach to the problem relies upon the conviction that structured information can be suitably encoded in terms of a vector, thus making possible the adoption of statistical/neural paradigms; it is so possible to use the large variety of

C. Arcelli et al. (Eds.): IWVF4, LNCS 2059, pp. 484–493, 2001.
© Springer-Verlag Berlin Heidelberg 2001

well-established and effective algorithms both for learning and for classifying patterns. The main disadvantage deriving from the use of these techniques is the impossibility of accessing the knowledge acquired by the system. In fact, after learning, the knowledge is implicitly encoded (e.g. within the weights of connections of the net) and its use, outside classification stage, is strongly limited. Examples of this approaches are [2,3,4].

Another approach, pioneered by [5], contains methods facing the learning problem directly in the representation space of the structured data [6,7,8]. So, if data are represented by graphs, the learning procedure generate graphs for representing the prototypes of the classes.

Some methods, ascribable to this approach, are based on the assumption that the prototypical descriptions are built by interacting with an expert of the domain [9,10]; the inadequacy of human knowledge to find a set of prototypes really representative of a given class significantly increases the risk of errors, especially in domains containing either many data or many different classes.

More automatic methods are those facing the problem as a symbolic machine learning problem [11,12], so formulated: "given a suitably chosen set of input data, whose class is known, and possibly some background domain knowledge, find out a set of optimal prototypical descriptions for each class". A formal enunciation of the problem and a more detailed discussion to related issues will be given in the next section. Dietterich and Michalski [8] provide an extensive review of this field, populated by methods which mainly differ in the adopted formalism [11,13], sometimes more general than that implied by the graphs.

The advantage making this approach really effective relies in the obtained descriptions which are explicit. Moreover, the property of being maximally general makes them very compact, i.e. containing only the minimum information for covering all the samples of a same class and for preserving the distinction between objects of different classes, as required for understanding the features driving the classification task. Due to these properties, the user can easily acquire knowledge about the domain by looking at the prototypes generated by the system, which appear simple, understandable and effective. Consequently he can validate or improve the prototypes or understand what has gone wrong in case of classification errors.

In this paper we propose a novel algorithm for learning structured descriptions, ascribable to the category of symbolic techniques. It faces the problem directly in the space of the graphs and obtains general and coherent prototypes with a low computational cost with respect to other symbolic learning systems. The proposed algorithm is tested with reference to a problem of handwritten character recognition from a standard database.

2. Rationale of the Method

The rationale of our approach is that of considering descriptions given in terms of Attributed Relational Graphs and devising a method which, inspired to basic machine learning methodologies, particularizes the inference operations to the case of graphs. To this aim, we first introduce a new kind of Attributed Relational Graph, devoted to represent prototypes of a set of ARG's. These graphs are called Generalized Attributed Relational Graphs (GARG's) as they contain generalized nodes, edges, and attributes. Then, we formulate a learning algorithm which builds such prototypes by means of a set of operations directly defined on graphs. The algorithm preserves the generality of the prototypes generated by classical machine learning algorithms. Moreover, similarly to most of machine learning systems [8,11,12], the prototypes obtained by our system are *consistent*, i.e. each sample is covered by only one prototype.

We assume that the objects are described in terms of Attributed Relational Graphs (ARG). An ARG can be defined as a 6-tuple $(N, E, A_N, A_E, \alpha_N, \alpha_E)$ where N and $E \subset N \times N$ are respectively the sets of the nodes and the edges of the ARG, A_N and A_E the sets of nodes and edge attributes and finally α_N and α_E the functions which associate to each node or edge of the graph the corresponding attributes.

We will suppose that the attributes of a node or an edge are expressed in the form $t(p_1, \ldots, p_{k_t})$, where t is a type chosen over a finite alphabet T of possible types and $(p_1, \ldots, p_{k_t})$ are a tuple of parameters, also from finite sets $P_1^t, \ldots, P_{k_t}^t$. Both the number of parameters (k_t, the *arity* associated to type t) and the sets they belongs to depend on the type of the attribute; for some type k_t may be equal to zero, so meaning that the corresponding attribute has no parameters. It is worth noting that the introduction of the type permits to differentiate the description of different kinds of nodes (or edges); in this way, each parameter associated to a node (or an edge) assumes a meaning depending on the type of the node itself. For example, we could use the nodes to represent different parts of an object, by associating a node type to each kind of part (see fig. 1).

A GARG is used for representing a prototype of a set of ARG's; in order to allow a prototype to match a set of possibly different ARG's (the samples covered by the considered prototype) we extend the attribute definition. First of all, the set of types of node and edge attribute is extended with the special type ϕ, carrying no parameter and allowed to match any attribute type, ignoring the attribute parameters. For the other attribute types, if the sample has a parameter whose value is within the set P_i^t, the corresponding parameter of the prototype belongs to the set $P_i^{*t} = \wp(P_i^t)$, where $\wp(X)$ is the power set of X, i.e. the set of all the subsets of X. Referring to the example of the geometric objects in Fig.1, a node of the prototype could have the attribute $rectangle(\{s, m\}, \{m\})$, meaning a rectangle whose width is small or medium and whose height is medium.

We say that a GARG $G^* = (N^*, E^*, A_N^*, A_E^*, \alpha_N^*, \alpha_E^*)$ covers a sample G and use the notation $G^* \models G$ (the symbol $\models$ denotes the relation here on called *covering*) iff there is a mapping $\mu : N^* \to N$ such that:

(1) μ is a *monomorphism*, and

(2)   the attributes of the nodes and of the edges of G^* are compatible with the corresponding ones of G.

The *compatibility* relation, denoted with the symbol $\succ$ is formally introduced as:

$$\forall t, \ \phi \succ t(p_1, \ldots, p_{k_t}) \ \text{ and } \ \forall t, \ t(p_1^*, \ldots, p_{k_t}^*) \succ t(p_1, \ldots, p_{k_t}) \Leftrightarrow p_1 \in p_1^* \wedge \ldots \wedge p_{k_t} \in p_{k_t}^*$$

$$(3)$$

Condition (1) requires that each primitive and each relation in the prototype must be present also in the sample. This allows the prototype to specify only the features which are strictly required for discriminating among the various classes, neglecting the irrelevant ones. Condition (2) constrains the monomorphism required by condition (1) to be consistent with the attributes of the prototype and of the sample, by means of the compatibility relation defined in (3): this latter simply states that the type of the attribute of the prototype must be either equal to ϕ or to the type of the corresponding attribute of the sample; in the latter case all the parameters of the attribute, which are actually sets of values, must contain the value of the corresponding parameter of the sample.

Another important relation that will be introduced is *specialization* (denoted by the symbol $\trianglelefteq$): a prototype G_1^* is said to be a *specialization* of G_2^* iff:

$$\forall G, \quad G_1^* \models G \Rightarrow G_2^* \models G \tag{4}$$

In other words, a prototype G_1^* is a specialization of G_2^* if every sample covered by G_1^* is also covered by G_2^*. Hence, a more specialized prototype imposes stricter

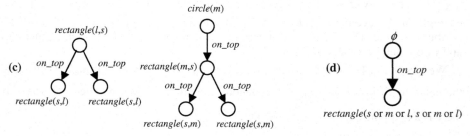

NODE TYPE ALPHABET

$T = \{\ triangle\ ,\ rectangle\ ,\ circle\ \}$

$k_{triangle} = 2 \qquad k_{rectangle} = 2 \qquad k_{circle} = 1$

$p_1^{triangle} = \text{base} = \{\ small\ ,\ medium\ ,\ large\ \} \cong \{s,m,l\}$

$p_2^{triangle} = \text{height} = \{\ small\ ,\ medium\ ,\ large\ \} \cong \{s,m,l\}$

$p_1^{rectangle} = \text{width} = \{\ small\ ,\ medium\ ,\ large\ \} \cong \{s,m,l\}$

$p_2^{rectangle} = \text{height} = \{\ small\ ,\ medium\ ,\ large\ \} \cong \{s,m,l\}$

$p_1^{circle} = \text{radius} = \{\ small\ ,\ medium\ ,\ large\ \} \cong \{s,m,l\}$

EDGE TYPE ALPHABET

$T = \{\ on\_top\ ,\ left\ \}$

$k_{on\_top} = 0 \qquad k_{left} = 0$

Fig. 1. (a) Objects made of three parts: circles, triangles and rectangles. (b) The description scheme introduces three types of nodes, each associated to a part. Each type contains a set of parameters suitable to describe each part. Similarly edges of the graph, describing topological relations among the parts, are associated to two different types. (c) The graphs corresponding to the objects in a) and (d) a GARG representing the two ARG's in c). Informally the GARG represents "any object made of a part on the top of a rectangle of any width and height".

requirements on the samples to be covered. In fig. 1d a prototype covering some objects is reported.

3. The Proposed Learning Algorithm

The goal of the learning algorithm can be stated as follows: there is a (possibly infinite) set S^* of all the patterns that may occur, partitioned into C different classes $S_1^*, \ldots, S_C^*$, with $S_i^* \cap S_j^* = \varnothing$ for $i \neq j$; the algorithm is given a finite subset $S \subset S^*$ (*training set*) of labeled patterns ($S = S_1 \cup \ldots \cup S_C$ with $S_i = S \cap S_i^*$), from which it tries to find a sequence of prototype graphs $G_1^*, G_2^*, \ldots, G_p^*$, each labeled with a class identifier, such that:

$$\forall G \in S^* \quad \exists i : G_i^* \models G \text{ (completeness)} \tag{5}$$

$$\forall G \in S^* \quad G_i^* \models G \Rightarrow \text{class}(G) = \text{class}(G_i^*) \text{ (consistency)} \tag{6}$$

where $\text{class}(G)$ and $\text{class}(G^*)$ refer to the class associated with samples G and G^*.

Of course, this is an ideal goal since only a finite subset of S^* is available to the algorithm; in practice the algorithm can only demonstrate that completeness and consistency hold for the samples in S. On the other hand, eq. (5) dictates that, in order to get as close as possible to the ideal case, the prototypes generated should be able to model also samples not found in S, that is they must be more *general* than the enumeration of the samples in the training set. However, they should not be too general otherwise eq. (6) will not be satisfied. The achievement of the optimal trade-off between completeness and consistency makes the prototypation a really hard problem.

A sketch of the learning algorithm is presented in Fig. 2; the algorithm starts with an empty list L of prototypes, and tries to cover the training set by successively adding consistent prototypes. When a new prototype is found, the samples covered by it are eliminated and the process continues on the remaining samples of the training set. The algorithm fails if no consistent prototype covering the remaining samples can be found. It is worth noting that the test of consistency in the algorithm actually checks whether the prototype is almost consistent, i.e. almost all the samples covered by G^* belongs to the same class:

$$Consistent(G^*) \Leftrightarrow \max_i \frac{\left| S_i(G^*) \right|}{\left| S(G^*) \right|} \geq \theta \tag{7}$$

where $S(G^*)$ denotes the sets of all the samples of the training set covered by a prototype G^*, $S_i(G^*)$ the samples of the class i covered by G^*, and $\bullet$ is a suitably chosen threshold, close to 1. Also notice that the attribution of a prototype to a class is performed after building the prototype.

FUNCTION *Learn(S) // Returns the ordered list of prototypes*
$L := [ ]$ *// L is the list of prototypes, initially empty*
WHILE $S \neq \varnothing$
   $G^* := FindPrototype(S)$
   IF NOT *Consistent*(G^*) THEN FAIL *// The algorithm ends unsuccessfully*
   // Assign the prototype to the class most represented
   class$(G^*) := \text{argmax}_i \, |S_i(G^*)|$
   $L := Append(L, G^*)$ *// Add G^* to the end of L*
   $S := S{-}S(G^*)$ *// Remove the covered samples from S*
END WHILE
RETURN *L*

Fig. 2. A sketch of the learning procedure.

According to (7) the algorithm would consider consistent a prototype if more than a fraction θ of the covered training samples belong to a same class, avoiding a further specialization of this prototype that could be detrimental for its generality.

The most important part of the algorithm is the *FindPrototype* procedure, illustrated in fig. 3. It performs the construction of a prototype, starting from a trivial prototype with one node whose attribute is ϕ (i.e. a prototype which covers any non-empty graph), and refining it by successive specializations until either it becomes consistent or it covers no samples at all. An important step of the *FindPrototype* procedure is the construction of a set Q of specializations of the tentative prototype G^*. The adopted definition of the heuristic function H, guiding the search of the current optimal prototype, will be examined later.

FUNCTION *FindPrototype(S)*
   // The initial tentative prototype is the trivial one, made of one node with attr. ϕ
   $G^* :=$ TrivialPrototype *// The current prototype*
   WHILE $|S(G^*)| > 0$ AND NOT *Consistent*(G^*)
      $Q := Specialize(G^*)$
      $G^* := \text{argmax}_{X \in Q} H(S, X)$ *// H is the heuristic function*
   END WHILE
   RETURN G^*

Fig. 3. The function *FindPrototype*

At each step, the algorithm tries to refine the current prototype definition, in order to make it more consistent, by replacing the tentative prototype with one of its specializations. To accomplish this task we have defined a set of specialization operators which, given a prototype graph G^*, produce a new prototype $\overline{G}^*$ such that $\overline{G}^* \trianglelefteq G^*$. The considered specialization operators are:

1. Node addition: $\overline{G}^*$ is augmented with a new node n whose attribute is ϕ.

2. Edge addition: a new edge (n_1^*, n_2^*) is added to the edges of G^*, where n_1^* and n_2^* are nodes of G^* and G^* does not contain already an edge between them. The edge attribute is ϕ.

3. Attribute specialization: the attribute of a node or an edge is specialized according to the following rule:

- If the attribute is ϕ, then a type t is chosen and the attribute is replaced with $t(P_1^t, \ldots, P_{k_t}^t)$. This means that only the type is fixed, while the type parameters can match any value of the corresponding type.

- Else, the attribute takes the form $t(p_1^*, \ldots, p_{k_t}^*)$, where each p_i^* is a (non necessarily proper) subset of P_i^t. One of the p_i^* such that $\left| p_i^* \right| > 1$ is replaced with $p_i^* - \{p_i\}$, with $p_i \in p_i^*$. In other words, one of the possible values of a parameters is excluded from the prototype.

The heuristic function H is introduced for evaluating how promising the provisional prototype is. It is based on the estimation of the consistency and completeness of the prototype (see eq. 5 and 6). To evaluate the consistency degree of a provisional prototype G^*, we have used an entropy based measure:

$$H_{\text{cons}}(S, G^*) = -\sum_i \frac{|S_i|}{|S|} \log_2 \frac{|S_i|}{|S|} - \left(-\sum_i \frac{|S_i(G^*)|}{|S(G^*)|} \log_2 \frac{|S_i(G^*)|}{|S(G^*)|} \right) \tag{8}$$

H is defined so that the larger is the value of $H_{\text{cons}}(S, G^*)$, the more consistent is G^*; hence the use of H_{cons} will drive the algorithm towards consistent prototypes.

The completeness of a provisional prototype is taken into account by a second term of the heuristic function, which simply counts the number of samples covered by the prototype:

$$H_{\text{compl}}(S, G^*) = \left| S(G^*) \right| \tag{9}$$

This term is introduced in order to privilege general prototypes with respect to prototypes which, albeit consistent, cover only a small number of samples.

The heuristic function used in our algorithm is the one described in [13]:

$$H(S, G^*) = H_{\text{compl}}(S, G^*) \cdot H_{\text{cons}}(S, G^*) \tag{10}$$

4. Experimental Results

In order to test the learning ability of the proposed method, preliminary tests have been carried out on digits belonging to the NIST database 19 [14].

Before describing them in terms of ARG's, characters have been decomposed in circular arcs by using the method presented in [15].

Fig. 4 illustrates the adopted description scheme in terms of ARG's; basically, we have two node types respectively used for representing the segments and their junctions, while the edges represent the adjacency of a segment to a junction. Node attributes are used to encode the size of the segments (normalized with respect to the size of the whole character) and their orientation; the edge attributes represent the relative position of a junction with respect to the segments it connects.

T_N = { *stroke* , *junction* }

k_{stroke} = 3

$k_{junction}$ = 0

p_1^{stroke} = size = { *very_short* , *short* , *medium* , *large* , *very_large* } = { *vs* , *s* , *m* , *l* , *vl* }

p_2^{stroke} = shape = { *straight* , *light_bent* , *bent* , *highly_bent* , *circle* } = { *s* , *lb* , *b* , *hb* , *c* }

p_3^{stroke} = orientation = { *n* , *nw* , *w* , *sw* , *s* , *se* , *e* , *ne* }

T_B = { *connection* }

$k_{connection}$ = 2

$p_1^{connection}$ = x-projection = { *leftmost* , *vertical* , *rightmost* } = { *l* , *v* , *r* }

$p_2^{connection}$ = y-projection = { *bottom* , *horizontal* , *above* } = { *b* , *h* , *a* }

Fig. 4. The alphabet of the types defined for the decomposition of the characters in circular arcs.

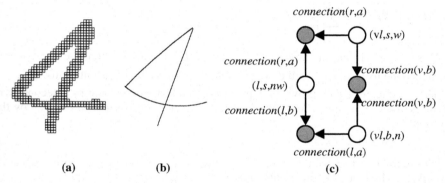

Fig. 5. (a) The bitmap of a character, (b) its decomposition in terms of circular arcs; (c) its representation in terms of ARG's.

In Fig. 5 an example of such a representation for a digit of the NIST database is shown.

No effort has been made to eliminate the noise from the *training set*: experimentally it has been observed that the chosen structural description scheme provides to the prototyper several isomorphic graphs belonging to different classes.

Some tests have been made by varying the composition of the *training set* and the consistency threshold; here only the results obtained by imposing a minimum consistency of the 90% are shown, as higher values are harmful for the generalization ability of the method. In particular, we present three different experiments: in the first one the *training set* is made up of 50 samples per class, randomly chosen by the database; in the second one 100 samples per class have been selected; while in the third one 200 samples per class.

From Fig. 6, it can be noted that the *training sets* are affected by noise: in fact, the recognition rates obtained on these sets range from 94% to 96%.

	0	1	2	3	4	5	6	7	8	9	R
0	98.21									1.79	
1		100									
2		1.92	92.31		1.92		1.92	1.92			
3			1.79	96.43	1.79						
4			1.96		88.24			1.96	3.92	3.92	
5				2.22		95.56	2.22				
6						1.89	98.11				
7								100			
8			1.89	1.89				1.89	94.34		
9				1.96				1.96		96.08	

(a)

	0	1	2	3	4	5	6	7	8	9	R
0	98.20					0.90	0.90				
1		97.50			0.83		0.83				0.83
2		0.95	96.19	0.95				0.95	0.95		
3			0.90	96.40	0.90				1.80		
4			0.98		97.06				0.98	0.98	
5			3.30	4.40		91.21	1.10				
6	3.81						94.29				1.90
7								100			
8		1.90		0.95				1.90	94.29	0.95	
9	0.97							6.80	1.94	90.29	

(b)

	0	1	2	3	4	5	6	7	8	9	R
0	96.85		0.90	0.45		0.45	1.35				
1		88.28	0.84		0.42		1.67			0.42	8.37
2			91.47	0.47	2.37	0.47		1.42	2.84		0.95
3	0.45	1.35	0.45	95.05	0.45	0.90	0.45	0.45	0.45		
4		0.49		0.49	97.54			0.99	0.49		
5		0.55	1.64	3.83	1.09	91.26	0.55		0.55		0.55
6		0.48	0.48	0.48			97.13				1.44
7								97.76		0.45	1.79
8		0.96	1.44	0.96		0.48	1.44	0.48	90.91	2.39	0.96
9	0.97		0.48		0.48	0.48		4.83	0.48	92.27	

(c)

Fig. 6. Misclassification tables on the *training set* as a function of its size: (**a**) 500 samples, (**b**) 1000 samples, (**c**) 2000 samples. The **R** column indicates the percentage of rejected samples.

However, the results obtained by the learning system, without taking into account the noise introduced by the description phase, are encouraging: the learning times for the above described tests were of 7, 12 e 22 minutes respectively, on a 600MHz Pentium III PC equipped with 128 MB of RAM. It is worth noting that multiplying by 2 the number of *training set* samples, the processing time grows by a factor slightly less than 2. This fact suggests that the better representativeness of the input data allows the generation of prototypes less affected by noise.

This hypothesis is confirmed by the total number of prototypes produced by the system: starting from 500 samples, 83 prototypes have been generated; with 1000 samples 119 prototypes have been built and with a training set of 2000 samples the number of generated prototypes was 203.

In other words, as the size of the training set grows, the performance of the system becomes better both in terms of generalization ability and of learning times.

5. Conclusions

In this paper we have presented a novel method for learning structural descriptions from examples, based on a formulation of the learning problem in terms of Attributed Relational Graphs. Our method, like learning methods based on first-order logic, produces general prototypes which are easy to understand and to manipulate, but it is based on simpler operations (graph editing and graph matching) leading to a smaller overall computational cost. A preliminary experimentation has been conducted, which seems to confirm our claims about the advantages of our method.

References

[1]   T. Pavlidis "Structural Pattern Recognition", Springer, New York, 1977.
[2]   P. Frasconi, M. Gori and A. Sperduti, "A general framework for adaptive processing of data structures, IEEE Trans. on Neural Networks, vol. 9, n. 5, pp. 768-785, 1998.
[3]   G.E. Hinton, "Mapping part-whole hierarchies into connectionist networks", Artificial Intelligence, vol. 46, pp. 47-75, 1990.
[4]   D.S. Touretzky, "Dynamic symbol structures in a connectionist network", In Artificial Intelligence, vol. 42, n. 3, pp. 5-46, 1990.
[5]   P.H. Winston, "Learning Structural Descriptions from Examples", Tech. Report MAC-TR-76, Dep. of Electrical Engineering and Computer Science - MIT, 1970.
[6]   N. Lavrac, S. Dzerosky, "Inductive Logic Programming: Techniques and Applications", Ellis Horwood, 1994.
[7]   A. Pearce, T. Caelly, and W. F. Bischof, "Rulegraphs for graph matching in pattern recognition", Pattern Recognition, vol. 27, n. 9, pp. 1231-1247, 1994.
[8]   T.G. Dietterich and R.S. Michalski, "A comparative review of selected methods for learning from examples", In R.S. Michalski et al. eds., Machine Learning: An Artificial Intelligence Approach, vol. 1, chap. 3, pp. 41-82. Morgan Kaufmann, 1983.
[9]   J. Rocha and T. Pavlidis, "A shape analysis model with applications to a character recognition system", IEEE Trans. on PAMI, vol. 16, n. 4, pp. 393-404, 1994.
[10]   H. Nishida, "Shape recognition by integrating structural descriptions and geometrical/statistical transforms", Computer Vision and Image Understanding, vol. 64, pp. 248-262, 1996.
[11]   R.S. Michalski, "Pattern recognition as rule-guided inductive inference", IEEE Trans. on Pattern Analysis and Machine Intelligence, vol. 2, n. 4, pp. 349-361, July 1980.
[12]   L.P. Cordella, P. Foggia, R. Genna and M. Vento, "Prototyping Structural Descriptions: an Inductive Learning Approach", Advances in Pattern Recognition, Lecture Notes in Computer Science, n. 1451, pp. 339-348, Springer-Verlag, 1998.
[13]   J.R. Quinlan, "Learning Logical Definitions from Relations", Machine Learning, vol. 5, n. 3, pp. 239-266, 1993.
[14]   P.J. Grother, NIST Special Database 19, Technical Report, National Institute of Standards and Technology, 1995.
[15]   A. Chianese, L.P. Cordella, M. De Santo and M. Vento, "Decomposition of ribbon-like shapes", Proc. 6th Scandinavian Conference on Image Analysis, pp. 416-423, 1989.

Training Space Truncation in Vision-Based Recognition

René Dencker Eriksen and Ivar Balslev

University of Southern Danmark,Odense,
The MaerskMc-Kinney Moller
Institute for Production Technology

Abstract. We report on a method for achieving a significant truncation of the training space necessary for recognizing rigid 3D objects
from perspective images. Considering objects lying on a table, the
configuration space of continuous coordinates is three-dimensional. In
addition the objects have a few distinct support modes. We show that
recognition using a stationary camera can be carried out by training
each object class and support mode in a two-dimensional configuration
space. We have developed a transformation used during recognition for
projecting the image information into the truncated configuration space
of the training. The new concept gives full flexibility concerning the
position of the camera since perspective effects are treated exactly. The
concept has been tested using 2D object silhouettes as image property
and central moments as image descriptors. High recognition speed and
robust performance are obtained.

Keywords: Computer vision for flexible grasping, recognition of 3D objects, pose estimation of rigid object, recognition from perspective images, robot-vision systems

1 Introduction

We describe here a method suitable for computer-vision-based flexible grasping
by robots. We consider situations where classes of objects with known shapes,
but unknown position and orientation are to be classified and grasped in a structured way. Such systems has many quality measures such as recognition speed,
accuracy of the pose estimation, low complexity of training, free choice of camera
position, generality of object shapes, ability to recognize occluded objects, and
robustness. We shall evaluate the properties of the present method in terms of
these quality parameters.

Recognition of 3D objects has been widely based on establishing correspondences between 2D features and the corresponding features on the 3D object
[1-3]. The features has been point like, straight lines or curved image elements.
The subsequent use of geometric invariants makes it possible to classify and pose
estimate objects [4-8]. Another strategy is the analysis of silhouettes. When perspective effects can be ignored, as when the objects are flat and the camera is

C. Arcelli et al. (Eds.): IWVF4, LNCS 2059, pp. 494–503, 2001.

remote, numerous well established methods can be employed in the search for match between descriptors of recorded silhouette and those of silhouettes in a data base [9-11]. Other method are based on stereo vision or structured light [1, 11-12].

In the present paper we face the following situation:

- The rigid objects do not have visual features suitable for 2D-3D correspondence search or 2D-2D correspondence search used in stereo vision.
- No structured light is employed.
- The camera is not necessarily remote, so that we must take perspective effects into account.

We propose here a 'brute force' method [13] in which a large number of images or image descriptors are recorded or constructed during training. Classification and pose estimation is then based on a match search using the training data base. A reduction of the configuration space of the training data base is desirable since it gives a simpler training process and smaller extent of the data bases. The novelty of the present method is the recognition based on training in a truncated configuration space.

The method is based on a nonlinear 2D transformation of the image to be recognized. The transformation corresponds to a virtual displacement of the object into an already trained position relative to the camera. As relevant for many applications we consider objects lying at rest on a table or conveyer belt. In Sect. 2 we describe the 3D geometry of the system. We introduce the concept, 'virtual displacement', and define the truncated training space. In Sect. 3 are described the relevant 2D transformation and the match search leading to the classification and pose estimation. We also specify our choice of descriptors and match criterion in the recognition. The method has been implemented by constructing a physical training setup and by developing the necessary software for training results and recognition. Typical data of the setup and the objects tested are given in Sect. 4. We also present a few representative test results.

The work does not touch upon the 2D segmentation on which the present method must rely. The 2D segmentation is known to be a severe bottleneck if the scene illumination and relative positions of the objects are inappropriate. In the test we used back-lighting and nonoccluded objects in order to avoid such problems. Therefore we can not evaluate the system properties in case of complex 2D segmentation.

2   The 3D Geometry and the Truncated Training Space

In the present work we consider a selection of physical objects placed on a horizontal table or a conveyer belt, see Fig.1. The plane of the table surface is denoted π. We consider gravity and assume object structures having a discrete number of ways - here called support modes - on which the surface points touch the table. This means that we exclude objects, which are able to roll with a constant height of the center-of-mass. Let i count the object classes and let j

count the support modes. With fixed j, each object's pose has three degrees of freedom, e.g. (x, y, ω), where (x, y) is the projection on the table plane π of its center-of-mass (assuming uniform mass density), and ω is the rotation angle of the object about a vertical axis through the center-of-mass.

A scene with one or more objects placed on the table is viewed by an ideal camera with focal length f, pin hole position H at a distance h above the table, and an optical axis making an angle α with the normal to the plane π, see Fig. 1. The origin of (x, y) is H's projection O on π, and the y-axis is the projection of the optical axis on π. Let (x, y, z) be the coordinates of a reference point of the object. We introduce the angle ϕ defined by:

$$\cos \phi = \frac{y}{\sqrt{x^2 + y^2}}, \quad \sin \phi = \frac{x}{\sqrt{x^2 + y^2}}. \tag{1}$$

Consider a virtual displacement of the object so that its new position is given by:

$$(x', y', \omega') = (0, \sqrt{x^2 + y^2}, \omega - \phi) \tag{2}$$

This displacement is a rotation about a vertical axis through the pinhole. Note that the same same points on the object surface are visible from the pin hole H in the original and displaced position. The inverse transformation to be used later is given by:

$$x = y' \sin \phi, \quad y = y' \cos \phi, \quad \omega = \omega' + \phi \tag{3}$$

The essential property of this transformation is that the corresponding 2D transformation is independent of the structure of the object. The truncation of the training space introduced in the present paper is based on this property.

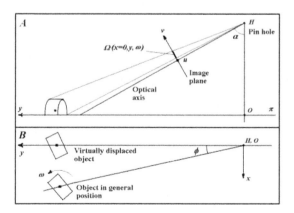

Fig. 1. Horizontal (A) and vertical (B) views of the system including the table, the camera, and the object before and after the virtual displacement.

We focus on image properties condensed in binary silhouettes. Therefore, we assume that the scene is arranged with a distinguishable background color so that each object forms a well defined silhouette $\Omega(i, j, x, y, \omega)$ on the camera image. Thus, $\Omega(i, j, x, y, \omega)$ is a list of coordinates (u, v) of set pixels in the image. We assume throughout that $(u, v) = (0, 0)$ is lying on the optical axis. The task in the present project is to determine (i, j, x, y, ω) from a measurement of an object's silhouette Ω_o and a subsequent comparison with the silhouettes $\Omega(i, j, x = 0, y, \omega)$ recorded or constructed in a reduced configuration space. In the data base the variables y and ω are suitably discretized. Silhouettes for the data base are either recorded by a camera using physical objects or constructed from a CAD representation.

3   The 2D Transformation and the Match Search

After the above mentioned virtual 3D displacement, an image point (u, v) of the object will have the image coordinates (u', v') given by:

$$u'(\phi, u, v) = \frac{f(u \cos \phi + v \sin \phi \cos \alpha - f \sin \phi \sin \alpha)}{u \sin \phi \sin \alpha + v(1 - \cos \phi) \sin \alpha \cos \alpha + f(\cos \phi \sin^2 \alpha + \cos^2 \alpha)} \tag{4}$$

$$v'(\phi, u, v) = \frac{f(-u \sin \phi \cos \alpha + v(\cos \phi \cos^2 \alpha + \sin^2 \alpha) + f(1 - \cos \phi) \sin \alpha \cos \alpha)}{u \sin \phi \sin \alpha + v(1 - \cos \phi) \sin \alpha \cos \alpha + f(\cos \phi \sin^2 \alpha + \cos^2 \alpha)} \tag{5}$$

This result can be derived by considering - in stead of an object displacement - three camera rotations about H: A tilt of angle $-\alpha$, a roll of angle ϕ, and a tilt of angle α. Then the relative object-camera-position is the same as if the object were displaced according to the 3D transformation described in Sect. 2. Note that the inverse transformation corresponds to a sign change of ϕ.

By transforming all points in a silhouette Ω according to (4-5), one obtains a new silhouette Ω'. Let us denote this silhouette transformation T_ϕ, so that

$$\Omega' = T_\phi(\Omega) \tag{6}$$

The 2D center-of-mass of Ω is $(u_{cm}(\Omega), v_{cm}(\Omega))$. The center-of-mass of the displaced silhouette Ω' is close to the transformed center-of-mass of the original silhouette Ω. In other words

$$u_{cm}(\Omega') \approx u'(\phi, u_{cm}(\Omega), v_{cm}(\Omega)) \tag{7}$$

$$v_{cm}(\Omega') \approx v'(\phi, u_{cm}(\Omega), v_{cm}(\Omega)) \tag{8}$$

This holds only approximately because the 2D transformation in (4-5) is nonlinear. In Fig 2 is shown a situation in which Ω' is a square. The black dot in Ω' is the transformed center-of-mass of Ω, which is displaced slightly from the center-of-mass of Ω'.

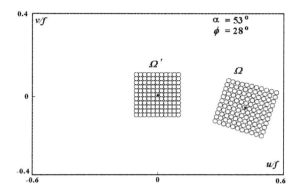

Fig. 2. An image component Ω and the transformed version Ω' in case that Ω' is a square. The values of α and ϕ are given in the upper right corner. The black dot Ω is the 2D center-of.mass. After transformation this point has a position shown as the black dot in Ω'.

Let $\Omega_{tr} = \Omega_{tr}(i, j, y, \omega)$ be the silhouettes of the training with $x = 0$. The training data base consist of descriptors of $\Omega_{tr}(i, j, y, \omega)$ with suitably discretized y and ω. In case of not too complex objects,

$$u_{cm}(\Omega_{tr}) \approx 0 \qquad (9)$$

The object to be recognized has the silhouette Ω_o. This silhouette defines an angle ϕ_o given by

$$\phi_o = \arctan(\frac{u_{cm}(\Omega_o)}{f \sin \alpha - v_{cm}(\Omega_o) \cos \alpha}) \qquad (10)$$

According to (4, 7, 10), the transformed silhouette $\Omega'_o = T_{\phi_o}(\Omega_o)$ has a 2D center-of-mass close to $u = 0$:

$$u_{cm}(\Omega'_o) \approx 0 \qquad (11)$$

We shall return to the approximations (7-9,11) later.

Eqs. (9) and (11) imply that Ω'_o is to be found among the silhouettes $\Omega_{tr}(i, j, y, \omega)$ of the data base. Because of the approximations (9,11), the similarity between Ω'_o and $\Omega_{tr}(i, j, y, \omega)$ is not exact with regards to translation, so one must use translational invariant descriptors in the comparison.

In the search for match between $\Omega_{tr}(i, j, y, \omega)$ and Ω'_o it is convenient to use that $v_{cm}(\Omega'_o) \approx v_{cm}(\Omega'_{tr}(i, j, y, \omega))$. It turns out, that $v_{cm}(\Omega'_{tr}(i, j, y, \omega))$ is usually a monotonous function of y, so - using interpolation - one can calculate a data base slice with a specified value v_{cm} and with i, j, ω as entries. This means that i, j, and ω can be determined by a match search between moments of Ω'_o

and moments in this data base slice. Note that the data base slice involves one continuous variable ω. With a typical step size of 3° the data base slice has only 120 records per support mode and per object class.

The result of the search are i_{match}, j_{match}, and ω_{match}. The value y_{match} can be calculated using the relation between y and $v_{cm}(\Omega_{tr})$ for the relevant values of i, j, and ω. The original pose (x, y, ω) can now be found by inserting $y' = y_{match}$, $\omega' = \omega_{match}$, and $\phi = \phi_o$ in Eq. (3).

If the approximation (9) brakes down, one must transform all the silhouettes of the data base, so that the match search takes place between Ω'_o and $\Omega'_{tr} = T_{\phi_{tr}}(\Omega_{tr})$ where

$$\phi_{tr} = \arctan(\frac{u_{cm}(\Omega_{tr})}{f \sin \alpha - v_{cm}(\Omega_{tr}) \cos \alpha}) \tag{12}$$

In this case $\phi = \phi_o - \phi_{tr}$ should be inserted in (3) in stead of ϕ_o.

We are left with the approximation (11), demonstrated in Fig. 2. This gives a slightly wrong angle ϕ_o. If the corresponding errors are harmful in the pose estimation, then one must perform an iterative calculation of ϕ_o, so that (11) holds exactly.

We conclude this section by specifying our choice of 1) image descriptors used in the data base, and 2) recognition criterion. In our test we have used as descriptors the 8-12 lowest order central moments, namely μ_{00}, μ_{20}, μ_{11}, μ_{02}, μ_{30}, μ_{21}, μ_{21}, and μ_{03}. The first order moments are absent since we use translational invarianat moments. In addition we used in some of the tests, the width and height of the silhouette. In the recognition strategy we minimized the Euclidean distance in a feature space of descriptors. The descriptors were normalized in such a way that the global variance of each descriptor was equal to one [14].

4   The Experiments

Fig. 3 shows the setup for training and test. We used a rotating table for scanning through the training parameter ω and a linear displacement of the camera for scanning through the parameter y. The pose estimation was checked by a grasping robot. In order to avoid 2D segmentation problems we used backlighting in both training and recognition.

The parameters for the setup and typical objects tested are shown in the Table 1.

We report here the result of a test of a single object, a toy rabbit manufactured by LEGO®, see Fig 4. Its 5 support modes are shown along with the support mode index used. After training using an angular step size of $\Delta\omega = 4^o$, we placed the rabbit with one particular support mode 100 random poses, i .e. values of (x, y, ω) in the field of view. The support modes detected by the vision system were recorded. This test was repeated for the remaining support modes. The results are shown in the Table 2. The two confused support modes were

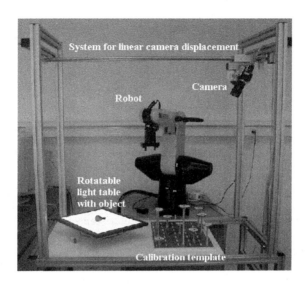

Fig. 3. Setup for training and test. The calibration template is used for calibrating the camera relative to a global coordinate system.

Table 1. Properties and parameters of the objects and the test setup.

Angle α	25°
Height h	800 mm
Field of view	400 mm x 300 mm
$\Delta\omega$ =angular step size during training	4°- 7.2^o
Δy =translational step size during training	50 mm
Camera resolution (pixels)	768 x 576
Number of support modes of objects	3-5
Typical linear object dimensions	25-40 mm
Typical linear dimensions of object images	40-55 pixels
Silhouette descriptors	8 lowest order centr. moments + width & height of silhouette
Number of data base records per object	900-1500 for $\Delta\omega = 7.2^o$
Training time per support mode	5 min.
Recognition time (after 2D segmentation)	5-10 ms

'support by four paws' and 'support by one ear and fore paws'. It can be understood from Fig. 4, that these two support modes are most likely to be mixed up in a pose estimation. We repeated the experiment with $\Delta\omega = 7.2^o$. In this case no errors were measured in the 500 tests.

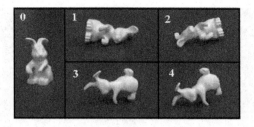

Fig. 4. The toy rabbit shown in its five support modes. The support mode indices used in Table 2 are written in the upper left corner.

Table 2. Statistics of the support mode detection when the toy rabit was placed at random positions (x, y, ω) in the field of view. Each support mode was tested 100 times and the experiments involved two different angular step sizes in the training.

Angular step size	\multicolumn{5}{c}{7.2^o}	\multicolumn{5}{c}{4^o}
True ↓, Detected →	0	1	2	3	4	0	1	2	3	4
0 standing	100					100				
1 lying on left side		100					100			
2 lying on right side			100					100		
3 on fore & hind paws				98	2				100	
4 on ear & fore paws				1	99					100

5   Discussion

A complete vision system for flexible grasping consists of two processes, one performing the 2D segmentation, and one giving the 3D interpretation. We have developed a method to be used in the second component only, since we used a illumination and object configuration giving very simple and robust segmentation.

The method developed is attractive with respect to the following aspects:

- High speed of the 3D interpretation.
- Generality concerning the object shape.
- Flexibility of camera position and object shapes, since tall objects, closely positioned cameras, and oblique viewing directions are allowed. In case of ambiguity in the pose estimation when viewed by a single camera, it is easy to use 2 or more cameras with independent 3D interpretation.
- Simple and fast training without assistance from vision experts.

The robustness and total recognition speed depends critically on the 2D segmentation, and so we can not conclude on these two quality parameters. The method in its present form is not suitable for occluded objects.

One remaining property to be discussed is the accuracy of the pose estimation. In our test the grasping uncertainty was about +/-2 mm and +/- 3°.

However, the origin of these uncertainties were not traced, so they may be reduced significantly by careful camera-robot co-calibration.

In our experiments we used a rather coarse descretization of y and ω, and only one object at a time was recognized. The recognition time in the experiment was typically 20 ms per object (plus segmentation time). This short processing times gives plenty of room for more demanding tasks involving more objects, more support modes, and higher accuracy through a finer discretizaion of y and ω.

6 Conclusion

We have developed and tested a computer vision concept appropriate in a brute force method based on data bases of image descriptors. We have shown that a significant reduction of the continuous degrees of freedom necessary in the training can be achieved by applying a suitable 2D transformation during recognition prior to the match search. The advantages are the reductions of the time and the storage used in the training process.

The prototype developed will be used for studying a number of properties and possible improvements. First, various types of descriptors and classification strategies will be tested. Here, color and gray tone information should be included. Second, the over-all performance with different 2D segmentation strategies will be studied, particularly those allowing occluded objects. Finally, the concept of training space truncation should be extended to systems recognizing objects of arbitrary pose.

References

1. B.K.P. Horn, Robot Vision, The MIT Press, Cambridge, Massachusetts, 1998.
2. R.M. Haralick, L.G. Shapiro, Computer and Robot Vision, Vol II, Addison Wesley, Reading, Massachusetts, 1993.
3. I. K. Park, K.M. Lee, S. U. Lee, Recognition and Reconstruction of 3D objects using Model-based Perceptual Grouping, Proc. Int. Conf Pattern Recognition 2000, Barcelona, Vol. 1 (A. Sanfeliu et. al., eds.) pp. 720-724, IEEE Computer Society, Los Alamitos
4. K. Nagao, B.K.P. Horn, Direct Object Recognition Using Higher Than Second Order Statistics of the Images, A.I.Memo#1526, MIT, 1995.
5. K. Nagao, W.E.L. Grimson, Using Photomtric Invariants for 3D Object Recognition, Computer Vision and Image Understanding 71, 1998, 74-93.
6. W.E.L. Grimson, Object Recognition by Computer: The Role of Geometric Constraints, The MIT Press, Cambridge, Massachusetts, 1990
7. J. L. Mundy and A. Zisserman, Repeated Structures: Image Corresponcence Constraints and 3D Structure Recovery, in Applications of Invariance in Computer Vision Springer-Verlag, Berlin (J.L. Mundy, A. Zisserman, and D. Forsyth, eds), 1994, pp. 89-106.
8. J. Ponce, D.J. Kriegman, Toward 3D Object Recognition from Image Contours, in 'Geometric Invariance in Computer Vision', (J.L. Mundy and A. Zisserman, eds.), The MIT Press, Cambridge, Massachusetts, 1992.

9. N. Götze, S. Drüe, G. Hartmann, Invariant Object Recognition with Discriminat Features based on Local Fast-Fourier Mellin Transformation, Proc. Int. Conf Pattern Recognition 2000, Barcelona, Vol. 1 (A. Sanfeliu et. al., eds.) pp. 948-951, IEEE Computer Society, Los Alamitos.
10. S. Abbasi, F. Mokhtarian, Automatic View Selection and Sigmal Processing, Proc. Int. Conf Pattern Recognition 2000, Barcelona, Vol. 1 (A. Sanfeliu et. al., eds.) pp.13-16, IEEE Computer Society, Los Alamitos
11. R.C.G. Gonzales, R.E. Woods, Digital Image Processing, Addison Wesley, Reading, Massachusetts, 1992.
12. M. Sonka, V. Hlavac, R. Boyle, Image Processing, Analysis and Computer Vision, PWS Publishing, Pacific Grove, California, 1999.
13. R.C. Nelson, A. Selinger, Learing 3D Recognition Models for General Objects from Unlabeled Imaginary: An Experiment in Intelligent Brute Force, Proc. Int. Conf Pattern Recognition 2000, Barcelona, Vol. 1 (A. Sanfeliu et. al., eds.) pp. 1-8, IEEE Computer Society, Los Alamitos
14. Balslev, Noise Tolerance of Moment Invariants in Pattern Recognition, Pattern Recognition Letters 19, 1998, 183-89.

Grouping Character Shapes by Means of Genetic Programming

Claudio De Stefano[1], A. Della Cioppa[2], A. Marcelli[2], and F. Matarazzo[2]

[1] Facoltà di Ingegneria
Università del Sannio
Piazza Roma, Palazzo Bosco Lucarelli
Benevento, Italy
Ph. +39 0824 305839
destefan@unina.it,
[2] Dipartimento di Ingegneria dell'Informazione e Ingegneria Elettrica
Università di Salerno
Via Ponte Don Melillo
I–84084 Fisciano (SA), Italy
Ph. +39 089 964254, +39 089 964274
dean@unina.it, marcelli@diiie.unisa.it

Abstract. In the framework of an evolutionary approach to machine learning, this paper presents the preliminary version of a learning system that uses Genetic Programming as a tool for automatically inferring the set of classification rules to be used by a hierarchical handwritten character recognition system. In this context, the aim of the learning system is that of producing a set of rules able to group character shapes, described by using structural features, into super–classes, each corresponding to one or more actual classes. In particular, the paper illustrates the structure of the classification rules and the grammar used to generate them, the genetic operators devised to manipulate the set of rules and the fitness function used to match the current set of rules against the sample of the training set. The experimental results obtained by using a set of 5,000 digits randomly extracted from the NIST database are eventually reported and discussed.

1 Introduction

The recognition of handwritten character involves identifying a correspondence between the pixels of the image representing the samples to be recognized and the abstract definitions of characters (models or prototypes). The prevalent approach to solve the problem is that of implementing a sequential process, each stage of which progressively reduces the information passed to subsequent ones. This process, however, requires that decisions about the relevance of a piece of information have to be taken since the early stages, with the unpleasant consequence that a mistake may prevent the correct operation of the whole system. For this reason, the last stage of the process is usually by far the most complex one, in that it must collect and combine in rather complex ways many piece of

C. Arcelli et al. (Eds.): IWVF4, LNCS 2059, pp. 504–513, 2001.
© Springer-Verlag Berlin Heidelberg 2001

information that, all together, should be able to recover, at least partially, the loss of information introduced in the previous ones. Moreover, due to the extreme variability exhibited by samples produced by a large population of writers, pursuing such an approach often requires the use of a large number of prototypes for each class, in order to capture the distinctive features of different writing styles. The combination of complex classification methods with large set of prototypes has a dramatic effect on the classifier: larger number of prototypes requires a finer discriminating power, which, in turn, requires more sophisticated methods and algorithms to perform the classification. While such complex methods are certainly useful for dealing with difficult cases, they are useless, and even dangerous, for simple ones; if the input sample happens to be almost noiseless, and its shape corresponds to the most "natural" human perception of a certain character, invoking complex classification schemes on large sets of prototypes results in overloading the classifier without improving the accuracy.

For this reason, we have investigated the possibility of using a preclassification technique whose main purpose is that of reducing the number of prototypes to be matched against a given sample while dealing with simple cases. The general problem of reducing a classifier computational cost has been faced since the 70's in the framework of statistical pattern recognition [1] and more recently within shape–based methods for character recognition [2,3]. The large majority of the preclassification methods for character recognition proposed in the literature belongs to one of two categories, depending on whether they differ in the set of features used to describe the samples or in the strategy adopted for labeling them [4]. Although they exhibit interesting performance, their main drawback is that a slight improvement in the performance results in a considerable increase of the computational costs.

In the framework of an evolutionary approach to character recognition we are developing [5], this paper reports a preliminary version of a learning system that uses Genetic Programming [6,7] as a tool for producing a set of prototypes able to group character shapes, described by using structural features, into super–classes, each corresponding to one or more actual classes. The proposed tool works in two different modes. During an off–line unsupervised training phase, rather rough descriptions of the shape of the samples belonging to the training set are computed from the feature set, and then allowed to evolve according to the Genetic Programming paradigm in order to achieve the maximum coverage of the training set. Then, each prototype is matched and labeled with the classes whose samples are matched by that prototype. At run time, after the feature extraction, the same shape description is computed for a given sample. The labels of the simplest prototype matching the sample represent the classes the sample most likely belongs to.

The paper is organized as follows: Section 2 describes the character shape description scheme and how it is reduced to a feature vector, which represent the description the Genetic Programming works on. In Section 3 we present our approach and its implementation, while Section 4 reports some preliminary experimental results and a few concluding remarks.

2   From Character Shape to Feature Vector

In the framework of structural methods for pattern recognition, the most common approach is based on the decomposition of an initial representation into elementary parts, each of which is simply describable. In this way a character is described by means of a structure made up by a set of parts interrelated by more or less complex links. Such a structure is then described in terms of a sentence of a language or of a relational *graph*. Accordingly, the classification, that is the assignment of a specimen to a class, is performed by parsing the sentence, so as to establish its accordance with a given grammar or by some graph matching techniques. The actual procedure we have adopted for decomposing and describing the character shape is articulated into three main steps: *skeletonization*, *decomposition* and *description* [8]. During the first step, the character skeleton is computed by means of a MAT–based algorithm, while in the following one it is decomposed in parts, each one corresponding to an arc of circle which we have selected as our basic elements. Eventually, each arc found within the character is described by the following features:

- *size of the arc*, referred to the size of its bounding box;
- *angle spanned by the arc*;
- *direction of the arc curvature*, represented by the oriented direction of the normal to the chord subtended by the arc.

The spatial relations among the arcs are computed with reference to arc projections along both the horizontal and vertical axis of the character bounding box. In order to further reduce the variability still presents among samples belonging to the same class, the descriptions of both the arcs and the spatial relations are given in discrete form. In particular, we have assumed the following ranges of values:

- *size*: small, medium, large;
- *span*: closed, medium, wide;
- *direction*: N, NE, E, SE, S, SW, W, NW;
- *relation*: over, below, to–the–right, superimposed, included.

Those descriptions are then encoded into a feature vector of 139 elements. The first 63 elements of the vector are used to count the occurrences of the different arcs that can be found within a character, the following 13 elements describe the set of possible relations and the remaining 63 ones count the number of the different relations that originates from each type of arc [9].

3   Learning Explicit Classification Rules

Evolutionary Learning Systems seem to offer an effective prototyping methodology, as they are based on Evolutionary Algorithms (EAs) that represent a powerful tool for finding solutions in complex high dimensional search space, when there is no *a priori* information about the distribution of the sample in the feature space [10]. They perform a parallel and adaptive search by generating new populations of individuals and evaluating their fitness while interacting with the environment. Since EAs work by directly manipulating an encoded representation of the problem, and because such a representation can hide relevant information, thus severely limiting the chance of a successful search, problem representation is a key issue in EAs. In our case, as in many others, the most natural representation for a solution is a set of prototypes or rules, whose genotype's size and shape are not known in advance, rather than a set of fixed–length strings. Since classification rules may be thought as computer programs, the most natural way to introduce them into our learning system is that of adopting the Genetic Programming paradigm [6,7]. Such a paradigm combines GAs and programming languages in order to evolve hierarchical computer programs of dynamically varying complexity (size and shape) according to a given defined behavior. According to this paradigm, populations of computer programs are evolved by using the Darwin's principle that evolution by *natural selection* occurs when a population of replicating entities possesses the *heritability* characteristic and are subject to *genetic variation* and *struggle to survive*.

Typically, Genetic Programming starts with an initial population of randomly generated programs composed of functionals and terminals especially tailored to deal with the problem at hand. The performance of each program in the population is measured by means of a fitness function, whose nature also depends on the problem. After the fitness of each program has been evaluated, a new population is generated by selection, recombination and mutation of the current programs, and replaces the old one. Then, the whole process is repeated until a termination criterion is satisfied.

In order to implement the Genetic Programming paradigm, the following steps has to be executed:

- definition of the structures to be evolved;
- choice of the fitness function;
- definition of the genetic operators.

3.1 Structure Definition

In order to define the individual structures that undergo to adaptation in Genetic Programming, one needs a program generator, providing syntactically correct programs, and an interpreter for the structured computer programs, in order to execute them.

The program generator is based on a grammar written for S–expressions. A grammar $\mathcal{G}$ is a quadruple $\mathcal{G} = (\mathcal{T}, \mathcal{V}, S, \mathcal{P})$, where $\mathcal{T}$ and $\mathcal{V}$ are disjoint finite alphabets. $\mathcal{T}$ is the *terminal alphabet*, $\mathcal{V}$ is the *non–terminal alphabet*, S is the *start symbol*, and $\mathcal{P}$ is the set of *production rules* used to define the strings belonging to our language, usually written as $v \rightarrow w$ where v is a string on $(\mathcal{T} \cup \mathcal{V})$ containing at least one non–terminal symbol, while w is an element of $(\mathcal{T} \cup \mathcal{V})^*$. For the problem at hand, the set of terminals is the following:

Table 1. The grammar for the random rules generator.

Production Rule No.	Production Rule	Probability
1	$S \longrightarrow A$	1.0
2	$A \longrightarrow CBC \mid (CBC) \mid (IMX)$	0.25 | 0.25 | 0.5
3	$I \longrightarrow a_1 \mid \dots \mid a_{139}$	uniform
4	$X \longrightarrow 0 \mid 1 \mid \dots \mid 9$	uniform
5	$M \longrightarrow < \mid \leq \mid = \mid \geq \mid >$	uniform
6	$C \longrightarrow A \mid \neg A$	uniform
7	$B \longrightarrow \vee \mid \wedge$	uniform

$$\mathcal{T} = \{a_1, a_2, \dots, a_{139}, 0, 1, \dots, 9\},$$

and the set $\mathcal{V}$ is composed as follows:

$$\mathcal{V} = \{\wedge, \vee, \neg, <, \leq, =, >, \geq, A, X, I, M, C, B\},$$

where a_i is a variable atom denoting the i–th element in the feature vector, and the digits $0, 1, \dots, 9$ are constant atoms used to represent the value of each element in the feature vector. It should be noted that the above sets satisfy the requirements of closure and sufficiency [6]. The adopted set of production rules is reported in Table 1.

Each individual in the initial random population is generated starting with the symbol S that, according to the above grammar, can be replaced only by the symbol A. This last, on the contrary, can be replaced by itself, by its opposite, or by any recursive combination of logical operators whose arguments are the occurrences of the elements in the feature vector. It is worth noticing that, in order to avoid the generation of very long individual, the clause IMX has a higher probability of being selected to replace the symbol A with respect to the other ones that appear in the second production rule listed in Table 1. As it is obvious, the set of all the possible structures is the set of all possible compositions of functions that can be obtained recursively from the set of predefined functions.

Finally, the interpreter is a simple model of a computer and is constituted by an automaton that computes Boolean functions, i.e. an acceptor. Such an automaton computes the truth value of the rules in the population with respect to a set of samples.

3.2   Fitness Function

The definitions reported in the previous subsection allow for the generation of the genotypes of the individuals. The next step to accomplish is the definition of a fitness function to measure the performance of the individuals. For this purpose, it should be noted that EAs suffer of a serious drawback while dealing with multi–peaked (multi–modal) fitness landscape, in that they are not able to deal with cases where the global solution is represented by different optima, i.e.

species, rather than by a single one. In fact, they do not allow the evolution of stable multiple species within the same population, because the fitness of each individual is evaluated independently on the composition of the population at any time [11]. As a result, the task of the evolutionary process is reduced to that of optimizing the average value of the fitness in the population. Consequently, the solution provided by such a process consists entirely of genetic variations of the best individual, i.e. of a stable single species distribution. This behaviour is very appealing whenever it is desirable the uniform convergence to the global optimum exhibited by the canonical EAs. On the other hand, there exist many applications of interest that require the discovering and the maintenance of multiple optima, such as multi–modal optimization, inference induction, classification, machine learning, and biological, ecological, social and economic systems [12].

In order to evolve several stable distributions of solutions, each represent-ing a given species, we need a mechanism that prevents the distribution of the species with highest selective value to replace the competing species inducing some kind of restorative pressure in order to balance the convergence pressure of selection. To this end, several strategies for the competition and cooperation among the individuals have been introduced. The natural mechanism to handle such cooperation and competition is *sharing* the environmental resources, i.e. similar individuals share such resources thus inducing *niching* or *speciation* [11, 12,13,14,15].

For the problem at hand, it is obvious that we deal with different species. In fact, our aim is to perform an unsupervised learning in such a way that the result of the evolution is the emergence of different rules (species) each of which covers different sets of samples. Thus, the global solution will be represented by the disjunctive–normal–form of the discovered species in the final population. More-over, in case of handwritten characters the situation gets an additional twist. In fact, it is indisputable that different writers may refer to different prototypes when drawing samples belonging to a given class. A classical example is that of the digit '7' that is written with an horizontal stroke crossing the vertical one in many European countries, while such a stroke is usually not used in North American countries.

Therefore, some kind of niching must be incorporated at different levels into the fitness function. In our case, we have adopted a niching mechanism based on *resource sharing* [14,15]. According to resource sharing the cooperation and competition among the niches is obtained as follows: for each finite resource s_i a subset P of prototypes from the current population is selected. They are let to compete for the resource s_i in the training set. Only the best prototype receives the payoff. In the case of a tie, the resource s_i is assigned randomly among all deserving individuals. The winner is detected by matching each individual in the sample P against the sample of the training set. In our case, a prototype p matches a sample s *if and only if* p covers s, i.e. the features present in the sample are represented also in the prototype and their occurrences satisfy the constraints expressed in the prototype. At the end of the cycle, the fitness of each individual is computed by adding all rewards earned. In formula:

$$\phi(p) = \sum_{i=1}^{m} c \cdot \mu(p, s_i),$$

where $\phi(p)$ is the the fitness of the prototype p, m is the number of samples in the training set, $\mu(p, s_i)$ is a function that takes into account if the prototype p is the winner for the sample s_i (it is 1 if p is the winner for the sample s_i and 0 otherwise) and c is a scaling factor.

3.3   Selection and Genetic Operators

The selection mechanism is responsible for choosing among the individuals in the current population the ones that are replicated, without alterations, in the new population. To this aim many different selection mechanisms have been proposed. Nevertheless, we have to choose a mechanism that helps the maintenance of the discovered niches by reducing the selection pressure and noise [16]. So, we have used in our Genetic Programming the well known Stochastic Universal Selection mechanism [16], according to which we have exactly the same expectation as Roulette Wheel selection, but lower variance in the expected number of individuals generated by the selection process.

As regards the variations operators, we have actually used only the mutation operator, performing both *micro–* and *macro–mutation*. The macro–mutation is activated when the point to be mutated in the genotype is a node, and it substitutes the relative subtree with another one randomly generated according to the grammar described in subsection 3.1. It follows from the above that the macro–mutation is responsible for modifying the structure of the decision tree corresponding to each prototype in the same general way with respect to that implemented by the classical tree–crossover operator. For this reason we have chosen not to implement the tree-crossover operator.

Eventually, the micro–mutation operator is applied whenever the symbol selected for mutation is either a terminal or a function, and it is responsible for changing both the type of the features and their occurrences in the prototype. Therefore, it resembles closely the classical point–mutation operator.

Finally, we have allowed also *insertions*, i.e. a new random node is inserted in a random point, along with the relative subtree if it is necessary, and *deletions*, i.e. a node in the tree is selected and deleted. Obviously, such kind of mutations are effected in a way that ensures the syntactic correctness of the newly generated programs.

4   Experimental Results and Conclusions

A set of experiments has been performed in order to evaluate the ability of Genetic Programming to generate classification rules in very complex cases, like the one at hand, and the efficiency of the preclassifier. The experiments were performed on a data set of 10,000 digits extracted from the NIST database and equally distributed among the 10 classes. This data set was randomly subdivided into a training and a test set, both including 5,000 samples. Each character was decomposed, described and eventually coded into a feature vector of 139 elements, as reported in Section 2. Starting from those feature vectors, we have considered only 76 features, namely the 63 features describing the type of the arcs found in the specimen and the 13 used for coding the type of the spatial relations among the arcs, and used that as descriptions for the preclassifier to work with.

It must be noted that the Genetic Programming paradigm is controlled by a set of parameters whose values affect in different ways the operation of the system. Therefore, before starting the experiments, suitable values should be assigned to the system parameters. To this purpose, we have divided this set of parameters into external parameters, that mainly affect the performance of the learning system and internal parameters, that are responsible for the effectiveness and efficiency of the search, As external parameters we have assumed the population size N, the number of generations G and the maximum depth of the trees representing the rules in the initial population D. The internal parameters are the mutation probability p_m, the number of rules n competing for environmental resources in the resourse sharing and the maximum size of the subtree generated by the macro–mutation operator d. The results reported in the sequel has been obtained with $N = 1000$, $G = 350$, $D = 10$, $p_m = 0.6$, $n = N$ and $d = 3$. Eventually, let us recall that, as mentioned in subsection 3.2, the fitness function requires to select the winner among the prototypes that match a given sample. We have adopted the Occam's razor principle of simplicity closely related to Kolmogorov Complexity definition [17,18], i.e. "*if there are alternative explanations for a phenomenon, then, all other things being equal, we should select the simplest one*". In the current system, it is implemented by choosing as winner the prototype whose genotype is the shortest one. With such a criterion we expect the learning to produce the simplest prototypes covering as many samples as possible, in accordance with the main goal of our work: to design a method for reducing the computational cost of the classifier while dealing with simple cases by reducing the number of classes to be searched for by the classifier.

As mentioned before, the first experiment was aimed at evaluating the capability of the Genetic Programming paradigm to deal with complex cases such as the one at hand. For this purpose, we have monitored the covering rate C, i.e. the percentage of the samples belonging to the training set covered by the set of prototypes produced by the system during the learning. Let us recall now that, in our case, a prototype p matches a sample s *if and only if p covers s*, i.e. the features present in the sample are represented also in the prototype and their occurrences satisfy the constraints. The learning was implemented by providing the system with the descriptions of the samples without their labels, so as to implement un unsupervised learning, and let it evolve the set of prototypes for achieving the highest covering rate. Despite the loss of the information due

Table 2. The experimental results obtained on the Test Set.

$\mathcal{C}$	$\mathcal{R}$	≤ 3	≤ 5	≤ 7	$\mathcal{E}$	≤ 3	≤ 5	≤ 7
87.00	67.16	74.18	24.72	1.10	19.84	81.65	17.64	0.10

to both the discrete values allowed for the features and the reduced number of features considered in the experiment, the system achieved a covering rate of 91.38% In other words, there were 431 samples in the training set for which the system was unable to generate or to maintain a suitable set of prototypes with the time limit of 350 generations.

In order to measure the efficiency of the classifier, we preliminarily labeled the prototypes obtained at the end of the learning phase. The labeling was such that each time a prototype matched a sample of the training set, the label of the sample, i.e. its class, was added to the list of labels for that prototype. At the end of the labeling, thus, each prototype had a list of labels of the classes it covered, as well as the number of samples matched in each class. A detailed analysis of these lists showed that there were many prototypes covering many samples of very few classes and very few samples of many other classes. This was due to "confusing" samples, i.e. samples belonging to different classes but having the same feature values, thus looking indistinguishable for the system. In order to reduce this effect, the labels corresponding to the classes whose number of samples covered by the prototype was smaller than 10% of the total number of matchings for that prototype were removed from the list. Finally, a classification experiment on the test set was performed, that yielded to the results reported in Table 2 in terms of covering rate $\mathcal{C}$, correct recognition rate $\mathcal{R}$ and error rate $\mathcal{E}$.

Such results were obtained by assuming that a sample was correctly preclassified if its label appeared within the list of labels associated to the prototype covering it. To emphasize the efficiency of the preclassifier, Table 2 reports the experimental results by grouping the prototypes depending on the number of classes they cover. For instance, the third column shows that 74.18% of the total number of samples correctly preclassified were covered by prototypes whose lists have at most 3 classes.

The experimental results reported above allow for two concluding remarks. The first one is that Genetic Programming represents a promising tool for learning classification rules in very complex and structured domains, as the one we have considered in our experiments. In particular, it is very appealing in developing hierarchical handwritten character recognition system, since it may produce prototypes at different level of abstraction, depending on the way the system is trained. The second one is that the learning system developed by us, although makes use of a fraction of the information carried by the character shape, is higly efficient. Since the classification is performed by matching the unknown sample against the whole prototype set to determine the winner, reducing the number of classes reduces the number of prototype to consider, thus resulting in an overall reduction of the classification time. Therefore, roughly speaking, the preclassifier proposed by us is able to reduce the classification time to 50% or less in more than 65% of the cases.

References

1. Fukunaga, K.; Narendra, P. M. 1975. *A Branch and Bound Algorithm for Computing K–Nearest Neighbors. IEEE Trans. on Computers* C–24:750–753.
2. Marcelli, A.; Pavlidis, T. 1994. Using Projections for Preclassification of Character Shape. In *Document Recognition*. L.M. Vincent, T. Pavlidis (eds.), Los Angeles: SPIE, 2181:4–13.
3. Marcelli, A.; Likhareva, N.; Pavlidis, T. 1997. *Structural Indexing for Character Recognition. Computer Vision and Image Understanding*, 67(1): 330–346.
4. Mori, S.; Yamamoto, k.; Yasuda, M. 1984. *Research on Machine Recognition od Handprinted Characters. IEEE Trans. on PAMI*, PAMI–6: 386–405.
5. De Stefano, C.; Della Cioppa, A.; and Marcelli, A. 1999. Handwritten numerals recognition by means of evolutionary algorithms. In *Proc. of the 3th Int. Conf. on Document Analysis and Recognition*, 804–807.
6. Koza, J. R. 1992. *Genetic Programming: On the Programming of Computers by Natural Selection*. Cambridge, MA: MIT Press.
7. Koza, J. R. 1994. *Genetic Programming II: Automatic Discovery of Reusable Programs*. Cambridge, MA: MIT Press.
8. Boccignone, G. 1990. Using skeletons for OCR. V. Cantoni et al. eds., *Progress in Image Analysis and Processing*, 235–282.
9. Cordella, L. P.; De Stefano, C.; and Vento, M. 1995. Neural network classifier for OCR using structural descriptions. *Machine Vision and Applications* 8(5):336–342.
10. Goldberg, D. E. 1989. *Genetic algorithms in search, optimization, and machine learning*. Reading, MA: Addison–Wesley.
11. Deb, K., and Goldberg, D. E. 1989. An investigation of niche and species formation in genetic function optimization. In Schaffer, J. D., ed., *Proc. of the 3th Int. Conf. on Genetic Algorithms*, 42–50. San Mateo, CA: Morgan Kaufmann.
12. Mahfoud, S. 1994. Genetic drift in sharing methods. In *Proceedings of the First IEEE Conference on Evolutionary Computation*, 67–72.
13. Booker, L. B.; Goldberg, D. E.; and Holland, J. H. 1989. Classifier Systems and Genetic Algorithms. In *Artificial Intelligence*, volume 40. 235–282.
14. Forrest, S.; Javornik, B.; Smith, R. E.; and Perelson, A. S. 1993. Using genetic algorithms to explore pattern recognition in the immune system. *Evolutionary Computation* 1(3):191–211.
15. Horn, J.; Goldberg, D. E.; and Deb, K. 1994. Implicit Niching in a Learning Classifier System: Nature's Way. *Evolutionary Computation* 2(1):37–66.
16. Baker, J. E. 1987. Reducing bias and inefficiency in the selection algorithm. In Grefenstette, J. J., ed., *Genetic algorithms and their applications : Proc. of the 2th Int. Conf. on Genetic Algorithms*, 14–21. Hillsdale, NJ: Lawrence Erlbaum Assoc.
17. Li, M.; Vitányi, P. 1997. *An Introduction to Kolmogorov Complexity and its Applications*. Series: Graduate text in computer Science. Springeri–Verlag, New York.
18. Conte, M.; De Falco, I.; Della Cioppa, A.; Tarantino, E.; and Trautteur, G. 1997. Genetic Programming Estimates of Kolmogorov Complexity. *Proc. of the Seventh Int. Conf. on Genetic Algorithms* (ICGA'97), 743–750. Morgan–Kaufmann.

Pattern Detection Using a Maximal Rejection Classifier

Michael Elad[1], Yacov Hel-Or[2], and Renato Keshet[3]

[1] Jigami LTD Israel, email: elad@jigami.com
[2] The Interdisciplinary Center, Herzlia, Israel, email: toky@idc.ac.il
[3] Hewlett-Packard Laboratories - Israel, email: renato@hpli.hpl.hp.com

Abstract. In this paper we propose a new classifier - the Maximal Rejection Classifier (MRC) - for target detection. Unlike pattern recognition, pattern detection problems require a separation between two classes, *Target* and *Clutter*, where the probability of the former is substantially smaller, compared to that of the latter. The MRC is a linear classifier, based on successive rejection operations. Each rejection is performed using a projection followed by thresholding. The projection vector is designed to maximize the number of rejected *Clutter* inputs. It is shown that it also minimizes the expected number of operations until detection. An application of detecting frontal faces in images is demonstrated using the MRC with encouraging results.

1   Introduction

In target detection applications, the goal is to detect occurrences of a specific *Target* in a given signal. In general, the target is subjected to some particular type of transformation, hence we have a set of target signals to be detected. In this context, the set of non-*Target* samples are referred to as *Clutter*. In practice, the target detection problem can be characterized as designing a classifier $C(z)$, which, given an input vector z, has to decide whether z belongs to the *Target* class $\mathbf{X}$ or the *Clutter* class $\mathbf{Y}$. In example based classification, this classifier is designed using two training sets - $\hat{\mathbf{X}} = \{x_i\}_{i=1..L_x}$ (*Target* samples) and $\hat{\mathbf{Y}} = \{y_i\}_{i=1..L_y}$ (*Clutter* samples), drawn from the above two classes.

Since the classifier $C(z)$ is usually the heart of a detection algorithm, and is applied many times, simplifying it translates immediately to an efficient detection algorithm. Various types of example-based classifiers are suggested in the literature [1,2,3]. The most simple and fast are the linear classifiers, where the projection of z is performed onto a projection vector u, thus, $C(z) = f(u^t z)$ where $f(*)$ is a thresholding operation (or some other decision rule). The Support Vector Machine (SVM) [2] and the Fisher Linear Discriminant (FLD) [1] are two examples of linear classifiers. In both cases the kernel u is chosen in some optimal manner. In the FLD, u is chosen such that the Mahalanobis distance of the two classes after projection will be maximized. In the SVM approach the

---

[4] This work has been done at HP Labs, Israel.

C. Arcelli et al. (Eds.): IWVF4, LNCS 2059, pp. 514–524, 2001.
© Springer-Verlag Berlin Heidelberg 2001

motive is similar, but the vector u is chosen such that it maximizes the margin between the two sets.

In both these classifiers, it is assumed that the two classes have equal importance. However, in typical target detection applications the above assumption is not valid since the a-priori probability of z belonging to $\mathbf{X}$ is substantially smaller, compared to that of belonging to $\mathbf{Y}$. Both the FLD and the SVM do not exploit this property. Moreover, in both of these methods, it is assumed that the classes are linearly separable. However, in a typical detection scenario the target class is surrounded by the clutter class, thus the classes are not linearly seperable (see, e.g. Figure 2). In order to be able to treat more complex, and unfortunately, more common scenarios, non-linear extensions of these algorithms are required [1,2]. Such extensions are typically at the expense of much more computationally intensive algorithms.

In this paper we propose the *Maximal Rejection Classifier* (MRC) that overcomes the above two drawbacks. While maintaining the simplicity of a linear classifier, it can also deal with non linearly separable cases. The only requirement is that the *Clutter* class and the convex hull of the *Target* class are disjoint. We define this property as two *convexly-separable* classes, which is a much weaker condition compared to linear-separability. In addition, this classifier exploits the property of high *Clutter* probability. Hence, it attempts to give very fast *Clutter* labeling, even if at the expense of slow *Target* labeling. Thus, the entire input signal is classified very fast.

The MRC is an iterative rejection based classification algorithm. The main idea is to apply at each iteration a linear projection followed by a thresholding, similar to the SVM and the FLD. However, as opposed to these two methods, the projection vector and the corresponding thresholds are chosen such that at each iteration the MRC attempts to maximize the number of rejected *Clutter* samples. This means that following the first classification iteration, many of the *Clutter* samples are already classified as such, and discarded from further consideration. The process is continued with the remaining *Clutter* samples, again searching for a linear projection vector and thresholds that maximizes the rejection of *Clutter* samples from the remaining set. This process is repeated iteratively until a small number or non of the *Clutter* samples remain. The remaining samples at the final stage are considered as *Targets*. The idea of rejection-based classifier was already introduced by [3]. However, in this work we extend the idea by using the concept of maximal rejection.

2   The MRC in Theory

Assume two classes are given in $\Re^n$, $\mathbf{X}$ (the *Target* class) and $\mathbf{Y}$ (the *Clutter* class). It is required to discriminate between these two classes, i.e., given an input z drawn from one of these classes, we would like to be able to label it correctly as either *Target* or *Clutter*. One important point, however, is that we know a-priori that for a typical input stream the vast majority of the inputs are

Clutters, i.e.:

$$P\{\mathbf{X}\} \ll P\{\mathbf{Y}\} \tag{1}$$

where $P\{\mathbf{X}\}$ is the a-priori probability that an input signal will be a *Target*, and $P\{\mathbf{Y}\}$ is defined similarly. Based on this knowledge, we would like the classifier to give a decision as fast as possible (i.e., with as few operations as possible). Thus, *Clutter* labeling should be performed fast, even if at the expense of slow *Target* labeling.

Similar to other linear classifiers [1,2], we suggest to first project the sample z onto a vector u, and label it based on the projected value $\alpha = u^T z$. Projecting the *Target* class and the *Clutter* class onto u results with a Probability Density Functions (PDF) $P\{\alpha|\mathbf{X}\}$ and $P\{\alpha|\mathbf{Y}\}$ respectively. We define the following intervals based on $P\{\alpha|\mathbf{X}\}$ and $P\{\alpha|\mathbf{Y}\}$:

$$
\begin{aligned}
C_t &= \{\alpha | P\{\alpha|\mathbf{X}\} > 0, P\{\alpha|\mathbf{Y}\} = 0\} \\
C_c &= \{\alpha | P\{\alpha|\mathbf{X}\} = 0, P\{\alpha|\mathbf{Y}\} > 0\} \\
C_u &= \{\alpha | P\{\alpha|\mathbf{X}\} > 0, P\{\alpha|\mathbf{Y}\} > 0\}
\end{aligned} \tag{2}
$$

(t-Target, c-Clutter and u-Unknown). After projection, z is labeled either as a *Target*, *Clutter*, or *Unknown*, based on the interval at which α belongs to.

Unknown classifications are obtained only in the C_u interval, where a decision cannot be made. Figure 1 presents an example for the construction of the intervals C_t, C_c and C_u and their appropriate decisions. The probability of the *Unknown* decision is given by:

$$P\{Unknown\} = \int_{\alpha \in C_u} P\{\mathbf{Y}\}P\{\alpha|\mathbf{Y}\}d\alpha + \int_{\alpha \in C_u} P\{\mathbf{X}\}P\{\alpha|\mathbf{X}\}d\alpha \tag{3}$$

The above term is a function of the projection vector u. We would like to find the vector u which minimizes the "Unknown" probability. However, since this is a complex minimization problem, an alternative minimization is developed here, using a proximity measure between the two PDF's.

If $P\{\alpha|\mathbf{Y}\}$ and $P\{\alpha|\mathbf{X}\}$ are far apart and separated from each other $P\{Unknown\}$ will be small. Therefore, an alternative requirement is to minimize the overlap between these two PDF's. We will define this requirement using the following expected distance between a point α_0 and a distribution $P\{\alpha\}$:

$$D(\alpha_0 \parallel P\{\alpha\}) = \int_\alpha \frac{(\alpha_0 - \alpha)^2 P\{\alpha\}}{\sigma^2} d\alpha = \frac{(\alpha_0 - \mu)^2 + \sigma^2}{\sigma^2}$$

where μ is the mean of $P\{\alpha\}$ and σ is the variance of $P\{\alpha\}$. The division by σ is performed in order to make the distance scale-invariant (or unit-invariant). Using this distance definition, the distance of $P\{\alpha|\mathbf{X}\}$ from $P\{\alpha|\mathbf{Y}\}$ can be defined as the expected distance of $P(\alpha|\mathbf{Y})$ from $P\{\alpha|\mathbf{X}\}$:

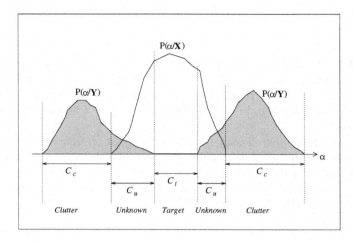

Fig. 1. The intervals C_t, C_c and C_u, for specific PDFs $P\{\alpha|\mathbf{X}\}$ and $P\{\alpha|\mathbf{Y}\}$.

$$D(P\{\alpha|\mathbf{Y}\} \| P\{\alpha|\mathbf{X}\}) = \int_{\alpha'} D(\alpha' \| P\{\alpha|\mathbf{X}\}) \, P\{\alpha'|\mathbf{Y}\}d\alpha' = \qquad (4)$$

$$= \int_{\alpha'} \frac{(\alpha' - \mu_x)^2 + \sigma_x^2}{\sigma_x^2} \, P\{\alpha'|\mathbf{Y}\}d\alpha' = \frac{(\mu_y - \mu_x)^2 + \sigma_x^2 + \sigma_y^2}{\sigma_x^2}$$

where $[\mu_x, \sigma_x]$ and $[\mu_y, \sigma_y]$ are the mean-variance pairs of $P\{\alpha|\mathbf{X}\}$ and $P\{\alpha|\mathbf{Y}\}$, respectively. Since we want the two distributions to have as small an overlap as possible, we would like to maximize this distance or minimize the *proximity* between $P\{\alpha|\mathbf{Y}\}$ and $P\{\alpha|\mathbf{X}\}$, which can be defined as the inverse of their mutual distance. Note, that this measure is asymmetric with respect to the two distributions, i.e the proximity defines the closeness of $P\{\alpha|\mathbf{Y}\}$ to $P\{\alpha|\mathbf{X}\}$, but not vice versa. Therefore, we define the overall proximity between the two distributions as follows:

$$Prox\,(P\{\alpha|\mathbf{Y}\}, P\{\alpha|\mathbf{X}\}) = \qquad (5)$$

$$= P\{\mathbf{X}\}\frac{\sigma_y^2}{\sigma_x^2 + \sigma_y^2 + (\mu_y - \mu_x)^2} + P\{\mathbf{Y}\}\frac{\sigma_x^2}{\sigma_x^2 + \sigma_y^2 + (\mu_y - \mu_x)^2}$$

Compared to the original expression in Equation 3, the minimization of this term with respect to u is easier. If $P\{\mathbf{X}\} = P\{\mathbf{Y}\}$, i.e. if there is an even chance to obtain *Target* or *Clutter* inputs, the proximity becomes:

$$Prox(P\{\alpha|\mathbf{Y}\} , P\{\alpha|\mathbf{X}\}) = \frac{\sigma_x^2 + \sigma_y^2}{\sigma_x^2 + \sigma_y^2 + (\mu_y - \mu_x)^2} \qquad (6)$$

which is associated with the cost function minimized by the Fisher Linear Discriminant (FLD)[1]. In our case $P\{\mathbf{X}\} \ll P\{\mathbf{Y}\}$ (Equation 1), thus, the first term is negligible in Equation 5 and can be omitted. Therefore, the optimal u should minimize the resulting term:

$$d(u) = \frac{\sigma_x^2}{\sigma_x^2 + \sigma_y^2 + (\mu_y - \mu_x)^2} \tag{7}$$

where $\sigma_y^2, \sigma_x^2, \mu_y$ and μ_x are all a function of the projection vector u.

There are two factors that control $d(u)$. The first factor is the distance between the two means μ_y and μ_x. Maximizing this distance will minimize $d(u)$. However, this factor is negligible when the two means are close to each other. This scenario is typical in detection cases when the target class in surrounded by the clutter class (see Figure 2). The other factor is the ratio between σ_x and σ_y. Our aim is to find a projection direction which results in a small σ_x and large σ_y. This means that the projection of *Target* inputs tend to concentrate in a narrow interval, whereas the *Clutter* inputs will spread with a large variance (see e.g. Fig 2).

For the optimal u, most of the *Clutter* inputs will be projected onto C_c, while C_t might even be an empty set. Subsequently, after projection, many of the *Clutter* inputs are usually classified, whereas *Target* labeling may not be immediately possible. This serves our purpose because there is a high probability that a decision will be made when a *Clutter* input is given. Since these inputs are more frequent, this means a faster decision for the vast majority of the inputs.

The method which we suggest follows this scheme: The classifier works in an *iterative* manner, projecting and thresholding with different parameters at each iteration sequentially. Since the classifier is asymmetric, the classification is based on *rejections*; *Clutter* inputs are classified and removed from further consideration while the remaining inputs are kept as suspected *Targets*. The iterations and the *rejection* approaches are both key concepts of the proposed scheme.

3   The MRC in Practice

Let us return to Equation 7 and find the optimal projection vector u. In order to do so, we have to express $\sigma_y^2, \sigma_x^2, \mu_y$ and μ_x as functions of u. It is easy to see that:

$$\mu_x = u^T \mathbf{M}_x \quad \text{and} \quad \sigma_x^2 = u^T \mathbf{R}_{xx} u \tag{8}$$

where we define:

$$\mathbf{M}_x = \int_z z P\{z|\mathbf{X}\} \, dz \quad ; \quad \mathbf{R}_{xx} = \int_z (z - \mathbf{M}_x)(z - \mathbf{M}_x)^T P\{z|\mathbf{X}\} \, dz \tag{9}$$

In a similar manner we express μ_y and σ_y^2. As can be seen, only the first and second moments of the classes play a role in the choice of the projection vector u.

In practice we usually do not have the probabilities $P\{z|\mathbf{X}\}, P\{z|\mathbf{Y}\}$, and inference on the *Target* or *Clutter* class is achieved through examples. For the two example-sets $\hat{\mathbf{X}} = \{x_k\}_{k=1}^{L_x}$ and $\hat{\mathbf{Y}} = \{y_k\}_{k=1}^{L_y}$, the mean-covariance pairs $(\mathbf{M}_x, \mathbf{R}_{xx}, \mathbf{M}_y,$ and $\mathbf{R}_{yy})$ are replaced with empirical approximations:

$$\hat{\mathbf{M}}_x = \frac{1}{L_x} \sum_{k=1}^{L_x} x_k \quad ; \quad \hat{\mathbf{R}}_{xx} = \frac{1}{L_x} \sum_{k=1}^{L_x} (x_k - \hat{\mathbf{M}}_x)(x_k - \hat{\mathbf{M}}_x)^T \qquad (10)$$

and similarly for $\hat{\mathbf{M}}_y$ and $\hat{\mathbf{R}}_{yy}$. The function we aim to minimize is therefore:

$$d(u) = \frac{u^T \hat{\mathbf{R}}_{xx} u}{u^T \left[\hat{\mathbf{R}}_{xx} + \hat{\mathbf{R}}_{yy} + \left(\hat{\mathbf{M}}_y - \hat{\mathbf{M}}_x \right) \left(\hat{\mathbf{M}}_y - \hat{\mathbf{M}}_x \right)^T \right] u} \qquad (11)$$

Similarly to [1,4,5], it is easy to show that u that minimizes the above expression satisfies:

$$\hat{\mathbf{R}}_{xx} u = \lambda \left[\hat{\mathbf{R}}_{xx} + \hat{\mathbf{R}}_{yy} + \left(\hat{\mathbf{M}}_y - \hat{\mathbf{M}}_x \right) \left(\hat{\mathbf{M}}_y - \hat{\mathbf{M}}_x \right)^T \right] u \qquad (12)$$

and should correspond to the smallest possible λ. A problem of the form $Au = \lambda Bu$, as in Equation 12, is known as the generalized eigenvalue problem [1,4,5], and has a closed form solution. Notice that given any solution u for this equation, βu is also a solution with the same λ. Therefore, without loss of generality, the normalized solution $\|u\| = 1$ is used.

After finding the optimal projection vector u, the intervals C_t, C_c, and C_u can be determined according to Equation 2. An input z is labeled as a *Target* or *Clutter* if its projected value $u^T z$ is in C_t or C_c, respectively. Figure 2 (left) presents this stage for the case where C_t is empty, i.e. there are no inputs which can be classified as *Target*.

Input vectors whose projected values are in C_u are not labeled. For these inputs we apply another step of classification, where the design of the optimal projection vector in this step is performed according to the following new distributions:

$$P\{z|\mathbf{Y} \ \& \ u_1^T z \in C_u\} \quad \text{and} \quad P\{z|\mathbf{X} \ \& \ u_1^T z \in C_u\}$$

We define the next projection vector u_2 as the vector which minimizes the proximity measure between the above two distributions. This minimization is performed in the same manner as described for the first step. Figure 2-right presents the second rejection stage, which follows the first stage shown in Figure 2-left.

Following the second step, the process continues similarly with projection vectors $u_3, u_4, \cdots$, etc. Due to the optimality of the projection vector at each

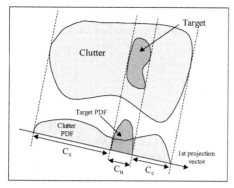 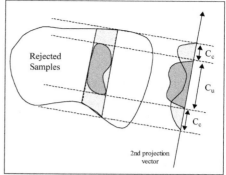

Fig. 2. Left: The first rejection stage for a 2D example. Right: The second rejection stage.

step, it is expected that a large portion of the input vectors will be labeled as *Clutter* at each step, while following steps will deal with the remaining input vectors. Applying the cascade of classifiers in such an iterative manner ensures a good performance of the classification with respect to an accurate labeling and a fast classification rate.

Since we exchanged the class probabilities with sets of points, it is impractical to define the intervals C_t, C_c, and C_u using Equation 2. This is because the intervals will be composed of many fragments each of which results from a particular example. Moreover, the domain of α cannot be covered by a finite set of examples. Therefore, it is more natural to define for each set, two thresholds bounding its projection values. As explained above, due to the functional that we are minimizing, in typical detection cases the *Target* thresholds define a small interval located inside the *Clutter* interval (see Figure 2). Therefore for simplicity, we define, for each projection vector, only a single interval $\Gamma = [T_1, T_2]$, which is the interval bounding the *Target* set. After projection we classify points projected outside Γ as *Clutter* and points projected inside Γ as *Unknown*.

In the case where the *Target* class forms a convex set, and the two classes are disjoint, it is theoretically possible to completely discriminate between them. This property is easily shown by noticing that we are actually extracting the *Target* set from the *Clutter* set by a sequence of two parallel hyper-planes, corresponding to the two thresholding operations. This constructs a parallelogram that bounds the *Target* set from outside. Since any convex set can be constructed by a set of parallel hyper-planes, exact classification is possible. However, if the *Target* set is non-convex, or the two classes are non-convexly separable (as defined in the Introduction), it is impossible to achieve a classification with zero errors; *Clutters* inputs which are inside the convex hull of the *Target* set cannot be rejected. Overcoming this limitation can be accomplished by a non-linear extension of the MRC, which is outside the scope of this paper. In practice, even if we deal with a convex *Target* set, false-alarms may exist due to the sub-optimal approach we are using, which neglects multi-dimensional moments higher than the second. However, simulations demonstrate that the number of false-alarms is typically small.

4   Face Detection Using the MRC

The face detection problem can be specified as the need to detect all instances of faces in a given image, at all spatial positions, all scales, all facial expressions, all poses, of all people, and under all lighting conditions. All these requirements should be met, while having few or no false alarms and mis-detections, and with as fast an algorithm as possible. This description reveals the complexity of the detection problem at hand. As opposed to other pattern detection problems, faces are expected to appear with considerable variations, even for the detection of frontal and vertical faces only. Variations are expected because of changes in skin color, facial hair, glasses, face shape, and more.

Several papers already addressed the face detection problem using various methods, such as SVM [2,6], Neural Networks [7,8,9], and other methods [10,11, 12,13]. In all of these studies, the above complete list of requirements is relaxed in order to obtain practical detection algorithms. Following [6,7,9,10,13], we deal with the detection of frontal and vertical faces only.

In all these algorithms, spatial position and scale are treated through the same method, in which the given image is decomposed into a Gaussian pyramid with near-unity (e.g., 1.2) resolution ratio. The search for faces is performed in each resolution layer independently, thus enabling the treatment of different scales. In order to be able to detect faces at all spatial positions, fixed sized blocks of pixels are extracted from the image at all positions (with full or partial overlap) for testing. In addition to the pyramid part, which treats varying scales and spatial positions, the core part of the detection algorithm is essentially a classifier which provides a *Face/Non-Face* decision for each input block.

We demonstrate the application of the MRC for this task. In the face-detection application, *Faces* take the role of targets, and *Non-Faces* are the clutter. The MRC produces very fast *Non-Face* labeling at the expense of slow *Face* labeling. Thus, on the average, it has a short decision time per input block.

The first stage in the MRC is to gather two example-sets, *Faces* and *Non-Faces*. As mentioned earlier, large enough sets are needed in order to guarantee good generalization for the faces and the non-faces that may be encountered in images. As to the *Face* set, the ORL data-base [1] was used. This database contains 400 frontal and vertical face images of 40 different individuals. By extracting the face portion from each of these images and scaling to 15×15 pixels, we obtained the set $\hat{\mathbf{X}} = \{x_k\}_{k=1}^{L_x}$ (with $L_x = 400$). The *Non-Face* set is required to be much larger, in order to represent the variability of *Non-Face* patterns in images. For this purpose we have collected from images with no faces more than 20 million *Non-Face* examples.

---

[1] http://www.cam-orl.co.uk/facedatabase.html: ORL database web-site

5   Results

We trained the MRC for detecting faces by computing 50 sets of kernels $\{u_k\}_{k=1}^{50}$ and associated thresholds $\{[T_1^k, T_2^k]\}_{k=1}^{50}$, using the above described databases of *Faces* and *Non-Faces*. Figures 3 and 4 show three examples of the obtained results. In these examples, the first stage rejected close to 90% of the candidates.

Fig. 3. An example for face detection with the MRC.

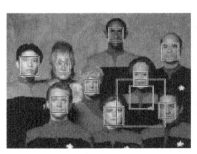

Fig. 4. Two examples for face detection with the MRC

This stage is merely a convolution of the input image (at every scale) with the first kernel, followed by thresholding. Successive kernels yield further rejection at about 50% for each projection. Thus, the complete MRC classification required an effective number of close to two convolutions per each pixel in each resolution layer. As can be seen from the examples, the MRC approach performed very well and was able to detect most of the existing faces. There are few false alarms, which typically correspond to blocks of pixels having a pattern which may resemble a face. In addition mis-detection occurs when a face is partially occluded or rotated too much. Generally speaking, the algorithm performs very well in terms of detection rate, false alarm rate, and most important of all, computational complexity. Due to space limitation we do not include more technical details in this paper. Comprehensive description of the results as well as comparative study with other face detection algorithms can be found in [14].

6   Conclusion

In this paper we presented a new classifier for target detection, which discriminates between *Target* and *Clutter* classes. The proposed classifier exploits the fact that the probability of a given input to belong to the *Target* class is much lower, as compared to its probability to belong to the *Clutter* class. This assumption, which is valid in many pattern detection applications, is exploited in designing an optimal classifier that detects *Target* signals as fast as possible. Moreover, exact classification is possible when the *Target* and the *Clutter* classes are convexly separable. The Fisher Linear Discriminant (FLD) is a special case of the proposed framework when the *Target* and *Clutter* probabilities are equal. In addition, the proposed scheme overcomes the instabilities arising in the FLD in cases where the mean of the two classes are close to each other. An improvement of the proposed technique is possible by rejecting *Target* patterns instead of *Clutter* patterns in advanced stages, when the probability of *Clutter* is not larger anymore. The performance of the MRC is demonstrated in the face detection problem. The obtained face detection algorithm is shown to be both computationally very efficient and accurate. Further details on the theory of the MRC and its application to face detection can be found in [15,14].

References

1. R. O. Duda and P. E. Hart, *Pattern Classification And Scene Analysis.* Wiley-Interscience Publication, 1973. 1st Edition.
2. V. N. Vapnik, *The Nature of Statistical Learning Theory.* Springer-Verlag, 1995. 1st Edition.
3. S. Baker and S. Nayar, "Pattern rejection," in *Proceedings 1996 IEEE Computer Society Conference on Computer Vision and Pattern Recognition - San Francisco, CA, USA, 18-20 June,* 1996.
4. G. H. Golub and C. F. V. Loan, *Matrix Computations.* Johns Hopkins University Press, 1996. 3rd Edition.
5. J. W. Demmel, *Applied Numerical Linear Algebra.* SIAM - Society for Industrial and Applied Mathematics, 1997. 1st Edition.

6. E. Osuna, R. Freund, and F. Girosi, "Training support vector machines: An application to face detection," in *IEEE Conference on Computer Vision and Pattern Recognition, Puerto-Rico*, 1997.

7. H. Rowley, S. Baluja, and T. Kanade, "Neural network-based face detection," *IEEE Trans. Pattern Analysis and Machine Intelligence*, vol. 20, pp. 23–38, 1998.

8. H. Rowley, S. Baluja, and T. Kanade, "Rotation invariant neural network-based face detection," in *Calnegie Mellon University Internal Report - CMU-CS-97-201*, 1997.

9. P. Juell and R. March, "A hirarchial neural network for human face detection," *Pattern Recognition*, vol. 29, pp. 781–787, 1996.

10. K. Sung and T. Poggio, "Example-based learning for view-based human face detection," *IEEE Trans. Pattern Analysis and Machine Intelligence*, vol. 20, pp. 39–51, 1998.

11. A. Mirhosseini and H. Yan, "Symmetry detection using attributed graph matching for a rotation-invariant face recognition system," in *Conference on Digital Image Computing: Techniques and Applications - Australia*, 1995.

12. A. Rajagopalan, K. Kumar, J. Karlekar, R. Manivasakan, and M. Patil, "Finding faces in photographs," in *International Conference of Computer Vision (ICCV) 97, Delhi, India*, 1997.

13. A. Tankus, H. Yeshurun, and N. Intrator, "Face detection by direct convexity estimation," in *Tel-Aviv University Technical Report*, 1998.

14. M. Elad, R. Keshet, and Y. Hel-Or, "Frontal and vertical face detection in images using the approximated invariant method (aim)," in *Hewlett-Packard Technical Report - HPL-99-5*, 1999.

15. M. Elad, Y. Hel-Or, and R. Keshet, "Approximated invariant method (AIM): A rejection based classifier," in *Hewlett-Packard Technical Report - HPL-98-160*, 1998.

Visual Search and Visual Lobe Size

Anand K. Gramopadhye* and Kartik Madhani

Department of Industrial Engineering,
Clemson University, Clemson, SC 29634-0920, USA

Abstract. An experiment was conducted to explore the transfer of training between visual lobe measurement tasks and visual search tasks. The study demonstrated that lobe practice improved the visual lobe, which in turn resulted in improved visual search performance. The implication is that visual lobe practice on carefully chosen targets can provide an effective training strategy in visual search and inspection. Results obtained from this research will help us in devising superior strategies for a whole range of tasks that have a visual search component (e.g., industrial inspection, military target acquisition). Use of these strategies, will ultimately lead to superior search performance.

1   Introduction

Visual search of extended fields has received continuing interest from researchers [1], [2], [3], [4]. It constitutes a component of several practically important tasks such as industrial inspection and military target acquisition [5], [6]. Because of its continuing importance in such tasks, training strategies are required which will rapidly and effectively improve search performance. The research presented here is a study focused on study of the most fundamental parameter, in visual search performance, the visual lobe size. Visual search has been modeled as a series of fixations, during which visual information is extracted from an area around the central fixation point referred to as the visual lobe [7], [8]. The visual lobe describes the decrease in target detection probability with increasing eccentricity angle from the line of sight [9]. The importance of the visual lobe as a determinant of visual search performance has been illustrated by Johnston [10] who showed that the size of the visual lobe had a significant effect on search performance. Subjects who had larger visual lobes exhibited shorter search times. Such a relationship between visual lobe and search performance has been confirmed by Erickson [11], Leachtenauer [12] and Courtney and Shou [13]. The concept of visual lobe is central to all mathematical approaches of extended visual search [14], [15], [16]. For a given search area the search time is dependent on the following [5]: the lobe size, the fixation time and the search strategy. The visual lobe area is affected by such factors as the adaptation level of the eye, target characteristics such as target size and target embeddedness, background characteristics, motivation, individual differences in peripheral vision and individual experience [17].

---

* Corresponding author. Tel: (864) 656-5540; fax: (864) 656-0795
E-mail address: agramop@ces.clemson.edu (A. K. Gramopadhye)

C. Arcelli et al. (Eds.): IWVF4, LNCS 2059, pp. 525–531, 2001.
© Springer-Verlag Berlin Heidelberg 2001

Studies in visual search have shown that both speed and accuracy can be improved with controlled practice [18], [14]. Training has consistently shown to be an effective way for improving various facets of visual inspection performance where the inspection task has a major search component [19], [20], [21], [22]. There is evidence to suggest that the visual lobe is amenable to training and best depicted by a negatively accelerated learning function [9], [12]. In a major field study of the role of photo-interpreter performance, Leachtenauer [12] found an increase in visual lobe size with training and a correlation between lobe size and interpreter search performance.

Thus drawing from the literature on visual lobe and visual search the question that needs answering is whether it is possible to train subjects to improve their visual lobe. Moreover does an improved visual lobe translate into improved search performance? To answer the above questions the following experiment was devised.

2   Experiment - Effects of Lobe Practice on Search Performance

2.1 Methodology

The experiment had as its objective the test of a transfer of training hypothesis. The objective was to determine whether lobe practice improved search performance. A criterion task was presented in the Pre-Training phase. Following this, a Training phase of practice was given. Finally the criterion task was presented once again in the Post-Training phase. In the Training phase, separate groups were given different training interventions, with subjects randomly assigned to the groups. These training intervention groups are characterized as Groups (G), while the criterion tasks are referred to as Trials (T).

2.2 Subjects

All subjects were undergraduate and graduate student volunteers, aged 20-30 years and were paid $ 5.00 per hour to participate. All were tested to ensure at least 20/20 vision, when wearing their normal corrective prescriptions if appropriate. Subjects in each experiment were briefed and given access to the criterion and training tasks to familiarize them with the tasks and computer interfaces. The eighteen subjects were randomly assigned to two groups, a Lobe Practice Group and a Control Group.

Lobe Practice Group. Practice on five trial blocks of the VL task, taking about 30 min for a total of 700 screens.

Search Practice Group. Practice on five trial blocks of the VS task, taking about 30 min for a total of 250 screens.

2.3 Equipment

The task was performed on an IBM PS2 (Model 70). Stimulus material consisted of search field of size 80 characters wide by 23 characters deep. The search field contained the following background characters: % ! | + - ( ) \ which were randomly located on the screen; with background character density of 10 % (i.e., only 10% of the spaces on the search field were filled, the rest of the search field being filled by blanks). The character '&' served as the target (Figure 1). The characters on the screen were viewed at 0.5 meters. The entire screen subtended an angle of 33.4° horizontally. The density of characters around the target was controlled to prevent any embedding affect, as in the task used by Czaja et al. [20] and Gallwey et al. [23]. Monk et al. [24] has shown that the number of characters in the eight positions adjacent to the target has a large effect on the performance in a search task. Hence this was kept constant at two out of eight positions for all screens.

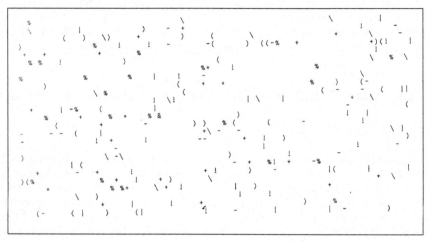

Fig. 1. Screen for experiment

2.4 Description of the VS and VL Tasks

The objective of the VL task was to estimate the horizontal extent of the visual lobe by determining how far into the periphery the subjects could see the target in a single eye fixation. In the VL task fixation cross was presented for 3s, followed by a VL test screen for 0.3s, and finally a repeat of the cross to prevent post-stimulus scanning. The purpose of the cross was to help the subjects follow their instructions to fixate at the central fixation point after each viewing of the VL screen. The target would appear randomly at any one of the six equally spaced, predetermined locations on the horizontal center line, with three positions on either side of the central fixation point. The subject's task was to locate the single target in VL test screen, which could only occur along the horizontal axis of the fixation cross. With such a brief duration of exposure, only a single fixation would be possible. Subjects using binocular vision indicated that they had found a target to the left of the fixation cross by pressing the key "Q" on the keyboard or the "P" key if the target was found on the right. If no target could be found, the space bar was pressed. No prior information concerning the position of the targets at the six predetermined positions was provided to the subjects before exposure to the VL screen. For equal number of screens of each VL experiment, the target was not present at any of the six target positions to discourage any guessing strategy.

In the visual search (VS) tasks, the subject's objective was to locate a target by self-paced search of the whole screen. The background characters and density were same as for the VL task. The entire screen was divided into four quadrants by the cross. When each VS screen appeared, timing began using the computers timing function, timing ended with the subject pressing one of four keys (L, A, X or M) appropriate to the quadrant in which the target was found.

2.5 Stimulus Materials

The stimulus materials used in this experiment comprised of the VL and VS task wherein each Trial block of the VL task comprised 360 screens with targets and 60 screens without targets. The criterion task were both the VS task, with a single Trial block of 150 screens and a single Trial block of the VL task administered before and after training. The training intervention comprised five trial blocks of the VL task.

2.6 Measurement and Analysis

The measurement in the VL tasks consisted of determining the probabilities of detection at each target distance (in degrees of visual angle) from the fixation cross. Left and right occurrences of the target were combined. On each trial block for each subject, the probabilities of detection were used to derive a single measure of visual lobe size. This was defined as the visual lobe horizontal radius. The probability of detection is plotted as a function of angular position of the target for two trial blocks to give one half of a visual lobe horizontal cross section. As the edges of the visual lobe have been found to be the form of a cumulative normal distribution [25], [26], the z values corresponding to each probability were regressed onto angular target position. The angular position giving a predicted z value of 0.00, corresponding to a detection probability of 0.50 was used as the measure of visual lobe size in the visual search tasks, search time, in seconds per screen, was measured.

3   Results

3.1 Effects of Training Intervention on the Criterion Tasks

With both VL and VS as criterion tasks, two analyses were undertaken. Visual lobes were calculated as before, and a mixed model ANOVA conducted on lobe sizes. There was no Group effect but a large Trial effect ($F(1,16) = 711.7$, $p < 0.0001$) and a large Group x Trial interaction ($F(1,16) = 10.6$, $p < 0.01$). An ANOVA conducted on the Pre-training Trial showed no group effect. For the Control group, the visual lobe horizontal radius increased after the training intervention from $10.1°$ to $11.52°$ while for the lobe practice group the change was from $9.13°$ to $13.06°$. A mixed model ANOVA of the mean search times for each subject showed significant Trial ($F(1, 16) = 166.1$, $p < 0.001$) and the Group x Trial interaction ($F(1,16) = 7.57$, $p < 0.001$)

effects. There was no difference between the groups on the Pre-training Trial. Mean search times decreased for the control group from 6.4 to 5.3 seconds, and for the lobe practice group from 6.4 to 4.5 seconds. Clearly, there was a good transfer from a lobe practice task to a search task as evidenced from the correlation between lobe size and visual search performance.

3.2 Effects During Practice

For the five practice trial blocks, a mixed model ANOVA on the lobe sizes showed a significant effect of Trials ($F(4,32) = 8.310$, $p < 0.001$) for the lobe practice group. Figure 2 plots this data, with the criterion lobe size trials also included.

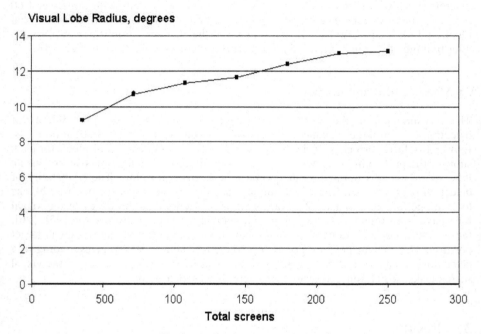

Fig. 2. Learning graph for lobe size

3.3 Search Time/Lobe Size Correlation

Both the correlations before practice ($r = 0.725$) and after practice ($r = 0.888$) were significant at $p < 0.001$.

4   Discussion and Conclusion

The experiment has reinforced the intimate relationship between visual lobe size and visual search performance, and extended the relationship to cover practice and transfer effects. Experiment showed the transfer effect, showing that lobe practice did indeed improve performance on a criterion search task. Additionally, the effect of lobe practice on lobe size was confirmed. The lobe practice effects on lobe size, found by Leachtenauer [12] for photo interpreters, were confirmed here by the experiment, and extended to the effects of search practice on lobe size. However, the slope of the learning function leveled off. This suggests that considerable practice is required to realize the full benefits of learning, so that directly improving a searcher's visual lobe size by practice at a lobe measurement task can be a practical training strategy. Given that lobe size and visual search performance were found to be highly correlated in these experiments, and that lobe practice transfers to search performance, this should be a powerful means of training in a difficult visual task such as inspection. Leachtenauer [12] had previously found an insignificant training effect on lobe size but a significant effect on search performance, using a training program aimed at a technique for expansion of the visual lobe. From our data, it appears that simple practice on the lobe measurement task is an effective training strategy.

The experiment confirmed that lobe task practice does improve performance on the criterion visual search task, showing this to be a viable and simple method for improving visual search performance. With the ability to construct computer-based simulators (such as the one used here) comes the ability to use digitally scanned images to develop a large library of faults, and incorporate these directly into a training program. Training can thus achieve a higher degree of realism and generality than was used here using available technology. Thus, many of the benefits of using computers in the knowledge-based and rule-based aspects of aircraft maintenance function (e.g. Johnson [27]) can be brought to the vital but more skill based functioning of visual inspection. From the results of this study the following conclusion can be drawn: Practice on a visual lobe detection task does transfer to a visual search task.

References

1. Lamar, E. S., 1960. Operational background and physical considerations relative to visual search problems. In A. Morris and P. E. Horne (Eds.) Visual search techniques. Washington, DC: National Research Council.
2. Morris, A. and Horne, E. P. (Eds), 1960. Visual search techniques. Washington, DC: National Academy of Sciences.
3. National Academy of Sciences, 1973. Visual search. Washington, DC: Author.
4. Brogan, D., 1990. Visual Search. Proceedings of the first international conference on visual search. University of Durham, England, Taylor and Francis.
5. Drury, C. G., 1992. Inspection Performance. In Gavriel Salvendy (Ed.), Handbook of Industrial Engineering, 2nd Edition, Chapter 88, John Wiley and Sons Inc., New York.
6. Greening, G. C., 1976. Mathematical modeling of air to ground target acquisition. Human Factors, 18: 111-148.
7. Engel, F. L., 1976. Visual conspicuity as an external determinant of eye movements and selective attention. Unpublished doctoral dissertation, T. H. Eindhoven.

8. Widdel, H., and Kaster, J., 1981. Eye movement measurements in the assessment and training of visual performance. In J. Moraal and K. F. Kraiss (Eds.) Manned systems design, methods, equipment and applications. pp 251-270 New York: Plenum.

9. Engel, F. L., 1977. Visual conspicuity, visual search and fixation tendencies of the eye. Vision Research, 17: 95-108.

10. Johnston, D. M., 1965. Search performance as a function of peripheral acuity. Human Factors, 7: 528-535.

11. Erickson, R. A., 1964. Relation between visual search time and peripheral acuity. Human Factors, 6: 165-178.

12. Leachtenauer, J. C., 1978. Peripheral acuity and photo-interpretation performance. Human Factors, 20: 537-551.

13. Courtney, A. J. and Shou, H.C., 1985(b). Visual lobe area for single targets on a competing homogeneous background. Human Factors, 27: 643-652.

14. Bloomfield, J. R., 1975. Theoretical approaches to visual search. In C. G. Drury and J. A. Fox (Eds.). Human reliability in quality control. Taylor and Francis, London.

15. Morawski, T., Drury, C. G. and Karwan, M. H., 1980. Predicting search performance for multiple targets. Human Factors, 22: 707-718.

16. Arani, T., Karwan, M. H. and Drury, C. G., 1984. A variable memory model of visual search. Human Factors, 26: 631-639.

17. Overington, I., 1979. The current status of mathematical modeling of threshold functions. In J. N. Clare and M. A. Sinclair (Eds.), Search and the human observer, 114-125, London: Taylor and Francis.

18. Parkes, K. R. and Rennocks, J., 1971. The effect of briefing on target acquisition performance, Loughborough, England: Loughborough University of Technology, Technical Report 260.

19. Martineck and Sadacca, R., 1965. Error keys as reference aids in image interpretation. US. Army Personnel Research Office, Washington, Technical Research Note 153, In Embrey, D. E. Approaches to training for industrial inspection. Applied Ergonomics, 1979, 10:139-144.

20. Czaja, S. J. and Drury, C. G., 1981. Training programs for inspection. Human Factors, 23, 473-484.

21. Micalizzi J. and Goldberg, J., 1989. Knowledge of results in visual inspection decisions: sensitivity or criterion effect? International Journal of Industrial Ergonomics, 4: 225-235.

22. Gramopadhye, A. K., 1992. Training for visual inspection. Unpublished Ph.D dissertation, State University of New York at Buffalo.

23. Gallwey, T. J. and Drury, C. G., 1986. Task complexity in visual inspection. Human Factors,   28 (5): 596-606.

24. Monk, T. M. and Brown, B., 1975. The effect of target surround characteristics on visual search performance. Human Factors, 17: 356-360.

25. Kraiss K., and Knaeuper A., 1982. Using visual lobe area measurements to predict visual search performance. Human Factors, 24(6): 673-682.

26. Bowler, Y. M., 1990. Towards a simplified model of visual search. In Visual Search D. Brogan (Ed.) Taylor and Francis, pp.3-19.

27. Johnson, W. B., 1990. Advanced technology training for aviation maintenance. Final report of the Third FAA Meeting on Human Factors Issues in Aircraft Maintenance and Inspection, 115-134.

Judging Whether Multiple Silhouettes Can Come from the Same Object

David Jacobs[1], Peter Belhumeur[2], and Ian Jermyn[3]

[1] NEC Research Institute
[2] Yale University
[3] New York University

Abstract. We consider the problem of recognizing an object from its silhouette. We focus on the case in which the camera translates, and rotates about a known axis parallel to the image, such as when a mobile robot explores an environment. In this case we present an algorithm for determining whether a new silhouette could come from the same object that produced two previously seen silhouettes. In a basic case, when cross-sections of each silhouette are single line segments, we can check for consistency between three silhouettes using linear programming. This provides the basis for methods that handle more complex cases. We show many experiments that demonstrate the performance of these methods when there is noise, some deviation from the assumptions of the algorithms, and partial occlusion. Previous work has addressed the problem of precisely reconstructing an object using many silhouettes taken under controlled conditions. Our work shows that recognition can be performed without complete reconstruction, so that a small number of images can be used, with viewpoints that are only partly constrained.

1 Introduction

This paper shows how to tell whether a new silhouette could come from the same object as previously seen ones. We consider the case in which an object rotates about a single, known axis parallel to the viewing plane, and is viewed with scaled orthographic projection. This is an interesting subcase of general viewing conditions. It is what happens when a person or robot stands upright as it explores a scene, so that the eye or camera is directed parallel to the floor. It is also the case when an object rests on a rotating turntable, with the camera axis parallel to the turntable.

It is easy to show that given any two silhouettes, and any two viewpoints, there is always an object that could have produced both silhouettes. So we suppose that two silhouettes of an object have been obtained to model it, and ask whether a third silhouette is consistent with these two images. We first characterize the constraint that two silhouettes place on an object's shape, and show that even when the amount of rotation between the silhouettes is unknown, this constraint can be determined up to an affine transformation. Next we show that for silhouettes in which every horizontal cross-section is one line segment, the

C. Arcelli et al. (Eds.): IWVF4, LNCS 2059, pp. 532–541, 2001.

question of whether a new silhouette is consistent with these two can be reduced to a linear program. Linear programming can also be used to test a necessary, but not sufficient condition for arbitrary silhouettes. We provide additional algorithms for silhouettes with cross-sections consisting of multiple line segments. We describe a number of experiments with these algorithms.

Much prior work has focused on using silhouettes to determine the 3D structure of an object. Some work uses a single silhouette. Strong prior assumptions are needed to make reconstruction possible in this case (eg., [10,1,4,9]). A second approach is to collect a large number of silhouettes, from known viewpoints, and use them to reconstruct a 3D object using differential methods (eg., [5], [3], [12], [11]) or volume intersection (eg., [8], [2]). These methods can produce accurate approximations to 3D shape, although interestingly, Laurentini[7] shows that exact reconstruction of even very simple polyhedra may require an unbounded number of images. Our current work makes quite different assumptions. We consider using a small number of silhouettes obtained from unknown viewpoints and ask whether the set of prior images and the new image are consistent with a single 3D shape without reconstructing a specific shape.

2   Constraints from Two Silhouettes

Let p, q, r denote the boundaries of three silhouettes. Let P, Q, R denote the filled regions of the silhouettes. When rotation is about the y axis, there will be two 3D points that appear in every silhouette, the points with highest and lowest y values. Denote the image of these points on the three silhouettes as: $p_1, p_2, q_1, q_2, r_1, r_2$. Let M denote the actual 3D object.

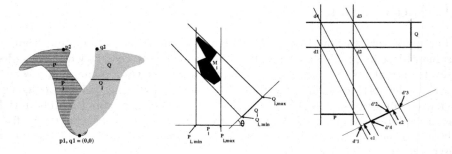

Fig. 1. Left: Two silhouettes with bottom points aligned. Middle: The $y = i$ plane. Right: Rectangular constraints project to a new image.

Given two silhouettes, p and q, we can always construct an object that can produce p and q, with a method based on volume intersection (eg., [8]). We may assume, without loss of generality (WLOG), that rotation is about the y-axis (when we consider three silhouettes this assumption results in a loss of generality). Also, WLOG assume that the silhouettes are transformed in the plane so that $p_1 = q_1 = (0,0)$ (see Figure 1, left), and the tangent to p and q at that point is the x axis. Assume the silhouettes are scaled so that p_2 and q_2 have the same y value. Assume also WLOG that M is positioned so that the point on M that projects to p_1 is placed at $(0,0,0)$. Moreover, we can assume that M is projected without scaling or translation. That is, in this setup, we can assume the object is projected orthographically to produce p, then rotated about the y axis, and projected orthographically to produce q.

If we cut through P and Q with a single horizontal line, $y = i$, we get two line segments, called P_i and Q_i. Denote the end points of P_i by $(P_{i,min}, i), (P_{i,max}, i)$. These line segments are projections of a slice through M, where it intersects the plane $y = i$. Call this slice M_i. The segment P_i constrains M_i. In particular, in addition to lying in the $y = i$ plane, all points in M_i must have $P_{i,min} \leq x \leq P_{i,max}$, and there must be points on M_i for which $P_{i,min} = x$ and for which $x = P_{i,max}$ (and there must be points on M_i that take on every intermediate value of x). We get constraints of this form for every i. Any model that meets these constraints will produce a silhouette p.

Now, suppose that Q has been produced after rotating M by some angle, θ (see Figure 1, middle). The constraints that Q places on M have the same form. In particular, P_i and Q_i provide the only information that constrains M_i. However, the constraints Q_i places are rotated by an angle θ relative to the constraints of P_i. Therefore, together, they constrain M_i to lie inside a parallelogram, and to touch all of its sides. Therefore, we can create an object that produces both silhouettes simply by constructing these parallelograms, then constructing an object that satisfies the constraints they produce.

We denote the entire set of constraints that we get from these two images by C_θ. We now prove that it is not important to know θ, because the set of constraints that we derive by assuming different values of θ are closely related. Let $C_{\frac{\pi}{2}}$ denote the constraints we get by assuming that $\theta = \frac{\pi}{2}$. Then

Lemma 1. $T_\theta C_\theta = C_{\frac{\pi}{2}}$ where: $T_\theta = \begin{pmatrix} 1 & 0 & 0 \\ 0 & 1 & 0 \\ -\cos\theta & 0 & \sin\theta \end{pmatrix}$

That is, C_θ consists of a set of parallelograms, and applying T_θ to these produces the set of rectangles that make up $C_{\frac{\pi}{2}}$.

We omit the proof of this lemma for lack of space. This shows that C_θ and $C_{\frac{\pi}{2}}$ are related by an affine transformation. Since the affine transformations form a group, this implies that without knowing θ we determine the constraints up to an affine transformation. This is related to prior results showing that affine structure of point sets can be determined from two images ([6]). However, our results are quite different, since they refer to silhouettes in which different sets of points generate the silhouette in each image, and our proof is quite different.

3   Comparing Three Silhouettes

3.1   Silhouettes with Simple Cross-Sections

Now we assume that the third silhouette r is generated by again rotating M about the same axis. This problem is easier than the case of general rotations, because each of the parallelograms constraining M project to a line segment. For any other direction of rotation, they project to parallelograms. To simplify notation we refer to $C_{\frac{\pi}{2}}$ as C. These constraints are directly determined by assuming that p constrains the x coordinates of M, and q constrains the z coordinates. The true constraints, C_θ, depend on the true angle of rotation between the first two images, which is not known.

We can translate and scale r so that the y coordinates of the tops and bottoms of the silhouettes are aligned. This accounts for all scaling, and all translation in y. We may then write the transformation that generates r in the form:

$$T_\phi M = \begin{pmatrix} 1 & 0 & 0 \\ 0 & 1 & 0 \end{pmatrix} \begin{pmatrix} \cos\phi & 0 & \sin\phi \\ 0 & 1 & 0 \\ -\sin\phi & 0 & \cos\phi \end{pmatrix} M + \begin{pmatrix} t \\ 0 \end{pmatrix}$$

which expresses x translation of the object, rotation about the y axis by ϕ, and then orthographic projection. We now examine how this transformation projects the vertices of the constraining parallelograms into the image. As we will see, the locations of these projected vertices constrain the new silhouette r. Since $C = T_\theta C_\theta$, we have $T_\theta^{-1} C = C_\theta$. Therefore, the projection of the true constraints, $T_\phi C_\theta = T_\phi T_\theta^{-1} C$, or:

$$T_\phi C_\theta = \begin{pmatrix} \cos\phi - \frac{\sin\phi\cos\theta}{\sin\theta} & 0 & \frac{\sin\phi}{\sin\theta} \\ 0 & 1 & 0 \end{pmatrix} C + \begin{pmatrix} t \\ 0 \end{pmatrix}$$

We will abbreviate this as: $\begin{pmatrix} a & 0 & b \\ 0 & 1 & 0 \end{pmatrix} C + \begin{pmatrix} t \\ 0 \end{pmatrix}$ with $a = \cos\phi - \frac{\sin\phi\cos\theta}{\sin\theta}, b = \frac{\sin\phi}{\sin\theta}$. a, b and t are unknowns, while the constraints, C and the new silhouette r are known. Our goal is to see if a, b and t exist that match the constraints and silhouette consistently.

Constraints on transformation parameters. We will be assisted by the fact that the projection of the constraints can be described by equations that are linear in a, b and t. However, because a and b are derived from trigonometric functions they cannot assume arbitrary values. So we first formulate constraints on these possible values.

We can show (derivation omitted) that a and b are constrained by:

$$-1 \le |a| - |b| \le 1; \qquad -(|a| + |b|)| \le 1 \le |a| + |b|$$

and that any a and b that meet these constraints lead to valid values for θ and ϕ. For any of the four possible choices of sign for a and b, these constraints are linear on a and b.

Constraints from new silhouette. Now consider again a cross-section of the constraints, C_i, and of the filled silhouette, R_i (see Figure 1). C_i is a rectangle, with known vertices $d_{i,1}, d_{i,2}, d_{i,3}, d_{i,4}$. Specifically: $d_{i,1} = (P_{i,min}, i, Q_{i,min})$ $d_{i,2} = (P_{i,max}, i, Q_{i,min})$ $d_{i,3} = (P_{i,min}, i, Q_{i,max})$ $d_{i,4} = (P_{i,max}, i, Q_{i,max})$.

Under projection, these vertices map to a horizontal line with $y = i$. We will consider constraints from just one $y = i$ plane, and drop i to simplify notation. Call the x coordinates of these projected points d'_1, d'_2, d'_3, d'_4. That is, $d'_j = (a, 0, b) \cdot d_j + t$. Notice that the sign of a and b determine which of these points have extremal values. For example, $a, b \geq 0 \Rightarrow d'_1 \leq d'_2, d'_4 \leq d'_3$. We continue with this example; the other three cases can be treated similarly.

R_i is a line segment, with end points whose x values we'll denote by e_1, e_2, with $e_1 \leq e_2$. Since M is constrained to lie inside C_i in the $y = i$ plane, we know that e_1 and e_2 must lie in between the two extremal points. That is: $d'_1 \leq e_1$, $e_2 \leq d'_3$. Furthermore, we know that M touches every side of C_i. This means that the projection of each side must include at least one point that is in R. This will be true if and only if: $e_1 \leq d'_2$, $e_1 \leq d'_4$, $d'_2 \leq e_2$, $d'_4 \leq e_2$.

These are necessary and sufficient constraints for r to be a possible silhouette of the shape that produced p and q. Finally, since $d'_j = (a, 0, b) \cdot d_j + t$ these constraints are linear in a, b, and t. As noted above, for $a, b \geq 0$ we also have linear constraints on a and b that express necessary and sufficient conditions for them to be derived from rotations. So we can check whether a new silhouette is consistent with two previous ones using linear programming.

Because of noise, the constraints might become slightly infeasible. It is therefore useful to specify a linear objective function that allows us to check how close we can come to meeting the constraints. We can write the constraints as, for example, $(a, 0, b)d_{i,1} + t \leq R_{i,min} - \lambda$. Then we run a linear program to satisfy these while maximizing λ. The constraints we have derived are all met if these constraints are met with $\lambda \geq 0$. If $\lambda < 0$ then λ provides a measure of the degree to which the constraints are violated.

3.2   Silhouettes with Complex Cross-Sections

Up to now, we have assumed that a horizontal cross-section of a silhouette consists of a single line segment. This will not generally be true for objects with multiple parts, holes, or even just concavities. These *multi-line* silhouettes complicate the relatively simple picture we have derived above. We wish to make several points about multi-line silhouettes. First, if we fill in all gaps between line segments we can derive the same straightforward constraints as above; these will be necessary, but not sufficient conditions for a new silhouette to match two previous ones. Second, if we merely require that the new silhouette have a number of lines that is consistent with the first two silhouettes, this constraint can be applied efficiently, although we omit details of this process for lack of space. Third, to exactly determine whether a new silhouette is consistent with previous ones becomes computationally more demanding, requiring consideration of either a huge number of possibilities, or an explicit search of the space of rotations.

But if we make a simple genericity assumption, the complexity can be reduced to a small size again.

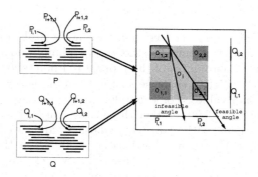

Fig. 2. Here we show two silhouettes, P and Q, that have cross-sections that consist of two line segments. On the right we show how the i cross-section leads to four parallelograms that must contain the object. Either the dark grey pair $(o_{1,1}, o_{2,2})$ or the patterned pair $(o_{1,2}, o_{2,1})$ must contain parts of the object. A third viewing angle, labeled "feasible angle" is shown for which this object may appear as a single line segment. An infeasible angle is also shown; from this viewpoint the object must produce two line segments in the image. Our system uses this constraint, though we omit details of how this is done. In some cases we make a continuity assumption across cross-sections. On the left, this means that, for example, if $P_{i,1}$ matches $Q_{i,1}$ ($o_{1,1}$ contains part of the object) then $P_{i+1,1}$ matches $Q_{i+1,1}$.

In the example shown in Figure 2, we can suppose either that $o_{1,1}$ and $o_{2,2}$ are occupied, or that $o_{1,2}$ and $o_{2,1}$ are occupied (there are other possibilities, which we can handle, but that we omit here for lack of space). When we assume, for example, that $o_{1,1}$ contains part of the object we can say that $P_{i,1}$ is matched to $Q_{i,1}$. Each of the two possible ways of matching $(P_{i,1}, P_{i,2})$ to $(Q_{i,1}, Q_{i,2})$ must be separately pursued, and gives rise to separate constraints that are more precise than the coarse ones we get by filling in the gaps in multi-line cross-sections. Suppose that for k consecutive cross-sections, the first two silhouettes each have two line segments. If we consider all possible combinations of correspondences across these cross-sections, we would have 2^k possibilities, a prohibitive number. But we can avoid this with a simple genericity assumption. We assume that in 3-D, it does not happen that one part of an object ends exactly at the same height that another part begins. This means that given a correspondence between line segments at one cross-section, we can typically infer the correspondence at the next cross-section.

3.3   Occlusion

The methods described above can also be applied to partially occluded silhouettes. To do this, something must be known about the location of the occlusion. For

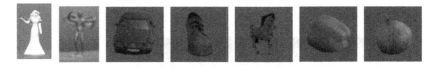

Fig. 3. Seven objects used in experiments.

example, if a cross-section is known to be occluded in one silhouette, that cross-section can be discarded. If a cross-section is known to be partially occluded in the third silhouette, the visible portion can be required to lie inside the projection of the constraining parallelogram derived from the other two. Occlusion may not only make it impossible to derive constraints from occluded cross-sections, it may also create uncertainty in determining which cross-sections correspond to each other. For example, if the bottom of an object is blocked in a third view, we will not know how many cross-sections are occluded. We can solve this by searching through different scalings of the silhouette, which imply different possible ways of matching its cross-section to the first two silhouettes. We can then select the scale or scales that allow the resulting constraints to be met.

3.4   Experiments

We test these ideas using the objects shown in Figure 3. Our experimental system varies in which approach we use to handle multi-lines.

Experiment 1: First, we experiment with coarse constraints that fill in any gaps present in a silhouette cross-section. Also, we heuristically throw away some constraints that may be sensitive to small misalignments between different silhouettes. In this experiment we use five silhouettes taken from a figure of Snow White (Figure 4) photographed after rotations of $20°$. First, all ten triplets of these silhouettes are compared to each other. In all cases they are judged consistent ($\lambda > 0$). Next, we compared each pair to 95 silhouettes, taken from the objects shown in Figure 3. About 6% of these other objects are also judged consistent with two Snow Whites (see Figure 4).

Fig. 4. Silhouettes of Snow White, numbers one to five from left to right. On the right, experimental results.

Fig. 5. On the left, silhouettes of Hermes. On the right, experiments comparing pairs of these to either a third silhouette of Hermes (first ten data points, shown as circles) or to silhouettes of other objects (shown as crosses). A horizontal line at $\lambda = -5$ separates correct answers with one false positive.

Experiment 2: Next, we performed a similar experiment using the silhouettes of Hermes in Figure 5. The axis of rotation was tilted slightly, so that the images do not exactly lie on a great circle on the viewing sphere. We heuristically compensate for this by searching for good in-plane rotations. For all 10 triples of silhouettes of Hermes, we obtain values of λ ranging from -4.4 to 1.7. However, when we compare randomly chosen pairs of Hermes silhouettes to randomly chosen silhouettes of the other five objects, we obtain only one case in twenty-five with λ larger than -4.4; other values are much smaller (see Figure 5).

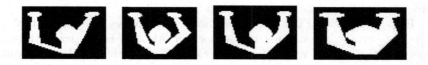

Fig. 6. The two figures on the left show the first and second silhouettes of the object used in experiment 4. The third silhouette shows this object from a new view. The fourth silhouette shows the same view with the object scaled so that its cross-sections are 1.8 times as big. This silhouette cannot be matched to the first two without greatly violating the constraints ($\lambda < -12$).

Experiment 3: We now show an experiment in which we search through possible correspondences between different object parts. We use a synthetic shape, vaguely like a human torso (Figure 6). Given three silhouettes, there are four possible correspondences between the "hands". We consider all four, then use the continuity constraint to determine correspondences at subsequent cross-sections. We compare two silhouettes to a third in which the shape has been "fattened", so that the cross-section of each part is scaled by a constant factor. When scale is 1, therefore, the third silhouette comes from the same object that produced the first two. In Figure 7 we show how λ varies with scale. We also show what happens if we do not hypothesize correspondences between the parts of the figure, but just fill in gaps in multiline cross-sections to derive a simple, conservative

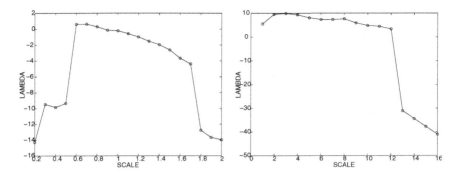

Fig. 7. On the left, we show how λ varies as the third silhouette scales. A scale of 1 means the third silhouette comes from the same object as the first two. On the right, we show this when individual parts are not matched, and only coarser constraints are derived by filling in all gaps in each cross-section of each silhouette. Note that the horizontal scale is approximately ten times larger on the right.

set of constraints. For an object like this, with small parts widely separated, this approach is much too conservative.

Fig. 8. On the left, the third Snow White silhouette, half occluded. The next two images show one hypothesized occlusion of the first two silhouettes that match this; the last two images show a second (false) match.

Experiment 4: Finally, we experiment with a case in which the first two Snow White silhouettes are matched to the third, but the bottom half of the third is occluded. In this case we try matching this half-silhouette to some top portion of the other two silhouettes, considering all possible top portions. We find two regions in the set of possible scales in which the third silhouette matches portions of the first two within one pixel of error; either when the first two are supposed about half occluded (the correct choice) or when they are supposed about 70% occluded (incorrect). Both are shown in Figure 8.

4 Conclusions

We have analyzed the problem of object recognition using silhouettes. We especially focus on the problem in which our knowledge of an object comes from seeing it from only a few viewpoints, under relatively unstructured viewing conditions, and in which we do not have a priori knowledge that restricts the model to belong to a special class of objects. This situation has not been much addressed, presumably because in this case it is not possible to derive a definite 3D model. However, we show that even though we may have considerable uncertainty about the 3D object's shape, there is still a lot of information that we can use to recognize the object from new viewpoints. This fits our general view that recognition can be done by comparing images, if our comparison method is based upon the knowledge that images are the 2D projections of the 3D world.

Our analysis has been restricted to the case where the objects or camera rotate about an axis that is parallel to the image plane. This is a significant restriction, but we feel that this case is worth analyzing for several reasons. First, it occurs in practical situations such as when a mobile robot navigates in the world, or in images generated to study human vision. Second, this analysis gives us insight into the more general problem, and provides a starting point for its analysis.

References

1. Brady, M. and Yuille, A., 1984, "An Extremum Principle for Shape from Contour," *IEEE Trans. PAMI* **6**(3):288-301.
2. Chien, C. and Aggarwal, J., 1989, "Model Construction and Shape Recognition from Occluding Contours," *IEEE Trans. PAMI* **11**(4):372–389.
3. Cipolla, R. and Blake, A., 1992, "Surface Shape from Deformation of Apparent Contours," *IJCV* **9**(2):83–112.
4. Forsyth, D., 1996. "Recognizing Algebraic Surfaces from their Outlines," *IJCV* **18**(1):21-40.
5. Giblin, P. and Weiss, R., 1987, "Reconstruction of Surfaces from Profiles," *IEEE Int. Conf. on Comp. Vis.*:136-144.
6. Koenderink, J. and van Doorn, A., 1991. "Affine structure from motion", *Journal of the Optical Society of America*, **8**(2): 377–385.
7. Laurentini, A., 1997, "How Many 2D Silhouettes Does it Take to Reconstruct a 3D Object," *Computer Vision and Image Understanding* **67**(1):81–87.
8. Martin, W. and Aggarwal, J., 1983, "Volumetric Descriptions of Objects from Multiple Views," *IEEE Trans. PAMI* **5**(2):150–158.
9. J. Ponce, D. Chelberg, W. B. Mann, 1989, "Invariant Properties of Straight Homogeneous Generalized Cylinders and Their Contours" *IEEE Trans. PAMI*, **11**(9) pp. 951-966.
10. Richards, W., Koenderink, J., and Hoffman, D., 1989, "Inferring 3D Shapes from 2D Silhouettes," in *Natural Computation*, edited by W. Richards.
11. Szeliski, R. and Weiss, R., 1998, "Robust Shape Recovery from Occluding Contours Using a Linear Smoother," *IJCV* **28**(1):27-44.
12. Vaillant, R. and Faugeras, O., 1992, "Using Extremal Boundaries for 3-D Object Modeling," *IEEE Trans. PAMI* **14**(2):157-173.

Discrete Deformable Boundaries for the Segmentation of Multidimensional Images

Jacques-Olivier Lachaud and Anne Vialard

LaBRI,* Université Bordeaux 1, 351 cours de la Libération, F-33405 Talence, France
(lachaud,vialard)@labri.u-bordeaux.fr

Abstract. Energy-minimizing techniques are an interesting approach to the segmentation problem. They extract image components by deforming a geometric model according to energy constraints. This paper proposes an extension to these works, which can segment arbitrarily complex image components in any dimension. The geometric model is a digital surface with which an energy is associated. The model grows inside the component to segment by following minimal energy paths. The segmentation result is obtained *a posteriori* by examining the energies of the successive model shapes. We validate our approach on several 2D images.

1 Introduction

A considerable amount of litterature is devoted to the problem of image segmentation (especially 2D image segmentation). Image components are determined either by examining image contours or by looking at homogeneous regions (and sometimes using both information). The segmentation problem cannot generally be tackled without adding to that information some *a priori* knowledge on image components, e.g., geometric models, smoothness constraints, reference shapes, training sets, user interaction. This paper deals with the segmentation problem for arbitrary dimensional images. We are interested in methods extracting an image component by deforming a geometric model. The following paragraphs present classical techniques addressing this issue.

Energy-minimizing techniques [4] have proven to be a powerful tool in this context. They are based on an iterative adaptation process, which locally deforms a parametric model. The model/image adequation is expressed as an energy, which is minimal when the model geometry corresponds to image contours. The continuity of the geometric model and tunable smoothness constraints provide a robust way to extract image components, even in noisy images. The adaptation process is sensitive to initialization since it makes the model converge on local minima within the image. The parametric definition of the model also restricts its topology to simple objects. Recent works now propose automated topology adaptation techniques to overcome this issue, both in 2D [10] and in 3D [6]. However, these techniques are difficult to extend to arbitrary dimensions.

---

* Laboratoire Bordelais de Recherche en Informatique - UMR 5800

C. Arcelli et al. (Eds.): IWVF4, LNCS 2059, pp. 542–551, 2001.

Front propagation techniques have been proposed to avoid the topology restriction induced by the model parameterization. Instead of deforming a geometric model in the image, they assign a scalar value to each point of the image space. The evolution of the points value is governed by partial differential equations, similar to the heat diffusion equation. The model is then implicitly defined as a level-set of this space, which is called a front. The equations are designed to make the front slow down on strong contours and to minimize its perimeter (or area in 3D) [9]. The implicit definition of the front ensures natural topology changes. However, this technique is not designed to integrate *a priori* knowledge on the image component (e.g., other geometric criteria, reference shape).

In region growing methods [1], the extraction of an image component follows two steps: (i) seeds are put within the component of interest and (ii) these seeds grow by iteratively adding pixels to them according to a merging predicate (homogeneity, simple geometric criterion). These methods are interesting because on one hand they have a simple dimension independent formulation and on the other hand they can segment objects of arbitrary topology. However, they are not well adapted to the extraction of inhomogeneous components.

This paper proposes an original approach based on a discrete geometric model that follows an energy-minimizing process. The discrete geometric model is the *digital boundary* of an object growing within the image. The model energy is distributed over all the boundary elements (i.e., the *surfels*). The energy of each element depends on both the local shape of the boundary and the surrounding image values. The number of possible shapes within an image grows exponentially with its size. Therefore, the following heuristic is used to extract components in an acceptable time. The model is initialized as an object inside the component of interest. At each iteration, a set of connected elements (i.e., a *voxel patch*) is locally glued to the model shape. The size and position of this set are chosen so that the object boundary energy be minimized. This expansion strategy associated with proper energy definitions induce the following model behavior: strong image contours forms "wells" in the energy that hold the model, the growth is at the same time penalized in directions increasing the area and local curvature of the object boundary, the model grows without constraints elsewhere.

This model casts the energy-minimizing principle in a discrete framework. Significant advantages are thus obtained: reduced sensibility to initialization, modeling of arbitrary objects, arbitrary image dimension. The paper is organized as follows. Section 2 recalls some necessary digital topology definitions and properties. Section 3 defines the model geometry and its energy. Section 4 presents the segmentation algorithm. Segmentation results on 2D images are presented and discussed in Section 5.

2   Preliminary Definitions

A *voxel* is an element of the discrete n-dimensional space $\mathbb{Z}^n$, for $n \geq 2$. Some authors [11] use the term "spel" for a voxel in an n-dimensional space; since we feel that no confusion should arise, we keep the term "voxel" for any dimension. Let M be a finite "digital parallelepiped" in $\mathbb{Z}^n$. An *image* I on $\mathbb{Z}^n$ is a tuple

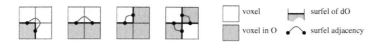

Fig. 1. Local computation of surfel adjacencies in the 2D case. The 4-adjacency (resp. 8-adjacency) has been chosen for the object elements (resp. background elements).

(M, f) where f is a mapping from the subset M of $\mathbb{Z}^n$, called the *domain* of I, toward a set of numbers, called the *range* of I. The *value* of a voxel $u \in M$ in the image I is the number $f(u)$. An *object* is any nonempty subset of the domain M. The *complement* of the object O in M is denoted by O^c.

Let ω_n be the adjacency relation on $\mathbb{Z}^n$ such that $\omega_n(u, v)$ is true when u and v differ of ± 1 on exactly one coordinate. Let α_n be the adjacency relation such that $\alpha_n(u, v)$ is true when $u \neq v$, and u and v may differ of either -1, 0, or 1 on any one of their coordinates. If ρ is any adjacency relation, a ρ-*path* from a voxel v to a voxel w on a voxel set A is a sequence $u_0 = v, \ldots, u_m = w$ of voxels of A such that, for any $0 \leq i < m$, u_i is ρ-adjacent to u_{i+1}. Its *length* is $m + 1$.

For any voxels u and v with $\omega_n(u, v)$, we call the ordered pair (u, v) a *surfel* (for "surface element" [11]). Any nonempty set of surfels is called a *digital surface*. For any given digital surface Σ, the set of voxels $\{v \mid (u, v) \in \Sigma\}$ is called the *immediate exterior* of Σ and is denoted by IE(Σ). The *boundary* ∂O of an object O is defined as the set $\{(u, v) \mid \omega_n(u, v)$ and $u \in O$ and $v \in O^c\}$.

Up to now, an object boundary is just viewed as a set. It is convenient to have a notion of surfel neighbors (i.e., a "topology") in order to define connected zones on an object boundary or to determine an object by tracking its boundary. In our case, this notion is compulsory to define a coherent model evolution. Besides, defining an object through its boundary is often faster.

Defining an adjacency between surfels is not as straightforward as defining an adjacency between voxels (especially for $n \geq 3$). The problem lies in the fact that object boundary components (through surfel adjacencies) must separate object components from background components (through voxel adjacencies). In this paper, we do not focus on building surfel adjacencies consistent with a given voxel adjacency. Consequently, given an object O considered with a voxel adjacency ρ, with either $\rho = \omega_n$ or $\rho = \alpha_n$, we will admit that it is possible to locally define a consistent *surfel adjacency* relation, denoted by β_O, for the elements of ∂O (3D case, see [3]; nD case, Theorem 34 of Ref. [7]). For the 2D case, Figure 2 shows how to locally define a surfel adjacency on a boundary.

The β_O-adjacency relation induces β_O-*components* on ∂O. β_O-*paths* on ∂O can be defined analogously to ρ-paths on O. The *length* of a β_O-path is similarly defined. The β_O-*distance* between two surfels of ∂O is defined as the length of the shortest β_O-path between these two surfels. The β_O-*ball* of size r around a surfel σ is the set of surfels of ∂O which are at a β_O-distance lesser or equal to r from the surfel σ. Let Σ be a subset of ∂O. We define the *border* $B(\Sigma)$ of Σ on ∂O as the set of surfels of Σ that have at least one β_O-neighbor in $\partial O \setminus \Sigma$. The k-*border* $B_k(\Sigma)$ of Σ on ∂O, $1 \leq k$, is the set of surfels of Σ which have a β_O-distance lesser to k from a surfel of $B(\Sigma)$.

3   The Discrete Deformable Boundaries Model

For the purpose of image segmentation, we introduce a geometric model to fit image components. The model *geometry* (i.e., its *shape*) is defined as an object in the image. It is equivalently defined as the boundary of this object. Note that the model is not necessarily connected. The model geometry is aimed to evolve from an initial shape toward an image component boundary. This evolution depends on an energy associated to each possible geometry of the model.

We identify the energy of an object to the energy of its boundary. The *energy* of an object boundary ∂O, denoted by $E(\partial O)$, is computed by summation of the energies of each of its surfels. To get finer estimates, the energy of a surfel may depend on a small neighborhood around it on the object boundary. That is why we use the notation $E(\sigma)_{\partial O}$ to designate the energy of the surfel σ with respect to ∂O (σ must be an element of ∂O). By definition, we set $E(\Sigma)_{\partial O} = \sum_{\sigma \in \Sigma} E(\sigma)_{\partial O}$, where Σ is any nonempty subset of an object boundary ∂O.

The surfel energy is the sum of several energy terms. Two types of surfel energies are distinguished: the surfel energies that only depend on the local geometry of the boundary around the surfel are called *internal energies*, the surfel energies that also depend on external parameters (e.g., local image values) are called *external energies*. The local geometric characteristics required for the surfel energy computation are based upon a neighborhood of the surfel on the object boundary: it is a β_O-ball on the object boundary, centered on the surfel. To simplify notations, we assume that this ball has the same size p for the computation of every surfel energy. The whole surfel energy is thus locally computed.

Internal energies allow a finer control on the model shape (e.g., smoothness). External energies express the image/model adequation or other external constraints (e.g., similarity to a reference shape). The following paragraphs present a set of internal and external energies pertinent to our segmentation purpose. This set is by no way restrictive. New energies can be specifically designed for a particular application to the extent that the summation property is satisfied.

Image features are generally not sufficient to clearly define objects: the boundary can be poorly contrasted or even incomplete. To tackle this problem, we use the fact that the shape which most likely matches an image component is generally "smooth". In our case, this is expressed by defining two internal energies. The *stretching energy* $E^s(\sigma)_{\partial O}$ of a surfel σ is defined as an increasing function of the area of the surfel σ. The *bending energy* $E^b(\sigma)_{\partial O}$ of a surfel σ is defined as an increasing function of the mean curvature of the surfel σ. Examples of area and mean curvature computations are given in Section 5. Note that these energies correspond to the internal energies of many deformable models [4], which regularize the segmentation problem.

In our context, we define a unique external energy, based on the image value information. Since the model evolution is guided by energy minimization, the *image energy* $E^I(\sigma)_{\partial O}$ of a surfel σ should be very low when σ is located on a strong contour. A simple way to define the image energy at a surfel $\sigma = (u, v)$ on an object boundary is to use the image gradient: $E^I(\sigma)_{\partial O} = -\|f(v) - f(u)\|^2$, if $I = (M, f)$. This definition is valid for arbitrary n.

The model grows by minimizing its energy at each step. Since the growing is incremental, an incremental computation of the energy would be pertinent.

More precisely, the problem is to compute the energy of an object O' given the energy of an object O included in O'. In our case, the digital surfaces ∂O and $\partial O'$ generally have much more common surfels than uncommon surfels (as the model is growing, this assertion is more and more true). The set of the common surfels is denoted by Φ. The p-border of Φ on ∂O is identical to the p-border of Φ on $\partial O'$. We can thus denote it uniquely by $B_p(\Phi)$. The energy of ∂O and $\partial O'$ is expressed by the two following equations:

$$E(\partial O) = \sum_{\sigma \in \partial O} E(\sigma)_{\partial O} = E(\Phi \setminus B_p(\Phi))_{\partial O} + E(B_p(\Phi))_{\partial O} + E(\partial O \setminus \Phi)_{\partial O},$$

$$E(\partial O') = \sum_{\sigma \in \partial O'} E(\sigma)_{\partial O'} = E(\Phi \setminus B_p(\Phi))_{\partial O'} + E(B_p(\Phi))_{\partial O'} + E(\partial O' \setminus \Phi)_{\partial O'}.$$

From the surfel energy computation, it is easy to see that the surfels of $\Phi \setminus B_p(\Phi)$, common to both ∂O and $\partial O'$, hold the same energy on ∂O and on $\partial O'$. However, the energy of the surfels of $B_p(\Phi)$ may (slightly) differ whether they are considered on ∂O or on $\partial O'$. We deduce

$$\underbrace{E(\partial O') - E(\partial O)}_{\text{variation}} = \underbrace{E(\partial O' \setminus \Phi)_{\partial O'}}_{\text{created surfels}} - \underbrace{E(\partial O \setminus \Phi)_{\partial O}}_{\text{deleted surfels}} + \underbrace{E(B_p(\Phi))_{\partial O'} - E(B_p(\Phi))_{\partial O}}_{\text{surfels close to displacement}}.$$

To get efficient energy computations at each step, each surfel of the model stores its energy. When a model grows from a shape O to a shape O', the energy of only a limited amount of surfels will have to be computed: (i) the energy of created surfels and (ii) the energy of the surfels nearby those surfels.

4   Segmentation Algorithm

In the energy-minimizing framework, the segmentation problem is translated into the minimization of a cost function in the space of all possible shapes. Finding the minimum of this function cannot be done directly in a reasonable time. Except for very specific problems (e.g., what is the best contour between two known endpoints), heuristics are proposed to extract "acceptable" solutions. For the snake, an "acceptable" solution is a local minimum. We propose a heuristic that builds a set of successive shapes likely to correspond to image components.

We first briefly outline the segmentation process. The model is initialized as a set of voxels located inside the object to be segmented. At each step, a voxel patch is added to the model. A *voxel patch* of radius k around a surfel σ on the boundary ∂O is the immediate exterior of the β_O-ball of size k around σ. To decide where a voxel patch is "glued" to O, its possible various locations are enumerated. Among the possible resulting shapes, the one with the smallest energy is chosen. Unlike most segmentation algorithms, this process does not converge on the expected image component. However, the state of the model at one step of its evolution is likely to correspond to the expected image component. The boundary of the object of interest is hence determined *a posteriori*. This technique is similar to the discrete bubble principle [2].

It is then possible to let the user choose the shape pertinent to his problem among the successive states of the model. A more automated approach can also be taken. Since the model growing follows a minimal energy path (among all possible shapes within the image), the image/model adequation can be estimated through the model energy at each step. Consequently, the shape of minimal energy often delineates an object which is a pertinent component of the image.

For now, k is a given strictly positive integer number. It corresponds to the radius of the voxel patch that is added on the model at each step. The following process governs the growing evolution of the model and assigns an energy to each successive model state (see Fig. 2):

1. Assign 0 to i. Let O_0 be equal to the initial shape. The initial shape is a subset of the image domain included in the component(s) to extract. Let E_0 be equal to $E(\partial O_0)$.
2. For all surfels $\sigma \in \partial O_i$, perform the following steps:
   a) Extract the β_{O_i}-ball V_σ of radius k around σ. Define O_σ as $O_i \cup \mathrm{IE}(V_\sigma)$.
   b) Incrementally compute $E(\partial O_\sigma)$ from $E(\partial O_i)$.
3. Select a surfel τ with minimal energy $E(\partial O_\tau)$.
4. Let O_{i+1} be equal to O_τ. Let E_{i+1} be equal to $E(\partial O_\tau)$. Increment i.
5. Go back to step 2 until an end condition is reached (e.g., $O = M$, user interaction, automated minimum detection).

In the experiments described in Section 5, the end condition is $O = M$, which corresponds to a complete model expansion.

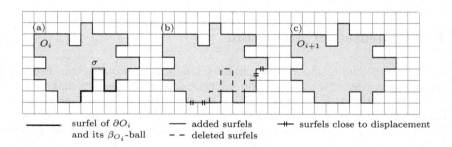

Fig. 2. A step in the model growing. (a) σ is a surfel of the model boundary at step i. (b) The voxel patch of radius 4 around σ is added to the object. (c) Model shape at step $i + 1$ if σ is chosen as the best place to make the model grow.

5 2D Experiments

In order to validate this segmentation approach, a 2D prototype has been implemented. The experiments emphasize the ability of our model to segment image components in various contexts: incomplete or weakly contrasted contours, inhomogeneous components. In 2D, voxels correspond to pixels and surfels are usually called pixel edges. We first consider how energies are computed and weighted. We then highlight the model capabilities on both synthetic and medical images.

5.1   Energy Computation

For our experiments, the energies for a surfel $\sigma = (u, v)$ are computed as follows:

- The stretching energy $E^s(\sigma)_{\partial O}$ is the contribution of the surfel σ to the model perimeter. The boundary ∂O can be viewed as one or several 4-connected paths. We adapt the Rosen-Profitt estimator to measure the contribution of σ to the boundary length. On the boundary path, σ has two end-points p_1 and p_2. These points can either be "corner" points or "non-corner" points. We set $E^s(\sigma)_{\partial O} = (\psi(p_1), \psi(p_2))/2$, where $\psi(p) = \psi_c = 0.670$ if p is a corner point and $\psi(p) = \psi_{nc} = 0.948$ otherwise. Note that other perimeter estimators could be used (e.g., see [5]).
- The bending energy $E^b(\sigma)_{\partial O}$ is computed as the ratio of the two distances l and D, defined as follows. The β_O-ball of size p around σ forms a subpath of ∂O, which has two endpoints e_1 and e_2. Its length l is computed by the same Rosen-Profitt estimator as above. The distance D is the Euclidean distance of e_1 and e_2. Many curvature estimation methods could be implemented (angular measurement, planar deviation, tangent plane variation, etc. [12]).
- The image energy $E^I(\sigma)_{\partial O}$ is defined as in Section 3 as $-\|f(v) - f(u)\|^2$. It is the only external energy used in the presented experiments.

The energy of a surfel σ on the boundary ∂O is the weighted summation of the above-defined energies:

$$E(\sigma)_{\partial O} = \alpha_s E^s(\sigma)_{\partial O} + \alpha_b E^b(\sigma)_{\partial O} + \alpha_I E^I(\sigma)_{\partial O},$$

where α_s, α_b and α_I are positive real numbers whose sum is one. To handle comparable terms, internal energies are normalized to $[0, 1]$. The image energy is normalized to $[-0.5, 0.5]$ to keep in balance two opposite behaviors: (i) should the image energy be positive, the shape of minimal energy would be empty, (ii) should the image energy be negative, long and sinuous shape would be favored. Note that most classical deformable models choose a negative image energy function. These coefficients allow us to tune more precisely the model behavior on various kinds of images. A set of coefficients pertinent to an image will be well adapted to similar images.

5.2   Results

In all presented experiments, we consider the model with the 4-connectedness. The surfel adjacency β_O is therefore defined as shown on Fig. 2. The parameter p is set to 4.

Since our model searches for contours in images, inhomogeneous components can efficiently be segmented. We illustrate this ability on the test image of Fig. 3a.

This test image raises another segmentation issue: contours are weak or even inexistant on some locations both between the disk and the ring and between the disk and the background. Fig. 3b emphasizes three steps in the model evolution: the initial shape, the iteration when the model lies on the disk–ring boundary (a local minimum in the energy function), the iteration when the model lies on the ring–background boundary (the minimum of the energy function). Fig. 3c displays the energy function. At each iteration, several patch radius sizes (i.e., k) have been tested for each surfel: k was between 0 and 6. The model successfully delineates the two boundaries during its evolution, although the first boundary induces a less significant minimum in the energy function than the second one: the first boundary is indeed not as well defined as the second one.

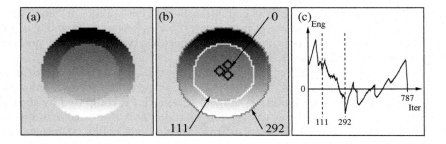

Fig. 3. Inhomogeneous components segmentation. (a) Test image: a circle filled with a shading from gray value 98 (top) to gray value 189 (bottom), a ring encircling it filled with a sharper shading from 0 (top) to 255 (bottom), a homogeneous background of gray value 215. (b) Three significant steps in the model evolution: initial shape, disk–ring boundary, ring–background boundary. (c) Energy curve for this evolution: the two extracted boundaries correspond to local minima of the energy function.

The second experiment illustrates the robustness of the segmentation process compared to the initial shape. The test image is a MR image[1] of a human heart at diastole (Fig. 4a). Our objective is to segment the right ventricle. This image component is inhomogeneous in its lower part and presents weak contours on its bottom side. The other contours are more distinct but are somewhat fuzzy. All these defects can be apprehended on Fig. 4b-c, which show the image after edge detection. The middle row (Fig. 4d-f) presents three evolutions, one per column, with three different initial shapes. Each image depicts three or four different boundaries corresponding to significant steps in the model evolution. The bottom row (Fig. 4g-i) displays the corresponding energy curve. For this experiment, only the patch radius size 3 is tested for each surfel (i.e., $k = 3$). Whichever is the initial shape, the model succeeds in delineating the right ventricle. The left ventricle may also be delineated in a second stage (near the end of the expansion),

[1] Acknowledgements to Pr. Ducassou and Pr. Barat, Service de Médecine Nucléaire Hôpital du Haut Levêque, Bordeaux, France.

but it is more hazardous: the proposed initial shapes are indeed extremely bad
for a left ventricle segmentation.

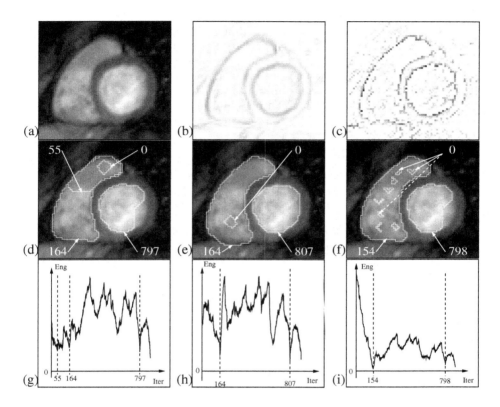

Fig. 4. Robustness of the segmentation to initialization. (a) Test image: a MR image
of a human heart at diastole. (b) Image after Sobel filtering. (c) Image after Laplace
edge detection. The bottom two rows depict the model behavior for three different
initial shapes (energy parameters are set to $\alpha_s = 0.25$, $\alpha_b = 0.25$, $\alpha_I = 1$). The middle
row (d-f) shows significant steps in the model evolution (initialization, important local
minima). The corresponding energy curves are drawn on the bottom row figures (g-i).

Our prototype does not include all the optimizations which could be implemented for the 2D case (e.g., efficient traversal of surfel adjacency graphs, various
precomputations). For the heart image, whose size is 68×63, the complete model
evolution takes 349s. The right ventricle is detected after 40s.

6   Conclusion

We have presented a discrete deformable model for segmenting image components. The segmenting process is carried out by expanding a digital surface

within the image under internal and external constraints. The external constraint (i.e., the image energy) stops the model expansion on strong contours. At the same time, the internal constraints regularize the model shape. The robust framework of energy-minimizing techniques is thus preserved. The significant advantages of this model are its dimension independence and its ability to naturally change topology. The first results on both synthetic and real-life data are very promising. They underline the model abilities to process images with poor contours and inhomogeneous components. Moreover, our segmentation technique is less sensitive to initialization than classical energy-minimizing techniques. Further information can be found in [8].

References

1. J.-P. Cocquerez and S. Philipp. *Analyse d'images : filtrage et segmentation.* Masson, Paris, Milan, Barcelone, 1995.
2. Y. Elomary. *Modèles déformables et multirésolution pour la détection de contours en traitement d'images.* PhD thesis, Univ. Joseph Fourier, Grenoble, France, 1994.
3. G.T. Herman and D. Webster. A topological proof of a surface tracking algorithm. *Computer Vision, Graphics, and Image Processing*, 23:162–177, 1983.
4. M. Kass, A. Witkin, and D. Terzopoulos. Snakes: Active contour models. *International Journal of Computer Vision*, 1(4):321–331, 1987.
5. J. Koplowitz and A. M. Bruckstein. Design of perimeter estimators for digitized planar shapes. *IEEE PAMI*, 11(6):611–622, 1989.
6. J.-O. Lachaud and A. Montanvert. Deformable meshes with automated topology changes for coarse-to-fine 3D surface extraction. *Medical Image Analysis*, 3(2):187–207, 1999.
7. J.-O. Lachaud and A. Montanvert. Continuous analogs of digital boundaries: A topological approach to iso-surfaces. *Graphical Models and Image Processing*, 62:129–164, 2000.
8. J.-O. Lachaud and A. Vialard. Discrete deformable boundaries for image segmentation. Research Report 1244-00, LaBRI, Talence, France, 2000.
9. R. Malladi, J. A. Sethian, and B. C. Vemuri. Shape Modelling with Front Propagation: A Level Set Approach. *IEEE PAMI*, 17(2):158–174, 1995.
10. T. McInerney and D. Terzopoulos. Medical Image Segmentation Using Topologically Adaptable Surfaces. In *Proc. of CVRMed-MRCAS, Grenoble, France*, volume 1205 of *LNCS*, pages 23–32. Springer-Verlag, 1997.
11. J. K. Udupa. Multidimensional Digital Boundaries. *CVGIP: Graphical Models and Image Processing*, 56(4):311–323, July 1994.
12. M. Worring and A. Smeulders. Digital curvature estimation. *Computer Vision, Graphics and Image Processing: Image Understanding*, 58(3):366–382, 1993.

Camera Motion Extraction Using Correlation for Motion-Based Video Classification

Pierre Martin-Granel[1], Matthew Roach[2], and John Mason[2]

[1] E.N.S.P.M.
Ecole Nationale Superieure de Physique de Marseille
13013 Marseille, FRANCE
pierre.martin-granel@ingenieur.cc
[2] University of Wales, Swansea
SA2 8PP, UK
{eeroachm, J.D.S.Mason}@swansea.ac.uk
http://galilee.swan.ac.uk

Abstract. This paper considers camera motion extraction with application to automatic video classification. Video motion is subdivided into 3 components, one of which, camera motion, is considered here. The extraction of the camera motion is based on correlation. Both subjective and objective measures of the performance of the camera motion extraction are presented. This approach is shown to be simple but efficient and effective. This form is separated and extracted as a discriminant for video classification. In a simple classification experiment it is shown that sport and non-sport videos can be classified with an identification rate of 80%. The system is shown to be able to verify the genre of a short sequence (only 12 seconds), for sport and non-sport, with a false acceptance rate of 10% on arbitrarily chosen test sequences.

1 Introduction

The classification of videos is becoming ever more important as the amount of multimedia material in circulation increases. Automatic classification of video sequences would increase usability of these masses of data by enabling people to search quickly and efficiently multimedia databases. There are three main sources of information in video: first the audio which has been extensively researched in the context of coding and recognition. Secondly, individual images which can be classified by their content. Thirdly, the dynamics of the image information held in the time sequence of the video and it this last attribute that makes video classification different to image classification. It is this dynamic aspect of video classification, at the highest level, we investigate here.

The most successful approaches to video classification are likely to use a combination of static and dynamic information. However in our approach we use only simple motion measures to quantify the contribution of the dynamics to high level classification of videos [1]. Approaches to motion extraction range from region-based methods such those proposed by Bouthemy *et al* [2] to tracking

C. Arcelli et al. (Eds.): IWVF4, LNCS 2059, pp. 552–562, 2001.
© Springer-Verlag Berlin Heidelberg 2001

objects as described by Parry *et al* [3]. It is generally accepted that motion present in videos can be classified into two kinds, foreground and background. The foreground motion is the motion created by one or multiple moving objects within the sequence. The background motion, usually the dominant motion in the sequence, is created by movement of the camera.

Those approaches that attempt to classify the background motion do so usually to label the video with the type of movement of the camera, as for example the work of Bouthemy [4] and Xiong [5]. Here we propose a method for separating the foreground object and background camera motions. Haering [6] uses foreground and background motion in combination with some static features to detect wildlife hunts in nature videos. We hypothesize that the separation of these two motion signals will benefit the classification of videos based on dynamics. Some preliminary results for the discriminating properties of camera motion alone in the context of a simple classification task of sport and non-sport are presented. The sports are chosen for their reasonably high pace during a game to make the illustrative example viable, e.g. Soccer, Rugby.

2   Video Dynamics

Three forms of video dynamics can be identified: two are motions within the sequence, namely foreground and background and the third which is of a different origin and is manually inserted in the form of shot and scene changes. Here the shot changes are automatically detected and discarded using pair-wise pixel comparisons. There are more robust approaches to shot or cut detection, for example Xiong [5]. Porter *et al* [7], use frequency domain correlation, however this is not currently integrated into the system and is the subject of further work. Interestingly Truong *et al* [8] suggest that the shot length has some discriminatory information and therefore this information is used in their genre classification of image sequences. We wish to separate all three of these motion signals in order to assess the classification potential of each individually. Here we consider the camera motion.

There can be some ambiguity in camera motion terminology, so here we clarify our interpretations of certain terms: *pan* is a rotation around a vertical axis, *tilt* is a rotation around an horizontal axis, *Z or image rotation* is a rotation around the optical axis. *X and Y translation* is caused by the camera moving left-right and up-down respectively normal to the optical axis, *zoom in/out* is a focal length change, and *Z translation* is a linear movement of the camera along the optical axis. It is actually difficult to discriminate automatically between pan, tilt, and X, Y translations because they introduce only subtly different perspective distortions. These distortions occur around object boundaries, and are most prominent where the difference in depth is large. It is the nature of the occlusions that is used to differentiate between the motions. For zoom and Z translation the objects in the image are magnified differently according to their depth position.

We can distinguish two main classes of models for camera movement: The first class aims at investigating the influence of the camera motion in the frames. This requires a 2D affine motion model extracted using general optical flow approaches. Some of these approaches use a 6-parameter camera model [9,10] describing the three rotations (pan, tilt, Z rotation) and three translations (X, Y, Z) of the camera. Generally, the model is reduced to 3-parameters [11,5], merging the terms pan and X translation, tilt and Y translation, disregarding the Z rotation, and replacing the Z translation by the notion of zoom.

The second class has the purpose to characterize the relative camera location in a 3D space. These claim to extract a 9-parameter camera model, for example [12]. In addition to the 6-parameter model with the three translations and three rotations, the X and Y focal lengths, and the zoom operation are included, also presented by Sudhi *et al* [10]. Some of these approaches use constraints on the epipole [12]. However this second class of approaches, although more accurate than the smaller model approaches, are obviously complicated, computationally expensive and the parameters are unlikely to be accurately determined from video analysis.

Here we use a simple three-parameters motion model to deal with camera motion X, Y and Z. Left and right, or X motion includes X translation and pan. Up and down, or Y motion includes Y translation and tilt, and Z motion includes Z translation and focal length modifications or zoom. This provides relative computational efficiency and is predicted to include most of the discriminatory information available for high-level classification.

3   The Correlation Approach

In this section, we describe the correlation approach used here to extract camera motion. In Section 3.1 we see how to obtain motion vectors to obtain an optical flow for blocks within the image using correlation. The first step consists of calculating a Pixel Translation and Deformation Grid (PTDG) between two consecutive frames using matching blocks. For each pixel in the image, the PTDG gives the translation and the deformation of its neighborhood.

In Section 3.2 we show how to extract the camera motion from the optical flow. A simple segmentation of this PTDG attempts to detect the different moving objects and to quantify their translations in the image. The dominant motion, over a certain threshold, is assumed to be caused by a mobile camera. Although a large rigid object in the foreground can create a similar effect this a rare occurrence and is not dealt with here. The translation of the background and the zoom in/out of the camera is deduced from this PTDG along a 3-parameter camera model. The rotation about the optical axis is not considered as it too is rare and is mainly found in music videos, a class we do not currently consider.

3.1   The NCS Principle

From the current frame, we extract a block of n×n pixels, which serves as a reference mask. The aim is to find the most similar block of pixels in the previ-

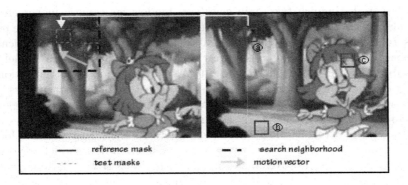

Fig. 1. Search for the matching mask from three different reference masks, ⓐ good correlation, ⓑ uniform area, ⓒ deformed area

ous frame by correlating the reference mask with all possible test masks in the previous frame. The test mask with the highest correlation coefficient, i.e the most similar block, is denoted the matching mask. It has been observed that the position of this matching mask is generally relatively close to the original position, therefore only a subregion of the image is searched; this is called the search neighborhood. The computation of the similarity indcies within a search neighborhood N×N is performed by comparing the reference mask with the different test masks. We obtain this comparison measure by using the normalized correlation coefficient $C_{k,l}$, for each pixel (k,l) in the search neighborhood, given by:

$$\mathfrak{C}_{k,l} = \frac{\sum\limits_{i,j,c} P_c(i,j) Q_c(i+k,j+l)}{\sqrt{\sum\limits_{i,j,c} P_c(i,j)} \sqrt{\sum\limits_{i,j,c} Q_c(i+k,j+l)}} \tag{1}$$

where (k,l) locates the N×N search neighborhood and defines the center of the test masks, (i,j) describe the mask n×n, $c \in \{R,G,B\}$ is the red, green or blue byte, P_c is the color value of the pixel in the current image, and Q_c is the color value of the pixel in the previous frame.

In figure 1 there are three example cases ⓐ, ⓑ and ⓒ illustrating different situations. The first, ⓐ, is a good correlation with a motion vector shown by the arrow. The second, ⓑ, is a uniform area. This gives many good correlations with resultant ambiguity. The third case, ⓒ, shows an example of a deformed area for which there are no good correlations. A 2D surface, called Neighborhood Correlation Surface (NCS) is obtained by plotting the correlation coefficients $\mathfrak{C}_{k,l}$ calculated over the search neighborhood. A value tending to 1 expresses an exact fitting between the reference mask from the current frame and the test mask from the previous frame.

In the Figure 1, the reference mask obtained ⓐ correlates almost exactly with a test mask in the search neighborhood. The reliability of this normalized correlation results from the fact that the coefficient decreases when the similarity decreases. The maximum of the NCS is usually represented by a peek. This gives us the previous position of the reference mask and therefore the evolution of this mask through time (from the matching mask to the reference mask). Figure 2 illustrates a typical successful correlation: the NCS shows confidently the position of the matching mask. This is the case of the reference mask ⓐ in Figure 1.

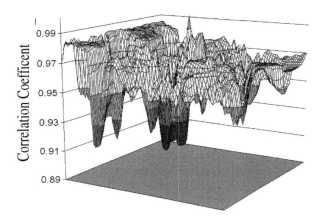

Fig. 2. NCS for the reference mask ⓐ: good similarity index

There are two cases where the NCS fails to indicate clearly the position of the matching mask in the previous frame. Figure 1 shows examples of each. Therefore it is necessary to introduce a limit beyond which the position of the matching mask, and therefore the subsequent motion vector, is regarded as unusable.

A first threshold is set to decide whether or not the reference mask and the matching mask are considered similar enough i.e. a good correlation. A maximum search correlation coefficient not reaching this threshold means that we cannot find a reliable matching mask: the reference mask is predicted to be included in a non-rigid object or in an area occluded or revealed by moving objects. This is referred to as deformation of the mask. This is the case of the reference mask ⓒ in Figure 1 for which no high peek in the NCS can be seen in Figure 3. The low maximum correlation coefficient indicates a deformation of the masks between the previous and the current frame.

A second threshold is set to prevent difficulties which occur when uniform zones are present. The correlation coefficients computed on an uniform search neighbourhood gives many high correlation coefficient values, and picking the

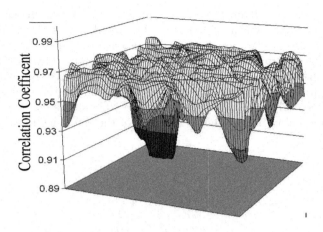

Fig. 3. NCS for the reference mask ©: bad similarity index, deformed area

maximum value may not ensure an accurate position of the matching mask. Therefore an NCS comprising of too many good correlations (coefficients over the first threshold) is considered as a uniform zone. This eliminates the need for further or apriori computation to ensure that the motion vector is valid, such as the variance analysis applied by [6]. The periodicity of peeks in the NCS can give an indication of the texture being uniform. This is the case of the reference mask ⓑ in Figure 1 the NCS for which is given in Figure 4.

For each frame a Pixel Translation and Deformation Grid (PTDG) is composed. The PTDG provides for each matching mask of the image the x-translation, y-translation, the maximum correlation coefficient and the number of successful correlations. The deformed masks (low maximum correlation coefficient: c.f. case ©) and uniform areas (high number of successful correlations: c.f. case ⓑ) are both rejected from further computations for camera motion.

3.2   Camera Motion Extraction

As stated previously, a relativly simple three-parameter model to deal with camera motion is used. The term zoom is used to encompass all Z motion. The Z rotation (about the optical axis) can be easily extracted from the PTDG, but is considered unnecessary for the classical scenes we propose to analyse. The first step is to compute a coarse global Z motion. The equation linking pixel coordinates and the camera focal length is given in Equation 2, as adopted by Xiong [5]. The focal length is assumed to be constant across the lens system. We consider only equal and opposite motion vector pairs from geometricaly opposite segments of the frame not rejected by the PTDG, as these are most likley to be part of the background. The relative segment length is not altered by a global

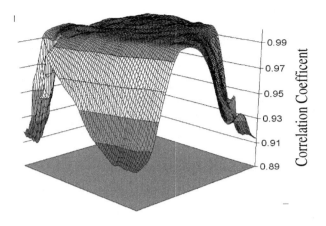

Fig. 4. NCS for the reference mask ⓑ: uniform area

translation or rotation; we assume that they are altered only by the zoom. Using the pairs of motion vectors to calculate the zoom improves the robustness to moving objects. We compute what we call the relative zoom average of the local zooms given by:

$$x = f_x \times \frac{X}{Z} \qquad y = f_y \times \frac{Y}{Z} \qquad \frac{l'}{l} = \frac{f'}{f} = z \qquad (2)$$

where $P = \{X, Y, Z\}$ designates the 3D coordinates of the object in a reference system which has (Oz) along the optical axis, where $p = \{x, y\}$ designates the image coordinates of the projection of this object in the image. l and l' are the segment lengths in the current and in the previous frame respectively. f_x and f_y are the x and y focal lengths in the current frame, with $f_x = f_y = f$, and f' is the focal length in the previous frame. z is the relative zoom.

Once this first estimation is complete, we adjust the PTDG to contain only the X and Y motion of the pixels. The motion vetors can segment the image into several groups. If we detect a large and coherent moving group (typically more than half the image), we assume that this is the moving background due to camera motion. If another coherent group is detected, we reject the pixels belonging to this group and then we reiterate the zoom computation to correct the PTDG. Obviously if a large rigid object is moving towards or away from the camera, the analysis of pairs of motion vectors on its area will succeed and it will affect the computation of the zoom. The correction of the PTDG will be skewed. However, we have observed that if the size of this rigid object is smaller than half the image, it is simple to separate from the background. After the background separation, a zoom computation considering only the background pixels, further improves accuracy.

4   Assessment

Results of this paper are divided into two parts. Primaraly, the experimental result relates to the assessment, both subjective and objective, of the camera compensation accuracy. Then the preliminary investigation in to whether the camera motion alone holds discriminatory dynamic information is presented. Features reflecting the camera motion signals, as in Figure 6, are statistically processed (inverse variance weighted, zero mean) and classified using a Bayesian based classifier in the form of a Gaussian Mixture Model (GMM).

4.1   Assessing the Camera Compensation

The first approach to assessing the camera compensation was based purely on subjective observation. A range of test scenes were observed to assess the compensation under fast and slow camera motions in the 3 parameters of the motion model. The observations were of thresholded differences of adjacent frames of the original sequences and compared with those differences after camera compensation. A difficult example of these comparison images is shown in Figure 5. The image created from the original sequence has motion in most areas of the image due to both object and camera motion as seen in 5(a). Conversely in 5(b), the compensated sequence shows a large reduction of the motion in the background in contrast to the motion around the foreground objects which does not show much reduction. With additional morphological noise filtering the background movement is further reduced as seen in 5(c).

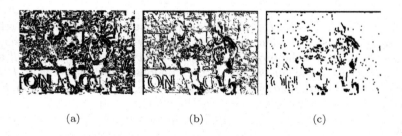

(a)   (b)   (c)

Fig. 5. Typical Examples of Camera Compensation (a) Original, (b) Compensated, (c) Noise Filtered

The second approach is an objective measure, using artificial test data. Here the accuracy in pixels of the predicted camera motion is measured. A test sequence is manually generated so that the ground truth of the camera motion is known. Static camera sequences are chosen and the frames are then perturbed in space through time to simialte camera motion. The sequences include moving forground objects to make sure the method was robust under various conditions, assuming that the dominant motion is the camera's motion. The Controlled test sequences are then put through the camera compensation algorithm and the measured camera motion compared with the ground truth. The average accuracy of the camera motion across the artificial test data was found to be in the order of 0.1 of a pixel.

4.2   Classification of Sport Using Camera Motion

The second part of the results show the potencial of the camera dynamics in classifying video sequences. The class of sport is chosen because this is predicted to be relatively different in terms of camera motions and is commonly quoted as having high discriminatable dynamics [2,8].

A total of 285 classifications were made, each approximately 12 seconds of video. The database used was made up of 28 scenes, 6 sport and 22 others including soap, news, cartoon, movies and nature. Each scene in the database comprises multiple shots and is about 40 seconds long giving a total of approximately 18 minutes. The short sequences used for classification are arbitrarily selected by automatically spliting a scene into 12 second sequences.

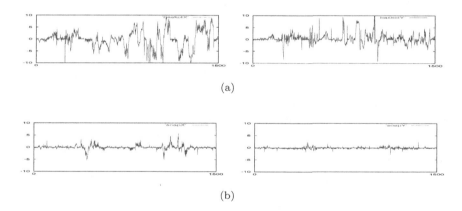

(a)

(b)

Fig. 6. Typical examples of second order camera motion (+10,-10 pixels) plotted against time (1500 frames). On the left is X motion on the right is Y motion. The scenes are (a) Basketball and (b) Soap.

Two motion signals of the camera, X and Y, for typical scenes can be seen in Figure 6. These motion signals are statistically modelled for sport and non-sport for classification. A "round-robin" technique is used to maximise the data usage. The GMM is trained on all scenes other than the one being tested i.e 27 scenes are split into short sequences of 12 seconds. The the test scene is changed and the training is done again. The feature extraction is that used for our holistic region-based approach described in [1].

Results are split into two: identification and verification. The identification error for sport and non-sport on short sequences on dynamics only was 20% based on just 12 second shots. Verification results were also obtained to further investigate the performance of the system for different applications. The system produced a false acceptance of 10% when false rejection was at 50%.

5   Conclusion

This proposed correlation approach tends to be simple but efficient and effective. The background search allows consideration of all kinds of scenes, and therefore ensures feasible cooperation with other methods for video content analysis. Moreover, the PTDG can easily be exploited to give more information about local features, and object annotations. The correlation is a robust estimator of block matching, is intensity independent, and based on optical flow. We have also shown that the camera motion itself has discriminatory information for video classification. The results show that if a search was made for sport the system would return 4 sport sequences and 1 non-sport. These results are encouraging when analysed because the test sequences were chosen randomly without vetting, and a proportion of them contain very little or no motion and the approach can not perform at all in these conditions as it is purely a dynamic based approach.

References

1. M.J. Roach and J.S.D. Mason. Motion-Based Classification of Cartoons. *accepted in ISIMP, Hong-Kong*, 2001.
2. R. Fablet and P. Bouthemy. Motion-based feature extraction and ascendant hierarchical classification for video indexing and retrieval. *3rd Int. Conf. on visual Information Systems, VISual'99, Amsterdam*, 1999.
3. H.S. Parry, A.D. Marshall, and K.C. Markham. Region Template Correlation for FLIR Target Tracking. *http://www.bmva.ac.uk/bmvc/1996/parry_1.ps.gz*.
4. F. Ganansia and P. Bouthemy. Video partioning and camrea motion characterization for content-based video indexing. *IEEE Int. Conf. on Image Processing*, 1996.
5. W. Xiong and J.C.M. Lee. Efficient Scence Change Detection and Camera Motion Annotation for Video Classification. *Computer Vision and Image Understanding, vol 71, No. 2*, pages 166–181, 1998.
6. N.C. Haering, R.J. Qian, and M.I. Sezan. A semantic event detection approach and its application to detecting hunts in wildlife video. *IEEE Trans. on Circuits and Systems for Video Tecnology*, 1999.
7. S. Porter, M. Mirmehdi, and B. Thomas. Video cut detection using frequency domain correlation. *Int. Conf. on Pattern Recognition*, 3:413–416, 2000.
8. B-T. Truong, S. Venkatesh, and C. Dorai. Automatic genre identification for content-based video categorization. *Int. Conf. Pattern Recgnition*, 4:230–233, 2000.
9. J.A. Fayman, O. Sudarsky, and E. Rivlin. Zoom tracking and its applications.
10. G. Sudhir and J.C.M. Lee. Video Annotation by Motion Interpretation using Optical Flow Streams. *HKUST-CS96-16*, 1997.

11. J.M. Martínez, Z. Zhang, and L. Montano. Segment-Based Structure from an Imprecisely Located Moving Camera. *http://www.cps.unizar.es/~josemari/iscv95.ps.gz*, 1995.
12. Y. Ma, R. Vidal, S. Hsu, and S. Sastry. Optical Motion Estimation from Multiple Images by Normalized Epipolar Constraint. In *IEEE Computer Society Conference on Computer Vision and Pattern Recognition*, 1999.

Matching Incomplete Objects Using Boundary Signatures

Adnan A.Y. Mustafa

Kuwait University, Department of Mechanical and Industrial Engineering,
P. O. Box 5969 - Safat, Kuwait 13060
symymus@kuc01.kuniv.edu.kw

Abstract. Object identification by matching is a central problem in computer vision. A major problem that any object matching method must address is the ability to correctly match an object to its model when parts of the object is missing due to occlusion, shadows, ... etc. In this paper we introduce boundary signatures as an extension to our surface signature formulation. Boundary signatures are surface feature vectors that reflect the probability of occurrence of a feature of a surface boundary. We introduce four types of surface boundary signatures that are constructed based on local and global geometric shape attributes of the boundary. Tests conducted on incomplete object shapes have shown that the *Distance Boundary Signature* produced excellent results when the object retains at least 70% of its original shape.

1 Introduction

Any reliable object recognition system must be able to analyze the world scene that the machine "sees" and correctly recognize all objects appearing in the scene. At the core of the object recognition subsystem is the matching process, where the system compares the result of its scene analyses to its object database, and hypothesizes about objects appearing in the scene. A major concern with matching is that most matching techniques fail at arriving at the correct match when partial information is missing. A typical example, is the case of object occlusion where only part of the object is visible to the camera (or vision sensor). Such a case usually results in object mismatch and incorrect object hypotheses and hence incorrect recognition.

In our previous work, we introduced surface signatures as robust feature descriptors for surface matching [1,2,3]. A surface signature is a feature vector that reflects the probability of occurrence of a surface feature on a given surface. Surface signatures are scale and rotation invariant. We showed that by using surface signatures correct identification was possible under partial occlusion and shadows. Previously we employed two types of surface signatures, surface curvature signatures and surface spectral signatures, which statistically represent surface curvature features and surface color features, respectively. In this paper we introduce surface boundary signatures as an extension to our surface signature formulation.

C. Arcelli et al. (Eds.): IWVF4, LNCS 2059, pp. 563–572, 2001.
© Springer-Verlag Berlin Heidelberg 2001

2   Related Literature

Shape analysis is divided into two categories, boundary-based shape analysis that is based on the analyses of the object's boundary and interior-based shape analysis that is based on the analyses of the object's interior region. Examples of boundary based shape analysis are Chain (or Freeman) codes, Polygonal Approximations and Shape Signatures. Examples of interior based shaped analysis include region moment methods and Medial Axis Transform (MAT) methods also known as skeleton techniques.

The literature on shape analyses methods is vast. In this section we only present a sample of recent papers that are relevant to our work. Hong [4] presented an indexing approach to 2-D object description and recognition that is invariant to rotation, translation, scale, and partial occlusion. The scheme is based on three polygonal approximations of object boundaries where local object structural features (lines and arcs) are extracted. Ozcan and Mohan [5] presented a computationally efficient approach which utilizes genetic algorithms and attributed string representation. Attributed strings were used to represent the outline features of shapes. Roh and Kweon [6] devised a contour shape signature descriptor in to the recognition of planar curved objects in noisy scenes. The descriptor consisting of five-point invariants was used to index a hash table. Nishida [7] proposed an algorithm for matching and recognition of deformed closed contours based on structural features. The contours are described by a few components with rich features. Kovalev and Petrou [8] extracted features from co-occurrence matrices containing description and representation of some basic image structures. The extracted features express quantitatively the relative abundance of some elementary structures. Mokhtarian et. al [9] used the maxima of curvature zero-crossing contours of curvature scale space image as a feature vector to represent the shapes of object boundary contours. For a recent survey on shape analysis techniques the reader is referred to [10].

3   Boundary Features and Signatures

We use the surface's (or object's) boundary (Ω) to extract local and global geometric shape attributes that are used to construct four boundary signatures. The boundary of any surface (or object) consists of a finite number (N) of an ordered sequence of points (λ) that defines the shape of the surface (see Fig. 1),

$$\Omega = \{\lambda_i = (x_i, y_i), i = 0, \ldots, N-1\} \tag{1}$$

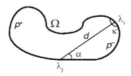

Fig. 1. The boundary and its feature vectors.

The following assumptions are made about Ω; Ω is closed (i.e. λ_0 follows λ_{N-1}), Ω has a single point thickness (i.e. Ω has been thinned), Ω is ordered in a counter-clockwise sense (the object is to the left) and Ω does not contain any internal holes.

3.1   The Boundary Intra-distance

The boundary intra-distance (d_{ij}) is defined as the Euclidean distance between two pair of points, λ_j and λ_i, on Ω (see Fig. 1),

$$d_{ij} = d(\lambda_i, \lambda_j) = \sqrt{\left(x_j - x_i\right)^2 + \left(y_j - y_i\right)^2} \tag{2}$$

Plots of the distance matrix ($\mathbf{d}$) for a *circle* ($\mathbf{d}_{circle}$) and an *ellipse* ($\mathbf{d}_{ellipse}$) are shown in Fig. 2. For illustrative purposes, each plot is shown from two view-points; a side view and a top view. Because of the symmetry inherit in the shape of a *circle*, $\mathbf{d}_{circle}$ has the unique feature of having the only profile containing diagonal lines with constant values. As a *circle* is stretched in a given direction, $\mathbf{d}_{circle}$ looses its symmetry producing $\mathbf{d}_{ellipse}$.

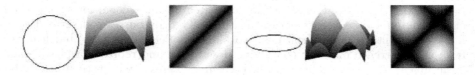

Fig. 2. Plots of $\mathbf{d}$: for a *circle*; $\mathbf{d}_{circle}$ (right) and an *ellipse* with eccentricity 0.995; $\mathbf{d}_{ellipse}$ (left)

When parts of an object boundary are missing $\mathbf{d}$ will also change from its original profile. However, if the amount of boundary missing is relatively small then $\mathbf{d}$ will not change by much. Let μ denote the percentage of boundary missing. Fig. 3 shows three variations of a *circle* with different amount of its original boundary missing along with their corresponding $\mathbf{d}$. We see that when the amount of boundary missing is small ($\mu = 0.1$) $\mathbf{d}_{circle\_0.9}$ is very similar to $\mathbf{d}_{circle}$. But as the amount of boundary missing increases the similarity of $\mathbf{d}$ to its original $\mathbf{d}$ decreases. Note that even when large amount of its boundary is missing, such as the case for the *circle* with 50% of its original boundary missing, we see that a large similarity still exists between $\mathbf{d}_{circle\_0.5}$ and $\mathbf{d}_{circle}$ that the shape can be identified as being (part of) a *circle*. It is this similarity in $\mathbf{d}$ and the other boundary features presented in this paper that we exploit in our work to arrive at successful matching when partial boundary information is missing.

3.2   The Boundary Intra-angle

The boundary intra-angle (α) is defined as the directional angle between boundary points (see Fig. 1). The boundary intra-angle of point λ_j with respect to λ_i

is given by,

$$\alpha_{ij} = \alpha(\lambda_i, \lambda_j) = \angle \mathbf{v}_{ij} = \arctan(y_j - y_i, x_j - x_i) \tag{3}$$

Values of α lie in the range $[0,2\pi[$. Note that α is skew symmetric, $\alpha_{ij} = -\alpha_{ji}$.

3.3   The Boundary Curvature Angle

The boundary curvature angle (κ°) represents the amount of local boundary bending as measured by the local boundary curvature angle (see Fig. 1). κ° at point λ_i is computed by,

$$\kappa_i^\circ = \cos^{-1}\left(\hat{\mathbf{v}}_{i-1} \cdot (-\hat{\mathbf{v}}_i)\right) \cdot \text{sign}\left(\hat{\mathbf{v}}_{i-1} \times \hat{\mathbf{v}}_i\right) \tag{4}$$

where the sign function is defined as, $\text{sign}(x) = -1$ if $x < 0, 1$ otherwise. $\hat{\mathbf{v}}_i$ is the unit direction vector from point λ_i to λ_{i+1}. Positive values of κ° between $0 < \kappa^\circ \leq \pi$ indicate shape concavities while negative values of κ° between $-\pi < \kappa^\circ < 0$ indicate shape convexities. For example, a *circle* has values of $\kappa^\circ = \pi$ everywhere. The profile of an *ellipse* is similar to that of a *circle*, but has smaller values of κ° at the ends of its major axis (dependent on eccentricity and profile resolution).

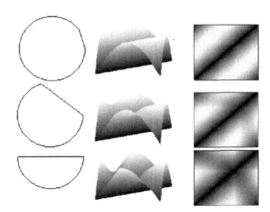

Fig. 3. Plots of the distance matrix (**d**) for a *circle* with different amount of its border missing (from top to bottom): $\mathbf{d}_{circle\_0.9}$ ($\mu = 0.1$), $\mathbf{d}_{circle\_0.7}$ ($\mu = 0.3$) and $\mathbf{d}_{circle\_0.5}$ ($\mu = 0.5$).

3.4   Chord Bending

Chord bending (γ), represents the amount of parameter bending between two points on Ω (see Fig. 1). Chord bending between two points, $\lambda_i, \lambda_j \in \Omega$ is defined as the ratio of the Euclidean distance between the two points (d_{ij}) to the parametrical distance between the two points (p_{ij}),

$$\gamma_{ij} = \gamma(\lambda_i, \lambda_j) = d_{ij}/p_{ij} \tag{5}$$

Since Ω is closed, two different values of p exist between any pair of points, one in the counter-clockwise direction and the other in the clockwise direction. We define the parameter distance between two points as the shortest of the two distances. A plot of the chord bending matrix (γ) for a *circle* and an *ellipse* is shown in Fig. 4. Note that the profile of an *ellipse* is distinctly different from that of a *circle*.

Fig. 4. Plot of the chord bending matrix (γ) for a *circle* (left) and an *ellipse* (right).

3.5   Boundary Signatures

A surface signature is a feature vector that reflects the probability of occurrence of the feature for a given surface. If $\mathbf{S}$ denotes a surface signature of size N, then by definition, $\Sigma S_i = 1$ for $0 \leq S_i \leq 1$ where $0 \leq i < N - 1$ [1,2,3].

   Given the boundary matrices, $\mathbf{d}$, α, κ° and γ, surface signatures can be constructed for these matrices. The *Distance Boundary Signature* ($\mathbf{S}_{DB}$) represents the frequency of occurrence of the normalized intra-distance between two points for a given intra-distance value. Recall that $d(\lambda_i, \lambda_j)$ represents the shape's intra-distance function between two points, λ_i, $\lambda_j \in \Omega$. Normalizing the distance function with respect to the maximum distance, produces the normalized distance function, $\hat{d}(\lambda_i, \lambda_j)$,

$$\hat{d}(\lambda_i, \lambda_j) = \frac{d(\lambda_i, \lambda_j)}{max(d(\lambda_i, \lambda_j))} \tag{6}$$

As a result, $0 \leq \hat{d} \leq 1$. The advantage of normalizing the intra-distance is that it produces a metric that is scale invariant. Let Ψ_d denote the inverse intra-distance function of $\hat{d}$, i.e.

$$\Psi_d(u) = f^{-1}(\hat{d}(\lambda_i, \lambda_j)) \tag{7}$$

where $u \equiv \hat{d}$. Ψ_d is defined only for $u \in [0, 1]$. Let $\tilde{\Psi}_d$ denote the normalized inverse intra-distance function obtained by normalizing the inverse intra-distance function with respect to the maximum inverse intra-distance value,

$$\tilde{\Psi}_d(u) = \frac{\Psi_d(u)}{max(\Psi_d(u))} \tag{8}$$

Values of $\tilde{\Psi}_d$ fall between $[0,1]$. Taking a closer look at u, we see that u is a continuous random variable representing the normalized intra-distance between two point pairs on Ω. Further examination into the normalized inverse intra-distance function, $\tilde{\Psi}_d$, reveals that it is actually the cumulative distribution function (cdf) of the random variable u,

$$\text{cdf}_d(u) = \tilde{\Psi}_d u) \tag{9}$$

Hence, the probability that the normalized distance (p_d) between two point pairs lie in the range $[u_1, u_2]$,

$$p_d(u_1 \le u \le u_2) = \int_{u_1}^{j_2} \text{pdf}_d(u)du = \text{cdf}_d(u_2) - \text{cdf}_d(u_1) = \tilde{\Psi}_d(u_2) - \tilde{\Psi}_d(u_1) \tag{10}$$

As stated earlier, $\mathbf{S}_{DB}$ represents the frequency of occurrence of the normalized Euclidean distance between two points for a given distance, i.e. $\mathbf{S}_{DB}$ is the pdf of u,

$$\mathbf{S}_{DB} = \text{pdf}_d(u) = \frac{d}{du}\left(\tilde{\Psi}_d(u)\right) \tag{11}$$

Similar analysis leads to the construction of the other boundary signatures. The *Angle* (or *Direction*) *Boundary Signature* ($\mathbf{S}_{AB}$) represents the frequency of occurrence of two boundary points oriented relative to each other at a given angle. The *Parameter Boundary Signature* ($\mathbf{S}_{PB}$) represents the frequency of occurrence of the chord bending between two points at a given value. The *Curvature Boundary Signature* ($\mathbf{S}_{CB}$) represents the frequency of occurrence of the curvature of a boundary point at a given curvature angle.

3.6   Measuring Signature Matching Performance

Matching observed objects to model objects is accomplished by comparing their signatures using the four error metrics described in [3]. These metrics compare the signature profiles based on signature distance, variance, spread and correlation. The four error metrics are then combined to form the *signature match error* (E), which gives a weighted signature error based on the four error metrics. We will refer to the correct model that an object should match to as the *model-match*.

Signature Recognition Rate and Recognition Efficiency. We define the *signature recognition rate* (Φ) as the percentage of correct hypotheses found for a given set using a particular signature type. The *signature recognition efficiency* (η) is defined as the efficiency of a signature in matching a surface to its model-match surface for a given surface and is calculated by,

$$\eta = (1 - AMCI) \cdot 100\% \tag{12}$$

where $AMCI$ is the average model-match percentile of a set.

Significance of Performance Measures. Φ and η complement each other in measuring the signature matching performance for a given set of surfaces. While Φ measures the percentage of surfaces that correctly produced a hypotheses, η measures the efficiency of matching by taking into consideration how "far off" -in terms of matching rank- the model-match is from the hypotheses found.

4 Experimental Results

Tests were conducted on objects of different shapes and sizes. The model set consisted of the 41 objects shown in Fig. 5. Shape signature files for these models were constructed off-line and stored in the model database. Testing was done on two sets; The first set consisted of objects that are completely visible while the second set consisted of incomplete objects. A sample of the test objects is shown in Fig. 6. Boundary signatures for a sample model are shown in Fig. 7.

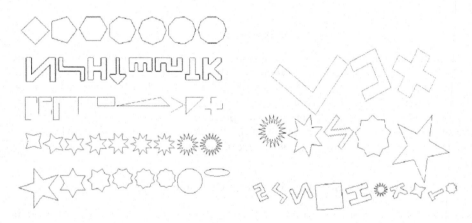

Fig. 5. Database models.   **Fig. 6.** Sample objects.

4.1 Matching Complete Objects

Testing was done on 129 instances of the models at random scales and rotations. In addition 6 instances of 2 new models not in the database were also tested to observe if the system can indeed identify the closest model match to that which is perceived visually.

Boundary signatures for these objects were constructed and subsequently matched to the model database. The results of matching are discussed in [11] and a summary follows:

- Matching using $\mathbf{S}_{DB}$ produced the best signature matching where 124 of the 135 objects were correctly hypothesized ($\Phi_{DB} = 91.9\%$). Matching using $\mathbf{S}_{PB}$

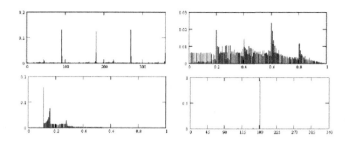

Fig. 7. Boundary signatures for model C. Top to bottom and left to right. $\mathbf{S}_{AB}$, $\mathbf{S}_{DB}$, $\mathbf{S}_{PB}$, $\mathbf{S}_{CB}$.

produced the second best matching rate ($\Phi_{PB} = 63.7\%$). Matching using $\mathbf{S}_{CB}$ and $\mathbf{S}_{AB}$ produced the worst results ($\Phi_{CB} = 15.6\%$ and $\Phi_{AB} = 8.1\%$).
- The signature efficiency using $\mathbf{S}_{DB}$ and $\mathbf{S}_{PB}$ are very high ($\eta_{DB} = 99.1\%$ and $\eta_{PB} = 91.1\%$). This indicates that using $\mathbf{S}_{DB}$ produced -on average- the model-match as the best match or the next best match, with a higher tendency for the former. The signature efficiency using $\mathbf{S}_{AB}$ and $\mathbf{S}_{PB}$ are very poor ($\eta < 60\%$).

4.2   Matching Incomplete Objects

In this section we present the results of matching incomplete objects. Forty-one incomplete objects were tested. For these objects μ varied from $\mu = 0.12$ to $\mu = 0.61$ with an average value of $\mu = 0.37$ (recall that μ denotes the amount of data missing). Fig. 8 shows a sample of these shapes. Matching results for these objects were as follows:

- The overall success rate of the signatures were all very weak, where $\Phi < 20\%$ for all four signatures. Matching using $\mathbf{S}_{DB}$ produced the highest success rate ($\Phi_{DB} = 19.5\%$) followed by $\mathbf{S}_{CB}$ ($\Phi_{PB} = 14.6\%$). Matching using either $\mathbf{S}_{AB}$ or $\mathbf{S}_{PB}$ produced the poorest results ($\Phi_{AB} = 2.4\%$ and $\Phi_{PB} = 7.3\%$). The performance of $\mathbf{S}_{DB}$ and $\mathbf{S}_{CB}$ are much better when $\mu < 0.3$ ($\Phi_{DB,\mu<0.3} = 66.7\%$ and $\Phi_{PB,,\mu,<0.3} = 16.6\%$).
- However, the signature efficiencies varied from $\eta = 53.6\%$ to $\eta = 78.4\%$. $\mathbf{S}_{PB}$ had the best performance with $\eta_{PB} = 78.4\%$ followed by $\mathbf{S}_{DB}$ with $\eta_{DB} = 76.0\%$. $\mathbf{S}_{CB}$ and $\mathbf{S}_{AB}$ both had poor performances with $\eta = 65.9\%$ and $\eta = 53.6\%$, respectively.
- When $\mu < 0.3$ the signature efficiencies for both $\mathbf{S}_{DB}$ and $\mathbf{S}_{CB}$ improve considerably ($\eta_{DB} = 92.3\%$ and $\eta_{CB} = 77.7\%$).
- From the points mentioned above, it is obvious that a correlation exists between μ and MCI (model-match percentile) for $\mathbf{S}_{DB}$ and $\mathbf{S}_{PB}$. The correlation coefficient calculated for the four signatures are 0.110, 0.764, 0.647 and 0.120 for $\mathbf{S}_{AB}$, $\mathbf{S}_{DB}$, $\mathbf{S}_{PB}$ and $\mathbf{S}_{CB}$, respectively. These correlation coefficients indicate, as earlier observed, that a strong correlation exists between the amount

of boundary missing and the model-match percentile for $\mathbf{S}_{DB}$ and $\mathbf{S}_{PB}$, and no correlation exists between the amount of boundary missing and the model-match percentile for $\mathbf{S}_{AB}$ and $\mathbf{S}_{CB}$.

Fig. 8. A sample of the incomplete objects.

5   Discussion and Conclusion

Table 1 gives a comparison of η for the four signatures as function of μ,

- The signature efficiencies (η) for both $\mathbf{S}_{DB}$ and $\mathbf{S}_{PB}$ are sensitive to μ as they degrade with increasing η while η for $\mathbf{S}_{AB}$ and $\mathbf{S}_{CB}$ are not sensitive to μ.
- Up to $\mu = 0.3$ using $\mathbf{S}_{DB}$ outperforms all other boundary signatures.
- Up to $\mu = 0.6$ the signature of choice is to use either $\mathbf{S}_{DB}$ or $\mathbf{S}_{PB}$.

The fact that Φ for both $\mathbf{S}_{DB}$ and $\mathbf{S}_{PB}$ are sensitive to μ while $\mathbf{S}_{DB}$ and $\mathbf{S}_{PB}$ are not, are due to the fact that the former pair of signatures are dependent on the distance between boundary points. When an object is incomplete, the object takes on a new shape defined as the boundary of the visible part of the object. The boundary of the object has smaller inter-distance between its boundary points. As the amount of the boundary missing (μ) increases the distortion in distance increases. On the other-hand, $\mathbf{S}_{CB}$ is a function of the boundary curvature, which retains its correct values on the visible portion of the original boundary. $\mathbf{S}_{AB}$ is a function of the inter-direction between boundary points. As

Table 1. Comparison of η

μ	$\mathbf{S}_{AB}$	$\mathbf{S}_{CB}$	$\mathbf{S}_{DB}$	$\mathbf{S}_{PB}$
$\mu = 0$	52.8%	59.9%	99.1%	91.1%
$0 < \mu < 0.3$	58.2%	77.7%	92.3%	76.8%
$0 < \mu < 0.7$	53.6%	65.9%	76.0%	78.4%

long as only a non-significant part of the object is missing the inter-direction between boundary points is not greatly distorted.

The high success rates obtained using the boundary signatures with the various shapes described above makes the use of signature matching acceptable for character recognition. Furthermore, the boundary signature error described in this paper can be combined with our previous work to define a *surface match error* based on curvature, spectral and boundary attributes. With this added feature, the excellent results previously obtained using curvature and boundary attributes can be further enhanced.

Acknowledgment. The author would like to thank the Kuwait University Research Administration for providing financial support for this research (KU-Grant EM-112).

References

1. Mustafa, A. A., Shapiro, L. G. and Ganter, M. A. 1996. "3D Object Recognition from Color Intensity Images". In the *13th International Conference on Pattern Recognition*, Vienna, Austria, pp. 627-631, August 25-30.
2. Mustafa, A. A., Shapiro, L. G. and Ganter, M. A. 1997. "Object Identification with Surface Signatures". The *7th International Conference on Computer Analysis of Images and Patterns*, Kiel, Germany, pp. 58-65, Sept. 10-12.
3. Mustafa, A. A., Shapiro, L. G. and Ganter, M. 1999. "3D Object Identification with Color and Curvature Signatures". *Pattern Recognition*, Elsevier Science. Vol. 32, No. 3, pg. 1-17.
4. Hong, D., Sarkodie-Gyan, T., Campbell, A. and Yan, Y. 1998. "A Prototype Indexing Approach To 2-D Object Description and Recognition". *Pattern Recognition*, Elsevier Science. Vol. 31, No. 6, pg. 699-725.
5. Ozcan, E. and Mohan, C. 1997. "Partial shape matching using genetic algorithms". *Pattern Recognition Letters*, Elsevier Science .V 18, pg. 987-992.
6. Roh, K. and Kweon, I. 1998. "2D Object Recognition Using Invariant Contour Descriptor And Projective Refinement" *Pattern Recognition*, Elsevier Science. V 31, N 4, pg. 441-455.
7. Nishida, Hirobumi. 1998. "Matching and Recognition of Deformed Closed Contours Based on Structural Transformation Models", *Pattern Recognition*, Elsevier Science. Vol. 31, No. 10, pg. 1557-1571.
8. Kovalev, V. and Petrou, M. 1996. "Multidimensional Co-occurrence Matrices for Object Recognition and Matching", *GMIP*, V58 (3), pp. 187-197.
9. Mokhtarian, F., Abbasi, S and Kittler, J 1996."Robust and Efficient Shape Indexing through Curvature Scale Space". In the *British Machine Vision Conference*, Edinburgh, pg. 53-62.
10. Loncaric, Sven 1998. "A survey of shape analysis techniques". *Pattern Recognition*, Elsevier Science. Vol. 31, No. 8, pg. 983-1001.
11. Mustafa, Adnan A. "Object Matching using Shape Surface Signatures". *SPIE conference on Machine Vision Applications in Industrial Inspection IX*, San Jose, Ca, Jan. 22-23, 2001.

General Purpose Matching of Grey Level Arbitrary Images

Francesca Odone[1], Emanuele Trucco[2], and Alessandro Verri[1]

[1] INFM-DISI, Università di Genova,
Via Dodecaneso 35, 16146 Genova, Italy
{odone,verri}@disi.unige.it,
[2] Department of Computing and Electrical Engineering,
Heriot–Watt University,
Edinburgh, EH14 4AS, UK
mtc@cee.hw.ac.uk

Abstract. In this paper we propose a method for measuring the similarity between two images inspired by the notion of Hausdorff distance. Given two images, the method checks pixelwise if the grey values of one are contained in an appropriate interval around the corresponding grey values of the other. Under certain assumptions, this provides a tight bound on the directed Hausdorff distance of the two grey-level surfaces. The proposed technique can be seen as an equivalent in the grey level case of a matching method developed for the binary case by Huttenlocher *et al.* [2]. The method fits naturally an implementation based on comparison of data structures and requires no numerical computations whatsoever. Moreover, it is able to match images successfully in the presence of severe occlusions. The range of possible applications is vast; we present preliminary, very good results on stereo and motion correspondence and iconic indexing in real images, with and without occlusion.

1 Introduction

This paper presents a general-purpose technique for matching arbitrary grey-level images, built around the concept of Hausdorff distance. The Hausdorff distance provides a useful measure for matching two sets of points, and its versatility is suggested by the diverse applications in which it appears, including defect detection [1], gesture recognition [4], robot localization [7], range image analysis [5], and content-based video and database indexing.

Hausdorff measures have been used in computer vision nearly exclusively to match binary patterns of contours or edges. The work by Huttenlocher *et al.* [2,3, 6] is the most representative here, and an apt springboard to introduce the main points and innovations of our technique. Huttenlocher *et al.* used the Hausdorff distance to implement an efficient search of a binary edge model M in a binary edge image I. They fixed a (small) threshold ρ, say $\rho = 1$ pixel, and a certain fraction f of the model edge points, say $f = 90\%$, and considered all possible translated versions M_t of M over I. Then, they built a dilated version of I, I_ρ,

C. Arcelli et al. (Eds.): IWVF4, LNCS 2059, pp. 573–582, 2001.

by setting equal to 1 all pixels within ρ from each edge location in I. Candidate matches were indicated by translations, if any, resulting in M_t with at least a fraction f of the edge points contained in the corresponding region of the dilated edge image I_ρ. A problem of this method is that, if the edge density in I is high, the likelihood of a successful match against *any* M_t is also high. To obviate this, the same procedure was applied a second time after swapping image and model (similar to the left-right consistency constraint of stereo matching). The authors also proposed extensions to allow for rotations and scaling of the model with respect to the image.

Our method, which includes Huttenlocher's in the special case of binary images, is a true grey-level matching technique. It achieves tolerance to occlusions and significant versatility. No explicit numerical computations are necessary, so that the method is also suitable for high-speed implementations. Finally, using the rich information of the grey levels makes the disambiguating second step (matching after swapping image and model) largely redundant.

The structure of this paper is as follows. Section 2 presents the necessary theory, Section 3 the method proposed, Section 4 experimental results with various applications. A discussion of our work is given in Section 5.

2   Theoretical Background

In this section we recall the main mathematical concepts behind our work, and discuss some geometric properties of Hausdorff distances. For simplicity, we restrict attention to the case of $\mathbb{R}^2$.

2.1   Hausdorff Distances

Given two finite point sets A and B of $\mathbb{R}^2$, the *directed Hausdorff distance*, $h(A, B)$, is measured in two steps. For a fixed point a of A, the first step computes the distance of a from each point b of B, and selects the distance between a and the closest point of B, d_a (see Figure 1, left). The second step takes the maximum of d_a for all a of A, $h(A, B)$ (see Figure 1, center).

Formally this can be written as

$$h(A, B) = \max_{a \in A} \min_{b \in B} ||a - b||. \tag{1}$$

Note that order matters, since in general $h(A, B) \neq h(B, A)$. In the case of Figure 1, for example, $h(B, A) \ll h(A, B)$, because all points of B are "close" to some points of A. *The directed Hausdorff distance*, which is not symmetric and thus not a "true" distance, *measures the degree of mismatch between A and B*. To obtain a distance in the mathematical sense, symmetry can be restored by taking the maximum between $h(A, B)$ and $h(B, A)$. This brings to the definition of *Hausdorff distance*, that is,

$$H(A, B) = \max(h(A, B), h(B, A)). \tag{2}$$

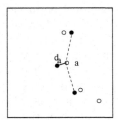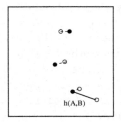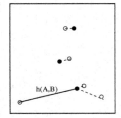

Fig. 1. Directed Hausdorff distance. Let the empty and filled dots be the elements of the sets A and B respectively. The distance of a point a of A from the set B is denoted by d_a (left); the maximum of d_a over all points of A is the directed Hausdorff distance and is denoted by $h(A, B)$ (center). The presence of a single outlier in A is sufficient to distort $h(A, B)$ significantly (right).

Being a distance in the mathematical sense, the Hausdorff distance is zero if and only if $A = B$. Instead, the directed Hausdorff distance is zero if and only if A is a subset of B. For our purposes, a useful property of both measures is their *ability to measure the distance between two sets with different number of points*, while an undesirable property is their sensitivity to outliers (see Figure 1, right). The next section shows how this can be countered effectively.

2.2   Geometric Interpretation

One way to gain intuition on Hausdorff measures is to think in terms of set inclusion, see Figure 2 (left). Let B_ρ be the set obtained by replacing each point of B with a disk of radius ρ, and taking the union of all of these disks. Effectively, B_ρ is obtained by dilating B by ρ. *The directed Hausdorff distance $h(A, B)$ is not greater than ρ if and only if $A \subseteq B_\rho$.* This follows easily from the fact that, in order for every point of A to be within distance ρ from some points of B, A must be contained in B_ρ.

This geometric interpretation suggests an interesting property of the directed Hausdorff distance, useful to counter the effect of outliers. Let us call $\hat{A}$ the subset of the points of A contained in B_ρ. Assume that, for a given value of ρ, $\hat{A}$ is *nearly* equal to A (in Figure 2 right, there is only one element of difference). While the directed Hausdorff distance between A and B can be distorted by the few points not in $\hat{A}$, $h(\hat{A}, B)$ is still not greater than ρ, which means that the potential outliers are defined and identified in one step.

3   The Method

In this section we present the method for finding a match of a grey level *model* M in a grey level *image* I, which can be classified among correlation techniques.

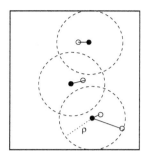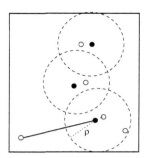

Fig. 2. Geometric interpretation of the directed Hausdorff distance. Let again the empty and filled dots be the elements of the sets A and B respectively. Let B_ρ be the union of the set of disks of radius ρ centered at the points of B. From the left figure it is clear that $A \subseteq B_\rho$ if $\rho = h(A, B)$ and *vice versa*. The right figure shows an example of how this property, through the concept of partial inclusion, can be used to overcome the sensitivity to outliers.

The idea is to search for a match by comparing all possible translated versions M_t of M with I and selecting the most appropriate translation $\hat{t}$, if any. The method consists of three steps.

1. Expand the model M into the 3D binary matrix $\mathcal{M}$, the third dimension being the grey value. That is, for i and j spanning the pixel locations and g the grey values:
$$\mathcal{M}(i, j, g) = \begin{cases} 1 \text{ if } M(i, j) = g; \\ 0 \text{ otherwise.} \end{cases}$$
Build the 3D binary matrix $\mathcal{I}$ from the image I in the same way.
2. Dilate the matrix $\mathcal{I}$ by growing its nonzero entries by a fixed amount in all three dimensions. Let $\mathcal{I}'$ be the resulting 3D binary matrix.
3. Finally, compute the size of the intersection between all possible translated versions $\mathcal{M}_t$ and $\mathcal{I}'$, call it $S(t)$. The candidate matches are identified by high values of $S(t)$, which can be thought of as a *similarity surface*.

In our examples, we simply assign the best match to the translation $\hat{t}$ for which $S(\hat{t}) \geq S(t)$ for all t, that is to the absolute maximum of $S(t)$, if the maximum is above a threshold τ, which can be chosen in accordance with the level of similarity required. As shown in the next section, even with this simplistic choice results are very good.

Three remarks are in order. First, the link between this method and the directed Hausdorff distance is as follows. If (a) the dilation of the matrix $\mathcal{I}$ is isotropic in an appropriate metric, and (b) $S(\hat{t})$ takes on the maximum possible value[1], then the directed Hausdorff distance between $\mathcal{M}_{\hat{t}}$ and $\mathcal{I}'$, $h(\mathcal{M}_{\hat{t}}, \mathcal{I}')$ is not greater than ρ.

---

[1] That is, $\mathcal{M}_{\hat{t}} \subseteq \mathcal{I}'$

Second, with an appropriate choice of the data structures, the algorithm is reduced to a set of entry-wise logical AND operations between $\mathcal{M}$ and $\mathcal{I}'$ and no numerical computations are required.

Third and finally, unlike the 2D binary image case, the reverse matching with swapped patches is not necessary to eliminate possible spurious matches in the grey-level case, since the dense grey-level patches contain enough information to make the directed Hausdorff distance unambiguous.

4   Experimental Results

In this section we show some applications of this method to stereo and motion correspondence, and object search.

4.1   Stereo and Motion

The proposed method can be used to find correspondences in stereo pairs and motion sequences. Figure 3 shows a well-known stereo pair with its disparity map. To compute the disparity map of the left image, at each iteration, the model M is a neighborhood of each pixel of the left image, the image I is a subset of the right image, selected with the help of the epipolar constraint. To obtain the right disparity map, it suffices to swap left and right in the above description. Left-right consistency is performed, followed by a post-processing to fill the holes caused by multiple matchings and occlusions.

Fig. 3. A stereo pair with its disparity map.

We have also successfully used our matching technique as a feature tracker. Features have been extracted from the first frame. Each feature can be seen as a model, and tracked along the sequence by applying the matching method to each frame: Figure 4 shows an example in which both the camera and the subject are moving. In the top row a region of the background, taken from frame 0 as model, is tracked for more than 40 frames (here we show frames 0, 20, and 40). In the bottom row the head of a walking person is tracked for 15 frames (the figure shows frame 0, 5, and 12); the target is lost at frame 14, when the subject walks into an area of shadow. Figure 5 shows an example where the image patch selected as a model in the first frame of a sequence is successfully tracked across the sequence, despite the appearance changes. The

Fig. 4. Stable region tracking of background (top row) and foreground motion (bottom row).

figure shows frames 0 (containing the model) 20, 40, 60, 70, 85 (the last frame of the sequence). The camera rotates around the statue, inducing a small image deformation (therefore, a dilation in all the three directions has been applied). The binary maps represent the distribution of positive scores (white pixels are the inclusion of $\mathcal{M}$ and $\mathcal{I}$). In the whole sequence (86 frames) there has been only one error, which occurred at frame 70 (see figure), because its background was too different from the real model (the binary map on the left represents the error, the one on the right a possible satisfactory matching).

4.2   Iconic Search

In this range of applications we are interested to search arbitrary models inside image databases. The model can be a region of interest of a specific image; for instance, a nose in a given image of a database of faces can be used to locate all the noses in the database. Here, the objects imaged are indeed human faces from the Olivetti face database[2]. Figure 6 shows an example. The pattern to be searched is a window around the subject's right eye in the top row (left image), the center image shows the location of the best match found, by applying the search on the same image from which the model has been extracted. The map at the right represent a (equalized) similarity grey level description of the quantities $S(\mathbf{t})$; each map pixel represents an image pixel: the whiter the map pixel intensity, the higher the value of the surface S, the closer is the neighbourhood of the image pixel to the model. The other two sections of Figure 6 show results of the eye search on various images of the face database. High surface values are scored systematically around the eyes. The absolute maximum misses the eyes

---

[2] A link to the Olivetti face database can be found at the Computer Vision Home Page.

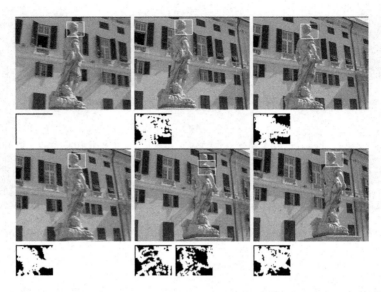

Fig. 5. Sparse samples of a sequence with binary maps describing the pixel positive scores of the obtained matches.

frequently when the subject wears spectacles, which is quite reasonable, since the presence of glasses modify the eyes pattern.

Figures 7 and 8 show occlusion experiments. Tolerance to occlusions is guaranteed by the tolerance to outliers previously discussed, which is related to the idea of partial inclusion. Figure 7 shows the image from which the model has been extracted and matching results with a different face image for increasing level of occlusion. Notice that changing the shape of the occlusion the results change and also the "breaking point" is different. This is because the results strongly depend on the amount of information contained in the non-occluded area. In Figure 7, for instance, a tiny model of about 100 pixels is enough to obtain good results, as long as the model contains the fold of a nostril.

Figure 8 shows a further example of how the "breaking point" changes with differently shaped occlusions. Of course, in practical applications it is more common to have occlusions in the image rather than in the model. Our choice of occluding parts of the model though, does not affect either the method or the results, but simplifies the synthetic manipulation of the occlusions.

5   Discussion

We have presented a method for measuring the similarity between two images inspired by the notion of Hausdorff distance. With an appropriate choice of the data structures, the algorithm can be implemented very efficiently. Reverse matching, necessary to disambiguate in the binary case, is redundant here.

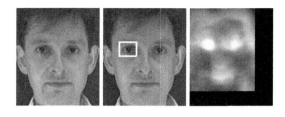

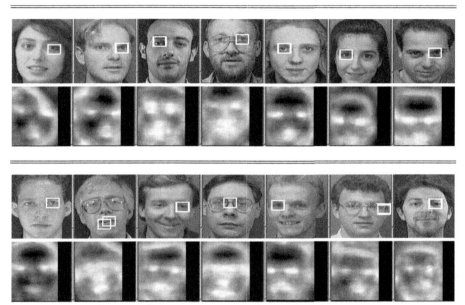

Fig. 6. Eye search in a database of faces (see text).

The excellent potential of the method has been illustrated in a number of experiments, even though the current selection of the best match on the similarity surface exploits only a fraction of the information actually provided by the surface itself. A better use of this information is the subject of current work along with a more systematic performance evaluation.

References

1. D. Chetverikov and Y. Khenokh. Matching for shape defect detection. In *Lecture Notes in Computer Science*, volume 1689, pages 367–374, 1999.
2. D. P. Huttenlocher, G. A. Klanderman, and W. J. Rucklidge. Comparing images using the hausdorff distance. *IEEE Trans. on Pattern Analysis and Machine Intelligence*, 9(15):850–863, 1993.
3. D. P. Huttenlocher and W. J. Rucklidge. Multi-resolution technique for comparing images using the Hausdorff distance. In *IEEE Int. Conf. on Computer Vision and Pattern Recognition*, pages 705–706, 1993.

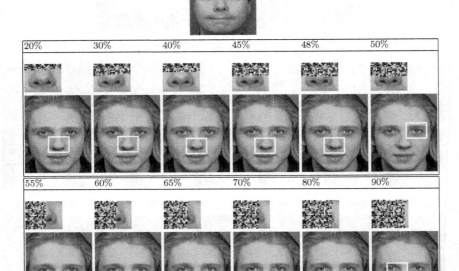

Fig. 7. Matching results with increasingly occluded models derived from the subject on top. Middle row: horizontal occlusions. Bottom row: vertical occlusions.

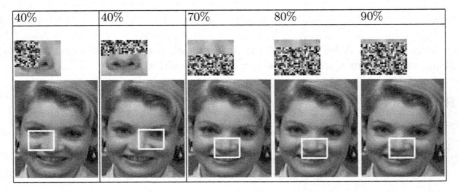

Fig. 8. Breaking points for differently shaped occlusions.

4. R. Kahn, M. Swain, P. Prokopowicz, and R. Firby. Gesture recognition using the Perseus architecture. In *Proceedings of the IEEE Conference on Computer Vision and Pattern Recognition*, pages 734–741, 1996.
5. C. Olson. Mobile robot self-localization by iconic matching of range maps. In *Proceedings of the Int. Conf. on Advanced Robotics*, pages 447–452, 1997.
6. W. J. Rucklidge. Efficiently locating objects using the hausdorff distance. *Int. Journ. of Computer Vision*, 24(3):251–270, 1997.
7. R. Sim and G. Dudek. Learning and evaluating visual features for pose estimation. In *ICCV99*, pages 1217–1222, 1999.

Many-to-many Matching of Attributed Trees Using Association Graphs and Game Dynamics

Marcello Pelillo[1], Kaleem Siddiqi[2], and Steven W. Zucker[3]

[1] University of Venice, Dept. of Computer Science
[2] McGill University, School of Computer Science
[3] Yale University, Depts. of Computer Science and Electrical Engineering

Abstract. The matching of hierarchical relational structures is of significant interest in computer vision and pattern recognition. We have recently introduced a new solution to this problem, based on a maximum clique formulation in a (derived) "association graph." This allows us to exploit the full arsenal of clique finding algorithms developed in the algorithms community. However, thus far we have focussed on one-to-one correspondences (isomorphisms), and many-to-one correspondences (homomorphisms). In this paper we present a a general solution for the case of many-to-many correspondences (morphisms) which is of particular interest when the underlying trees reflect real-world data and are likely to contain structural alterations. We define a notion of an ϵ-morphism between attributed trees, and provide a method of constructing a weighted association graph where maximal weight cliques are in one-to-one correspondence with maximal similarity subtree morphisms. We then solve the problem by using *replicator* dynamical systems from evolutionary game theory. We illustrate the power of the approach by matching articulated and deformed shapes described by shock trees.

1 Introduction

The matching of relational structures is a classic problem in computer vision and pattern recognition, instances of which arise in areas as diverse as object recognition, motion and stereo analysis (see, e.g., [3]). A well-known approach to solving this problem consists of transforming it into the equivalent problem of finding a maximum clique in an auxiliary graph structure, known as the *association graph* [3]. The idea goes back to Ambler *et al.* [1], and has since been successfully employed in a variety of different problems. This framework is attractive because it casts relational structure matching as a pure graph-theoretic problem, for which a solid theory and powerful algorithms have been developed. Although the maximum clique problem is known to be NP-hard, powerful heuristics exist which efficiently find good approximate solutions [5].

In many computer vision problems, relational structures are organized in a hierarchical manner [6], i.e., are *trees*. However, in standard association graph formulations, the solutions are not constrained to preserve this partial order. Hence, the extension of such techniques to tree matching problems is of considerable interest. We have recently introduced a solution to this problem by providing a

C. Arcelli et al. (Eds.): IWVF4, LNCS 2059, pp. 583–593, 2001.

novel way of deriving an association graph, based on the graph-theoretic notions of connectivity and the distance matrix [7]. We have proved that in this new formulation there is a one-to-one correspondence between maximum cliques in the derived association graph, and maximum subtree isomorphisms. The framework has also been extended to handle the matching of trees whose nodes have one or more associated attributes, by casting the attributed tree matching problem as an equivalent problem of finding a maximum weight clique in a weighted association graph.

Whereas thus far we have focussed on one-to-one correspondences (isomorphisms) and many-to-one correspondences (homomorphisms) between the structures being matched [7,4] it is clear that these maybe overly restrictive assumptions for problems where structural alterations may occur in both of the underlying trees, i.e., nodes have been deleted or additional nodes are present. In this paper we provide a generalization of the association graph framework to handle many-to-many correspondences (morphisms). We define a notion of an ϵ-morphism (a many-to-one mapping) between weighted attributed trees, and provide a method of constructing an association graph where maximal weight cliques are in one-to-one correspondence with maximal similarity subtree morphisms. We then solve the problem by using *replicator* dynamical systems from evolutionary game theory. We illustrate the approach by matching articulated and deformed 2D shapes described by shock trees.

2   Many-to-many Tree Matching

Formally, an *attributed tree* is a triple $T = (V, E, \alpha)$, where (V, E) is the "underlying" rooted tree and $\alpha : V \to \mathcal{A}$ is a function which assigns an attribute vector $\alpha(u)$ to each node $u \in V$. Two nodes $u, v \in V$ are said to be *adjacent* (denoted $u \sim v$) if they are connected by an edge. We shall also consider a function $\delta : \mathcal{A} \to \Re^*_+$ which assigns to each set of attributes (and therefore to each node in the tree) a real positive number. This will be interpreted as the negligibility of the corresponding node in the tree. Specifically, a node will be declared "negligible" if the value of the function δ corresponding to its attributes is smaller than a fixed threshold ϵ. This will allow us to associate a *cluster* of nodes (defined in the following) in the first subtree to a single node in the other one, thereby defining a many-to-one mapping from the first to the second tree.

For technical simplicity, we shall transform our node-weighted tree into an edge-weighted *derived attributed tree*, by simply moving the δ-value associated to every node to the edge connecting it to its parent. The root of the original tree will become a child of a newly created dummy root, and its weight will be moved to the corresponding edge. In the derived tree, we shall speak of a "negligible" edge, when its weight is less than threshold ϵ.

Given a fixed threshold $\epsilon \geq 0$, we define the distance $d_\epsilon(u, v)$ between two nodes u and v in an attributed tree, as the number of non-negligible edges (i.e. with weight less than ϵ) on the (unique) path from u to v. We define the function

$\text{lev}_\epsilon(u)$ of a node in an attributed tree, as the distance d_ϵ from the root of the tree and node u.

Given $\epsilon \geq 0$, we define an ϵ-*cluster* in a derived attributed tree T as a subset of nodes of T such that for every two nodes u and v in it, we have:

$$d_\epsilon(u, v) = 0 .$$

A set with only one node is a particular case of cluster that we call a *singleton*. It can easily be proved that given an ϵ-cluster C, for all $u, v \in C$, and $z \notin C$, we have:

$$d_\epsilon(u, z) = d_\epsilon(v, z) .$$

From this observation, the next proposition follows [8]:

Proposition 1. *Given two ϵ-clusters C' and C'', for any pair of nodes, one in C' and the other in C'', their ϵ-distance is constant.*

This allows us to extend the notion of ϵ-distance to pairs of clusters, which we shall denote by $d_\epsilon(C', C'')$, and in turn to generalize the notion of adjacency to clusters. Specifically, two disjoint ϵ-clusters C' and C'' in a derived attributed tree are said to be ϵ-*adjacent* (denoted $C' \sim_\epsilon C''$) if:

$$d_\epsilon(C', C'') = 1 .$$

It is clear that when C' and C'' are singletons, this is equivalent to the traditional notion of adjacency between nodes.

Finally, we are in a position to introduce the notion of a parent-child relationship between pairs of clusters. Let C' and C'' be two disjoint ϵ-clusters in a derived attributed tree T. We say that C' is an ϵ-*parent* of C'' when:

$$C' \sim_\epsilon C''$$

and

$$\text{lev}_\epsilon(C') \leq \text{lev}_\epsilon(C'')$$

where $\text{lev}_\epsilon(C)$ is defined as $d_\epsilon(root(T), C)$.

Now, let $T_1 = (V_1, E_1, \alpha_1)$ and $T_2 = (V_2, E_2, \alpha_2)$ be two attributed trees, and let $M \subseteq V_1 \times V_2$ be any relation on V_1 and V_2. Define the sets H_1 and H_2 as follows:

$$H_1 = \{u \in V_1 \ : \ \exists w \in V_2 \text{ such that } (u, w) \in M\}$$

and

$$H_2 = \{w \in V_2 \ : \ \exists v \in V_1 \text{ such that } (u, w) \in M\} .$$

Moreover, for each $u \in H_1$ let $M[u]$ be defined as:

$$M[u] = \{w \in V_2 \ : \ (u, w) \in M\}$$

and, similarly, for each $w \in H_2$ let $M[w]$ be defined as

$$M[w] = \{u \in V_1 \; : \; (u, w) \in M\} \; .$$

A relation $M \subseteq V_1 \times V_2$ is called a *subtree ϵ-morphism* if the following properties hold:

The subgraphs induced by H_1 and H_2 are connected (i.e. are trees) $\qquad$ (1)

$$\forall u \in H_1, \; \forall w \in H_2 \; : \; M[u] \text{ and } M[w] \text{ are } \epsilon\text{-clusters} \qquad (2)$$

$$\begin{array}{l} \forall u, v \in H_1 \; : \; u \text{ is an } \epsilon\text{-parent of } v \Leftrightarrow M[u] \text{ is an } \epsilon\text{-parent of } M[v] \\ \forall w, z \in H_2 \; : \; w \text{ is an } \epsilon\text{-parent of } z \Leftrightarrow M[w] \text{ is an } \epsilon\text{-parent of } M[z] \end{array} \qquad (3)$$

Clearly, in realistic applications, it would be desirable to find a morphism which pairs nodes having "similar" attributes. To this end, let σ be any similarity measure on the attribute space, i.e., any (symmetric) function which assigns a positive number to any pair of attribute vectors. If M is a subtree ϵ-morphism between two attributed trees $T_1 = (V_1, E_1, \alpha_1)$ and $T_2 = (V_2, E_2, \alpha_2)$, the overall similarity between the matched structures can be defined as follows:

$$S(M) = \sum_{(u,w) \in M} \sigma(\alpha_1(u), \alpha_2(w))$$

The ϵ-morphism M is called a *maximal similarity subtree ϵ-morphism* if we cannot add further matchings to M, while retaining the morphism property. It is called a *maximum similarity subtree ϵ-morphism* if $S(M)$ is the largest among all ϵ-morphisms between T_1 and T_2.

3 The Tree Association Graph

Let u and v be two nodes of an attributed tree T, joined by path $u = x_0 x_1 \ldots x_n = v$, and let x^* be the node in the path nearest to the root of T; formally: $x^* = \mathrm{argmin}_{x \in \{x_0, \ldots, x_n\}} \mathrm{lev}(x)$. We define:

$$n_\epsilon(u, v) = d_\epsilon(u, x^*)$$
$$p_\epsilon(u, v) = d_\epsilon(x^*, v) \; .$$

The weighted ϵ-tree association graph (ϵ-TAG) of two attributed trees $T_1 = (V_1, E_1, \alpha_1)$ and $T_2 = (V_2, E_2, \alpha_2)$ is the graph $G_\epsilon = (V, E, \omega)$ where

$$V = V_1 \times V_2$$

such that for any two nodes (u, w) and (v, z) in V

$$(u, w) \sim (v, z) \Leftrightarrow \begin{cases} n_\epsilon(u, v) = n_\epsilon(w, z) \\ p_\epsilon(u, v) = p_\epsilon(w, z) \end{cases} \qquad (4)$$

and ω is a function which assigns a positive weight to each node $(u, w) \in V$ as follows:

$$\omega(u, w) = \sigma(\alpha_1(u), \alpha_2(w)) . \tag{5}$$

Intuitively, condition (4) imposes that the number of negligible ascent edges between u and v must be equal to the number of negligible ascent edges between w and z. The same applies to descent edges. Notice that, when $\epsilon = 0$ we obtain the same association graph structure as originally studied for the tree isomorphism case [7].

Given a subset of nodes C of V, the total weight assigned to C is simply the sum of all the weights associated with its nodes. A *maximal weight* clique in G is one which is not contained in any other clique having larger total weight, while a *maximum weight* clique is a clique having largest total weight. The maximum weight clique problem is to find a maximum weight clique of G [5].

The following result establishes a one-to-one correspondence between the attributed tree morphism problem and the maximum weight clique problem (see [8] for the proof).

Theorem 1. *Any maximal (maximum) similarity subtree ϵ-morphism between two attributed trees induces a maximal (maximum) weight clique in the corresponding weighted ϵ-TAG, and vice versa.*

Once the tree morphism problem has been formulated as a maximum weight clique problem, any clique finding algorithm can be employed to solve it (see [5] for a recent review). In the work reported in this paper, we used an approach recently introduced in [7,9], which is summarized in the next section.

4   Matching via Game Dynamics

Let $G = (V, E, \omega)$ be an arbitrary weighted graph of order n, and let S_n denote the standard simplex of $\mathbb{R}^n$:

$$S_n = \{ \mathbf{x} \in \mathbb{R}^n : \mathbf{e}'\mathbf{x} = 1 \text{ and } x_i \geq 0, \ i = 1 \ldots n \}$$

where $\mathbf{e}$ is the vector whose components equal 1, and a prime denotes transposition. Given a subset of vertices C of G, we will denote by $\mathbf{x}^c$ its *characteristic vector* which is the point in S_n defined as

$$x_i^c = \begin{cases} \omega(u_i)/\Omega(C), & \text{if } u_i \in C \\ 0, & \text{otherwise} \end{cases}$$

where $\Omega(C) = \sum_{u_j \in C} \omega(u_j)$ is the total weight on C.

Now, consider the following quadratic function

$$f(\mathbf{x}) = \mathbf{x}'A\mathbf{x} \tag{6}$$

where $A = (a_{ij})$ is the $n \times n$ symmetric matrix defined as follows:

$$a_{ij} = \begin{cases} \frac{1}{2\omega(u_i)} & \text{if } i = j \,, \\ 0 & \text{if } i \neq j \text{ and } u_i \sim u_j \,, \\ \frac{1}{2\omega(u_i)} + \frac{1}{2\omega(u_j)} & \text{otherwise} \,. \end{cases} \quad (7)$$

The following theorem allows us to formulate the maximum weight clique problem as a quadratic program, thereby switching from the discrete to the continuous domain (see [9] for proof).

Theorem 2. *Let $G = (V, E, \omega)$ be an arbitrary weighted graph and consider a matrix $A \in \mathcal{M}(G)$. Then, the following hold:*

(a) A vector $\mathbf{x} \in S_n$ is a local minimizer of f on S_n if and only if $\mathbf{x} = \mathbf{x}^c$, where C is a maximal weight clique of G.
(b) A vector $\mathbf{x} \in S_n$ is a global minimizer of f on S_n if and only if $\mathbf{x} = \mathbf{x}^c$, where C is a maximum weight clique of G.

Moreover, all local (and hence global) minimizers of f on S_n are strict.

We now turn our attention to a class of simple dynamical systems that we use for solving our quadratic optimization problem. Let W be a non-negative real-valued $n \times n$ matrix, and consider the following dynamical system:

$$\dot{x}_i(t) = x_i(t) \left[(W\mathbf{x}(t))_i - \mathbf{x}(t)'W\mathbf{x}(t) \right], \quad i = 1 \ldots n \quad (8)$$

where a dot signifies derivative w.r.t. time t, and its discrete-time counterpart

$$x_i(t+1) = x_i(t) \frac{(W\mathbf{x}(t))_i}{\mathbf{x}(t)'W\mathbf{x}(t)}, \quad i = 1 \ldots n \,. \quad (9)$$

It is readily seen that the simplex S_n is invariant under these dynamics, which means that every trajectory starting in S_n will remain in S_n for all future times. Both (8) and (9) are called *replicator equations* in evolutionary game theory, since they are used to model evolution over time of relative frequencies of interacting, self-replicating agents [14].

Theorem 3. *If $W = W'$ then the function $\mathbf{x}(t)'W\mathbf{x}(t)$ is strictly increasing with increasing t along any non-stationary trajectory $\mathbf{x}(t)$ under both continuous-time (8) and discrete-time (9) replicator dynamics. Furthermore, any such trajectory converges to a stationary point. Finally, a vector $\mathbf{x} \in S_n$ is asymptotically stable under (8) and (9) if and only if $\mathbf{x}$ is a strict local maximizer of $\mathbf{x}'W\mathbf{x}$ on S_n.*

The previous result is known in mathematical biology as the fundamental theorem of natural selection [14] and, in its original form, traces back to R. A. Fisher. Motivated by this result, we use (as in [7]) replicator equations as a simple heuristic for solving our attributed tree matching problem. Indeed, note that replicator equations are maximization procedures while ours is a minimization problem. However, it is straightforward to see that the problem of minimizing the quadratic form $\mathbf{x}'A\mathbf{x}$ on the simplex is equivalent to that of maximizing $\mathbf{x}'(\gamma\mathbf{ee}' - A)\mathbf{x}$, where γ is an arbitrary constant. Let $T_1 = (V_1, E_1, \alpha_1)$ and $T_2 = (V_2, E_2, \alpha_2)$ be two attributed trees, and let $G = (V, E, \omega)$ be the corresponding ϵ-TAG. By letting

$$W = \gamma\mathbf{ee}' - A \qquad (10)$$

where A is defined in (7) and $\gamma = \max a_{ij}$, we know that the replicator dynamical systems (8) and (9), starting from an arbitrary initial state, which is usually taken to be the simplex barycenter, will iteratively maximize the function $\mathbf{x}'W\mathbf{x}$ (and hence minimize $\mathbf{x}'A\mathbf{x}$) over the simplex and will eventually converge to a strict local optimizer which will then correspond to the characteristic vector of a maximal weight clique in the ϵ-TAG. This will in turn induce a maximal similarity subtree ϵ-morphism between T_1 and T_2.

5 An Example: Matching Shock Trees

We illustrate our framework with numerical examples of shape matching. We use a *shock graph* representation based on a coloring of the shocks (singularities) of a curve evolution process acting on simple closed curves in the plane [13]. Results on shape matching using edit operations between trees obtained from a related representation have also been recently reported in [11].

Shocks are grouped into distinct types according to the local variation of the radius function along the medial axis. Intuitively, the radius function varies monotonically at a type 1, reaches a strict local minimum at a type 2, is constant at a type 3 and reaches a strict local maximum at a type 4. The shocks comprise vertices in the graph, and their formation times direct edges to form a basis for tree matching. Each graph can be reduced to a unique attributed rooted tree, providing the requisite hierarchical structure for our matching algorithm. An illustrative example appears in Figure 5. The vector of attributes assigned to each node $u \in V$ of the attributed shock tree $T = (V, E, \alpha)$ is given by $\alpha(u) = (x_1, y_1, r_1, v_1, \theta_1; ...; x_m, y_m, r_m, v_m, \theta_m)$. Here m is the number of shocks in the group, and $x_i, y_i, r_i, v_i, \theta_i$ are, respectively, the x coordinate, the y coordinate, the radius (or time of formation), the speed, and the direction of each shock i in the sequence, obtained as outputs of the shock detection process.

In order to apply our framework we must define a measure of significance for each node (related to the ϵ parameter) as well as a similarity measure between the attributes of two nodes u and v. The significance measure used here is the amount of boundary reconstructed by the shocks in a particular node, expressed as a fraction of the total boundary length. Intuitively, nodes are deemed less significant when they correspond to ligature-like regions and their associated shocks reconstruct only small portions of the boundary [2]. The similarity measure used is the same as the one described in [7], but scaled by the average significance of the two nodes being matched. The measure provides a number between 0 and

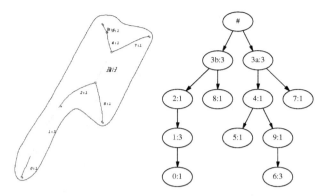

Fig. 1. LEFT: Shock groups are identified on the skeleton using the notation identifier:shocktype. RIGHT: The corresponding shock tree [13].

1, which represents the overall similarity between the geometric attributes of the two nodes being compared. The measure is designed to be invariant under rotations and translations of two shapes, and to satisfy the requirements of the weight function discussed above.

We selected 24 silhouettes representing seven different object classes; the tools shapes were taken from the Rutgers Tools database. For each object class one or more prototype objects were chosen and were matched against *all* entries in the database using the clique-finding replicator equations described in the previous section.

The selection of a proper value for the parameter ϵ is clearly crucial for the performance of the matching process. Our intuition is that the larger the value of ϵ, the larger the cluster formed by the matching process. In the experiments presented here ϵ was set by trial and error to be equal to 0.004.

We ranked the results using a score given by the quantity:

$$W \times \frac{1}{2}\{ \sum_{u \in H_1} (l(u)/L_1) + \sum_{v \in H_2} (l(v)/L_2)\}, \tag{11}$$

where H_1 and H_2 are the sets of matched nodes in the first and in the second tree, respectively, W is the overall similarity between matched nodes (i.e., the weight of the maximal clique found), L_1 and L_2 the total boundary lengths of each shape and $l(u), l(v)$ the lengths of boundaries reconstructed by nodes u and v, respectively. The score represents the weight of the maximal clique scaled by the average of the total (relative) length reconstructed by the nodes in each tree that participates in the match. The top 8 matches are shown for each query shape, in Table 1. It is evident that the best matches are typically to instances from the same object class. These results represent an improvement with respect to those obtained for the isomorphism case [7], and a slight improvement over those obtained for homomorphism on the same database [4]. Specifically, the many-to-

Table 1. A tabulation of the top 8 matches for several prototype shapes, with $\epsilon = 0.004$. The scores indicate the value of index (11).

Prot.	Top 8 matches							
	1	2	3	4	5	6	7	8
	1.000	0.764	0.711	0.615	0.499	0.459	0.430	0.421
	1.000	0.711	0.668	0.607	0.506	0.428	0.423	0.421
	1.000	0.788	0.784	0.689	0.643	0.622	0.528	0.474
	1.000	0.835	0.784	0.722	0.577	0.513	0.496	0.489
	1.000	0.819	0.749	0.652	0.593	0.496	0.474	0.464
	1.000	0.762	0.749	0.681	0.665	0.447	0.429	0.423
	1.000	0.819	0.762	0.691	0.670	0.458	0.449	0.433
	1.000	0.508	0.430	0.402	0.388	0.354	0.295	0.295
	1.000	0.722	0.715	0.643	0.545	0.492	0.407	0.400
	1.000	0.904	0.888	0.427	0.343	0.330	0.329	0.321
	1.000	0.888	0.843	0.435	0.347	0.325	0.311	0.310
	1.000	0.535	0.503	0.446	0.435	0.428	0.427	0.395
	1.000	0.780	0.535	0.399	0.300	0.298	0.293	0.243

many matching algorithm tends to provide a sharper distinction between object classes.

However, more experiments have to be done to assess the overall impact of the many-to-many matching algorithm, particularly when the database of shapes is much larger. We are currently carrying out such experiments, using a database of shock trees built using a novel algorithm for computing skeletal graphs [10].

6   Conclusions

We have expanded our earlier work on matching hierarchical relational structures to handle many-to-many mappings. This has been done by introducing a notion of ϵ-morphism between attributed trees, and by constructing an association graph whose maximal weight cliques are shown to be in one-to-one correspondence with maximal similarity subtree morphisms. Computational examples of matching articulated and deformed shapes give improved results over our earlier work [7,4], because the framework now allows for negligible nodes in either one of the two trees to be ignored during the matching process.

References

1. A. P. Ambler, H. G. Barrow, C. M. Brown, R. M. Burstall, and R. J. Popplestone. A Versatile Computer-controlled Assembly System. In *Proc. 3rd Int. J. Conf. Artif. Intell.*, pages 298–307, Stanford, CA, 1973.
2. J. August, K. Siddiqi and S. W. Zucker. Ligature Instabilities in the Perceptual Organization of Shape. *Computer Vision and Image Understanding*, 76(3), 1999, pages 231–243.
3. D. H. Ballard and C. M. Brown. *Computer Vision.* Prentice-Hall, Englewood Cliffs, N.J, 1982.
4. M. Bartoli, M. Pelillo, K. Siddiqi, and S. W. Zucker. Attributed Tree Homomorphism Using Association Graphs. *Proceedings of the International Conference on Pattern Recognition*, Vol. 2, 2000, pages 133–136.
5. I. M. Bomze, M. Budinich, P. M. Pardalos, and M. Pelillo. The Maximum Clique Problem. In *Handbook of Combinatorial Optimization (Supplement Volume A)*, D.-Z. Du and P. M. Pardalos (Eds.), Kluwer Academic Publishers, Boston, MA, 1999.
6. D. Marr and K. H. Nishihara. Representation and Recognition of the Spatial Organization of Three Dimensional Structure. *Proceedings of the Royal Society of London*, B, 1978, pages 269–294.
7. M. Pelillo, K. Siddiqi, and S. W. Zucker. Matching Hierarchical Structures Using Association Graphs. *IEEE Trans. Pattern Anal. Machine Intell.*, 21(11), 1999, pages 1105–1120.
8. M. Pelillo, K. Siddiqi, and S. W. Zucker. Atributed Tree Morphism Using Association Graphs. *In preparation.*
9. I. M. Bomze, M. Pelillo, and V. Stix. Approximating the Maximum Weight Clique Using Replicator Dynamics. *IEEE Trans. Neural Networks*, 11(6), Nov. 2000.
10. P. Dimitrov, C. Phillips and K. Siddiqi. Robust and Efficient Skeletal Graphs. *Proceedings of the Conference on Computer Vision and Pattern Recognition*, 2000, pages 417–423.

11. P. Klein, D. Sharvit, and B. Kimia. A tree-edit-distance algorithm for comparing simple, closed shapes. *Proceedings, ACM-SIAM Symposium on Discrete Algorithms (SODA)* 2000, pages 696–704.
12. M. Pelillo. Replicator dynamics in combinatorial optimization, In *Encyclopedia of Optimization*, C. A. Floudas and P. M. Pardalos (Eds.). Kluwer Academic Publishers, Boston, MA (in press).
13. K. Siddiqi, A. Shokoufandeh, S. Dickinson, and S. W. Zucker. Shock Graphs and Shape Matching, *Int'l J. Computer Vision*, 35(1), 1999, pages 13–32.
14. J. Weibull. *Evolutionary Game Theory*. MIT Press, Cambridge, MA, 1995.

An Expectation-Maximisation Framework for Perceptual Grouping

Antonio Robles-Kelly* and Edwin R. Hancock

Department of Computer Science
University of York
York, Y01 5DD, UK
email: arobkell,erh@cs.york.ac.uk

Abstract. This paper casts the problem of perceptual grouping into an evidence combining setting using the apparatus of the EM algorithm. We are concerned with recovering a perceptual arrangement graph for line-segments using evidence provided by a raw perceptual grouping field. The perceptual grouping process is posed as one of pairwise relational clustering. The task is to assign line-segments (or other image tokens) to clusters in which there is strong relational affinity between token pairs. The parameters of our model are the cluster memberships and the pairwise affinities or link-weights for the nodes of a perceptual relation graph. Commencing from a simple probability distribution for these parameters, we show how they may be estimated using the apparatus of the EM algorithm. The new method is demonstrated on line-segment grouping problems where it is shown to outperform a non-iterative eigenclustering method.

1 Introduction

Perceptual grouping is an important process which permits low-level features to be organised into higher level relational structures that can be subsequently used for scene understanding and object recognition. Broadly speaking, the available literature on perceptual grouping can be divided into three areas according to the level of abstraction at which they operate. At the lowest level of abstraction the available algorithms are concerned with computation of the grouping field. There are several contributions which deserve special mention Heitger and von der Heydt [4] have shown how to model the line extension field using directional filters whose shapes are motivated by studies of the visual field of monkeys. Williams and his co-workers [16][15] have taken a different approach using the stochastic completion field. Here the completion field of curvilinear features is computed using Monte Carlo simulation of particle trajectories between the endpoints of contours. At the intermediate level of abstraction, several authors have investigated the use of iterative relaxation style operators for edgel grouping. This approach was pioneered by Shashua and Ullman [12] and later refined by

---

* Supported by CONACYT, under grant No. 151752.

C. Arcelli et al. (Eds.): IWVF4, LNCS 2059, pp. 594–605, 2001.

Guy and Medioni [3] among others. Parent and Zucker have shown how co-circularity can be used to gauge the compatibility of neighbouring edges [8]. Matas and Kittler [6] have shown how Waltz's dictionary of line-label configurations can be used for junction re-enforcement. At the highest level of abstraction, the available algorithms pose the grouping problem as that of recovering a graph which represents the relational arrangement of segmental entities previously extracted from raw image data. One of the most popular methods here is to use ideas from spectral graph theory to locate the salient relational structure. For instance, both Sarkar and Boyer [11] and Perona and Freeman [10] have used the eigenstructure of a perceptual affinity matrix to find disjoint subgraphs that represent the main arrangements of segmental entities. Finally, it is worth mentioning that several authors have used similar algorithms based on eigenvalues of an affinity matrix to iteratively segment image data. One of the best known is the normalised cut method of Shi and Malik [13]. Recently, Weiss [14] has shown how this and other closely related methods can be improved using a normalised affinity matrix.

The observation underpinning this paper is that although considerable effort has been expended at intermediate level to develop algorithms for combining evidence from a raw grouping field, the higher-level graph-based methods use static affinity relationships or relational abstractions as input. The aim in this paper is to develop a different approach to the problem which poses the recovery of the perceptual arrangement graph in an evidence combining framework. We pose the problem as one of pairwise clustering which is parameterised using two sets of indicator variables. The first of these are cluster membership variables which indicate to which perceptual cluster a segmental entity belongs. The second set of variables are link weights which convey the strength of the perceptual relations between pairs of nodes in the same cluster. Our contribution is to show how to estimate both sets of indicator variables using the apparatus of the EM algorithm.

The outline of this paper is as follows. In Section 2 we develop a mixture model for the grouping problem. Section 3 shows how the parameters of this mixture model, namely the cluster membership probabilities and the pairwise link-weights can be estimated using the EM algorithm. In Section 4 we describe a simple model which can be used to initialise the link-weights. Section 5 provides a sensitivity study on synthetic data and also furnishes some examples on real world images. Finally, Section 6 concludes the paper by summarising our contributions and offering directions for future research.

2   Maximum Likelihood Framework

We pose the problem of peceptual grouping as that of finding the pairwise clusters which exist within a set of objects segmented from raw image data. These objects may be point-features such as corners, lines, curves or regions. However, in this paper we focus on the problem of grouping line-segments. The process of pairwise clustering is somewhat different to the more familiar one of central clustering. Whereas central clustering aims to characterise cluster-membership using the cluster mean and variance, in pairwise clustering it is link-weights between nodes which are used to establish cluster membership. Although less well studied than central clustering, there has recently been renewed interest in pairwise clustering aimed at placing the method on a more principled footing using techniques such as mean-field annealing [5].

We abstract the problem in the following way. The raw peceptual entities are indexed using the set V. Our aim is to assign each node to one of a set of pairwise clusters which are indexed by the set Ω. To represent the state of organisation of the perceptual relation graph, we introduce some indicator variables. First, we introduce a cluster membership indicator which is unity if the node i belongs to the perceptual cluster $\omega \in \Omega$ and is zero otherwise, i.e.

$$s_{iw} = \begin{cases} 1 & \text{if } i \in \omega \\ 0 & \text{otherwise} \end{cases} \tag{1}$$

The second model ingredient is the link-weight $A_{i,j}$ between pairs of nodes $(i,j) \in V \times V - \{(i,i|i \in V\}$. When the link-weights become binary in nature, they convey the following meaning

$$A_{ij} = \begin{cases} 1 & \text{if } i \in \omega \text{ and } j \in \omega \\ 0 & \text{otherwise} \end{cases} \tag{2}$$

When the link-weights satisfy the above conditionm, then the different clusters represent disjoint subgraphs.

Our aim is to find the cluster membership variables and the link weights which partition the set of raw perceptual entities into disjoint pairwise clusters. We commence by assuming that there are putative edges between each pair of nodes (i,j) belonging to the Cartesian self-product $\Phi = V \times V - \{(i,i)|i \in V\}$. Further suppose that $p(A_{ij})$ is the probability density for the link weight appearing on the pair of nodes $(i,j) \in \Phi$. Our aim is to locate disjoint subgraphs by updating the link weights untill they are either zero or unity. Under the assumption that the link-wieghts on different pairs of nodes are independent of one-another, then the likelihood function for the observed arragement of perceptual entities can be factorised over the set of putative edges as

$$P(G) = \prod_{(i,j) \in \Phi} P(A_{ij}) \tag{3}$$

We are interested in paritioning the set of perceptual entities into pairwise clusters using the link weights between them. We must therefore entertain the possibility that each of the Cartisian pairs appearing under the above product, which represent putative perceptual relations, may belong to each of the pairwise clusters indexed by the set Ω. To make this uncertainty of association explicit, we construct a mixture model over the perceptual clusters and write

$$P(A_{ij}) = \sum_{\omega \in \Omega} P(A_{ij}|\omega)P(\omega) \qquad (4)$$

According to this mixture model, $P(A_{ij}|\omega)$ is the probability that the nodes i and j are connected by an edge with link weight A_{ij} which falls within the perceptual cluster indexed ω. The total probability mass associated with the cluster indexed ω is $P(\omega)$. In most of our experiments, we will assume that there are only two such sets of nodes; those that represent a foreground arrangement, and those that represent background clutter. However, for generality we proceed under the assumption that there are an arbitrary number of perceptual clusters. As a result, the probability of the observed set of perceptual entities is

$$P(G) = \prod_{(i,j)\in|Phi} \sum_{\omega \in \Omega} P(A_{ij}|\omega)P(\omega) \qquad (5)$$

To proceed, we require a model of probability distribution for the link-weights. Here we adopt the Bernoulli model

$$p(A_{ij}|\omega) = A_{i,j}^{s_{i\omega}s_{j\omega}}(1 - A_{i,j})^{1-s_{i\omega}s_{j\omega}} \qquad (6)$$

This distribution takes on is largest values when either the link weight A_{ij} is unity and $s_{i\omega} = s_{j\omega} = 1$, or if the link weight $A_{i,j} = 0$ and $s_{i\omega} = s_{j\omega} = 0$.

3   Expectation-Maximisation

Our aim is to find the cluster-membership weights and the link-weights which maximize the likelihood function appearing in Equation (5). One way to locate the maximum likelihood perceptual relation graph is to update the binary cluster and edge indicators. This could be effected using a number of optimisation methods including simulated annealing and Markov Chain Monte Carlo. However, here we use the apparatus of the EM algorithm originally developed by Dempster, Laird and Rubin [1]. Our reason for doing this is that the cluster-membership variables $s_{i\omega}$ must be regarded as hidden data whose distribution is governed by the link weights A_{ij}. Since at the outset we know neither the associations between nodes and clusters nor the strength of the link weights within clusters, this information must be treated as hidden data. In other words, we must use the EM algorithm to estimate them.

The idea underpinning the EM algorithm is to recover maximum likelihood solutions to problems involving missing or hidden data by iterating between two computational steps. In the E (or expectation) step we estimate the a posteriori probabilities of the hidden data using maximum likelihood parameters recovered in the preceding maximisation (M) step. The M-step in-turn aims to recover the parameters which maximise the expected value of the log-likelihood function. It is the available a posteriori probabilities from the E-step which allows the weighting of log-likelihood required in the maximisation-step.

3.1   Expected Log-Likelihood Function

For the likelihood function appearing in Equation (5), the expected log-likelihood functions is defined to be

$$Q(G^{(n+1)}|G^{(n)}) = \sum_{w \in \Omega} \sum_{(i,j) \in |Phi} P(w|A_{ij}^{(n)}) \ln p(A_{ij}^{(n+1)}|w) \qquad (7)$$

where $p(A_{ij}^{(n+1)}|w)$ is the probability distribution for the link-weights at iteration $n+1$ and $P(w|A_{ij}^{(n)})$ is the a posteriori probability that the pair of nodes with link weight $A_{ij}^{(n)}$ belong to the cluster indexed w at iteration n of the algorithm. When the probability distribution function from equation (6) is substituted, then the expected log-likehood function becomes

$$Q(G^{(n+1)}|G^{(n)}) = \sum_{w \in \Omega} \sum_{(i,j) \in \Phi} \zeta_{i,j,w}^{(n)} \left\{ s_{iw}^{(n+1)} s_{jw}^{(n+1)} \ln A_{ij}^{(n+1)} + \right.$$
$$\left. (1 - s_{iw}^{(n+1)} s_{jw}^{(n+1)}) \ln(1 - A_{i,j}^{(n+1)}) \right\} \qquad (8)$$

where we have used the shorthand $\zeta_{i,j,w}^{(n)} = P(w|A_{ij}^{(n)})$ for the a posteriori cluster membership probabilities. After some algebra to collect terms, the expected log-likelihood function simplifies to

$$Q(G^{(n+1)}|G^{(n)}) = \sum_{w \in \Omega} \sum_{(i,j) \in \Phi} \zeta_{i,j,w}^{(n)} \left\{ s_{iw}^{(n+1)} s_{jw}^{(n+1)} \ln \frac{A_{ij}^{(n+1)}}{1 - A_{ij}^{(n+1)}} + \ln(1 - A_{i,j}^{(n+1)}) \right\} \qquad (9)$$

3.2   Maximisation

In the maximisation step of the algorithm, we aim to recover the cluster and edge parameters s_{iw} and A_{ij}. The edge parameters are found by computing the derivatives of the expected log-likelihood function. As a result the updated link-weights are given by

$$A_{ij}^{(n+1)} = \frac{\sum_{w \in \Omega} \zeta_{i,j,w}^{(n)} s_{iw}^{(n+1)} s_{jw}^{(n+1)}}{\sum_{w \in \Omega} \zeta_{i,j,w}^{(n)}} \qquad (10)$$

In other words, the link-weight for the pair of nodes (i, j) is simply the average of the product of individual node cluster memberships over the different perceptual clusters.

We use the soft-assign ansatz of Bridle [2] to update the cluster membership assignment variables. This involves exponentiating the partial derivatives of the expected log-likelihood function in the following manner

$$s_{i\omega}^{(n+1)} = \frac{\exp\left[\frac{\partial Q(G^{(n+1)}|G^{(n)})}{\partial s_{i\omega}^{(n+1)}}\right]}{\sum_{\omega\in\Omega}\exp\left[\frac{\partial Q(G^{(n+1)}|G^{(n)})}{\partial s_{i\omega}^{(n+1)}}\right]} \tag{11}$$

As a result the update equation for the cluster membership indicator variables is

$$s_{i\omega}^{(n+1)} = \frac{\exp\left[\sum_{j\in V}\zeta_{i,j,\omega}^{(n)}s_{j\omega}^{(n)}\ln\frac{A_{i,j}^{(n+1)}}{1-A_{ij}^{(n+1)}}\right]}{\sum_{\omega\in\Omega}\exp\left[\sum_{j\in V}\zeta_{i,j,\omega}^{(n)}s_{j\omega}^{(n)}\ln\frac{A_{i,j}^{(n+1)}}{1-A_{ij}^{(n+1)}}\right]}$$

$$= \frac{\prod_{j\in V}\left\{\frac{A_{i,j}^{(n+1)}}{1-A_{ij}^{(n+1)}}\right\}^{\zeta_{i,j,\omega}^{(n)}s_{j\omega}^{(n)}}}{\sum_{\omega\in\Omega}\prod_{j\in V}\left\{\frac{A_{ij}^{(n+1)}}{1-A_{ij}^{(n+1)}}\right\}^{\zeta_{i,j,\omega}^{(n)}s_{j\omega}^{(n)}}} \tag{12}$$

3.3   Expectation

The a posteriori probabilities are updated in the expectation step of the algorithm. The current estimates of the parameters $s_{i\omega}^{(n)}$ and $A_{ij}^{(n)}$ are used to compute the probability densities $p(A_{ij}^{(n)}|\omega)$ and the a posteriori probabilities are updated using the formula

$$P(\omega|A_{ij}^{(n)}) = \frac{p(A_{ij}^{(n)}|\omega)\alpha^{(n)}(\omega)}{\sum_{\omega\in\Omega}p(A_{ij}^{(n)}|\omega)\alpha^{(n)}(\omega)} \tag{13}$$

where $\alpha^{(n)}(\omega)$ is the available estimate of the class-prior $P(\omega)$. This is computed using the formula

$$\alpha^{(n)}(\omega) = \frac{1}{|V|^2}\sum_{(i,j)\in\Phi}P(\omega|A_{ij}^{(n)}) \tag{14}$$

4   Initial Line-Grouping Field

We are interested in locating groups of line-segments that exhibit strong geometric affinity to one-another. In this section we provide details of a probabilistic linking field that can be used to gauge geometric affinity. To be more formal suppose we have a set of line-segments $\mathcal{L} = \{\Lambda_i; i = 1, ..., n\}$. Consider two lines Λ_i and Λ_j drawn from this set. Their respective lengths are l_i and l_j. Our model of the linking process commences by constructing the line $\Gamma_{i,j}$ which connects

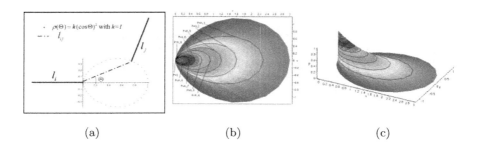

(a)   (b)   (c)

Fig. 1. (a) Geometric meaning of the parameters used to obtain P_{ij}; (b) Plot showing the level curves; (c) 3D plot showing A_{ij} on the z axis

the closest pair of endpoints for the two lines. The geometry of this connecting line is represented using the polar angle θ_{ij} of the line $\Gamma_{i,j}$ with respect to the base-line Λ_i and its length ρ_{ij}. We measure the overall scale of the arrangement of lines using the length of the shorter line $\hat{\rho}_{i,j} = \min[l_i, l_j]$. The relative length of the gap between the two line-segments is represented in a scale-invariant manner using the dimensionless quantity $\xi_{i,j} = \frac{\rho_{i,j}}{\hat{\rho}_{i,j}}$.

Following Heitger and Von der Heydt [4] we model the linking process using an elongated polar grouping field. To establish the degree of geometric affinity between the lines we interpolate the end-points of the two lines using the polar lemniscate $\xi_{i,j} = k \cos^2 \theta_{i,j}$.

The value of the constant k is used to measure the degree of affinity between the two lines. For each linking line, we compute the value of the constant k which allows the polar locus to pass through the pair of endpoints. The value of this constant is

$$k = \frac{\rho_{i,j}}{\hat{\rho}_{i,j} \cos^2 \theta_{i,j}} \tag{15}$$

The geometry of the lines and their relationship to the interpolating polar lemniscate is illustrated in Figure 1a. It is important to note that the polar angle is defined over the interval $\theta_{ij} \in (-\pi/2, \pi/2]$ and is rotation invariant.

We use the parameter k to model the linking probability for the pair of line-segments. When the lemniscate envelope is large, i.e. k is large, then the grouping probability is small. On the other hand, when the envelope is compact, then the grouping probability is large. To model this behaviour, we assign the linking probability using the exponential distribution

$$A_{ij}^{(0)} = \exp[-\lambda k] \tag{16}$$

where λ is a constant whose best value has been found empirically to be unity. As a result, the linking probability is large when either the relative separation of the endpoints is small i.e. $\rho_{i,j} << \hat{\rho}_{i,j}$ or the polar angle is close to zero or π, i.e. the two lines are colinear or parallel. The linking probability is small when either the relative separation of the endpoints is large i.e. $\rho_{i,j} >> \hat{\rho}_{i,j}$ or the polar angle is close to $\frac{\pi}{2}$, i.e. the two lines are perpendicular. In Figures 1 b and c we show a plot of the linking probability as a function of $\frac{\rho_{i,j}}{\hat{\rho}_{i,j}}$ and $\theta_{i,j}$.

5   Experiments

In this Section we provide some experiments to illustrate the utility of our new perceptual grouping method. There are two aspects to this study. We commence by providing some examples for synthetic images. Here we investigate the sensitivity of the method to clutter and compare it with an eigen-decomposition method. The second aspect of our study focusses on real world images.

The first sequence of synthetic images is shown in Figure 2. Here the foreground structure is an approximately circular arrangement of line-segments. In the first row of Figure 2 we show the arrangement of lines with increasing amounts of added clutter. In the subsequent rows we show the results of grouping. In each row the first image is the pattern of foreground line segments extracted by applying the eigendecomposition method described in [7] to the grouping field already detailed in Section 4. The second, third and fourth images show the results obtained with the first, second and third iterations of the EM algorithm. Here the line segments are coded according to the value of the cluster membership weights $s_{i\omega_f}$ where ω_f is the foreground cluster label. As the EM algorithm iterates, then so the foreground cluster weights tend to unity. In each case, the foreground cluster located by the EM algorithm contains less noise contamination than the result delivered by eigendecomposition. Moreover, none of the line segments leaks into the background.

We have repeated the experiments described above for a sequence of synthetic images in which the density of distractors increases. For each image in turn we have computed the number of distractors merged with the foregound pattern and the numnber of foregound line-segments which leak into the background. Figures 3 a and b respectively show the fraction of nodes merged with the foreground and the number of nodes which leak into the background as a function of the number of distractors. In both cases, the shoulder response curve for the EM algorithm occurs at a significanlty higher error rate than that for the eigen-decomposition method.

Finally, we present results on real-world images. We have used the airplane and turtle images shown in Figures 4a and e. The edges shown in Figures 4b and f have been extracted from the raw images using the Canny edge-detector. Straight-line segments have been extracted using the method of Yin [9]. The resulting clusters obtained with the EM algorithm can be seen in Figures 4c and g. For comparison, in Figures 4d and h we show the results obtained with the eigendecomposition method. In both cases the groiping obtained by the EM algorithm is cleaner and contains less spurious clutter.

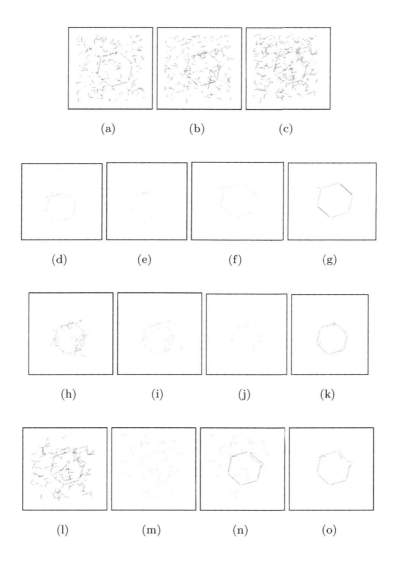

Fig. 2. Top row: test patterns with 180, 200 and 250 random lines added respectively; second row: result of the applying eigendecomposition to the patterns in the top row; third, fourth and fifth rows: each row shows the first, second and third iterations of the EM approach to one of the patterns in the top row.

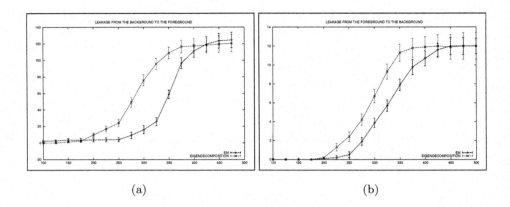

(a)   (b)

Fig. 3. Comparison between the EM algorithm and eigendecomposition.

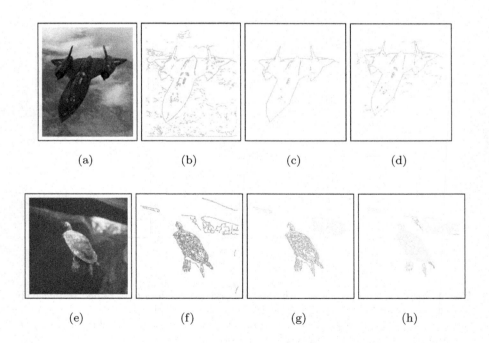

(a)   (b)   (c)   (d)

(e)   (f)   (g)   (h)

Fig. 4. First column: real world images; second column: output of the Canny edge detector; third column: result of the EM algorithm; fourth column: result of eigendecomposition.

6 Conclusions

In this paper, we have presented a new perceptual clustering algorithm which uses the EM algorithm to estimate link-weights and cluster membership probabilities. The method is based on a mixture model over pairwise clusters. The cluster membership probabilities are modelled using a Bernoulli distribution for the link-weights. We apply the method to the problem of line-segment grouping. Here the method appears robust to severe levels of background clutter. It is also relatively insensitive to the gap length and opening angles of the line-segments.

There are a number of ways in which the method proposed in the paper can be improved. Presently, we use a soft-assign method to update the cluster membership variables. We are currently investigating whether this step can be rendered more efficient using matrix factorisation method along the lines suggested by Perona and Freeman [10].

References

1. N. Laird A. Dempster and D. Rubin. Maximum-likehood from incomplete data via the em algorithm. *J. Royal Statistical Soc. Ser. B (methodological)*, 39:1–38, 1977.
2. J. S. Bridle. Training stochastic model recognition algorithms can lead to maximum mutual information estimation of parameters. In *NIPS 2*, pages 211–217, 1990.
3. G. Guy and G. Medioni. Inferring global perceptual contours from local features. *International Journal of Computer Vision*, 20(1/2):113–133, 1996.
4. F. Heitger and R. von der Heydt. A computational model of neural contour processing. In *IEEE CVPR*, pages 32–40, 1993.
5. T. Hofmann and M. Buhmann. Pairwise data clustering by deterministic annealing. *IEEE Tansactions on Pattern Analysis and Machine Intelligence*, 19(1), 1997.
6. Matas J and Kittler J. Junction detection using probabilistic relaxation. *Image and Vision Computing*, 11(4):197–202, 1993.
7. A. Robles Kelly and E. R. Hancock. Grouping-line segments using eigenclustering. In *Proceedings of the British Machine Vision Conference*, 2000.
8. P. Parent and S. Zucker. Trace inference, curvature consistency and curve detection. *IEEE Transactions on Pattern Analysis and Machine Intelligence*, 11(8):823–839, 1989.
9. Yin Peng-Yeng. Algorithms for straight line fitting using k-means. *Pattern Recognition Letters*, 19:31–41, 1998.
10. P. Perona and W. T. Freeman. Factorization approach to grouping. In *ECCV*, pages 655–670, 1998.
11. S. Sarkar and K. L. Boyer. Quantitative measures of change based on feature organization: Eigenvalues and eigenvectors. *Computer Vision and Image Understanding*, 71(1):110–136, 1998.
12. A. Shashua and S. Ullman. Structural saliency: The detection of globally salient structures using a locally connected network. In *Proc. 2nd Int. Conf. in Comp. Vision*, pages 321–327, 1988.
13. J. Shi and J. Malik. Normalized cuts and image segmentations. In *CVPR*, pages 731–737, 1997.

14. Y. Weiss. Segmentation using eigenvectors: A unifying view. In *IEEE International Conference on Computer Vision*, pages 975–982, 1999.
15. L. R. Williams and D. W. Jacobs. Local parallel computation of stochastic completion fields. *Neural Computation*, 9(4):859–882, 1997.
16. L. R. Williams and D. W. Jacobs. Stochastic completion fields: A neural model of illusory contour shape and salience. *Neural Computation*, 9(4):837–858, 1997.

Alignment-Based Recognition of Shape Outlines

Thomas B. Sebastian, Philip N. Klein, and Benjamin B. Kimia

Brown University, Providence RI, USA

Abstract. We present a 2D shape recognition and classification method based on matching shape outlines. The correspondence between outlines (curves) is based on a notion of an *alignment curve* and on a measure of similarity between the intrinsic properties of the curve, namely, length and curvature, and is found by an efficient dynamic-programming method. The correspondence is used to find a similarity measure which is used in a recognition system. We explore the strengths and weaknesses of the outline-based representation by examining the effectiveness of the recognition system on a variety of examples.

1   Introduction

The representation of the shape of objects can have a significant impact on the effectiveness of a recognition strategy. Shapes have been represented as curves [11, 21,2,6,22], point sets [1,15,20], feature sets [3,7], and by medial axis [23,17,18,14, 12,10,9], among others. This paper develops an approach to object recognition based on a curve-based representation of shape outline using the proposed concept of an *alignment curve*, and identifies the strengths and weaknesses of using curves to represent shapes for object recognition and for indexing into image databases by shape context.

In many object recognition and content-based image indexing applications, the object outlines are represented as curves and matched. The matching relies on either aligning feature points using an optimal similarity transformation [1, 15,20] or on a deformation-based approach to aligning the properties of the two curves [11,21,2,6,22]. *Transformation-based* methods rely on matching feature points by finding the optimal rotation, translation, and scaling parameters [1, 15,20]. *Deformation-based* methods typically involve finding a mapping from one curve to the other that minimizes an "elastic" performance functional, which penalizes "stretching" and "bending" [4,19,2,22]. The minimization problem in the discrete domain is transformed into one of matching shape signatures with curvature, bending angle, or orientation as attributes [5,13,6,11,21]. The curve-based methods in general typically suffer from one or more of the following drawbacks: asymmetric treatment of the two curves, sensitivity to sampling, lack of rotation and scaling invariance, and sensitivity to articulations and deformations of parts. We address some of these issues in the proposed method.

Another type of shape representation models the shape outline as point sets and matches the points using an assignment algorithm. Gold *et al.* [7] use graduated assignment to match image boundary features. In a recent approach, Belongie *et al.* [3] use the Hungarian method to match unordered boundary points,

C. Arcelli et al. (Eds.): IWVF4, LNCS 2059, pp. 606–618, 2001.
© Springer-Verlag Berlin Heidelberg 2001

using a coarse histogram of the relative location of the remaining points as the feature. These methods have the advantage of not requiring ordered boundary points, but do not necessarily preserve the coherence of shapes in matching.

Shapes have also been represented by medial axis or its variants and then matched. Shock graph matching have been used in [17,18,14] for object recognition and image indexing tasks. Zhu and Yuille [23] have proposed a framework (FORMS) for matching animate shapes by comparing their skeletal graphs. These approaches do not explicitly model the instabilities of the symmetry-based representations, which can be problematic when dealing with visual transformations like occlusion, view-point variation, and articulation. Liu and Geiger [12] use the A* algorithm to match shape axis trees. Their algorithm does not preserve ordering of edges at nodes which can result in matches that do not preserve coherence of the shapes. Klein *et al.* [10,9] have recently proposed an edit-distance based approach to shape matching, which is very effective, but like other graph matching techniques can in general be computationally intensive. This gives rise to the question whether the additional effort required in skeletal matching is justified by the improvements in recognition rates for particular applications. A goal of this paper is to examine the effectiveness of outline-based matching techniques in general.

In this paper, we present an outline-based recognition method, which relies on finding the optimal correspondence between 2D outlines (curves) by comparing their *intrinsic* properties, namely, length and curvature. The basic premise of the approach is that the goodness of the optimal correspondence can be expressed as the sum of the goodness of matching subsegments. This allows us to cast the problem of finding the optimal correspondence as an energy minimization problem, which is solved by an efficient dynamic-programming algorithm. We introduce the notion of an *alignment curve* to ensure a symmetric treatment of the two curves being matched. The problem formulation and the mathematics underlying the matching process is described in Section 2. In Section 3 we discuss the proposed curve matching framework in application to shape classification and handwritten character recognition. In Section 4, we discuss some of the shortcomings and limitations of curve-based representation for recognition.

2   Curve Matching

This section first discusses the problem of matching and aligning two curve segments followed by a discussion pertaining to closed curves. Denote the curve segments to be matched by $C(s) = (x(s), y(s))$, $s \in [0, L]$ and $\bar{C}(\bar{s}) = (\bar{x}(\bar{s}), \bar{y}(\bar{s}))$, $\bar{s} \in [0, \bar{L}]$, where s is arc length, x and y are coordinates of each point, L is length, and where each is similarly defined for $\bar{C}$. A central premise of this approach is that the "goodness" of the overall optimal match is the sum of "goodness" of the optimal matches between two corresponding subsegments. This allows an energy functional to convey the goodness of a match as a function of the correspondence or alignment of the two curves as proposed earlier in [4,22]. Let a mapping $g : [0, L] \to [0, \bar{L}]$, $g(s) = \bar{s}$, represent an alignment of the two curves.

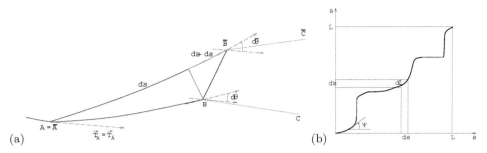

(a)   (b)

Fig. 1. (a) The cost of deforming an infinitesimal segment AB to segment $\bar{A}\bar{B}$, when the initial points and the initial tangents are aligned ($A = \bar{A}$, $T_A = T_{\bar{A}}$), is related to the distance $B\bar{B}$, and is defined by $|d\bar{s} - ds| + R|d\bar{\theta} - d\theta|$. (b) The alignment curve allows for a finite segment from one curve to be aligned with a single point on one curve, thus allowing for the curve segment deletion or addition.

Cohen *et al.* [4] use "bending" and "stretching" energies in a physical analogy similar to the one used in formulating active contours [8] in the form of

$$\mu[g] = \int_{\mathcal{C}} |\frac{\partial}{\partial s}(\bar{\mathcal{C}}(\bar{s}) - \mathcal{C}(s))|^2 ds + R \int_{\mathcal{C}} (\kappa_{\mathcal{C}}(s) - \kappa_{\bar{\mathcal{C}}}(\bar{s}))^2 ds,$$

where κ is the curvature, R is a parameter, and $\bar{s} = g(s)$. Younes [22] uses a similar functional. A key drawback of these approaches for recognition is that they are not invariant to the rotation of one curve with respect to the other, as the cost functional is a function of the absolute orientation of the curves. In addition, the issue of invariance to sampling has not been addressed. We now formulate the problem in an intrinsic manner which addresses both issues:
Definition: Let $\mathcal{C}|_{[s_1,s_2]}$ denote the portion of the curve from s_1 to s_2 and $g|_{([s_1,s_2],[\bar{s}_1,\bar{s}_2])}$ the restriction of the mapping g to $[s_1, s_2]$, where $\bar{s}_1 = g(s_1)$ and $\bar{s}_2 = g(s_2)$. Define a measure μ on this alignment function,

$$\mu[g]|_{([s_1,s_2],[\bar{s}_1,\bar{s}_2])} : g|_{([s_1,s_2],[\bar{s}_1,\bar{s}_2])} \rightarrow \mathbf{R}^+,$$

constructed such that it is inversely proportional to the goodness of the match, *i.e.*, it denotes the cost of deforming $\mathcal{C}|_{[s_1,s_2]}$ to $\bar{\mathcal{C}}|_{[\bar{s}_1,\bar{s}_2]}$.

We restrict this measure μ to one which satisfies an *additivity property*, *i.e.*, $\mu[g]|_{([s_1,s_3],[\bar{s}_1,\bar{s}_3])} = \mu[g]|_{([s_1,s_2],[\bar{s}_1,\bar{s}_2])} + \mu[g]|_{([s_2,s_3],[\bar{s}_2,\bar{s}_3])}$, where $\bar{s}_i = g(s_i)$. This property implies that the match process can be decomposed into a number of smaller matches, which in turn implies that it can be written as a functional $\mu[g]|_{([0,L],[0,\bar{L}])} = \int_0^L \mu[g]|_{([s,s+ds][g(s),g(s+ds)])} ds$. Then, the optimal match is given by $g^* = \underset{g}{\operatorname{argmin}} \mu[g]|_{([0,L],[0,\bar{L}])}$.

Consider two infinitesimal curve segments $\mathcal{C}|_{[A,B]}$ and $\bar{\mathcal{C}}|_{[\bar{A},\bar{B}]}$ of lengths ds, $d\bar{s}$, and curvatures κ, $\bar{\kappa}$, respectively. In our approach we only compare the intrinsic aspects of the curves. Thus, we can align the curves such that the points A and $\bar{A}$, and their tangents $\boldsymbol{T}_A$ and $\boldsymbol{T}_{\bar{A}}$ coincide, Figure 1(a). The cost of matching the infinitesimal curve segments is the degree by which B and $\bar{B}$ and their respective tangents differ, namely,

$$\mu[g]\Big|_{([s_1,s_1+ds],[\bar{s}_1,\bar{s}_1+d\bar{s}])} = |d\bar{s} - ds| + R|d\bar{\theta} - d\theta|, \tag{1}$$

where R is a constant. Then, the resulting functional is given by

$$\mu[g] = \int_C \left[\left| \frac{d\bar{s}}{ds} - 1 \right| + R \left| \frac{d\bar{\theta}(\bar{s})}{d\bar{s}} \frac{d\bar{s}}{ds} - \frac{d\theta(s)}{ds} \right| \right] ds \tag{2}$$

The functional penalizes "stretching" and "bending". However, this formulation of the curve matching problem is inherently asymmetric. This is precisely the objection raised by Tagare et al. [19] to algorithms which are based on differentiable function of one curve to the other. They instead propose a "bimorphism", which diffeomorphically maps a pair of curves to be matched, and which corresponds to a closed curve in space of $\mathcal{C}_1 \times \mathcal{C}_2$. Specifically, they formulate a cost function that minimizes differences in local orientation change $|d\bar{\theta} - d\theta|$ along each differential segment of this curve, and seek a pair of functions ϕ_1 and ϕ_2, elements of the bimorphism, which optimize this cost functional.

We approach this asymmetry issue in a somewhat similar fashion. Observe that the formulation allows for mapping an entire segment of the first curve to a single point in the second curve, but it is not possible to map a single point in the first curve to a segment in the second curve. This is because the notion of an alignment is captured by a (uni-valued) function g. To alleviate this difficulty we adopt a view where an alignment between two curves is represented as a pairing of two particles, one on each curve traversing their respective paths monotonically, but with finite stops allowed. Let the alignment be specified in terms of two functions h and $\bar{h}$ relating arc length along $\mathcal{C}$ and $\bar{\mathcal{C}}$ to the newly defined curve parameter ξ, i.e., $s = h(\xi)$, and $\bar{s} = \bar{h}(\xi)$. In cases where h is invertible, we have $\bar{s} = \bar{h}(h^{-1}(s)) = \bar{h} \circ h^{-1}(s)$, which allows for the use of an alignment function, $g = \bar{h} \circ h^{-1}$, as before. However, when h is not invertible, i.e., when the first particle stops along the first curve for some finite time, g is not defined. While this formulation allows for a symmetric treatment of the curves, note that a superfluous degree of freedom is introduced, as in [19], because different traversals h and $\bar{h}$ may give rise to the same alignment. While Tagare et al. [19] treat this degree of redundancy in the optimization involving two functions, we remove this additional degree of redundancy by proposing the notion of an *alignment curve*, α, with coordinates h and $\bar{h}$

$$\alpha(\xi) \triangleq (h(\xi), \bar{h}(\xi)), \quad \xi \in [0, \tilde{L}], \quad \alpha(0) = (0,0), \quad \alpha(\tilde{L}) = (L, \bar{L}),$$

where ξ is the arc length along the alignment curve and $\tilde{L}$ is its length. The alignment curve can now be specified by a *single function*, namely, $\psi(\xi)$, $\xi \in [0, \tilde{L}]$, where ψ denotes the angle between the tangent to the curve and the x-axis, Figure 1(b). The coordinates h and $\bar{h}$ can then be obtained by integration

$$h(\xi) = \int_0^\xi \cos(\psi(\eta))d\eta, \qquad \bar{h}(\xi) = \int_0^\xi \sin(\psi(\eta))d\eta, \quad \xi \in [0, \tilde{L}].$$

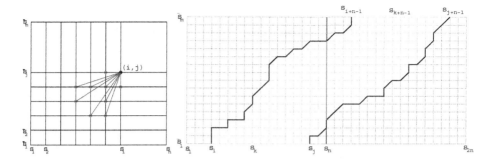

Fig. 2. (a) This figure illustrates the template that is used to in the Dynamic Programming implementation of Equation 4. The entry at (i, j) is the cost to match the curve segments $x_1, x_2, \ldots, x_i$ and $y_1, y_2, \ldots, y_j$ $d(i, j)$. To update the cost at (i, j) (blue dot) we limit the choices of the k and l, so that only costs at the a limited set of points (green dots) are considered. (b) This figure illustrates the grid used by the dynamic-programming method to compute the optimal alignment curve for closed curves. Discrete samples along the curves are the axes. The first curve $\mathcal{C}$ is repeated. If the blue curves are optimal alignment curves from $(s_i, \bar{s}_1)$ to $(s_i + n - 1, \bar{s}_m)$ and $(s_j, \bar{s}_1)$ to $(s_j + n - 1, \bar{s}_m)$, then the alignment curve from $(s_k, \bar{s}_1)$ to $(s_k + n - 1, \bar{s}_m)$ for $i < k < j$ does not cross the blue lines, so the search can be restricted to the green area. Full details are discussed in [16].

Note that ψ is constrained by monotonicity ($h' \geq 0$ and $\bar{h}' \geq 0$) to lie in $[0, \frac{\pi}{2}]$. The alignment between $\mathcal{C}$ and $\bar{\mathcal{C}}$ is then fully represented by single function ψ.

The goodness of the match corresponding to the alignment curve can now be rewritten in terms of ψ. First, if $h' \neq 0$ and $\bar{h}' \neq 0$ for $\xi \in [\xi_1, \xi_2]$, then $g = \bar{h} \circ h^{-1}$ is well defined and we rewrite $\mu[\psi]$ in terms of g using Equation 1, which after some simplification results in

$$\mu(\psi)\big|_{[\xi_1, \xi_2]} = \int_{\xi_1}^{\xi_2} \left[|\cos(\psi) - \sin(\psi)| + R|\kappa(h) \cos(\psi) - \bar{\kappa}(\bar{h}) \sin(\psi)| \right] d\xi \quad (3)$$

Second, consider that one of h' or $\bar{h}'$ is zero at a point, say $h'(\xi) = 0$, implying that this point maps to a corresponding interval $[\bar{h}(\xi), \bar{h}(\xi + d\xi)]$. The cost of mapping the point $h(\xi)$ to the interval $[\bar{h}(\xi), \bar{h}(\xi + d\xi)]$ is defined by enforcing continuity of the cost with deformations: consider the cost of aligning the interval $[h(\xi), h(\xi + d\xi)]$ to the interval $[\bar{h}(\xi), \bar{h}(\xi + d\xi)]$ as the first interval shrinks to a point, i.e., as $\psi \to \frac{\pi}{2}$, $\cos(\psi) \to 0$, Similarly, the case where an interval in the first curve is mapped to a point in the second curve, should be the limiting case of $\psi \to 0$ or $\sin(\psi) \to 0$. Thus, the overall cost of the alignment ψ is well defined

Fig. 3. Examples of the optimal alignment between curves obtained using the curve matching algorithm. The alignment is indicated by arbitrarily coloring portions of the aligned curves by identical colors with a number indicating the each portion's end point. Observe that the alignment is intuitive for both open and closed curves.

in all cases of Equation 3, and is found by minimizing

$$
\begin{cases}
\mu[\psi] = \int_0^{\tilde{L}} \big[\,|\cos(\psi) - \sin(\psi)| + R|\kappa(h)\cos(\psi) - \bar{\kappa}(\bar{h})\sin(\psi)|\,\big]d\xi, \\[2mm]
0 \le \psi \le \dfrac{\pi}{2}, \int_0^{\tilde{L}} \cos(\psi)d\xi = L, \text{ and } \int_0^{\tilde{L}} \sin(\psi)d\xi = \bar{L}.
\end{cases}
\tag{4}
$$

Then, the optimal alignment is given by $\psi^* = \operatorname*{argmin}_{\psi} \mu(\psi)\big|_{[0,\tilde{L}]}$.

Definition: Let the *edit distance* between two curve segments $\mathcal{C}$ and $\bar{\mathcal{C}}$ be defined as the cost of the optimal alignment of the two curves, $d(\mathcal{C}, \bar{\mathcal{C}}) = \mu(\psi^*)$. It is straightforward to show the following [16].

Lemma 1. *If h^* and $\bar{h}^*$ specify the optimal alignment given by ψ^*, the distance function satisfies the following suboptimal property for $\xi_1 < \xi_2 < \xi_3, s_i = h^*(\xi_i), \bar{s}_i = \bar{h}^*(\xi_i), i = 1, 2, 3,$*

$$
d(\mathcal{C}\big|_{[s_1,s_3]}, \bar{\mathcal{C}}\big|_{[\bar{s}_1,\bar{s}_3]}) = d(\mathcal{C}\big|_{[s_1,s_2]}, \bar{\mathcal{C}}\big|_{[\bar{s}_1,\bar{s}_2]}) + d(\mathcal{C}\big|_{[s_2,s_3]}, \bar{\mathcal{C}}\big|_{[\bar{s}_2,\bar{s}_3]}).
\tag{5}
$$

Matching Closed Curves: The edit distance between two closed curves is the minimum cost of the matching the open curve segments starting at s_1 and $\bar{s}_1$, and terminating at s_1^* and $\bar{s}_1^*$ having traversed the entire curve.

$$
d(\mathcal{C}_{closed}, \bar{\mathcal{C}}_{closed}) = \min_{[s_1,\bar{s}_1]} d(\mathcal{C}\big|_{[s_1,s_1^*]}, \bar{\mathcal{C}}\big|_{[\bar{s}_1,\bar{s}_1^*]}).
$$

When matching closed curves, we do not have to find the alignment for all pairs of start point correspondences. It is sufficient to choose a start point s_1 on curve $\mathcal{C}$, and the find the optimal alignments for all possible start points on the curve $\bar{\mathcal{C}}$. If we choose another point s_2, instead of s_1, we will get the same optimal alignment using Lemma 1.

The curve matching is implemented using a fast[1] dynamic-programming method, as outlined in Figure 2 and described in detail in [16]. Figure 3 illustrates the alignment for two simple cases. In all the examples, we set $R = 10$.

---

[1] The complexity of the algorithm to match curve segments and closed curves is $O(n^2)$ and $O(n^2 log(n))$, respectively, where n is the number of samples along the curves. It takes 0.04 secs and 1.6 secs to match curve segments and closed curves with 50 samples respectively on an SGI INDIGO^2 (195MHz).

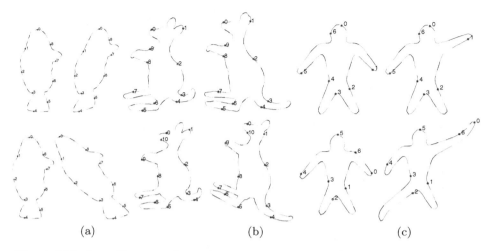

Fig. 4. This figure illustrate the performance of the curve matching algorithm in presence of an affine transformation (a), view-point variation (b) and articulation and deformation of parts (c). The alignment is indicated by arbitrarily coloring portions of the aligned curves by identical colors with a number indicating the each portion's end point. Observe that the matches are intuitive, *e.g.*, hands, legs and head of the dolls correspond in the presence of articulation, stretching and bending. Note that the different views of the kangaroo were obtained by taking snapshots of a 3D model.

We have also seen experimentally that the alignment is relatively insensitive to the choice of R.

3   Recognition Using Shape Outline Alignment

In this section, we examine the effectiveness of curve matching for recognizing shape outlines and characters. The curve alignment framework gives a correspondence between two curves, which is then used to measure the similarity between two curves. One can either use edit distance or normalized Euclidean distance between corresponding points [11,6]. For curve matching to be effective in object recognition, it has to perform well under a variety of visual transformations such as occlusion, articulation and deformation of parts, and view-point variation, which we examine now. Figure 4 shows that the curve matching algorithm works well in the presence of commonly occurring visual transformations, affine transformations, modest amounts of view-point variation, and under some articulation and deformations like stretching and bending of parts.

Object Recognition: We illustrate the use of curve matching for shape classification on a database of 36 shapes. The database consists of shapes from six different categories (fishes, tools, planes, rabbits, "greebles", and hands) with six different shapes in each category. Comparisons are made between every pair of shapes. The five nearest neighbors for each shape are highlighted. Observe that

Table 1. Costs of matching pairs of shape outlines in a database of 36 shapes, 6 samples of each of 6 categories. The five nearest neighbor are from the "correct" category 36/36, 35/36, 35/36, 33/36, 27/36 cases.

```
       tool           |      pencil         |       A          |      duck        |      droplet     |      hand
----------------------+---------------------+------------------+------------------+------------------+------------------
 0  5  9 ..  7 10 | .. .. 12 .. .. .. | .. .. .. .. .. .. | .. .. .. .. .. .. | .. .. .. .. .. .. | .. .. .. .. .. ..
 5  0 .. 10  6  9 | .. .. .. .. .. .. | .. .. 13 .. .. .. | .. .. .. .. .. .. | .. .. .. .. .. .. | .. .. .. .. .. ..
 9 ..  0  6  6  7 | .. .. .. .. .. .. | .. .. 10 .. .. .. | .. .. .. .. .. .. | .. .. .. .. .. .. | .. .. .. .. .. ..
.. 10  6  0  7  6 | .. .. .. .. 12 .. | .. .. .. .. .. .. | .. .. .. .. .. .. | .. .. .. .. .. .. | .. .. .. .. .. ..
 7  6  6  7  0  7 | .. .. .. .. .. .. | .. .. .. .. .. .. | .. .. .. .. .. .. | .. .. .. .. .. .. | .. .. .. .. .. ..
10  9  7  6  7  0 | .. .. .. .. .. .. | .. .. .. .. .. .. | .. .. .. .. .. .. | .. .. .. .. .. .. | .. .. .. .. .. ..
----------------------+---------------------+------------------+------------------+------------------+------------------
.. .. .. .. .. .. |  0  2 12 10  6  3 | .. .. .. .. .. .. | .. .. .. .. .. .. | .. .. .. .. .. .. | .. .. .. .. .. ..
.. .. .. .. .. .. |  2  0 12  5  4  3 | .. .. .. .. .. .. | .. .. .. .. .. .. | .. .. .. .. .. .. | .. .. .. .. .. ..
12 .. .. .. .. .. | ..  12  0  3  4  4 | .. .. .. .. .. .. | .. .. .. .. .. .. | .. .. .. .. .. .. | .. .. .. .. .. ..
.. .. .. .. .. .. | 10  5  3  0  5  7 | .. .. .. .. .. .. | .. .. .. .. .. .. | .. .. .. .. .. .. | .. .. .. .. .. ..
.. .. .. 12 .. .. |  6  4  4  5  0  9 | .. .. .. .. .. .. | .. .. .. .. .. .. | .. .. .. .. .. .. | .. .. .. .. .. ..
.. .. .. .. .. .. |  3  3  4  7  9  0 | .. .. .. .. .. .. | .. .. .. .. .. .. | .. .. .. .. .. .. | .. .. .. .. .. ..
----------------------+---------------------+------------------+------------------+------------------+------------------
.. .. .. .. .. .. | .. .. .. .. .. .. |  0 .. ..  9  9  9 | .. .. .. 13 .. .. | .. .. .. 10 .. .. | .. .. .. .. .. ..
.. .. .. .. .. .. | .. .. .. .. .. .. | 14  0  7 10 11 .. | 13 .. .. .. .. .. | .. .. .. .. .. .. | .. .. .. .. .. ..
.. 13 10 .. .. .. | .. .. .. .. .. .. | ..  7  0  8 12 .. | .. .. .. .. .. .. | .. .. .. .. .. .. | .. .. .. .. .. ..
.. .. .. .. .. .. | .. .. .. .. .. .. |  9 10  8  0  9 .. | .. .. .. 17 .. .. | .. .. .. .. .. .. | .. .. .. .. .. ..
.. .. .. .. .. .. | .. .. .. .. .. .. |  9 11 12  9  0 13 | .. .. .. .. .. .. | .. .. .. .. .. .. | .. .. .. .. .. ..
.. .. .. .. .. .. | .. .. .. .. .. .. |  9 .. .. .. 13  0 | .. .. 13 13 .. .. | .. .. 13 .. .. .. | .. .. .. .. .. ..
----------------------+---------------------+------------------+------------------+------------------+------------------
.. .. .. .. .. .. | .. .. .. .. .. .. | .. 13 .. .. .. .. |  0  3  4  3  5  5 | .. .. .. .. .. .. | .. .. .. .. .. ..
.. .. .. .. .. .. | .. .. .. .. .. .. | .. .. .. .. .. .. |  3  0  5  8  5  8 | .. .. .. .. .. .. | .. .. .. .. .. ..
.. .. .. .. .. .. | .. .. .. .. .. .. | .. .. .. .. .. 13 |  4  5  0  4  5  7 | .. .. .. .. .. .. | .. .. .. .. .. ..
.. .. .. .. .. .. | .. .. .. .. .. .. | 13 .. .. 17 .. 13 |  3  8  4  0 11  3 | .. .. .. .. .. .. | .. .. .. .. .. ..
.. .. .. .. .. .. | .. .. .. .. .. .. | .. .. .. .. .. .. |  5  5  5 ..  0  4 | .. .. 11 .. .. .. | .. .. .. .. .. ..
.. .. .. .. .. .. | .. .. .. .. .. .. | .. .. .. .. .. .. |  5  8  7  3  4  0 | .. .. .. .. .. .. | .. .. .. .. .. ..
----------------------+---------------------+------------------+------------------+------------------+------------------
.. .. .. .. .. .. | .. .. .. .. .. .. | .. .. .. .. .. .. | .. .. .. .. .. .. |  0  3  3  4  3  2 | .. .. .. .. .. ..
.. .. .. .. .. .. | .. .. .. .. .. .. | .. .. .. .. .. .. | .. .. .. .. .. .. |  3  0  4  2  2  2 | .. .. .. .. .. ..
.. .. .. .. .. .. | .. .. .. .. .. .. | .. .. .. .. .. 13 | .. .. .. .. 11 .. |  3  4  0  4  3  3 | .. .. .. .. .. ..
10 .. .. .. .. .. | .. .. .. .. .. .. | .. .. .. .. .. .. | .. .. .. .. .. .. |  4  2  4  0  4  3 | .. .. .. .. .. ..
.. .. .. .. .. .. | .. .. .. .. .. .. | .. .. .. .. .. .. | .. .. .. .. .. .. |  3  2  3  4  0  3 | .. .. .. .. .. ..
.. .. .. .. .. .. | .. .. .. .. .. .. | .. .. .. .. .. .. | .. .. .. .. .. .. |  2  2  3  3  3  0 | .. .. .. .. .. ..
----------------------+---------------------+------------------+------------------+------------------+------------------
.. .. .. .. .. .. | .. .. .. .. .. .. | .. .. .. .. .. .. | .. .. .. .. .. .. | .. .. .. .. .. .. |  0  1  0  6  5  6
.. .. .. .. .. .. | .. .. .. .. .. .. | .. .. .. .. .. .. | .. .. .. .. .. .. | .. .. .. .. .. .. |  1  0  1  6  5  7
.. .. .. .. .. .. | .. .. .. .. .. .. | .. .. .. .. .. .. | .. .. .. .. .. .. | .. .. .. .. .. .. |  0  1  0  6  5  6
.. .. .. .. .. .. | .. .. .. .. .. .. | .. .. .. .. .. .. | .. .. .. .. .. .. | .. .. .. .. .. .. |  6  6  6  0  5 10
.. .. .. .. .. .. | .. .. .. .. .. .. | .. .. .. .. .. .. | .. .. .. .. .. .. | .. .. .. .. .. .. |  5  5  5  5  0 11
.. .. .. .. .. .. | .. .. .. .. .. .. | .. .. .. .. .. .. | .. .. .. .. .. .. | .. .. .. .. .. .. |  6  7  6 10 11  0
```

the shapes are categorized intuitively, *i.e.*, the nearest neighbors of the "tool" shapes are in the "tool" category and similarly for others.

Handwritten character recognition: As another example we have selected handwritten character recognition which due to its inherently one-dimensional nature, is well suited to this approach. As in the case of recognizing shape outlines, the optimal alignment between two characters is found and then used to compute a distance measure between the two. We have used a database of 88 digits consisting of 6 different characters, to perform recognition experiments. Matching is done between every pair of characters in the database, and the top 25 matches of a few sample characters are shown in Table 2. Observe that in

Table 2. The top 20 matches for a few handwritten digits. The database used in this experiment consists of 88 handwritten digits. The number below the matching character is the computed distance. Observe that most of the top matches of a character are samples of the same character, *i.e.*, the top matches of "2" are samples of "2".

query																				
2	4	5	8	8	8	9	9	10	11	11	11	13	14	14	14	15	15	15	15	18
3	5	6	6	6	7	7	7	7	7	8	8	10	11	15	16	16	16	16	17	17
6	4	4	4	4	4	5	6	6	7	7	7	8	8	9	22	23	24	25	26	26
8	7	7	7	7	8	9	9	10	10	11	11	11	12	14	25	26	26	27	28	28

Table 3. The top five matches for a few sample characters are shown. The number below each match is the computed distance measure between the two characters.

q	1	2	3	4	5		q	1	2	3	4	5		q	1	2	3	4	5
0	4	7	10	13	14		*L*	4	11	15	16	18		*9*	6	8	9	11	15
1	2	4	5	6	10		*m*	5	15	17	21	22		*G*	5	12	19	19	19
2	4	11	11	13	15		*8*	12	17	18	20	21		*V*	5	8	11	12	13
3	5	11	16	16	19		*P*	4	17	18	18	19		*W*	6	16	17	18	21
6	2	10	13	14	15		*S*	4	10	11	11	12		*Z*	3	7	10	13	14

most cases, the top matches are other samples of the same character, indicating the potential of this approach in handwritten character recognition.

Prototype formation: Prototypes have been used to improve the efficiency of object and character recognition [5] and indexing into image databases. Typically, a representative sample is used as the prototype. Instead, an "average" curve can be used. The curve alignment framework allows us to generate the average of set of curves [16]. Figure 5 shows the average curve for a set of fish outlines, and handwritten digits. The average outline of handwritten characters are used in the handwritten character recognition experiments with excellent results. For 327 digits and alphabets (34 categories) written by one subject, 323 characters (98.8%) were correctly recognized. The top five matches for a few sample characters are shown in Table 3.

Morphing: Morphing a shape to another has a variety of applications in computer graphics and animation. The proposed curve matching framework can be used to generate a sequence of images morphing a shape to another when the shapes are not very dissimilar. Figure 5 shows the morphing of the outline of a cat to that of a kangaroo. Curve matching has also been used in a variety of other applications including tracking objects in a video sequence, comparing

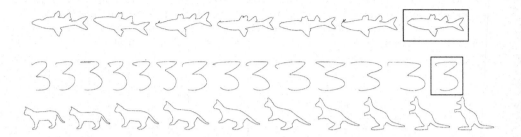

Fig. 5. Top and middle rows: A collection of curves (blue) and their average (red). Bottom row: Sequence of deforming the outline of a cat to that of a kangaroo.

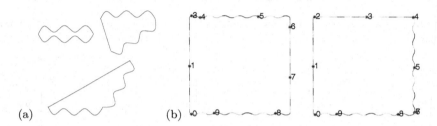

Fig. 6. (a) Curves do not represent the interior of the shape, and hence cannot adequately distinguish between perceptually distinct shapes whose outlines have similar features. (b) This also implies that in relating two shapes by curve matching, outline features take precedence over matching the shape interior! Curve matching aligns the wavy sides of the two squares, ignoring the spatial configuration as a square.

medical structures, registering 3D volume datasets by aligning characteristic 3D space curves like ridges. Thus, our proposed scheme can be useful in numerous applications.

4   Discussion and Conclusion

We have presented a computational framework to find the optimal correspondence between two 2D curves. The main contribution of this paper is to propose a new scheme for curve matching that is symmetric in its treatment of the two curves, is highly efficient, and works well in a variety of computer vision applications including shape classification, hand-written character recognition, prototype formation, and morphing. The optimal correspondence is computed by using the concept of an alignment curve and due to the use of intrinsic properties is invariant to rotations and translations and gives the intuitive matches in the presence of visual transformations like viewpoint variation, articulation and occlusions of limited extent.

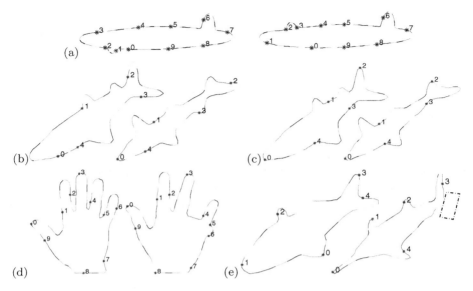

Fig. 7. This figure illustrates the sensitivity of the curve matching to spatial arrangement of parts (a, b, c, d) and to occlusion (e). Top row: We perceive two ellipses with protrusions. The larger protrusion is matched correctly, as it lies on the same side of the ellipse. However, the smaller protrusion is matched incorrectly as it lies on opposite sides of the ellipse. Middle row: The correspondence of the fishes in (b) is incorrect, as a fin on the fish on the left is matched to the head of the fish on the right. This incorrect match is because there is an extra fin in the fish on the left. (c) illustrates that the correspondence is intuitive when the fin on the "correct" side is removed. Bottom row: The missing finger and the small bump of the hand on the right causes the curve matching to give the un-intuitive match (d). (e) shows a case where curve matching gives the un-intuitive correspondence for similar shapes in presence of occlusion. Part of the tail of fish on the right (shown by the box) is occluded in this case.

We have studied the effectiveness of curve matching for shape matching and classification, especially in comparison to our group's work on shock graph-based methods [10,9], and evaluated its strengths and weaknesses. While the full comparison is beyond the scope of this paper, we summarize the main points of differences below [16]. The major advantage of curve matching is its computational efficiency. We have shown that with our proposed curve matching method we can achieve acceptable recognition rates for shape matching even under a range of visual transformations while maintaining computational efficiency. However, we have identified a number of areas where curve matching fails for 2D shape recognition. The first shortcoming of curve-based representation is that they do not represent interior of the shape. Hence, curve matching cannot easily distinguish between some perceptually distinct shapes when the local curve-based features are in conflict with the global shape percept, Figure 6. Another drawback of curve representation and hence curve matching is the sensitivity to the presence and spatial arrangement of parts. Figure 7 shows examples where curve matching gives the un-intuitive correspondence when the parts are arranged differently. Curve matching works well in the presence of occlusion, if it does not affect the overall part structure of the object. When the occlusion adds or deletes a part, curve matching can fail, as shown in Figure 7(e).

In conclusion, curve-based representation is the natural choice in handwritten character recognition and in other applications where the data is inherently one-dimensional. Also, for shape recognition, prototype formation and morphing where the variation in shape does not alter the part structure, curve matching works well. However, in the presence of large scale variations of the outline resulting in changes in the part structure, curve matching can fail, and more comprehensive representations which explicitly represent the shape interior, such as skeletal graphs are necessary despite their relatively higher computational cost.

Acknowledgements. Philip Klein acknowledges the support of NSF Grant CCR-9700146 and Benjamin Kimia acknowledges the support of NSF grants IRI-9700497 and BCS-9980091. We are grateful to Farzin Mokhtarian for providing the fish outlines, Jayashree Subrohmania for the handwritten character data, and Michael Tarr for the "greebles".

References

1. N. Ayache and O. Faugeras. HYPER: A new approach for the recognition and positioning of two-dimensional objects. *PAMI*, 8(1):44–54, January 1986.
2. R. Basri, L. Costa, D. Geiger, and D. Jacobs. Determining the similarity of deformable shapes. *Vision Research*, 38:2365–2385, 1998.
3. S. Belongie and J. Malik. Matching with shape contexts. *CBAIVL*, 2000.
4. I. Cohen, N. Ayache, and P. Sulger. Tracking points on deformable objects using curvature information. *ECCV*, pages 458–466, 1992.
5. S. Connell and A. Jain. Learning prototypes for on-line handwritten digits. *ICPR*, page PRP1, August 1998.
6. Y. Gdalyahu and D. Weinshall. Flexible syntactic matching of curves and its application to automatic hierarchical classification of silhouettes. *PAMI*, 21(12):1312–1328, December 1999.
7. S. Gold and A. Rangarajan. A graduated assignment algorithm for graph matching. *PAMI*, 18(4):377–388, 1996.
8. M. Kass, A. Witkin, and D. Terzopoulos. Snakes: Active contour models. *IJCV*, 1:321–331, 1988.
9. P. Klein, T. Sebastian, and B. Kimia. Shape matching using edit-distance: an implementation. *SODA*, pages 781–790, 2001.
10. P. Klein, S. Tirthapura, D. Sharvit, and B. Kimia. A tree-edit distance algorithm for comparing simple, closed shapes. *SODA*, pages 696–704, 2000.
11. H. Liu and M. Srinath. Partial shape classification using contour matching in distance transformation. *PAMI*, 12(11):1072–1079, November 1990.
12. T. Liu and D. Geiger. Approximate tree matching and shape similarity. *ICCV*, pages 456–462, September 1999.

13. E. Milios and E. Petrakis. Shape retrieval based on dynamic programming. *Trans. Image Proc.*, 9(1):141–146, 2000.
14. M. Pelillo, K. Siddiqi, and S. Zucker. Matching hierarchical structures using association graphs. *PAMI*, 21(11):1105–1120, November 1999.
15. J. Schwartz and M. Sharir. Identification of partially obscured objects in two and three dimensions by matching noisy characteristic curves. *Intl. J. Rob. Res.*, 6(2):29–44, 1987.
16. T. Sebastian, P. Klein, and B. Kimia. Curve matching using alignment curve. Technical Report LEMS 184, LEMS, Brown University, June 2000.
17. D. Sharvit, J. Chan, H. Tek, and B. B. Kimia. Symmetry-based indexing of image databases. *JVCIR*, 9(4):366–380, December 1998.
18. K. Siddiqi, A. Shokoufandeh, S. Dickinson, and S. Zucker. Shock graphs and shape matching. *IJCV*, 35(1):13–32, November 1999.
19. H. Tagare, D. O'Shea, and A. Rangarajan. A geometric correspondence for shape-based non-rigid correspondence. *ICCV*, pages 434–439, 1995.
20. S. Umeyama. Parameterized point pattern matching and its application to recognition of object families. *PAMI*, 15(2):136–144, February 1993.
21. H. Wolfson. On curve matching. *PAMI*, 12(5):483–489, May 1990.
22. L. Younes. Computable elastic distance between shapes. *SIAM J. Appl. Math.*, 58:565–586, 1998.
23. S. C. Zhu and A. L. Yuille. FORMS: A flexible object recognition and modeling system. *IJCV*, 20(3), 1996.

Behind the Image Sequence: The Semantics of Moving Shapes

Guido Tascini, A. Montesanto, R. Palombo, and P. Puliti

Istituto di Informatica, Università di Ancona,
via Brecce Bianche, 60131 Ancona,
tascini@inform.unian.it

Abstract. The paper describes a method for analysing a sequence of images by building static images, representing the environment on which shapes move. From the background and moving objects it is possible to reconstruct the original image sequence as well as to generate new ones. The analysis uses a linguistic interface that allows to express the semantics of video's. Both background and movement analysis allows to extract the shapes contained in a video. The description of video shapes and of their spatio-temporal properties is performed by a Prolog program; so the program using facts describes the Syntax of the video's, while the layout of predicates contains the description of the semantics. Then the content of a 'video-base' may be extracted: the approach uses a prototype film, whose description is used as a dynamic query for automatic extraction of other film semantics.

1 Introduction

Visual representation is characterised by richness of semantics[2] and in particular filmic representation corresponds to long and complex linguistic description. Video's stored in large data bases have problems for retrieving their content[4] . This content is hard to retrieve when the images are static and becomes a challenge in a dynamical context and when the image number is various hundred. In this case the involved information is complex and the access to it is normally hard and slow.In fact various researchers attempted to extract from a sequence of frames the content. The aim of such a research is to associate each film to a global description that constitutes a way to index[15] it in a database. The problem of build an interface for multimedia content description is related to standard MPEG. In fact the Moving Picture Expert Group[13] actually is working to a new standard, named *MPEG-7* [14], which deals with a 'multimedia content description interface'. This standard concerns especially multimedia descriptions, schema of descriptions and description definition languages. From this standard a series of characteristics born like the followings: a description must be apt both to user and application; different people may produce different analysis of the same scene, as well as may give different interpretation of the same event. A description has a proper granularity: too general description results in detail loss, while too detailed one losses important general concepts. Besides some details may interest some peoples and not other one. MPEG-7 document not includes an automatic extraction of descriptions and says nothing about a browser for showing

C. Arcelli et al. (Eds.): IWVF4, LNCS 2059, pp. 619–629, 2001.

these descriptions. The paper focuses on two aspects of the interface. Firstly it takes care of the extraction of a compact representation [7]of a frame series and on a description based on little number of features[16]. Secondly it defines a query language constituted by a set of predicates and facts; these last are derived by a series of features automatically extracted from the frame-sequence.

2   Background-Based Video Analysis

Mosaic technique and movement analysis allow to extract all the Video Objects (VO) contained in a film. The description of VO and of their movements is made through the automatic layout of a Prolog program. So Prolog programs using facts describes the Syntax of the video's, while the layout of predicates contains the description of the semantics. It is possible to refer a dynamic shape to background and to define a mobile point of view that in simulates the TV-Camera movement. A background may be built by putting, side by side, different frames in order to build a panoramic scene [10]. The union of frames allows to reconstruct the environment on which the film was run. The information on a video is redundant because each point of the scene is repeatedly shown in consecutive frames. The data shared by all frames constitute the data of global scene. Mosaicing allows to transform a frame-based representation in a background-based representation. The spatial information and the temporal information constitute the base of analysis. Besides while the visual field of a single image is restricted the camera moving extends the field that will be present in the background.

If the background includes the spatial information the moving objects must be removed from the scene when we are searching for spatial information because they cover the background. Some objects may obscure the background during the the film leaving the mosaic not complete. On the contrary in some frames the cover background is shown and then promptly captured. Similar considerations may be valid for temporal information that is not linked to movement. The camera movement is basic for frame join The assumptions for movement analysis are the following: - input is constituted by a series of frames; - each frame represents an image of the scene at a given instant; - each image is a projection of the scene depending on camera position[19]. - the change in a scene is due to the camera movement and to the displacement of shapes, that are almost rigid. The system reveals the change in a scene and from these detect the movement of the observer and of the shapes. They are three dynamic situations: 1) Stationary Camera, Moving Object - SCMO; 2) Moving Camera, Stationary Object-MCSO; 3) Moving Camera, Moving Object - MCMO; SCMO is valid for surveillance applications, while the other two are more versatile and are used in navigation systems and recently in the video indexing. There is also, in literature, the subdivision of the dynamic scene analysis in two steps [12]: Perceptive; Cognitive.

3 Background Extraction

The background extraction needs four steps: 1)movement detection; 2) movement model definition and computation of dynamic parameters; 3)cancellation of moving objects respect to background; 4)alignment and building of the background.

Movement detection. Movement detection using gradients is accurate only if the difference between frames is a pixel fraction so the implicit Taylor approximation is acceptable. Besides the derivatives become critic in case of fast movement and often it will be necessary pyramidal multi-resolution [1]. Methods based on features[16,18] often depend on type of images. For instance the corner method applied to smooth and circular shapes. In addition often in these method the correlation is used for corrispondence detection [5]. For these reason we have chose for movement detection a correlation method. We consider here the image divided in blocks and the movement is analysed through the correlation between blocks; in particular we use a variant of the BMA[13] algorithm. The correlation here is computed inside a window including the block and this solution overcomes the speed problem: the film speed may increase as the window size increase. Besides the BMA algorithm allows to consider images distant 3 or 4 frames. The reference-blocks are built in greed as in figure 1.

Fig. 1. The reference-blocks. Each point represents the centre of a block.

In this algorithm each point is candidate for being followed and they are not searched point with particular characteristics as in other approach[17,22]. The used search method is the FSA with a similarity method MSE . In the example of figure 1 they are considered blocks of 5x5 pixels and a maximum size of the speed vector equal to 20 points in vertical and in horizontal direction. The movement vector is obtained by searching, with the similarity measure MSE, the current block in a window 20x20 in the reference-frame. The FSA algorithm is computationally simple and requires 40^2 analysis of matching criterion. The MSE measure is computed with the formula:

$$MSE = \sum_{\substack{i=0 \ j=0}}^{M-1, N-1} \left(I_n(k+i, l+j) - I_{n-1}(k+i+u, l+j+v) \right)^2$$

To couple points of two images we will search the minimum of MSE. In some cases the method gives results ambiguous. In general points where the matching is not clear are refused. After the correspondences, the detection of *optical flow* [8], that is the speed **v** of the blocks, is computed by the following formula:

$$v \propto \sqrt{(x_2 - x_1)^2 + (y_2 - y_1)^2}$$

where (x_1, y_1) and $(x_2 y_2)$ are the co-ordinates of the same block in different frames. The optical flow will be represented by vectors departing from the source-frame and with orientation, versus and length capable of connecting the destination block in the destination frame.

Movement Model. The choice of the model implies using pre-defined knowledge, or models[3,20], to obtain the right interpretation. Normally we introduce the affine model[20] with 6 parameters. By considering three points we obtain six equations. The solutions may be influenced by: - the quantisation noise and imperfect-matching noise; - the 2D nature of the affine model; - the presence of autonomously moving objects.

If n are the corresponding points we will write n systems with 2n equations. Unfortunately the previous reported problem produces not sound equations. A solution is obtained by minimising the error between real position in the destination frame and computed position, using movement parameters. If μ is the mean of error distribution and σ the variance, the blocks that obey to the formula: $Err > \mu + \sigma$ will be considered not background blocks. All points with error higher of the previous value will be abandoned in the next calculation. The discarded point will be considered as part of those objects with proper movement, or point with bad value of speed due to accumulation of error. The algorithm detects the dominant movement and the areas with discordant movement. Figure 2 shows the computed optical flow. The circled points are the points whose movement is discord (south-east) with the dominant one (east). After the system computes the difference between actual and preview frame. Since this operation compensate the background movement, the result is constituted only by the discordant points.

Cancellation of objects with autonomous movement. Some methods of object cancellation [11,21] attempt to group frame zones with coherent movements. Here we attempt: - to detect the background, which represents the environment where the scenes are realised; -to detect the objects with movement independent from the background. The difference criterion, for this aim, used by M. Irani, B. Russo e S. Peleg [11], is not enough robust and they are revealed zones without movement. In the method of J.Y.A. Wang and E.H. Adelson [21] the surface are detected, in optical flow images, only by similarity. In the previous images we can see that points are miss-classified as background: for instance some point of the guardrail. Our choice classify as not background blocks detected by both methods: – by dissimilarity of

Fig. 2. The optical flow.

Fig. 3. Masked objects that move respect to background.

movement from that one of tv-camera: – by difference between images, after realignment of actual with preview frame. A final computation concerns the grouping of blocks in a single object: this is obtained by neighbouring and connection criteria: in fact sometime parts belonging the same object have the same movement. We hypothesise that point of the same object are neighbouring or connected : the 8-connection is used. The mosaic construction requires masking the blocks of moving objects, by black zones, as in figure 3. Given two consecutive frames the mosaic will be constituted by the pixels of the second frame; if some pixels are masked we go back to the first frame. If they are still masked pixels, they are supplied by the old mosaic, and so on. This method allows up-to-date information; during the time new zones will be included in the background.

4. Linguistic Semantic Interface

The analysis of image sequences pass trough a linguistic interface that allows to express in simple manner the semantics of the video's. This semantic linguistic interface allows to analyse a reference film and compare it with other films in order to find those with similar semantics. In practice a given film may constitute a video-query on a film DB. The possibility of referring all objects in a scene to a background allows to describe the scenes in terms of movement parameters and in terms of spatial relations.

The extractor of video descriptions. We can now extract a series of features concerning the background and the moving shapes. These features are converted in a database of facts. The video is analysed by extracors that extract some characteristics and describe them by means of facts in a database. Figure 4 shows the schema of an extractor. Prolog-facts represents features of the video and in general they are more facts for the same particular. The set of all facts represents the *syntactic description* of the video. They will be added to the database concerning all analysed video. Some procedures including predicates will specify the *semantics* of relationships among elements derived by the extractor. Particular instances obtained by inferential engine of Prolog will constitute the description of the video.

The choice of Prolog as description language depends on the facility in defining new predicates and on its commercial diffusion. The inferential engine of the language allows in addition interpreting the descriptions. At the end the possibility for the user to define new predicates and new rules allows to introduce personal primitives, interpretations and the right description granularity. Submitting proper queries that in our case assume the aspect of a 'goal' may test the Prolog knowledge base so built. All the 'unification's' that satisfy the posed goal will be the descriptions that is the result of the navigation in the knowledge base. The movement may be checked by distinguishing the displacement of the camera and by separating from the background the shapes with autonomous movement. This separation allows to show characteristics of the detected shapes. For instance we can axaminate the path by computing frame by frame the mass-centre and refer it to co-ordinates of the frame. But if we know the movement of each frame respect to next one we can refer the mass-centre to the panoramic background. For each piece of video the fact descriptor produces a predicate of the type:

video(Number,Name,Frames).
Where number is an integer that identify the video in the data base, name is the name given to the video and frames is the number of frames that constitute the piece. For each frame a descriptor produces facts of the type:

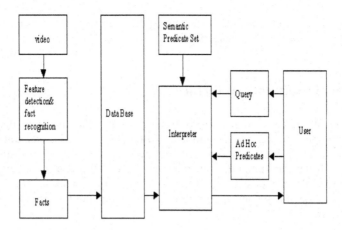

Fig. 4. Extractor of video descriptions

mass-centre (Video,Frame,Shape,X,Y)
where video stand for video under analysis, frame for number of current frame, X and
Y are the co-ordinate of the reference point in the frame. Another predicate describes
the transformation from a frame to another and reconstructs the camera movement:
affine(A,B,C,D,E,F,Frame,Video)
where the letters from A to F allow to identify and reconstruct Frame from its
preceding one. Predicates that describe frame-by-frame the object movement may be
the followings:
fsouth(V,F,S,X,Y,F1,X1,Y1):- mass-centre(V,F,S,X,Y),
            mass-centre (V,F1,S,X1,Y1),
                Y>Y1,F<F1.
fnorth(V,F,S,X,Y,F1,X1,Y1):- mass-centre (V,F,S,X,Y),
 mass-centre (V,F1,S,X1,Y1),
 Y<Y1,F<F1.
The following predicates express the position of a shape respect to the mosaic and the
affine-transformation from mosaic and frame under analysis:
position(S,X,Y,T,V):-mass-centre(Video,Frame,Shape,X,Y),
                video(Video,Name,Mosaic),
                globalaffine(A,B,C,D,E,F,Mosaic,Frame,V),
                X is integer(X1+A+B*X1+C*Y1),
        Y is integer(Y1+D+E*X1+F*Y1).
globalaffine(A,B,C,D,E,F,End,End,V):-A is S, B is S, C is S,
                D is S, E is S, F is S.
globalaffine(A,B,C,D,E,F,End,Start,V):-End\=Start,
            T2 is End-1,
                globalaffine(A1,B1,C1,D1,E1,F1,T2,Start,V),
        affine(A2,B2,C2,D2,E2,F2,T2,V),
        A is A1+A2, B is B1+B2, C is C1+C2,
        D is D1+D2, E is E1+E2, F is F1+F2.
We can now write predicates to describe the movement inside the panoramic image.

south(V,F,S,X,Y,F1,X1,Y1):-position(S,X,Y,F,V),
position(S,X1,Y1,F1,V),
X<X1,T<T1.

The temporal relations of movement may be described by predicate like **come** and **go**:

come(Ogg,Ogg1,F,F1,V):-distance(Obj,Obj1,D,F,V),
distance(Obj,Obj1,D1,F1,V),
F<F1,D>D1.

go(Ogg,Ogg1,F,F1):-distance(Obj,Obj1,D,F),
distance(Obj,Obj1,D1,F1),
F<F1,D<D1.

Other temporal relations are those between events, like before, after, contemporary, that are described by order relations of the type: **F>F1, F<F1, F=F1**.

The predicate path describes in detail, using previous facts, the journey of objects.

The camera movement is described by predicates of the type: **pan** and **tilt**.

Results from queries. The video of the example previous analysed that is named 'car', is composed by 56 frames of 300x200 corresponding to 2 sec. Of vision, sampled each 3 frame. It show a car that run on a road moving high-low and left-right. The camera has a similar movement. In the mass-centre path the numbers represent the frame numbers and give a temporal reference. There is a loop absent in practice. Figure 5 instead shows the same path situated on the mosaic.

Fig. 5. Movement of car respect to background.

The path is more understandable: it proceeds from north to south and from east to west. As we see the different paths depend on different points of view. The facts automatically extracted from the system areof the type:

video(1,car,18).
mass-centre(1,1,1,202,211).
mass-centre(1,2,1,203,212).
...
mass-centre(1,18,1,195.4,238.3).

affine(-1.961,-0.0007111,0.0002773,0.8482,-0.0007373,-0.0003905,18,1).
affine(-2.901,-0.005026,-0.004864,-9.16,-0.01003,-0.0008144,17,1).
...

affine(-2.388,0.003019,-0.0002567,5.572,0.001408,0.002383,2,1).
affine(0,0,0,0,0,0,1,1).
If we pose the goal:
fsud(1,_,_,_,_,1,Frame,1),
we will have the answers: **Frame = 2   Frame = 4,** that means that in frame-intervals
1-2 and 1-4 the car move respect to screen border through south. Differently when
we pose the query:
sud(1,_,_,_,_,1,Frame,1)
   The writing of predicate *move* is strictly related to the semantics and may be
submitted to any database that is for any video. The possible instances of he variable
Movement instead give the description of those video in which satisfy the predicate
move. We can build predicates for searching trough the database. For instance with
the predicate:
move(Object,[south],Start,End,Video).

Dynamic Video Query. The semantic approach, embedded in the Prolog predicates,
may be of course defined by means the consultation of an expert. For instance by
consulting an art director. Then the art director may analyse a film and choose this as
a typical one, that is it choose a film containing some, given, technical properties, like
pan,   point-of-view movement, etc.. The technical content may be automatically
analised by the semantic extractor previous defined. If we have a video base, in which
the set of Prolog predicates and facts, are condensed in a semantic index: that is the
indexing is realised by means of semantic description of the video's, we can use the
chose video as dynamic query. This will automatically extract from the video base all
video with similar semantics.

5. Conclusion

The paper describes a method for analysing a sequence of images by means of a
background on which shapes   move. The   approach reduces the video analysis
complexity and allows to capture the moving objects with their shape and their
dynamic features. The analysis is performed in two directions: 1-detection of spatio-
temporal relationships among objects and between objects and background, 2-
inquiring of knowledge base to discover the video content. The mean is a robust
extraction of a mosaic from a series of images. Then the movement of objects,
independently from the background, and the camera movement are analysed. From
the background and moving objects it is possible to reconstruct the original image
sequence as well as to generate new ones. A linguistic interface allows to express in
simple manner the semantics of the video's. Mosaicing technique and movement
analysis allow to extract all the Video Shapes (VS) contained in a film. The
description of VS and of their movements is made through the automatic layout of a
Prolog program. So Prolog programs using facts describes the Syntax of the video's,
while the layout of predicates contains the description of the semantics. A reference
film may be analysed and compared with other films to find those one with similar
semantics. The linguistic interface, based on a fact-data-base that is automatically

built by the system, constitutes an interesting research sector. In particular the future work has to be appointed on the fact interpretation, on the 'intelligent' video analysis, as well as on the video indexing and retrieval by dynamic properties of shapes and on the automatic knowledge extraction from unknown video's.

References

1. F. Ackermann, M. Hahn: Image Pyramids for Digital Photogrammetry – Digital Photogrammetric Systems – Wichmann Ed.– 43-57, Stuttgart;
2. Anil K. Jain: Fundamentals of Digital image Processing – Cap. 9 Image Analysis and Computer Vision — Ed. Prentice Hall Information and System Sciences Series;
3. R. Bergen, P.J. Burt: A Three-Frame Algorithm for Estimating Two-Component Image Motion – IEEE Transaction on Pattern Analysis and Machine Intelligence –V14 N9 886-896 (1992)
4. M. De Marsicoi, L. Cinque, S. Levialdi: Indexing pictorial documents by their content: a survey of current techniques – Image and Vision Computing –V.15, 119-141 (1997).
5. P.R. Giaccone. G.A. Jones: Segmentation of Global Motion using Temporal Probabilistic Classification — BMVC '98,;
6. E. Gülch : Automatic Extraction of Geometric Features from Digital Imagery — Digital Photogrammetric System – Ebner, Fritsch, Heipke Ed., Wichmann;
7. Harpreet S. Sawhney, Serge Ayer: Compact Representations of Video Through Dominant and Multiple Motion Estimation – IEEE Transactions On Pattern Analisys and Machine Intelligence – Vol. 18, N. 8 (1996).
8. B.K.P. Horn, B.F. Schunck: Determinig Optical Flow — Artificial Intelligence, V17, 185-203 (1981).
9. M. Irani, P. Anadan, J. Bergen, R. Kumar, S. Hsu: Efficient Representations of Video Sequences and Their Applications – Signal Processing: image Comunication, V8 N4 (1996).
10. M. Irani, P. Anandan: Video Indexing Based on Mosaic Representations – Proceedings of IEEE, may 1998;
11. M. Irani, Benny Rousso, Shmuel Peleg: Computing Occluding and Trasparent Motions — International Journal of Computer Vision, V12, 5-16 (1994).
12. Ramesh Jain, Rangachar Kasturi, Brian G. Schunck: Machine Vision - Cap.14 – Dynamic Vision;– Mc Graw Hill Series in Computer Science.
13. Rob Koenen: MPEG-4 Overview - (Melbourne Version) - - International Organisation for Standardisation ISO/IEC JTC1/SC29/WG11 Coding of Moving Pictures and Audio ISO/IEC JTC1/SC29/WG11 N2995 – October 1999;
14. Rob Koenen: MPEG-7: Context, Objectives and Technical Radmap, V.12 - (Vancouver Version) - - International Organisation for Standardisation ISO/IEC JTC1/SC29/WG11 Coding of Moving Pictures and Audio ISO/IEC JTC1/SC29/WG11 N2861 – July 1999;
15. Emile Sahouria : Video Indexing Based on Object Motion –- May 1997 - WWW;
16. J. Shi, C. Tomasi: Good Features to Track — IEEE Conference on Computer Vision and Pattern Recognition – Seattle 1994 June;
17. S. M. Smith: Reviews of Optical Flow, Motion segmentation, Edge finding and Corner Finding — Technical Report TR97SMS1 – Oxford University (1997).
18. T. Tommasini, A. Fusiello, V. Roberto: Robust Feature Tracking — Dipartimento di Matematica e Informatica Università di Udine;
19. R.C. Gonzales, R.E. Woods: Digital Image Processing — Ed. Addison Wesley Pubblishing Company;

20. Newman Sproull: Principles of Interactive Computer Graphics 2° Edition — Computer Science Series;
21. J.Y.A. Wang, E.H. Adelson: Representing Moving Image with Layers – IEEE Transaction on Image Processing, Special Issue : Image Sequence Compression –V3 N5, 625-638 (1994).
22. Han Wang, Michael Brady: Real-time corner detection algorithm for motion estimation — Image and Vision Computing –V13 N9, 695-703 (1995).

Probabilistic Hypothesis Generation for Rapid 3D Object Recognition

June-Ho Yi

School of Electrical and Computer Engineering
Sungkyunkwan University
Suwon 440-746, Korea
email: jhyi@ece.skku.ac.kr

Abstract. A major concern in practical vision systems is how to retrieve the best matched models without exploring all possible object matches. This research presents probabilistic hypothesis generation based on indexing approach for the rapid recognition of three dimensional objects. We have defined the discriminatory power of a feature for a model object is defined in terms of *a posteriori* probability. This measure displays belief that a model appears in the scene after a feature is observed. We compute off-line the discriminatory power of features for model objects from CAD model data using computer graphic techniques. In order to speed up the indexing or selection of correct objects, we generate and verify the object hypotheses for features detected in a scene in the order of the discriminatory power of these features for model objects. Experimental results on synthetic and real range images show the effectiveness of our probabilistic method for hypothesis generation.

Keywords: 3D, object recognition, probabilistic, indexing

1   Introduction

The fundamental issue in model-based recognition is how to rapidly narrow down the number of candidate models without actually searching through all the models. This problem has motivated the use of indexing or hashing for efficient retrieval of correct object model objects. In indexing, the feature correspondence and search of model database are replaced by a table look-up mechanism and this indexing table is computed off-line [1-2]. Recently, there have been some research works based on probabilistic indexing [3-4] where not only correspondence hypotheses but also the probability of each one being a correct interpretation is provided.

Wheeler and Ikeuchi [4] compiled statistical information about image features and object features off-line from a large set of ray-traced images of each object. They represented the likelihood of hypotheses and their inter-dependencies using MRF(Markov random field) to select a set of hypotheses with strong supporting evidence. Their system only considers polyhedral model and does not handle

C. Arcelli et al. (Eds.): IWVF4, LNCS 2059, pp. 630–639, 2001.

situations where only one surface is visible, while our system can handle objects with curved surfaces and single surface view situations. Beis and Lowe [3] also employed a probabilistic approach. They used 4-straight-line-segment chains (three angles and the ration of the interior edge lengths) as indexing vector and trained an indexing function (a linear combination of Gaussian centered on the indexing vectors) from synthetic images taken from various viewpoints. Their indexing vectors can not handle objects with curved edges. From the indexing function computed, they obtain the probability of each hypothesis being a correct interpretation of the data. Performance of these systems cannot be compared directly because they have been developed based on different assumptions and they perform in different scenarios using different features to generate object hypotheses.

We have employed a formal probabilistic solution for efficient indexing of correct model objects using a Bayesian framework. We define a decision-theoretic measure of the discriminatory power of a feature for a model object in terms of *a posteriori* probability. We estimate this measure off-line using computer graphic techniques. This measure allows us to employ salient features of model objects first for object recognition. In our system design, a measure of how well a feature can be detected, called "the detectability of a feature" is defined as a function of the feature itself, the viewpoint, sensor characteristics, and the feature detection algorithm. The detectability of a feature is incorporated into the formulation of the discriminatory power of a feature for a model object by considering model dependent information and sensing dependent information separately based on their conditional independence. In order to speed up the indexing or selection of the correct objects, we generate and verify the object hypotheses for the features detected in the scene, in the order of the discriminatory power of these features for model objects. By considering the object hypotheses in this order, we verify only a few correct hypotheses of the scene objects, resulting in the acceleration of recognition.

The following section gives a brief overview of our vision system. In section 3, we define a decision-theoretic measure of the discriminatory power of a feature for a model object. In section 4, we describe how object features for indexing are automatically compiled using our example feature, LSG (Local Surface Group). Section 5 presents experimental results on the effectiveness of our probabilistic indexing scheme.

2   System Overview

Let us briefly overview the entire object recognition system proposed. The system is divided into two parts. One is off-line compilation of model information and the other is on-line recognition.

The first component is concerned with the automatic computation of object representations from a CAD model database for recognition of model objects. The second component is the range image simulator. One module of this component is for simulating the sensing process to estimate the detectability of features.

The other module renders each model object for a viewpoint sampling. From the rendered images, knowledge of all objects in the domain of interest is compiled. After renderings are done for all model objects, the third component computes *a posteriori* probability that a model object appears in a scene given a detected feature, i.e., the discriminatory power of a feature for a model object. As a result, a feature indexing table is constructed where features are linked to the models with *a posteriori* probabilities. This indexing table is loaded at recognition time.

The on-line process consists of feature extraction, matching, and verification modules. The input to the feature extraction process is a dense range (depth) map from a single viewpoint. The feature extraction module detects features for generating object hypotheses. During the matching phase, features extracted from the scene are indexed by means of the precomputed indexing tables. A set of hypotheses are created and ordered in decreasing order of the probabilities associated with them. We validate these hypotheses applying a series of filters using geometric constrains. Finally, we obtain a list of valid hypotheses that will enter the verification stage. At the verification stage, the valid hypotheses are verified in the order they appear in the list (i.e., in the order of their probability).

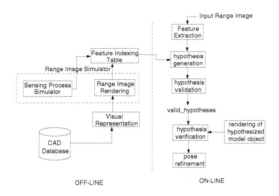

Fig. 1. System Overview

3   Discriminatory Power of a Feature for a Model Object

We exploit the discriminatory power of a feature for a particular model object for efficient indexing of the best matched models. In order to define the discriminatory power of a feature for a model object in terms of *a posteriori* probability, we start with the joint probability $P(m_k, M_i, viewpoint_j)$, $k = 1, \cdots, f$, $i = 1, \cdots, N$, and $j = 1, \cdots, v$ where m_k, M_i, and $viewpoint_j$ denote a feature for indexing, the i-th model object, and the j-th viewpoint of a set of viewpoint samplings, respectively. f, N, and v represent the numbers of features, models, and viewpoints, respectively. This joint probability encodes the information conveyed by a feature of a model object. The same feature may occur in several different models. If a feature to be used is viewpoint independent, we can ignore $viewpoint_j$ in $P(m_k, M_i, viewpoint_j)$.

3.1 Definitions and Notations

$P(m_k, M_i, viewpoint_j)$: joint probability of m_k (a feature for indexing), M_i (i-th model object), and $viewpoint_j$ (j-th viewpoint of a set of viewpoint samplings), where $k = 1, \cdots, f$, $i = 1, \cdots, N$, and $j = 1, \cdots, v$.

$P(M_i)$: The probability that a given object in a scene is M_i. Then we have $\sum_{i=1}^{N} P(M_i) = 1$.

$P(m_k/M_i)$: a likelihood function, that is,
$P(m_k/M_i) > P(m_k/M_j)$ means that the model M_i is more "likely" to be the model object that the feature m_k belongs to than the model M_j, in that m_k would be a more plausible instance of the features of the model M_i than the model M_j.

$P(M_i/m_k)$: This *a posteriori* probability reflects the updated belief that model M_i appears in the scene after the feature m_k is observed.

D_{m_k} : Detectability of a feature m_k. It measures how well a feature m_k can be detected.

Definition :

$P(M_i/m_k)$ is the discriminatory power of the feature m_k for a particular model object M_i.

The detectability of a feature is considered in the computation of the discriminatory power of a feature for a model object. The detectability of a feature, D_{m_k}, depends on the feature m_k itself (i.e. feature class). For example, a vertex feature may be less reliably detectable than a surface feature. D_{m_k} changes as the viewpoint varies. For example, when a planar surface is detected in various viewpoints, it is more difficult to detect in a viewpoint involving a very high sloped appearance of the planar surface than would be the case in a viewpoint giving a flat appearance of the planar surface. The sensor's capability is also important for a feature to be reliably detectable. Finally, the detectability of a feature can vary according to the feature detection algorithm used. Therefore, we represent the detectability of a feature m_k, D_{m_k} as a function of the above four factors:

$$0 \leq D_{m_k} = f(m_k, \text{ viewpoint, sensor, feature detection algorithm}) \leq 1 \quad (1)$$

3.2 Computation of Discriminatory Power

In the following, we will describe how to estimate *a posteriori* probability, $P(M_i/m_k)$. Let us denote estimates of quantities defined in the previous section by a hat above the symbol. $\hat{P}(M_i)$ and $\hat{P}(m_k, viewpoint_j/M_i)$ can be calculated once we know the specific application domain and determine which feature to use. $\hat{P}(M_i)$ can be computed by observing the frequency of the appearance of the model object M_i in the scene and normalizing it by the total number of observations of all models. Once the decision to use feature m_k for the indexing of model objects is made, we compute $\hat{P}(m_k, viewpoint_j/M_i)$ by counting the number of appearances of the feature m_k in model object M_i for the $viewpoint_j$. However, the feature m_k is often not perfectly detectable, i.e., D_{m_k} is not 1.0. To incorporate feature detectability into the computation of the discriminatory power, we consider model dependent information and sensing dependent information separately. That is, we estimate the model dependent information, $\hat{P}(m_k, viewpoint_j/M_i)$, assuming perfect detectability of the feature m_k and incorporate the sensing dependent information by multiplying these two terms as follows:

$$
\hat{P}(m_k, viewpoint_j/M_i) \cdot \hat{D}_{m_k}
$$
$$
= \left(\frac{\text{\# occurrences of the feature } m_k \text{ in } M_i \text{ for } viewpoint_j}{\text{\# occurrences of all features } m_{l,l=1,\cdots,f_m} \text{ in } M_i \text{ for all viewpoints}} \right) \cdot \hat{D}_{m_k}
$$
$$
(2)
$$

This way, feature detectability can be incorporated into the computation of the discriminatory power when CAD model data is used. Therefore, the likelihood $\hat{P}(m_k/M_i)$ and a $posteriori$ probability $\hat{P}(M_i/m_k)$ are computed as

$$
\hat{P}(m_k/M_i) = \sum_{j=1}^{v} \hat{P}(m_k, viewpoint_j/M_i) \cdot \hat{D}_{m_k} \tag{3}
$$

and

$$
\hat{P}(M_i/m_k) = \frac{\hat{P}(M_i) \sum_{j=1}^{v} \hat{P}(m_k, viewpoint_j/M_i) \cdot \hat{D}_{m_k}}{\sum_{i=1}^{N} \hat{P}(M_i) \sum_{j=1}^{v} \hat{P}(m_k, viewpoint_j/M_i) \cdot \hat{D}_{m_k}}. \tag{4}
$$

Note that if a feature m_k does not exist in the model object M_i, $\hat{P}(m_k/M_i) = 0.0$. As previously stated, the same formulation (3) and (4) can be applied to viewpoint independent features by ignoring the $viewpoint_j$ term.

Given a particular feature and viewpoint, estimating D_{m_k} amounts to determining how different feature detection algorithms behave under different sensor characteristics (for example, signal/noise ratio) [5]. For the case of edge detection in which the feature is an edge, the Sobel operator is known to perform better in noisy situation than the Robert's cross. Therefore D_{m_k} for edge features would be higher for the Sobel operator than for the Robert's cross. In fact, it is possible to analytically determine the probability of detecting an edge using either algorithm with a given signal/noise ratio. In the current prototype system, D_{m_k} is assumed a constant.

4   Construction of Indexing Table

In this section, we describe our example feature, LSG, for object hypothesis generation and present how to construct an indexing table.

4.1   Model Features for Object Hypothesis Generation

Our object recognition system can employ a wide class of features for object hypothesis generation as long as the discriminatory power, $P(M_i/m_k)$ can be computed. In the current prototype system, we use the LSG (local surface group). LSG is not a simple feature but a viewpoint dependent feature structure that contains several attributes. Figure 2 shows an example of a LSG that consists of a visible surface patch C_1 and its two adjacent surface patches, P_1 and P_2, that are simultaneously visible for the given viewpoint. Once we know the adjacent surfaces that are simultaneously visible, we access the node attribute set of the attribute-relational graph corresponding to the model object and can extract the information shown in the LSG. The most popular surface types used in computer vision are quadric surfaces because the majority of man-made objects can be modeled by them. Among the quadrics, *planar*, *cylindrical* (*ridge* and *valley*), and *spherical* (*peak* and *pit*) surfaces are supported in the current prototype system. The last entry of each surface patch in the attribute <simultaneously-visible-adjacent-surfaces: < list of surfaces >> is the angle between the surface orientation of the seed surface and the adjacent surface. This angle is only applicable only when two surface types are either planar or cylindrical (*ridge, valley*). Surface orientation is defined for planar surfaces as the direction of surface normal and for cylindrical surfaces as the direction of the axis, respectively. For pairs of other surface types, the angle value is set to NIL which indicates an undefined value. Note that the LSG is a viewpoint dependent feature structure and that the number of LSGs for a viewpoint is theoretically at most the number of visible surface patches for this given viewpoint. LSG can be extended to incorporate other feature attributes such as color and texture information.

Indexing involves a tradeoff between the complexity of the indexing feature and efficiency of indexing using the feature. If the indexing feature is complex (for example, a whole object as an indexing feature), indexing will not be efficient but will result in only a few candidate models. On the other hand, if a simple feature is used for indexing (for example, a surface patch as the indexing feature), indexing will be easy while many model objects are indexed for a feature detected in the scene. We will use a subset of the LSG as an indexing feature because the use of the complete LSG as an indexing feature makes the indexing of model objects complex and computationally expensive. Choosing an indexing feature of optimal complexity in the sense of recognition performance is a topic for further work. We will call our indexing feature "Indexing_LSG". An example of the Indexing_LSG that is used in our current system for the indexing of model objects is shown in Figure 2. In an Indexing_LSG, only the surface type information of simultaneously visible adjacent surface patches and the sum of the angles listed in the LSG are encoded without distinguishing respective adjacent surface patches. Different instances of this Indexing_LSG are the m_k's in section 3.

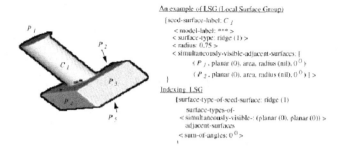

Fig. 2. An example of LSG

4.2   Automatic Compilation of LSGs

A model object is taken from our model database, rendered using z buffer algorithm along with surface labels for a viewpoint sampling, and LSGs are compiled. To obtain a viewpoint sampling, we use the dual of a geodesic polyhedron with frequency Q (default 4) of geodesic division based on icosahedron. This dual polyhedron generates $10Q^2 + 2$ (default 162) viewpoints on a unit sphere. The process of compiling LSGs can be summarized as:

For $i = 1, 2, \cdots, N$
    For $j = 1, 2, \cdots, v$
        Render the range image of the object, M_i, for $viewpoint_j$
        along with surface labels.
        Scan the range image to collect LSGs.
    end for j
end for i
Compute $\hat{P}(M_i/m_k)$'s and return the indexing table.

5   Experimental Results

5.1   Extracting and Ordering Scene LSGs

To compute LSGs from an input scene image, we segment the image first and characterize feature attributes of surface patches such as primitive surface type, area/radius of surface, surface normal direction, and so on.

5.2   Probabilistic Hypothesis Generation

We have experimented our approach using the 20 object model database shown in Figure 3. We have visualized in Figure 4 the distribution of $\hat{P}(M_i/m_k)$. Let us make several comments about the information displayed in Figure 4. If the discriminatory power of a feature (m_k) for a model object (M_j) is 1.0 (i.e. $P(M_j/m_k) = 1.0$), the feature, m_k, is unique to the model object. In other words, if the feature, m_k, is detected in the scene, it is certain that the model M_j is in the scene. On the other hand, suppose that feature, m_2, is detected in the scene. Then, $\left[\hat{P}(M_3/m_2) = 0.425\right] > \left[\hat{P}(M_0/m_2) = 0.401\right] > \left[\hat{P}(M_5/m_2) = 0.174\right]$ indicates the belief that appearance of model objects in the scene is plausible in the order of M_3, M_0 and M_5. Model object, M_3, is hypothesized first and then M_0 and M_5. Similarly, when several features are detected in the scene, object hypotheses are generated in the order of the discriminatory power, $P(M_i/m_k)$.

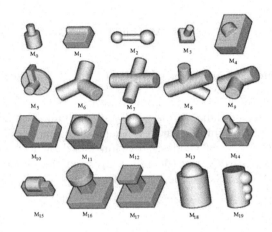

Fig. 3. Model database

5.3   Hypothesis Verification

We verify the hypotheses listed in the *valid_hypotheses*, one by one, in the order in which they appear. In order to verify an object hypothesis, we first find what model surfaces, other than the initially matched model surfaces in the hypothesis, should appear in the scene image. We compute a candidate view of the hypothesized model object from the initially matched pairs of scene and model surfaces in the object hypothesis. We render the hypothesized model object for the computed view and compute a list of neighboring surface pairs. Then the verification routine checks whether the model surface pairs can be found in the list of scene surface pairs. Compatibility between a model surface and the corresponding scene surface is determined based on the geometric constraints such as surface area, radius, and angle between two surfaces.

5.4   Indexing Efficiency

To experimentally determine the effectiveness of our indexing scheme, we define a measure of capability to index correct objects for our technique. We name this measure *indexing-efficiency-measure* and it is defined as:

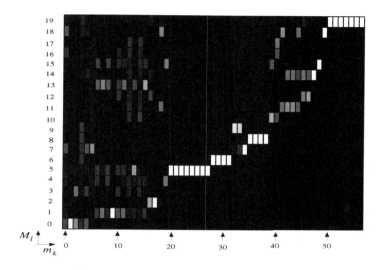

Fig. 4. Distribution of $P(M_i/m_k)$'s for 20 object model database shown in Figure 3. $P(M_i/m_k) = 0.0$ (black) and $P(M_i/m_k) = 1.0$ (white)

Definition:
indexing-efficiency-measure = position of the successfully verified hypothesis in the list of hypotheses initially generated.

We have experimented using a set of synthetic and real range images. A tabular summary of the results is shown in Table 1. We have generated synthetic range images of all objects in the model database for 10 randomly selected views for each object (total of 200 experiments). For real range images, we have built four objects, M_0, M_3, M_5 and M_{15}. 13 range images of these objects for several different poses were scanned. The average value of *indexing-efficiency-measure* was 2.80 and 2.68 for the synthetic and the real range images, respectively. That is, correct hypotheses were located near the third position in the list of hypotheses. This proves the effectiveness of our indexing scheme for the current model database although we adopted a simplified version of a LSG as an Indexing_LSGs. The average number of hypothesis verifications leading to successful recognition was 1.7 for the synthetic range images and 1.8 for the real range images, respectively, because hypothesis validation using geometric constraints served as an extra filter before each hypothesis entered actual verification procedure.

6   Summary and Conclusions

We have proposed a probabilistic method for efficient generation of object hypotheses, based on indexing approach. We achieve rapid recognition by generating the object hypotheses for the features detected in the scene in the order of the discriminatory power of these features for model objects. The discriminatory

Table 1. Experimental results

number of images	recognition accuracy	indexing-efficiency-measure
200 synthetic images	89.0% (178/200)	2.80
13 real images	84.6% (11/13)	2.68

power of an indexing-LSG in favor of an object model is computed off-line by compiling statistics from the rendered images of the model objects in the model database.

We experimentally proved the effectiveness of our indexing scheme using a feature structure called LSG (Local Surface Group) for generating the object hypotheses. The novelty of our approach is in the use of a formal probabilistic solution for efficient indexing of correct model objects, resulting in a speeding up of recognition.

Acknowledgements. This work was supported by grant number 1999-2-515-001-5 from the Basic Research program of the Korea Science and Engineering Foundation.

References

[1] Y. Lamdan, J. Schwartz, and H. Wolfson. Object recognition by affine invariant matching. In *Proc. IEEE Conf. on Computer Vision and Pattern Recognition*, June 1988.

[2] P. Flynn and A. K. Jain. Object recognition using invariant feature indexing of interpretation tables. *CVGIP: Image Understanding*, March 1992. February 1992.

[3] J. Beis and D. Lowe. Learning indexing functions for 3-D model-based object recognition. In *AAAI Workshop*, April 1993.

[4] M. Wheeler and K. Ikeuchi. Sensor modeling, probabilistic hypothesis generation, and robust localization for object recognition. *IEEE Trans. Patt. Anal. Machine Intell.*, 17(3):252–265, March 1995.

[5] T. Kanungo, M. Y. Jaisimha, J. Palmer, and R. M. Haralick. A methodology for quantitative performance evaluation of detection algorithms. *IEEE Trans. Image Processing*, 4(12):1667–1674, December, 1995.

Efficient Shape Description Using NURBS

Djordje Brujic[1], Iain Ainsworth[1], Mihailo Ristic[1], and Vesna Brujic[2]

[1] Imperial College of Science, Technology & Medicine, Dept. of Mechanical
Engineering, London SW7 2BX, United Kingdom
{d.brujic, i.ainsworth, m.ristic}@ic.ac.uk
[2] University of Surrey, School of Mechanical & Materials Engineering,
Surrey GU2 7XH, United Kingdom
mes1vb@surrey.ac.uk

Abstract. In this paper we present an efficient method for smooth surface generation from unorganised points using NURBS. This is a preferred alternative to using triangular meshes, which are expensive to store, transmit, render and are difficult to manipulate. The proposed method does not require triangulation prior to surface fitting because it generates NURBS directly. Two fundamental problems must be addressed to accomplish this task: parameterisation of measured data and overcoming ill-conditioning of the least squares surface fitting. We propose to solve the parameterisation problem by employing a suitable base surface, automatically generated from the data points, or provided as a CAD model if available. Ill-conditioning was solved by introducing additional fitting criteria in the minimisation functional, which constrain the fitted surface in the regions with insufficient number of data points. Surface fitting is performed by treating the surface as a whole without the need to either identify or re-measure the regions with insufficient data. The accuracy of fitting is dictated by the number of control points. The improvements in data compression, shape analysis and rendering are presented. The realised computational speed and the quality of the results were found to be highly encouraging.

1 Introduction

The problem of approximating a surface from a cloud of unordered 3D points appears in many areas including computer vision, computer aided design and object recognition. With the advancements in technology allowing fast digitisation of a large number of points on an object, there is a clear need for new methods that can handle large amounts of data in acceptable time and memory space.

At the same time, in today's industrial practice many products are designed using free-form surfaces. The principal modelling entity is NURBS (Non-Uniform Rational B-spline). Trimmed NURBS are also widely used, because they largely overcome the limitations imposed by the strictly rectangular domain of tensor product surfaces and provide additional flexibility. NURBS are especially suitable for a web applications as they allow data compression and are compatible

C. Arcelli et al. (Eds.): IWVF4, LNCS 2059, pp. 643–653, 2001.

with OpenGL. They are also supported by a large number of CAD programs. It is therefore natural to try to reconstruct measured objects using NURBS and trimmed NURBS.

Terzopoulos and Qin [14] propose dynamic NURBS for scattered data fitting, based on a computational physics framework for shape modelling, in which the fitted surface results from the numerical integration of a set of non-linear equations. However, they report that in their implementation matrix assembly and matrix vector multiplication quickly become too costly, so in practice they are limited to using surfaces of the order of 10 by 10 control points. Also, blended deformable models of Snyder [13] and generative modelling of Ramamoorthi and Arvo [11] might also be seen as tools for surface fitting. They use the measured data cloud with a user defined class of models which are a generalisation of swept surfaces. A shortcoming of this approach is that it is limited in representing local detail.

The main work on surface updating was done by Ma, Kruth and He [6,7] who present methods for least squares fitting of B-splines to unorganised points. It is well known that methods based on least squares fitting have a potential problem with rank deficient matrices, which is a direct result of an insufficient coverage of certain regions by the measurements [1]. The main idea proposed by Ma and He [6] is to avoid the singularity problem by excluding from fitting those control points for which the position is not defined by the data. After applying this solution it is still possible that the system is ill-conditioned due to sparsity of data in some regions, so it was suggested that those regions should be re-measured. However, this is not very practical for shape refinement since fitting, knot insertion and measuring may need to be re-iterated a number of times.

In this paper we present a novel method to generate or update the model based on a set of unorganised measured points in three-dimensional space. Our approach features the following:

- Eliminates the need for pre-processing/triangulation and directly generates NURBS,
- Starts with a CAD model if available, or automatically generates initial surface,
- Solves the problem of ill-conditioning through regularisation, and
- Exploits banded-matrix-based algorithms to gain in computational efficiency

2   Overview of the Algorithm

The algorithm is based on least squares fitting. To define a fitting problem the measurement data set, the degree of the curve or surface to be constructed and the error bound specification are required as input data. It is usually not known in advance how many control points are required to obtain the desired accuracy, hence approximation methods are generally iterative.

Least squares fitting requires parameterisation of the input data. The assignment of u, v parameters is crucial as it has a strong effect on the shape of the fitted surface. A number of methods to parameterise points have been published but the majority of them make the assumption that the data is ordered. Since our work aims to deal with both ordered and unordered data, an alternative method was developed following the suggestion by Ma and Kruth [7], where the parameterisation can be achieved by projecting the points onto a base surface, from which the u_k and v_k values are obtained. The base surface might be seen as the first approximation of the final surface. Consequently, there are three main parts to this algorithm. The first is to initialise the fitting surface and the related parameters, the second to fit the surface to the data points, and the third to insert additional knots if necessary.

2.1   Base Surface

There is no general solution to the problem of how to generate the base surface. Various authors suggest different strategies depending on the complexity of the object. In the approach adopted in this work a CAD model of the object is used as a base surface, if available. Otherwise, we devised an algorithm to generate the base surface automatically within our method, for the case of single-valued data and for the case where the object is of tubular shape. If data is single-valued, such as, for example, a single-view range image of an object, then a single-valued surface may be used to represent the model, in which case a rectangle is automatically generated. If, however, the surface is closed and of a tubular geometry a generalised cylinder is automatically generated. In both cases the initial number of control points may be defined by the user, with uniform knot vectors and weights set to 1. The flexibility of the surface can be further increased by knot refinement during the fitting procedure, where in each step a number of knots is inserted in areas with largest deviation until required accuracy is achieved.

Planar Base Surface Generation. We compute the centre of mass and principal axes for the cloud of data. Measured points are then projected onto a plane defined by the centre of mass and least principal axis. A rectangle containing all projected points is constructed in that plane. Finally, we generate a NURBS surface as bilinear interpolation of the rectangle's edges.

Generalised Cylinder Generation. Using effective shape control techniques for generalised cylinders it is possible to model natural shapes such as trees, arms, legs and bodies.

The description of objects with cylindrical shapes has been extensively used in the computer vision community. A generalised cylinder is obtained by sweeping a 2D cross-section along a trajectory space curve (called spine), in which the 2D cross section may change its shape dynamically while moving along the trajectory curve. In contrast to voxelised data, we start with an unorganised cloud of points sampled on a surface. A 3D curve must be specified inside the cloud of points in order to use a cylindrical model and to represent the surface. Although essential, the issue of finding a suitable axis in order to design a generalised cylinder has rarely been discussed.

We have implemented the method suggested by Nazarian, Chedot and Sequeira [5] where recursive subdivision of the set of data points is used. At each step the subsets of data points are split by a plane perpendicular to their main axes of inertia and passing through their barycenters. The resulting axes is a polyline composed of main axes of parts of the initial set of data points. The authors recognise that their method does not always work. Moreover, in some cases, it is even impossible to find the spine automatically due to the shape and non-uniform density of measured points. For those cases, we have implemented a new method to interactively define the spine. The details of the interactive method are outside the scope of this paper.

Using the spine we generate a constant radius cylinder as a swept surface [9], which is then subject to least squares minimisation, the variables being the positions of control points. We interpret the generalised cylinder as a smooth deformation of a cylinder.

2.2   Iterative NURBS Fitting

Least squares fitting of NURBS through a set of points would lead to a non-linear optimisation problem if the unknowns are the control points, parameters (u, v), knots, and the weights. However, the non-linear nature of the problem can be avoided and the optimisation can be greatly simplified, if the weights and the knot vectors are set *a priori*. We propose using the weights and knot vectors obtained from the initial model and then optimising only the positions of control points.

By denoting the measured points as $Q_1, ..., Q_M$, we set up a linear least squares problem for the unknown control points. The functional to be minimised is:

$$f = \sum_{k=1}^{M} \left| Q_k - \mathbf{S}(u_k, v_k) \right|^2 \tag{1}$$

where u_k and v_k are the parametric co-ordinates corresponding to the closest point to each measured point.

The new position of the N control points are obtained as a solution to the system of normal equations [1]:

$$\mathbf{A}^T \mathbf{A} \mathbf{a} = \mathbf{A}^T \mathbf{b} \tag{2}$$

where $\mathbf{A}$ is an $M \times N$ matrix of B-spline coefficients corresponding to the closest points on the base surface for all measured points, $\mathbf{a}$ is the vector of control points and $\mathbf{b}$ is the vector of measured points [9].

Normal equations can be solved using a number of methods. We have implemented the iterative method of Gauss-Seidel, as well as Cholesky decomposition.

Rank Deficiency and Ill-Conditioning of the Least Squares Problem.
The matrix $\mathbf{A}^T\mathbf{A}$ in the normal equations is prone to rank deficiency. As the size of the system increases it is also likely that the set of equations becomes ill-conditioned. The problem of rank deficiency arises from the localised nature of the basis functions and its detailed presentation is provided by Dierckx [2]. The main reasons for this are:

- Incomplete data sets due to inaccessibility (most often close to the edges)
- Incomplete data sets due to the use of trimmed NURBS in modelling
- Knot insertion

Fig. 1. Surface fitting and knot insertion. (a) Initial surface and unevenly distributed measurement cloud; (b) low flexibility fitted surface; (c) high flexibility fitted surface

As an illustration, Fig. 1 presents an example in which a base surface is updated using unevenly distributed measurements (Fig. 1a). The results of surface fitting (Fig. 1b) show that the available number of control points provide insufficient flexibility for the surface. This is a clear case for employing knot insertion which provides additional flexibility (Fig. 1c). Consequently, fitting is significantly improved in the central area, but this leads to a situation where empty knot segments start to appear in the sparsely measured outer regions, causing ill-conditioning of matrix $\mathbf{A}^T\mathbf{A}$ and corruption of the shape.

Proposed Solution: Regularisation. If the problem is ill-posed there is no way to overcome this unless additional *a priori* information about the solution is available. The majority of authors use smoothness as such additional criterion to restore stability and construct efficient numerical algorithms, but this choice proves costly in terms of computational speed and memory space. We have adopted a different, novel approach introducing new constraints, which proved computationally highly efficient.

In developing the solution for the regularisation problem, it was noted that when the system becomes unstable, the control points associated with the areas with insufficient data move in an uncontrollable fashion, away from the surface. Our principal concept is based on the fact that control points do approximate the surface and it seems natural to keep them as close to the surface as possible. This is achieved by introducing an additional criterion ("α-criterion"), which is to minimise the sum of the squared distances between the control points and their corresponding points on the fitted surface. We expected this criterion to smoothen the surface and to provide an equivalent of energy minimisation.

Mathematically, this is reflected in expanding the functional of Equation (1) to include an additional "α-criterion" term, as follows:

$$f = \sum_{k=1}^{M} \left| Q_k - \mathbf{S}(u_k, v_k) \right|^2 + \alpha \sum_{i=0}^{n} \sum_{j=0}^{m} \left| \mathbf{P}_{i,j} - \mathbf{S}(u_{i,j}, v_{i,j}) \right|^2 \tag{3}$$

where $\mathbf{P}_{i,j}$ are the control points and $\mathbf{S}(u_{i,j}, v_{i,j})$ are their corresponding points on the surface, while coefficient $\alpha \geq 0$ provides the required trade-off flexibility. Naturally, the question arises as to how to define the corresponding surface points. We adopted a solution using Greville abscissae [3,4], because they are obtained using knot averaging and therefore provide high regularity of a matrix.

After implementing the concept, extensive experiments have been conducted with a variety of shapes and the inclusion of α-criterion proved highly beneficial, particularly in case of large deformations of the initial model.

The use of Greville points improves conditioning of the system but does not guarantee uniqueness of the solution. Thus, in the knowledge that matrix $\mathbf{A}^T \mathbf{A}$ is possibly degenerate, a new criterion and a corresponding weight need to be introduced in order to set up a well-posed problem. We labelled it "β-criterion" and introduced it with the aim to limit the displacement of the control points relative to their original positions. This constraint minimises the combined movement of all control points [9].

The minimisation problem still remains linear and the cost function to be minimised now becomes:

$$f = \sum_{k=1}^{M} \left| Q_k - \mathbf{S}(u_k, v_k) \right|^2 + \alpha \sum_{i=0}^{n} \sum_{j=0}^{m} \left| \mathbf{P}_{i,j} - \mathbf{S}(u_{i,j}, v_{i,j}) \right|^2 + \beta \sum_{i=0}^{n} \sum_{j=0}^{m} \left| \mathbf{P}_{i,j} - \mathbf{P}_{i,j}^o \right|^2$$
$$\tag{4}$$

where $\mathbf{P}_{i,j}^o$ is the original position of the control point $\mathbf{P}_{i,j}$, and $\beta \geq 0$ is a weighting factor.

A necessary condition for a minimum is that the gradient of f equals to zero, which gives the following generalised normal equations:

$$\left[\mathbf{A}^T \mathbf{A} + \mathbf{a} \left(\mathbf{B} - \mathbf{I} \right)^T \left(\mathbf{B} - \mathbf{I} \right) + \beta \mathbf{I} \right] \mathbf{a} = \mathbf{A}^T \mathbf{b} + \beta \mathbf{a}^o \tag{5}$$

where $\mathbf{B}$ is the matrix of B-spline coefficients corresponding to Greville points, I is a unity matrix and $\mathbf{a}^o$ is the vector of original control points.

The system of equations (5) is solved using the same methods as those used for solving (2). The only additional task is to compute $(\mathbf{B-I})^T (\mathbf{B-I})$, the computational cost of which is negligible since, in general, the number of control points is far fewer than the number of measured points.

It is worth noting that the two parts of the minimisation will have comparable weights [10] by choosing:

$$\alpha = \frac{\text{Tr}(\mathbf{A}^T\mathbf{A})}{\text{Tr}((\mathbf{B\text{-}I})^T(\mathbf{B\text{-}I}))} \tag{6}$$

In the proposed regularisation method, the quality of the fitted surface may be controlled by a trade-off between the weights of α and β. It is therefore instructive to analyse the fitting results for different values of these parameters.

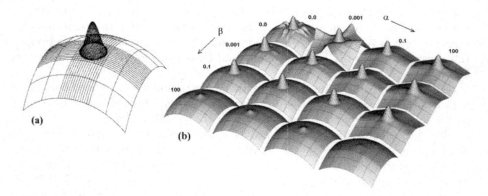

Fig. 2. (a) Measured data and the base surface; (b) Effect of varying α and β on the fitting results

In Fig. 2 the updating of the model is presented. Note that the data is available only in the small part of the object (cone-like deformation in the centre). The parameter values in Fig. 2b represent relative weights according to Equation (6). From these results the following conclusions may be drawn: Sufficiently small α and β have negligible effect in the regions with enough data but stabilise the surface in regions with no data. In particular, small positive β will introduce the effect of removing instability from a system. Increasing α or β will ultimately begin to affect even the regions covered by the data, which is undesirable.

Apart from removing instability, the β-criterion also acts to preserve the original shape. It is worth noting that this effect is unwanted on the measured regions and can be effectively eliminated by iterative fitting.

The α-criterion has the effect of flattening the regions unpopulated, or sparsely populated by the data points.

In all our experiments we were using $\alpha = 0.1$ and $\beta = 1e - 9$. The users are, however, encouraged to make their own selection of weights to control the shape of the unmeasured regions. This involves setting a balance between maintaining the original shape of the unmeasured regions on the one hand, and reducing the energy contained in them, on the other.

2.3   Surface Trimming

In Fig. 3b it is shown that the generated surface contains superfluous areas that do not correspond to the measured cloud. These are due to the rectangular shape of the initial surface. Hence, these areas must be trimmed and an appropriate automatic trimming routine has been developed. The procedure is as follows:

- For each measured point find the closest point on the fitted surface
- Find the convex hull of these points using Graham's algorithm [8]
- Fit a trimming curve through the convex hull in a parametric space

2.4   Computational Efficiency

Fitting complex surfaces through large measurement data sets can be expensive in terms of both computational time and memory, a significant proportion of which can be attributed to the matrix multiplication $\mathbf{A}^T\mathbf{A}$. The time and memory requirements can be drastically reduced by exploiting the sparse and banded nature of $\mathbf{A}$. This means that only non-zero elements of $\mathbf{A}$ are stored and directly multiplied. As a result it is possible to reduce both the number of computations and the memory requirements.

The key features of our technique can be summarised as follows:

- $\mathbf{A}^T\mathbf{A}$ is computed without storing $\mathbf{A}$ or $\mathbf{A}^T$ so that the memory space needed is linear with a number of control points.
- Time to compute $\mathbf{A}^T\mathbf{A}$ is linear with the number of measured points and does not depend on a number of control points.
- Time to solve $\mathbf{A}^T\mathbf{A}$ is linear with the number of control points and does not depend on the number of measured points.

A full discussion of the algorithm implementation is beyond the scope of this paper, and it will be a subject of a separate publication. Nevertheless, the results in Table 1 are included in order to provide an indication of the performance in terms of computational speed and memory requirements (Pentium III, 400MHz).

Table 1. Computational performance with regards to time and memory

Number of data points	Number of Control Points 10	100	1,000	10,000
100	0.03s 42.6KB	0.09s 146KB	-	-
1,000	0.08s 42.6KB	0.14s 146KB	0.78s 1054KB	-
10,000	0.62s 42.6KB	0.69s 146KB	1.39s 1054KB	7.2s 9126KB
100,000	6.17s 42.6KB	6.20s 146KB	6.97s 1054KB	13s 9126KB
1,000,000	61.8s 42.6KB	61.5s 146KB	63.0s 1054KB	69s 9126KB

3   Results

The reconstruction of the Igea artefact (Fig. 3) demonstrates the quality of the results obtained using the rectangular base surface. Courtesy of University of Thessalonica, measured data is available from www.cyberware.com. The measured cloud and the automatically generated NURBS rectangle are visualised in the Fig. 3a. The fitted and trimmed surfaces are presented in the Figs. 3b–c.

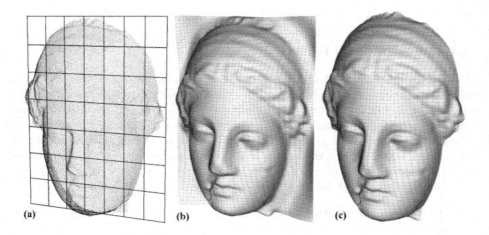

(a)   (b)   (c)

Fig. 3. Igea artefact fitting stages using planar base surface

The reconstruction of the scanned knee, Fig. 4 demonstrates the quality of the results obtained using the cylindrical base. Data and base cylinder are presented in Fig. 4a. The smoothing effects of the fitting can be seen when comparing the NURBS fitted surface of Fig. 4b with the triangulation (courtesy of Cyberware) in Fig. 4c.

In order to demonstrate the quality of the results obtained using CAD model, we present a car windscreen involving a trimmed NURBS surface as shown in Fig. 5. The grey area of Fig. 5a represents the digitised region corresponding to the untrimmed surface, whilst the isoparametric curves show the base surface. The modified shape after fitting is shown in Fig. 5b.

4   Conclusions

The paper has presented a new method for modelling the shape of an object from a cloud of digitised points using NURBS. No special assumptions are made about the data distribution. Least squares fitting, which is the basis of the method, frequently suffers from the problems of rank deficient and ill-conditioned matrices. This was overcome by the regularisation of the least squares problem

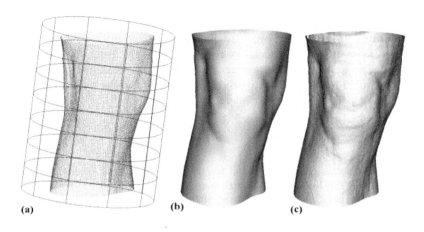

Fig. 4. (a) Measured points and base surface, (b) fitted surface, (c) triangulated points

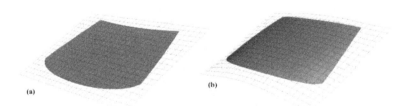

Fig. 5. Fitting of a trimmed NURBS CAD model surface

through adoption of additional criteria in the minimising functional, producing good results in both the measured and the unmeasured surface regions. The method requires an initial approximation of the object shape and this may be generated automatically, from the point cloud. Alternatively, in industrial applications, the initial shape may be provided by a pre-defined CAD model. The implemented algorithms utilise the sparse structure of the matrices and achieve a considerable improvement over previously reported implementations in terms of computational speed and memory requirements. The method was successfully applied to a wide range of shapes, art objects as well as engineering parts, leading us to conclude that the proposed approach is feasible and attractive for a wide range of potential applications.

References

1. Bjorck, A Numerical methods for least squares problems, Society for Industrial and Applied Mathematics, Philadelphia (1996).
2. Dierckx, P Curve and surface fitting with splines, Oxford, Clarendon (1993).
3. Farin, G E Curves and surfaces for computer-aided geometric design: a practical guide, 4th ed., Academic Press, San Diego; London (1997).
4. Gordon, W and Riesenfield,. R 'B-spline curves and surfaces' Computer Aided Geometric Design ed. Barnhill and Riesefield, Academic Press, New York (1974) 95–125.
5. Nazarian, B, Chedor, C and Sequeira, J 'Automatic reconstruction of irregular tubular structures using generalised cylinders' MICAD 96 - Revue Internationale de CFAO at d'Infographie 11:11-20
6. Ma, W and He, P R 'B-spline surface local updating with unorganised points' Computer Aided Design 30 (11) (1998) 853–862.
7. Ma, W and Kruth, J P 'Parameterization of randomly measured points for least squares fitting of B-spline curves and surfaces' Computer-Aided Design 27 (9) (1995) 663–675.
8. O'Rourke, J Computational geometry in C, Cambridge University Press (1998).
9. Piegl, L A and Tiller, W The NURBS book, 2nd edn. Springer (1997).
10. Press, W H, Teukolsky, S A, Vetterling W T and Flannery, B P Numerical Recipes in C : The Art of Scientific Computing, 2nd edn, Cambridge University Press (1993).
11. Ramamoorthi, R. and Arvo, J 'Creating Generative models from Range Images' Computer Graphics proceedings (1999) 195–204.
12. Ristic, M, Brujic, D and Ainsworth, I. 'Precision Reconstruction of Manufactured Free-Form Components' 12th Annual International Symposium SPIE Electronic Imaging 2000, San Jose, California (2000).
13. Snyder, J and Kajiya, J 'Generative Modelling: A symbolic system for geometric modelling' Computer Graphics (SIGGRAPH 92 proceedings) (1992) 369–378.
14. Terzopoulos, D and Qin, H 'Dynamic NURBS with Geometric Constraints for interactive Sculpting' ACM Transactions on Graphics, 13 (2) (1994) 103–136.

Non-manifold Multi-tessellation: From Meshes to Iconic Representations of Objects

Leila De Floriani, Paola Magillo, Franco Morando, and Enrico Puppo

Department of Computer and Information Sciences (DISI),
University of Genova, Via Dodecaneso 35, 16146 Genova - Italy
{deflo,magillo,morando,puppo}@disi.unige.it
http://www.disi.unige.it/research/Geometric_modeling/

Abstract. This paper describes preliminary research work aimed at obtaining a multi-level iconic representation of 3D objects from geometric meshes. A single-level iconic model describes an object through parts of different dimensions connected to form a hypergraph. The multi-level iconic model, called Non-manifold Multi-Tessellation, incorporates decompositions of an object into parts at different levels of abstraction, and permits to refine an iconic representation selectively.

1 Introduction

Polygonal meshes, in particular triangle meshes, are widely used representations of three-dimensional shapes in computer graphics, virtual reality, and simulation. As devices and systems for 3D object reconstruction become more and more common and reliable [3], meshes increase their relevance in applications. For instance, meshes are a suitable input to *model databases* [11,16] within systems for generic 3D shape recognition and classification.

Triangle meshes can approximate arbitrarily well the shape of an object, but they do not provide information on either its structure, or its morphological features. On the contrary, *iconic models*, intended as concise, part-based representations of an object, provide more structured descriptions, even if sometimes less accurate, thus giving a valid support for many application tasks.

This paper describes key ideas and some preliminary results of our ongoing work on multiresolution iconic representations of 3D objects. We consider an object initially described as a triangle mesh, and we devise its progressive decomposition into parts, of different dimensions, that leads to an iconic representation of the object. This approach reflects the intuitive idea that some parts of an object can be perceived as lower-dimensional, depending on the level of abstraction Each part of an object is represented as a geometric complex of the proper dimension and is characterized by some geometrical and topological shape features. The different parts arising from the decomposition are connected at non-manifold junctions.

An iconic model can be provided at different levels of abstraction, depending on the number and dimension of its parts, and on their connection structure. We

C. Arcelli et al. (Eds.): IWVF4, LNCS 2059, pp. 654–664, 2001.
© Springer-Verlag Berlin Heidelberg 2001

propose next a model that can encode a whole range of levels of detail, allowing us to refine the representation locally, in those parts of the object that are more interesting for a given application.

In this paper, we introduce the single- and multi-level iconic models, and we address their design, and related data structures. More specific issues concerning their construction, and applications of such model are just briefly outlined, as they are and will be the subject of our current and future research.

1.1   Related Work

There exists a large body of literature on the segmentation and representation of objects based on part decomposition (see, e.g., [1] for a fairly complete set of references. Typical approaches are based on a set of elementary shapes (generalized cylinders, geons, superquadrics) [13,15,5]. Siddiqi and Kimia [17] developed a framework for partitioning schemes for decomposing 2D shapes and presented a hierarchical scheme which combines a boundary-based and a part-based approach. There exist, however, very few proposals working on mesh-based representations. An example is the recent work by Cutzu [4], who uses a triangular mesh as object representation in finding a part decomposition that takes into account the perceptual similarities among several of the object views.

A method for subdividing non-manifold geometric complexes into manifold parts has been proposed in [9] for an application to triangle mesh compression for transmission. Their approach is somehow similar to our method for identifying manifold components in a non-manifold complex.

Mesh simplification is popular in computer graphics to produce descriptions of objects at different levels of detail and several proposals exist in recent literature (see, e.g., [8] for a survey). Many methods are based on *edge collapse*, i.e., on a local operator which contracts an edge to a vertex.

Multiresolution models based on meshes have also been extensively studied in computer graphics in order to provide compact ways of describing several levels of details in a single data structure [8]. However, existing models are mainly oriented to visualization. A general multiresolution model for representing 3D shapes described by triangle meshes was proposed in [6]. Such model was, however, restricted to describe manifold shapes. The model we propose in this paper is somehow inspired to that work.

1.2   Preliminaries

In the following, we briefly and informally review some standard concepts of algebraic topology. See, e.g., [2] for a more formal treatment.

Simplices of dimension 0, 1, and 2 are points (vertices), straight-line segments (edges), and triangles, respectively. A *simplicial complex* Σ is a set of *simplices* such that no two simplices intersect, except when either a simplex is a facet of another simplex of higher dimension, or two simplices of the same dimension share some facet. The *carrier* of a simplicial complex is the subset of space covered by the union of all its simplices. A complex having a uniformly dimensional carrier is called a *mesh*.

The *star* σ^* of a simplex $\sigma \in \Sigma$ is the set of all simplices for which σ is a facet. A vertex v said to be a *manifold* vertex if and only if either v^* contains no triangle and at most two edges, or it consists of a single fan of triangles (i.e., an either open or cyclic sequence of adjacent triangles all sharing one vertex). An edge e is said to be a *manifold* edge if and only if its star e^* contains at most two triangles. A simplicial complex Σ is said to be a *manifold* complex if and only if all its vertices and edges are manifold.

The boundary of a k-dimensional manifold mesh is the union of all its $(k-1)$-simplices that contain less than two k-simplexes in their star. If the boundary is empty, then the mesh is said *without boundary*. The carrier of two-dimensional manifold mesh without/with boundary is a closed/open surface; the carrier of a one-dimensional manifold mesh without/with boundary is a closed/open line.

A two-manifold mesh is characterized, from a topological point of view, by the number of boundaries (more precisely, the number of connected components of its boundary), and the topological type of the surface obtained by closing each boundary with a disc.

2   An Iconic Object Model

Non-manifold, mixed-dimensional simplicial complexes are suitable to represent the shape of a 3D object as an aggregation of parts, where non-manifold edges and vertices represent junctions among different parts. In order to identify and explicitly represent the parts, we decompose a complex into maximal manifold components, each having a certain dimension, and we organize such components into a hypergraph representing their assembly structure.

In the following, we give a constructive definition of the *iconic model*. We refer to Fig. 1 as running example. Let Σ be a non-manifold, mixed-dimensional simplicial complex. First, we find all non-manifold edges of Σ, i.e. those edges having three or more triangles in their star. We replace each such edge e with as many copies as there are triangles in e^* (in Fig. 1b, edges v_1v_2 and v_2v_3 are replaced with three copies each). Then, we find all non-manifold vertices, by checking the star of each vertex. For each such vertex v, we decompose v^* into maximal subsets, such that each subset is either a single edge or a fan of triangles. We replace v with as many copies of v as there are maximal subsets in v^* (see Fig. 1c). Note that the replication of vertices and edges is a purely topological operation since each copy maintains its position.

The above process decomposes Σ into a set of parts $\{\Gamma_1 \ldots \Gamma_m\}$, where each part is either a one- or a two-manifold simplicial complex. We call these parts the *manifold components* of Σ. Each component Γ_i is characterized by its dimension, number of boundaries and, if two-dimensional, by its orientability and genus.

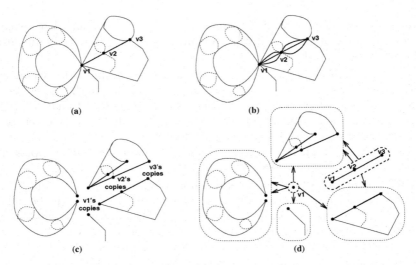

Fig. 1. (a) A complex with three non-manifold vertices v_1, v_2, v_3 and two non-manifold edges v_1v_2 and v_2v_3. (b) Splitting non-manifold edges. (c) Splitting non-manifold vertices. (d) The iconic hypergraph.

The iconic model we propose consists of a hypergraph, that we call the *Iconic Hypergraph* (IH). Its nodes are the manifold components of Σ, and its hyperarcs correspond to (groups of) simplices that connect different components.

The hyperarcs of the IH are obtained as follows. We initially set one hyperarc for each non-manifold vertex v and for each non-manifold edge e of Σ, where the hyperarc connects all components that contain a copy of v and e, respectively. Two hyperarcs connecting the same manifold components are said to be *similar*. Each maximal set of similar hyperarcs, such that the union of their corresponding simplices is connected, is then replaced with a single hyperarc. Now, the set of simplices associated with each hyperarc is either a 0-dimensional complex (i.e., a single vertex), or a 1-dimensional complex which is not necessarily manifold. In the latter case, we decompose such 1-dimensional complex into manifold parts (in the same way as we did for Σ), and we replace its corresponding hyperarc with as many hyperarcs as there are manifold parts. We call *junction* the simplicial complex associated with a hyperarc since it represents a point or line where two or more components touch each other.

The iconic hypergraph for the mesh of Figure 1c is depicted in Figure 1d. For the sake of simplicity, we will draw manifold components through graphical symbols, namely blobs, sticks, and dots, depending on dimension. Hyperarcs will correspond to points where different symbols meet (see Figure 3).

The input complex Σ is encoded through a standard data structure for representing non-manifold, mixed-dimensional complexes [18], which provides information sufficient to run the construction procedure efficiently. Each component Γ_i (node of the IH) is encoded in a standard data structure for manifold triangular meshes, in which every triangle is described by its three vertices and is linked to its three adjacent triangles. For each component Γ_i we also store synthetic information (dimension, number of boundaries, orientability, genus), and the list of its incident hyperarcs. A hyperarc stores pointers to the nodes it connects, a representation of the associated junction (0- or 1-dimensional complex), and synthetic information for it. Further details on the definition and construction of the IH, as well as on data structures that we use to encode it are given in [12].

3 Non-manifold Simplification

Initially, a two-manifold triangle mesh Σ_0 describing an object is given. The iconic model associated with such mesh is trivial and consists of just one node. We introduce *non-manifold simplification* as a way to modify Σ_0 by contracting the dimension of some of its parts, thus decomposing the object into parts. Non-manifold simplification is an iterative and progressive process, which incrementally produces local changes on a current mesh (initialized with Σ_0). The iconic hypergraph associated with the current mesh is affected as a consequence.

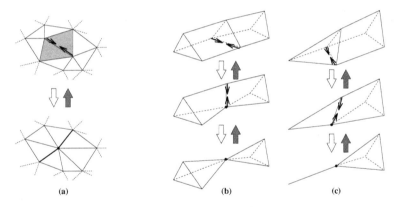

Fig. 2. Transformations of a simplicial complex through edge collapses (white arrows) and vertex splits (dark arrows).

Non-manifold simplification is based on the iterative application of a local modification operator, called *edge collapse:* an edge $e = v_1v_2$ shrinks to one point v along its length; each triangle in the star of e collapses to an edge, while the other simplices in v_1^* and v_2^* are deformed into simplices incident in v. Figure 2a shows an example of an edge collapse in a manifold mesh.

An edge collapse can produce non-manifold configurations and thus it modifies the iconic model corresponding to the current mesh. In Figure 2b, two successive collapses create first a non-manifold edge and then a non-manifold vertex; in Figure 2c, we have a similar situation in which, in addition, a part of the mesh becomes one-dimensional. Collapses, that lead to new non-manifold configurations, can be detected by performing a local analysis of the portion of complex modified by the operation [7].

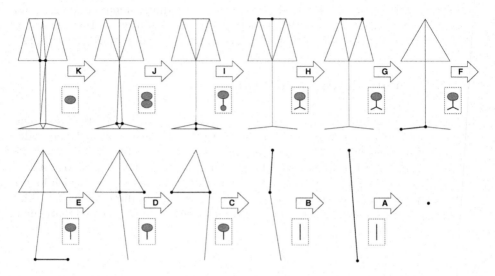

Fig. 3. A sequence of edge collapses simplifying an initial mesh into a single point. For simplicity, a 2D shape is considered. Arrows denote edge collapses. The small icon made of blobs and sticks depicts the iconic model associated with each simplicial complex in the sequence.

A sequence of edge collapses may lead to a representation in which the different parts of an object are described by subcomplexes of different dimension. Figure 3 shows a two-dimensional example of a sequence of edge collapses applied to a mesh representing a lamp. For instance, collapse K identifies the cap and the base as distinct parts; collapse J further splits the base by revealing the one-dimensional nature of the stem; etc.; eventually, the whole shape is reduced to a single point.

On the other hand, not all sequences are suitable to produce an iconic model. Figure 4 shows an example of a "bad" sequence. Our aim is to find sequences that can identify *meaningful* parts during the simplification process. This can be achieved by finding a suitable *cost function* that assigns to each edge e of the current complex a non-negative value, i.e., the estimated cost of collapsing e. At each step, the edge of minimum cost is collapsed.

The cost function must consider a weighted combination of parameters measuring the changes in shape geometry and topology caused by an edge collapse. Cost functions proposed in the literature are aimed essentially at rendering and take into account only geometry. Examples are: length of the edge to be collapsed, variation in surface area, variation in differential properties, smoothness, preservation or destruction of sharp features. Simplification of mixed dimensional

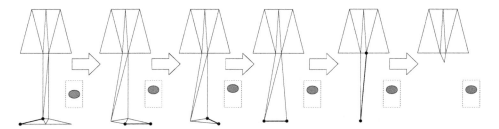

Fig. 4. A simplification sequence that is not able to reveal the part-based structure: the base of the lamp is merged into the cap, instead of being identified as a separate part.

complexes must take into account also changes in topology and in the structure of the iconic model. Exemples are: creation or elimination of non-manifold entities, generation of lower-dimensional parts, etc. Such issues are the subject of our ongoing research. We remark that, however, the whole simplification scheme is completely parametric over the cost function adopted.

4   Non-manifold Multi-tessellation

In this section, we introduce a multi-level iconic model, called *Non-manifold Multi-Tessellation (NMT)*. This model is obtained by building on top of the sequence of edge collapses $[c_1, c_2, \ldots, c_m]$ described in the previous section. The basic idea of organizing simplification steps in a partial order is inherited from our previous multi-resolution model for the manifold case [6].

We define *top simplices* in a complex Σ those simplices of Σ that have an empty star (i.e., they are not facets of other simplices). In the following, Σ is characterized just as the collection of its top simplices, all other simplices being implicitly represented as facets. We see the collapse of an edge $e = v_1 v_2$ into a vertex v as an operation that removes the top simplices of v_1^* and v_2^*, and replaces them with the top simplices of v^*).

We define a *partial order* over the set of edge collapses $\{c_1, c_2, \ldots, c_m\}$ as the transitive closure of the following *dependency relation*: a collapse c_i *depends on* another collapse c_j if and only if c_i creates some top simplices removed by c_j. The partially ordered set of edge collapses is represented graphically as a Directed Acyclic Graph (DAG) where the nodes are the collapses, and the arcs denote dependency links. Figure 5a shows the DAG corresponding to the simplification sequence of Figure 3.

The above DAG encodes resolution changes at the granularity level of a single edge collapse, which is eccessively fine-grained for our purpose. Indeed, we wish to consider larger resolution changes, corresponding to sets of edge collapses that produce *meaningful modifications* in the underlying iconic model.

We assume to have some criterion to select a subset $R \subset \{c_1, \ldots, c_m\}$ of *relevant edge collapses*. We group the DAG nodes into clusters, each cluster

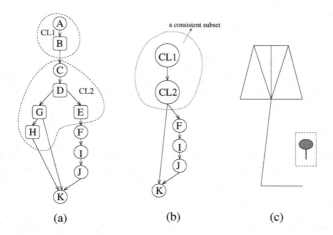

Fig. 5. (a) The DAG representing the partial order among the vertex splits of Figure 3; nodes that change the iconic model are drawn round; dotted circles enclose nodes forming one cluster. (b) The resulting NMT after clustering; a consistent subset of nodes is highlighted. (c) The mesh and iconic model associated with such consistent subset.

containing one relevant node c_i and all the non-relevant nodes lying on paths that start from c_i and end either at another relevant node, or at the common descendant of c_i and another relevant node. For each cluster, we form a *macronode*, and call *Non-manifold Multi-Tesselation* (NMT) the DAG resulting from clustering. Each node of the NMT corresponds to several collapses which are seen as a single, atomic modification of the simplicial complex and its iconic model.

As an example, we may assume that any collapse modifying the topology of the complex is relevant (see Figure 5). Note, however, that this may be still not enough selective, since it might encode several intermediate step that do not necessarily correspond to meaningful modifications. We are currently investigating several definitions of a relevant edge collapse, and related clustering techniques. Note, however, that the above definition of NMT is completely parametric over the definition of a relevant edge collapse.

The NMT can be used for obtaining iconic models at various levels of abstraction, possibly different over the various object parts. Each such iconic model corresponds to selecting a subset S of nodes of the NMT and simplifying Σ_0 through all the edge collapses that are not contained in such nodes. The selected set of macro-nodes must, however, be consistent with the partial order defined in the DAG.

We say that a subset S of nodes of a NMT is *consistent* if, for every node $u \in S$, all the nodes w, that are predecessors of u in the partial order, are also in S. Any consistent set S has an associated simplicial complex Σ_S, and a related iconic hypergraph, which corresponds to an intermediate level of detail (see, e.g.,

Figure 5). Σ_S is obtained by starting from the single point and undoing all edge collapses contained in set S.

Undoing an edge collapse means recovering the original edge $e = v_1 v_2$ from the vertex v on which it was collapsed: such operation is called a *vertex split*. Note that consistency of set S with respect to the partial order implies that, when splitting v, the star of v is exactly the same as it was just after collapsing e during the simplification process. Vertex v is replaced by edge e; some of the edges in v^* are expanded into triangles of e^*; each remaining top-simplex $\sigma \in v^*$ either replaces v with v_1, or replaces v with v_2, or becomes two simplices of the same dimension as σ, one incident in v_1 and the other one incident in v_2.

The NMT supports a dynamic algorithm to modify a mesh Σ_S and its iconic hypergraph representation while adding and removing macro-nodes to and from S. The algorithm performs a traversal of the DAG describing the NMT, starting at the given consistent set S, and involves two phases: a *contraction* phase, in which nodes are subtracted from set S (and Σ_S is coarsened through the corresponding edge collapses) in parts of the complex where resolution is too high; and an *expansion* phase, in which nodes are added to S (and Σ_S is refined through the corresponding vertex splits) in parts where the resolution is not sufficient (see [12] for details).

In order to support the dynamic algorithm, we use the data structure described in Section 2 for encoding the iconic hypergraph associated with the current mesh Σ_S, a standard data structure for the DAG, and a data structure to represent the updates contained in each macro-node. To this aim, we store a reference to vertex v, two references to the endpoints v_1 and v_2 of e in the vertex table, an offset vector used to update vertex positions, and a bit-mask of *split-codes* that specifies how to transform each top-simplex $\sigma \in v^*$ when splitting v (see [12] for details).

5   Conclusions and Future Work

In this paper, we have presented an iconic model for three-dimensional objects which is based on the explicit representation of object parts through meshes, and of their assembly structure. We have introduced a multi-level model which spans a whole range of object iconic representations, and allows also for selectively extracting different iconic representations for different object parts. A construction algorithm for the single-level model has been described, and a parametric construction technique for the multi-level model through progressive simplification and clustering has been outlined. Specific simplification and clustering strategies to reveal the part structure of an object starting at a detailed yet unstructured manifold mesh are the subject of our current investigations.

Our aim is to build a model database supporting, for instance, queries by similarity and shape recognition with a hierarchical approach. The basic observation is that objects with a similar structure will have similar iconic representations at a low level of detail. Thus, we can design a database organized as an AND/OR DAG in which every object is represented by a subgraph, corresponding to its

NMT description. Objects which have common representations, up to a certain level of detail, share the same data structure, up to that level. Data structures for encoding such a database, techniques for its construction and update, and techniques for queries by similarity and model-driven object recognition will be the subjects of our future research.

Acknowledgments. This work has been partially supported by the European Research Training Network "MINGLE - Multiresolution in Geometric Modelling", contract number HPRN-CT-1999-00117.

References

1. Three Dimensional Object Description and Computation Techniques, *Annotated Computer Vision Bibliography,* 11,
   http://iris.usc.edu/Vision-Notes/bibliography/contentsdescribe.html
2. M.K. Agoston, Algabraic Topology: a First Course, in *Pure and Applied Mathematics,* M. Dekker (ed.), New York, 1976.
3. F. Bernardini and H. Rushmeyer, The 3D Model Acquisition Pipeline, *Eurographics 2000 State of the Art Reports,* Eurographics Association, pp.41-62, 2000.
4. F. Cutzu, Computing 3D Object Parts from Similarities among Objcet Views, *Proc. Int. Conf. on Computer Vision and Pattern Recognition,* IEEE Computer Society Press, 2000.
5. S.J. Dickinson, R. Bergevin. I. Biederman, J.O. Eklundh, R. Munck-Fairwood, A.K. Jain, A.P. Pentland, Panel Report: The Potential of Geons for Generic 3-D Object Recognition", *Image and Vision Computing ,* 15(4): 277-292, 1997.
6. L. De Floriani, P. Magillo, E. Puppo. Multiresolution representation and reconstruction of triangulated surfaces. In *Advances in Visual Form Analysis,* World Scientific, pp.140-149, 1997.
7. T. K. Dey, H. Edelsbrunner, S. Guha and D. Nekhayev. Topology preserving edge contraction. *Publications de l'Institut Mathematique (Beograd),* 60(80):23–45, 1999.
8. M. Garland, Multiresolution Modeling: Survey & Future Opportunities, *Eurographics '99 – State of the Art Reports,* Eurographics Association, 1999, pp. 111–131.
9. A. Gueziec, G. Taubin, F. Lazarus, W. Horn, Converting Sets of Polygons to Manifold Surfaces by Cutting and Stitching, *Proc. IEEE Visualization,* 1998, pp. 383–390.
10. D. Luebke, C. Erikson, View-dependent simplification of arbitrary polygonal environments, *Proc. ACM SIGGRAPH,* 1997, pp. 199–207.
11. G.G. Medioni, A.J.R. Francois, 3-D Structures for Generic Object Recognition, *Proc. Int. Conf. on Pattern Recognition,* Barcelona, Spain, 2000, pp. 30–37.
12. F. Morando, L. De Floriani, P. Magillo, E. Puppo. Decomposing a Non-Manifold Mesh into Manifold Components *in preparation.*
13. A. Pentland, Recognition by Parts, *Proc. First International Conference on Computer Vision,* IEEE Computer Society Press, London, June 1987.
14. J. Popovic, H. Hoppe. Progressive simplicial complexes. *Proc. ACM SIGGRAPH,* 1997, pp. 217–224.
15. E. Rivlin and S.J. Dickinson and A. Rosenfeld, Recognition by Functional Parts, *Image Understanding Workshop,* Monterey, CA, 1994, pp. 1531–1539.

16. K. Sengupta, K.L. Boyer, Organizing large structural model bases, *IEEE Trans. on Patterns Analysis and Machine Learning*, 17(14):321–332, 1995.
17. K. Siddiqi, B.B. Kimia, Parts of Visual Form: Computational Aspects, *PAMI*, 17(3): 239-251, 1995. Corrections in: *PAMI*, 17(5): 544-544, 1995.
18. K. Weiler, Topological Structures for Geometric Modeling, *PhD Thesis*, Rensselaer Polytechnic Institute, Troy, NY, August 1986.

Image Indexing by Contour Analysis: A Comparison

Riccardo Distasi[1], Michele Nappi[1], Maurizio Tucci[1], and Sergio Vitulano[2]

[1] Dipartimento di Matematica e Informatica, Università di Salerno, 84084 Baronissi (SA), Italy. {ricdis,micnap,mautuc}@unisa.it
[2] Dipartimento di Scienze Mediche, Facoltà di Medicina, Via S. Giorgio 12, 09124 Cagliari, Italy. vitulano@vaxca1.unica.it

Abstract. This paper describes three systems for image indexing and retrieval based on contour analysis. The systems compared are F-Index for Contours (FIC), Hierarchical Entropy-based Representation (HER) and Sketch-based query by Dialogue (SQD). The first system has been modified for contour-matching, since it was originally designed for a different purpose. The choice of these specific systems has been made because of their similar conception, aim and computational complexity. An experimental and conceptual comparison has been carried out in order to assess retrieval precision, efficiency and usability. The results show that FIC and HER have similar performance in the high end of the spectrum, while SQD has less precise retrieval and less efficient search.

1 Introduction

The days where computers were text-only systems are long gone. Nowadays, even low-end personal computers are able to display, employ and process images—at least in the form of icons, but usually in much more complex ways. The typical user has several image files in his storage devices; the high-end user or the graphic specialist may well have thousands. Furthermore, there are computer systems that are entirely devoted to the archival or treatment of images: many multimedia databases are made in great part by images, and several applications in a wide range of specific fields rely on digital images for many of their functions. As the number of images available to a system increases, the need for an automatic image retrieval system of some kind becomes more stringent.

For a human being, it is very easy to recognize shapes and textures independently from their position and orientation; however, it is much harder to specify exactly the steps involved in such recognition, and this makes it difficult to devise an algorithm that can be programmed into a computer. Indeed, the general problem of image classification and retrieval by content, that is, based only on the actual content of the pictorial scene without the aid of textual labels or other metadata, is a rather hard one. Although no general solution to this problem has been found yet, there are a number of results that solve, at least partially, particular problems in specific areas. Indeed, it is much easier to devise an image retrieval system if the designer has *a priori* information on the type of images

C. Arcelli et al. (Eds.): IWVF4, LNCS 2059, pp. 665–676, 2001.
© Springer-Verlag Berlin Heidelberg 2001

involved in the queries. The resulting indexing system might not be suitable for general application, but this kind of ad hoc solution can be useful as long as it works in a specific environment.

The techniques proposed in the scientific literature fall almost invariably in the category of feature extraction methods. The key idea is that of analyzing the pictorial scene in order to obtain n numerical features. By doing so, an image is mapped from image (or pixel) space into a single point in n-dimensional feature space, where traditional—and exact—spatial access methods may be used to retrieve points (i.e., images) that are close to the query image. This type of user interaction paradigm is called 'query by example,' because the user supplies an example (the query image) and the system looks for images that are close to it in feature space—and hopefully also from a perceptive point of view. Another possible paradigm for user interaction is called 'query by sketch.' In this case, the user draws a sketch of some object which should appear in the retrieved images. Depending on the nature of the feature extraction engine, either or both paradigms might be applicable.

The underlying assumption with all feature extraction based methods is that proximity in feature space implies some kind of proximity in image space. Under certain hypotheses about the feature extraction process and the definition of distance in feature space, this is indeed the case. However, the type of features utilized and the organization of the information have a considerable influence over the usability of the system and the subjective perceived quality of the end result, which may vary widely from system to system. Also, the computational requirements can vary from lightweight through taxing to nearly infeasible.

This paper describes three systems for image indexing and retrieval that extract their features from an analysis of the contour. The pros and cons of each system are pinpointed and compared. The methods we will discuss are: (a) F-Index [1], a Discrete Fourier Transform based index, originally devised for time series and adapted to work with object contours specifically for the present comparison(FIC); (b) Hierarchical Entropy-based Representation for object contours (HER) [6]; (c) Sketch-based Query by Dialog (SQD) [4]. The F-Index and HER were originally designed with a query by example interface, while SQD had a sketch-oriented interface with a particular attention to user interaction. However, all of these methods can be queried by both paradigms. As we shall see in the next section, all of these systems should be at least robust to certain types of image transformations such as scaling, rotations and reflections.

Other methods, which might be regarded as similar under several aspects, have been excluded from the comparison because their computational cost is significantly higher. Such is the case, for instance, of Elastic Matching [5].

The outline of the paper is as follows. Section 2 describes the methods one by one, discussing the relative strengths and weaknesses from the point of view of their design. The experiments that have been done in order to asses the methods from the practical point of view of their actual performance are presented and discussed in Section 3. Finally, Section 4 draws a few concluding remarks.

2   The Methods

The choice of the methods considered has been aimed at making the comparison as meaningful as possible. In fact, they all share several characteristics:

1. The data that make up the index are derived from the contour of the most prominent object in the pictorial scene. This implies that all of these methods are intended to be applied to images of the 'one object against a background' type, so they work best with this kind of images. Another implication is that segmentation is a necessary step in order to separate the object from the background. The exact nature of the segmentation algorithm needs not concern us here, as long as it is the same for all the methods being compared.
2. The execution time is in the low range of the spectrum of available techniques. As a consequence, all these methods can be effectively implemented on low end workstations.
3. The performance of these methods is very similar as long as the index size is the same. However, two of these methods, namely HER and FIC, provide a way to tune index size, while the third (SQD) produces indices of a fixed size, independently of both image size and user choice of parameters.
4. The ability to offer feedback to the user as to the actual look of the query shape after feature extraction, so that it is possible to refine subsequent queries. This ability is integrated into the user interface of SQD, while for the other two methods the user must iterate the querying process until satisfied with the result.

One aspect where they do differ is in the ability that the user has to adjust the operation of the system. FIC and HER allow the user to trade off speed for accuracy and index size by changing the cutoff frequency (FIC) or the number of maxima that end up in the index (HER). In contrast, SQD has its feature space fixed at 4 dimensions, but it is only used for a filtering phase before 'real' search takes place in another space. On the other hand, SQD has several adjustable thresholds that may be tweaked in order to influence the size of the answer set. However, adjusting these thresholds does not often yield predictable results. The designers of SQD point this out: in their implementation, the values are hard-wired into the system based on experimental results [4].

Another difference between FIC and HER on one side and SQD on the other is that while the former systems use a spatial access method for searching, SQD recurs to simple sequential search.

2.1   F-Index for Contours

The F-Index was first introduced by Agrawal, Faloutsos and Swami to perform similarity search in time series databases [1]. Among the methods we consider, it might well be the simplest to understand and implement. It has also been the first to appear in the literature, although the modification that allows it to be used for image contours (FIC) has been purposely made for the present paper.

In order to obtain a time series from our contour data, we scan the contour clockwise starting from its top left pixel, recording the distance between each pixel and the center of mass, as shown in Figure 1. This yields a periodic time series with as many points as there are pixels in the object contour.

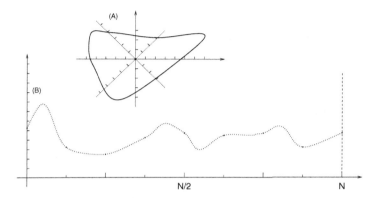

Fig. 1. Representing an N-pixel 2-D contour (A) as a periodical 1-D time series (B)

Once we have a time series $x(\cdot)$ with N points, the steps performed to obtain its FIC representation are the following:

1. Obtain the coefficients of its DFT (Discrete Fourier Transform).

$$X(f) = \frac{1}{\sqrt{N}} \sum_{k=0}^{N-1} x(k) \exp\left(\frac{-j2\pi fk}{N}\right), \tag{1}$$

where $j = \sqrt{-1}$ is the imaginary unit. This yields N complex numbers.
2. Discard all DFT coefficients but the first M. In other words, the parameter M is the *cutoff frequency*.
3. Construct a multidimensional index in $2M$-space using an appropriate spatial access method—in this case, an R*-tree [2].

The F-Index and its derivation FIC use the discrete form of the Fourier Transform [7], which has several nice well-known mathematical properties, most importantly linearity. Therefore, by invoking Parseval's theorem, it can be proven that searching by the F-Index we can expect no *false dismissals*. In other words, images that lie within the specified distance from the query image will never fail to appear in the answer set. The reason is that since many Fourier coefficients are discarded, the distance between two items in feature space is less than the original distance in pixel space. This indeed makes sure that there are no false dismissals, but on the other hand it might introduce some *false alarms* that must be filtered out in a postprocessing step.

Although the DFT and the closely related DCT (Discrete Cosine Transform), used by JPEG, are able to capture a good deal of information about images, sharply straight lines can't be effectively represented unless we are willing to use enough coefficients. As shown by Zahn and Roskies [11], an adequate approximation of a polygon requires 15–30 coefficients. When the object has highly irregular or jagged contours, even 30 coefficients are not enough to characterize the shape adequately for accurate reconstruction. The increasingly good approximation of a shape by its DFT is shown graphically in Figure 2.

FIC allows a reconstruction of the query shape based on the data from its DFT. However crude it might be, even a 3-coefficient Fourier approximation of the shape is something that the user can refer to in order to evaluate the effectiveness of the query. This undoubtedly enhances the usability of the system.

$M = 1$ $M = 4$ $M = 10$ $M = 60$

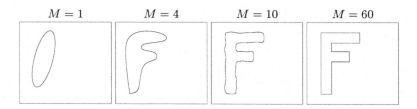

Fig. 2. Approximation of a shape by the first M coefficients of its DFT

As for the spatial access method used in the FIC technique, it employs Euclidean distance as its metric in both image space and feature space. The data structure of choice, R*-trees, have been shown experimentally to perform well for dimensions up to 20 [8], which means 10 complex Fourier coefficients.

The user can adjust the behavior of FIC by varying the cutoff frequency M, thus trading off search time and index size for accuracy. The designers of the F-Index experimentally found that 3 is a good value for the cutoff [1], but they were dealing with 'real' 1-D data with spectrum patterns not unlike pink or brown noise, in which most of the relevant information involves medium- or long-term trends. In this case, we are dealing with 1-D encodings of 2-D data, which often makes it appropriate to increase the cutoff frequency slightly.

2.2 Hierarchical Entropy-Based Representation

As in FIC's case, also the Hierarchical Entropy-based Representation (HER) technique [6] was originally designed for the indexing and retrieval of time series. However, it has been shown to work well whenever the data, no matter what their origin, could be meaningfully made into a time series by some kind of transformation. In the case of image contours, this is done by scanning the contour in the same way as for FIC.

Supposing we have a time series $x(\cdot)$ with N points, let us define the energy of the i-th sample as $E(i) = |x(i)|^2$. The total energy of $x(\cdot)$ is simply $E = \sum_{i=0}^{N-1} E(i)$, while the relative energy of $x(i)$ is $E_r(i) = E(i)^2/(E - E(i))$.

The HER representation vector $\widehat{\mathbf{y}}$ of the sequence $x(\cdot)$ is obtained as follows:

1. Find the signal maxima and put them in a queue Q in decreasing magnitude order, along with their x-axis positions. If the number of signal maxima is m, the queue Q now contains $\big((i_1, x(i_1)), \ldots, (i_m, x(i_m))\big)$;
2. Compute the relative energy $E_r(t)$ of the first (largest) maximum in Q, say $x(t)$.
3. Compute the standard deviation $\sigma(t)$ relative to the current maximum $x(t)$ using

$$\sigma(t) = \frac{1}{\sqrt{2\pi \left(E_r(t)\right)^2}}. \tag{2}$$

In other words, we are considering $x(t)$ to be the midpoint of a Gaussian distribution. Compute the entropy relative to $x(t)$ as

$$S(t) = \frac{1}{x(t)} \sum_{k=-\sigma(t)}^{\sigma(t)} |x(t+k)|. \tag{3}$$

4. Concatenate the values $\big(x(t), S(t), t\big)$ to the end of the HER output vector $\widehat{\mathbf{y}}$. Remove $x(t)$ from Q.
5. Go back to Step 2 until we have removed a predefined number M of maxima from the queue Q.

An alternative form for Step 5 keeps on iterating until the fraction of the total energy remaining in the signal $x(\cdot)$ falls below a given threshold. In many cases, the alternate test offers more control on index accuracy at the expense of unpredictable index size. In order to perform our comparison with preset index sizes, we have preferred the simpler 'number of maxima' test.

Differently from Fourier-based methods, this representation was never meant for reconstructing the signal. However, it is indeed possible to reconstruct the contour if one feature is added to the HER representation: the angle made by the current maximum and some reference line—say, the positive X axis. This feature might be employed to enhance the system's usability by providing the user with feedback about the actual appearance of the query shape, at the cost of a 33% increase in index size. In this case, Figure 3 shows how a HER reconstruction changes when increasing the number M of maxima. In Figure 3, it is assumed that all interpolation between maxima is done by straight line segments. In principle it is possible to use curves as to fit the position i where the maximum $x(i)$ occurs, but in practice the final effect is usually not worth the extra effort.

The spatial access method used by HER when processing queries does not use Euclidean distance in feature space. Rather, the distance between two representation vectors $\widehat{\mathbf{y_1}}$ and $\widehat{\mathbf{y_2}}$ is defined as

$$D(\widehat{\mathbf{y_1}}, \widehat{\mathbf{y_2}}) = \sum_{k=0}^{\infty} |\widehat{\mathbf{y_1}}_k - \widehat{\mathbf{y_2}}_k|. \tag{4}$$

The data structure employed for the spatial organization of feature vectors is based on k-d-trees [3].

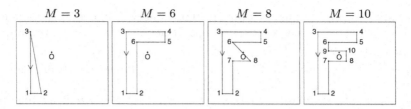

Fig. 3. Approximation of a shape by its M largest maxima using HER

2.3   Sketch-Based Query by Dialog

The Sketch-based Query by Dialog method (SQD) [4] stands out in this group
for two peculiarities. First, it was conceived from the start as an interactive,
'query, browse, refine the query' method. Second, because of how the contour is
sampled, the index has a fixed size independent of the original contour. Index
size can't be changed by tweaking any parameter, either.

In order to obtain its SQD representation $\hat{z}$ (called 'signature' by its designers), a contour is sampled starting from the top left pixel at fixed $1°$ intervals,
thus obtaining $N = 360$ samples at $\theta_i = 2i\pi/N$ for $i = 0, \dots, N-1$. The distance
from the contour's center of mass is recorded for each sample. If the contour is
concave, there might be more than one point corresponding to a single angle;
in this case, the maximum distance is recorded. An example of this process is
illustrated in Fig. 4. sort, but it distorts the concavities, as shown in Fig. 5.
Compare Fig. 4 with Fig. 6, where the distance is recorded for each single pixel
in the contour.

In order to achieve scaling and rotation invariance, SQD normalizes the distances with respect to the maximum distance and shifts the whole sequence of
samples to put the farthest contour pixel in the first position.

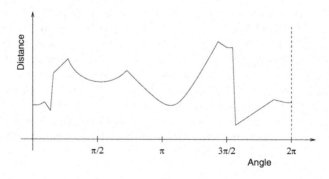

Fig. 4. Converting the 'F' contour into a time series by sampling at fixed angle incre-
ments (SQD signature)

Fig. 5. SQD reconstruction

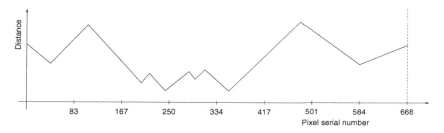

Fig. 6. Converting the 'F' contour into a time series pixel by pixel (FIC, HER)

The contour, and therefore the image, is represented by the whole signature $\hat{\mathbf{z}}$. In order to perform the search, 4 features are extracted from $\hat{\mathbf{z}}$:

1. The sum of all N distances. This feature is larger for elongated objects and smaller for nearly circular shapes.
2. The variance of the distances. This helps distinguish jagged, sharply varying contours from smooth ones.
3. The ratio between minimum and maximum distance. A low value in this feature might point to a concave contour segment.
4. The integral of the Fourier spectrum of the sequence of distances. Given the N complex DFT coefficients $X(f)$, $f = 0, \ldots, N$, the integral spectrum is given by the sum of their magnitudes:

$$S = \sum_{f=0}^{N} |X(f)|. \tag{5}$$

This feature, like variance, is greater for sharply varying contours than it is for smooth ones, but it is more sensitive to local variations between pixels in the same neighborhood, that is, high frequencies in the DFT.

The system was originally designed for query by sketch, but query by example is also possible and it has indeed been used for our tests. There is no real spatial access method and distance in feature space is a highly nonlinear combination of individual feature-by-feature distances and inter-signature distance, all defined by absolute differences.

Here is how searching works. When the user submits a query shape, the system goes through two steps: first it evaluates 4 distances for each item in the database—one for each feature. These distances are compared against predefined thresholds and each time a distance falls below the threshold the candidate database item scores a point. At the end of first step, each database item has therefore a score between 0 and 4. The second step is restricted to items that have scored over another 'score threshold.' Here, the system computes the distance between the item's and the query's signature, defined as

$$D(\widehat{\mathbf{y}}, \widehat{\mathbf{z}}) = \sum_{k=0}^{N-1} |\widehat{\mathbf{y}}(k) - \widehat{\mathbf{z}}(k)|, \tag{6}$$

which is the final ranking factor for the presentation of the answer set.

A remark about this searching scheme is in order. The first step involves a sequential scan of the whole database, which can result in long searching times if the number of items is substantial. However, in principle it would be possible to implement this phase in a smarter and more efficient way utilizing a spatial data structure, especially considering that feature space dimensionality is low by design (only 4 features).

3   The Experiments

The systems under examination have been tested by means of a series of experiments aimed mainly at assessing effectiveness, efficiency and usability. All the tests have been performed on a heterogeneous database of about 400 images that includes tools, animals and pasta. This database consists of the Scaroff database [9, 10] plus a few additions. The results are averaged over 20 queries for each system.

Effectiveness, that is, precision of the retrieval, has been measured by the quantity known as Normalized Recall (NR), defined as follows. Consider a database D of $|D|$ objects where the number of objects relevant to the query is $N < |D|$. Suppose that the objects are sorted so that the most relevant object is X_1 and the least relevant is X_N. Let A be the ordered answer set returned by a query, and let r_i be the rank of i-th most relevant object X_i in A. The *ideal rank* (IR) of the query is then defined as

$$\text{IR} = \frac{1}{N} \sum_{i=1}^{N} i = \frac{N+1}{2}. \tag{7}$$

Note that the ideal rank does not depend on A. The *average rank* (AR) of A is then

$$\text{AR} = \frac{1}{N} \sum_{i=1}^{N} r_i. \tag{8}$$

The difference $\text{AR} - \text{IR}$ gives a measure of the precision achieved by the query. This quantity is usually normalized in order to obtain a value between 0 and 1— the Normalized Recall.

$$\text{NR} = \frac{\text{AR} - \text{IR}}{|D| - N}. \tag{9}$$

As Fig. 7 shows, FIC and HER have very similar performance, while SQD does not achieve the best results.

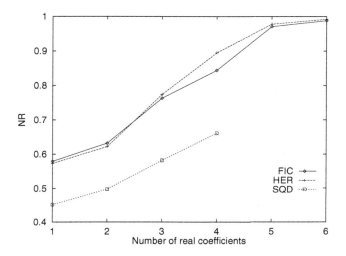

Fig. 7. Normalized Recall as a function of the number of coefficients

Invariance various kinds of image transformations is also an issue that influence the quality of retrieval. These systems all include some method for ensuring at least robustness, if not exact invariance, to image rotations, reflections and scaling. Indeed, the representations used by FIC, HER and SQD are intrinsically invariant to rotations and nearly invariant to scaling—if we neglect the discreteness of pixels as opposed to continuous Cartesian space.

The next illustration (Fig. 8) shows how FIC, HER and SQD respond to the same sample query. All the results have the query image in the upper left, while the answer set, sorted best match first, is shown on the right. In order to make the comparison as fair as possible, in this case FIC and HER have been set up to use 4 real numbers for indexing as SQD does. These results are fairly representative of the standard behavior exhibited by these systems in that SQD returns a significantly higher number of false alarms. The rabbits returned by FIC and SQD can be justified by noting how similar these rabbit pictures are to a fish: the snout and ears can be mistaken for a fish tail, while the fore legs bear a definite resemblance to a ventral fin.

The efficiency of the systems has been assessed by considering index size and response time. The three systems have indices of similar size, at least for the parameter values that have been considered. Both FIC and HER allow the user to adjust the dimension of feature space—and consequently the size of the index—by varying the cutoff frequency (FIC) or the number of maxima (HER). On the other hand, SQD has its dimension fixed to 4. Predictably, FIC and HER's indices are slightly larger.

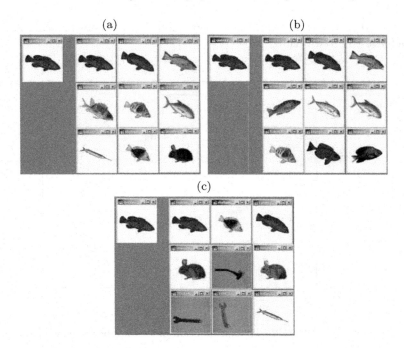

Fig. 8. Sample queries: FIC (a); HER (b) and SQD (c)

As for the response time, arguably more important, SQD's is perceivably longer because of the lack of a spatial access method in the search phase. In fact, FIC's and HER's searching times are about an order of magnitude shorter than SQD's. This result should not be a surprise: for these moderate feature space dimensions (1 to 6) the asymptotic performance of FIC and HER is basically logarithmic in the size of the database, while SQD's is linear.

Finally, a few words about usability. With FIC and HER, the user can see what the index data 'looks like.' In other words, it is possible to evaluate the quality of the index data a priori by looking at the contour reconstructed by means of the very same data that will be used to perform the actual search. This is shown in Figs. 2 and 3. On the other hand, SQD reconstructs the contour based on the whole 'signature,' but the search—at least in the first phase—is performed by utilizing only 4 numerical features extracted from the signature. It is not possible to attempt any contour reconstruction from these 4 numbers, so it is harder for the user to predict the quality of the results in a reliable way.

4   Conclusion

This paper has described three systems for image index and retrieval based on contour analysis: the F-Index for Contours (FIC), the Hierarchical Entropy-based Representation (HER) and the Sketch-based Query by Dialogue (SQD).

The three systems have been selected on the basis of their similar concept, aim and computational complexity. The F-Index for Contours was initially designed for time series matching [1] and modified specifically for this paper.

Experiments were made with the intent of assessing the effectiveness (retrieval precision), efficiency and usability of the systems. The experimental results show that FIC and HER have similar performance, with a Normalized Recall well above 0.9 if enough coefficients are extracted from the data. SQD, on the other hand, has less precise retrieval but emphasizes and supports user intervention to iterate the search until the answer set 'converges' to the desired results.

References

1. R. Agrawal, C. Faloutsos, A. Swami, "Efficient similarity search in sequence databases," Proc. *FODO: 4th Int'l Conf. on Foundations of Data Organization and Algorithms*, pp. 69–84, Evanston, IL, Oct. 1993.
2. N. Beckmann, H. P. Kriegel, R. Schneider, B. Seeger, "The R^*-tree: an efficient and robust access method for points and rectangles," *Proc. ACM SIGMOD*, pp. 322–331, May 1990.
3. J. L. Bentley, "Multidimensional binary search trees used for associative searching," *Comm. ACM*, Vol. 18, No. 9, pp. 509–517, Sept. 1975.
4. A. Del Bimbo, M. De Marsico, S. Levialdi, G. Peritore, "Query by dialog: and interactive approach to pictorial querying," *Image and Vision Computing* 16, pp. 557–569, Elsevier, 1998.
5. A. Del Bimbo, P. Pala, "Visual Image Retrieval by Elastic Matching of User Sketches," *IEEE Trans. Pattern Analysis and Machine Intelligence*, vol. 18, n. 2, Feb. 1997, pp. 121–132.
6. R. Distasi, D. Vitulano, S. Vitulano, "A hierarchical representation for content based image retrieval," *Journal of Visual Languages and Computing*, Special Issue on Multimedia Databases and Image Communication, Vol. 5, n. 8, Aug. 2000.
7. A. V. Oppenheim and R. W. Schafer, *Digital Signal Processing*, Prentice-Hall, Englewood Cliffs, NJ, USA, 1975.
8. M. Otterman, "Approximate Matching with High Dimensionality R-trees," M.Sc. scholarly paper, Dept. of Computer Science, Univ. of Maryland, MD, USA, 1992.
9. S. Sclaroff, "Distance to Deformable Prototypes: Encoding Shape Categories for Efficient Search," In A. W. M. Smeulders, R. Jain, Eds., *Image Databases and Multi-Media Search*, Series on Software Engineering and Knowledge Engineering, vol. 8, pp. 25–37, World Scientific.
10. S. Sclaroff, Image database used in shape-based retrieval experiments available via ftp at `ftp://cs-ftp.bu.edu/sclaroff/pictures.tar.Z`.
11. C. T. Zahn and R. Z. Roskies, "Fourier descriptors for plane closed curves," *IEEE Trans. Computers*, Vol. C21, 1972.

Shape Reconstruction from an Image Sequence

Atsushi Imiya and Kazuhiko Kawamoto

Computer Science Division
Dept. of Information and Image Sciences, Chiba University
1-33 Yayoi-cho, Inage-ku, 263-8522, Chiba, Japan
imiya@ics.tj.chiba-u.ac.jp

Abstract. This paper clarifies a sufficient condition for the reconstruction of an object from its shadows. The objects considered are finite closed convex regions in three-dimensional Euclidean space. First we show a negative result that a series of shadows measured using a camera moving along a circle on a plane is insufficient for the full reconstruction of an object even if the object is convex. Then, we show a positive result that a series of pairs of shadows measured using a general stereo system with some geometrical assumptions is sufficient for full reconstruction of a convex object.

1 Introduction

The reconstruction of three-dimensional shapes from measured data such as range data, photometric information, and stereo image pairs, is called "Shape from X." In this paper, we deal with "Shape from shadows." This problem is also called 'Shape from counter," [1] and "Shape from profile" [2] in computer vision and "Shape from plane probing," [3], in computational geometry. In computer vision, theoretical analysis of reconstruction algorithms is paid little attention.

This paper proves that a series of shadows measured using a camera moving along a circle on a plane is insufficient for the full reconstruction of the visible hull of an object. Although this type of measuring system is sometimes used in computer vision, our results show that we cannot reconstruct full profiles of objects using such a camera system even if the objects are convex. Next, we prove a positive result that a series of pairs of shadows measured using a general stereo system with some simple geometric assumptions fully reconstructs convex objects. Then, using the same mathematical ideas with "Shape from shadows," we prove similar sufficient condition for "Shape from range data" and "Shape from photometric data." We also clarify the relation between "Shape from shadows" and image reconstruction from line integrals using the characteristic functions of line integrals. We prove similar properties between the two problems for the orbits of source points. The orbits are spatial curves on which the eye center of the camera system and the x-ray source move for "Shape from shadows" and image reconstruction from line integrals, respectively.

The illumination problem [4] estimates the minimum and maximum numbers of view points for the reconstruction of a convex body from their views from an

C. Arcelli et al. (Eds.): IWVF4, LNCS 2059, pp. 677–686, 2001.
© Springer-Verlag Berlin Heidelberg 2001

appropriate set of view points. The illumination problem is equivalent to shape reconstruction from silhouettes or shadows. However, it is in general difficult to answer the configuration of view points for a given object. There are many results for the reconstruction of a convex polygon from their shadows. For example see [5], and [6]. Laurentini [7,8] was concerned with geometric properties of silhouette-based shape reconstruction for polyhedrons, and clarified the relation among the visible hull and the convex hull of a polyhedron.

Tuy [11] proved that for a positive functions defined in a finite closed convex region in three-dimensional Euclidean space, it is possible to reconstruct the function from line integrals measured by cone beams, if source of line integrals moves on a pair of circles with the same radius lying on a mutually perpendicular planes which encircle the region. Shape from perspective projections has relations with the cone-beam reconstruction problem since it is possible to determine boundary from line integrals.

2   Shape Reconstruction from Support Planes

In this section, we summarized the result of convex geometry in three-dimensional Euclidean space $\mathbf{R}^3$ [12], since the analytical relations between a convex object and its tangent planes were sometimes dealt in a well-described form for computer vision by several authors [10,13]. Let $x - y - z$ be an orthogonal coordinate system in $\mathbf{R}^3$. We call the system the world coordinate system. We denote a vector in the world coordinate $\boldsymbol{x} = (x, y, z)^\top$, where $\top$ means the transpose of a vector. Setting $\boldsymbol{x}^\top \boldsymbol{y}$ to be the inner product of $\boldsymbol{x}$ and $\boldsymbol{y}$, we define the length of a vector as $|\boldsymbol{x}| = \sqrt{\boldsymbol{x}^\top \boldsymbol{x}}$. Thus, $|\boldsymbol{x} - \boldsymbol{y}|$ is the Euclidean distance between $\boldsymbol{x}$ and $\boldsymbol{y}$. Furthermore, setting $0 \leq \theta < \pi$, and $0 \leq \phi < 2\pi$, we define a rotation matrix

$$\boldsymbol{R} = \begin{pmatrix} \cos\phi\sin\theta & \sin\theta\sin\phi & \cos\theta \\ \cos\phi\cos\theta & \sin\phi\cos\theta & -\sin\theta \\ -\sin\phi & \cos\phi & 0 \end{pmatrix}. \tag{1}$$

Next, for the basis of the world coordinate system $\boldsymbol{e}_1 = (1, 0, 0)^\top$, $\boldsymbol{e}_2 = (0, 1, 0)^\top$, and $\boldsymbol{e}_3 = (0, 0, 1)^\top$, we define a set of orthogonal basis vectors

$$\hat{\boldsymbol{e}}_i = \boldsymbol{R}^\top \boldsymbol{e}_i, \quad i = 1, 2, 3. \tag{2}$$

Thus, we obtain the relation $\boldsymbol{R}^\top = (\hat{\boldsymbol{e}}_1, \hat{\boldsymbol{e}}, \hat{\boldsymbol{e}}_3)$. Setting $\boldsymbol{D}$ to be the vector gradient operator on the unit sphere

$$\boldsymbol{D}\boldsymbol{x} = \left(\frac{\partial}{\partial\theta}\boldsymbol{x}, \frac{1}{\sin\theta}\frac{\partial}{\partial\phi}\boldsymbol{x} \right), \tag{3}$$

where θ and ϕ are the polar angle and the longitude on the sphere, the relation $\boldsymbol{D}\hat{\boldsymbol{e}}_1 = (\hat{\boldsymbol{e}}_2, \hat{\boldsymbol{e}}_3)$, holds. This equation leads to the equation $\boldsymbol{R} = (\hat{\boldsymbol{e}}_1, \boldsymbol{D}\hat{\boldsymbol{e}}_1)$. Setting K to be a bounded closed convex set in $\mathbf{R}^3$, we denote the boundary of K as ∂K. If a plane touches K at a point on ∂K, this plane is called a support plane of K. We set that $h(\theta, \phi)$ is the Euclidean distance between the origin of the world coordinate system and the support plane of K, the normal vector of which is $\hat{e}_1$. $h(\theta, \phi)$ is a function on the unit sphere. K exists in a half space,

$$\mathbf{H}(\theta, \phi) = \{x \mid x^\top \hat{e}_1 \leq h(\theta, \phi)\}. \tag{4}$$

Therefore, we call a plane

$$x^\top \hat{e}_1 = h(\theta, \phi), \tag{5}$$

a support plane of K.

Furthermore, let $\hat{\nabla}$ be the scaler gradient operator on the unit sphere $\mathbf{S}^2$; that is, for function $h(\theta, \phi)$ defined on the unit sphere,

$$\hat{\nabla} h(\theta, \phi) = \left(\frac{\partial}{\partial \theta} h(\theta, \phi), \frac{1}{\sin \theta} \frac{\partial}{\partial \phi} h(\theta, \phi) \right)^\top. \tag{6}$$

The following proposition is a well-known result in convex geometry[12].

Proposition 1 *Let* $h = \left(h(\theta, \phi), \hat{\nabla} h(\theta, \phi) \right)^\top$. *Then,* $x \in \partial K$ *is obtained by*

$$x = Rh. \tag{7}$$

Equation (7) is called the support plane expression of a convex body. From eq. (5) if the normal vectors of support planes are defined, we can obtain the support plane expression of an object.

Several authors in the computer vision and image processing field re-found the eqs. (7), and its two-dimensional version [10,13]. However, the support function is the fundamental results in convex geometry [12]. Figure 1 (a) shows the relation between a convex object and the support plane.

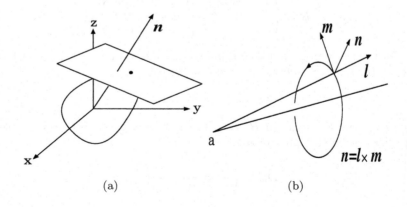

(a)   (b)

Fig. 1. The configurations of a convex object and its support plane (a) and the three orthogonal vectors (b).

3   Shape Reconstruction from Perspective Projections

For a point $\boldsymbol{a} = (a, b, c)^{\top}$, setting $-\boldsymbol{a}$ to be the positive direction of the ξ axis, and $\boldsymbol{a}$ to be the origin of a coordinate system, we define a right-handed orthogonal coordinate system $\xi - \eta - \zeta$. We call the system the observation system and denote a vector in this coordinate system as $\boldsymbol{\vartheta} = ((\xi, \eta, \zeta))^{\top}$. Thus, the world coordinate system and the observation coordinate system are related by

$$\boldsymbol{x} = \boldsymbol{U}\boldsymbol{\vartheta} - \boldsymbol{a} \tag{8}$$

for an appropriate rotation matrix $\boldsymbol{U}$, where $\boldsymbol{U}$ depends on the directions of the ξ and η axes. Setting $\boldsymbol{a}$ to be the camera center and $((\xi, \eta, f))^{\top}$ such that $|f| < |\boldsymbol{a}|$ and $f|\boldsymbol{a}| > 0$ to be an imaging plane, we denote the shadow of K on an imaging plane as $\tilde{K}(\boldsymbol{a})$ and denote the boundary of shadow as $\partial \tilde{K}(\boldsymbol{a})$. Furthermore, we denote the boundary curve of $\partial \tilde{K}(\boldsymbol{a})$ as

$$\boldsymbol{r}(t, \boldsymbol{a}) = ((\xi(t, \boldsymbol{a}), \eta(t, \boldsymbol{a}), f))^{\top}, \ 0 \leq t \leq T(\boldsymbol{a}), \tag{9}$$

such that $\boldsymbol{r}(t, \boldsymbol{a}) = \boldsymbol{r}(t + T(\boldsymbol{a}), \boldsymbol{a})$ where $T(\boldsymbol{a})$ is the total length of $\partial \tilde{K}(\boldsymbol{a})$, since $\partial \tilde{K}(\boldsymbol{a})$ is a closed curve on an imaging plane. The vector

$$\boldsymbol{l}(t, \boldsymbol{a}) = \frac{\boldsymbol{U}\boldsymbol{r}(t, \boldsymbol{a})}{|\boldsymbol{r}(t, \boldsymbol{a})|} \tag{10}$$

is the N-vector of $\boldsymbol{r}(t, \boldsymbol{a})$ in the world coordinate system [14]. $\boldsymbol{l}(t, \boldsymbol{a})$ moves on a closed curve $\partial L(\boldsymbol{a})$ on the unit sphere for each $\boldsymbol{a}$. We call these closed curve $\partial L(\boldsymbol{a})$ the N-curve of $\boldsymbol{r}(t, \boldsymbol{a})$. Next, setting

$$\dot{\boldsymbol{l}}(t, \boldsymbol{a}) = \frac{\partial}{\partial t} \boldsymbol{l}(t, \boldsymbol{a}), \ \boldsymbol{m}(t, \boldsymbol{a}) = \frac{\dot{\boldsymbol{l}}(t, \boldsymbol{a})}{\left| \dot{\boldsymbol{l}}(t, \boldsymbol{a}) \right|}, \tag{11}$$

we obtain the relation

$$\boldsymbol{l}(t, \boldsymbol{a})^{\top} \boldsymbol{m}(t, \boldsymbol{a}) = 0. \tag{12}$$

Furthermore, $\boldsymbol{m}(t, \boldsymbol{a})$ is the normalized tangent vector of $\partial \tilde{K}(\boldsymbol{a})$. From eq. (12), setting

$$\boldsymbol{n}(t, \boldsymbol{a}) = \boldsymbol{l}(t, \boldsymbol{a}) \times \boldsymbol{m}(t, \boldsymbol{a}) = \boldsymbol{l}(t, \boldsymbol{a}) \times \frac{\dot{\boldsymbol{l}}(t, \boldsymbol{a})}{\left| \dot{\boldsymbol{l}}(t, \boldsymbol{a}) \right|}, \tag{13}$$

we obtain a moving orthogonal basis $\{\boldsymbol{l}(t, \boldsymbol{a}), \boldsymbol{m}(t, \boldsymbol{a}), \boldsymbol{n}(t, \boldsymbol{a})\}$. Figure 1 (b) shows relations of these three mutually orthogonal vectors. Moreover, $\boldsymbol{n}(t, \boldsymbol{a})$ is the normal vector of a plane which touches K and passes through point $\boldsymbol{a}$. Thus, a support plane of K is given by

$$P(\boldsymbol{a}) = \left\{ \boldsymbol{x} | \boldsymbol{x}^{\top} \boldsymbol{n}(t, \boldsymbol{a}) = \boldsymbol{a}^{\top} \boldsymbol{n}(t, \boldsymbol{a}) \right\}. \tag{14}$$

From eq. (14), setting

$$\hat{\boldsymbol{e}}_1 = \boldsymbol{l}(t, \boldsymbol{a}) \times \frac{\dot{\boldsymbol{l}}(t, \boldsymbol{a})}{\left| \dot{\boldsymbol{l}}(t, \boldsymbol{a}) \right|}, \tag{15}$$

we can obtain $\boldsymbol{R}$ and $h(\theta, \phi)$, using the relations

$$\boldsymbol{R} = \left(\boldsymbol{l}(t, \boldsymbol{a}) \times \frac{\dot{\boldsymbol{l}}(t, \boldsymbol{a})}{\left| \dot{\boldsymbol{l}}(t, \boldsymbol{a}) \right|}, \boldsymbol{D} \left(\boldsymbol{l}(t, \boldsymbol{a}) \times \frac{\dot{\boldsymbol{l}}(t, \boldsymbol{a})}{\left| \dot{\boldsymbol{l}}(t, \boldsymbol{a}) \right|} \right) \right) \tag{16}$$

and

$$h(\theta, \phi) = \boldsymbol{a}^\top \left(\boldsymbol{l}(t, \boldsymbol{a}) \times \frac{\dot{\boldsymbol{l}}(t, \boldsymbol{a})}{\left| \dot{\boldsymbol{l}}(t, \boldsymbol{a}) \right|} \right). \tag{17}$$

Thus, eqs. (15) and (17) imply that we can reconstruct a convex body from its shadows, if we obtain $\boldsymbol{n}(t, \boldsymbol{a})$, which is defined by eqs. (11) and (13). Equations (15) and (17) yield the following theorem.

Theorem 1 *If vector $\boldsymbol{n}(t, \boldsymbol{a})$ is measured all over the unit sphere, it is possible to obtain full reconstruction of a convex body from its shadows.*

Theorem 1 is a modification of proposition 1, which is a classical result of convex geometry. It is, however, important from the standpoint of computer vision because the theorem shows a sufficient condition for shape reconstruction from shadows obtained by perspective projections. Also, from eqs. (15) and (17), we can reconstruct a convex object using only boundary information of shadows.

4   A Sufficient Condition for Reconstruction

4.1   Shape from Shadows

For a point $\tilde{\boldsymbol{x}} \in \tilde{K}(\boldsymbol{a})$, setting $\boldsymbol{x}(\boldsymbol{a})$ to be the N-vector of $\tilde{\boldsymbol{x}}$, we define a cone

$$C(\boldsymbol{a}) = \{ \boldsymbol{x} | \boldsymbol{x} = \lambda \boldsymbol{x}(\boldsymbol{a}), \lambda \geq 0 \} . \tag{18}$$

We call $C(\boldsymbol{a})$ the view-cone at $\boldsymbol{a}$. The boundary of $C(\boldsymbol{a})$ is

$$\partial C(\boldsymbol{a}) = \{ \boldsymbol{x} | \boldsymbol{x} = \lambda \boldsymbol{l}(t, \boldsymbol{a}), \boldsymbol{l}(t, \boldsymbol{a}) \subset \partial L(\boldsymbol{a}), \lambda \geq 0 \} . \tag{19}$$

If a pair of view-cones which have the same vertex satisfy the relation

$$C_1(\boldsymbol{a}) \subseteq C_2(\boldsymbol{a}), \tag{20}$$

we write $\partial L_1(\boldsymbol{a}) \preceq \partial L_2(\boldsymbol{a})$, where $\partial L_i(\boldsymbol{a})$ is called the associated N-curve of a view-cone $C_i(\boldsymbol{a})$.

If $\boldsymbol{l}(t, \boldsymbol{a})$ moves on $\partial L(\boldsymbol{a})$, $\boldsymbol{n}(t, \boldsymbol{a})$ moves on the unit sphere and forms a closed curve $\partial N(\boldsymbol{a})$, which we call the orthogonal N-curve. From geometrical consideration, it is obvious that if $\partial L_1(\boldsymbol{a}) \preceq \partial L_2(\boldsymbol{a})$, then $\partial N_1(\boldsymbol{a}) \succeq \partial N_2(\boldsymbol{a})$. Furthermore, setting

$$\partial C(\boldsymbol{a})^\perp = \{ \boldsymbol{x} | \boldsymbol{x} = \lambda \boldsymbol{n}(t, \boldsymbol{a}), \lambda \geq 0 \} \tag{21}$$

if eq. (20) holds, then the relation $C_1(\boldsymbol{a})^\perp \supseteq C_2(\boldsymbol{a})^\perp$ holds. Setting $C_1(\boldsymbol{a})$ and $C_2(\boldsymbol{a})$ to be view-cones of K and B to be a sphere which encircles K, respectively, we obtain the following theorem, where the origin of the world coordinate system is at the center of B.

Theorem 2 *For a bounded closed set A in* $\mathbf{R}^3$, *if*

$$\bigcup_{a \in A} C_2(a)^\perp \supseteq \mathbf{R}^3, \tag{22}$$

then we can reconstruct K from shadows which are obtained by perspective projection.

(Proof) $\forall a \in A \subset \mathbf{R}^3$, the relation $C_1(a)^\perp \subseteq \mathbf{R}^3$ holds. If $\bigcup_{a \in A} C_2(a)^\perp \supseteq \mathbf{R}^3$, then we obtain

$$\mathbf{R}^3 \subseteq \bigcup_{a \in A} \mathbf{C}_2(a)^\perp \subset \bigcup_{a \in A} \mathbf{C}_1(a)^\perp \subseteq \mathbf{R}^3. \tag{23}$$

This relation concludes the relation $\bigcup_{a \in A} C_1(a)^\perp = \mathbf{R}^3$. Furthermore, this equation leads to $\bigcup_{a \in A} n_1(t, a) = \mathbf{S}^2$.   (Q. E. D)

In the following, we show some examples of a bounded closed set A.

Example 1 *Let* P_1 *and* P_2 *be a pair of perpendicular planes which pass through the center of B. Setting* a_1 *and* a_2 *to be circles on* P_1 *and* P_2 *of the center of which are at the center of B with radii a and b, respectively, if*

$$a^{-2} + b^{-2} > r^{-2}, \tag{24}$$

where r is the radius of B, then $a_1 \cup a_2$ *is an example of A.*

Example 2 *Let* P_1 *and* P_2 *be a pair of parallel planes which touch B. Setting* a_1 *and* a_2 *to be circles with the radius d, the center of which are on B, if* $d > r$, $d_1 \cup d_2$ *is an example of A.*

In Figures 2(a) and 2(b), we show the spatial configurations of K, which is enclosed in B, B, and A for examples 1 and 2, respectively. It is clear from the figures that the configurations of examples 1 and 2 satisfy the condition of theorem 2. These examples show that for the reconstruction of a convex object from its shadows the orbit on which the camera center of perspective projection moves is a one-dimensional manifold, or a collection of curves in $\mathbf{R}^3$.

The camera orbit configuration of example 2 has the significant property that if the regions of interest of two cameras overlap for each a, we can apply the method of shape from stereo pairs. In this case, the stereo images are measured using a general stereo system, two optional axes of which are not parallel. The most interesting camera configuration is that in which the epipolar line of the stereo system is perpendicular to each circle on which the camera moves.

4.2   Shape from Range Data and Photometric Data

The main idea used in the analysis of the problem is that shadows of convex objects are functions on the unit sphere. This property yields a sufficient condition for the reconstruction of objects from the boundaries of their shadows. The

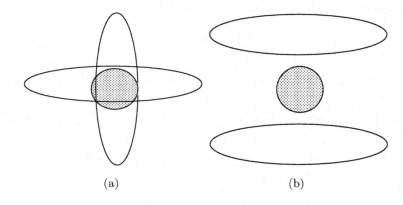

(a)   (b)

Fig. 2. The Configuration of the camera motions of examples.

range date and the photometric data of a convex object are also functions on the unit sphere. Using this property we also derive a sufficient geometric condition for full recovery of convex objects from range data and photometric data. Here, we assume that the range data and photometric data are also obtained using perspective projection geometry.

Setting $d(\tilde{\boldsymbol{x}}, \boldsymbol{a})$ to be range data or photometric data at $\tilde{\boldsymbol{x}} \in \tilde{K}(\boldsymbol{a})$, for a fixed $\boldsymbol{a}$, $d(\tilde{\boldsymbol{x}}, \boldsymbol{a})$ is a function of $\boldsymbol{x}(\boldsymbol{a})^{\perp} \in C(\boldsymbol{a})^{\perp}$. This geometric property implies that for a bounded closed set $\boldsymbol{A}$ in $\mathbf{R}^3$ if

$$\bigcup_{\boldsymbol{a} \in \boldsymbol{A}} \boldsymbol{x}(\boldsymbol{a})^{\perp} = \mathbf{S}^2, \tag{25}$$

then we can reconstruct objects from the pair $(\boldsymbol{x}(\boldsymbol{a})^{\perp}, \boldsymbol{d}(\boldsymbol{a}, \tilde{\boldsymbol{x}}))$. Since eq. (25) is equivalent to

$$\bigcup_{\boldsymbol{a} \in \boldsymbol{A}} C(\boldsymbol{a})^{\perp} = \mathbf{R}^3, \tag{26}$$

we obtain the following theorem as in the case of shape reconstruction from shadows.

Theorem 3 *For a bounded closed set $\boldsymbol{A}$ which satisfies eq. (26) if $d(\tilde{\boldsymbol{x}}, \boldsymbol{a})$ is a function on $\mathbf{S}^2$, we can reconstruct ∂K.*

Theorem 3 is valid for non-convex objects if $d(\tilde{\boldsymbol{x}}, \boldsymbol{a})$ is a function on $\mathbf{S}^2$. Thus we obtain the following theorem

Theorem 4 *For a bounded closed set $\boldsymbol{A}$ in $\mathbf{R}^3$, we can reconstruct ∂K from $d(\tilde{\boldsymbol{x}}, \boldsymbol{a})$, when $C(\boldsymbol{a})$ is the view-cone of a sphere B which encloses K.*

This theorem also leads to the same configurations of camera orbits as examples 1 and 2 for range date and photometric data obtained using perspective projection geometry.

5   Tomography and Voting Method

Setting $f(\boldsymbol{x})$ to be a positive, and integrable and square integrable function defined in K, for a $\boldsymbol{a} \in \mathbf{R}^3$ and $\boldsymbol{\omega} \in \mathbf{S}^2$,

$$g(\boldsymbol{a}, \boldsymbol{\omega}) = \int_0^\infty f(\boldsymbol{a} + t\boldsymbol{\omega})dt \qquad (27)$$

is the divergent x-ray transform [15]. Reconstruction of $f(\boldsymbol{x})$ from $g(\boldsymbol{a}, \boldsymbol{\omega})$ is the mathematical model for the reconstruction of volume distributions form cone beams for the x-ray computerized tomography. If $\boldsymbol{a}$ moves on the beam-source orbits which is same with the view-points orbit defined in example 1 for $a = b$, it is possible to reconstruct fully $f(\boldsymbol{x})$ from $g(\boldsymbol{a}, \boldsymbol{\omega})$. Therefore, from $g(\boldsymbol{a}, \boldsymbol{\omega})$, we can reconstruct ∂K in the same condition with example 1.

On the other hand our data are shadows of $f(\boldsymbol{x})$ measured by perspective projections. Therefore, denoting the characteristic function of $g(\boldsymbol{a}, \boldsymbol{\omega})$ and the ray cone as

$$\chi(\boldsymbol{a}, \boldsymbol{\omega}) = \begin{cases} 1 \text{ , if } g(\boldsymbol{a}, \boldsymbol{\omega}) > 0 \\ 0 \text{ , otherwise,} \end{cases} \qquad (28)$$

and

$$C(\boldsymbol{a}, \boldsymbol{\omega}) = \{(\boldsymbol{a}, \boldsymbol{\omega}) \,|\, \chi(\boldsymbol{a}, \boldsymbol{\omega}) = 1, \, \boldsymbol{a} \in \mathbf{R}^3, \boldsymbol{\omega} \in \mathbf{S}^2\}, \qquad (29)$$

respectively, we obtain the following relations

$$K = \bigcap_{\boldsymbol{a} \in A} C(\boldsymbol{a}, \boldsymbol{\omega}), \ \partial K = \bigcap_{\boldsymbol{a} \in A} \partial C(\boldsymbol{a}, \boldsymbol{\omega}), \qquad (30)$$

where $\partial C(\boldsymbol{a}, \boldsymbol{\omega})$ is the boundary of $C(\boldsymbol{a}, \boldsymbol{\omega})$.

The support plane method for shape reconstruction is an algebraic expression of the second equation of eq. (30). These geometric properties show a mathematical relationship between "Shape from shadows" and the image reconstruction form projections of the x-ray computerized tomography since "Shape from shadows" focuses to shape reconstruction.

Setting the characteristic function in the view cone to be

$$c(\boldsymbol{x}; \boldsymbol{a}, \boldsymbol{\omega}) = \begin{cases} 1, \, \boldsymbol{x} \in C(\boldsymbol{a}, \boldsymbol{\omega}) \\ 0, \text{ otherwise,} \end{cases} \qquad (31)$$

if we vote $c(\boldsymbol{x}; \boldsymbol{a}, \boldsymbol{\omega})$ in to the space, we have a function

$$k(\boldsymbol{x}) = \sum_{\boldsymbol{a} \in A} c(\boldsymbol{x}; \boldsymbol{a}, \boldsymbol{\omega}). \qquad (32)$$

as the results of voting. For a positive integer τ, a set of points

$$K_\tau = \{\boldsymbol{x}|, k(\boldsymbol{x}) \geq \tau\} \qquad (33)$$

defines an object. The construction of shape by K_τ is called shape reconstruction by voting.

If an object is a collection of points in a space, projection $\tilde{K}(\boldsymbol{a})$ is a collection of points on a plane for each view point $\boldsymbol{a}$. Furthermore, if an object is a collection of line segments in a space, projection $\tilde{K}(\boldsymbol{a})$ is a collection of line segments on a plane for each view point $\boldsymbol{a}$. A point and a line segment on an imaging plane determine a half line and a fan whose origins are vector $\boldsymbol{a}$, respectively. We can reconstruct the position of a point and a line segment in a space as the common sets of a many half lines and many fans, respectively which are measured from verious directions. A point and a line segment are convex objects whose dimensions are one and two, respectively. A polyhedron is a collection of vertices (points) and edges (line segments) on a closed surface. These geometric properties conclude that the voting method permits us to reconstruct a class of nonconvex polyhedrons without holes from a series of images, if each vertex and each edge of a polyhedron are measured in several images [16]. In Figures 3 (a) and (b), we show an image of a nonconvex polyhedron, which are collection of points on a plane, and the reconstructed polyhedron by voting of half lines in a space, respectively.

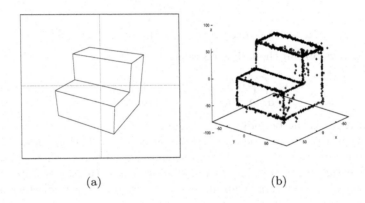

(a)   (b)

Fig. 3. An example of the reconstruction of a nonconvex polyhedron by voting of half lines in a space.

6   Conclusions

Our results have mathematically clarified that for full recovery of convex objects, two cameras each of which moves on a circle are sufficient. We also clarified the relation between shape reconstruction from shadows and shape reconstruction by voting, which is used for the model generation for mixed reality [17].

We also showed that the voting method permits us to reconstruct a class of nonconvex polyhedrons form a series of images using projections of vertices and edges which determine half lines and fans, respectively.

References

1. J. Aloimonos, "Visual shape computation," Proceedings of IEEE, Vol.76, pp.899-916, 1988.
2. S.S. Skiena, "Interactive reconstruction via geometric probing," IEEE Proceedings, Vol.80, pp.1364-1383, 1992.
3. D.P. Dobkin, H. Edelsbrunner, and C.K. Yap, "Probing convex polytopes," Proc. 18th ACM Symposium on Theory of Computing, pp.424-432. 1986.
4. V. Boltyanski, H. Martin, and P.S. Soltan, *Excursions into Combinatorial Geometry,* Springer-Verlag; Berlin, 1997.
5. R.S.-Y. Li, "Reconstruction of polygons from projections," Information Processing Letters, Vol.28, pp.235-240, 1988.
6. M. Lindembaum and A. Bruckstein, "Reconstructing a convex polygon from binary perspective projections," Pattern Recognition, Vol.23, pp.1343-1350, 1990.
7. A. Laurentini, "The visual hull concept for silhouette-bases image understanding," IEEE PAMI Vol.16, pp.150-163, 1994.
8. A. Laurentini, "How for 3D shape can be understood from 2D silhouettes," IEEE PAMI Vol.17, pp.188-195, 1995.
9. R.J. Gardner, *Geometric Tomography,* Cambridge University Press, Cambridge, 1995.
10. J.-Y. Zheng, "Acquiring 3-D models from sequences of contours," IEEE Pattern Analysis and Machine Intelligence, Vol.16, pp.163-178, 1994.
11. H.K. Tuy, "An inversion formula for cone-beam reconstruction," SIAM J. Applied Mathematics, Vol.43, pp.546-552, 1983.
12. H.W. Guggenheimer, *Applicable Geometry,* Robert E. Kniegen Pub. Inc, New York 1977.
13. R. Vaillant and O.D. Faugeras, "Using external boundaries for 3-D object modeling," IEEE Pattern Analysis and Machine Intelligence, Vol. 14, pp. 157-173, 1992.
14. K. Kanatani, *Geometric Computation for Machine Vision,* Oxford University Press, Oxford, 1993
15. C. Hammaker, K.T. Smith, D.C. Solomon, and L. Wagner, "The divergent beam x-ray transform," Rocky Mountain Journal of Mathematics, Vol.10, pp.253-283, 1980.
16. A. Imiya and K. Kawamoto, "Performance analysis of shape recovery by random sampling and voting, " pp.227-240, in R. Klette, H. Siegfried Stiel, M. A. Viergever and K.L. Vincken eds. *Performance Characterization in Computer Vision,* Kluwer Academic Publishers, Dordrecht, 2000.
17. K. Kutulakos and S. M. Seitz, "A theory of shape by space carving," Proceedings of 7th ICCV, Vol. 1, pp.307-314, 1999.

3D Shape Reconstruction from Multiple Silhouettes: Generalization from Few Views by Neural Network Learning

Itsuo Kumazawa and Masayoshi Ohno

Department of Computer Science, Tokyo Institute of Technology,
Meguro-ku Ookayama 2-12-1,
Tokyo 152-8552 Japan
kumazawa@cs.titech.ac.jp

Abstract. In this report, we present a 3D shape modeling method using the shape's silhouettes from multiple views to determine the model (polyhedron) parameters. The polyhedron parameters are determined by neural networks, each of which represents the model's silhouette observed from a view point, and determines the polyhedron parameters by the back propagation algorithm so that the model's silhouette from each view approximates the corresponding silhouette of the target shape. By conducting basic experiments, we verified the effectiveness of the method.

1   Introduction

In many cases, a 3D shape model is manually designed by specifying parameters of a polyhedron so that the polyhedron provides a good approximation of the target shape. However, in order to reduce the amount of time and labor required for designing, interactive design tools with user friendly graphical interface[7] and automated systems with range measuring devices have been developed.

To save measurement cost, instead of directly measured 3D data, multiple 2D views have been used for disparity-based 3D reconstruction[3], or generating new views by image-based rendering methods without a 3D model[11].

In this paper, as a part of attempt to simplify measurement and automate the 3D modeling, we present a method to construct 3D shape model for a target shape by deforming the 3D shape model so that each of its silhouettes from multiple viewpoints fit to a corresponding silhouette of the target shape. Although some shapes with concave surface cannot be modeled due to the limitation of the silhouette, accurate shape modeling is expected for many practical shapes.

2   Analytical Silhouette Representation by a Neural Network

A number of deformable shape models have been developed[1][2][4][5][8][10]. Among them, the shape representation neural network[9] has the following benefits.

C. Arcelli et al. (Eds.): IWVF4, LNCS 2059, pp. 687–695, 2001.
© Springer-Verlag Berlin Heidelberg 2001

(i)   It provides fully analytical representation and partial derivatives with re-
      spect to model parameters are analytically computed for gradient descent
      algorithms to update model parameters iteratively.
(ii) Represented shapes are blurred just by changing gain parameters of sigmoid
      functions. Coarse-to-fine matching is easily implemented by changing gain
      parameters during matching process.
(iii) Shape approximation ability of multi-layer network of nonlinear units is
      higher than that of linear combination of base functions.
(iv) A number of powerful learning algorithms for multi-layer neural networks
      such as back propagation can be used for shape deformation.

In this paper, we use a polyhedron to represent a 3D shape. The positions of
vertices are parameters to specify the polyhedron. Each face of the polyhedron
is a triangle and its projection onto an observation plane is also a triangle. According to the shape representation neural network, a silhouette of a polyhedron
is represented as follows.

Each triangular face of the polyhedron is projected onto the observation
plane. This projected face is described by taking an AND operation over the
half plane regions with borders corresponding to the three edges of the triangle.
Each half plane is represented by a neuron as shown in Fig. 2. The neuron in
this figure outputs 1 for inputs (x, y) inside the half plane: $ax + by + c > 0$
and outputs 0 for inputs (x, y) inside the half plane: $ax + by + c < 0$. When
the sigmoid function is used for the neuron's activation function. The edge of
the half plane is blurred by decreasing σ_{edge} in Eq.(2) as the sigmoid function
changes its graph as shown in Fig. 1. The whole silhouette of the polyhedron is

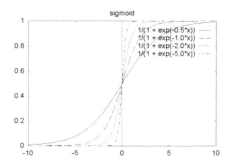

Fig. 1. Sigmoid functions for various gain values.

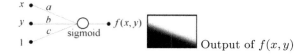

Fig. 2. Representation of a half plane by a single neuron.

represented by taking an OR operation over the projected triangular faces. As the AND operation and the OR operation are computed by neuron functions, this whole computation is executed by a multi-layer neural network.

2.1   Representation of a Half Plane Region

The function $f(x, y)$ which outputs 1 for inputs (x, y) inside the half plane: $ax + by + c > 0$ and outputs 0 for inputs (x, y) inside the half plane: $ax + by + c < 0$ is realized by a sigle neuron as follows.

$$sigmoid(s, \sigma) = \frac{1}{1 + e^{-\sigma s}} \tag{1}$$

$$f(x, y) = sigmoid(ax + by + c, \ \sigma_{edge}) \tag{2}$$

2.2   Representation of a Triangular Region

A triangular region is obtained as an intersection of three half planes. The intersection is obtained by AND operation over three functions $f_i(x, y)(i = 1, 2, 3)$ where $f_i(x, y)$ represents the $i - th$ half region. The sigmoid function with a sufficiently large gain value with an appropriate threshold performs the AND operation in the following equation.

$$F(x, y) = sigmoid(f_1(x, y) + f_2(x, y)$$
$$+ f_3(x, y) - \epsilon, \ \sigma_{AND}) \tag{3}$$

Thus the triangular region is represented by a neural network shown in Fig.4 where the Translation part performs conversion (Fig.3) between three vertices of the triangle $v_i(i = 1, 2, 3)$ and the line parameters $\alpha_i = (a_i, b_i, c_i)^T (i = 1, 2, 3)$ by the following equations.

$$\alpha'_i = \begin{pmatrix} a'_i \\ b'_i \\ c'_i \end{pmatrix} = \begin{pmatrix} v_{y_i} - v_{y_j} \\ v_{x_j} - v_{x_i} \\ v_{x_i} v_{y_j} - v_{x_j} v_{y_i} \end{pmatrix} \tag{4}$$

$$\alpha_i = \frac{1}{\sqrt{a'^2_i + b'^2_i}} \alpha'_i \tag{5}$$

2.3   Representation of Perspective Projection of a Triangular Region

Each triangular face of the polyhedron in the 3D space is projected onto an observation space as shown in Fig. 5. According to this projection, Vertices $V_i = (V_{x_i}, V_{y_i}, V_{z_i})^T (i = 1, 2, 3)$ in the 3D space is transformed to vertices $v_i(i = 1, 2, 3)$ on the observation plane. This transformation is dependent on the camera

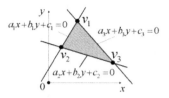

Fig. 3. A triangular region with vertices v_1, v_2, v_3

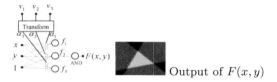

Output of $F(x, y)$

Fig. 4. Representation of a triangular region by a neural network.

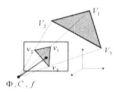

Fig. 5. Perspective projection of a triangular surface.

position, the view angles and the focus distance. These camera parameters are denoted by $\boldsymbol{\Phi} = (\phi, \theta, \psi)^T, \boldsymbol{C} = (C_x, C_y, C_z)^T$, and f Using these parameters, the perspective projection is formulated as follows.

$$\boldsymbol{R_z}(\phi) = \begin{bmatrix} \cos\phi & -\sin\phi & 0 \\ \sin\phi & \cos\phi & 0 \\ 0 & 0 & 1 \end{bmatrix} \tag{6}$$

$$\boldsymbol{R_y}(\theta) = \begin{bmatrix} \cos\theta & 0 & \sin\theta \\ 0 & 1 & 0 \\ -\sin\theta & 0 & \cos\theta \end{bmatrix} \tag{7}$$

$$\boldsymbol{R_x}(\psi) = \begin{bmatrix} 1 & 0 & 0 \\ 0 & \cos\psi & -\sin\psi \\ 0 & \sin\psi & \cos\psi \end{bmatrix} \tag{8}$$

$$R(\boldsymbol{\Phi}) = \boldsymbol{R_x}(\psi)\boldsymbol{R_y}(\theta)\boldsymbol{R_z}(\phi) \tag{9}$$

$$\boldsymbol{V}_i' = (V_{x_i}', V_{y_i}', V_{z_i}')^T \tag{10}$$

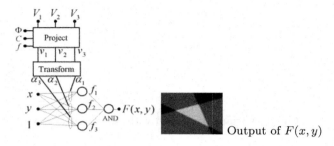

Output of $F(x, y)$

Fig. 6. Representation of perspective projection of a triangular region by a neural network.

$$= R(\Phi)V_i + C \tag{11}$$

$$v_i = \begin{pmatrix} f\frac{V'_{x_i}}{V'_{z_i}} \\ f\frac{V'_{y_i}}{V'_{z_i}} \end{pmatrix} \qquad (i = 1, 2, 3) \tag{12}$$

Fig.6 shows a way of representing a perspective projection of a triangular region by a neural network. In this figure, Project Part performs the projection formulated by Eq.(6)-(12).

2.4   Representation of a Silhouette of a Polyhedron

The whole projection of a polyhedron, that is observed as a silhouette of the polyhedron from a viewpoint, is obtained by taking OR operation of the projected triangular regions as shown in Fig. 7. This combination of projected triangular regions is computed by the neural network shown in Fig. 8. In this figure, a switching operation is introduced instead of OR operation to reduce the computation cost. When computing $F(x, y)$, a triangular region far from the point (x, y) does not affect the result as the sigmoid function decays so fast as shown in Fig.1. From this reason, we can exclude the triangular regions far from (x, y) for computing $F(x, y)$. The switching operation is used to choose sufficiently near

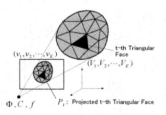

Fig. 7. Perspective projection of a polyhedron

triangular regions. Due to this switching operation, the function $F(x,y)$ is no longer analytical. However, as the result by switching provides an approximation for the analytical case of using OR operation, the benefit of the shape representation network still exists. An example of polyhedron silhouette is illustrated in Fig. 8. In this example, edges of triangular regions are indicated by thin lines to understand each triangular shape.

3   Shape Reconstruction by Training Neural Networks

Silhouettes of a polyhedron model are computed for multiple viewpoints as shown in Fig. 9. These silhouettes of a model are compared with the corresponding silhouettes of the target shape. The vertices of the polyhedron $V_k (k = 1, 2, \cdots, K)$ are modified for the model to provide better approximation. This modification is performed by the gradient descent algorithm as follows.

(i)   The energy function to evaluate the difference between the model silhouettes $F^{(m)}(x,y)(m = 1, 2, \cdots, M)$ and the corresponding target silhouettes $G^{(m)}(x,y)(m = 1, 2, \cdots, M)$ is defined by

$$e^{(m)}(x,y) = \left| F^{(m)}(x,y) - G^{(m)}(x,y) \right|^2, \tag{13}$$

$$\varepsilon^{(m)} = \sum_{x,y} e^{(m)}(x,y), \tag{14}$$

$$E = \sum_m \varepsilon^{(m)}. \tag{15}$$

(ii)   Compute partial derivatives $\partial e / \partial V_k$ by the back propagation algorithm. As a switching operation is introduced instead of the OR operation, the backward signal flow occurs as indicated by a mesh region in Fig. 10.

(iii)   Update iteratively the vertex positions of the polyhedron model by

$$V_k = V_k - \eta \frac{\partial E}{\partial V_k} \qquad (k = 1, 2, \cdots, K) \tag{16}$$

where η is a small positive number.

4   Experiment

Two artificial target shapes, Fish and Airplane, were used for the experiments. These shapes were numerically produced using polyhedrons with sufficiently large number of vertices. Their silhouettes were computed numerically for various views. The size of polyhedron (the number of vertices) used as a shape model was 62 for Fish and 114 for Airplane. Silhouettes from 5 views were used in the training for Fish and silhouettes from 3 views were used in the training for Airplane. The silhouette from View No.0 was not used for the training and used

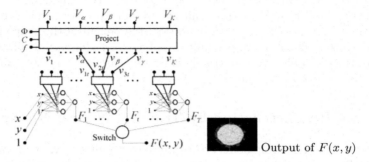

Fig. 8. Representation of a perspective projection of a polyhedron by a neural network

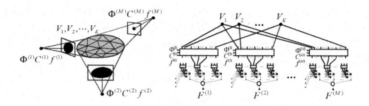

Fig. 9. Polyhedron silhouettes observed from multiple views with camera parameters $\boldsymbol{\Phi}^{(m)}, \boldsymbol{C}^{(m)}, f^{(m)}(m = 1, 2, \cdots, M)$.

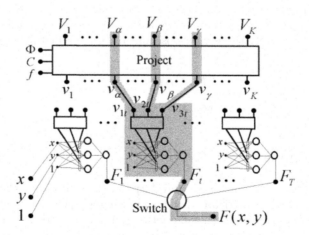

Fig. 10. Backward signal route of the back propagation algorithm.

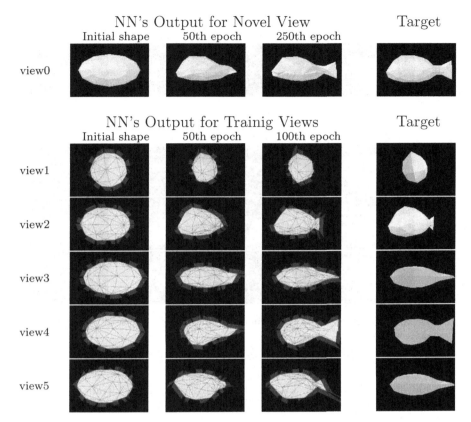

Fig. 11. Exp.1 Shape modeling experiment for a fish shape. A polyhedron with 62 vertices was used.

to test the generalization ability of the learning algorithm. The result for Fish was good but some errors still remained for Airplane even after 2000 times of parameter updating. The three views seemed to be insufficient for reconstructing airplanes.

References

1. Bardinet, E., Cohen, L.D., 1998. A Parametric Deformable Model to Fit Unstructured 3D Data. Computer Vision and Image Understanding 71,(1),39-54.
2. Barr, A., 1984. Global and Local Deformations of Solid Primitives, Computer Graphics, 18, 21-30.
3. Chan, M., Metaxas, D., 1994 Physics-Based Object Pose and Shape Estimation from Multiple Views, Proc. International Conference on Pattern Recognition, Vol.1, IEEE Computer Society Press, 326-330.
4. Chung, P.C., Tsai, C.T, Y.N. Sun., 1994. Polygonal Approximation Using a Competitive Hopfield Neural Network. Pattern Recognition 27, 1,505-1,512.

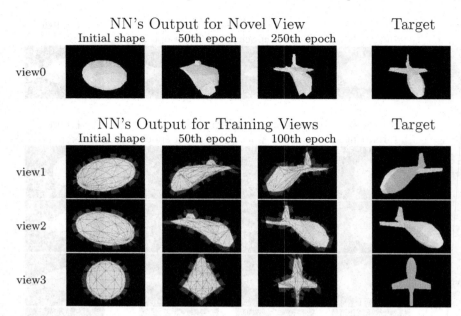

Fig. 12. Exp.2 Shape modeling experiment for an airplane shape. A polyhedron with 114 vertices was used.

5. Cootes, T.F., Taylor, C.J., Cooper, D.H., Graham, J., 1992 Training Models of Shape from Sets of Examples, Proc. British Machine Vision Conference, Springer-Verlag, 9-18.
6. Hartley, R.I., 1997. In defense of the eight-point algorithm, IEEE Trans. Pattern Analysis and Machine Intelligence 19,(6), 580-593.
7. Igarashi, T., Matsuoka, S., Tanaka, H., 1999. Teddy:A Sketching Interface for 3D Freeform Design, ACM SIGGRAPH'99, Los Angels, 409-416.
8. Jain, A.K., Zhong, Y., Lakshmanan, S., 1996. Object Matching Using Deformable Templates, IEEE Trans. Pattern Analysis and Machine Intelligence 18, (3), 267-278.
9. Kumazawa, I., 2000. Compact and parametric shape representation by a tree of sigmoid functions for automatic shape modeling, Pattern Recognition Letters 21 651-660.
10. Shum, H.Y., Hebert, M., Ikeuchi, K., Reddy, R., 1997. An Integral Approach to Free-Form Object Modeling, IEEE Trans. Pattern Analysis and Machine Intelligence 19, (12), 1,366-1,375.
11. Zhang, Z., 1998. Image-based Geometrically-Correct Photorealistic Scene/Object Modeling(IBPhM):A Review, Proc. 3rd Asian Conference on Computer Vision (ACCV'98), 340-349.
12. Zhang, Z., 1998 Determining the epipolar geometry and its uncertainty:A review, International Journal of Computer Vision.

Robust Structural Indexing through Quasi-Invariant Shape Signatures and Feature Generation

Hirobumi Nishida

Ricoh Software Research Center, 1-1-17 Koishikawa, Bunkyo-ku, Tokyo 112-0002, Japan
`hirobumi.nishida@nts.ricoh.co.jp`

Abstract. A robust method is presented for retrieval of model shapes that have parts similar to the query shape presented to the image database. Structural feature indexing is a potential approach to efficient shape retrieval from large databases, but it is sensitive to noise, scales of observation, and local shape deformations. To improve the robustness, shape feature generation techniques are incorporated into structural indexing based on quasi-invariant shape signatures. The feature transformation rules obtained by an analysis of some particular types of shape deformations are exploited to generate features that can be extracted from deformed patterns. Effectiveness is confirmed through experimental trials with databases of boundary contours, and is validated by systematically designed experiments with a large number of synthetic data.

1 Introduction

Efficient and robust retrieval from large image databases by shape [1] is an important problem. In particular, shapes observed in natural scenes are often occluded, corrupted by noise, and partially visible. It is challenging to develop a robust method for efficient retrieval of model shapes that have parts similar to the query shape presented to the image database.

Shape retrieval from image databases has been studied recently for improving robustness against noise and shape deformations. For structural organization [2,3] of databases composed of boundary contours of objects, Del Bimbo [4] and Mokhtarian *et al.* [5] apply the curvature scale-space approach to feature indexing, and Sclaroff [6] proposes a method for image indexing with the modal matching. However, these methods assume that the query shape is presented as a *closed* contour. Structural feature indexing is efficient, but is sensitive to noise, scales of observation, and local shape deformations. The correct model does not necessarily receive as many votes as expected for the ideal case, and the performance is degraded drastically. Stein *et al.* [7] cope with this problem by extracting features from several versions of polygonal approximations of boundary contours. This method can also be applied when the query shape is presented as part of a boundary contour, but the efficiency is degraded.

Efficiency and robustness are important, but sometimes incompatible criteria for performance evaluation. The improvement of robustness implies that the scheme for classification and retrieval should tolerate certain types of variations and deformations for images. Obviously, it may lead to inefficiency if some brute-force methods are

C. Arcelli et al. (Eds.): IWVF4, LNCS 2059, pp. 696–705, 2001.
© Springer-Verlag Berlin Heidelberg 2001

employed such as a generate-and-test strategy by generating various images with a number of different parameters.

An idea to achieving both efficiency and robustness is to handle deformations in the *feature domain* by generating features that can be extracted from deformed patterns [8]. *Feature generation models* can be obtained through an analysis of feature transformations due to some particular types of shape deformation. Robustness can be improved by incorporating generated features in the indexing process. This approach has been applied successfully when queries are presented as closed contours [8]. It is attractive to extend and generalize this approach to the problem being considered.

In this paper, based on the structural indexing with feature generation models, an efficient, robust method is presented for retrieval of model shapes that have parts similar to the query shape presented to the image database. This paper is organized as follows: In Section 2, a structural representation of curves by quasi-convex/concave features along with quantized-directional features [8] is outlined. In Section 3, based on the shape representation outlined in Section 2, we describe the quasi-invariant shape signature, the model database organization through feature indexing, and the shape retrieval through voting. In Section 4, the transformation rules of shape signatures are introduced to generate features that can be extracted from deformed patterns caused by noise and local shape deformations. The proposed algorithm is summarized in Section 5. In Section 6, the proposed method is validated by systematically designed experiments with a large number of synthetic data. Section 7 concludes this paper.

2   Shape Representation

The structural representation of curves [8] is outlined in this section, based on quasi-convex/concave structures incorporating $2N$ quantized-directional features (N is a natural number). As shown in Fig. 1a, the curve is first approximated by a series of line segments. On a 2-D plane, we introduce N-axes together with $2N$ quantized-direction codes. For instance, when $N = 4$, eight quantized-directions are defined along with the four axes as shown in Fig. 1b. Based on these N-axes together with $2N$ quantized-direction codes, the analysis is carried out hierarchically.

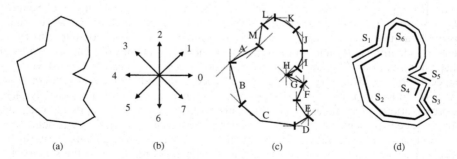

(a)   (b)   (c)   (d)

Fig. 1.   (a) A closed contour with a polygonal approximation, (b) quantized-directional codes when $N = 4$, (c) sub-segments when $N = 4$, (d) segments when $N = 4$.

A curve is decomposed into *sub-segments* at extremal points along each of the N-axes. Fig. 1c illustrates the decomposition of a contour shown in Fig. 1a into sub-segments when $N = 4$. For adjacent sub-segments a and b, suppose that we turn counterclockwise when traversing them from a to b, and the joint of a and b is an extremal point along the axes toward the directions $(j, j + 1(\mathrm{mod}2N), \ldots, k)$. Then, we write the concatenation of these two sub-segments as $a \xrightarrow{j,k} b$. For instance, the joint of sub-segments H and G in Fig. 1c is an extremal point along the three axes toward the directions 3, 4, and 5. Therefore, the concatenation of H and G is written as $H \xrightarrow{3,5} G$. In this way, we obtain the following concatenations for the sub-segments illustrated in Fig 1c.

$$L \xrightarrow{3,3} M, \quad K \xrightarrow{2,2} L, \quad J \xrightarrow{1,1} K, \quad I \xrightarrow{0,0} J,$$

$$H \xrightarrow{7,7} I, \quad H \xrightarrow{3,5} G, \quad F \xrightarrow{0,1} G, \quad F \xrightarrow{4,4} E,$$

$$D \xrightarrow{7,0} E, \quad C \xrightarrow{6,6} D, \quad B \xrightarrow{5,5} C, \quad A \xrightarrow{3,4} B, \quad A \xrightarrow{7,7} M$$

By linking local features around joints of adjacent sub-segments, some sequences of the following form can be constructed:

$$a_0 \xrightarrow{j(1,0),j(1,1)} a_1 \xrightarrow{j(2,0),j(2,1)} \cdots \xrightarrow{j(n,0),j(n,1)} a_n \tag{1}$$

A part of the curve corresponding to a sequence of this form is called a *segment*. When a segment is traversed from a_0 to a_n, one turns counterclockwise around any joints of sub-segments. The following segments, as shown in Fig. 1d, are generated from the 13 sub-segments:

$$S_1 : A \xrightarrow{7,7} M, \quad S_2 : A \xrightarrow{3,4} B \xrightarrow{5,5} C \xrightarrow{6,6} D \xrightarrow{7,0} E,$$

$$S_3 : F \xrightarrow{4,4} E, \quad S_4 : F \xrightarrow{0,1} G, \quad S_5 : H \xrightarrow{3,5} G,$$

$$S_6 : H \xrightarrow{7,7} I \xrightarrow{0,0} J \xrightarrow{1,1} K \xrightarrow{2,2} L \xrightarrow{3,3} M.$$

A segment is characterized by a pair of integers $\langle r, d \rangle$, *characteristic numbers*, representing the angular span of the segment and the direction of the first sub-segment:

$$r = \sum_{i=1}^{n} (j(i,1) - j(i,0))_{\mathrm{mod}2N} + \sum_{i=1}^{n-1} (j(i+1,0) - j(i,1))_{\mathrm{mod}2N} + 2, \quad d = j(1,0) \tag{2}$$

The characteristic numbers are given by $\langle 2,7 \rangle$, $\langle 7,3 \rangle$, $\langle 2,4 \rangle$, $\langle 3,0 \rangle$, $\langle 4,3 \rangle$, and $\langle 6,7 \rangle$, respectively, for the six segments shown in Fig. 1d.

Adjacent segments are connected by sharing the first sub-segments or last ones of the corresponding sequences. These two types of connection are denoted by $S \xrightarrow{h} T$ and $S \perp T$, respectively, for two adjacent segments S and T. For instance, connections are denoted by $S_1 \xrightarrow{h} S_2 \perp S_3 \xrightarrow{h} S_4 \perp S_5 \xrightarrow{h} S_6 \perp S_1$ for the six segments shown in Fig. 1d.

3 Quasi-Invariant Shape Signature, Indexing, and Voting

Based on the shape representation outlined in Section 2, we describe the quasi-invariant shape signature, the model database organization through feature indexing, and the shape retrieval through voting. For the model database organization, we assume that each model shape is presented to the system as line drawings or boundary contours of objects. In the shape retrieval, we assume that line drawings or parts of some model shape can be given as a query to the system.

3.1 Quasi-Invariant Shape Signatures

In order to retrieve images for a query given as a partially visible shape, the shape signature is required to tolerate rotation, scaling, and translation. Therefore, features depending on orientation, size, and location cannot be employed as shape signatures. Based on the characteristic numbers and connections of segments extracted from model shapes or query shapes, the shape signature is constructed to satisfy this requirement. We assume that a series of n segments S_i $(i = 1,2,...,n)$ have been extracted with characteristic numbers $\langle r_i, d_i \rangle$ and lengths l_i. The angular span r_i does not depend on orientation, size, or location. Furthermore, the lack of information due to dropping orientation, size, and location can be compensated by employing a triplet of the angular spans of two consecutive segments and their length ratio as the shape signature. From two consecutive segments S_i and S_{i+1} connected as $S_i \overset{c}{-} S_{i+1}$, $c \in \{h,t\}$, the quasi-invariant shape signature is constructed as follows:

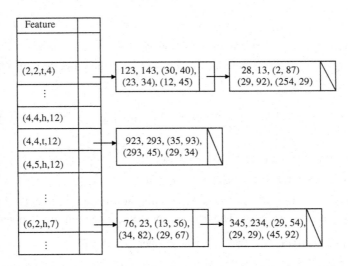

Fig. 2.   Model database organization by structural indexing. Each table item stores a list whose element is composed of the model identifier, length, location of the center of gravity, and locations of the two end points of the curve segment.

$$\left(r_i, r_{i+1}, c, \left\lfloor Q \cdot \frac{l_{i+1}}{l_i + l_{i+1}} \right\rfloor \right),$$

where Q is the number of quantization levels for length-ratio parameters.

3.2   Indexing

From each model shape, shape signatures are extracted from all pairs of consecutive segments. A large table, as illustrated in Fig. 2, is constructed for a model set by assigning a table address to a shape signature and by storing there a list whose item is composed of the following elements: the model identifier that has the corresponding shape signature, and shape parameters of the curve segment corresponding to the shape signature, namely length, location of the center of gravity, and locations of the two endpoints, computed on the model shape.

3.3   Voting

Classification of the query shape is carried out by voting for the transformation space associated with each model. For each model, voting boxes are prepared for the quantized transformation space $(\sigma, \theta, x_T, y_T)$, where σ is the scaling factor, θ is the rotation angle, and (x_T, y_T) is the translation vector. Shape signatures are extracted from the curve segment given as a query to the shape database. For each extracted shape signature, model identifiers and shape parameters are retrieved from the table by computing the table address. By comparing the shape parameters of the extracted shape signature with the registered parameters, the transformation parameters $(\sigma, \theta, x_T, y_T)$ can be computed for each model and the voting box corresponding to the transformation parameters associated with the model is incremented by one. In the implementation, transformation parameters σ and θ are computed from the line segment connecting the two endpoints, and (x_T, y_T) is computed from the location of the center of gravity.

4   Feature Generation Models

Shape signatures extracted from the curve are sensitive to noise and local shape deformations, and therefore, the correct model does not necessarily receive as many votes as expected for the ideal case. Furthermore, when only one sample pattern is available for each class, techniques of statistical or inductive learning from training data cannot be employed for obtaining *a priori* knowledge and feature distributions of deformed patterns. To cope with these problems, we analyze the feature transformations caused by some particular types of shape deformations, constructing feature transformation rules. Based on the rules, we generate segment features that can be extracted from deformed patterns caused by noise and local shape deformations. In both processes of model database organization and classification, the generated features by the transformation rules are used for structural indexing and voting, as well as the features actually extracted from curves.

The following two types of feature transformations are considered in this work:

- Change of convex/concave structures caused by perturbations along normal directions on the curve and scales of observation, along with transformations of characteristic numbers (the angular span of the segment and the direction of the first sub-segment).
- Transformations of characteristic numbers caused by small rotations.

We describe these two types of transformation in the rest of this section.

4.1   Transformations of Convex/Concave Structures

The convex/concave structures along the curve are changed by noise and local deformations, and also depend on scales of observations. For instance, two parts of curves shown in Fig. 3a are similar to one another in terms of global scales, but their structural features are different. When $N = 4$, the curve shown on left is composed of three segments connected as $S_1 \perp S_2 \overset{h}{=} S_3$ with characteristic numbers $\langle 6,6 \rangle$, $\langle 2,6 \rangle$, and $\langle 3,2 \rangle$, whereas the one shown on right is composed of five segments connected as $S_1' \perp S_2' \overset{h}{=} S_3' \perp S_4' \overset{h}{=} S_5'$ with characteristic numbers $\langle 6,6 \rangle$, $\langle 2,6 \rangle$, $\langle 2,2 \rangle$, $\langle 2,6 \rangle$, and $\langle 3,2 \rangle$. To cope with such deformations, structural features on the two curves are edited so that their features can become similar to one another. For instance, the structural features illustrated in Fig. 3a can be edited by merging the two segment blocks $\{S_1, S_2, S_3\}$ and $\{S_1', S_2', S_3', S_4', S_5'\}$ to two segments S and S' as shown Fig. 3b. In the structural indexing and voting processes, for an integer M specifying the

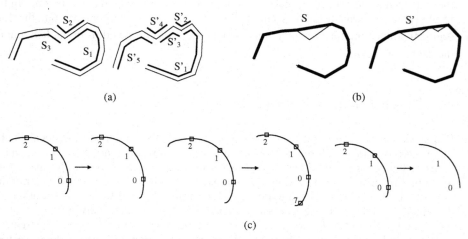

(a)   (b)

(c)

Fig. 3.   (a) Part of curves similar to one another in terms of global scales, (b) editing structural features by merging segment blocks, (c) transformations of characteristic numbers of segments by small rotations.

maximum number of segments to be merged, shape signatures are generated by applying RULE 1 described below to segment blocks.

Rule 1: Let $\langle r_i, d_i \rangle$ be the characteristic number of the segment s_i, and $\lambda(s_j s_{j+1} \cdots s_{j+k-1})$ be length of the curve composed of k consecutive segments $s_j s_{j+1} \cdots s_{j+k-1}$. From a segment block

$$\Sigma(s_j, m, n) = \left\{ s_{j+k} \middle| k = 0, 1, \ldots, m, \ldots, m+n-1; s_j \overset{c_j}{\frown} s_{j+1} \cdots s_{j+m-1} \overset{*}{\frown} s_{j+m} \cdots s_{j+m+n-1} \right\}$$

where m and n are odd such that $1 \le m, n \le M$, a shape signature

$$\left(\sum_{k=0}^{m-1} (-1)^k r_{j+k}, \sum_{k=0}^{n-1} (-1)^k r_{j+m+k}, C_j, \left\lfloor Q \cdot \frac{\lambda(s_j s_{j+1} \cdots s_{j+m-1})}{\lambda(s_j s_{j+1} \cdots s_{j+m-1} s_{j+m} \cdots s_{j+m+n-1})} \right\rfloor \right),$$

is generated if $\displaystyle\sum_{k=0}^{m-1} (-1)^k r_{j+k} \ge 2$, $\displaystyle\sum_{k=0}^{n-1} (-1)^k r_{j+m+k} \ge 2$, $r_{j+2k-2} - r_{j+2k-1} + r_{j+2k} \ge 2$

for $k = 1, \ldots, (m-1)/2$, and $r_{j+m+2k-2} - r_{j+m+2k-1} + r_{j+m+2k} \ge 2$ for $k = 1, \ldots, (n-1)/2$.

For instance, when $N = 4$ and $M = 3$, from the six segments illustrated in Fig. 1d with characteristic numbers $\langle 2, 7 \rangle$, $\langle 7, 3 \rangle$, $\langle 2, 4 \rangle$, $\langle 3, 0 \rangle$, $\langle 4, 3 \rangle$, and $\langle 6, 7 \rangle$, the following shape signatures (length-ratio omitted) are generated by *Rule 1*: (2, 7, h), (2, 8, h), (7, 2, t), (7, 3, t), (8, 4, t), (2, 3, h), (2, 5, h), (3, 6, h), (3, 11, h), (3, 4, t), (5, 2, t), (4, 6, h), (4, 11, h), (6, 2, t), (11, 2, t), (11, 3, t). In total, at most $n \cdot \lceil M/2 \rceil$ shape signatures are generated from n segments.

4.2   Transformations of Characteristic Numbers by Small Rotations

The characteristic number $\langle r, d \rangle$ ($r \ge 2$) can be transformed by rotating the shape. Rules can be introduced for generating characteristic numbers by rotating the shape slightly (see Fig. 4c).

Rule 2: When the curve composed of the two consecutive segments S_1 and S_2 with characteristic numbers $\langle r_1, d_1 \rangle$ and $\langle r_2, d_2 \rangle$ is rotated by angle ψ ($-\pi/N \le \psi \le \pi/N$), the angular spans r_1 and r_2 can be transformed into one of the 9 cases: (1) (r_1, r_2), (2) $(r_1, r_2 - 1)$, (3) $(r_1, r_2 + 1)$, (4) $(r_1 - 1, r_2)$, (5) $(r_1 - 1, r_2 - 1)$, (6) $(r_1 - 1, r_2 + 1)$, (7) $(r_1 + 1, r_2)$, (8) $(r_1 + 1, r_2 - 1)$, (9) $(r_1 + 1, r_2 + 1)$. Note that the cases (4—6) are applicable only if $r_1 \ge 3$, and that the cases (2), (5), and (8) are applicable only if $r_2 \ge 3$.

For instance, when $N = 4$ and $M = 3$, the 16 shape signatures have been generated by *Rule 1* from the six segments illustrated in Fig. 1d. Then, by applying *Rule 2* to these generated ones, 120 shape signatures, in total, are further generated.

5   Algorithm

In the model database organization step by structural indexing, shape signatures are generated from each model shape by *Rules 1* and *2*, and the model identifier with the shape parameters is appended to the list stored at the table address corresponding to each generated shape signature. For each model i ($i = 1,2,\ldots,n$), let c_i be the number of shape signatures generated by *Rules 1* and *2*. For instance, $c_i = 120$ for the contour shown in Fig. 1a when $N = 4$ and $M = 3$. In the classification and retrieval by voting for models and the transformation space, from shape signatures extracted from the query shape, shape signatures are generated by *Rules 1* and *2*. Model identifier lists are retrieved from the tables by using the addresses computed from the generated shape signatures, and the transformation parameters are computed for each model on the lists. The voting box is incremented by one for the model and the computed transformation parameters. Let v_i ($i = 1,2,\ldots,n$) be the maximum votes among the voting boxes associated with the model i. The query shape is classified by selecting out some models according to the descendant order of v_i / c_i. Examples of shape retrieval are given in Fig. 4, where query shapes are presented at top along with retrieved model shapes.

6   Experiments

In this section, the proposed algorithm is evaluated quantitatively in terms of the robustness against noise and shape deformations, based on the systematically designed, controlled experiments with a large number of synthetic data [8]. We examined the probability that the correct model is included in top $t\%$ choices for various values of the deformation parameter β [8] when curves composed of $r\%$ portions of a model shape are given as queries. For given values of r and β, a sub-contour of $r\%$ of length is randomly extracted from the model shape, and then, it is deformed by the deformation process as described in Nishida [8].

The main contribution of this work is to incorporate the shape feature generation into the structural indexing for coping with shape deformations and feature transformations. Therefore, the performance was compared with a naive method extracting features from several versions of piecewise linear approximations of the curve with a variety of error tolerances for approximations.

We carried out several experimental trials by changing the number of models from 200 to 500, examining the classification accuracy in terms of the deformed portions of model shapes given as queries to image databases. In the implementation of the naive method, by changing the error tolerance with Ramer's method from 1% to 20%, with a step of 1%, of the widest side of the bounding box of the curve, twenty versions of approximations were created for each model shape and the query shape.

Table 1 presents the average classification rates for top 1%, 2%, 3%, 5%, and 10% choices when $\beta \in [0.0,0.5]$, $\beta \in [0.5,1.0]$, and $\beta \in [1.0,1.5]$. For instance, when a curve segment composed of 80% portions of a model shape subjected to the deformation process with $\beta \in [0.5, 1.0]$ is given as a query to shape databases of 500 models, the correct models are included in top 15 choices (3%) with probability 98.2% for proposed algorithm and with probability 83.9% for the naive method. The computation time of the proposed method is comparable to that of the naive method. Clearly, significant improvements of robustness against noise and local shape deformations can be observed for the proposed algorithm in terms of classification accuracy without a degradation of efficiency. Through the experiments, the effectiveness has been verified through the experiments for the shape signature and the shape feature generation models.

7 Conclusion

Structural feature indexing is a potential approach to efficient shape retrieval from large image databases, but the indexing is sensitive to noise, scales of observation, and local shape deformations. It has now been confirmed that efficiency of classification and robustness against noise and local shape transformations can be improved at the same time by the feature indexing approach incorporating shape feature generation techniques [8]. In this paper, based on this approach, an efficient, robust method has been presented for retrieval of model shapes that have parts similar to the query shape presented to the image database. The effectiveness has been confirmed by experimental trials with a large database of boundary contours and has been validated by systematically designed experiments with a large number of synthetic data.

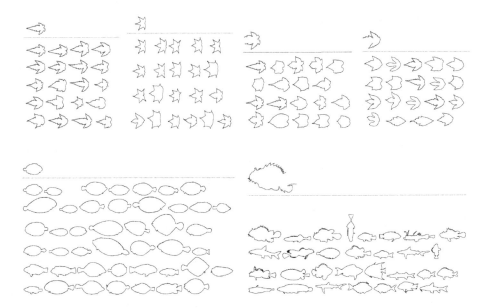

Fig. 4.   Examples of shape retrieval.

References

1. R. Mehrotra and J.E. Gary, Similar-shape retrieval in shape data management, *Computer* **28**(9), 1995, 57—62.
2. A. Califano and R. Mohan, Multidimensional indexing for recognizing visual shapes, *IEEE Transactions on Pattern Analysis and Machine Intelligence* **16**(6), 1994, 373—392.
3. W.I. Grosky and R. Mehrotra, Index-based object recognition in pictorial data management, *Computer Vision, Graphics, and Image Processing* **52**, 1990, 416—436.
4. A. Del Bimbo and P. Pala, Image indexing using shape-based visual features, *Proc. 13th Int. Conf. Pattern Recognition, Vienna, August 1996*, vol. C, pp. 351—355.
5. F. Mokhtarian, S. Abbasi, and J. Kittler, Efficient and robust retrieval by shape content through curvature scale space, *Proc. First International Workshop on Image Database and Multimedia Search, Amsterdam, August 1996*, pp. 35—42.
6. S. Sclaroff, Deformable prototypes for encoding shape categories in image databases, *Pattern Recognition*, **30**(4), 1997, 627—641.
7. F. Stein and G. Medioni, Structural indexing: efficient 2-D object recognition, *IEEE Trans. Pattern Analysis & Machine Intelligence* **14**(12), 1992, 1198—1204.
8. H. Nishida, Structural shape indexing with feature generation models, *Computer Vision and Image Understanding* **73**(1), 1999, 121—136.

Table 1. Average classification rates (%) of deformed patterns by the proposed algorithm in terms of the portion of model shapes ($r\%$) presented as queries, in comparison with the naive method.

β	Portion (r)	Method	Classification rates (%) for top $t\%$ choices				
			$t = 1$	$t = 2$	$t = 3$	$t = 5$	$t = 10$
0.0— 0.5	100%	**Nishida**	**100.0**	**100.0**	**100.0**	**100.0**	**100.0**
		naive	92.8	96.8	98.3	98.8	99.7
	80%	**Nishida**	**99.6**	**99.7**	**99.7**	**99.8**	**99.9**
		naive	86.3	92.3	94.8	97.3	98.9
	60%	**Nishida**	**92.4**	**95.0**	**95.9**	**96.7**	**98.0**
		naive	61.6	73.1	79.0	84.7	91.6
0.5— 1.0	100%	**Nishida**	**99.8**	**99.8**	**99.8**	**99.9**	**99.9**
		naive	80.6	87.4	89.9	92.8	96.5
	80%	**Nishida**	**96.0**	**97.8**	**98.2**	**99.1**	**99.5**
		naive	69.5	79.8	83.9	88.1	93.1
	60%	**Nishida**	**75.1**	**82.7**	**85.7**	**89.0**	**93.4**
		naive	39.5	51.2	58.3	66.8	78.1
1.0— 1.5	100%	**Nishida**	**91.1**	**94.3**	**96.2**	**97.3**	**98.2**
		naive	55.1	65.8	71.0	76.2	84.2
	80%	**Nishida**	**68.4**	**75.9**	**79.8**	**84.9**	**90.1**
		naive	40.7	52.0	57.7	66.0	76.5
	60%	**Nishida**	**36.6**	**46.0**	**51.7**	**59.0**	**68.5**
		naive	17.5	27.3	33.3	41.3	55.5

Fast Reconstruction of 3D Objects from Single Free-Hand Line Drawing

Beom-Soo Oh and Chang-Hun Kim

Department of Computer Science and Engineering, Korea university, 1, 5ka, Anam-dong, sungbuk-ku, SEOUL 136-701, KOREA
{obs, chkim}@cgvr.korea.ac.kr

Abstract. This paper proposes an efficient algorithm that not only can narrow down the search domain of face identification but also can reconstruct various 3D objects from a single free-hand line drawing. The algorithm is executed in two stages. In the face identification stage, we generate and classify potential faces into implausible, basis and minimal faces by using geometrical and topological constraints to reduce search space. The proposed algorithm searches the space of minimal faces only to identify actual faces of an object fast. In the object reconstruction stage, we introduce 3D regularities and quadric face regularities to reconstruct 3D object accurately. Furthermore, the proposed method can be applied to a wide scope of general objects containing flat and quadric faces. The experimental results show that the proposed method identifies faces much faster than previous ones and efficiently reconstructs various objects from a single free-hand line drawing.

1 Introduction

During the conceptual design stage of mechanical parts, designers tend to draw their basic ideas of the mechanical parts mainly on papers with pencil. The method of representing 3D information by using a line drawing is easy to input geometrical information. Once the 3D model is obtained, it can be manipulated/modified, and further detail can be sketched in to obtain more detailed and accurate object. This approach provides designers with the means to convey their ideas to a CAD system. Therefore, it is necessary to develop an algorithm for automatically reconstructing 3D objects from a free-hand line drawing.

Much work has been studied on the reconstruction of 3D objects from a line drawing. Marti et al [1] expanded junction library by using line-labeling techniques. However, he relied on the line font (dashed/solid) to extract spatial information. Marill [2] suggested an optimization-based reconstruction for depth information of vertices using MSDA. Leclerc et al [3] identified all non-self-intersecting closed circuits of edges. However, his method does not applied to the case of concave faces and the case with

C. Arcelli et al. (Eds.): IWVF4, LNCS 2059, pp. 706–715, 2001.
© Springer-Verlag Berlin Heidelberg 2001

ambiguities in a sketch. In addition, he amended Marill [2]'s method using face planarity; however, their methods limited object types. Sphitalni et al [4] identified faces
efficiently by using maximum rank equation and face adjacency theory. However, it
requires large search domain of faces including numerous implausible faces. Lipson et
al [5] reconstructed 3D object containing flat and cylindrical faces based on optimization method that formalizes various image regularities. However, the reconstruction
results tend to produce a somewhat distorted 3D object, and they are limited to objects
containing flat and cylindrical faces.

Despite many methods are proposed, it is difficult to develop a practical reconstruction system for the following three reasons: (1) Because 2D line drawing corresponds to multiple 3D objects and contains tremendous potential faces, previous
methods require large combinatorial searches of face identification. (2) The reconstruction results tend to produce a somewhat distorted 3D object due to the inherent
inaccuracies in line drawing. (3) In addition, the error of reconstruction of a curved
object is significantly increased because most of 2D image regularities are derived
from planar configuration of 2D entities.

In this paper, we describe a novel algorithm for identifying 2D actual faces of an
object fast and reconstructing 3D objects efficiently from a single free-hand line
drawing. Conventional methods classify potential faces into implausible and minimal
faces to identify actual faces through the combinatorial search of minimal faces. This
paper proposes a new method for minimizing the number of minimal faces by efficiently classifying potential faces into implausible, basis and minimal faces. By introducing constraints for considering relation between line drawing and an object, we
recognize basis faces that can be determined to actual faces without searches, implausible faces that can't be actual faces and undetermined minimal faces. 2D actual faces
of an object are identified fast by searching reduced minimal faces only. Furthermore,
the proposed algorithm reconstructs various objects containing flat and quadric faces
by introducing constraints of 3D regularities and quadric face regularities.

2   Overview of the Reconstruction Process

In this paper, the input is a single free-hand line drawing [4, 5]. 2D sketch represents a
general object in wireframe. The projection reveals all edges and vertices uniquely. In
addition, all drawn lines represent real edges, silhouette curves or intersections of
faces in the 3D object. The algorithm supports general (manifold and non-manifold)
objects containing flat and quadric faces.

Reconstruction of 3D objects from 2D line drawing consists of two stages: face
identification and object reconstruction. In the face identification stage, the algorithm
first analyzes line drawing to obtain edge-vertex graph [6], then it restore topological
information of an object using topological/geometrical constraint of edge-vertex
graph. In the object reconstruction stage, the algorithm reconstructs geometrical information of an object by using various constraints of regularities. The system concept
is illustrated in Fig. 1.

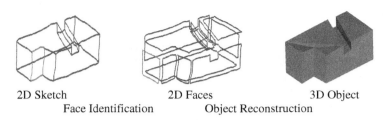

2D Sketch   2D Faces   3D Object
       Face Identification   Object Reconstruction

Fig. 1. System overview

3   Identification of Faces

In this section, we will discuss the method for reducing search space of face identification by classifying potential faces based on geometrical and topological constraints.

3.1 Face Classification

Because there are numerous potential faces that potentially correspond to faces of the depicted object in a line drawing, it is necessary to reduce search space of face identification. We describe several constraints to cut down search space by classifying potential faces into implausible, minimal and basis faces as in Table 1. We refer each face set to *PF*, *IF*, *MF*, and *BF*, respectively.

Table 1. Classification of potential faces

Face class	Description	Symbol
Potential	All candidate faces	PF
Implausible	Implausible faces of all object	IF
Minimal	Candidate of actual faces of an object	MF
Basis	Actual faces of all objects	BF

If we set the actual faces of an object be *AF*, we can drive Equations (1) - (3), which mean that we can identify actual faces by searching minimal faces only.

$$IF \cup MF \cup BF = PF .$$
(1)

$$IF \cap MF = MF \cap BF = BF \cap IF = \varnothing .$$
(2)

$$BF \subseteq AF, AF \subseteq (BF \cup PF) .$$
(3)

3.2 Face Classification Step

We define rank $R(v)$ and $R(e)$ as the number of faces whose boundary contains that entity, and the upper bound of the ranks are denoted by $R^+(v)$ and $R^+(e)$ [3]. In addition, we define $RF(v)$ and $RF(e)$ as the sets of faces whose boundary contains that entity.

There are 6 steps for classification of potential faces.

Step 1. Generate all potential faces using n edges[4], i.e., PF. Initially, $IF = MF = BF = \phi$.

$$PF = make\_potential\_face\{e_1, \cdots, e_n\}. \tag{4}$$

Step 2. Find implausible faces, IF, contain internal edge(s).

$$\{f \mid f, f_1, f_2 \in PF, [f = (f_1 \cup f_2) - (f_1 \cap f_2), if\ \forall e \in (f_1 \cap f_2), \tag{5}$$
$$e\ is\ the\ internal\ edge\ of\ f\ \}$$

Step 3. Find basis faces, BF.

$$\{f \mid f \in (PF - IF) = F, \tag{6}$$
$$[Connected\ edges\ e_1, e_2, n[RF(e_1) \cap RF(e_2)] = 1, f \in RF(e_1) \cap RF(e_2)]\}$$

Step 4. Find implausible faces by using maximum rank.

$$\{f \mid f \in (PF - BF - IF), [\exists e, RBF(e) = R^+(e), f \in (RF(e) - RBF(e))]\}. \tag{7}$$

Step 5. Recover over-reduced minimal faces.

$$\{f \mid f \in IF, F = (PF - IF), f \in makeface\{e \mid (R + (e) - n(RF(e)) \geq 1\}\}. \tag{8}$$

Step 6. Repeat Step 3 ~ Step 5 until no change of the face class. All faces in (PF-IF-BF) are undetermined minimal faces.

For example, in Step 1, we can generate 15 potential faces from 2D sketch of itself in Fig. 2.

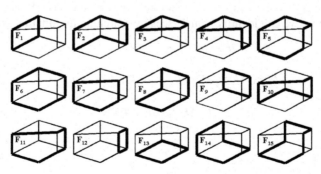

Fig. 2. 15 all potential faces that could be used in 3D reconstruction from 2D sketch

In Step 2, we can find 7 implausible faces, $f_2, f_4, f_5, f_6, f_8, f_{11}$, and f_{14}. Applying Equation (5) we can find 6 basis faces, $f_1, f_3, f_7, f_9, f_{12}, f_{13}$. However, according to the face adjacency theorem [3], faces f_7, f_{13} cannot coexist. Therefore, some constraints must be added into Step 3.

$$\{f \mid f_1, f_2 \in BF, \forall e \in (f_1 \cap f_2) \text{ are smooth}[4].\} . \tag{9}$$

By applying Equation (9), two faces f_7 and f_{13} remains in potential faces. In Step 4, we find implausible faces f_{10} and f_{15} as shown Fig. 3. Step 5~6 do not affect on this example. Finally, we extract 4 basis faces f_1, f_3, f_9 and f_{12}, and two minimal faces f_7 and f_{13}. By searching minimal faces only, we can identify actual faces of an object fast.

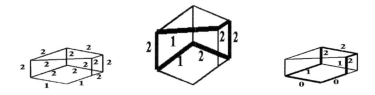

Fig. 3. Implausible faces complying with Equation (7)

Identifying faces of an object in sketch can be formulated as a selection problem, i.e., selecting k faces among the m potential faces such that the k faces represent a valid object by combinatorial searches 2^m.

We can identify actual faces by using minimizing Equation (10) [4]. By minimizing the number of minimal faces in combinatorial search, we identify actual faces fast.

$$\mid R^+(e) - R(e) \mid + \mid R^+(v) - R(v) \mid . \tag{10}$$

4   Reconstruction of 3D Objects

In this section, we introduce constraints of 3D regularities and quadric regularities to minimize distortion of the reconstructed 3D object resulted from inaccurate 2D sketch.

4.1 Basic Reconstruction Algorithm

To reconstruct geometrical information of 3D object, Lipson et al [5] proposed 13 image regularities. A 3D configuration can be represented a compliance function by summing the contributions of the regularity terms. The final compliance function to be optimized takes the form

$$W^T \sum [\alpha_{regularity}] . \tag{11}$$

However, reconstruction results tend to produce a somewhat distorted 3D object due to the inherent inaccuracies in the sketch and 2D image regularities.

4.2 3D Regularities and Quadric Face Regularities

We introduce some geometrical constraints of 3D regularities and quadric face regularities that are used into Equation (11) with 2D image regularities to reconstruct 3D objects more accurately.

[Face parallelism] : A parallel pair of planes in the sketch plane reflects parallelism in space. The term used to evaluate is

$$\alpha_{\substack{face \\ parallelism}} = \sum_{i=1}^{n} [\cos^{-1}(n_1 \cdot n_2)]^2 \cdot \tag{12}$$

where, n_1 and n_2 denote all possible pairs of normal of parallel faces, and n is the number of pairs of parallel faces.

[Face orthogonality] : An orthogonal pair of faces in the sketch plane reflects orthogonality in space. The term used to evaluate is

$$\alpha_{\substack{face \\ orthogonality}} = \sum_{i=1}^{n} [\sin^{-1}(n_1 \cdot n_2)]^2 \cdot \tag{13}$$

where, n_1 and n_2 denote all possible pairs of normal of orthogonal faces, and n is the number of pairs of orthogonal faces.

It is simple to find parallel or orthogonal faces by using angular distribution graph that identifies prevailing axis system. First, we find each edge's prevailing axis. Then, all faces contain at most two prevailing axes. If two faces containing two axes have the same axes, and then they are parallel faces, else they are orthogonal faces.

In addition, we introduce simple radius regularities affecting quadric faces.

[Radius equality]

$$\alpha_{\substack{radius \\ equality}} = \sum_{i=1}^{n} (d_1 - d_2)^2 \cdot \tag{14}$$

where, d_1 and d_2 are distance from center of curve to the end-vertices, and n is the number of quadric faces.

In addition, we assign high weight to the regularity of face planarity to reconstruct the most plausible solution with reality.

5 Experiments

5.1 Results

To estimate the efficiency of the proposed algorithm, we applied the method to various objects as shown in Fig. 5. The experiment is done on a PC with Pentium III processor (450MHz).

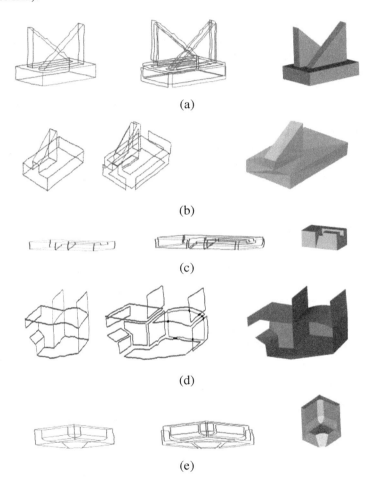

(a)

(b)

(c)

(d)

(e)

Fig. 4. Experimental Results (left: 2D sketch; center: 2D faces; right: 3D object)

5.2 Discussions

Table 2 shows that our method efficiently narrows down the search space of face identification to a manageable size. And, the total time is dramatically reduced in most cases by using the proposed method.

To evaluate the effect of 3D regularities, we check the 3D error and 2D error. We define 3D error as the distance between the depth of reconstructed object's vertices and the real depth of synthetic object's vertices, and we define 2D error as the sum of regularities proposed by Lipson et al [5].

When 3D regularities and quadric face regularities are used to improve the model (after 20 iteration), they can perturb the error curve as demonstrated by the sudden spike (Fig.5 and Fig. 6). However, they improve significantly the shape of an object with more iteration.

Fig. 5 shows that although 2D error is not improved, the constraints of 3D regularities improve the shape of reconstructed polyhedral object significantly. Fig. 6 shows the effect of quadric face regularities. When 3D regularities and quadric faces regularities are used in curved object, they reduces 2D error as well as 3D error significantly in the case of quadric object. However, the error in curved objects still more significant than that in polyhedral objects because most of regularities are derived from 2D planar configuration of a line drawing.

We apply the proposed method to real tower scenes to evaluate the efficiency of the proposed algorithm (Fig. 7).

Table 2. Evaluation of face identification

Fig. 4.	Method	PF	IF	BF	MF	SOL	Time(sec)
(a)	A	33	17	16	0	1	30
	B	33	14	-	19	1	142
(b)	A	37	25	8	4	2	60
	B	37	18	-	19	2	60
(c)	A	279	265	14	0	1	138
	B	279	159	-	120	1	1200
(d)	A	205	193	12	0	1	551
	B	205	164	-	41	1	1091
(e)	A	896	882	14	0	1	4420
	B	896	679	-	202	1	7283

A: proposed method, B: [4]'s method

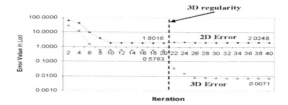

Fig. 5. Evaluation of 3D regularities in the case of polyhedral object: Fig. 4(c)

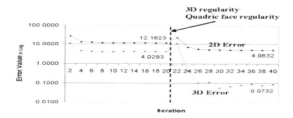

Fig. 6. Evaluation of 3D regularities and quadric face regularities in the case of quadric object: Fig. 4(e)

6   Conclusions

We have presented an efficient algorithm for reconstructing a 3D object from a single free-hand line drawing. We cut down the search domain by classifying potential faces into implausible, basis and minimal faces. As a result, we can identify actual faces of an object fast by searching minimal faces only. In addition, we introduced the 3D regularities and quadric regularities to reconstruct various 3D objects more accurately.

Future works will be focused on the regularities of curved faces and on the reduction of reconstruction errors.

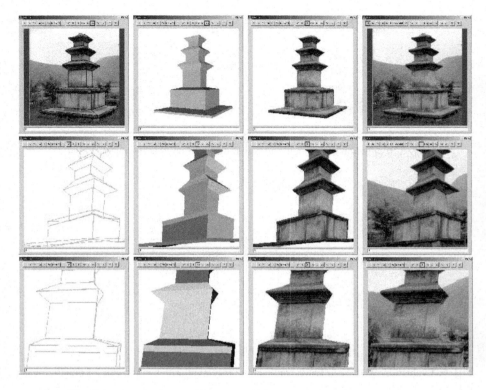

Fig. 7. Real tower scene : 3D reconstruction & navigation from one image

References

1. E. Martí, J., Regincós, J., López-Krahe., J.J. Villanueva.: Hand line drawing interpretation as three-dimensional object. Vol.32. Signal Processing (1993) 91-110
2. B T. Marill.: Emulating the Human Interpretation of Line Drawings as Three Dimensional Objects. Vol. 6, No. 2. Int. J. Comput. Vision (1991) 147-161
3. Y. G. Leclerc., M. A. Fiscler.: An optimization based approach to the interpretation of single line drawings as 3D wire frames. Vol. 9, No 2. Int. J. of Computer Vision (1992) 113-136
4. M. Shpitalni., H. Lipson.: Identification of faces in a 2D Line Drawing Projection of a Wireframe Object. Vol. 18, No. 10. IEEE Trans. Pattern Analysis & Machine Intell (1996) 1000-1012
5. H. Lipson., M. Shpitalni.: Optimization-Based Reconstruction of a 3D Object From a Single Freehand Line Drawing. Journal of Computer Aided Design (1995)
6. M. Shpitalni., H. Lipson.: Classification of Sketch Strokes and Corner Detection using Conic Sections and Adaptive Clustering. Vol. 119, No. 2. Trans of the ASME, Journal of Mechanical Design (1997)

Color and Shape Index for Region-Based Image Retrieval

B.G. Prasad[1*], S.K. Gupta[2], and K.K. Biswas[2]

[1] Department of Computer Science and Engineering,
P.E.S.College of Engineering, Mandya, 571402, INDIA.
[2] Department of Computer Science and Engineering,
Indian Institute of Technology, New Delhi, 110016, INDIA.
{bgprasad,skg,kkb}@cse.iitd.ernet.in,
WWW home page: http://www.cse.iitd.ernet.in/~ skgupta/

Abstract. Most CBIR systems use low-level visual features for representation and retrieval of images. Generally such methods suffer from the problems of high-dimensionality leading to more computational time and inefficient indexing and retrieval performance. This paper focuses on a low-dimensional color and shape based indexing technique for achieving efficient and effective retrieval performance. We propose a combined index using color and shape features. A new shape similarity measure is proposed which is shown to be more effective. Images are indexed by dominant color regions and similar images form an image cluster stored in a hash structure. Each region within an image is further indexed by a region-based shape index. The shape index is invariant to translation, rotation and scaling. A JAVA based query engine supporting query-by-example is built to retrieve images by color and shape. The retrieval performance is studied and compared with a region-based shape indexing scheme.

1 Introduction

The past few years have seen many advanced techniques evolving in Content-Based Image Retrieval (CBIR). Applications like medicine, entertainment, education, manufacturing, etc. make use of vast amount of visual data in the form of images. This envisages the need for fast and effective retrieval mechanisms in an efficient manner. A major approach directed towards achieving this goal is the use of low-level visual features of the image data to segment, index and retrieve relevant images from the image database. Recent CBIR systems based on features like color, shape, texture, spatial layout, object motion, etc., are cited in [1],[2]. Of all the visual features, color is the most dominant and distinguishing one in almost all applications.

---

* This work is partly supported by the AICTE Young Teachers Career Award

C. Arcelli et al. (Eds.): IWVF4, LNCS 2059, pp. 716–725, 2001.

1.1   Previous Work in CBIR

Current CBIR systems such as IBM's QBIC [3],[4] allow automatic retrieval based on simple characteristics and distribution of color, shape and texture. But they do not consider structural and spatial relationships and fail to capture meaningful contents of the image in general. Also the object identification is semi-automatic. The Chabot project [5] integrates a relational database with retrieval by color analysis. Textual meta-data along with color histograms form the main features used. VisualSEEK [6] allows query by color and spatial layout of color regions. Text based tools for annotating images and searching is provided. A new image representation which uses the concept of localized coherent regions in color and texture space is presented by Carson et al. [7]. Segmentation based on the above features called "Blobworld" is used and query is based on these features.

Some of the popular methods to characterize color information in images are color histograms [8],[9], color moments [10] and color correlograms [11]. Though all these methods provide good characterization of color, they have the problem of high-dimensionality. This leads to more computational time, inefficient indexing and performance. To overcome these problems, use of SVD [9], dominant color regions approach [12],[13] and color clustering [14] have been proposed.

1.2   Recent Work in Shape-Based CBIR

Shape also is an important feature for perceptual object recognition and classification of images. It has been used in CBIR in conjunction with color and other features for indexing and retrieval.

Shape description or representation is an important issue both in object recognition and classification. Many techniques, including chain code, polygonal approximations, curvature, fourier descriptors and moment descriptors have been proposed and used in various applications [15]. Recently, techniques using shape measure as an important feature have been used for CBIR. Features such as moment invariants and area of region have been used in [3],[16], but do not give perceptual shape similarity. Cortelazzo [17] used chain codes for trademark image shape description and string matching technique. The chain codes are not normalized and string matching is not invariant to shape scale. Jain and Vailaya [18] proposed a shape representation based on the use of a histogram of edge directions. But these are not scale normalized and computationally expensive in similarity measures. Mehrotra and Gary [19] used coordinates of significant points on the boundary as shape representation. It is not a compact representation and the similarity measure is computationally expensive. Jagadish [20] proposed shape decomposition into a number of rectangles and two pairs of coordinates for each rectangle are used to represent the shape. It is not rotation invariant.

A region-based shape representation and indexing scheme that is translation, rotation and scale invariant is proposed by Lu and Sajjanhar [21]. It conforms to human similarity perception. They have compared it to Fourier descriptor model and found their method to be better. But, the images database consists of only 2D planar shapes and they have considered only binary images. Moreover, shapes with similar eccentricity but different shapes are retrieved as matched images. Our aim is to extend this method to represent color image regions and augment the color index of our previous work [13] with the shape features. Our shape indexing feature and similarity measure is different and shown to be effective in retrieval. A combined index based on color and shape has been implemented to improve retrieval efficiency and effectiveness.

The paper is organised as follows: Section 2 describes the color and shape features used for indexing. The indexing scheme, querying and similarity measure are explained in section 3. Section 4 highlights the results of our approach and sample output. A comparision of performance with the scheme employed in [21] is also covered in this section.

2   Color and Shape Features

The initial step in our approach is to index images based on dominant color regions [13]. Image regions thus obtained after segmentation and indexing are used as input to the shape module. The region-based shape representation proposed in [21] is modified to calculate the shape features required for our proposed shape indexing technique and similarity measure. It is simple to calculate and robust. We show that the retrieval effectiveness is better compared to their method.

2.1   Color Indexing Approach

To segment images based on dominant colors, a color quantization in RGB space using 25 perceptual color categories is employed as is done in [13]. From the segmented image we find the enclosing minimum bounding rectangle (MBR) of the region, its location, image path, number of regions in the image, etc., and all these are stored in a metafile for further use in the construction of an image index tree. An index tree for the entire database is constructed when the query engine is initiated. At query time, similar images matched on the basis of color and spatial location are retrieved. To the above color index, we have included a region-based shape index similar to the one in [21] which is invariant to rotation, scaling and translation. Since their representation is suited only for 2-D binary image regions, we have used a different shape feature to index the color regions and also a suitable shape similarity measure. A comparison of the two techniques has been carried out for an image database consisting of flags, flowers, fruits, vegetables and simulated shape regions.

2.2   Shape Representation

Definitions of terminology:

- **Major axis**: it is the straight line segment joining the two points on the boundary farthest away from each other (in case of more than one, select any one).

- **Minor axis**: it is perpendicular to the major axis and of such length that a rectangle with sides parallel to major and minor axes that just encloses the boundary can be formed using the lengths of the major and minor axes.
- **Basic rectangle**: the above rectangle formed with major and minor axes as its two sides is called basic rectangle.
- **Eccentricity**: the ratio of the major to the minor axis is called eccentricity of the region.
- **Centroid or Center of gravity**: a single point of an object/region towards which other objects/regions are gravitationally attracted. For 2D shapes, the coordinates (X_c, Y_c) of the centroid are defined as:

$$X_c = \sum_x \sum_y f(x,y)x / \sum_x \sum_y f(x,y)$$
$$Y_c = \sum_x \sum_y f(x,y)y / \sum_x \sum_y f(x,y)$$

where (x,y) are pixel coordinates and f(x,y) is set to 1 for points within or on the shape and set to 0 elsewhere.

Basic idea. Given a shape region, a grid space consisting of fixed-size square cells is placed over it so as to cover the entire shape region as shown in figure 1. We assign a "1" to cells with at least 25% of pixels covered and "0" to each of the other cells. A binary sequence of 1's and 0's from left to right is obtained as the shape feature representation. For example, the shape in the above figure can be represented by a binary sequence 11111111 01111111 00110110 00000110 00000010 00000000.

The smaller the grid size, the more accurate the shape representation is and more the storage and computation requirements. The representation is compact, easy to obtain and translation invariant. Hence, a scale and rotation normalization is carried out to make it invariant to scale and rotation.

Rotation normalization. The purpose of rotation normalization is to place shape regions in a unique common orientation. Hence the shape region is rotated such that its major axis is parallel to the x-axis.

There are still two possibilities as shown in figure 1 and 2, caused by 180^0 rotation. Further, two more orientations are possible due to the horizontal and vertical flips of the original region as shown in figures 3 and 4 respectively. Two binary sequences are needed for representing these two orientations. But only

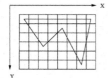
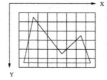
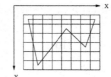
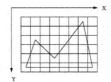

Fig. 1. Grid on region overlayed.

Fig. 2. Region rotated by 180°.

Fig. 3. Horizontal flip of region.

Fig. 4. Vertical flip of region.

one sequence is stored and at the time of retrieval we can account for these two sequences.

Scale normalization. To achieve scale normalization, we proportionally scale all the shape regions so that their major axes have the same length of 96 pixels.

Shape index. Once the shape region has been normalized for scale and rotation invariance, using a fixed size of grid cells (say 8x8), we obtain a unique sequence for each shape region. The grid size in our proposed method is kept as 96 x 96 pixels. Each sub-grid cell is of size 12x12 pixels giving a binary sequence of length 64 bits per shape region. Using this sequence, we find both the row and column totals of the 8x8 grid and store them as our shape index, which is more robust and gives a better perceptual representation to the coverage of the shape. A suitable shape similarity measure using this index is employed for matching images at query time.

3 Indexing Scheme and Querying

3.1 Color Index

A composite index based on a color look-up table is formed consisting of 25 colors. The index is unique and given by the equation below:

$$Index = \sum_{i=1}^{No\_RGN} C_i * 25^{n-i}$$

Suppose (C1, C2, C3) are the color indices of three dominant regions found within an image, where C1 represents index of the first dominant region, C2 represents index of the second dominant region and C3 represents index of the third dominant region.

Then, the index is given by

$$Index = C1 * 25^2 + C2 * 25^1 + C3 * 25^0$$

Images with similar indices are stored in same hash entry of the hash table structure. Each entry also stores the color region features such as location, area, percentage of color, etc., associated to each region of the image which is used in the matching criteria.

3.2 Shape Index

For each color region processed above, we compute the shape descriptor as follows:

1. Compute the major and minor axes of each color region.

2. Rotate the shape region to align the major axis to X-axis to achieve rotation normalization and scale it such that major axis is of standard fixed length (96 pixels).
3. Place the grid of fixed size (96x96 pixels) over the normalized color region and obtain the binary sequence by assigning 1's and 0's accordingly.
4. Using the binary sequence, compute the row and column total vectors. These along with the eccentricity form the shape index for the region.

3.3 Querying

Given a query image, we apply the same process on the query image to obtain the color and shape features. Our implementation supports both Query-by-example and Query-by-feature for color matching. The shape matching module supports only Query-by-example. Based on the color index of the query image, a list of matching images are retrieved from the hash structure. Then the shape descriptors are used to find matching images from this initial set to retrieve the final images matched on both color and shape.

The query process is as follows:

1. The query image is processed to obtain a list of matching images based only on color features.
2. For each color region in the query image, the shape representation of each region is evaluated. To take care of the problem of 180^0 rotation and vertical and horizontal flips, we need to store 4 sets of the shape index.
3. Compare the shape index of regions in the query image to those in the list of images retrieved on color.
4. Regions with only matching eccentricity within a threshold (t) are compared for shape similarity.
5. The matching images are ordered depending on the difference in the sum of the difference in row and column vectors between query and matching image.

3.4 Similarity Measure

Let R and R' represent the row vectors of test image and query image respectively. Similarly, C and C' represent the column vectors of the test image and query image respectively. The similarity measure is computed as follows:

1. Calculate the row and column vectors of all the regions in the query image.
2. Find the row and column difference between query image regions and regions in the image to be tested using the equation:

$$R_d = \sum_i \left(|R_i - R'_i| \right)$$
$$C_d = \sum_i \left(|C_i - C'_i| \right)$$

where R_d and C_d are the row and column differences between the test image and query image region, R_i and C_i are the ith bit of row and column vectors in image and R'_i and C'_i are the ith bit of row and column vectors in the query image.
3. If $(R_d + C_d) < T$ (threshold), then the images match.

4   Experimental Results and Performance

The experimental database consists of about 200 images of flags and 120 images of fruits, flowers and simulated objects(squares, rectangles, triangles, circles, etc). Each image in the database is indexed on color and shape features. A hash table stores images of similar index based on the features extracted. Images are retrieved first based on the color index and displayed. Then shapes of all regions in the query image are compared to the region shapes in the displayed set to find images similar on the basis of shape index.

An example output for retrieval for the image database of flowers, fruits and simulated shape regions is shown in figure 5 for matching on color and in figure 6 for matching on shape. It can be observed that images non-similar in shape are eliminated. The image on the left of the screen is the query image. Figures 7 and 8 show the corresponding results for the image database of flags.

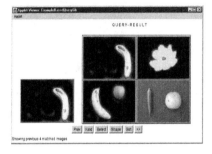

Fig. 5. Retrieval of images based on matching color regions

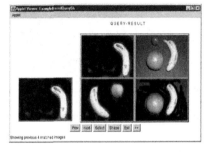

Fig. 6. Retrieval of images based on matching shape regions

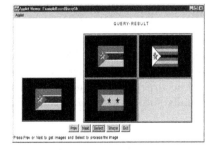

Fig. 7. Retrieval of images based on matching color regions

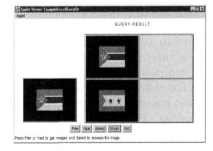

Fig. 8. Retrieval of images based on matching shape regions

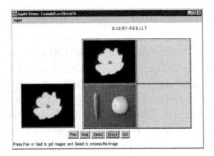

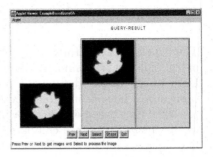

Fig. 9. Retrieval results based on similarity measure of [21].

Fig. 10. Retrieval results based on our proposed similarity measure.

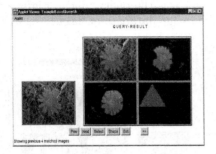

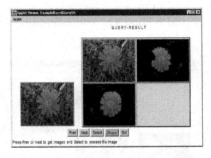

Fig. 11. Retreival results based on similarity measure of [21].

Fig. 12. Retrieval results based on our proposed similarity measure.

We have compared the results of our technique with that proposed in [21]. The output for the two comparative techniques are shown in figures 9 and 10 for the image database of flowers, fruits and simulated shape regions. Figures 11 and 12 show the corresponding difference in retrieval results for the flag image database. The outputs show that there is better pruning of the matched images using the row and column vector based technique for matching images.

The retireval performance is measured using recall and precision, as is standard in all CBIR systems. Recall measures the ability of retrieving all relevant or similar items in the database. It is defined as the ratio between the number of relevant or perceptually similar items retrieved and the total relevant items in the database. Precision measures the retrieval accuracy and is defined as the ratio between the number of relevant or perceptually similar items and the total number of items retrieved.

The graph in figure 13 shows the retrieval performance for the image database of flowers, fruits and simulated shapes in terms of color and shape curves. Similar analysis is done for the flag image database and shown in figure 14.

A comparative study of the two different techniques of indexing and retrieval is shon in figures 15 and 16 respectively. It is seen that our indexing method and similarity measure provides better retrieval effectiveness.

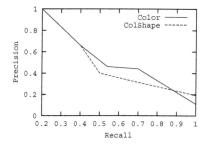

Fig. 13. Recall-precision graph for image database of flowers etc.

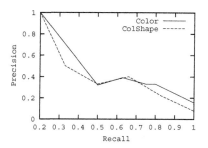

Fig. 14. Recall-precision graph for image database of flags.

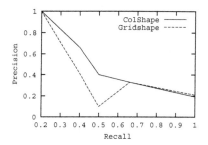

Fig. 15. Recall-Precision comparing the two similarity measures for image database of flowers, etc.

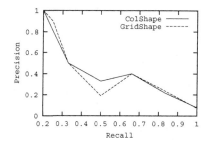

Fig. 16. Recall-Precision comparing the two similarity measures for image database of flags.

5   Conclusions

A combined color and shape based low-dimensional indexing technique has been implemented. Images are segmented into dominant regions based on perceptually similar color regions using a color quantized indexing method. Such segmented out regions are stored in a hash structure as similar image clusters. Shape features of these regions are used to further prune the retrieval of images from a sample image database. The shape representation is based on a grid-based coverage of the region which is normalized to achieve invariance in scale, rotation and size. The index is a robust one. Our proposed index based on row and column vectors and the related similarity measure is shown to provide an efficient and effective retrieval performance. A JAVA based search engine using query-by-example has been developed on Windows-NT platform. The results and performance analysis of our method shows that it is effective and efficient. It can be further enhanced by including texture features.

References

1. V.N.Gudivada and V.V.Raghavan.: Special issue on content-based image retrieval systems - guest eds. IEEE Computer. **28(9)** (1995) 18–22
2. M.De Marsicoi, L.Cinque, and S.Levialdi.: Indexing pictorial documents by their content: a survey of current techniques. Image and Vision Computing. **15(2)** (1997) 119–141
3. M.Flickner etal.,: Query by image and video content: the qbic system. IEEE Computer. **28(9)** (1995) 23–32
4. W.Niblack etal.,: The qbic project: Querying images by content using color, texture and shape. In Storage and Retrieval for Image and Video Databases (SPIE). **1908** 173–187
5. V.E.Ogle and M.Stonebaker.: Chabot: Retrieval from a relational database of images. IEEE Computer. **28(9)** (1995) 40–48
6. J.R.Smith and S.F.Chang.: Visualseek: A fully automated content-based image query system. ACM Multimedia. (1996) 87–98
7. C.Carson etal.,: Region-based image querying.: In CVPR'97 Workshop on Content-based Access to Image and Video libraries (CAIVL'97) (1997)
8. M.J.Swain and D.H.Ballard.: Color indexing. Intl. Journal of Computer Vision. **Vol. 7 No. 1** (1991) 11-32
9. J.Hafner etal.,: Efficient color histogram indexing for quadratic form distance functions. PAMI. **Vol. 17 No. 7** July. (1995) 729-736
10. M.Stricker and A.Dimai.: Color indexing with weak spatial constraints. Proceedings of SPIE Storage and Retrieval of Still Image and Video Databases IV. **Vol. 2670** (1996) 29-40
11. J.Huang etal.,: Image indexing using color correlograms. Proceedings of CVPR. (1997) 762-768
12. H.Zhang etal,.: Image retrieval based on color features: an evaluation study. Proceedings of SPIE. **Vol. 2606** (1995) 212-220
13. K.C.Ravishankar, B.G.Prasad, S.K.Gupta and K.K.Biswas.: Dominant Color Region Based Indexing Technique for CBIR. In proceedings of the International Conference on Image Analysis and Processing (ICIAP'99). Venice. Italy. Sept. (1999) 887-892
14. X.Wan and C.J.Kuo.: A multiresolution color clustering approach to image indexing and retrieval. Proceedings of ICASSP. (1998)
15. R.C.Gonzalez and P.Wintz.: Digital Image Processing. 2nd Edition. Addison-Wesley. Reading. Mass. (1987)
16. D.Mohamad, G.Sulong and S.S.Ipson.: Trademark Matching using Invariant Moments. Second Asian Conference on Computer Vision. 5-8 Dec, Singapore. (1995)
17. G.Cortelazzo etal.,: Trademark Shapes Description by String-Matching Techniques. Pattern Recognition. **27(8)** (1994) 1005-1018
18. A.K.Jain and A.Vailaya.: Image Retrieval using Color and Shape. Second Asian Conference on Computer Vision. 5-8 Dec. Singapore. (1995) 529-533
19. R.Mehrotra and J.E.Gary.: Similar-Shape Retrieval in Shape Data Management. IEEE Computer. **28(9)** (1995) 57-62
20. H.V.Jagadish.: A Retrieval Technique for Similar Shapes. Proceedings of ACM SIGMOD. Colorado. ACM. New York. May (1991) 208-217
21. G.Lu and A.Sajjanhar.: Region-Based Shape Representation and Similarity Measure Suitable for Content-Based Image Retrieval. Multimedia Systems. **7** (1999) 165-174

Virtual Drilling in 3-D Objects Reconstructed by Shape-Based Interpolation

Adrian G. Borş[1], Lefteris Kechagias[2], and Ioannis Pitas[2]

[1] Department of Computer Science, University of York
York YO10 5DD, U.K. - E-mail: adrian.bors@cs.york.ac.uk
[2] Department of Informatics, University of Thessaloniki, Box 451,
54006 Thessaloniki, Greece - E-mail: {lkechagi,pitas}@zeus.csd.auth.gr

Abstract. In this paper we propose a virtual drilling algorithm which is applied on 3-D objects. We consider that initial we are provided with a sparse set of parallel and equi-distant slices of a 3-D object. We propose a volumetric interpolation algorithm for recovering the 3-D shape from the given set of slices. This algorithm employs a morphology morphing transform. Drilling is simulated on the resulting volume as a 3-D erosion operation. The proposed technique is applied for virtual drilling of teeth considering various burr shapes as erosion elements.

1 Introduction

3-D object representation and processing simulation is required in many fields such as medicine, architecture, computer aided design, etc. [1,2,3]. Very often we are not provided with the complete information about the object to be modeled and processed. In this paper we show how virtual processing operations can be simulated on volumes described by means of a group of slices representing parallel sections of its structure. Depending on the type of the object we represent such images can be acquired by Computer Tomography (CT), Magnetic Resonance Imaging (MRI) or by mechanical slicing and digitization. Usually, the pixel size within a slice is different from the spacing between two adjacent slices. In such situations it is necessary to interpolate additional slices in order to obtain an accurate volumetric description of the object. In this paper we employ mathematical morphology for reconstructing a full 3-D shape from a group of slices and afterwards for modeling 3-D drilling in the resulting shape.

There are two main categories of interpolation techniques for reconstructing objects from sparse sets: grey-level and shape-based. Grey-level interpolation methods employ nearest-neighbor, splines, linear [2], or polynomial interpolation. Other algorithms employ feature matching [4] or homogeneity similarity [5] for determining the direction of interpolation. Interpolation of additional cross-sections from shape contours in vector form is described in [6]. A distance function from each pixel to the object boundary is considered for interpolation in [1]. Extensions of this algorithm are proposed in [7,8]. An algorithm which uses the elastic matching interpolation, spline theory and surface consistency is

C. Arcelli et al. (Eds.): IWVF4, LNCS 2059, pp. 729–738, 2001.
© Springer-Verlag Berlin Heidelberg 2001

considered in [9]. In [10] each slice is eroded by a morphological operator until its number of pixels reaches the mean of those from the slices to be interpolated. A mixed shape and grey level based interpolation method is proposed in [11].

A mathematical morphology based function interpolation algorithm called the skeleton by influence zones transform (SKIZ) has been employed in [12]. The SKIZ transform interpolates by employing dilations of the intersection and of the complementary of the union of two sets [12]. However, such an approach does not correspond to a natural morphing of one set into the next one. In this paper we propose a morphing procedure for estimating the intermediary slices of the two given sets. The morphing transforms two neighboring sets by using combinations of dilations and erosions. The interpolated set corresponds to the idempotency of the two morphed sets after a certain number of iterations. This produces a new set sharing similarities in shape with both initial sets. The morphing transformation is applied repeatedly onto the new stack of interpolated sets until we recover an appropriate object shape.

After reconstructing the 3-D volumes we simulate a drilling operation. Virtual surgery using 3-D data visualization has lately attracted a lot of attention due to its potential use in surgical intervention planning and training [14]. We propose a morphological algorithm using 3-D structural elements for simulating drilling. Virtual drilling is modeled as a succession of volumetric erosions oriented along the chosen direction. The paper is organized as follows. Section 2 describes the interpolation algorithm and Section 3 the simulation of drilling using a 3-D operator. In Section 4 we provide experimental results when applying the proposed virtual drilling tool on a set of teeth reconstructed by interpolation, while the conclusions of this study are drawn in Section 5.

2   Geometrically Constrained 3-D Interpolation

Let us assume that we have two sets X_i and X_{i+1}, which are sharing at least one common point $X_i \cap X_{i+1} \neq \emptyset$. We align these sets according to an $(n-1)$-dimensional hyperplane (axis for 2-D sets) using matching or a centering operation. Let us consider $X_{i,m}$ an element (pixel in 2-D or voxel in 3-D) contained into the set X_i, where m denotes an ordering number and denote the complement (background) of the set X_i by $X_i^c = E - X_i$. After alignment, each element $X_{i,m}$ in one set will have a corresponding element which may be a member of the other set $X_{i+1,m} \in X_{i+1}$, or may be part of its background $X_{i+1,m}^c \in X_{i+1}^c$.

Our morphing transformation ensures a smooth transition from one shape set to the other one by means of generating several intermediary sets. Let us consider the elements located on the boundary set, denoted by C_i :

$$C_i = \{X_{i,m} \in X_i | \; \exists X_{i,l}^c \; l \in \mathcal{N}_{B_1}(X_{i,m})\} \tag{1}$$

where $\mathcal{N}_{B_1}(X_{i,m})$ denotes the neighborhood of the location $X_{i,m}$, having the same size and shape as the structuring element B_1. In our morphing operation, the elements of a boundary set C_i are changed differently according to their correspondences from the second given set X_{i+1}. These changes are defined in terms of mathematical morphology basic operations such as dilations and erosions [13]. The dilation of a set A by the structuring element B is given by :

$$A \oplus B = \bigcup_{b \in B} A_b \qquad (2)$$

where $\oplus$ denotes dilation and A_b represents a structuring element centered onto an element of the set A. The erosion of a set A by using the structuring element B is given by :

$$A \ominus B = \bigcap_{b \in B} A_b \qquad (3)$$

where $\ominus$ denotes erosion. These operations correspond to the Minkowski set addition and subtraction. The most commonly used structuring element is the elementary ball of dimension n. The dilation with the elementary ball, *i.e.* for $n = 1$, expands the given set with a uniform layer of elements while the erosion operator takes out such a layer from the given set. The structuring element considered in this paper consists of a pixel and its horizontal and vertical immediate neighbors. We can identify three possible correspondence cases for the elements of the two aligned sets. One situation occurs when the border region of one set corresponds to the interior of the other set. In this case we dilate the border elements :

$$\text{If} \quad X_{i,m} \in C_i \wedge X_{i+1,m} \notin C_{i+1}$$
$$\text{then} \qquad \text{perform } X_{i,m} \oplus B_1 \qquad (4)$$

where B_1 is the structuring element for the set X_i. A second case occurs when the border region of one set corresponds to the background of the other set. In this situation we have erosions of the boundary elements :

$$\text{If} \quad X_{i,m} \in C_i \wedge \exists\, X_{i+1,m}^c$$
$$\text{then} \quad \text{perform } X_{i,m} \ominus B_1 \qquad (5)$$

No modifications are performed when both corresponding elements are members of their sets boundary :

$$\text{If} \quad X_{i,m} \in C_i \wedge X_{i+1,m} \in C_{i+1}$$
$$\text{then} \qquad \text{perform no change} \qquad (6)$$

The last situation corresponds to the regions where the two sets coincide locally and no change is necessary, while (4) and (5) correspond to morphing transformations.

By including all these local changes we define the following morphing transformation applied on the set X_i depending onto the set X_{i+1} and on the structuring element B_1 :

$$f(X_i | X_{i+1}, B_1) = [(X_i \ominus B_1) \bigcup ((X_i \bigcap X_{i+1}) \oplus B_1)] \bigcap (X_i \bigcup X_{i+1}) \qquad (7)$$

A similar morphing operation $f(X_{i+1} | X_i, B_2)$ is defined onto the set X_{i+1} depending on the set X_i and on the structuring element B_2. The proposed morphing transformation is illustrated in Figure 1.

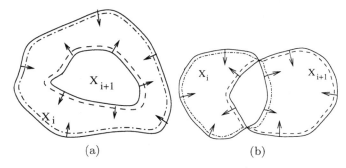

Fig. 1. Exemplification of mathematical morphology morphing. The result produced by equation $f(X_{i+1}|X_i, B_2)$ is represented with dashed lines while the result produced by equation $f(X_i|X_{i+1}, B_1)$ is represented with dot-dashed lines: (a) $(X_i \ominus B_1) \supset (X_{i+1} \oplus B_2)$; (b) $X_i - X_{i+1} \neq \emptyset$ and $X_{i+1} - X_i \neq \emptyset$.

The morphing operation defined by $f(X_i|X_{i+1}, B_1)$ and by $f(X_{i+1}|X_i, B_2)$ is applied iteratively onto the sets resulted from the previous morphings. For isotropic interpolation we use identical structuring elements, $B_1 = B_2 = B$, when morphing the two sets. The succession of morphing operations creates new sets starting from the two initial extremes. With each iteration these sets are closer in shape and size to each other. Eventually, the morphological transformations processing each slice will lead to the idempotency of the resulting sets. This set will represent the resulting interpolation.

This procedure can be easily extended for gray scale objects. In our approach we employ bilinear gray-level interpolation in the overlapping area of the two sets $(X_i \cap X_{i+1})$. In the regions where only one of the sets is defined (*i.e.* $X_i - X_{i+1} \neq \emptyset$ and $X_{i+1} - X_i \neq \emptyset$) we replicate the gray level values of the existing set for the interpolated slice.

3   Simulating Drilling by Volumetric Erosion

Shape-based interpolation provides us with the 3-D object reconstruction. We need the reconstructed volume in order to simulate virtual processing operations. In the following we propose a mathematical morphology based system for simulating drilling. Let us consider that the volume is made up of isotropic material and that the effects of drilling do not depend on the direction. The drilling of the 3-D volume proceeds along a certain direction. Let us consider a parametric spherical coordinate system in which the direction is shown by two angles (θ, ϕ), where θ represents the angle made by the drilling direction with the image plane (x, y) and ϕ represents the angle made by the projection of the drilling direction on the image plane with the horizontal axis x. The drilling of a volume $\mathcal{O}$ produces a drilled volume, denoted as $\breve{\mathcal{O}}$, and can be represented as a succession of erosions with a volumetric structuring element denoted as $B^{(3)}$:

$$\breve{\mathcal{O}} = \mathcal{O}(x, y, z) \ominus B^{(3)} \tag{8}$$

where (x,y,z) denotes the point where the drilling is applied. The eroded voxels have assigned the value of the background while the object voxels corresponding to the surface of the structuring element become part of the boundary of the new object $\breve{O}$. Drilling is simulated by successively eroding volumes of the 3-D structuring element size in the given direction. The drilling direction is shown by the following parametric equations :

$$x(i) = x(i-1) - d\cos(\theta)\cos(\phi) \tag{9}$$

$$y(i) = y(i-1) - d\cos(\theta)\sin(\phi) \tag{10}$$

$$z(i) = z(i-1) - d\sin(\theta) \tag{11}$$

where d is the width of the drilling element in the direction of the drilling. and where the starting drilling point has the coordinates $(x(0),y(0),z(0))$. The number of times the 3-D erosion is applied depends on the speed and duration of the drilling.

We can significantly speed up the erosion process by changing the reference system from being centered on the user position into considering the processed 3-D object as the reference. In this case we replace the heavy burden of calculating the directional erosion with rotating the entire volume such that the direction of drilling becomes parallel with z axis. We consider three shapes for modeling the volumetric erosion element: spherical, cylindrical and conical. At each erosion we extract a volume with the shape given by the corresponding structural element $B^{(3)}$. The corresponding region of the volume O has assigned the same grey level as its background, denoted as O^c. In this case the erosion with the spherical element at the iteration i is modeled by :

$$\{(x,y,z) \in \breve{O},\ (x - x_{IM})^2 + (y - y_{IM})^2 + (z - z(i))^2 < d^2\} \in \breve{O}^c \tag{12}$$

where d is the radius of the spherical erosion element and (x_{IM}, y_{IM}) are the coordinates of the 3-D object point projection on the image plane, where the erosion takes place. The direction of erosion is considered perpendicular onto the image plane in this case. When employing the cylindrical erosion element, the local drilling effect is given by :

$$\{(x,y,z) \in \breve{O},\ z(i) > z > z(i) - d, (x - x_{IM})^2 + (y - y_{IM})^2 < R^2\} \in \breve{O}^c \tag{13}$$

where R is the radius and d is the height of the cylindrical erosion element. The depth is conventionally considered as a negative number. For the conical erosion element, the local drilling effect is simulated by :

$$\{(x,y,z) \in \breve{O},\ z(i) > z > z(i) - d, (x - x_{IM})^2 + (y - y_{IM})^2 < [(z(i) - z)\tfrac{R}{d}]^2\} \in \breve{O}^c \tag{14}$$

where R and d are the conical erosion element radius and height, respectively. In this case the drilling direction is identical with that of the projection ray used for the volumetric visualization (we have used a parallel ray tracing algorithm). The changes in the volume rendering are localized only in an area around (x_{IM}, y_{IM}) depending on the drilling tool size. This contributes to a significant computational complexity reduction.

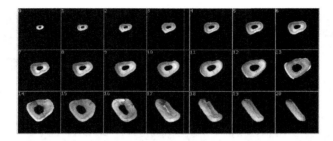

Fig. 2. Segmented and aligned set of slices for an incisor

4   Simulation Results

We have applied this algorithm in virtual dentistry. We employ the interpolation algorithm to reconstruct teeth. Afterwards, we apply the virtual drilling algorithm on the reconstructed 3-D teeth. In our simulations we consider three different types of teeth: an incisor (single root), a premolar (two roots) and a molar (three roots). Teeth from each of these categories have been mechanically sliced and digitized. The teeth boundaries as well as the root canals are segmented in each slice and the resulting object slices are aligned using a semi-automatic procedure. Initial slices after alignment are displayed in Figure 2 for an incisor. We have used the morphological interpolation algorithm described in Section 2 in order to reconstruct the tooth from the given initial group of slices. Tooth cross-sections are interpolated between each two consecutive slices. In the case of the incisor, the interpolation algorithm is applied recursively four times. Thus we obtain 336 interpolated slices from 22 original slices. The 3-D reconstruction from two different viewing angles are shown in Figures 3a, 3b, for the incisor, in Figures 3c, 3d, for the premolar and in Figures 3e, 3f for the molar, respectively. This result shows a smooth transition interpolating well even between slices having large geometrical shape variations. The morphology of the reconstructed tooth is accurate despite the fact that most of the slices have been reconstructed by interpolation.

We have compared the mathematical morphology interpolation algorithm with a linear interpolation algorithm. The linear interpolation algorithm takes the midpoints of the line segments between pixels on object contours of the two slices, in both x (horizontal) and y (vertical) directions as the interpolated slice contour. We have applied the linear interpolation algorithm on the incisor sequence displayed in Figure 2. For assessing the performance of the interpolation algorithms we have devised the following objective measure. Let X_i, X_{i+1} and X_{i+2} be three original tooth slices and $\hat{X}_{i+1}$ be the result of interpolating X_i and X_{i+2}. Let $|X|$ denote set cardinality. The ratio $|XOR(X_{i+1}, \hat{X}_{i+1})|/|X_{i+1}|$, representing the percentage of wrongly estimated pixels, can be used as a performance measure. In Table 1 we provide the results for reconstructing three different slices from the incisor group of sets as well as the average result for re-

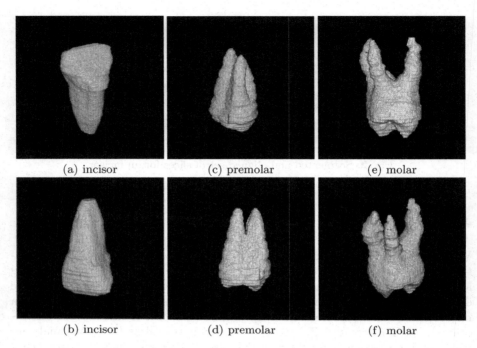

(a) incisor   (c) premolar   (e) molar

(b) incisor   (d) premolar   (f) molar

Fig. 3. Two different 3-D views for each of the three teeth.

constructing any intermediary slice $\hat{X}_{i+1}$ from the given group of sets X_i, X_{i+2}, for any $i \in \{1, ..., N-2\}$, where N is the number of initial sets. We can observe from Table 1 that the interpolated slice obtained by morphing is closer to the original slice than that interpolated by linear interpolation. The 3-D molar reconstructed by linear interpolation is displayed in Figure 4a, while in Figure 4b we show the same molar reconstructed by morphological morphing as described in this paper.

Table 1. Objective comparison measure between morphological morphing and linear interpolation when reconstructing an incisor.

Frames $i,i+1,i+2$	Frame Difference (%) $\frac{\|XOR(X_{i+2},X_i)\|}{\|X_i\|}$	Morphological Morphing (%) $\frac{\|XOR(\hat{X}_{i+1},X_{i+1})\|}{\|X_{i+1}\|}$	Linear Interpolation (%) $\frac{\|XOR(\hat{X}_{i+1},X_{i+1})\|}{\|X_{i+1}\|}$
4,5,6	62.9	5.9	11.925
10,11,12	26.8	6.84	9.46
18,19,20	27.2	7.5	14.28
Entire volume	51.5	9.25	11.46

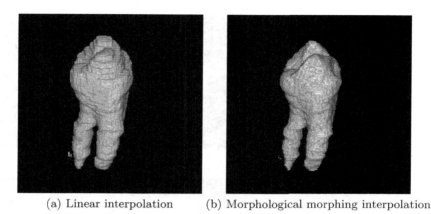

(a) Linear interpolation   (b) Morphological morphing interpolation

Fig. 4. Reconstruction of a molar in 3-D.

After rendering and displaying the volume of a tooth, we simulate drilling by using a volumetric erosion element for the dental burr. We have chosen three different geometrical shapes for modeling the dental burr: spherical, conical and cylindrical. We associate the action of a drilling tool (dental burr) with the repetition of erosions done with various 3-D structural elements.

The 3-D structural elements employed in our experiments for simulating different types of dental burrs are shown in Figures 5a, 5b, 5c. The elementary drilling operation consists of eroding the 3-D object with the structuring element corresponding to the shape of the drilling burr. Effects of drilling on a tooth when using spherical, cylindrical and conical erosion elements are displayed in Figures 5d, 5e and 5f respectively. We have applied the proposed morphological drilling tool for virtual dentistry by testing the drilling algorithm on teeth reconstructed by 3-D interpolation. Dentists have used an entire set of virtual drilling burrs with several different geometric parameter values. A 3-D tooth after being drilled by a dentist is displayed in Figure 6a. A set of sections through the drilled tooth is shown in Figure 6b. Two dental operations that have a particular treatment significance can be observed in these figures.

5   Conclusions

We simulate a processing operation such as drilling on a 3-D volume given a set of sparse sets which represent parallel and equidistant object sections. We have proposed an algorithm for reconstructing the 3-D object from the given group of slices. This algorithm interpolates between each two adjacent slices by means of morphological shape-based interpolation. Virtual drilling is simulated by a succession of erosions. The proposed algorithms are applied on teeth that have been mechanically sliced and digitized. The morphological drilling algorithm can be used as a tool in a virtual sculpturing environment.

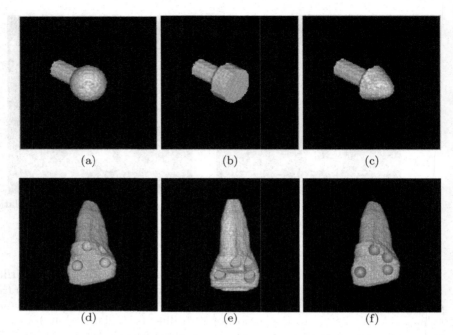

(a)   (b)   (c)

(d)   (e)   (f)

Fig. 5. Shapes of volumetric erosion elements employed as burrs for tooth drilling:
(a) spherical; (b) cylindrical; (c) conical; (d), (e), (f) volumetric erosion results on an
incisor for the given set of tools, assuming only one erosion.

References

1. S. P. Raya, J.K. Udupa, "Shape-based interpolation of multidimensional objects,"
   IEEE Trans. on Medical Imaging, vol. 9, pp. 32-42, 1990.
2. A. Goshtasby, D. A. Turner, L.V. Ackerman, "Matching of Tomographic Slices for
   Interpolation," *IEEE Trans. on Medical Imaging,* vol. 11, no. 4, pp. 507-516, Dec.
   1992.
3. N. Nikolaidis, I. Pitas, *3-D Image Processing Algorithms.* J. Wiley & Sons, 2000.
4. M. Moshfeghi, "Directional Interpolation for Magnetic Resonance Angiography
   Data," *IEEE Trans. on Medical Imaging,* vol. 12, no. 2, pp. 366-379, June 1993.
5. W. E. Higgins, C. J. Prlick, B. E. Ledell, "Nonlinear Filtering Approach to 3-D
   Grey-Scale Image Interpolation," *IEEE Trans. on Medical Imaging,* vol. 15, no. 4,
   pp. 580-587, 1996.
6. J. D. Boissonnat, "Shape reconstruction from planar cross-sections," *Computer
   Vision, Graphics and Image Processing,* vol. 44, no. 1, pp. 1-29, 1988.
7. P.N. Werahera, G.J. Miller. G.D. Taylor, T. Brubaker, F. Daneshgari, E.D. Craw-
   ford, "A 3-D Reconstruction Algorithm for Interpolation and Extrapolation of
   Planar Cross Sectional Data," *IEEE Trans. on Medical Imaging,* vol 14, no. 4, pp.
   765-771, Dec 1995.

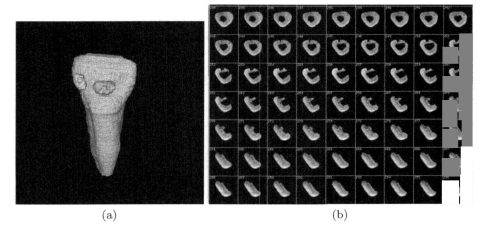

(a)   (b)

Fig. 6. Incisor treatment simulation: (a) 3-D view of the drilled incisor; (b) drilling effects on cross-sectional slices through incisor.

8. W.E. Higgins, C. Morice, E. L. Ritman, "Shape-based interpolation of tree-like structures in three-dimensional images," *IEEE Trans. on Medical Imaging,* vol. 12, no. 3, pp. 439-450, Sep. 1993.

9. S.-Y. Chen, W.-C.. Lin, C.-C. Liang, C.-T. Chen, "Improvement on Dynamic Elastic Interpolation Technique for Reconstructing 3-D Objects from Serial Cross Sections," *IEEE Trans. on Medical Imaging,* vol. 9, no. 1, pp. 71-83, March 1990.

10. M. Joliot, B. M. Mazoyer, "Three-Dimensional Segmentation and Interpolation of Magnetic Resonance Brain Images," *IEEE Trans. on Medical Imaging,* vol. 12, no. 2, pp. 269-277, June 1993.

11. G. J. Grevera, J.K. Udupa, "Shape-Based Interpolation of Multidimensional Grey-level images," *IEEE Trans. on Medical Imaging,* vol. 15, no. 6, pp. 881-892, 1996.

12. S. Beucher, "Sets, partitions and functions interpolations," *International Symposium on Mathematical Morphology and its Applications to Image and Signal Processing IV,* Amsterdam, Netherlands, June 3-5, 1998, pp. 307-314.

13. J. Sera, *Image Analysis and Mathematical Morphology.* New York: Academic Press, 1982.

14. R. Yoshida, T. Miyazawa, A. Doi, T. Otsuki, "Clinical Planning Support System - CliPSS," *IEEE Computer Graphics and Applications,* vol 13, no. 6, pp. 76-84, Nov. 1993.

Morphological Image Processing for Evaluating Malaria Disease

Cecilia Di Ruberto[1], Andrew Dempster[2], Shahid Khan[3], and Bill Jarra[3]

[1] Department of Mathematics, University of Cagliari, Cagliari, Italy
dirubert@vaxca1.unica.it
[2] Department of Electronic Systems, University of Westminster, London, UK
dempsta@cmsa.wmin.ac.uk
[3] National Institute for Medical Research, London, UK

Abstract. This work describes a system for detecting and classifying malaria parasites in images of Giemsa stained blood slides in order to evaluate the parasitaemia of the blood. The first aim of our system is to detect the parasites by means of an automatic thresholding based on a morphological approach. Then we propose a morphological method to cell image segmentation based on grey scale granulometries and openings with disk-shaped elements, flat and hemispherical, that is more accurate than the classical watershed-based algorithm. The last step of the system is classifying the parasites by morphological skeleton.

1 Introduction

In malarial blood the red corpuscles of vertebrates are infected by malaria parasites. The parasite develops in a highly regulated manner through distinct cycles in the vertebrate host [8]. The parasite attacks red corpuscles, in which it first appears as minute speck of chromatin surrounded by scanty protoplasm, and gradually becomes ring-shaped and is known as a *ring* or *immature trophozoite*. It grows at the expense of the red cell and assumes a form differing widely with the species but usually exhibiting active pseudopodia, i.e. projections of the nuclei. Pigment granules appear early in the growth phase and the parasite is known as a *mature trophozoite*. As the nucleus begins to divide and take up peripheral positions, the parasite is known as a *schizont*. The infected red blood cell ruptures. Some parasites on entering red cells become round sexual *gametocytes*, instead of asexual schizonts.

The aim of our system is to detect the parasites using a scan of a colour photograph of stained malarial rodent blood from a microscope in order to evaluate the parasitaemia of the blood i.e. counting the number of parasites per number of red blood cells [1]. A manual analysis of slides is tiring, time-consuming and requires a trained operator. So our task is to automate the process. The image processing system is made of three main steps: detection of parasites, cell segmentation and classification of parasites. In Section 2 we describe the different phases of the image analysis, beginning from parasites detection. We propose a method to automatically separate the parasites from the rest of an infected

C. Arcelli et al. (Eds.): IWVF4, LNCS 2059, pp. 739–748, 2001.
© Springer-Verlag Berlin Heidelberg 2001

blood image using colour and size information [2]. Then we introduce an efficient morphological method to segment cell images improving the accuracy of the classical watershed-based algorithm [3]. The last aim of the analysis is classifying the detected parasites using the morphological skeletons. In Section 3, the experimental results are presented, providing a comparison with the classical watershed-based algorithm and some numerical results. Finally, Section 4 draws the conclusions.

2   A Morphological Approach to Malarial Blood Analysis

Mathematical morphology is well suited for biological and medical image analysis. In fact it offers a powerful tool for extracting image components that are useful in the representation and description of region shape, size and colour. Our proposed technique is based on morphological methods, using *granulometries* to evaluate the size of the red cells and the nuclei of parasites and the *regional maxima* to mark relevant bright image objects, i.e. to detect the nuclei of parasites. We are also interested in morphological techniques for pre- or post-processing, such as *morphological filters* to suppress or smooth some areas, *thinning* and *gradient* to improve cells contours and *reconstruction by dilation* to recover objects of interest (schizonts, infected red cells) [7,9].

At the moment the processed images are taken from a microscope, developed on photographic paper and then scanned. This process would in future be automated. Our source images are taken at a range of magnifications, with some variation in stain colour and lighting conditions (see the sample 465x702 pixels image in Figure 1). We analyse the hue-saturation-value colour space. In both the hue component, H, and the saturation one, S, the bright objects are of interest. This is because the Giemsa staining solution stains up nucleic acids i.e. highlights the DNA in these objects, showing it as a dark, saturated purple. Therefore the white cells and the parasites, which contain DNA, are much brighter than the other objects in the image.

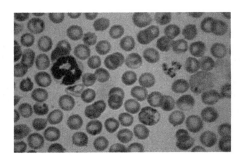 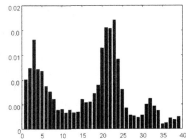

Fig. 1. A sample malarial blood image   **Fig. 2.** The pattern spectrum of the filtered and closed saturation image

2.1   Automatic Thresholding Using Granulometry and Regional Maxima

To detect the bright objects we convert the H and S images into binary images using simple thresholding and the product of these binary images is then the marker image for the parasites and the white blood cells. So, our aim is to automate the selection of these two thresholds in order to separate the objects of interest, i.e. those containing DNA, from the rest of the image. The images to process contain noise from the sample, from the microscope light, from the chemical development process or from the scanner. The noise is smoothed by a median filtering using a 5x5 window, on both H and S images. In order to enhance the bright image objects and make flatter, darker and cleaner the image background, we apply a morphological area closing on both the H and S components. This is especially useful to better estimate the cell size from the pattern spectrum.

Bright image objects can be detected by looking for the regional maxima in the image, in other words the parasites and the white blood cells correspond to the regional maxima of the H and S images. As all the objects of interest are roundish, it is useful to choose a disk-shaped structuring element for the connectivity of the maxima. The choice of its size is crucial for the effectiveness of the markers. A small disk locates too many regional maxima and by choosing too large a disk relevant image objects could be missed. But we are looking for parasite locations, i.e. regional extrema inside the red cells. So, the size of the connected components in the image, location of regional maxima, must be smaller than the size of the red cells. Therefore, the size of the structuring element defining the connectivity of the regional extrema could be chosen as equal to the size of the red cells. In order to evaluate the red cell sizes we apply a granulometric analysis on the image based on opening operations with disk-shaped structuring elements. Figure 2 shows the resulting size distribution for the saturation image (note the use of the disk-shaped SE). The histogram indicates the presence of two predominant particle sizes in the input image, relative to the nuclei of the trophozoites (3-8 pixels) and to the red blood cells (20-25 pixels). Let us denote by c the greatest size of the red cells estimated from the pattern spectrum. In the sample image c is equal to 25. At this point it is possible to look for all the regional maxima in the H and S images, according to the connectivity of a disk-shaped structuring element with radius c. Let us denote by MH and MS the marker images, containing the regional maxima of H and S, respectively. The marker image of the nuclei of the parasites, MHS, is then the intersection of the two marker images, MH and MS, after dilating both the images by a disk-shaped structuring element whose radius is equal to the nuclei size (equal to 5 in our sample image), i.e.:

$$MHS = MH \cap MS. \tag{1}$$

The MHS image locates the regional maxima in the image, i.e. connected component set of constant grey level. But the parasite nuclei are not all of constant grey level and so the MHS image marks only a subset of the pixels within the nuclei. To detect all the parasites we need a grey level characterising the nuclei. Let us denote by μH and μS the average grey level of the nuclei marked by MHS, computed on the H and S images, respectively, i.e.:

$$\mu H = \frac{1}{sum(MHS)} \sum_{p \in MHS} H(p) \tag{2}$$

and

$$\mu S = \frac{1}{sum(MHS)} \sum_{p \in MHS} S(p) \tag{3}$$

where $sum(MHS)$ is the number of pixels marked as regional maxima in MHS, $H(p)$ and $S(p)$ are the grey levels of the pixel p in the H and S images, respectively. The values μH and μS are chosen as threshold values to detect the objects of interest in the H and S images. The image THS identifying all the parasites and the white blood cells is then the intersection of TH and TS:

$$THS = TH \cap TS \tag{4}$$

where TH and TS are obtained thresholding H and S, respectively, by means of μH and μS (in Figure 3 the THS contours overlaid the original image).

2.2   Detection of White Blood Cells and Schizonts

The objects present in the thresholded image THS are parasites of all types and white blood cells. We can isolate the white cells by means of a morphological erosion with a disk-shaped structuring element whose size is achieved by the granulometric analysis. The size of white blood cells can be inferred from the average dimension of the red cells and hence the size of the structuring element can be selected as equal to this dimension since the chromatin spots are surely smaller than a red cell. In our sample case this size is 22. The white cell marked by this erosion is then reconstructed by dilation.

   In order to identify the schizonts in the THS image we look at how clustered the remaining objects are. This means we need a measure of separation of the objects in the image plane, in other words we need to compute the distance between sets in a discrete space. The distance between sets can be defined using the notion of *Hausdorff distance* [4]. Two sets are within Hausdorff distance λ from each other iff any point of one set is within distance λ from some point of the other set. Let us denote by A and B two sets. By a morphological approach the Hausdorff distance between these sets is the minimum of the radius λ of the disks S such that A dilated by S_λ contains B or B dilated by S_λ contains A. All the objects whose distance is smaller than the average (or maximum) size of red cells make up a schizont. All the other remaining objects identify nuclei of parasites. Applying a morphological reconstruction by dilation of the mask THS image from the marker schizont image by a 3x3-cross structuring element, we are able to localise the schizonts in the input image (see Figure 4).

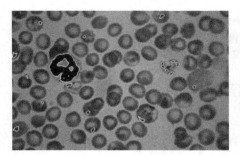

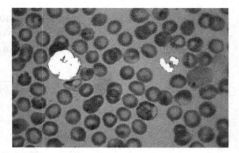

Fig. 3. The contours of parasites and white blood cell

Fig. 4. White cell and schizonts detected and depicted in "white"

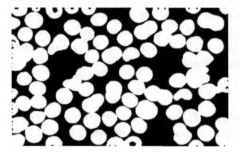

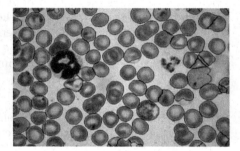

Fig. 5. The marker image used in the segmentation

Fig. 6. The contours of the red cells, single and composite

At this point all the remaining spots in the THS image identify nuclei of trophozoites, immature or mature, and gametocytes. The evaluation of parasitaemia requires also counting the red cells and so locating the red cells they infect. This requires the segmentation of the red cells. Before segmentation we remove from the input image the objects we have already identified (white cells and schizonts) so that the image to analyse contains only red cells, some of which are infected by parasites.

2.3 Segmentation of Red Blood Cells

The aim is to isolate each individual red blood cell, especially when they are overlapping and partially occluded and form clusters in the viewing field of the microscope, and so to locate and recognize the cell contours. To isolate the red blood cells we use the green component image because it is cleanest. However, thresholding the input image, we have observed that cells disappear into the background at the sides as noise comes up in the centre. Therefore we correct this non-uniform illumination using a paraboloid model of the illumination.

Thresholding the green image at this stage would leave 'holes' in the middle of the red blood cells. After filling the holes in the red cells by a morphological area opening, the last step of this pre-segmentation phase consists in making the background flat and clean. So, we threshold the image, setting to zero all the pixels belonging to the background, and remove items smaller than a red blood cell from it by means of a morphological area-open filter. At this point segmenting the image means retrieving the red cell blocks. In order to identify red cell bodies we use the granulometric analysis on the image, already done in the previous phase of the processing. The image consists of objects of two main different sizes, cells and trophozoites. Some objects are overlapping and they also are too cluttered to enable detection of individual particles. Therefore the first step of the segmentation process consists in estimating the size distribution to evaluate the smallest size s of the red cells and applying a morphological opening by a hemispherical structuring element of radius s on the image. In our sample image s is equal to 20.

The morphological gradient on the result of the opening produces a first rough localisation of the red cell contours. We binarize the gradient image and close the holes in order to get a binary image we can use as marker image in the classical watershed-based segmentation [9] (see Figure 5) to find the contours of the red blood cells (see Figure 6). This gives us a partial segmentation with some compound cells still to deal with. In fact the red blood cells can be overlapping or partially occluded. Therefore each cell body area detected by segmentation can identify both individual and composite cells. So cell parameters measuring the roundness are calculated. If a cell has a large ratio of the major axis over the minor axis (greater or equal to 1.3), i.e. it is elongated, it is treated as a composite cell (see Figure 7). The composite cells are then separated into the individual contributing cells applying a morphological opening by a flat disk-shaped structuring element of size s on the composite cells. The morphological gradient on the result of the opening produces a first rough localisation of the individual red cell bodies: the binarization of the gradient image produces rough individual contours. After closing the small holes between adjacent contours by a morphological area closing, a morphological thinning leaves contours around the individual cells (in Figure 8 the contours of all the individual red cells).

2.4   Identification of the Infected Red Cells and the Parasites' Nuclei

Finding the trophozoites identifies the infected red cells in the image. Each trophozoite, both immature and mature, presents a nucleus we have already isolated in the THS image after having removed the white red cells and the schizonts. So, applying a morphological reconstruction by dilation of this mask binary image conditioned to the segmented image using a disk-shaped structuring element [9], we are able to identify and isolate the infected red cells in the input image.

2.5   Classification of Parasites

The last step of the system is to classify the parasites in the other three classes of objects, immature trophozoites, mature trophozoites and gametocytes. In the paper we present a method still based on a morphological approach.

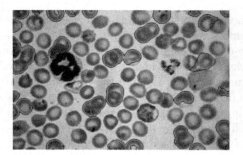

Fig. 7. The contours of the composite red cells

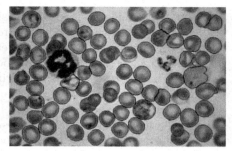

Fig. 8. The contours of the individual red cells

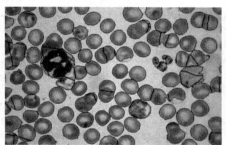

Fig. 9. The classical watershed-based algorithm segmentation

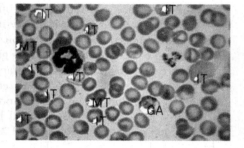

Fig. 10. The parasites classification by morphological skeleton

Each trophozoite, both immature and mature, is characterised by a nucleus, a circular spot of chromatin, particularly evident in the hue and saturation components of the input image. A mature trophozoite differs from an immature one because of the presence of active pseudopodia, the presence of pigment granules around the nucleus that become more numerous in case of a gametocyte. So, the classification is solved analysing the shape of the parasite automatically detected in the first step of the malarial processing using morphological operators.

In all pattern recognition problems we need to extract the features of an object to classify it. In order to classify the parasites we try to simplify the shape as much as possible in order to make the classification, that is a topological analysis, as simple as possible. One possibility consists in creating a version of the pattern that is as thin as possible, i.e thinning the object to a set of idealised thin lines which summerise the information of the original object while preserving its topology. The resulting thin lines are called the *skeleton* or *medial axis* of the input pattern and they are the thinnest representation of the original pattern that preserves the topology. The detection of the endpoints is important for classifying our objects, being strictly related to the shape of a parasite. As we have already observed an immature trophozoite is characterised by a nucleus, a circular spot of chromatin, and so the skeleton does not present many endpoints.

While a skeleton of a mature trophozoite presents more endpoints, because of the presence of pigment granules around the nucleus. The number of endpoints increases in case of gametocytes. Many algorithms that generate digital skeletons have been proposed. But most of them produce a non-connected skeleton, that is useless for shape description application since homotopy is not preserved and characteristic points such as multiple points or endpoints in the continuous case are lost. Digital skeletons can be generated by thinning algorithms. In [7] the thinning process has been analysed, including the proof of convergence, the condition for one-pixel thick skeletons and the connectivity of skeletons. A digital set can be skeletonised so as to preserve these important properties by thinning the set with SEs preserving homotopy, i.e. homotopic SEs. The skeleton is obtained by thinning the input image with homotopic or a series of homotopic SEs and their rotations until stability has been reached.

The skeleton of a binary object contains useful information about it and the endpoints of skeleton are an interesting shape descriptor that we have used for our pattern recognition purpose. The more endpoints of a skeletonised object more different the object is from a circular one. All the immature parasites are disk-shaped so the number of endpoints is small, while for mature parasites the shape is more irregular and rough, leaving the disk shape, and the irregularity increases in the gametocytes. We have used this morphological feature to recognize our parasites' sample and the successful classification is showed in Figure 10.

3   Experimental Results

In Figures 1-10 the different steps of the procedure are presented. In Figure 1 the initial sample image, scanned from a colour photograph of stained malarial rodent blood, is presented. In the image there exist all the different kinds of parasites, a polymorph white blood cell and red cells. In Figure 2 the pattern spectrum of the filtered and closed saturation component and in Figure 3 in white colour the contours of the parasites and white blood cell marked by the regional maxima, using a structuring element with radius 25. In Figure 4 the isolated schizonts and blood cell are depicted in white on the input image. From Figures 5 to 8 the main steps of segmentation phase are showed and in Figure 9 the contours detected by the classical watershed-based algorithm is presented. Finally, after having isolated the infected red cells, the classification of the extracted parasites is illustrated in Figure 10. The classificaton is obtained by using the morphological skeletons: the label IT is for immature trophozoite, the label MT for immature trophozoite while GA is for gametocyte. All the parasites have been correctly identified and classified.

In Figure 11 we present the numerical results obtained processing 12 images of malarial blood. Each image has been analysed by two biologists (called A and B in the table) and by our method (C). For each image we have counted the immature trophozoites (IT), the mature trophozoites (MT), the schizonts (SC), the total parasites (PAR), the total red blood cells (RBC) and the percentage

Images	I1	I2	I3	I4	I5	I6	I7	I8	I9	I10	I11	I12
IT A	38	33	32	35	37	24	38	40	36	31	39	30
IT B	35	36	42	33	23	28	30	23	39	33	28	26
IT C	38	36	39	39	40	33	31	36	39	42	38	28
MT A	9	4	5	2	4	1	1	5	3	4	4	5
MT B	9	5	3	3	10	3	5	5	3	5	10	7
MT C	9	5	4	3	7	3	3	4	3	4	8	7
SC A	1	2	1	4	2	3	1	1	0	1	1	2
SC B	0	2	0	2	0	0	1	0	0	0	1	0
SC C	0	2	1	3	0	0	1	0	0	0	1	0
PAR A	48	39	38	41	43	28	40	46	39	36	44	37
PAR B	44	43	45	38	33	31	36	28	42	38	39	33
PAR C	47	43	44	45	47	36	35	40	42	46	46	35
RBC A	170	165	164	163	153	136	155	155	157	155	172	165
RBC B	171	164	168	166	154	138	154	154	154	154	170	191
RBC C	173	165	171	170	160	138	160	159	158	153	173	173
% PAR A	28.2	23.6	23.2	25.1	28.1	20.6	25.8	29.7	24.8	23.2	25.6	22.4
% PAR B	25.7	26.2	26.8	22.9	21.4	22.5	23.5	18.2	27.3	22.1	22.9	17.3
% PAR C	27.1	26.0	25.7	26.4	29.3	26.0	21.8	25.1	26.5	30.0	26.5	20.2
% IT A	79.2	84.6	84.2	84.5	86.0	85.7	95.0	87.0	92.3	86.1	88.6	81.1
% IT B	79.6	83.7	93.3	86.8	69.7	90.3	83.3	82.1	92.9	86.8	71.8	78.8
% IT C	80.8	83.7	88.6	8.6	85.1	91.6	88.5	90.0	92.8	91.3	82.6	90.0
% MT A	18.7	10.3	15.2	4.9	9.4	3.6	2.5	10.8	7.7	11.1	2.3	13.5
% MT B	20.4	11.6	6.7	7.9	30.3	9.7	13.9	17.9	7.1	13.2	25.6	21.2
% MT C	19.1	11.6	9.0	6.6	14.8	8.3	8.5	10.0	7.1	8.6	17.3	20.0
% SC A	2.1	5.1	2.6	9.4	4.6	10.7	2.5	2.2	0	2.8	2.3	5.4
% SC B	0	4.7	0	5.3	0	0	2.8	0	0	0	2.6	0
% SC C	0	4.6	2.2	6.6	0	0	2.8	0	0	0	2.1	0
WBC A	0	0	0	1	0	0	0	0	1	0	0	0
WBC B	0	0	0	1	0	0	0	0	1	0	0	0
WBC C	0	0	0	1	0	0	0	0	1	0	0	0

Fig. 11. The counting results on some malarial images

of parasites on red blood cells (% PAR), the percentage of trophozoites that are immature (% IT), the percentage of trophozoites that are mature (% MT), the percentage of parasites that are schizonts (% SC) and the white blood cells (WBC). The counting results of the two biologists are more different from each other than our counts are from either of them, i.e. the A-B comparisons are worse than both A-C and B-C. As it's possible to observe from the numerical results, our automatic method seems to be a good compromise between the two experienced users. The experimental results have turned out to be very encouraging and we hope to test our method on a larger database of images.

4 Concluding Remarks

In this paper we have presented a morphological method to analyse malarial rodent blood images. The aim of malarial blood image processing is to detect the parasites infecting the red cells in order to evaluate the number of parasites per number of red blood cells. The proposed method identifies automatically the parasites using colour and size information, extracted by a morphological approach. We have used the regional maxima to detect the nuclei of the parasites, according to a connectivity of a disk-shaped structuring element whose radius is the greatest size of the red blood cells. The latter is obtained by a granulometric analysis of a filtered component of the image in the hue-saturation-value space. We have also presented a morphological system to segment blood images. The granulometric analysis and the opening by non-flat disk-shaped structuring element are the main steps of the segmentation process. Granulometry is used to capture information about objects of particular size and shape. In our case the objects of interests are cells, i.e. bright blobby image parts. We have used grey scale granulometries based on opening with disk-shaped elements. We have chosen a non-flat (hemisphere) disk-shaped structuring element to enhance the roundness and the compactness of the red cells before applying the watershed algorithm as these features could be lost or be too weak to produce an accurate segmentation if directly evaluated on the input image. In Figure 9 the classical watershed-based algorithm segmentation applied on the input image of Figure 1 is showed. On the contrary we have used a flat disk-shaped structuring element to detect the points of contact in composite cells. Openings with flat disks capture information about the height of overlapping objects, allowing the separation of composite structures into the individual composing parts. Finally we have presented a method for classification of parasites based on morphological skeleton, using endpoints as features for recognition. The sample images we have presented show that the proposed morphological analysis achieves very good and accurate performance. At the moment the method is dependent on the exposure and light conditions. So our future research will be focused on the automated choice of the morphological parameters in different exposure and magnification situation and on the examination of extending the proposed morphological technique to analyse malarial human blood images.

References

1. Dempster, A., Di Ruberto, C.: Morphological Processing of Malarial Slide Images. Proceedings Matlab DSP Conference 1999, Espoo, Finland (1999) 82–90
2. Di Ruberto, C., Dempster, A., Khan, S., Jarra, B.: Automatic Thresholding of Infected Blood Images using Granulometry and Regional Extrema. Proceedings ICPR'2000 - 15th International Conference on Pattern Recognition, Barcelona, Spain (2000) 445–448
3. Di Ruberto, C., Dempster, A., Khan, S., Jarra, B.: Segmentation of Blood Images using Morphological Operators. Proceedings ICPR'2000 - 15th International Conference on Pattern Recognition, Barcelona, Spain (2000) 401–404
4. Edgar, G.A.: Measure, Topology and Fractal Geometry. Springer-Verlag (1995)
5. Harms, H., Gunzer, U., Aus, H.M.: Combined Local Colour and Texture Analysis of Stained Cells. Computer Vision, Graphics and Image Processing **33** (1986) 364–376
6. Jang, B., Chen, R.T.: Analysis of Thinning Algorithms using Mathematical Morphology. IEEE Trans. on Pattern Analysis and Machine Intelligence **v. 12 n. 6** (1990) 541–551
7. Serra, J.: Image Analysis and Mathematical Morphology, v. 2. Academic Press (1992)
8. Smyth, J.D.: Introduction to Animal Parasitology. Cambridge University Press, Cambridge (1994)
9. Soille, P.: Morphological Image Analysis, Principles and Applications. Springer-Verlag (1999)
10. Volkova, S.E., Ilyasova, N.Y., Ustinov, A.V., Khramov, A.G.: Methods for Analysing the Images of Blood Preparations. Optics & Laser Technology **v. 27 n. 4** (1995) 255–261

A Binocular License Plate Reader for High Precision Speed Measurement

Giovanni Garibotto

Elsag, Via Puccini, 2 Genova, Italy
Giovanni.garibotto@elsag.it

Abstract. The paper describes a vision processing system for traffic speed computation. Vehicle tracking is performed by using high-level features, that are clusters of license plate characters to achieve increased robustness of the system. By using a binocular arrangement of the Computer Vision system, the spatial and temporal range of image capture between consecutive views is extended, which leads to an improved accuracy in speed computation. Multiple views can be collected, from the same vehicle in transit, depending on the amount of speed. The effectiveness of the proposed solution is demonstrated through a geometric simulation model, where almost all operating conditions and constraints are fully exploited and tested. An error sensitivity analysis is carried out, to identify the most critical components of the system and the possible sources of errors in speed computation. This simulation approach is proved quite useful in this complex scenario, where it is commonly very difficult to collect true speed measures from the vehicles in transit. Beside the simulation analysis, a series of experimental results have been collected by a first prototype that is available since the end of year 2000.

1. Introduction

All public administrations all over the world are heavily involved in the identification of reliable solutions for traffic control. Most required functions are license plate reading and speed violation detection systems, to provide traffic flow and access control to restricted areas in downtown or historical centers. Speed enforcement is definitely one of the most important applications, since speed limits violation is one of the most relevant causes of accidents everywhere.

Most established solutions currently in use are based on radar or laser technologies with an increasing interest in using optical systems [1]. In the last few years there has been an increasing exploitation of Computer Vision technology for vehicle tracking and speed measurement, due to the recent improvements in the algorithms and the availability of more processing power. Some results have moved from academic research labs to applied solutions [2]. Interesting solutions have been recently proposed [3] to compute the travel time of vehicles and the network level origin-destination matrix based on Computer Vision. It is based on the use of separate License Plate Readers placed along a road at a suitable distance (from hundred meters up to a few kilometers apart). The identification of the same license plate in the two sites allows the system to compute the average speed of the vehicle in this space

C. Arcelli et al. (Eds.): IWVF4, LNCS 2059, pp. 749–758, 2001.
© Springer-Verlag Berlin Heidelberg 2001

interval with high precision. The resulting measure can be successfully used both for monitoring purposes (to provide information about the average transit time along selected routes) as well as for enforcement (to detect speed violation).

Our proposed approach is aimed to compute the instantaneous speed of the vehicles by using a binocular system placed on the side of the road, to frame the image of the vehicle at two consecutive times, very closed together. By tracking the vehicle in the two images it is possible to get a quite accurate estimation of the vehicle speed. Actually, at lower speed, a number of views can be collected from each individual acquisition camera. The features used for matching and tracking are high level features (clusters of matched license plate characters) to minimize matching errors. An error sensitivity analysis is also referred, by using a simulation model of the binocular configuration. In this way it is possible to identify potential sources of errors and predict the behavior of the system in the different operating conditions.

The first section of the paper describes the proposed approach of speed estimation with a binocular vision system. Geometric constraints are discussed and the main mathematical relations used for speed measure. The second section shortly recalls the License Plate Recognition system by Elsag O^2CR reading technology, which is used for high-level feature detection and tracking. The following error sensitivity analysis allows to point out potential sources of errors as well as to strengthen weak and strong properties of the processing chain, and the consequences of perspective and calibration constraints.

2. Speed Estimation by a Binocular Vision System

The proposed approach is based on the computation of speed from the geometric features of License Plates acquired by a binocular configuration as depicted in fig.1. The License Plate LP is supposed to be framed from the first sensor S1 with a time stamp t_1=ndt, and from the second sensor S2 at time t_2=mdt where the time sampling dt is the same (i.e. 20 msec. at 50 Hz) since both cameras are synchronized together.

Fig.1 shows the different parameters involved in the process, as will be later explained in the text. As clearly shown in the figure, all the analysis is carried out in a reference trajectory plane which is supposed to contain both the optical axes of the two cameras and the trajectory of the license plates. This model does not take into account the different heights of the license plates of the vehicles. Such approximation has been proved to have negligible consequences in the measurement process. The common reference system is centered into the acquisition unit in between the two imaging sensors. In general the orientation of the license plates may be unpredictable but it is quite likely to be orthogonal to the current trajectory, and this can be used as a system constraint. Landmarks L_1, L_2, L_3, the nominal trajectory TN, and the distance, dbs, from the road-side, of the acquisition system, are calibration parameters and will be discussed later. Fig.2 summarizes the binocular speed measuring process. A vehicle transit event TE$\{Lp_1,Lp_2\}$ consists in the matching of two consecutive views (Lp_{1_i},Lp_{2_j}) of the same license plate, along the current trajectory. Of course there could be many instances of such correspondence, depending on the actual speed of the vehicle and the efficiency of the License Plate reader.

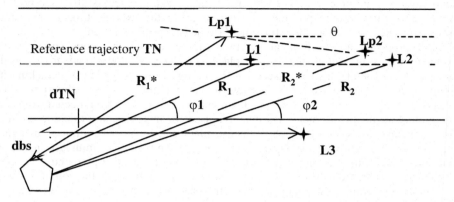

Fig. 1. Geometric configuration of the binocular system installed on the road side, The acquisition system is identified by the two camera unit CO1 and CO2.

Once a match has been found, the essential imaging features are extracted from the cluster of recognized characters, namely the center of mass of the clusters, $b_1 = (ub_1, vb_1)$, $b_2 = (ub_2, vb_2)$, and the average size of the characters, $sz_1 = (dh_1, dl_1)$ and $sz_2 = (dh_2, dl_2)$. The next step is the computation of the position of the license plates Lp_1, Lp_2 in the reference plane, i.e. the vectors R_1^* and R_2^* as shown in fig.1.

The centers of mass (b_1, b_2) are the projections of the license plate onto the imaging sensors and provide information about the direction of the vectors R_1^*, R_2^*. The module of the vector (i.e. the distance of the license plate from the sensor) is obtained from the average vertical size of the characters (dh), as:

$$|R^*| \ [mm] = f \ [pixels] * HT \ [mm] \ / \ dh \ [pixels] \qquad (1)$$

Where HT is the height (in mm.) of the actual character, f is the focal length of the imaging sensor, in pixel units, to take into account intrinsic parameters of the sensor, and dh is the average height of the matched characters (which can be computed with subpixel precision).

When the two vectors R_1^*, R_2^*, from the two matched views (LP_1, LP_2) of the same license plate (at time stamps $t_1 = ndt$, $t_2 = mdt$), have been computed, the speed estimation is quite straightforward:

$$v\text{-}_{est} = (\ ds \ / \ (m - n) \ dt \ ) \ / \cos \delta \qquad (2)$$

where $tg \ \delta = ns \ / \ ds$
$$ns = |R_2^*| \sin(\varphi_2 - \beta_{2est}) - |R_1^*| \sin(\varphi_1 - \beta_{1est}) + BT \cos(\alpha)$$
$$ds = |R_2^*| \cos(\varphi_2 - \beta_{2est}) - |R_1^*| \cos(\varphi_1 - \beta_{1est}) + BT \sin(\alpha)$$
and $\alpha = (\varphi_1 + \varphi_2) / 2$

φ_1 , φ_2 are the angles of the two optical axes of the sensors w.r.t. the nominal trajectory, BT is the telemetric basis of the binocular system. β_{1est} , β_{2est} are the angles of orientation of the vectors R_1*, R_2*, w.r.t. the optical axes of the sensors, (in the reference plane) and are obtained from the centers of mass b_1, b_2 as:

$$\text{tg } \beta_{1est} = ub_1 \, / \, f_1 \text{ [pixels]}$$
$$\text{tg } \beta_{2est} = ub_2 \, / \, f_2 \text{ [pixels]}$$

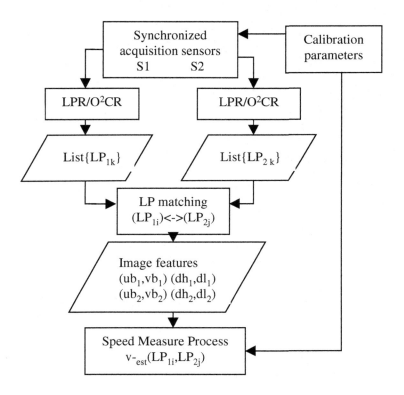

Fig. 2. Flow diagram of the binocular speed measurement process

In the above formula (3) the angle δ is an important cue on the reliability and consistency of the measure. Moreover, a bias of δ, in many consecutive speed measures, may be a serious hint of an error in the installation of the sensor, as briefly discussed in the following.

3. Experimental Set-Up of the Binocular System

A prototype version of a binocular vision system for license plate reading from the road side has been integrated and tested. Fig.3 shows the transportable system in a recent experimental test along the highway. It consists of three components: a binocular acquisition unit, a processing box and a battery to provide the required autonomy in the different applications.

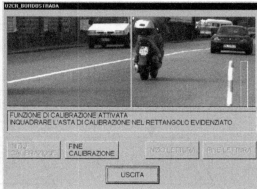

Fig. 3. Experimental binocular system   **Fig. 4.** Landmark L3 installation procedure

The selected imaging units are high-sensitivity cameras. For each specific application it is possible to select the most suitable optical parameters (from 12 to 75 mm focal length) and control convergence/divergence of the optical axes for each individual camera, during the calibration phase. Fig.3 shows the acquisition head, with the two cameras combined with an IR illumination source (based on halogen lamps) to achieve maximum efficiency to operate at long distance (over 24 meters). A TFT-LCD interface unit has been integrate in the rear part of the acquisition unit, to optimize human-machine interaction, simplify installation procedures (fig.4) and provide a visual feedback on the behavior of the system.

Actually the system has been developed primarily for road side License Plate reading in security applications where the transportable unit from the field can be connected to an operative supervisory unit for data collection and decision making. The proposed application for speed measurement is one of the additional features of the system. Other potential applications are data collection on the traffic flow within urban areas or along selected ways.

3.1   Calibration and Installation Procedures

A very simple and effective calibration procedure has been developed in the lab environment, to fix the acquisition geometry, according to the selected operating conditions (i.e. the distance from the road side)

Two license plates are placed along the nominal trajectory in positions L1 and L2 (as in fig.1) to be acquired by the two cameras, along the optical axes, with a nominal scale factor (image resolution) corresponding to the selected focal lengths at the predicted distance (R_1, R_2).

A third vertical landmark L3 is placed at a pre-selected distance (as depicted in fig.1), to control the correct orientation (pan angle) of the whole acquisition system w.r.t. the nominal trajectory. The vertical landmark L3 must be acquired by the sensor CO2 at a predefined column of the image. When the system is mounted in the field it is sufficient to verify this correspondence on the image plane to accept the correct installation as shown in fig.4. Actually it is possible to build a table of correspondences between the position of the landmark L3 and the distance from the side of the road (dsb in fig.1). It is worth to remark the basic assumption of no slope on the road and the acquisition head is supposed to be placed almost at the same height of the license plates to be read (from 50 to 90 cm above the ground).

4. License Plate Recognition

The key feature of the process is the availability of an extremely effective License Plate Recognition system able to detect the maximum number of instances of the LP's in transit with high accuracy. The LPR system should be able to work continuously on the video sequence at the maximum rate (50 Hz for a standard CCIIR) to pick-up the license plates from the video flow, without the help of external triggering devices. It is required an extremely accurate precision of character segmentation and localization since any error at this level would propagate to the range estimate R* and the following speed measure in equation (3).

Elsag has developed its first LPR (License Plate Reading) systems since the beginning of the 90's in a variety of applications [4], from pay-toll gate control (Telepass) along the Highways, parking access control, urban traffic monitoring and Road Pricing applications. This technology is the result of a long experience in Elsag on the subject of Intelligent Character Recognition applied to document and form processing [5] as well as for address understanding in mail sorting, including cursive handwriting [6]. A wide variety of solutions have been experimented using both statistical and neural network solutions. The adopted solution is based on a fairly classical pattern matching approach, and it represents a satisfactory compromise between accuracy and computational efficiency.

The task of license plate recognition involves many system-engineering issues and multidisciplinary competence of electro-optics, and computer vision. Infrared lighting is widely used especially at night, to minimize external uncontrolled conditions [7]. The position of the sensor and its orientation are other essential issues, to minimize perspective deformations of the characters and achieve the necessary resolution in the scene. In general, whatever kind of camera is used, it should be placed to avoid occlusion of the license plate, and allow the acquisition of the plate field as much as squared as possible. Angled views of up to 30°, in the horizontal or vertical plane are usually manageable. It is good practice to ensure that the license-plate characters appear as near vertical as possible in the digitized image.

The following section summarizes the real-time video processing approach which is used by O^2CR/LPR Elsag proprietary solution.

4.1 The Plate Recognition Process

The recognition process consists in three main blocks [8]. A first pre-processing block performing a screening of the video sequence in order to select images or regions of interest which are affected by motion.

An image processing engine performs license plate location, character segmentation, OCR and context analysis and validation.

A temporal post-processing block follows, to perform data fusion and tracking among different images of the video sequence. The context verification process exploits both spatial and syntactic information in order to select the best hypothesis for the numberplate.

The final temporal post-processing stage aims to extract a single numberplate for a given vehicle. The presence of a vehicle is detected by the observation of a group of video frames affected by image motion and showing a sufficient number of recognized characters. Such definition allows the detection of a vehicle even if the contextual process does not extract any numberplate hypothesis.

All the number plate hypothesis, with both spatial and syntactic information, are gathered from the image belonging to such group and a clustering approach is used in order to evaluate the best choice. Main characteristics of the O2CR LPR system are briefly summarized as follows [8]:

- Adaptive control of the exposure time of the camera to optimize the contrast image of the license plate
- Data temporal integration and feedback process to recover character segmentation errors.
- Continuous processing on the video sequence (free-running) and simultaneous multiple license reading, without the need of external triggering systems.
- Standard PC processing environment.
- Integration of high sensitivity imaging sensors with IR-LED integrated illumination system.
- Recognition performance (certified by Italian standards UNI10772)
- Between 90 and 95% according to operating conditions, with reading tolerance over 30°
- Over 97% in favourable conditions
- Processing speed at video frequency (50 fields/sec) to detect high speed transit.
- Multinational license plate (multifont and training to deal with various contextual information)

5. Simulation Results (Error Sensitivity Analysis)

To be able to predict the performance and the precision which can be reached by the system, a simulation model has been implemented, trying to generate all possible situations which will appear in the real field. Following the block diagram of fig.1., a first input section has been arranged. It allows to select a wide range of operating conditions, including optical parameters of the imaging sensors, the supposed trajectory (near and far from the road side) the amount of speed of the supposed vehicle, relative orientations, etc. It is possible to take into account also the orientation of the plate in the vertical plane, to verify their consequences in the speed estimation process. A second module has been implemented to simulate the image processing section, to generate the same amount of information as it will be provided by the License Plate reader, i.e. position and size of the recognized characters. The final section is devoted to implement the actual speed measurement process.

The main limitation of this approach is the use of a purely geometric simulation model. As such all simulation results are based on the assumption to use an ideally fast imaging acquisition system with an indefinitely short integration time and with an optimal intensity response. On the other hand, by separate testing the image acquisition and processing system, it has proved that the required performance can be successfully achieved in the real situation.

Actually the primary source of errors has been proved to be the precision of the image features localization and in particular the size of the matched characters in the two image planes.

The following table refers the computed errors in the speed measures by comparing the results obtained with an integer quantization (Np=1) of the character heights (dh_1, dh_2), a subpixel precision of 10% (Np=10), and a precision of 1% (Np=100). The other main simulation parameters are:

☐ Focal lengths: f_1= 25 mm, f_2= 50 mm,
☐ Dbs= 2 m, Nominal trajectory TN= 5 m,
☐ Displacement error of landmark3 = 47 cm.
☐ Vertical slant of the license plate = 5°
☐ Trajectory deviation θ = 2°.

Table 1. Speed measurement results (err %), as a function of character size precision (Np) for a reference speed of 120 km/h

Np	1	10	100
v-est [km/h]	133,12	121,01	120,51
err %	10,93%	0,845%	0,424%
dh_1	13	13,2	13,28
dh_2	13	13,8	13,87

Actually the size of the characters is by definition an average measure from a number of characters (typically 7) and Np=10 precision is always experimentally achieved. Moreover improved results can be achieved by time integration (averaging multiple estimates from the possible matching pairs of the same license plates on both image sensors). The following table refers a comparison of results (with Np=10 precision), for different speed values. V-est refer the computed speed value, err% is the computed error, (dh_1, dh_2) are the computed vertical sizes of the characters, (Nr_1, Nr_2) are the maximum number of readings in the two image planes CO1 and CO2 respectively.

Table 2. Simulation results at different speed values

Actual speed [km/h]	75	95	120	150	200
v-est [km/h]	74,85	94,84	120,64	149,56	201,39
err %	-0,19%	-0,168%	0,533%	-0,296%	0,694%
delta-est[°]	2,30	2,31	2,42	2,29	2,45
ht1	14,20	14,20	14,20	14,20	14,20
ht2	13	12,9	12,9	12,8	13
Nreading1	6	5	4	3	2
Nreading2	6	5	4	3	2

In this case an installation error of the landmark L3 (-30 cm) has been simulated, but no relevant errors have been introduced in the measures. The actual trajectory has been supposed to have a slant direction of 3° to the right, with respect to the lane direction. The speed measurement error is always below 0,7% and is quite constant, irrespective of the amount of speed. As it was anticipated, this is mainly due to fact that the simulation results do not take into account possible radiometric problems of the acquisition system. An estimate of slant orientation is computed (delta-est) and is quite in accordance with the actual simulated direction slant of 3°. The number of readings (in the assumption to work at maximum frequency of 50 fields/sec) is obviously higher at lower speed values. Even if in the real situations not all such readings will be usable for speed measurement computation, there is a significant possible improvement in precision by averaging multiple estimates.

Moreover, the simulation model allows to check the correct configuration.. For instance it is proved that the system is able to detect possible errors in the displacement of the installation landmark L3, and provide the necessary information to correct them. Moreover it is possible to see the effects of other critical situations like the slant of the license plate in the vertical plane or the direction of trajectory (θ).

6. Experimental Results

The prototype speed measurement system (as in fig.3) has been evaluated in two different conditions. A first analysis has been carried out in a fully controlled environment in the lab, to map the critical function R*(dh) of eq.1. The obtained results are found within the range of precision required (Np=10). The second test has been performed in the field to prove the detection capability of the binocular imaging system when dealing with a wide range of speed, from slow trucks up to very high speed cars. A series of experiments have been performed along a 2-lane highway to read the license plates of the vehicles in transit. To provide a ground truth on the computed speed a commercial laser-based speed measurement system has been used.

The obtained results (although yet limited in number) have been satisfactory for measured speed above 180 km/h. Further experiments are planned in the near future and will be extensively referred and discussed at the conference.

7. Conclusions

The paper refers a proposed approach for speed measurement based on a binocular configuration which has been already experimented successfully for license plate reading of vehicles at high speed along the highways. The proposed approach has some interesting and promising features:

➤ Joint speed and LP reading to support vehicle identification with a unique measure
➤ High precision, due to a possible time integration and average of multiple readings, virtually limited only by shutter speed and light intensity
➤ Potential of increased performance with increased resolution of visual sensors
➤ Wide measurement range (more than a single lane) even with conventional resolution digital cameras
➤ Self-awareness to control the installation conditions, to discard noisy ambiguous data
➤ Many possible configurations (short and long distance) depending on the range of maximum speed detectable
➤ Possibility to use partial readings of the license plate in order to perform license plate matching and speed measurement

The referred results are also supported by a simulation model which provides a nice control of the main parameters of the system. The next steps of the development will involve a thorough investigation and experimentation of the system with different sensor configuration (narrow IR and high speed sensors). Moreover the system will be evaluated against certification requirements to get a qualified speed measurement system for speed enforcement purposes.

References

1. Traffic Technology International 2000, UK and International Press, 2000.
2. "The Computer Vision Homepage", http://www.cs.cmu.edu/~cil/vision.html
3. P.W.Shuldiner, J.B.Woodson, "Acquiring Travel Time and Network Level origin-destination data by Machine Vision analysis of Video License Plate Images", 1996 National Traffic Data Acq. Conf., Albuquerque, New Mexico, May 5-9, 1996.
4. E.Ottaviani et al., Real-time vehicle number plate recognition from video sequences, to be published in *Proc. CSCC 99, Image Processing for intelligent transport systems*, Athens 1999
5. E. Appiani, L. Boato, S. Bruzzo, A.M. Colla, M. Davite and D. Sciarra, "STRETCH": A System for Document Storage and Retrieval by Content, *Proc. DEXA '99, Workshop W07: DAUDD'99, Florence, Italy*, Sep. '99,
6. G.Nicchiotti, C.Scagliola, "Generalised Projections: a Tool for Cursive Handwriting Normalisation", *Proc. 5th Int.Conference on Document Analysis and Recognition (ICDAR'99)*, Bangalore, India, September '99, Kluwer Academic Publisher Boston/Dordrecht/London, 1999,
7. Y.Cui, Q.Huang, Extracting characters of license plates from video sequences, Machine Vision Ap. Vol.10, pp.308-320, 1998.
8. G.Garibotto, P.Castello, E. Del Ninno, " Dinamic Vision for License Plate Recognition", chap. 6.6 "Multimedia Video-Based Surveillance Systems" Kluwer Academic Publisher Boston/Dordrecht/London, 1999.

Volume and Surface Area Distributions of Cracks in Concrete

George Nagy[1], Tong Zhang[1], W.R. Franklin[1], Eric Landis[2], Edwin Nagy[2], and Denis T. Keane[3]

[1] Rensselaer Polytechnic Institute, Troy, NY, USA
[2] University of Maine, Orono, ME, USA
[3] Northwestern University, Evanston, IL, USA

Abstract. Volumetric images of small mortar samples under load are acquired by X-ray microtomography. The images are binarized at many different threshold values, and over a million connected components are extracted at each threshold with a new, space and time efficient program. The rapid increase in the volume and surface area of the foreground components (cracks and air holes) is explained in terms of a simple model of digitization. Analysis of the data indicates that the foreground consists of thin, convoluted manifolds with a complex network topology, and that the crack surface area, whose increase with strain must correspond to the external work, is higher than expected.

1 Objectives and Scope of the Paper

Many attempts to model or recognize shape and form are based on a bi-level representation of relatively simple objects. In contrast, we are faced with an engineering problem characterized by sequences of large, complex, volumetric gray-scale images. This data was produced by a unique imaging instrument designed for observing the internal structure of dense, heterogeneous materials. The resulting measurements will ultimately be used in multiscale modeling of the microstructure for improved understanding of the macroscopic mechanical properties of concrete [1,2,3,4]. So far we can report only some observations from which we attempt to deduce aggregate and individual shape properties of a large collection of objects and to separate material properties from image processing artifacts. Our work raises far more questions than it answers. We present it here in the hope of gaining assistance from the segment of the image processing community dedicated to allied pursuits.

More specifically, we propose to analyze thin, warped, interconnected volumetric entities in sequences of density images of samples of mortar. The data is obtained by high-resolution 3-D microtomographic imaging using an X-ray imager at the National Synchrotron Light Source at Brookhaven[5,6,7]. The images show crack formation in mortar and concrete under increasing strain. Other applications with similar filiform and quasi-manifold configurations are membranes, plant and animal vasculature, nerve fibers, and polymers, all of which can be

C. Arcelli et al. (Eds.): IWVF4, LNCS 2059, pp. 759–768, 2001.
© Springer-Verlag Berlin Heidelberg 2001

imaged through soft-tissue tomography, magnetic resonance imaging, 3-D ultrasound, or confocal microscopy. The topological complexity of such data precludes 2-D analysis.

The detection, quantification and further analysis of structural changes in cement-based materials offers the potential for a more rational approach to the design, testing, repair, or replacement of concrete structures. The total replacement value of concrete structures in the US has been estimated to be over six trillion dollars. While continuum approaches based on plasticity and linear elastic fracture mechanics have led to considerable success in predicting failure in fine-grained materials such as metals, non-linear effects have resisted analysis in heterogeneous and quasi-brittle materials such as concrete. The study of microstructure coupled with traditional stress-strain measurements offers the most promising approach [8,9,10]. In concrete, cracks are thought to originate from one or more porous voids, and they may even spread preferentially through voids and pre-existing cracks. We hope that the detailed mechanisms of crack origination and propagation may be revealed by 3-D X-ray imaging.

2   Data Collection

Microtomography yields a 3-D map of absorptivity from hundreds of through-transmission radiographs of the specimen taken from different angles. It is similar to medical CAT scans, except for much higher beam intensity and detector resolution. The specimen is rotated on a stage designed to allow the application of a load while minimizing X-ray absorption [11,12,13](Figure 1).

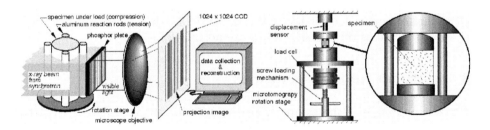

Fig. 1. (left) The system for x-ray microtomography; (right) the load cell for holding the composite material specimens under calibrated loads.

The specimens are small mortar cylinders under axial compressive load. The stress is continuously monitored by a conventional load cell, and the platen-to-platen displacement by a linear-voltage displacement microprobe. The data is collected and preprocessed by the University of Maine team at Beamline X2B. The X-ray source is synchrotron radiation with a highly collimated narrow-band beam monochromated to 32keV. The detector is a phosphor plate from which light is captured by a high-resolution CCD camera. The specimens are exposed to the beam at 720 different angles over a 180 degree range. Each exposure lasts 8-12 seconds, depending on the synchrotron beam current. This combination results in a very high resolution (2-6 μm) 3-D array, but the capture cross section is limited to about 6mm by the beam width.

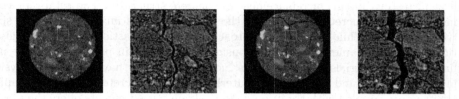

Fig. 2. Two 2-D circular slices of concrete, gray-scale, one each from successive loads, at corresponding heights. Enlargement of 100×100 squares across the large crack.

The resulting data (Figure 2) consists of sequences of 3-D integer arrays representing localized X-ray absorption. The dimensions of each array, reconstructed by the EXXON Direct Fourier Reconstruction algorithm on site, are typically 1024×1024×800 voxels. A new array is generated from the specimen after each of 5 or 6 load-and-release cycles with progressively greater loads. The last image is intended to capture the state of the specimen after the peak of the stress-strain curve, when any further load would cause it to crumble. Each sample requires about 6 hours of "beam-time". The images are originally recorded as 32-bit floating-point numbers, and then scaled to eight bit integers. In our representation, high gray values (shown as white or light gray) indicate high X-ray absorption, and low values (shown as dark gray or black) indicate voids consisting of cracks and air holes. When the data is binarized to 0-1, a higher threshold increases the number of foreground (black, i.e., 0) voxels.

3   Methods and Observations

The surroundings of the sample concrete cylinder are transparent to X-rays just as are the voids inside the cylinder. However, many cracks reach the external surface. In order to apply connected-components analysis, it is necessary to separate the exterior volume from the crack volume. This is accomplished by "shrink-wrapping" the cylinder. The resulting cross-sections are neither truly circular nor convex, and change along the axis.

Our 3-D processing relies heavily on connected components (CC) analysis [14,15,16]. We have developed and tested a robust algorithm that is space- and time-efficient because it is implemented as a Find-Union on 1-D runs, with path compression. In a test array of size 800×800×765 (489,600,000 voxels), it finds six million six-connected components in 200 seconds (400 MHz Pentium with 640Mbytes of RAM). In addition to listing all of the connected components with their constituent voxels, the program reports the volume, surface area (number of free faces) and the number of fore-ground runs in each CC. The code and test cases are freely available on `http://www.ecse.rpi.edu/Homepages/wrf/research/connect/`.

We use VTK, the 3D Visualization Toolkit, to visualize the cracks. VTK is an open source, surface-based rendering software system from kitware.com. Its rendering support is based on triangulating the gray-scale isosurfaces using Marching Cubes [17,18]. Other routines were developed to analyze the volume distribution, free surface histograms and merge graph of the connected components.

The remainder of the paper presents the observations in detail and attempts to explain them in terms of the characteristics of the sample and of a simple model of the digitization process.

3.1   Effects of Amplitude Quantization

The radiographic quality of the image data is quite consistent. There is little variation in grayscale from sample to sample, because fluctuation in the electron beam intensity is compensated for by periodic recording of a blank picture (without the sample). The gray scale does not capture the full dynamic range. The high absorption regions of mortar have a uniform value of 255 and some cracks and voids are saturated at 0. Nevertheless, there is sufficient contrast to discriminate the structure.

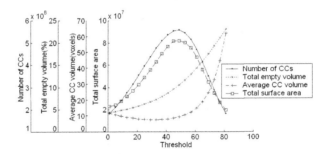

Fig. 3. Number of CCs, foreground (empty) volume, average CC volume, and surface area against binarization threshold for the whole shrink-wrapped volume of a sample.

The aggregate (larger particles) is even more opaque to X-rays than the mortar, while air is transparent. Given the high-contrast nature of the object under study, any threshold in a wide range should be equally satisfactory for isolating the voids (cracks and air holes). It turns out, however, that the choice of threshold has a very significant effect on the characteristics of the binarized image.

According to Figure 3, the total foreground volume increases gradually from 4% of the sample to 24%. The apparent crack-volume changes by a factor of more than three with threshold in the operational range of threshold from 40 to 60. This is a much larger change than that due to increased load at a constant threshold. It can be explained by the model of digitization presented below. At the same time, the number of CC's rises from nearly 1.8 million to nearly 5.5 million, then decreases to 1.8 million again. We conjecture (see below) that at the lower thresholds only thick voids are revealed. The eventual drop is expected: if the threshold is raised above the value of all voxels, then the entire sample will consist of a single connected component. Figure 4 shows representative 2-D cross sections at two thresholds at successive loads.

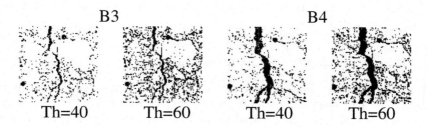

Fig. 4. 2-D bilevel X-section at thresholds of 40 and 60 at successive loads.

Because the work performed by the external force is expected to equal the work required to stretch the internal surfaces, the crack surface area is an important parameter. Furthermore, the ratio A/V of surface area to volume (akin to perimeter/area in 2-D) is a useful measure of rotundity that may separate cracks from air holes. The ratio $A^{\frac{3}{2}}/V$ is a *shape invariant*.

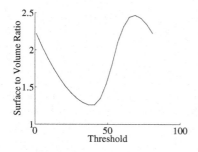

	Cube	One-voxel thick slab
Free face	Voxel Number	Voxel Number
0	5832	0
1	1944	0
2	216	7744
3	8	352
4	0	4
Total	8000	8100

Fig. 5. Surface area to volume ratio of the largest CC in a 200×200×200 voxel block.

Fig. 6. Free surface distribution

The Area/Volume ratio of the largest CC in a 200×200×200 block (Figure 5) falls as expected to a threshold of 40, then rises as even thinner and more tortuous cracks are merged to it. The maximum Area/Volume ratio is only 2.5.

In compact objects, most of the voxels would be either interior voxels, with no free face, or surface voxels with exactly one free face. The number of free faces is shown in Figure 6 for a 20×20×20 cube and a one-pixel thick 90×90 slab. In concrete, there are many voxels with several free surfaces, as seen in Figure 7, indicating highly irregular, tortuous surfaces. The skew increases with threshold, because although the larger CCs have more interior voxels, we are adding many smaller cracks.

The logarithmic scatter plots of Figure 7 show that the CCs range from flat or filiform (upper envelope: A/V constant) to filled-out shapes (lower envelope: $A^{\frac{3}{2}}/V$ constant). The larger the components, the thinner they are.

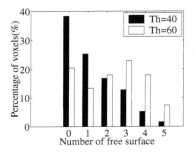 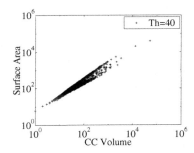

Fig. 7. (left) Bar chart of distribution of free faces at two thresholds; (right) scatter plot of surface area vs. CC volume at two thresholds.

3.2   Effects of Spatial Quantization

Because of the presence of so much boundary surface (between foreground and background), reducing the resolution by subsampling by 8 the data has a very different effect from averaging it over 2×2×2 volumes. Figure 8 compares histograms resulting from subsampling and interpolation. This result indicates that we can expect radically different results as the resolution of the imager is enhanced.

3.3   Crack Size and Connectivity

Figure 9 shows that the distribution of crack size is qualitatively similar at different thresholds. About half of the CCs are smaller than 1000 voxels, while the largest CC accounts for one third to one half of the total foreground volume. The CCs span six orders of magnitude in size.

The presence of a huge number of tiny foreground CCs at any threshold is certainly suggestive of noise, but may also be a property of the material. We will soon obtain multiple images of the same sample to resolve this question. We

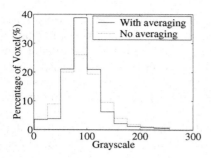

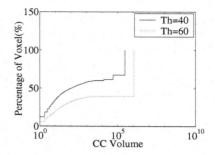

Fig. 8. Grayscale histogram of reduced samples.

Fig. 9. The empty voxel distribution versus CC volume of a $200 \times 200 \times 200$ voxel block.)

have also noted patterns of horizontal circular caused by the reconstruction of irregularities in the phosphor or spread of the X-ray beam.

It is possible to trace the merging of the largest components as the threshold changes (Figure 10). Each merger results in an abrupt increase in the volume and surface area of the resulting composite CC. The mergers are caused by the emergence of thin "bridge" cracks as the threshold is increased. A complete merge graph has millions of nodes. Ultimately we are interested in tracing the merger of cracks with increasing load.

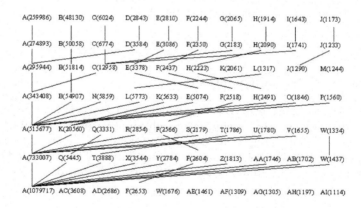

Fig. 10. Merge graph of the largest cracks(the numbers being the volume of cracks).

3.4 Point Spread Function

Most of the above observations can be explained using a simple model of digitization. We present the model in one dimension in order to be able to graph the functions. In the model, the cracks have a constant (one) density. The width of the cracks is distributed exponentially. The space between cracks is also distributed exponentially. The point spread function is modeled with a Gaussian. The 'analog' crack signal is convolved with the Gaussian, then thresholded and sampled. (The relative order of thresholding and sampling is immaterial.) The left of Figure 11 shows the original crack distribution, the convolved signal, and the binarized distribution at two different thresholds. As the threshold is decreased, nearby cracks are merged and new, thinner cracks appear. The width of the point spread function is more than 2 voxel diameters, as indicated by the cross section of cracks in the right of Figure 11.

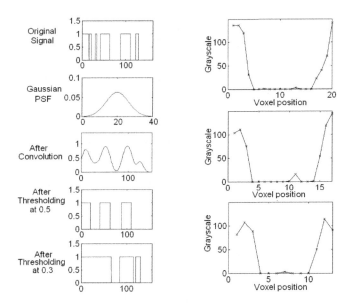

Fig. 11. (left) Model of Digitization; (right) sample profiles of grayscale across larger cracks.

4   Discussion

The samples consist of a huge number of very thin cracks, and a few wide cracks. The entire volume, except for the aggregate, appears to be traversed by cracks (*craquelure*). It is possible that most of these cracks are connected, but the point-spread function of the imager is too large to yield convincing proof.

A fast CC program is essential for studying the image at a wide range of threshold settings because the large point-spread function of the imaging system, compared to the spatial sampling interval, obscures the intrinsically high contrast between mortar and voids. This also accounts for the rapid growth in the volume of black pixels as the threshold is increased. For now, we can only speculate whether the point-spread function is dominated by the phosphor or the granularity of the reconstruction algorithm (the cooled CCD camera is not a likely culprit).

The thin cracks have very convoluted boundaries, which result in a high surface-area to volume ratio. Therefore, visualization software yields very little insight into the structure, and a quantitative approach is required. With increasing threshold, the number of boundary voxels increases faster than the number of interior voxels, resulting in an overall increase in the surface-area to volume ratio. The results on subsampling indicate that the current spatial sampling resolution is insufficient to give a true measure of the (possibly fractal) surface area of the crack boundaries. Such a measure is necessary to compare the increase in total crack surface area from load to load with theoretical predictions based on the loading and relaxation stress-strain curves. Although the resolution of the X-ray beam and of the optical system allows us to increase linear resolution by at least a factor of four (at the cost of a 64-fold increase in acquisition time and data volume, and a corresponding decrease in sample volume), electron micrographs would be useful here.

Under load, the volume of the largest connected components grows more quickly than the total volume of black pixels, as observed at any threshold. Equivalently, as the load increases, the volume of small connected components relative to the total black volume decreases because their expansion under load results in the small cracks being connected to the larger cracks. This effect is superficially similar to apparent crack growth with increasing threshold.

5   Future Work

We have much work ahead of us. We don't yet have any effective measures of crack shape and crack topology. Before modeling crack propagation under load, it will be necessary to model both the cracks and the density variations in the material itself. Furthermore, current 3-D image registration techniques will have to be extended to the compound problem of bringing into correspondence objects exhibiting both global distortion (motion of the sample) and local changes due to crack growth. After separating cracks from noise specks and air holes on the basis of volume and surface area, we can verify that cracks grow from load to load, air holes remain the same, and noise specks appear and disappear randomly.

Another important issue is the pore structure connectivity that governs water penetration from the surface. From the CC analysis we can determine which cracks open to the surface. By tracing these cracks, we can compute what fraction of the volume is a given distance from the surface. The change of permeability with crack growth affects long-term durability of concrete. An open question is the rate at which hairline cracks merge to form macro-cracks under load. Our long-term goal is the parametrization of crack growth for multiscale finite element modeling and analysis.

References

1. J.G.M. van Mier, Fracture Processes of Concrete, CRC Press, New York, 1997.
2. Z.P. Bazant, Analysis of Work-of-Fracture Method for Measuring Fracture Energy of Concrete. Journal of Engineering Mechanics, **122**(2), 138-144, 1996.
3. S.P. Shah, ed., Toughening mechanisms in quasi-brittle materials, Kluwer-Academic, Dordrcht 1990.
4. S.P. Shah, S.E. Swartz, C. Ouyang, Fracture mechanics of concrete: applications of fracture mechanics to concrete, rock, and other quasi-brittle materials, Wiley, New York, (1995).
5. F.O.Slate, S.Olsefski, X-ray study of internal structure and microcracking of concrete, J.Am. ConcreteInst.60,5,pp.575-588,1963
6. Flannery, B. P., Deckman, H. W., Roberge, W. G., and D'Amico, K. L. (1987) Three-Dimensional X-ray Microtomography. Science, **237**, 1439-1444.
7. H. W. Deckman, J. H. Dunsmuir, K.L. D'Amico, S.R. Ferguson, B.P. Flannery, Development of Quantitative X-ray Microtomography. Materials Research Society Symposium Proceedings, **217**, 97-110, 1991.
8. E. N. Landis, E. N. Nagy, D. T. Keane and N. Huynh, Observations of Internal Fracture in Mortar using X-Ray Microtomography, Proceedings of the ASCE Engineering Mechanics Specialty Conference, Ft. Lauderdale, FL, May, 637-640, 1996.
9. E. N. Landis, E. N. Nagy, D. T. Keane and S. P. Shah, Observations of Internal Crack Growth in Mortar using X-ray Microtomography, in Proceedings of the 2nd International Conference on Nondestructive Testing of Concrete in the Infrastructure, June 12-14, Nashville, TN, Society for Experimental Mechanics, 54-59, 1996.
10. E. N. Landis, E. N. Nagy and D. T. Keane, Microtomographic Measurements of Internal Damage in Portland Cement-Based Composites, Journal of Aerospace Engineering, V. 10, No. 1, 2-6, 1997.
11. E. N. Nagy and E. N. Landis, Analysis of Microtomographic Images to Measure Work of Fracture in Concrete, in Review of Progress in Quantitative Nondestructive Evaluation 16, D. O. Thompson and D. E. Chimenti, Eds., Plenum Press, New York, 1761-1766, 1997.
12. E. N. Nagy and E. N. Landis, Energy-Microcrack Growth Measurements for Mortar Cylinders in Compression, in Nondestructive Characterization of Materials in Aging Systems, R. Crane, Eds., Materials Research Society, 1998.
13. E.N. Landis, E.N. Nagy, D.T. Keane, G. Nagy, A technique to measure three-dimensional work-of-fracture in concrete, Journal of Engineering Mechanics 126, 6, pp. 599-605, June 1999.
14. C. Ronse and P.A. Devijver, Connected components in binary images: the detection problem, Research Studies Press, Letchworth, England, 1984.
15. H. Samet, Applications of Spatial Data Structures, Addison-Wesley 1989.
16. G. Borgefors, I. Nystrom, G. Sanniti de Baja, Connected components in 3D neighborhoods, Procs. 10th Scandinavian Conf. on Image Analysis (SCIA'97), 557-570, Helsinki 1997.
17. W.J. Schroeder, Geometric triangulation with application to fully 3-D automatic mesh generation. PHD dissertation, Rensselaer Polytechnic Institute, Troy, NY May 1991.
18. L.L. Schumaker, Triangulations in CAGD, IEEE Computer Graphics and Applications 13, 1, 47-52, 1993.

Integration of Local and Global Shape Analysis for Logo Classification*

Jan Neumann, Hanan Samet, and Aya Soffer**

Computer Science Department
Center for Automation Research
Institute for Advanced Computer Studies
University of Maryland
College Park, Maryland 20742
jneumann@cfar.umd.edu, hjs@cs.umd.edu and ayas@il.ibm.com

Abstract. A comparison is made of global and local methods for the shape analysis of logos in an image database. The qualities of the methods are judged by using the shape signatures to define a similarity metric on the logos. As representatives for the two classes of methods, we use the negative shape method which is based on local shape information and a wavelet-based method which makes use of global information. We apply both methods to images with different kinds of degradations and examine how a given degradation highlights the strengths and shortcomings of each method. Finally, we use these results to combine information from both methods and develop a new method which is based on the relative performances of the two methods.

Keywords: shape representation, shape recognition, image databases, symbol recognition, logos

1 Introduction

We examine three different approaches for classifying images with several components in an image database. One approach uses local methods to represent the image, the second uses global methods, while the third combines both using an adaptive weighting scheme based on relative performance. The local method uses so-called negative symbols, as described in [8], to compute a number of statistical and perceptual shape features for each connected component of an image and its background. The global method uses a wavelet decomposition of the horizontal and vertical projections of the global image as described in [5]. As a sample application of well-defined multi-component images, we use logos.

Several studies have reported results on some form of logo recognition. Each study used either global or local methods. These include local invariants [4,7],

---

* The support of the National Science Foundation under Grants CDA-95-03994, IRI-97-12715, EIA-99-00268, and IIS-00-86162 is gratefully acknowledged.
** Currently at IBM Research Lab, Haifa 31905, Israel.

C. Arcelli et al. (Eds.): IWVF4, LNCS 2059, pp. 769–778, 2001.
© Springer-Verlag Berlin Heidelberg 2001

wavelet features [5], neural networks [3], and graphical distribution features [6]. The performance in case of certain degradations was examined.

In this paper we compare the local and global methods under the influence of several image degradations. The performance measure is the ranking of the original logo after inputing a degraded version of it into the classifier. The results exhibit the advantages and disadvantages of local methods, based on shape features, in contrast to global methods, rooted in signal processing. Finally, we present an algorithm that combines both methods into a single, robust framework by adaptively weighting the contributions of each method according to an estimate of their relative performance.

2   Preprocessing: Normalization of the Images

The classification methods should be scale, translation, and rotation invariant. To achieve this, we apply some preprocessing steps to the input images before we start the computation of any features. The logos contained in the UMD-Logo-Database are gray-scale images that are scanned versions of black and white logos. Using an empirically determined preset threshold, we transform the input image into a binary image for which we compute its centroid. After shifting the image so that the centroid is located at the image center, which gives us translational invariance, we rotate the image around the centroid so that the major principal axis is aligned with the horizontal. This gives us rotational invariance. Finally, we resize the logo component so that its bounding box is a given percentage of the image size. This accounts for changes in scale of the input logos. These transformations make it possible to perform the following computations without reference to orientation, position, and scale.

3   The Wavelet Method

Given a normalized image we compute the horizontal and vertical projections of this binary image which are defined as $P(y) = \sum_{x=1}^{m} I(x,y)$ and $P(x) = \sum_{y=1}^{n} I(x,y)$. This means that we are counting the number of white pixels for each column and row. Next, we use a wavelet transform to apply a low-pass filter to the projections. In our experiments we used the Haar wavelet and the Daubechies wavelet s8 as implemented in the MATLAB wavelet toolbox and described in [9]. We do a 4-level Haar wavelet decomposition and for the 256x256 images that we used we get 16 low-pass coefficients per projection. In the case of the Haar wavelet this amounts to a repeated process of averaging and down-sampling. Finally, we end up with a 32-dimensional vector describing the logo as there are 16 coefficients for each of the two coordinate axes. This process is illustrated in Figure 1. These coefficient vectors, called *signatures*, are now used to compare different logos among each other. We use the L_1-Norm to compute the difference between their signatures, because the L_1-Norm is known to be robust against outliers and very fast to compute [9].

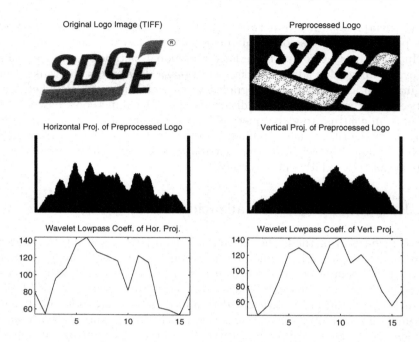

Fig. 1. The Wavelet signatures (from top-down, left-right): original image, normalized image, horizontal projection, vertical projection, low-pass wavelet coefficients of horizontal projection, low-pass wavelet coefficients of vertical projection (x-axis: index of coefficient, y-axis: coefficient magnitude).

4   The Negative Shape Method

The novel idea of the negative shape method as defined in [8] for the representation of symbol-like data such as found in logos is that we compute the shape features not just on the components of the foreground that constitute the symbol itself, but also on the components that make up the background of the image containing the symbol.

4.1   Choice of Shape Features

We start with the normalized images and do a connected component labeling of the image. For each component of the labeled image, we compute the following shape features:

1. **F1: Invariant moment:** The trace of the covariance matrix of the positions of the pixels that make up the logo, that is the sum of its diagonal entries.
2. **F2: Eccentricity:** The ratio between the length and width of the axis-aligned bounding box of the component after the normalization described in

Section 2. This gives us information about the extent of the elongation of a component.

3. **F3: Circularity:** The ratio between perimeter of the component and the perimeter of a circle of equivalent area: $CIRC = \frac{Perimeter^2}{4 \cdot \pi \cdot Area}$.

4. **F4: Rectangularity:** The ratio between the area of the component and the area of its bounding box.

5. **F5: Hole Area Ratio:** The ratio between the area of the holes inside the component and the area of the solid part of the component.

6. **F6,F7: Horizontal (Vertical) Gap Ratio:** The ratio of the square of the gap count to the area of the component where the gap count is defined as the number of pixels inside the component that have a right (bottom) neighbor that does not belong to the component.

4.2   The Classification Procedure

For the negative shape method we define the distance measure between two logos $Logo_1$ and $Logo_2$ as follows:

1. Normalize the value range for each element of the feature vector over all the logos of all the images in the dataset.
2. For each component of $Logo_1$ find the component of $Logo_2$ that has the smallest distance(L_2-norm) in feature space to it.
3. The average of these minimal distances over all the components of $Logo_1$ yields a measure for the distance between the two logos.

5   Comparison between the Methods

All methods were implemented in Matlab^{TM} [1] and were applied to the logos contained in the UMD-Logo-Database (123 logos) [2]. The system was tested by providing it with an input logo and ranking the logos in the database based on their similarity to this logo. All methods always found the matching logo in the database. In particular, they ranked it first when the input logo is an uncorrupted version of one of the logos in the database. Below, we investigate the robustness of the methods when the logos are corrupted using four different image degradation methods as described in Figures 2a, 3a, 4a, and 5a. For each method, we degrade the images in the database to a varying degree, input them into the classifier, and then examine the rank (in terms of feature space distance) of the original, uncompromised logo. Here we examine the median of the rankings of the original logo over all the input logos (part b of all the figures) and how often in terms of the percent of all logos the original logo was ranked among the closest five of all logos (part c of all the figures). Each graph consists of three curves: the dashed curve corresponds to the negative shape method, the gray curve corresponds to the wavelet method, and the solid curve corresponds to the combined method which has not yet been described. The combined method was devised based on the results of these experiments and thus we defer its explanation and the analysis of the results using this method to the next section (i.e. Section 6) once we understand the pros and cons of the two methods.

5.1   Additive Random Noise

To model the image degradation that is caused by processes such as fax transmissions or photo copying, we add Gaussian noise of zero mean and varying standard deviation (varying from 0.1 % to 50 % of the maximum possible pixel value of the image as indicated on the x-axis) to the gray-scale input images (e.g., Figure 2a).

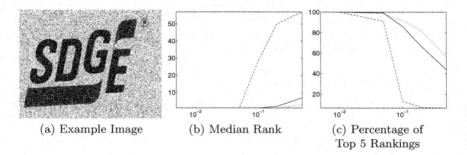

(a) Example Image   (b) Median Rank   (c) Percentage of
                                                Top 5 Rankings

Fig. 2. Gaussian Noise: The x-axis denotes the standard deviation of the Gaussian noise with respect to the maximal pixel value of the original image. The dashed curves in (b) and (c) correspond to the negative shape method, the gray curves to the wavelet method, and the solid curves to the combined method.

All the methods perform very well for small amounts of noise, but the wavelet method outperforms the negative shape noticeably (Figures 2b and 2c) for higher amounts of noise. Even when applying much noise (e.g., a standard deviation which is 20% of the possible pixel value), the average rank of the original logo is close to the top 10 (Figure 2b) and about 80% of the logos are ranked in the top 5 (Figure 2c). If we apply the negative shape method to such a heavily degraded image, the original logo is ranked in a nearly random manner (median rank 40th out of 123 logos as seen in Figure 2b) and the percentage of top 5 classifications is below 10% (Figure 2c).

It is to be expected that the wavelet method outperforms the negative shape method when adding random noise since the use of isotropic noise with an equal probability for adding or subtracting pixels should have only a small effect on the global histogram used in the wavelet method. We use noise of zero mean. Consequently, on the average, the distribution of white and black pixels in a row or column should not change much, and thus neither should the projection change much. On the other hand, in the negative shape method, we compute the feature vectors only on a small subset of pixels of each component. In this case, the noise will change the spatial distribution of the pixels more drastically because of the smaller number of pixels involved. Thus the negative shape method is less robust towards zero-mean Gaussian noise than the wavelet method.

5.2   Reduced Resolution

To see how the methods handle differences in image resolution, which is obviously not offset by the scaling invariance since we work on digitized images, we reduce the size of the input images through sub-sampling using bilinear interpolation (e.g., Figure 3). The parameter value is the size ratio between the original and the sub-sampled image as indicated on the x-axis.

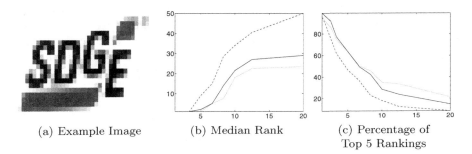

| (a) Example Image | (b) Median Rank | (c) Percentage of Top 5 Rankings |

Fig. 3. Reduced Resolution: The x-axis denotes the ratio between the size of the original and sub-sampled images. The dashed curves in (b) and (c) correspond to the negative shape method, the gray curves to the wavelet method, and the solid curves to the combined method.

As in Section 5.1, the wavelet method outperforms the negative shape method, although the negative shape method does not exhibit the same breakdown in performance as in the case of random noise. Since we use the low-pass wavelet coefficients for the classifier, the reduced resolution does not influence the performance of the wavelet method drastically. This is because sub-sampling an image by bilinear interpolation has a similar effect as low-pass filtering the image. The low-pass wavelet coefficients of a low-pass filtered image are in general very similar to the low-pass coefficients computed on the original image due to the fact that the low frequency components of the image are not affected noticeably by the sub-sampling operation. As before, the negative shape method is affected by this degradation because even when large scale changes are hardly visible, local shape features such as circularity, rectangularity and gap ratios are more susceptible to local changes due to a loss of detail.

5.3   Occlusion

To model the occlusion of parts of a logo, we add a component to the logo image which in this case is a black rectangle of varying size. The parameter here is the percentage of the image that is occluded by the rectangle (e.g., Figure 4a).

The performance graphs show that occlusion has a greater effect on the wavelet method than the negative shape method (Figures 4b and 4c) although both methods are able to handle small occlusions well.

Since the addition of an extra object or the omission of parts of the image causes global changes to the distribution of pixels in each row or column, the projections are strongly affected and thus so are the wavelet coefficients. Because of the local structure of the shape features, the components that are not occluded are not degraded at all and their feature values are unchanged. In the classifier we average the best feature vector matches for all the components in the input image. Since an occlusion is more likely to combine components into larger aggregates than to break them into many new ones, these few new components which do not have a corresponding component in the original image, are influencing the feature distance only to a small degree. Except for very degenerate configurations, the influence of the new components is averaged out by the continuing good matches of the feature vectors of the remaining uninfluenced components.

5.4   Swirling the Image

Swirling is a smooth deformation of an image which can be used to model a non-isotropic stretching of a logo. The relative position of each row is shifted to the left or right by an offset given by a smooth function, where the offset is limited to a certain percentage of the image width which is given as a parameter. This deforms the logo as if we would stretch a rubber sheet in different directions (e.g., Figure 5a).

This degradation has very different effects on the two methods. The performance of the wavelet method worsens rapidly with increasing swirl until we basically get a random ranking (our test size is 123, therefore, an average ranking of around 50 is nearly the expected median ranking for a logo that is not in the database). In contrast, the median rank of the original logo when using the negative shape method is lower than 10 (Figure 5b). It is possible to locally approximate this deformation as a combination of translations and rotations. The local features used by the negative shape method are rotation and translation

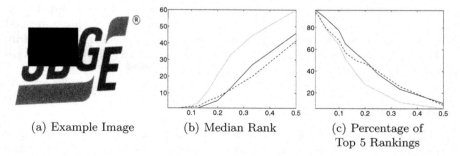

(a) Example Image   (b) Median Rank   (c) Percentage of
                                        Top 5 Rankings

Fig. 4. Occlusion of part of the image: The x-axis denotes the percentage of image area that is occluded. The dashed curves in (b) and (c) correspond to the negative shape method, the gray curves to the wavelet method, and the solid curves to the combined method.

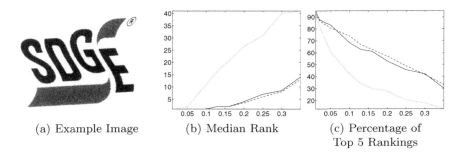

(a) Example Image   (b) Median Rank   (c) Percentage of
                                            Top 5 Rankings

Fig. 5. Swirl of the image: The x-axis denotes the maximum horizontal displacement of an image row in percentage of image width. The dashed curves in (b) and (c) correspond to the negative shape method, the gray curves to the wavelet method, and the solid curves to the combined method.

invariant due to the component normalization. Therefore, it is much less affected by this degradation than the wavelet method. Recall that the wavelet method is only globally rotation and translation invariant due to the global preprocessing, but not locally.

6   Combination of Both Methods

In Section 5 we saw that the wavelet and the negative shape methods perform very differently if the input logo is corrupted by either local or global degradations. To take advantage of the respective strengths of both methods we devised the following performance-dependent weighting scheme. First, for each undegraded logo l in the dataset we compute the average feature space distance of l to all other logos for both the wavelet and the negative shape methods. This is followed by calculating the average of these average distances for the two methods which we denote by A_w for the wavelet method and A_s for the negative shape method. We define the ratio between these two averages (i.e. $\frac{A_w}{A_s}$) to be the expected ratio E for the two methods. We determined how this ratio changed when we applied both methods to degraded inputs. The understanding of this relationship between the change in ratio and the relative performance of the two methods when applied to degraded images enabled us to adaptively weight the respective contributions of the two methods when combining them into a single distance measure. The relative weights are based on the change in the ratio because a a large increase of the feature space distance for one method compared to the other indicates a breakdown in its performance.

When classifying an input logo which has been degraded using one of the processes described in Section 5, we first compute the feature distances of this logo to all the other logos for the wavelet method which we denote by W and for the negative shape method which we denote by S. In addition, we define the averages of W and S over the whole dataset by D_w and D_s, respectively. Next, we compare the ratio between D_w and D_s (i.e., $\frac{D_w}{D_s}$) to the expected ratio between W and S which we assume to be similar to the precomputed value E. If the difference in the ratios indicates that one of the two methods is performing worse than expected, we decrease its weight in the final classification and increase the weight of the other method. The combined feature distance C for a single degraded input logo is a weighted sum of the wavelet method feature distance W and negative shape method feature distance S:

$$C = \frac{E \cdot D_s}{D_w} \cdot \frac{W}{E} + S \qquad (1)$$

The factor E, that describes the average ratio between W and S, is only included in order to facilitate understanding the rationale behind the final weighting method. If we divide W by E, then we effectively normalize W, so that its magnitude is equal to the magnitude of S. Thus, if the ratio $\frac{D_w}{D_s}$ equals the expected ratio E, then we believe that both methods will perform well and we use an approximately equal weighting of the two feature distances W and S. If now the ratio $\frac{D_w}{D_s}$ either grows larger (smaller) than E because the degradation of the input logo causes the wavelet method to compute feature distances larger (smaller) than the negative shape method (up to the expected ratio E), then the contribution of W in equation 1 will be reduced (increased) because we have less (more) confidence in the wavelet method's ability to classify the input logo correctly.

This adaptive weighting scheme increases the robustness of the classification noticeably. When we examine the performance criteria in Section 5, we see that the combined method is able to capture the different behavior of the methods and adapts its weights accordingly. Comparing the performance of the combined method on images degraded as described in Section 5.1 and Section 5.4 where the wavelet and the negative shape method exhibit very different performances, we see that our weighting scheme is able to detect the change in relative performance and adjust the weights to mimic the classification of the better performing method. For the degradations described in Sections 5.2 and 5.3 where the performance difference between the two basic methods is not as pronounced, the combined method lags slightly behind the better performing method in the median rank criterion (part (b) of all the Figures), but equals or surpasses the performance of the better method in terms of the other criterion (part (c) of all the Figures). This shows that our combined scheme is effective in capturing global as well as local shape information and is thus able to deal well with the image degradations of the kind that we described.

7   Summary and Future Work

Both the wavelet as well as the negative shape method are well-suited for certain kinds of image degradations but are very sensitive to others. This discrepancy in performance can be explained by the difference between local shape feature-based and global, filter-based methods. On the one hand, we have the wavelet method that operates on the global image and computes features that are relatively invariant to degradations that are isotropic. On the other hand, we have the negative shape method which operates on local image regions. Thus its features are relatively invariant to changes that leave the image at other locations mostly intact such as occlusions or preserve the local image structure such as the swirl deformation. We take advantage of the fact that both basic methods perform very differently on images that exhibit degradations of either local or global nature by devising a performance-dependent weighting scheme that combines the results of both methods. Our combined algorithm shows a noticeable improvement in the robustness of the classification by combining the strengths and avoiding the weaknesses of the respective methods. This weighting scheme performs the better the more different the performances of the underlying methods are because this makes it easier to detect if one method is performing poorly with respect to the other method. Therefore, the wavelet and the negative shape methods are very well-suited to be combined by a performance-dependent weighting scheme.

For future work it is planned to improve the synergy between the two methods by using local image information to estimate how much an image region is degraded and then use this locality information to adaptively weigh the feature vectors on the component level.

References

1. Matlab - the language of technical computing. www.mathworks.com.
2. Umd logo database.
   http://documents.cfar.umd.edu/resources/database/UMDlogo.html.
3. F. Cesarini, E. Fracesconi, M. Gori, S. Marinai, J. Q. Sheng, and G. Soda. A neural-based architecture for spot-noisy logo recognition. In *Proceedings of the Fourth International Conference on Document Analysis and Recognition*, pages 175–179, Ulm, Germany, August 1997.
4. D. Doermann, E. Rivlin, and I. Weiss. Applying algebraic and differential invariants for logo recognition. *Machine Vision and Applications*, 9(2):73–86, 1996.
5. M. Y. Jaismha. Wavelet features for similarity based retrieval of logo images. In *Proceedings of the SPIE, Document Recognition III*, volume 2660, pages 89–100, San Jose, CA, January 1996.
6. T. Kato. Database architecture for content-based image retrieval. In *Proceedings of the SPIE, Image Storage and Retrieval Systems*, volume 1662, pages 112–123, San Jose, CA, February 1992.
7. M. Kliot and E. Rivlin. Shape retrieval in pictorial databases via geometric features. Technical Report CIS9701, Technion - IIT, Computer Science Department, Haifa, Israel, 1997.
8. A. Soffer and H. Samet. Using negative shape features for logo similarity matching. In *Proceedings of the 14th International Conference on Pattern Recognition*, pages 571–573, Brisbane, Australia, August 1998.
9. G. Strang and T. Nguyen. *Wavelets and Filter Banks*. Wellesley-Cambridge Press, Wellesley, MA, 1996.

Motion Tracking of Animals for Behavior Analysis

Petra Perner

Institute of Computer Vision and Applied Computer Sciences, Arno-Nitzsche-Str. 45, 04277
Leipzig
ibaiperner@aol.com http://www.ibai-research.de

Abstract. In this paper, we are presenting our results for motion tracking
animals in stabling. This system was used in order to record the behavior of
pigs in stabling. We used an object-oriented method for our application instead
of a block-oriented method. First of all, we calculated a reference image. This
image was used in order to separate the objects from the background. Then,
object pixels were grouped into an object by the line-coincidence method.
Movement parameters are calculated for each object. Finally, an object
correction is done for those objects that were occluded by the boundary of the
stabling. The resulting tracking path and the movement parameters are
displayed on screen for the user.

1 Introduction

Researchers and farmers record the movements of animals on video in order to
observe animal behavior. Afterwards, they analyze these videos by looking up each
image sequence and taking notes about the spatial position of each animal. This is a
very time consuming process. However, such an observation of animals is important
in order to understand the behavior of animals in stabling and other environments
[1][2]. The resulting knowledge can help to improve animal welfare as well as meat
quality.

The task of behavior analysis is not only important for the study of animal welfare,
it also becomes important for many other tasks such as group behavior analysis in
public traffic areas, soccer game reporting and pharmacological studies.

An automatic system has to recognize the object and to track the object before it is
possible to describe the semantic concepts of the behavior of the observed object such
as object "stand", "moves" or more complex concepts such as object under "nervous
excitement". In this paper, we are presenting our results for motion tracking of pigs in
stabling.

Visual object tracking has become an important subject in computer vision. We
can identify block-oriented and object-oriented methods [3]. An object-oriented
method is described in Mae et. al [4]. They combined the optical flow with edge
detection in order to separate objects from background. The intention of this work is
contour detection of moving objects in a highly structured background. In Hoetner et.
al [5] a block-oriented approach is described that uses the signal difference and the
texture for the detection of moving objects. In Iketani et. al [6] the segmentation of

C. Arcelli et al. (Eds.): IWVF4, LNCS 2059, pp. 779–786, 2001.
© Springer-Verlag Berlin Heidelberg 2001

the images is done by partitioning image into blocks and determining the value of the optical flow for each block. Regions with the same value for the optical flow were grouped into one object. That allows us to detect objects on an in-stationary background. For our application, we chose an object-oriented method to prevent objects getting separated or combined together by the blocks.

2   Image Acquisition

The movements of the pigs in stabling were recorded by a video camera when the pigs came into stabling for the first time. The length of the video was from 20s to one minute. The camera had a fixed position that allowed us to look at the stabling diagonally from the top. In each video, we could see the boundary of the stabling. First, images from the empty stabling were taken for 0.5 s. The spatial resolution of the image sequence is 768 x 576 pixels for each image. It is possible to see the time in each video at the lower right corner.

3   Outline of Our Method

The outline of our method is shown in Figure 1. The image sequence of the empty stabling is extracted from the video and given to the pre-processing unit. This unit calculated a reference image, a threshold and new matrix which contains the boundary of the stabling. Afterwards, objects and background were separated in each image of the video sequence based on the reference image. Then, objects were determined in the image and the motion parameters were calculated. Objects may be occluded by the boundary of the stabling. The objects and motion parameters were corrected for this reason. Finally, the tracking path and the motion parameters were displayed on screen.

4   Image Pre-processing

The image sequence from the empty stabling (see Figure 2) was extracted from the video sequence and a reference background image was calculated from these images:

$$\{ anf_k(i, j) : k = 1,.., K \} \quad ref(i, j) = \frac{1}{k}\sum_{k=1}^{K} anf_k(i, j) \cdot \tag{1}$$

In addition to that, we extracted the boundary of the stabling from the reference image $ref(i,j)$ and stored it in an image matrix called $gitter(i,j)$. For that reason, we calculated the histogram of $ref(i,j)$ and determined the grey level threshold which allowed us to separate the background from the boundary of the stabling. A pixel in the matrix $gitter(i,j)$ is one if the pixel belongs to the boundary of the stabling and it is zero if the pixel does not belong to the boundary.

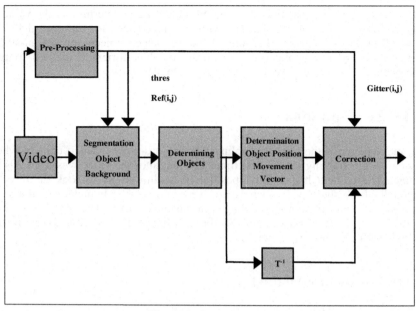

Fig. 1. Overall Structure of our Algorithm

5   Separation of Object and Background

A threshold was determined from the initial images and the reference images in order to separate objects and the background. We determined the variance from the difference of the initial images and the reference image:

$$sq = \frac{1}{k-1} \sum \left(anf_k(i, j) - ref(i, j) \right) \tag{2}$$

Afterwards, we calculated the histogram over all difference pixel. The threshold for the object segmentation was determined:

$$\sum_{s=0}^{grenz} h(sq) = 0.95 \qquad thresh = 2 \bullet grenz \tag{3}$$

Now, the reference image *ref(i,j)* was subtracted from the actual image *act(i,j)*. The resulting image was segmented into object pixel and background pixel based on the threshold described above. The resulting binary image has the name *arb(i,j)*. The object pixels were grouped into separate objects by the line coincidence method [7]. Each object was labelled by an object number *objnr*. Objects smaller than a predefined threshold were interpreted as image noise and eliminated from the list of objects.

Fig. 2. Empty stabling

After the first real object was found the determination of the object position and the movement vector were started.

6   Determination of the Object Position and the Movement Vector

The center of gravity mi and mj were determined for each object. Then, the movement vector of each object was determined by the following equation:

$$bewi(objnr,k) = mi(objnr,k+n+1) - mi(objnr,k)$$
$$bewj(objnr,k) = mj(objnr,k+n+1) - mj(objnr,k)$$

(4)

Unfortunately, this method for the determination of the movement vector has some disadvantages. The method can not determine rotations of the object according to its inner axis and 3D movements. However, it was sufficient for our problem.

Another parameter that is determined is the area of each object anz at each time t. That means that for each object we determined $objnr, mi, mj, bwei, bewj$ and anz. These parameters are the basis for further determination.

7   Correction of the Objects and the Object Parameters

The boundary of the stabling can sometimes partially occlude the pigs. For instance, objects may be separated by the boundary. The computed movement parameters were used to correct these disturbances and deformations. The binary matrix $arb(i,j,k)$ at the time k were taken in order to prove, for each object, if the object was behind or in front of the boundary. In the case where the object was in front of the boundary then the object occluded the boundary and no correction was necessary. Only in the case, where the object was behind the boundary was a correction necessary. For that the object was extracted from $arb(i,j)$ so that we obtained a new matrix $arb1(i,j)$ that only contained the pixel of this object. The matrix $arb1(i,j)$ was added up with the matrix

Fig. 3. Original image (first pig comes into the stabling) and result of motion analysis

gitter(i,j). In the case that the resulting matrix had the value "2" inside then no correction was necessary. If this case did not occur then a dilation was done on *arb1(i,j)*

$$arb1(i, j) \oplus M = \left\{(i, j) : M_{i,j} \cap arb1(,) \neq 0\right\} \tag{5}$$

M is a 3 x 3 mask containing the value "1" and $M_{i,j}$ is the mask M that was shifted to the pixel *(i,j)*. The resulting matrix was computed with the mask *gitter (i,j)* by the *logical AND* function. If the result was zero, then the object was not occluded by the boundary of the stabling and correction was not necessary. A correction was made if the resulting values of this operation were "1". The object was isolated from the matrix *arb(i,j,k-1)*. Then, it was shifted by the calculated motion vector *(=(trans,i,j))* and combined with the matrix *arb1(i,j)* by the logical AND function. For the resulting corrected object the proof was carried out as to whether has a nonempty intersection with the object in *arb(i,j,k)*. Then, these objects were combined to a single object. Afterwards, the *parameter mi(,), mj(,), bewi(,), bewj(,),* and *anz* were calculated for the corrected object.

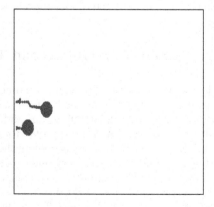

Fig. 4. Image number 82 and tracking path of pigs

8   Output of the System

The resulting tracking path of each object is displayed on screen. Figures 3-8 show images at different times t and the tracked path of the objects. It shows that we are able to track the objects by our algorithm.

However, on the recorded path we can see that the movement of the pig is not a smooth line. The pigs are tripping a bit back and forth at the same place. They rotate on their own axis.

Our system records for each pig the coordinates, motion parameters and the time. It gives us the basic information needed for behavior analysis. The next step must be the mapping of the information extracted from the image to the semantic concepts that the veterinarian needs for his analysis. Only when this task is solved do we have a fully automatic system. However, it also shows the complexity of the task of behavior analysis. The tracking of the objects is not simple and it is even harder in a real world environment. The next step, the mapping of the image information to the semantic concepts needs a clear understanding of what the concepts are and how we can describe them by the image content.

Recently, the veterinarians have shown they are happy with a listing of the coordinates, the motion parameters and the associated time for each pig.

 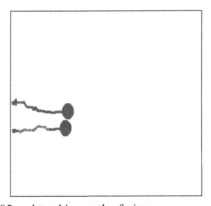

Fig. 5. Image number 105 and tracking path of pigs

 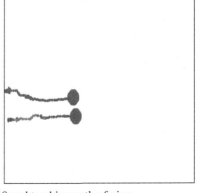

Fig. 6. Image number 118 and tracking path of pigs

9   Conclusion

We have presented our system for motion tracking of animals in stabling. Our system was used for the analysis of movements of pigs when they enter new stabling.

Furthermore, we have investigated other methods for motion tracking. However, we found that our method has several advantages over these methods for our application. First of all, it is easy to calculate. Secondly, it can correct occluded object parts, which helps to improve the determination of the object position and the motion parameters. Finally, it takes into account the changing shape of the objects that, for example, occurs by rotation of an object.

 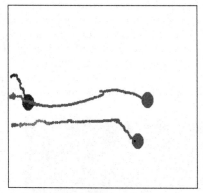

Fig. 7. Image number 205 and tracking path of pigs

 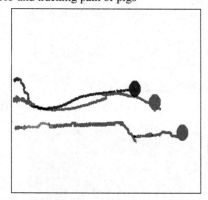

Fig. 8. Image number 327 and tracking path of pigs

Acknowledgement. We thank Dr. Laube from the Albrecht-Daniel-Thaer-Institute in Leipzig for drawing our attention to this application and for providing the videos.

References

1. Oliver, G.W.; Visualizing the tracking and diving behavior of marine mammals: a case study; Proceedings. Visualization '95, Los Alamitos, CA, USA: IEEE Comput. Soc. Press, 1995. p.397-399.
2. Stabenow, B.; Schon, P.-C, Brussow, K.-P.,; Biotelemetry in physiologic research of farm animals; In Proceedings of the 18th Annual International Conference of the IEEE Engineering in Medicine and Biology Society. Editor(s): Boom, H.; Robinson, C.; Rutten, W.; Neuman, M.; Wijkstra, H. New York, NY, USA: IEEE, 1997, vol.1, p.288-9.
3. Kummer, G., Perner, P.; Motion Analysis; IBaI Report January 1999, ISSN 1431-2360.
4. Mae, Y., Shirai, Y.; Tracking moving objects in 3D space based on optical flow and edges; Proc. 14th Intern. Conf. on Pattern Recognition, Brisbane, Australia 1998, vol. II, pp. 1439-1441.
5. Hötter, M., Mester, R., Meyer, M.; Detection of moving objects using a robust displacement estimation including a statistical error analysis, Proc. 13th Intern. Conf. on Pattern Recognition, Vienna, Austria, 1996, vol. IV, pp. 249-255.
6. Iketani, A., Nagai, A., Detecting persons on changing background, Proc. 14th Intern. Conf. on Pattern Recognition, Brisbane, Australia, 1998, vol. I, pp. 74-76.
7. Simon, H; Automatic Image Analysis in Medicine and Biology; Steinkopff-Verlag, Dresden 1975, p. 100-110.

Head Model Acquisition from Silhouettes

Kwan-Yee K. Wong, Paulo R.S. Mendonça, and Roberto Cipolla

Department of Engineering
University of Cambridge
Cambridge, CB2 1PZ, UK
[kykw2|prdsm2|cipolla]@eng.cam.ac.uk
http://svr-www.eng.cam.ac.uk/research/vision

Abstract. This paper describes a practical system developed for generating 3D models of human heads from silhouettes alone. The input to the system is an image sequence acquired from circular motion. Both the camera motion and the 3D structure of the head are estimated using silhouettes which are tracked throughout the sequence. Special properties of the camera motion and their relationships with the intrinsic parameters of the camera are exploited to provide a simple parameterization of the fundamental matrix relating any pair of views in the sequence. Such a parameterization greatly reduces the dimension of the search space for the optimization problem. In contrast to previous methods, this work can cope with incomplete circular motion and more widely spaced images. Experiments on real image sequences are carried out, showing accurate recovery of 3D shapes.

1   Introduction

The reconstruction of 3D head models has many important applications, such as video conferencing, model-based tracking, entertainment and face modeling [13]. Existing commercial methods for acquiring such models, such as laser scans, are expensive, time-consuming and cannot cope with low-reflectance surfaces. Image based systems can easily overcome these difficulties by tracking point features along video sequences [5]. However, this can be remarkably difficult for human faces, where there are not many reliable landmarks with long life span along the sequence.

In this paper we present a practical system for generating 3D head models from silhouettes alone. Silhouettes are comparatively easy to track and provide useful information for estimating the camera motion [1,10] and reconstruction [12,2,15]. Since they tend to concentrate around regions of high curvature, they provide a compact way of parameterizing the reconstructed surface. In our system, images are acquired by moving the camera along a circular path around the head. This imposes constraints on the fundamental matrix relating each pair of images, simplifying the motion estimation. The system does not require the motion to be a full rotation and the images can be acquired at more widely spaced positions around the subject, an advantage over the technique introduced in [10].

C. Arcelli et al. (Eds.): IWVF4, LNCS 2059, pp. 787–796, 2001.
© Springer-Verlag Berlin Heidelberg 2001

Section 2 presents the theoretical background of motion estimation from silhouettes. The algorithms for model building are described in Section 3, and Section 4 shows the experimental results. Conclusions are given in Section 5.

2   Theoretical Background

The fundamental difficulty in solving the problem of structure and motion from silhouettes is that, unlike point or line features, the silhouettes do not readily provide image correspondences that allow for the computation of the epipolar geometry, summarized by the fundamental matrix. The usual solution to this problem is the use of *epipolar tangencies* [11,3], as shown in Fig. 1. An epipolar tangent point is the projection of a *frontier point* [3], which is the intersection of two contour generators. If 7 or more epipolar tangent points are available, the

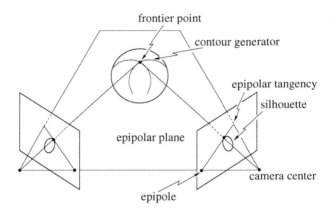

Fig. 1. A frontier point is the intersection of two contour generators and is visible in both views. The frontier point projects onto a point on the silhouette which is also on an epipolar tangent

epipolar geometry can be estimated. The intrinsic parameters of the cameras can then be used to recover the motion [6,4]. However the unrealistic demand for a large number of epipolar tangent points makes this approach impractical. By constraining the motion to be circular, a parameterization of the fundamental matrix with only 6 degrees of freedom (dof) is possible [14,5,9]. This parameterization explicitly takes into account the main image features of circular motion, namely the image of the rotation axis, the horizon and a special vanishing point, which are fixed throughout the sequence. This makes it possible to estimate the epipolar geometry by using only 2 epipolar tangencies [9].

In [10], a practical algorithm has been introduced for the estimation of motion and structure from silhouettes of a rotating object. The image of the rotation axis and the vanishing point are first determined by estimating the harmonic homology associated with the image of surface of revolution spanned by the object. In order to obtain such an image, a dense image sequence from a complete circular motion is required. In this paper, the parameters of the harmonic homology and other motion parameters are estimated simultaneously by minimizing the reprojection errors of epipolar tangents. This algorithm does not require the image of such a surface of revolution and thus can cope with incomplete circular motion and more widely spaced images, an advantage over the algorithm presented in [10].

2.1 Symmetry and Epipolar Geometry in Circular Motion

Consider a pinhole camera undergoing circular motion. If the camera intrinsic parameters are kept constant, the projection of the rotation axis will be a line l_s which is pointwise fixed on each image. This means that, for any point x on l_s, the equation $x^T F x = 0$ is satisfied, where F is the fundamental matrix related to any image pair in the sequence. For circular motion, all the camera centers lie on a common plane. The image of this plane is a special line l_h, the *horizon*. Since the epipoles are the images of the camera centers, they must lie on l_h. In general, l_s and l_h are not orthogonal. Another feature of interest is the vanishing point v_x which corresponds to the normal direction of the plane defined by the camera center and the axis of rotation. The vanishing point and the horizon satisfy $v_x^T l_h = 0$. A detailed discussion of the above can be found in [14,5,9].

Consider now a pair of cameras, denoted as P_1 and P_2, related by a rotation about an axis not passing through their centers, and let F be the fundamental matrix associated with this pair. It has been shown that corresponding epipolar lines associated with F are related to each other by a harmonic homology W [9], given by

$$W = \mathbb{I} - 2 \frac{v_x l_s}{v_x^T l_s}. \tag{1}$$

Note that W has 4 dof: 2 corresponding to the axis and 2 corresponding to the vanishing point. If P_1 and P_2 point towards the axis of rotation, v_x will be at infinity and W will be reduced to a skew symmetry S with only 3 dof. Besides, if the cameras also have zero skew and aspect ratio 1, the transformation will be further specialized to a bilateral symmetry B with only 2 dof. A pictorial description of these transformations can be seen in Fig. 2

In [16], an algorithm has been presented for estimating the camera intrinsic parameters from 2 or more silhouettes of surfaces of revolution. For each silhouette, the associated harmonic homology W is estimated and this provides 2 constraints on the camera intrinsic parameters:

$$v_x = K K^T l_s, \tag{2}$$

where K is the 3×3 camera calibration matrix. Conversely, if the camera intrinsic parameters are known, (2) provides 2 constraints on W and as a result W has only 2 dof.

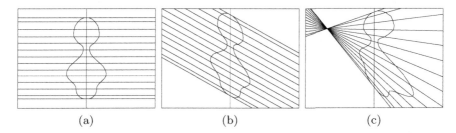

$$\text{(a)} \qquad\qquad\qquad \text{(b)} \qquad\qquad\qquad \text{(c)}$$

Fig. 2. (a) A curve displaying bilateral symmetry. The horizon is orthogonal to the axis. (b) Same curve, distorted by an affine transformation. The horizon is no longer orthogonal to the axis, and each side of the curve is mapped to the other by a skew symmetry transformation. (c) The curve is now distorted by a special projective transformation (harmonic homology), and the lines of symmetry intersect at a point corresponding to the vanishing point

2.2 Parameterization of the Fundamental Matrix

In [8,17], it has been shown that any fundamental matrix $\mathbf{F}$ can be parameterized as $\mathbf{F} = [\mathbf{e}_2]_\times \mathbf{M}$, where $\mathbf{M}^{-T}$ is any matrix that maps the epipolar lines from one image to the other, and $\mathbf{e}_2$ is the epipole in the second image. In the special case of circular motion, it follows that

$$\mathbf{F} = [\mathbf{e}_2]_\times \mathbf{W}. \qquad (3)$$

Note that $\mathbf{F}$ has 6 dof: 2 to fix $\mathbf{e}_2$, and 4 to determine $\mathbf{W}$. From (2), if the camera intrinsic parameters are known, 2 parameters are enough to define $\mathbf{W}$ and thus $\mathbf{F}$ will have only 4 dof.

An alternative parameterization for the fundamental matrix in the case of circular motion [14,5,9] is given by

$$\mathbf{F} = [\mathbf{v}_x]_\times + \kappa \tan\frac{\theta}{2}(\mathbf{l}_s\mathbf{l}_h^T + \mathbf{l}_h\mathbf{l}_s^T), \qquad (4)$$

where θ is the angle of rotation between the cameras. The constant κ can be determined from the camera intrinsic parameters [9] if $\mathbf{l}_s$, $\mathbf{v}_x$ and $\mathbf{l}_h$ are properly normalized. θ is the only parameter which depends on the particular pair of cameras being considered, while the other 4 terms are common to all pairs of images in the sequence. When the camera intrinsic parameters are known, 2 parameters are enough to fix $\mathbf{l}_s$ and $\mathbf{v}_x$. Since $\mathbf{v}_x$ must lie on $\mathbf{l}_h$, only 1 further parameter is needed to fix $\mathbf{l}_h$. As a result, the fundamental matrix has only 4 dof.

3 Algorithms

Before the 3D model can be reconstructed from the silhouettes of the head, the motion of the camera has to be estimated. By using the parameterization shown in (4), the $\binom{N}{2} = N(N-1)/2$ fundamental matrices relating all possible pairs of cameras in a sequence of N images, taken by a rotating camera with known intrinsic parameters, can be defined with the 3 parameters which fix $\mathbf{l}_s$, $\mathbf{v}_x$ and $\mathbf{l}_h$, together with the $N-1$ angles of rotation between adjacent cameras. By enforcing the epipolar constraint on the corresponding epipolar tangent points, these $N+2$ motion parameters can be estimated by minimizing the reprojection errors of corresponding epipolar tangents (see Fig. 3). Since a silhouette has at least two epipolar tangencies (one at its top and another at its bottom), there will be totally $2\binom{N}{2} = N(N-1)$ measurements from all pairs of images. Due to the dependence between the associated fundamental matrices, however, these $N(N-1)$ measurements only provide $2N$ (or 2 when $N=2$) independent constraints on the $N+2$ parameters. As a result, a solution will be possible if $N \geq 3$.

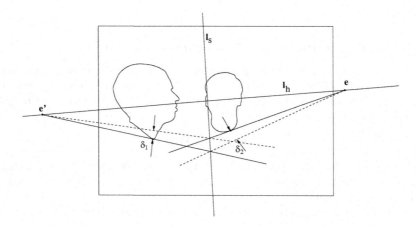

Fig. 3. The parameters of the fundamental matrix associated with each pair of images in the sequence can be estimated from the reprojection errors of epipolar tangents. The solid lines are tangents to the silhouettes passing through the epipoles, and the dashed lines are the epipolar lines corresponding to the tangent points

The minimization of the reprojection errors will generate a consistent set of fundamental matrices, which, together with the camera intrinsic parameters, can be decomposed into a set of camera matrices describing a circular motion compatible with the image sequence. The algorithm for motion estimation is summarized in Algorithm 1. Having the motion of the camera estimated, the 3D model can then be reconstructed from the silhouettes using the simple triangulation technique introduced in [15].

Algorithm 1 Estimation of the motion parameters from silhouettes

track the silhouettes of the head using cubic B-spline snakes;
initialize l_s, l_h and the $N-1$ angles between the N cameras;
while not converged **do**
  for each image pair **do**
    form fundamental matrix;
    locate epipolar tangents;
    compute reprojection errors;
  end for
  update parameters to minimize the sum of reprojection errors;
end while

---

4   Experiments and Results

In order to evaluate the performance of the algorithm described in Section 3, 2 human head sequences, each with 10 images, were acquired using the setup shown in Fig. 4. The camera is mounted to the extensible rotating arm of the tripod, whose height can be adjusted according to the height of the subject. Each image in the sequence was taken after rotating the arm of the tripod roughly by 20°, with the subject standing close to the tripod. The intrinsic parameters of the camera are obtained from an offline calibration process. The silhouettes of the heads are tracked using cubic B-spline snakes [2] (see Fig. 5 and 6).

Fig. 4. Experimental setup used to acquire image sequences around human heads. The camera is mounted to the rotating arm of the tripod with the subject standing close to the tripod. Although the camera motion is constrained to be circular, the camera orientation and rotation angle are unknown

The initial guess for the horizon and the image of the rotation axis was picked by observation, and the angles of rotation were initialized as 10° respectively. The

Fig. 5. Image sequence (I) used in the experiment, with the silhouettes of the head tracked using cubic B-spline snakes

Fig. 6. Image sequence (II) used in the experiment, with the silhouettes of the head tracked using cubic B-spline snakes

sum of the reprojection errors was minimized using the Levenberg-Marquardt algorithm [7].The reconstructed 3D head models can be found in Fig. 7 and 8. The shapes of the ears, lips, noise and eyebrows demonstrate the quality of the 3D models recovered.

5   Conclusions

In this paper we have presented a simple and practical system for building 3D models of human heads from image sequences. No prior model is assumed, and in fact the system can be applied to a variety of objects. The only constraint on the camera is that it must perform circular motion, though the exact camera orientations and positions are unknown. Besides, the camera is not required to perform a full rotation and there is no need for using a dense image sequence. The silhouettes of the head are the only information used for both motion estimation and reconstruction, circumventing the lack, instability and occlusion of landmarks on faces. The silhouettes also provide a natural and compact way of parameterizing the head model, concentrating contours around regions of high curvature. The experimental results show the accuracy of the acquired model.

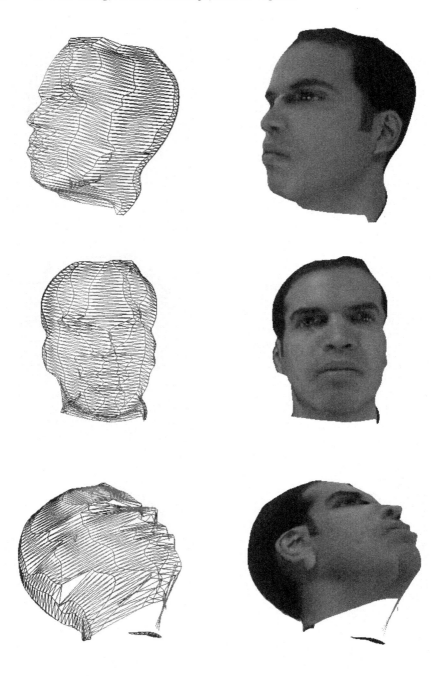

Fig. 7. Different views of the VRML model from the model building process using the 10 images in Fig. 5

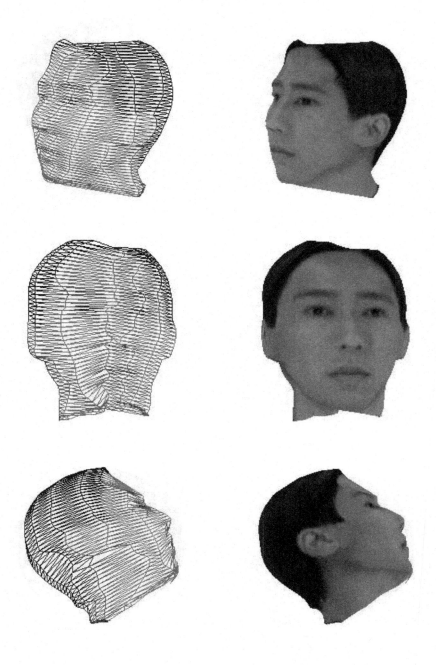

Fig. 8. Different views of the VRML model obtained from the model building process using the 10 images in Fig. 6

Acknowledgements. P. R. S. Mendonça would like to acknowledge CAPES/Brazilian Ministry of Education for the grant BEX 1165/96-8 that partially funded this research.

References

1. R. Cipolla, K. Åström, and P.J. Giblin. Motion from the frontier of curved surfaces. In *Proc. 5th Int. Conf. on Computer Vision*, pages 269–275, 1995.
2. R. Cipolla and A. Blake. Surface shape from the deformation of apparent contours. *Int. Journal of Computer Vision*, 9(2):83–112, 1992.
3. R. Cipolla and P. J. Giblin. *Visual Motion of Curves and Surfaces*. Cambridge University Press, Cambridge, 1999.
4. O. Faugeras. Stratification of three-dimensional vision: Projective, affine and metric representations. *J. Opt. Soc. America A*, 12(3):465–484, 1995.
5. A. W. Fitzgibbon, G. Cross, and A. Zisserman. Automatic 3D model construction for turn-table sequences. In *3D Structure from Multiple Images of Large-Scale Environments, European Workshop SMILE'98*, Lecture Notes in Computer Science 1506, pages 155–170, 1998.
6. H. C. Longuet-Higgins. A computer algorithm for reconstructing a scene from two projections. *Nature*, 293:133–135, 1981.
7. D. G. Luenberger. *Linear and Nonlinear Programming*. Addison-Wesley, USA, second edition, 1984.
8. Q.-T. Luong and O. Faugeras. The fundamental matrix: Theory, algorithm, and stability analysis. *Int. Journal of Computer Vision*, 17(1):43–75, 1996.
9. P. R. S. Mendonça, K.-Y. K. Wong, and R. Cipolla. Recovery of circular motion from profiles of surfaces. In B. Triggs, A. Zisserman, and R. Szeliski, editors, *Vision Algorithms: Theory and Practice*, Lecture Notes in Computer Science 1883, pages 151–167, Corfu, Greece, Sep 1999. Springer–Verlag.
10. P. R. S. Mendonça, K.-Y. K. Wong, and R. Cipolla. Camera pose estimation and reconstruction from image profiles under circular motion. In D. Vernon, editor, *Proc. 6th European Conf. on Computer Vision*, volume II, pages 864–877, Dublin, Ireland, Jun 2000. Springer–Verlag.
11. J. Porrill and S. B. Pollard. Curve matching and stereo calibration. *Image and Vision Computing*, 9(1):45–50, 1991.
12. R. Vaillant and O. D. Faugeras. Using extremal boundaries for 3D object modelling. *IEEE Trans. on Pattern Analysis and Machine Intell.*, 14(2):157–173, 1992.
13. T. Vetter. Synthesis of novel views from a single face image. *Int. Journal of Computer Vision*, 28(2):103–116, June 1998.
14. T. Vieville and D. Lingrand. Using singular displacements for uncalibrated monocular visual systems. In *Proc. 4th European Conf. on Computer Vision*, volume II, pages 207–216, 1996.
15. K.-Y. K. Wong, P. R. S. Mendonça, and R. Cipolla. Reconstruction and motion estimation from apparent contours under circular motion. In T. Pridmore and D. Elliman, editors, *Proc. British Machine Vision Conference*, volume 1, pages 83–92, Nottingham, UK, Sep 1999. British Machine Vision Association.
16. K.-Y. K. Wong, P. R. S. Mendonça, and R. Cipolla. Camera calibration from symmetry. In R. Cipolla and R. Martin, editors, *The Mathematics of Surfaces IX*, pages 214–226, Cambridge, UK, Sep 2000. Institute of Mathematics and its Applications, Springer–Verlag.
17. Z. Zhang. Determining the epipolar geometry and its uncertainty: A review. *Int. Journal of Computer Vision*, 27(2):161–195, 1998.

Author Index

Printed in the United States
By Bookmasters